LONDON MATHEMATICAL SOCIETY LECTURE NOTE SERIES

Managing Editor: Professor N.J. Hitchin, Mathematical Institute,
University of Oxford, 24–29 St Giles, Oxford OX1 3LB, United Kingdom

The titles below are available from booksellers, or, in case of difficulty, from Cambridge University Press at www.cambridge.org.

148 Helices and vector bundles, A.N. RUDAKOV *et al*
149 Solitons, nonlinear evolution equations and inverse scattering, M. ABLOWITZ & P. CLARKSON
150 Geometry of low-dimensional manifolds 1, S. DONALDSON & C.B. THOMAS (eds)
151 Geometry of low-dimensional manifolds 2, S. DONALDSON & C.B. THOMAS (eds)
152 Oligomorphic permutation groups, P. CAMERON
153 L-functions and arithmetic, J. COATES & M.J. TAYLOR (eds)
155 Classification theories of polarized varieties, TAKAO FUJITA
158 Geometry of Banach spaces, P.F.X. MÜLLER & W. SCHACHERMAYER (eds)
159 Groups St Andrews 1989 volume 1, C.M. CAMPBELL & E.F. ROBERTSON (eds)
160 Groups St Andrews 1989 volume 2, C.M. CAMPBELL & E.F. ROBERTSON (eds)
161 Lectures on block theory, BURKHARD KÜLSHAMMER
163 Topics in varieties of group representations, S.M. VOVSI
164 Quasi-symmetric designs, M.S. SHRIKANDE & S.S. SANE
166 Surveys in combinatorics, 1991, A.D. KEEDWELL (ed)
168 Representations of algebras, H. TACHIKAWA & S. BRENNER (eds)
169 Boolean function complexity, M.S. PATERSON (ed)
170 Manifolds with singularities and the Adams-Novikov spectral sequence, B. BOTVINNIK
171 Squares, A.R. RAJWADE
172 Algebraic varieties, GEORGE R. KEMPF
173 Discrete groups and geometry, W.J. HARVEY & C. MACLACHLAN (eds)
174 Lectures on mechanics, J.E. MARSDEN
175 Adams memorial symposium on algebraic topology 1, N. RAY & G. WALKER (eds)
176 Adams memorial symposium on algebraic topology 2, N. RAY & G. WALKER (eds)
177 Applications of categories in computer science, M. FOURMAN, P. JOHNSTONE & A. PITTS (eds)
178 Lower K- and L-theory, A. RANICKI
179 Complex projective geometry, G. ELLINGSRUD *et al*
180 Lectures on ergodic theory and Pesin theory on compact manifolds, M. POLLICOTT
181 Geometric group theory I, G.A. NIBLO & M.A. ROLLER (eds)
182 Geometric group theory II, G.A. NIBLO & M.A. ROLLER (eds)
183 Shintani zeta functions, A. YUKIE
184 Arithmetical functions, W. SCHWARZ & J. SPILKER
185 Representations of solvable groups, O. MANZ & T.R. WOLF
186 Complexity: knots, colourings and counting, D.J.A. WELSH
187 Surveys in combinatorics, 1993, K. WALKER (ed)
188 Local analysis for the odd order theorem, H. BENDER & G. GLAUBERMAN
189 Locally presentable and accessible categories, J. ADAMEK & J. ROSICKY
190 Polynomial invariants of finite groups, D.J. BENSON
191 Finite geometry and combinatorics, F. DE CLERCK *et al*
192 Symplectic geometry, D. SALAMON (ed)
194 Independent random variables and rearrangement invariant spaces, M. BRAVERMAN
195 Arithmetic of blowup algebras, WOLMER VASCONCELOS
196 Microlocal analysis for differential operators, A. GRIGIS & J. SJÖSTRAND
197 Two-dimensional homotopy and combinatorial group theory, C. HOG-ANGELONI *et al*
198 The algebraic characterization of geometric 4-manifolds, J.A. HILLMAN
199 Invariant potential theory in the unit ball of C^n, MANFRED STOLL
200 The Grothendieck theory of dessins d'enfant, L. SCHNEPS (ed)
201 Singularities, JEAN-PAUL BRASSELET (ed)
202 The technique of pseudodifferential operators, H.O. CORDES
203 Hochschild cohomology of von Neumann algebras, A. SINCLAIR & R. SMITH
204 Combinatorial and geometric group theory, A.J. DUNCAN, N.D. GILBERT & J. HOWIE (eds)
205 Ergodic theory and its connections with harmonic analysis, K. PETERSEN & I. SALAMA (eds)
207 Groups of Lie type and their geometries, W.M. KANTOR & L. DI MARTINO (eds)
208 Vector bundles in algebraic geometry, N.J. HITCHIN, P. NEWSTEAD & W.M. OXBURY (eds)
209 Arithmetic of diagonal hypersurfaces over infite fields, F.Q. GOUVEA & N. YUI
210 Hilbert C^*-modules, E.C. LANCE
211 Groups 93 Galway / St Andrews I, C.M. CAMPBELL *et al* (eds)
212 Groups 93 Galway / St Andrews II, C.M. CAMPBELL *et al* (eds)
214 Generalised Euler-Jacobi inversion formula and asymptotics beyond all orders, V. KOWALENKO *et al*
215 Number theory 1992–93, S. DAVID (ed)
216 Stochastic partial differential equations, A. ETHERIDGE (ed)
217 Quadratic forms with applications to algebraic geometry and topology, A. PFISTER
218 Surveys in combinatorics, 1995, PETER ROWLINSON (ed)
220 Algebraic set theory, A. JOYAL & I. MOERDIJK
221 Harmonic approximation, S.J. GARDINER
222 Advances in linear logic, J.-Y. GIRARD, Y. LAFONT & L. REGNIER (eds)
223 Analytic semigroups and semilinear initial boundary value problems, KAZUAKI TAIRA
224 Computability, enumerability, unsolvability, S.B. COOPER, T.A. SLAMAN & S.S. WAINER (eds)
225 A mathematical introduction to string theory, S. ALBEVERIO, J. JOST, S. PAYCHA, S. SCARLATTI
226 Novikov conjectures, index theorems and rigidity I, S. FERRY, A. RANICKI & J. ROSENBERG (eds)
227 Novikov conjectures, index theorems and rigidity II, S. FERRY, A. RANICKI & J. ROSENBERG (eds)
228 Ergodic theory of Z^d actions, M. POLLICOTT & K. SCHMIDT (eds)

229 Ergodicity for infinite dimensional systems, G. DA PRATO & J. ZABCZYK
230 Prolegomena to a middlebrow arithmetic of curves of genus 2, J.W.S. CASSELS & E.V. FLYNN
231 Semigroup theory and its applications, K.H. HOFMANN & M.W. MISLOVE (eds)
232 The descriptive set theory of Polish group actions, H. BECKER & A.S. KECHRIS
233 Finite fields and applications, S.COHEN & H. NIEDERREITER (eds)
234 Introduction to subfactors, V. JONES & V.S. SUNDER
235 Number theory 1993–94, S. DAVID (ed)
236 The James forest, H. FETTER & B. GAMBOA DE BUEN
237 Sieve methods, exponential sums, and their applications in number theory, G.R.H. GREAVES *et al*
238 Representation theory and algebraic geometry, A. MARTSINKOVSKY & G. TODOROV (eds)
239 Clifford algebras and spinors, P. LOUNESTO
240 Stable groups, FRANK O. WAGNER
241 Surveys in combinatorics, 1997, R.A. BAILEY (ed)
242 Geometric Galois actions I, L. SCHNEPS & P. LOCHAK (eds)
243 Geometric Galois actions II, L. SCHNEPS & P. LOCHAK (eds)
244 Model theory of groups and automorphism groups, D. EVANS (ed)
245 Geometry, combinatorial designs and related structures, J.W.P. HIRSCHFELD *et al*
246 p-Automorphisms of finite p-groups, E.I. KHUKHRO
247 Analytic number theory, Y. MOTOHASHI (ed)
248 Tame topology and o-minimal structures, LOU VAN DEN DRIES
249 The atlas of finite groups: ten years on, ROBERT CURTIS & ROBERT WILSON (eds)
250 Characters and blocks of finite groups, G. NAVARRO
251 Gröbner bases and applications, B. BUCHBERGER & F. WINKLER (eds)
252 Geometry and cohomology in group theory, P. KROPHOLLER, G. NIBLO, R. STÖHR (eds)
253 The q-Schur algebra, S. DONKIN
254 Galois representations in arithmetic algebraic geometry, A.J. SCHOLL & R.L. TAYLOR (eds)
255 Symmetries and integrability of difference equations, P.A. CLARKSON & F.W. NIJHOFF (eds)
256 Aspects of Galois theory, HELMUT VÖLKLEIN *et al*
257 An introduction to noncommutative differential geometry and its physical applications 2ed, J. MADORE
258 Sets and proofs, S.B. COOPER & J. TRUSS (eds)
259 Models and computability, S.B. COOPER & J. TRUSS (eds)
260 Groups St Andrews 1997 in Bath, I, C.M. CAMPBELL *et al*
261 Groups St Andrews 1997 in Bath, II, C.M. CAMPBELL *et al*
263 Singularity theory, BILL BRUCE & DAVID MOND (eds)
264 New trends in algebraic geometry, K. HULEK, F. CATANESE, C. PETERS & M. REID (eds)
265 Elliptic curves in cryptography, I. BLAKE, G. SEROUSSI & N. SMART
267 Surveys in combinatorics, 1999, J.D. LAMB & D.A. PREECE (eds)
268 Spectral asymptotics in the semi-classical limit, M. DIMASSI & J. SJÖSTRAND
269 Ergodic theory and topological dynamics, M.B. BEKKA & M. MAYER
270 Analysis on Lie Groups, N.T. VAROPOULOS & S. MUSTAPHA
271 Singular perturbations of differential operators, S. ALBEVERIO & P. KURASOV
272 Character theory for the odd order function, T. PETERFALVI
273 Spectral theory and geometry, E.B. DAVIES & Y. SAFAROV (eds)
274 The Mandelbrot set, theme and variations, TAN LEI (ed)
275 Computational and geometric aspects of modern algebra, M. D. ATKINSON *et al* (eds)
276 Singularities of plane curves, E. CASAS-ALVERO
277 Descriptive set theory and dynamical systems, M. FOREMAN *et al* (eds)
278 Global attractors in abstract parabolic problems, J.W. CHOLEWA & T. DLOTKO
279 Topics in symbolic dynamics and applications, F. BLANCHARD, A. MAASS & A. NOGUEIRA (eds)
280 Characters and Automorphism Groups of Compact Riemann Surfaces, T. BREUER
281 Explicit birational geometry of 3-folds, ALESSIO CORTI & MILES REID (eds)
282 Auslander-Buchweitz approximations of equivariant modules, M. HASHIMOTO
283 Nonlinear elasticity, R. OGDEN & Y. FU (eds)
284 Foundations of computational mathematics, R. DEVORE, A. ISERLES & E. SULI (eds)
285 Rational points on curves over finite fields: Theory and Applications, H. NIEDERREITER & C. XING
286 Clifford algebras and spinors 2nd edn, P. LOUNESTO
287 Topics on Riemann surfaces and Fuchsian groups, E. BUJALANCE, A. F. COSTA & E. MARTINEZ (eds)
288 Surveys in combinatorics, 2001, J. W. P. HIRSCHFELD (ed)
289 Aspects of Sobolev-type inequalities, L. SALOFF-COSTE
290 Quantum groups and Lie theory, A. PRESSLEY
291 Tits buildings and the model theory of groups, K. TENT
292 A quantum groups primer, S. MAJID
293 Second order partial differential equations in Hilbert spaces, G. DA PRATO & J. ZABCZYK
294 Introduction to operator space theory, G. PISIER
295 Geometry and integrability, L. MASON & Y. NUTKU (eds)
296 Lectures on invariant theory, I. DOLGACHEV
297 The homotopy category of simply connected 4-manifolds, H.-J. BAUES
298 Higher operads, higher categories, T. LEINSTER
299 Kleinian groups and hyperbolic 3-manifolds, Y. KOMORI, V. MARKOVIC & C. SERIES (eds)
300 Introduction to Möbius differential geometry, U. HERTRICH-JEROMIN
301 Stable modules and the D(2)-problem, F. E. A. JOHNSON
302 Discrete and continuous nonlinear Schrödinger systems, M. ABLOWITZ, B. PRINARI & D. TRUBATCH
304 Groups St Andrews 2001 in Oxford v1, C. M. CAMPBELL, E. F. ROBERTSON & G. C. SMITH (eds)
305 Groups St Andrews 2001 in Oxford v2, C. M. CAMPBELL, E. F. ROBERTSON & G. C. SMITH (eds)
306 Peyresq lectures on geometric mechanics and symmetry, J. MONTALDI & T. RATIU (eds)
307 Surveys in combinatorics 2003, C. D. WENSLEY (ed)

London Mathematical Society Lecture Note Series. 311

Groups
Topological, Combinatorial and Arithmetic Aspects

Edited by

T. W. Müller
Queen Mary, University of London

SEP/AE
math

PUBLISHED BY THE PRESS SYNDICATE OF THE UNIVERSITY OF CAMBRIDGE
The Pitt Building, Trumpington Street, Cambridge, United Kingdom

CAMBRIDGE UNIVERSITY PRESS
The Edinburgh Building, Cambridge CB2 2RU, UK
40 West 20th Street, New York, NY 10011-4211, USA
477 Williamstown Road, Port Melbourne, VIC 3207, Australia
Ruiz de Alarcón 13, 28014 Madrid, Spain
Dock House, The Waterfront, Cape Town 8001, South Africa

http://www.cambridge.org

First published 2004

Printed in the United Kingdom at the University Press, Cambridge

Typeface Times. 10/13 pt. *System* LaTeX 2_ε [TB]

A record for this book is available from the British Library

ISBN 0 521 54287 1 paperback

Contents

Authors and participants

H. Abels, Fakultät für Mathematik, Universität Bielefeld, POB 100131, D-33501 Bielefeld, Germany (abels@mathematik.uni-bielefeld.de)

P. Abramenko, Department of Mathematics, University of Virginia, POB 400 137 (Kerchof Hall), Charlottesville, VA 22904, USA (pa8e@virginia.edu)

S. I. Adian, Steklov Mathematical Institute, 42 ul. Vavilova, 117966 Moscow GSP-1, Russia (adian@log.mian.su)

H. Behr, Fachbereich Mathematik, J. W. Goethe-Universität, POB 111932, 60054 Frankfurt a. M., Germany (helmut.behr@math.uni-frankfurt.de)

R. Bieri, Fachbereich Mathematik, J. W. Goethe-Universität, POB 111932, 60054 Frankfurt a. M., Germany (bieri@math.uni-frankfurt.de)

M. Bridson, Department of Mathematics, Imperial College, 180 Queen's Gate, London SW7 2BZ, UK (m.bridson@ic.ac.uk)

K.-U. Bux, Department of Mathematics, Cornell University, Malott Hall 310, Ithaka, NY 19853-4201, USA (bux_math@kubux.net)

P. J. Cameron, School of Mathematical Sciences, Queen Mary, University of London, Mile End Road, London E1 4NS, UK (p.j.cameron@qmul.ac.uk)

I. M. Chiswell, School of Mathematical Sciences, Queen Mary, University of London, Mile End Road, London E1 4NS, UK (i.m.chiswell@qmul.ac.uk)

D. J. Collins, School of Mathematical Sciences, Queen Mary, University of London, Mile End Road, London E1 4NS, UK (d.j.collins@qmul.ac.uk)

A. Dress, Fakultät für Mathematik, Universität Bielefeld, POB 100131, D-33501 Bielefeld, Germany (dress@mathematik.uni-bielefeld.de)

R. Geoghegan, Department of Mathematical Sciences, SUNY, Binghamton, NY 13902-6000, USA (ross@math.binghamton.edu)

R. I. Grigorchuk, Steklov Mathematical Institute, Gubkina Street 8, Moscow 117966, Russia (grigorch@mi.ras.ru) and Department of Mathematics, Texas A&M University, College Station, Texas 77843-3368, USA (grigorch@math.tamu.edu)

F. Grunewald, Mathematisches Institut, Heinrich-Heine Universität, D-40225 Düsseldorf, Germany (fritz@math.uni-duesseldorf.de)

H. Helling, Fakultät für Mathematik, Universität Bielefeld, POB 100131, D-33501 Bielefeld, Germany (helling@mathematik.uni-bielefeld.de)

W. Imrich, Institute of Applied Mathematics, Montanuniversität Leoben, A-8700 Leoben, Austria (imrich@unileoben.ac.at)

R. Kaplinsky, Jerusalem ORT College, Givat Ram, PB 39161, Jerusalem 91390, Israel (rkaplins@mail.ort.org.il)

I. Lysionok, Steklov Mathematical Institute, 42 ul. Vavilova, 117966 Moscow GSP-1, Russia (lysionok@euclid.mi.ras.ru)

A. Mann, Institute of Mathematics, The Hebrew University, Givat Ram, Jerusalem 91904, Israel (mann@vms.huji.ac.il)

J. Mennicke, Fakultät für Mathematik, Universität Bielefeld, POB 100131, D-33501 Bielefeld, Germany (mennicke@mathematik.uni-bielefeld.de)

T. W. Müller, School of Mathematical Sciences, Queen Mary, University of London, Mile End Road, London E1 4NS, UK (t.w.muller@qmul.ac.uk)

V. Nekrashevych, Faculty of Mechanics and Mathematics, Kyiv Taras Shevchenko University, vul. Volodymyrska, 60, Kyiv, 01033, Ukraine (nazaruk@ukrpack.net)

J. R. Parker, Department of Mathematical Sciences, University of Durham, Durham DH1 3LE, UK (j.r.parker@durham.ac.uk)

L. Reeves, Mathematical Institute, University of Oxford, 24–29 St Giles', Oxford OX1 3LB, UK (reeves@maths.ox.ac.uk)

U. Rehmann, Fakultät für Mathematik, Universität Bielefeld, POB 100131, D-33501 Bielefeld, Germany (rehmann@mathematik.uni-bielefeld.de)

B. Remy, Institut Fourier – UMR 5582, Universite Grenoble 1 – Joseph Fourier, 100 rue des maths, BP 74 – 38402 Saint-Martin d'Heres, France (bertrand.remy@ujf-grenoble.fr)

D. Segal, All Souls College, Oxford OX1 4AL, UK
(dan.segal@all-souls.oxford.ac.uk)

C. M. Series, Mathematics Institute, University of Warwick, Coventry
CV4 7AL, UK (cms@maths.warwick.ac.uk)

S. N. Sidki, Departamento de Matemática, Universidade de Brasília,
Brasília-Df, 70.910-900, Brazil (sidki@mat.unb.br)

E. B. Vinberg, Department of Mechanics and Mathematics, Moscow State
University, Leninskie gory, 119899 Moscow, Russia
(vinberg@ebv.pvt.msu.su)

J. S. Wilson, School of Mathematics and Statistics, University of Birmingham,
Edgbaston, Birmingham B15 2TT, UK (jsw@for.mat.bham.ac.uk)

Preface

In 1999 a number of eminent mathematicians were invited to Bielefeld, to present papers at a one-week conference devoted to interactions between (mostly) infinite groups on the one hand and topological, combinatorial and arithmetic structure on the other. The present volume consists of articles invited from participants in this conference.

A glance at the table of contents gives an immediate impression of the breadth and depth of the contributions included here. The study of topological finiteness properties, a beautiful field of research inhabiting the fertile region between group theory and geometry, is the subject matter of papers by Abramenko, Behr, and Bux, while the article by Bieri and Geoghegan extends this theme towards a theory of group actions on non-positively curved spaces. Another exciting topic of somewhat similar geometric flavour is the theory of Kac-Moody groups, of which Remy's article gives a timely and masterly exposition, incorporating much of his own research. The paper by Chiswell, almost a monograph on its own provides a surprisingly accessible and highly readable account of the theory of Euler characteristics, an account which had been sorely missed in the literature.

The papers by Nekrashevych and Sidki and by Parker and Series both explore the fruitful connection between groups and inherent geometric structure on the one hand, and formal languages and automata on the other. In recent years, automorphisms of regular 1-rooted trees of finite valency have been the subject of vigorous research as a source of remarkable groups, whose structure reflects the recursiveness of these trees; cf. for instance the article by Sergiescu surveying the construction due to Gupta and Sidki of infinite Burnside groups (in: Group Theory from a Geometrical Viewpoint, World Scientific, 1991). This recursiveness in turn can be interpreted in terms of automata; in fact, each automorphism of the tree has a natural interpretation as input-output automaton, where the states, finite or infinite in number, are themselves automorphisms of the tree. This is the context of the article by Nekrashevych and Sidki, which presents a penetrating and original study of state-closed groups of automorphisms of the binary tree, with emphasis on m-dimensional affine groups, that is, groups of the form $\mathbb{Z}^m \cdot \mathrm{GL}(m, \mathbb{Z})$.

For a closed 2-manifold M, the mapping class group $\mathcal{M}(M)$ is defined as the group of all autohomomorphisms of M modulo the subgroup of those deformable to the identity. By a theorem of Nielsen (Acta Math. 50, 189–358), $\mathcal{M}(M)$ is isomorphic to $\mathrm{Out}(\pi_1(M))$, the group of all automorphisms of the fundamental group $\pi_1(M)$ of M modulo inner automorphisms. This important result opens up the possibility of an algebraic approach to the investigation of mapping class groups, which until recently was standard. By way of contrast, the fundamental paper by Parker and Series in this volume studies the mapping class group of the twice punctured torus Σ_2 from a topological and dynamical point of view. Their approach rests on the analogy between Fuchsian groups acting on the hyperbolic plane and the mapping class group $\mathcal{M}(\Sigma_2)$ acting on the corresponding Teichmüller space $\mathcal{T}(\Sigma_2)$. In this analogy, the boundary S^1 of the hyperbolic plane is replaced by the Thurston boundary of $\mathcal{T}(\Sigma_2)$, the space $\mathcal{PML}(\Sigma_2)$ of projective measured laminations on Σ_2, on which $\mathcal{M}(\Sigma_2)$ also acts. There is a well-known relationship between the modular group $\mathrm{PSL}(2, \mathbb{Z})$, thought of as a Fuchsian group acting on hyperbolic 2-space, and continued fractions, thought of as points of the limit set of $\mathrm{PSL}(2, \mathbb{Z})$. The celebrated Bowen-Series construction generalizes this relationship to a large class of Fuchsian groups Γ; it gives rise to a Markov map defined on the boundary at infinity (that is, the limit set $\Lambda(\Gamma)$ of Γ), generating continued fraction expansions for points in $\Lambda(\Gamma)$, whose admissible sequences simultaneously give an elegant solution to the word problem in Γ. A crucial step in the authors' program is the construction of an analogous Markov map on $\mathcal{PML}(\Sigma_2)$, which has the same relation to the action of $\mathcal{M}(\Sigma_2)$ on $\mathcal{PML}(\Sigma_2)$ as the Bowen-Series map has to the action of Γ on $\Lambda(\Gamma) = S^1$. The construction of this Markov map in turn gives rise to an explicit automatic structure on $\mathcal{M}(\Sigma_2)$ in the sense of Epstein et al. In order to illustrate their ideas and techniques, Parker and Series begin by discussing in some detail the (well-known) situation for the once punctured torus, explaining along the way relevant background material and important definitions. Having given this illuminating example, each step is then generalized to the much more demanding situation for Σ_2. While the authors concentrate on the case of the twice punctured torus, their methods in principle appear general and powerful enough to tackle even more advanced situations, and one can hope for important future research along similar lines.

Combinatorial group theory in the classical spirit of Nielsen and Magnus is the context for Collins' contribution; it combines a careful and clear introduction to Magnus' method of investigation for one-relator groups with the solution of a long-standing problem in the field. Situated in the same area, but in a rather different spirit, the paper by Grigorchuk and Wilson explores the intimate connection between atomic groups (in the sense of Pride's pre-order on groups) and just infinite groups. Their main result provides a sufficient condition for branch groups to be atomic, leading to a number of beautiful and highly non-trivial new examples of atomic groups.

In a lecture at Oberwolfach in the 1980s, Helmut Wielandt defended the proposition that topology in permutation groups is of no use, and came to the conclusion that (with the addition of the word "almost") this was indeed the case. Starting with an account of Wielandt's own work on the subject (the connection of non-Hausdorff topologies preserved by a permutation group and the concepts of primitivity and strong primitivity), Cameron's article sets out to qualify this verdict. He gives a detailed account of a spectacular theorem by Macpherson and Praeger, according to which a primitive group preserving no non-trivial topology is highly transitive. The original proof uses substantial machinery from model theory, and Cameron provides a new argument, subtle but essentially elementary, to replace part of their proof. Other topics include Peter Neumann's work on automorphism groups of filters, Mekler's theorem providing a necessary and sufficient condition for a countable permutation group to act on the rationals by homeomorphisms, and the natural topology (of pointwise convergence) on permutation groups.

The theory of subgroup growth, an exciting and fast developing part of what has become known in recent years as 'asymptotic group theory', studies the number theory of arithmetic functions counting (various types of) finite index subgroups in a group, and the connection of such arithmetic information with the algebraic structure of the underlying group. This theory has grown, over the last two decades, out of the work of Grunewald, Lubotzky, Mann, Segal, and others including the present editor, and has already given rise to a number of spectacular results, most of which are described in the recent book by Lubotzky and Segal (Birkhäuser, 2003). The papers by Mann, Müller, and Segal all deal with various aspects of this theory: the article by Avinoam Mann explores the use of probabilistic arguments to prove assertions concerning the subgroup growth of profinite groups; Müller's paper derives detailed congruences for the number of index n subgroups and the number of free subgroups of given index in Hecke groups, that is, groups of the form $C_2 * C_q$ for some $q \geq 3$. The results and methods of the latter paper have opened up a new and thriving chapter of subgroup growth theory dealing with modular properties of subgroup counting functions. Segal's article revolves around the so-called *Gap Conjecture*, which is concerned with the question whether, for finitely generated groups, there is a gap in the possible types of subgroup growth just above polynomial growth. His main result constructs counter-examples to this conjecture possessing a number of remarkable additional properties.

The three articles by Abels, Helling, and Vinberg and Kaplinsky deal with various arithmetic aspects of groups. Hyperbolic lattices in dimension 3, that is, discrete cofinite subgroups of $SL(2, \mathbb{C})$, appear to exhibit the tendency to have integer-valued character functions. Indeed, until recently, the only example in the literature of a lattice with non-integral character occurred in Vinberg's seminal 1967 paper on discrete hyperbolic groups generated by reflections. The article by Helling starts out with a discussion in geometric language of Vinberg's example, complementing Vinberg's results by determining the trace field of this lattice. Helling's main contribution is the construction of a whole series of cocompact lattices, which appear as the result of Dehn surgery along the figure eight knot, at the same time providing strong evidence that an infinite sub-series of these lattices should not admit an integer-valued character. This paper represents a major advance in the study of what appears to be a very difficult problem, and is likely to inspire further research.

By a generalized triangle group one means a group Γ with a fixed presentation of the form

$$\Gamma = \langle x, y \mid x^k = y^\ell = \big(w(x, y)\big)^m = 1\rangle,$$

where $k, \ell, m \geq 2$ and $w(x, y)$ is a word of the form

$$w(x, y) = x^{k_1} y^{\ell_1} x^{k_2} y^{\ell_2} \cdots x^{k_s} y^{\ell_s}$$

with $s \geq 1$, $0 < k_j < k$, and $0 < \ell_j < \ell$. It is also required that $w(x, y)$ is not itself a power of a shorter word. Since their introduction around 1987 by Baumslag, Morgan, Shalen and others, generalized triangle groups have been the object of intense research. One of the main tools in their study are so-called essential homomorphisms, that is, homomorphisms $\rho : \Gamma \to PSL(2, \mathbb{C})$ such that

$$|\rho(x)| = k, \quad |\rho(y)| = \ell, \quad \text{and} \quad |\rho(w(x, y))| = m.$$

Generalized triangle groups are known to admit such an essential homomorphism to $PSL(2, \mathbb{C})$, and, for most Γ, the image of ρ is in fact infinite. There are, however, infinite generalized triangle groups all of whose essential images are finite. These are the so-called pseudo-finite generalized triangle groups, whose classification appears to be difficult, and is one of the major open problems in the field. The paper by Vinberg and Kaplinsky in this volume represents a substantial advance in this direction; making use of essential homomorphisms in sufficiently large finite cyclic groups as well as methods from number theory, the authors completely cover the cases $m \geq 3$ and $s \leq 3$. Despite these impressive results however, the problem to classify pseudo-finite generalized triangle groups still has to be considered wide open; but ideas and methods employed in this paper will, it is hoped, turn out to be useful in the study of other aspects of this attractive and important problem.

Let G be a reductive group over a local field, Γ an S-arithmetic subgroup of G, and let $\mathfrak{S}$ be a fundamental domain for Γ in G (a so-called Siegel domain). In his 1959 Japan lectures on reduction theory, Carl Ludwig Siegel asked, whether, in modern terminology, the natural map $\mathfrak{S} \to \Gamma\backslash G$ is a coarse isometry. More precisely, Siegel asked this question only for the case where $G = \mathrm{SL}(n, \mathbb{R})$, $\Gamma = \mathrm{SL}(n, \mathbb{Z})$, and with respect to the pseudo-metric on G coming from the standard Riemannian metric on the symmetric space of G (the space of positive-definite real symmetric $n \times n$-matrices). Joint work of Abels and Margulis has recently led to a positive answer in full generality, for arbitrary reductive groups G, S-arithmetic subgroups Γ, and for pseudo-metrics on G which are norm-like. Here, a pseudo-metric is called norm-like, if it is coarsely isometric to a metric coming from the operator norm of a rational representation, or, equivalently, coming from a norm on a maximal split torus. This spectacular result in turn leads to the question of determining which pseudo-metrics are norm-like. The paper by Abels contains foundational material on these two topics. More specifically, it provides four descriptions of one and the same quasi-isometry class of pseudo-metrics on a reductive group G over a local field: the word metric corresponding to a compact set of generators of G, the pseudo-metric given by the isometric action of G on a metric space, the pseudo-metric coming from the operator norm for a representation of G, and the pseudo-metric given on a split torus over a local field K by valuations of the K^*-factors. As a result, one has four intriguingly different descriptions of one and the same natural and distinguished quasi-isometry class of pseudo-metrics. This paper will be indispensable for everyone wishing to study in depth the work of Abels and Margulis, as well as that of other authors, on Siegel's fundamental problem.

When setting out (rather naively) in 1998 to organize the conference which forms the background for this volume, and to edit resulting articles, little did I foresee how formidable and time consuming this task would turn out to be (a feeling no doubt shared by most inexperienced editors). The aim was to produce a stimulating book full of ideas, but also introducing sound technique, comprehensible to good students, which would open up future development in each of the various areas under discussion. All in all, the outcome, in my opinion, surpasses even these rather ambitious expectations; and, in retrospect, I find that I have learned immensely, and have thoroughly enjoyed this task. My gratitude extends to the individual authors; each of them has shared exciting ideas and important insights, but has also made a considerable effort to communicate them. Thanks are also due to Heinz Helling and Barbara Schulten for help with the organization and running of the original conference, and to Jan-Christoph Puchta for conveying to me some of his advanced TeX wisdom.

1

Reductive groups as metric spaces

H. Abels

1. Introduction

In this paper four descriptions of one and the same quasi-isometry class of pseudo-metrics on a reductive group G over a local field are given. They are as follows. The first one is the word metric corresponding to a compact set of generators of G. The second one is the pseudo-metric given by the action of G by isometries on a metric space. That these two pseudo-metrics on a group G are quasi-isometric holds in great generality. The third pseudo-metric is defined using the operator norm for a representation ρ of G. This pseudo-metric depends very much on the representation. But for a reductive group over a local field it does not up to quasi-isometry. The fourth pseudo-metric is given on a split torus over a local field K by valuations of the K^*–factors. The main result is that these four pseudo-metrics on a reductive group over a local field coincide up to quasi-isometry. We thus have four different descriptions of one and the same very natural and distinguished quasi-isometry class of pseudo-metrics.

This paper contains foundational material for joint work in progress with G. A. Margulis on the following two topics. One is work on the following question of C. L. Siegel's. Given a reductive group G over a local field and an S–arithmetic subgroup Γ of G, it is one of the main results of reduction theory to describe a fundamental domain R for Γ in G, a so called *Siegel domain*. Siegel asked in his Japan lectures [8, Sect. 10] on reduction theory of 1959, if – in our terminology, see Section 2.3 – the natural map $R \to \Gamma\backslash G$ is a coarse isometry. He asked this question only for the special case $G = SL(n, \mathbb{R})$, $\Gamma = SL(n, \mathbb{Z})$ and d the pseudo-metric on G coming from the standard Riemannian metric on the symmetric space of G, the space of positive definite real symmetric $n \times n$–matrices. We now have a positive answer in full generality, for arbitrary reductive groups G over local fields, S–arithmetic subgroups Γ and for pseudo-metrics d on G which are norm-like. We call a pseudo-metric on G *norm–like* if it is coarsely isometric to a metric coming from the operator norm of a rational representation, or, equivalently, coming from a norm on a maximal split torus, see Sections 5 and 6. This raises of course the question which pseudo-metrics are norm-like. Note that coarse isometry is a much stricter equivalence relation among pseudo-metrics than quasi-isometry. We show in this paper that the three last types of pseudo-metrics on reductive groups are norm-like. It is an open question whether the first one, namely the word metric, or, more generally (Section 3.8), any coarse path pseudo-metric, gives a norm-like pseudo-metric. In joint work in progress with G. A. Margulis we show that this is the case if G is a torus or if the rank r of a maximal split torus in the semi-simple part of G is equal to one. This is probably even true for $r = 2$.[1]

The question of Siegel has an interesting history. A first positive answer was given by Borel in [1]. It was discovered much later (see the notes to paper [1] in Borel's Oeuvres vol. IV) that the proof contains a gap. It occurs on pp. 550 – 552, (12) does not imply (14), but (14) is essential to prove (5), the main inequality. There are now proofs for Siegel's conjecture, in its original form [2] and more generally for real reductive groups G, ordinary arithmetic subgroups and the pseudo-metric d coming from the symmetric space [4, 6].

Here are some more details about our approach to Siegel's question. For the sake of exposition we restrict ourselves to the case $G = SL(n, \mathbb{R})$ and $\Gamma = SL(n, \mathbb{Z})$. Let T be the subgroup of $SL(n, \mathbb{R})$ of diagonal matrices $t = \operatorname{diag}(t_1, \ldots, t_n)$ of determinant one, a maximal $\mathbb{R}$–split torus. The *negative Weyl chamber* is by definition the subset $C^- = \{\operatorname{diag}(t_1, \ldots, t_n) \in T \mid 0 < t_1 \leq t_2 \leq \cdots \leq t_n\}$. A *Siegel set* R in $SL(n, \mathbb{R})$ is, by definition, a subset of G of the form $K \cdot C^- \cdot L$, where K and L are compact subsets of G. The main result of reduction theory for this case states that for appropriate sets K and L the Siegel set R is a set of representatives for G/Γ. So the natural map $G \to G/\Gamma$ restricts to a surjection $\pi : R \to G/\Gamma$. It has other nice properties, e.g., $\pi \mid_R$ is a proper map. The question of Siegel mentioned above asked about the metric properties of π. Let d be a right invariant pseudo-metric on G. Define a pseudo-metric $\overline{d}$ on G/Γ in the natural way, i.e., $\overline{d}(g\,\Gamma, h\,\Gamma) = \inf\{d(g\gamma, h) \mid \gamma \in \Gamma\}$. Now Siegel's question was: is $\pi : R \to G/\Gamma$ a coarse isometry? In other words: is there a constant C such that

$$\overline{d}(g\,\Gamma, h\,\Gamma) \leq d(g, h) \leq \overline{d}(g\,\Gamma, h\,\Gamma) + C$$

for every pair g, h of points of R? Siegel himself showed in [8, Sect. 10] that this is the case if we fix one variable, that is, for every $g \in G$ there is a constant

[1] Note added in proof: this is even true in general; see forthcoming joint work with G. A. Margulis.

$C = C(g)$ such that the right inequality holds for every $h \in R$. It suffices to show this for one point $g \in G$.

Here are the main steps of our proof that the answer is yes. We may assume that g and h are in the negative Weyl chamber C^- and that $d = d^{\rho}_{\mathrm{op}}$ is the metric coming from a rational representation, see Section 5. We prove that, for $\gamma \in \Gamma$,

$$d(g\gamma, h) \underset{\text{(I)}}{\to} \geq d(a(g\gamma), h) \underset{\text{(II)}}{\to} \geq d(w^{-1}g, h) \underset{\text{(III)}}{\to} \geq d(g, h)$$

up to constants, where $G = K \cdot A \cdot N$, $g = k(g) \cdot a(g) \cdot n(g)$, is the Iwasawa decomposition and $\gamma \in B\, w\, B$ in the Bruhat decomposition with w an element of the Weyl group S_{n-1}. Note that (III) is a very special property of reflection groups. It does for example not hold for g, h in the fundamental domain of a finite rotation group and w in this group. An important step in the proof of (II) is

$$\text{(II}'\text{)}\ a(g\gamma) = w^{-1} g\, w + r$$

where r is up to a compact error term the exponential of a positive linear combination of $\Sigma_{w^{-1}}$ where $\Sigma_{w^{-1}} = \{\alpha \in \Phi^+ \mid w^{-1}\alpha\, w \in \Phi^+\}$ and Φ^+ is the set of positive roots. That we found (II′) is due to discussions with Alex Eskin who showed us a geometric picture of this fact.

Let us point out the following features of this proof. It is different from both Ding's [2] which is by induction on n, and from Leuzinger's [6] which uses Tits buildings and facts about the geometry of symmetric and locally symmetric spaces, in particular their geometry at infinity. Our proof works in full generality, for arbitrary local fields and arbitrary S–arithmetic subgroups. Also, we admit arbitrary norm-like metrics, not only those coming from the symmetric space or the Bruhat–Tits building. Finally it gives further information concerning reduction theory, namely the inequalities stated above.

2. Metrics

We first need to recall some concepts concerning metric spaces.

2.1. Let X be a set. A function $d : X \times X \to \mathbb{R}$ is called a *pseudo-metric* (on X) if d is non-negative, zero on the diagonal, symmetric and fulfills the triangle inequality, i.e., if

$$d(x, y) \geq 0 \text{ for every } x, y \text{ in } X$$
$$d(x, x) = 0 \text{ for every } x \text{ in } X$$
$$d(x, y) = d(y, x) \text{ for every } x, y \text{ in } X$$
$$d(x, y) + d(y, z) \geq d(x, z) \text{ for every } x, y, z \text{ in } X.$$

So a pseudo-metric on X is a *metric* on X if and only if $d(x, y) = 0$ implies

$x = y$. A pair (X, d) consisting of a set X and a (pseudo-) metric d on X is called a *(pseudo-) metric space*. In a pseudo-metric space (X, d) the ball of radius r with center x is denoted $B_d(x, r)$ or $B(x, r)$. So

$$B_d(x, r) = \{y \in X : d(x, y) \leq r\}.$$

2.2. Let (X, d) and (X', d') be pseudo-metric spaces. A map $f : X \to X'$ is called a *quasi-isometry* if there are real numbers $C_1 > 0$ and C_2 such that

$$C_1^{-1} \cdot d(x, y) - C_2 \leq d'(f(x), f(y)) \leq C_1 \cdot d(x, y) + C_2$$

and $X' = \bigcup_{x \in X} B_{d'}(f(x), C_2)$. Thus, for every point $x' \in X'$ there is a point $x \in X$ such that $d(x', f(x)) \leq C_2$. Define a map $g : X' \to X$ by choosing for every $x' \in X$ a point $x = g(x')$ with this property. Then $g : X' \to X$ is a quasi-isometry, actually with the same multiplicative constant C_1, and we have $d(x, gf(x)) \leq C_2$ and $d'(x', fg(x')) \leq C_2$ for every $x \in X$ and $x' \in X'$.

2.3. A map $f : X \to X'$ between pseudo-metric spaces (X, d) and (X', d') is called a *coarse isometry* if f is a quasi-isometry and the multiplicative constant C_1 can be chosen to equal 1. Equivalently, the function $(x, y) \mapsto d'(f(x), f(y)) - d(x, y)$ is bounded on $X \times X$ and every point of X' is at bounded distance from $f(X)$. Finally, $f : X \to X'$ is called an *isometry* if both these bounds are zero, i.e., if f is surjective and $d'(f(x), f(y)) = d(x, y)$ for every x, y in X. If f is a (coarse) isometry, then so is any map $g : X' \to X$ considered above. It follows that if there is a (quasi-, coarse) isometry from X to X' then there is one from X' to X. Two pseudo-metrics on the same set are called *(quasi-, coarsely) isometric* if the identity map is a (quasi-, coarse) isometry. It follows that these relations are equivalence relations between pseudo-metric spaces and also between pseudo-metrics on the same set.

2.4. We will mainly be interested in pseudo-metrics on groups. So let G be a group. A pseudo-metric d on G will be called *left invariant (right invariant)* if every left translation (right translation) is an isometry. So d is left invariant on G if and only if $d(gh_1, gh_2) = d(h_1, h_2)$ for every g, h_1, h_2 in G. Define a function f on G by $f(g) = d(e, g)$. If d is a left (right) invariant pseudo-metric on G, then f is non-negative, zero at the identity element, symmetric and fulfills the triangle inequality, i.e.,

$$\begin{aligned} f(g) &\geq 0 && \text{for every } g \in G, \\ f(e) &= 0 && \text{for the identity element } e, \\ f(g) &= f(g^{-1}) && \text{for every } g \in G \text{ and} \\ f(gh) &\leq f(g) + f(h) && \text{for every } g, h \text{ in } G. \end{aligned}$$

Conversely, given a function f with these properties then $d(g, h) := f(g^{-1}h)$, resp. $d(g, h) = f(hg^{-1})$, defines the unique left (right) invariant pseudo-metric d on G such that $d(e, g) = f(g)$ for every $g \in G$. A function f on G with these properties is sometimes called a norm on G. But we want to reserve the term "norm" for a more special situation.

3. The word metric

Let G be a group and let Σ be a set of generators of G. Then the *word length* $\ell_\Sigma(g)$ of an element $g \in G$ with respect to Σ is defined as

$$\ell_\Sigma(g) = \inf\left\{r : g = a_1^{\varepsilon_1} \dots a_r^{\varepsilon_r},\ a_i \in \Sigma,\ \varepsilon_i \in \{+1, -1\}\right\}.$$

The function ℓ_Σ has the properties stated above and furthermore $\ell_\Sigma(g) = 0$ implies $g = e$. So $d_\Sigma(g, h) := \ell_\Sigma(g^{-1}h)$ defines a left invariant metric d_Σ on G, which is called the *word metric* associated with Σ. The ball of radius r with center e is

$$B_{d_\Sigma}(e, r) = (\Sigma \cup \Sigma^{-1})^r = \left\{a_1^{\varepsilon_1} \dots a_r^{\varepsilon_r} : a_i \in \Sigma\ ,\ \varepsilon_i \in \{+1, -1\}\right\},$$

and thus consists of all words of length at most r with respect to the alphabet $\Sigma \cup \Sigma^{-1}$. The word metric d_Σ depends of course on Σ. But if Σ and Σ' are both *finite* sets of generators of G then d_Σ and $d_{\Sigma'}$ are quasi-isometric, since if $\ell_\Sigma(\Sigma')$ is bounded by C_1 then $d_\Sigma \leq C_1 \cdot d_{\Sigma'}$. Similarly:

3.1. Lemma. *Let G be a locally compact topological group and let Σ and Σ' be compact sets of generators of G. Then the word metrics d_Σ and $d_{\Sigma'}$ on G are quasi-isometric. They are actually Lipschitz equivalent, that is, the additive constant C_2 in the definition of quasi-isometry may be chosen equal to zero.*

By the preceding argument it suffices to show the following.

3.2. Lemma. *Let G be a locally compact topological group and let Σ be a compact set of generators of G. Then every compact subset of G has bounded word length ℓ_Σ.*

Proof. The sequence of compact subsets $A_n = B_{d_\Sigma}(e, n) = (\Sigma \cup \Sigma^{-1})^n$ of G covers the locally compact space G. So one of them contains a non-empty open subset U of G by the Baire category theorem, say $U \subset A_n$. Then A_{2n} is a neighbourhood of the identity element e, since A_{2n} contains $U \cdot U^{-1}$. If now K is a compact subset of G there is a finite subset M of K such that $M \cdot A_{2n}$ contains K. Thus $\ell_\Sigma(K) \leq \ell_\Sigma(M) + 2n$. □

3.3. Remark. Both Lemmas 3.1 and 3.2 remain true if Σ and Σ' are relatively compact sets of generators of G which contain a non-empty open subset of G, as follows from the second part of the proof of Lemma 3.2. But Lemma 3.2, and hence Lemma 3.1, is not true for an arbitrary relatively compact set of generators of G; e.g., let G be the additive group $\mathbb{R}$. The word length $\ell_{\Sigma'}$ corresponding to the set of generators $\Sigma' = [0, 1]$ is $\ell_{\Sigma'}(x) = \lceil |x| \rceil$, the smallest integer $\geq |x|$. Consider the following set of generators Σ. There is a basis B of the $\mathbb{Q}$–vector space $\mathbb{R}$ such that $B \subset [0, 1]$ and B contains for every $n \in \mathbb{N}$ an element b_n with $0 \leq b_n \leq \frac{1}{n}$. Such a basis can be obtained from a given basis of $\mathbb{R}$ over $\mathbb{Q}$ by multiplying every basis element with an appropriate rational number. Put $\Sigma = \{q \cdot b : \ b \in B, \ q \in \mathbb{Q} \cap [0, 1]\} \subset [0, 1]$. Then Σ is a set of generators of $\mathbb{R}$, contained in $[0, 1]$ but ℓ_Σ is unbounded on $[0, 1]$, since $\ell_\Sigma(n\, b_n) = n$. In fact, for every real number $x = \sum_{b \in B} q_b \cdot b$ with $q_b \in \mathbb{Q}$, we have $\ell_\Sigma(x) = \sum_{b \in B} \lceil |q_b| \rceil$.

Here is a geometric approach to the word metric.

3.4. Definition. A pseudo-metric d on a set X is called a *coarse path pseudo-metric* if there is a real number C such that for every pair of points x, y in X there is a sequence $x = x_0, x_1, \ldots, x_t = y$ for which $d(x_{i-1}, x_i) \leq C$ for $i = 1, \ldots, t$ and

$$d(x, y) \geq \sum_{i=1}^{t} d(x_{i-1}, x_i) - C.$$

In other words, the triangle inequality $d(x, y) \leq \sum_{i=1}^{t} d(x_{i-1}, x_i)$ is in fact an equality up to a bounded error.

3.5. A left invariant pseudo-metric d on a group G is a coarse path pseudo-metric if and only if the function f with $f(g) = d(e, g)$ has the following property. There is a real number C such that for every $g \in G$ there is a sequence $g_1, \ldots, g_t$ of elements of G such that $g = g_1 \cdot \cdots \cdot g_t$, $f(g_i) \leq C$ for $i = 1, \ldots, t$ and $f(g) \geq \sum_{i=1}^{t} f(g_i) - C$. The equivalence is seen as follows. Starting with $g \in G$ take a sequence $x_0 = e, x_1, \ldots, x_t = g$ as above and put $g_i = x_{i-1}^{-1} \cdot x_i$. Conversely, for x, y in G take a sequence $g_1, \ldots, g_t$ as above for $g = x^{-1} y$ and put $x_i = x \cdot g_1 \cdot \cdots \cdot g_i$.

3.6. Example. A word metric d_Σ on a group is a coarse path metric, since by definition $C = 1$, $B(e, 1) = \Sigma \cup \Sigma^{-1} \cup \{e\}$ and the error in the triangle inequality is zero with notation as in 3.4.

3.7. One can generalize this example as follows. Given a set of generators Σ of G and a bounded function $\omega : \Sigma \to [0, \infty)$ on Σ we can define a *weighted* *word length* on G by

$$\ell_{\Sigma,\omega}(g) = \inf \left\{ \sum_{i=1}^{t} \omega(g_i) : \ t \in \mathbb{N} \cup \{0\}, \ g = g_1^{\varepsilon_1} \dots g_t^{\varepsilon_t}, \right.$$

$$\left. g_i \in \Sigma, \ \varepsilon_i \in \{+1, -1\} \right\}.$$

Then $\ell_{\Sigma,\omega}$ has all the properties of 2.4 so that $d_{\Sigma,\omega}(g, h) := \ell_{\Sigma,\omega}(g^{-1}h)$ defines a left invariant pseudo-metric on G which is in fact a coarse path pseudo-metric, as is readily seen. Furthermore, $d_{\Sigma,\omega}$ is the supremum of the pseudo-metrics d on X with the property that $d(e, g) \leq \omega(g)$ for $g \in \Sigma$.

3.8. The importance of this generalization lies in the following fact: every left invariant coarse path pseudo-metric is a weighted word pseudo-metric up to coarse isometry. More precisely, let d be a left invariant coarse path pseudo-metric on G and let C be as in 3.4. Then $\Sigma := B_d(e, C)$ is a set of generators of G and we have

$$d_{\Sigma,\omega} - C \leq d \leq d_{\Sigma,\omega},$$

where $\omega : \Sigma \to [0, C]$ is defined by $\omega(g) = d(e, g)$ for $g \in \Sigma$.

3.9. Note that a metric d' that is coarsely isometric to a coarse path metric d need not be a coarse path metric itself; e.g., on $G = \mathbb{Z}$ the metric

$$d'(x, y) = \begin{cases} 0 & \text{if } x = y \\ |x - y| + 1 & \text{if } x \neq y \end{cases}$$

is left invariant and coarsely isometric to the Euclidean metric on $\mathbb{Z}$ which is a left invariant coarse path metric, in fact the word metric for the set of generators $\Sigma = \{1\}$. But d' is not a coarse path metric. For a given $C > 0$ the error term $d_{\Sigma,\omega}(e, n) - d'(e, n)$ grows linearly in $|n|$, where $\Sigma = B(0, C)$ and $\omega(m) = d'(0, m)$. If we consider coarse path pseudo-metrics up to quasi-isometry only, we do not need a weight function ω by the following lemma.

3.10. Lemma. *Let d be a left invariant coarse path pseudo-metric on a group G. Then d is quasi-isometric to a word metric, namely to d_Σ with $\Sigma = B_d(e, C)$, where C is as in* 3.4.

Proof. Let C be a constant as in Definition 3.4, put $f(g) = d(e, g)$ and $\Sigma = B(e, C)$. Then for every $g \in G$ there are $g_1 \dots g_t \in \Sigma$ such that $g = g_1 \dots g_t$ and

$$f(g) \geq \sum_{i=1}^{t} f(g_i) - C. \qquad (*)$$

We may assume that for every $i = 1, \ldots, t-1$ we have $f(g_i) + f(g_{i+1}) > C$. Since if this is not the case we combine successive factors $g_i, g_{i+1}, \ldots, g_j$ into one factor $g_i \cdot \cdots \cdot g_j$ such that $f(g_i \cdot \cdots \cdot g_j) \leq C$ but $f(g_i \cdot \cdots \cdot g_j \cdot g_{j+1}) > C$, starting with $i = 1$. This does not destroy the validity of $(*)$ by the triangle inequality. Then

$$f(g) \geq \frac{t-1}{2} \cdot C - C.$$

This together with the obvious inequality $t \geq \ell_\Sigma(g)$ shows one of the inequalities of the desired quasi-isometry. The other one is seen as follows. The group G is generated by $\Sigma = B(e, C)$. Let $s = \ell_\Sigma(g)$, $g = g_1 \cdot \cdots \cdot g_s$, $g_i \in \Sigma$. Then $f(g) \leq \sum_{i=1}^{s} f(g_i) \leq C \cdot \ell_\Sigma(g)$. □

As a corollary we obtain the following uniqueness result.

3.11. Proposition. *Let G be a locally compact topological group. Let d and d' be two left invariant coarse path pseudo-metrics on G with the following two properties:*

(C) *Compact sets have bounded diameter.*
(P) *Balls of bounded radius are relatively compact.*

Then d and d' are quasi-isometric. Furthermore, if such a pseudo-metric d exists on G, then G has a compact set Σ of generators and d is quasi-isometric to the corresponding word metric d_Σ.

The letter C alludes to "compact" or "continuous". Note that C holds if d is continuous, but that the pseudo-metrics we consider need not be continuous, e.g., d_Σ is not, in general. The letter P alludes to "proper" since (P) holds if and only if inverse images of compact sets have compact closure for the map $x \mapsto d(e, x)$ (or for the map $x \mapsto d(y, x)$ for some (any) point $y \in G$).

Proof. d is quasi-isometric to the word metric d_Σ with $\Sigma = B_d(e, C)$, where C is as in Definition 3.4 for d. Then Σ is relatively compact by (P) for d. Similarly for d' which is quasi-isometric to $d_{\Sigma'}$, with Σ' relatively compact. Then Σ' is of bounded diameter for d, by (C), hence for the quasi-isometric word metric d_Σ, too. Thus $\ell_\Sigma(\Sigma') \leq C_1$, say, which implies $d_\Sigma \leq C_1 \cdot d_{\Sigma'}$ and similarly for the converse. Finally, Σ is a relatively compact set of generators for G, so G has a compact set Σ'' of generators. It follows as above that $d_\Sigma \leq C_2 \cdot d_{\Sigma''}$ for some $C_2 > 0$ and the converse $d_{\Sigma''} \leq C_3 d_\Sigma$ for some $C_3 > 0$ by the Baire category argument of the proof of 3.2. □

4. A geometric pseudo-metric

4.1. Let (X, d) be a pseudo-metric space and let the group G act on X by isometries. Let x_0 be a point of X. Then

$$d_{X,x_0}(g, h) := d(g\,x_0, h\,x_0) \tag{4.1}$$

defines a left invariant pseudo-metric d_{X,x_0} on G. For another point $x_1 \in X$ the pseudo-metrics d_{X,x_0} and d_{X,x_1} on G are coarsely isometric, see 2.3. There are many examples of this type. Here are two of them. Another one is 4.5.

4.2. Let G be a connected Lie group. There is a left invariant Riemannian tensor on G which gives rise to a left invariant path metric d_{Riem} on G. Any two such metrics d_{Riem} are quasi-isometric, in fact Lipschitz equivalent, since any two norms on the finite dimensional real vector space T_eG are equivalent. This metric on G can be regarded as an example of a geometric pseudo-metric as above, if we let G act on (G, d_{Riem}) by left translations.

4.3. Let G be a Lie group, which is connected or, more generally, has a finite group of connected components, and let K be a maximal compact subgroup of G. Then the homogeneous space $X = G/K$ carries a Riemannian tensor invariant against the action of G on X. Thus, for the corresponding path metric d_X on X the group G acts by isometries. Again, d_X is unique up to Lipschitz equivalence. Hence the corresponding pseudo-metrics d_{X,x_0} on G are unique up to Lipschitz equivalence if we fix x_0, e.g. $x_0 = e \cdot K$, and are unique up to quasi-isometry for arbitrary $x_0 \in X$.

We ask if the pseudo-metrics 4.2 and 4.3 are quasi-isometric to each other and quasi-isometric to the word metric for a compact set of generators. The answer is yes by the following proposition in view of Lemma 3.1.

4.4. Proposition. *Let (X, d) be a locally compact space X with a coarse path pseudo-metric d having the properties (C) and (P) of 3.11. Suppose the locally compact group G acts properly on X by isometries such that $G\backslash X$ is compact. Then G has a compact set Σ of generators and the pseudo-metric d_{X,x_0} on G is quasi-isometric to the word metric d_Σ. It follows that any two pseudo-metrics of the form d_{X,x_0} for spaces X as above are quasi-isometric. (Recall that an action of a locally compact group G on a locally compact space X is called proper if for every compact subset K of X the subset $\{g \in G;\ g\,K \cap K \neq \emptyset\}$ of G is compact.)*

Proof. Let K be a compact subset of X such that $X = GK$. We may assume that $x_0 \in K$. Let D be the diameter of K. Then every point of X is of distance $\leq D$ from some point of the orbit $G\,x_0$. Thus the embedding of the orbit $G\,x_0$

into X is a coarse isometry from $(G\,x_0, d\,|_{G\,x_0})$ to (X, d). It follows that the G–map $G \to X$, $g \mapsto g\,x_0$, is a coarse isometry since it is the composition of the isometry $G \to G\,x_0$, $g \mapsto g\,x_0$, and the coarse isometry $G\,x_0 \to X$. Note that the pseudo-metric d_{X,x_0} on G also has the properties (C) and (P). To see (P) use that the action of G on X is proper. Let C be a constant as in the Definition 3.4 of a coarse path pseudo-metric. Let $\Sigma = B(e, C + 2D)$ with respect to the pseudo-metric d_{X,x_0} on G. Then Σ is relatively compact. We may assume that Σ is a neighbourhood of e by taking a larger constant C or D if necessary. We claim that Σ generates G and that d_{X,x_0} and d_Σ are quasi-isometric which implies our claim in view of Remark 3.3. To prove this let $g \in G$. There is a coarse path $x_0, x_1, \ldots, g\,x_0 = x_t$ such that $d(x_{i-1}, x_i) \le C$ for $i = 1, \ldots, t$ and

$$d(x_0, g\,x_0) \ge \sum_{i=1}^{t} d\,(x_{i-1}, x_i) - C.$$

For every $i = 0, \ldots, t$ there is an element $g_i \in G$ such that $d(x_i, g_i\,x_0) \le D$. Here we put $g_0 = e$ and $g_t = g$. Then $d_{X,x_0}(g_{i-1}, g_i) = d(g_{i-1}x_0, g_i x_0) \le C + 2D$, so $g_{i-1}^{-1} \cdot g_i \in \Sigma$ and hence g is in the group generated by Σ. Furthermore, we may assume that in our coarse path we have $d(x_{i-1}, x_i) + d(x_i, x_{i+1}) > C$ for $i = 1, \ldots, t-1$, since otherwise we leave out some points of our coarse path. It follows that

$$d_{X,x_0}(e, g) \ge \sum_{i=1}^{t} d(x_{i-1}, x_i) - C \ge \frac{t-1}{2} \cdot C - C$$

and thus

$$d_\Sigma(e, g) \le t \le \frac{2}{C} d_{X,x_0}(e, g) + 3.$$

The inverse inequality is easy to see, as follows. Let $g = g_1 \ldots g_t$, $g_i \in \Sigma$ with $t = \ell_\Sigma(g)$. Note that $\Sigma = \Sigma^{-1}$, so we can avoid factors of the form g_i^{-1}. Then

$$d_{X,x_0}(e, g) = d(x_0, g\,x_0) \le \sum_{i=1}^{t} d(h_{i-1}x_0, h_i x_0)$$

where $h_i = g_1 \ldots g_i$. Thus $h_t = g$, $h_0 = e$ and $h_{i-1}^{-1} h_i = g_i \in \Sigma$ and hence

$$d(h_{i-1}x_0, h_i\,x_0) = d(x_0, h_{i-1}^{-1} h_i\,x_0) = d(x_0, g_i\,x_0) \le C + 2D$$

which implies $d_{X,x_0}(e, g) \le (C + 2D)t = (C + 2D)\ell_\Sigma(g) = (C + 2D) d_\Sigma(e, g)$. □

4.5. Example. For a non-archimedian local field K the group $G = \underline{G}(K)$ of K–points of a simple algebraic group $\underline{G}$ defined over K acts on the corresponding

Bruhat–Tits building X. There is a G–invariant metric d on X for which every apartment of X is isometric to a Euclidean space. We thus obtain a geometric metric d_{X,x_0} on G, unique up to quasi-isometry, and quasi-isometric to the word metric d_Σ for a compact set Σ of generators of G, by 4.4, since the action of G on X is proper and $G \setminus X$ is compact.

5. The operator norm

For a local field K we define a function on the group $GL_n(K)$ by

$$|g|_{\mathrm{op}} = \sup(|\log \|g\|\,|\,,\,|\log \|g^{-1}\|\,|),$$

where $\|\cdot\|$ is the operator norm associated with the normed vector space K^n. According to 2.4 the function $|\cdot|_{\mathrm{op}}$ gives a left invariant pseudo-metric $d_{\mathrm{op}}(g,h) = |g^{-1}h|_{\mathrm{op}}$ on $GL_n(K)$, which is unique up to coarse isometry since any two norms on K^n are Lipschitz equivalent.

We thus can endow every closed subgroup of $GL_n(K)$ with a left invariant pseudo-metric which is unique up to quasi-isometry. But note that if G is a locally compact topological group and ρ_1 and ρ_2 are two faithful representations of G into groups $GL_n(K)$ as above with closed image, then the two induced pseudo-metrics $d_{\mathrm{op}}^{\rho_i}$ defined by $d_{\mathrm{op}}^{\rho_i}(g,h) := d_{\mathrm{op}}(\rho_i(g), \rho_i(h))$, $g, h \in G$, on G need not be quasi-isometric; e.g., for $G = \mathbb{R}$ let $\rho_1 : \mathbb{R} \to GL_1(\mathbb{R})$, $\rho_1(t) = e^t$, and $\rho_2 : \mathbb{R} \to GL_2(\mathbb{R})$, $\rho_2(t) = \left(\begin{smallmatrix} 1 & t \\ 0 & 1 \end{smallmatrix}\right)$. Then $d_{\mathrm{op}}^{\rho_1}$ is coarsely isometric to the Euclidean metric on $\mathbb{R}$ but $d_{\mathrm{op}}^{\rho_2}(e,t)$ is coarsely isometric to $\log(1+|t|)$.

If G is the group $\underline{G}(K)$ of K–points of a reductive K–group $\underline{G}$, though, then the operator pseudo-metric d_{op}^{ρ} does not depend on ρ up to quasi-isometry, as will follow from our uniqueness result 6.6, which we will prove after yet another description of the quasi-isometry class of the pseudo-metrics we are interested in.

6. The norm on a torus

Let K be a local field, i.e., a locally compact non-discrete topological field. There is a proper continuous homomorphism

$$v : K^* \to \mathbb{R}$$

from the multiplicative group K^* of non-zero elements of K to the additive group $\mathbb{R}$, given by the logarithm of the valuation. Every continuous homomorphism from K^* to $\mathbb{R}$ is a multiple of v. The image of v is closed and co-compact in $\mathbb{R}$, namely equal to $\mathbb{R}$ if the field K is archimedean and a non-trivial cyclic subgroup of $\mathbb{R}$, if K is non-archimedean. More generally, let T be a topological group which is isomorphic to a direct product of n groups K_i^*, where K_i

are – possibly different – local fields. Then there is a proper continuous homomorphism

$$v : T \to \mathbb{R}^n,$$

namely $v(x_1, \dots, x_n) = (v_1(x_1), \dots, v_n(x_n))$, where $v_i : K_i^* \to \mathbb{R}$ are as above. Every other proper continuous homomorphism $v' : T \to \mathbb{R}^n$ is of the form $v' = \alpha \circ v$ with $\alpha \in GL(n, \mathbb{R})$. Its image is closed and co-compact in $\mathbb{R}^n$. If $\|\cdot\|$ is a norm on the $\mathbb{R}$–vector space $V = \mathbb{R}^n$ we obtain an invariant pseudo-metric d_{norm} on T by putting

$$d_{\text{norm}}(t_1, t_2) = \|v(t_1) - v(t_2)\| = \|v(t_1 t_2^{-1})\|$$

with corresponding function f given by $f(t) = d_{\text{norm}}(e, t) = \|v(t)\|$.

6.1. Definition. Any function f on T obtained in this way will be called a *norm* on T.

Here is a handy characterization of norms.

6.2. Proposition. *Let G be a topological group isomorphic to a direct product of a finite number of K_i^*, K_i local fields. Let d be an invariant pseudo-metric on G and let f be the corresponding function. Then d is a norm on G if and only if f has the properties (C) and (P) and*

$$f(g^2) = 2f(g) \qquad ((*)_2)$$

holds for every $g \in G$.

We will prove a slightly more general result in a moment, namely 6.3 (2).

One may ask which locally compact topological groups G admit such a function f which has the properties above. The answer is, these G are as close to split tori as one can expect. The precise answer is as follows.

6.3. Proposition. *Let G be a locally compact topological group. Suppose there is a left invariant pseudo-metric d on G such that the corresponding function f has the properties (C), (P) and $(*)_2$. Then*

(1) *G contains a unique largest – and hence normal – compact subgroup K and G/K is isomorphic to a group $V \times D$, where V is a finite-dimensional real vector space and D is a discrete torsion free abelian group.*
(2) *If G has a compact set of generators then f has the following form:*

$$f(g) = \|\pi(g)\|$$

for every $g \in G$, where $\pi : G \to G/K$ is the natural projection and $\|\cdot\|$ is a norm on the finite-dimensional real vector space $(G/K) \otimes \mathbb{R} \cong V \oplus (D \otimes \mathbb{R})$. The norm $\|\cdot\|$ on $V \oplus (D \otimes \mathbb{R})$ is uniquely determined by f.

We are thus lead to the following definition. This definition of a norm coincides with 6.1 if G is a direct product of a finite number of K_i^*, K_i a local field, by 6.2.

6.4. Definition. Let G be a locally compact topological group. A *norm* on G is a function $f : G \to \mathbb{R}$ with the following properties:

- f is non-negative, zero at the identity element, symmetric and fulfills the triangle inequality (see 2.4);
- (C) f is bounded on compact sets;
- (P) Sets on which f is bounded have compact closure;
- $(*)_2$ $\quad f(g^2) = 2f(g)$.

6.5. Corollary. *Let G be a locally compact topological group having a compact set of generators.*

a) *Any two norms on G are quasi-isometric, in fact Lipschitz equivalent.*
b) *Two norms on G are coarsely isometric if and only if they coincide.*

One may ask if a norm on G gives a coarse path pseudo-metric. This is true if G/K is a vector space but is not true otherwise. For a complete answer see the next section.

Proof of Proposition 6.3 and Corollary 6.5. Let $f : G \to \mathbb{R}$ have the properties of Proposition 6.3. Using Definition 6.4 we call f a norm on G. We first show that

$$f(g^n) = |n| f(g) \qquad ((*)_n)$$

for every $n \in \mathbb{Z}$ and $g \in G$. We may assume that $n \in \mathbb{N}$ since f is symmetric, i.e., $f(g) = f(g^{-1})$. We have $f(g^n) \leq nf(g)$ by the triangle inequality and $f(g^{2^m}) = 2^m f(g)$ by $(*)_2$. Hence if $n \leq 2^m$, say $n = 2^m - \ell$, we obtain $f(g^n) = f(g^{2^m} \cdot g^{-\ell}) \geq f(g^{2^m}) - f(g^\ell) \geq 2^m f(g) - \ell f(g) = nf(g)$, the converse inequality.

Next we show that the set $K = \{x \in G; f(x) = 0\}$ is the largest compact subgroup of G. First of all, K is a subgroup of G, by the triangle inequality, symmetry of f and since $f(e) = 0$. And K has compact closure, by (P). So $\overline{K}$ is a compact subgroup of G. On the other hand, if x is an element of G for which the cyclic subgroup $< x >$ generated by x has compact closure, then $f(x) = 0$, by $(*)_n$ and (C). Thus every compact subgroup of G is contained in K.

It follows that K is a normal subgroup of G and G/K has no non-trivial compact subgroup. Furthermore f is constant on the cosets of $G \mod K$ by the triangle inequality and hence induces a function $\overline{f} : G/K \to \mathbb{R}$, which is again a norm and has the additional property that

$$\overline{f}(x) = 0 \text{ implies } x = e,$$

equivalently, that the associated pseudo-metric is actually a metric. We now claim that

$$f(xyx^{-1}) = f(y) \text{ for every } x, y \text{ in } G.$$

We have

$$f(xyx^{-1}) \leq 2f(x) + f(y)$$

by the triangle inequality. Applying this to y^n instead of y we obtain for $n \in \mathbb{N}$:

$$\begin{aligned} nf(xyx^{-1}) &= f((xyx^{-1})^n) = f(xy^nx^{-1}) \leq 2f(x) + f(y^n) \\ &= 2f(x) + nf(y). \end{aligned}$$

Dividing by n and letting $n \to \infty$ yields

$$f(xyx^{-1}) \leq f(y)$$

for every $x, y \in G$ which of course implies our claim.

It follows from (P) that the group G has the following property, called $[FC]^-$. For every element $y \in G$ the conjugacy class $\{xyx^{-1}; x \in G\}$ has compact closure. The structure of these groups is known: if G has no non-trivial compact subgroup then G is isomorphic to $V \times D$, where V is a finite dimensional real vector space and D is a discrete torsion free abelian group, see [3, Theorem 3.16].

It thus remains to show part (2) of 6.3. Suppose that G has a compact set of generators. Then D is finitely generated since it is the image of G under a continuous homomorphism. So D, being torsion free abelian, is a lattice in the finite dimensional $\mathbb{R}$–vector space $D \otimes \mathbb{R}$. To prove (2) we may assume that $K = \{e\}$ by what we proved already and hence that $G = V \times D$. We thus have to show that $f : V \times D \to \mathbb{R}$ is the restriction of a (unique) norm on $W := V \oplus (D \otimes \mathbb{R})$. Recall that f has the following properties:

$$f(a) \geq 0 \qquad \text{for every } a \in G, \tag{1}$$

$$f(a) = 0 \qquad \text{iff } a = 0, \tag{2}$$

$$f(a+b) \leq f(a) + f(b) \quad \text{for } a, b \in G, \tag{3}$$

$$f(na) = |n| f(a) \qquad \text{for } a \in G \text{ and } n \in \mathbb{Z}. \tag{4}$$

f has the properties (C) and (P). Here we have written the group law in G as addition and correspondingly the identity element as 0.

Let C be a compact convex subset of W such that $C = -C$ and $C \cap G$ contains a basis B of W. Then $f(C \cap G)$ is bounded by (C), say $f(C \cap G) \subset [0, \rho]$. It follows from (3) and (4) that

$$f(a) \leq \rho \cdot \ell_B(a) + C_1$$

for every $a \in G$, where ℓ_B is the following norm on the real vector space W

$$\ell_B\left(\sum_{b \in B} \alpha_b b\right) = \sum_{b \in B} |\alpha_b|,$$

and $C_1 = \frac{1}{2}\rho \dim W + \sup\{f(g), g \in G \cap \frac{1}{2} \dim W \cdot C\}$. Using the homogeneity (4) we can actually conclude that

$$f(a) \leq \rho \cdot \ell_B(a), \tag{5}$$

since we have $f(na) \leq \rho\, \ell_B(na) +$ constant for every $n \in \mathbb{N}$ and both f and ℓ_B fulfill (4). The function f extends to a unique function f_1 on $G \otimes \mathbb{Q}$ with the property (4). Then f_1 has the properties (1) – (5). In the next step f_1 extends to a unique continuous function F on W, by (5) for f_1. It follows that F has the properties (1) and (3) – (5). It remains to show (2) for F. Let C be a compact subset of W such that $W = C + G$, e.g. our compact convex subset above will do. If $F(w) = 0$ for $w \neq 0$ in W, then F is bounded on $\mathbb{R}w + C$, by (3) and since F is continuous, but $\mathbb{R}w + C$ intersects G in an unbounded set, which contradicts (P) for f. This proves 6.3. Concerning 6.5, a) follows from the fact that any two norms on a finite dimensional real vector space are equivalent, b) is clear in view of positive homogeneity. □

The main result of this paper is the following theorem, which states that all the pseudo-metrics we considered are quasi-isometric for a reductive group G over a local field.

6.6. Theorem. *Let $\underline{G}$ be a reductive group over a local field k and let $G = \underline{G}_k$.*

(1) *Then the following pseudo-metrics on G are quasi-isometric:*
 - *d_Σ for any compact set Σ of generators of G,*
 - *d_{geom} for the symmetric space G/K if k is archimedian or the Bruhat–Tits building of G if k is non–archimedian,*
 - *d_{op} for any faithful representation of $\underline{G}$ defined over k.*

(2) *Any invariant pseudo-metric d on G is uniquely determined up to coarse isometry by its restriction to $T = \underline{T}_k$ where $\underline{T}$ is a maximal k-split torus in $\underline{G}$.*

(3) *Finally, $d\,|_T$ is quasi-isometric to any norm on T for any of the pseudo-metrics on G considered in* (1).

The theorem holds more generally for any pseudo-metric d_{geom} on G which comes from an action of G as in Proposition 4.4.

Proof. The pseudo-metrics of the types d_Σ and d_{geom} are quasi-isometric among themselves and to each other by 3.1 and 4.4. We have a Cartan decomposition of G: there is a compact subset K of G such that $G = K \cdot T \cdot K$. This implies (2), since all the pseudo-metrics we consider have the property (C), and by the way also (P) of 3.11. Also, all our claims are trivial or easily seen, if $\underline{G}$ is a torus. It thus remains to prove that $d_G\,|_T$ and d_T are quasi-isometric for the pseudo-metrics d_G and d_T on G and T, respectively, of the three types considered in (1). This is obvious for d_{op}. For d_{geom} it follows from the following facts: for the archimedean case the orbit of a maximal $\mathbb{R}$–split torus is a totally geodesic flat sub-manifold of the symmetric space G/K (see [7, § 5]). For the non–archimedean case, every minimal gallery in the Bruhat–Tits building is contained in one apartment. Finally, for the word metric d_Σ corresponding to a compact set Σ of generators of G we have

$$d_\Sigma\,|_T \leq \frac{1}{r} d_\Theta,$$

if $r \in \mathbb{N}$ is such that $(\Sigma \cup \Sigma^{-1})^r$ contains a compact set Θ of generators of T. On the other hand, Σ is bounded for the operator pseudo-metric d_{op} on G (for a given faithful representation of $\underline{G}$ defined over k), say $d_{\text{op}}(e, \Sigma) \subset [0, R]$ hence $d_{\text{op}}(e, x) \leq C^{-1} d_\Sigma(e, x)$ for every $x \in G$. The quasi-isometry of $d_{\text{op}}\,|_T$ with d_Θ now implies a converse inequality between d_Θ and $d_\Sigma\,|_T$. Thus d_Θ and $d_\Sigma\,|_T$ are quasi-isometric. □

So Theorem 6.6 says that all the interesting pseudo-metrics on G are quasi-isometric. And the proof was not difficult. It is quite a different matter to determine which of them are coarsely isometric. For the application to Siegel's question mentioned in the introduction we are particularly interested in the following class of pseudo-metrics on G which we call norm–like.

6.7. Definition. A left invariant pseudo-metric on a reductive group $G = \underline{G}_k$ over a local field k is called *norm-like* if its restriction to one, equivalently every, maximal split torus is coarsely isometric to a norm.

The claimed equivalence is implied by the fact that every inner automorphism is a coarse isometry with respect to every left invariant pseudo-metric.

6.8. Remarks. Let $G = \underline{G}_k$ be a reductive group over a local field.

(a) Every norm-like pseudo-metric has the properties (C) and (P) of 3.11.
(b) Any two norm-like pseudo-metrics on G are quasi-isometric.

(c) The metrics of type d_{op} are norm-like.
(d) The metric d_{geom} is norm-like for the symmetric space G/K if k is archimedian or the Bruhat–Tits building of G if k is non-archimedian.
(e) There is no left invariant pseudo-metric on $SL_2(\mathbb{R})$ whose restriction to every split torus is a norm.

Proof. The Cartan decomposition implies (a), and (b) follows since any two norms on T are quasi-isometric, see 6.5 a).

(c) Let ρ be a faithful representation of $\underline{G}$ defined over k in the vector space V. If $e_1, \ldots, e_n$ is a common eigenbasis of V for T with corresponding characters $\lambda_1, \ldots, \lambda_n$, then for the sup-norm $\|\Sigma\alpha_i\ell_i\| = \sup_i |\alpha_i|$ on V we have $d_{\text{op}}(e, t) = \sup_i |\log v(\lambda_i(t))|$ for $t \in T$. So $t \mapsto d_{\text{op}}(e, t)$ is a norm on T.

(d) Follows from the facts concerning d_{geom} mentioned in the proof of Theorem 6.6.

(e) Note that two coarsely isometric norms on a split torus are in fact equal, by 6.5 b). Hence if the restriction of d to every split torus is a norm, every inner automorphism Int_g induces an isometry $\text{Int}_g : T \to \text{Int}_g(T)$, since Int_g induces a coarse isometry of norms. Thus the function $x \mapsto d(e, x)$ is constant on every conjugacy class $C(x)$ of $x \in T$. But the closure of $C(x)$ is not compact for $x \neq e$. We thus get a contradiction to property (P) of 3.11. □

7. Norm versus coarse path metrics

A norm $\|\cdot\|$ on a vector space W gives a coarse path metric on W. But it is not true that it induces a coarse path metric on every lattice in W. Here is a description of those norms which induce a coarse path metric on a given closed subgroup of W. The following proposition concerning the general case of an arbitrary closed subgroup of W may be hard to digest at first sight. The subsequent Corollary 7.3 describing the case of a lattice in W gives a criterion that is much easier to state. Also, the Example 7.2 of the Euclidean norm may help to see the point of the proposition.

7.1. Proposition. *Let $\|\cdot\|$ be a norm on the finite-dimensional real vector space $W = V \oplus (D \otimes \mathbb{R})$, where D is a finitely generated torsion free abelian group. The restriction of $\|\cdot\|$ to $V \oplus D$ is a coarse path metric if and only if the norm one ball $B_1 = \{w \in W : \ \|w\| \leq 1\}$ has the following property: there is a finite subset S of D, not containing 0, and for every $d \in S$ a compact subset K_d of $(B_1 \cap (V + \mathbb{R}d)) \smallsetminus (B_1 \cap V)$ such that B_1 is the convex hull of $\bigcup_{d\in S} K_d \cup (B_1 \cap V)$.*

7.2. Example. The Euclidean norm on W does not give a coarse path metric on $V \oplus D$ unless $V = W$.

7.3. Corollary. *If D is a finitely generated torsion-free abelian group, then a norm $\|\cdot\|$ on $D \otimes \mathbb{R}$ induces a coarse path metric on D if and only if the norm one ball $B_1 = \{w \in W : \|w\| = 1\}$ has a finite number of extremal points and all of them are real multiples of elements of D.*

Proof. We first show necessity. Let $\|\cdot\|$ be a norm on W which induces a coarse path metric on $V \oplus D$. Let C be the constant in the definition of a coarse path metric and let S be the finite set of elements $d \neq 0$ in D such that $d + V$ contains a point of norm at most C. The case $W = V$ is trivial. So we assume $D \neq \{0\}$. Then S contains a basis of D and hence every point of W has distance at most $C_1 := \frac{C}{2} \cdot \operatorname{rank} D$ from some point of $V + D$. It follows that for every $w \in W$ and every $n \in \mathbb{N}$ there are integers $\beta_d(n) \in \mathbb{N}$ and vectors $v_d(n) \in V$ for $d \in S_0 = S \cup \{0\}$ such that

$$\|d + v_d\| \leq C,$$

$$\|\textstyle\sum_{d \in S_0} \beta_d(n) \cdot (d + v_d(n)) - nw\| \leq C + C_1,$$

$$\text{and} \qquad \|nw\| - C_1 \leq \textstyle\sum_{d \in S_0} \beta_d(n) \|d + v_d(n)\| \leq \|nw\| + C + C_1.$$

Then, for $d \neq 0$, the sequence $\frac{\beta_d(n)}{n}$ is bounded since $\|d + v_d(n)\| \geq \inf\{\|d + v\|, v \in V\}$ and for $d = 0$ we may and do choose $\|v_0\| \geq \frac{C}{2}$ at the expense of another constant $\frac{C}{2}$ in the inequalities above. Then $\frac{\beta_d(n)}{n}$ and $d + v_d(n)$ converge for $d \in S_d$ and n in some subsequence of $\mathbb{N}$, to β_d and $d + v_d$, say. We thus obtain

$$\|d + v_d\| \leq C, \ \|v_0\| = C, \ \beta_d \geq 0,$$

$$\textstyle\sum_{d \in S_0} \beta_d (d + v_d) = w,$$

$$\textstyle\sum_{d \in S_0} \beta_d \|d + v_d\| = \|w\|.$$

Note that we have a positive lower bound ε as follows:

$$\|d + v_d\| \geq \varepsilon > 0 \ \text{ for } \ d \in S_0.$$

For $\|w\| = 1$ we rewrite the equations above: put $\alpha_d = \beta_d \cdot \|d + v_d\|$. Then

$$\textstyle\sum_{d \in S_0} \alpha_d \cdot w_d = w,$$

$$\alpha_d \geq 0, \ \textstyle\sum \alpha_d = ||w||,$$

$$\text{where} \qquad w_d = \frac{d + v_d}{\|d + v_d\|}$$

for $d \in S$, and w_d is contained in the compact subset $K_d = B_1 \cap (V + [\varepsilon, \infty)\cdot d)$ for $d \neq 0$ and $w_0 \in B_1 \cap V$. This proves necessity.

To prove sufficiency, suppose $\|\cdot\|$ is as in the claim of the proposition. We may assume that the set S is symmetric, i.e., $-s \in S$ if $s \in S$ and then assume that every point w_d of K_d is of the form $w_d = r_d(d + v_d)$ with positive r_d. There is a positive number ε such that if $w_d = r_d(d + v_d) \in K_d$ then $r_d \geq \varepsilon$ for every $d \in S$, since all the K_d are compact and do not intersect V. Every $w \in W$ can be written as $w = \|w\| \sum_{d \in S_0} \alpha_d w_d$ with $S_0 = S \cup \{0\}$, $w_d \in K_d$, $w_0 \in K_0 := B_1 \cap V$, $\alpha_d \geq 0$ and $\Sigma \alpha_d = 1$. We have $\|w_d\| = 1$ if $\alpha_d > 0$ and hence $\|d + v_d\| = \frac{1}{r_d} \leq \frac{1}{\varepsilon}$ if we set $w_d = r_d(d + v_d)$. For every $d \in S$ there is a non-negative integer n_d such that $|n_d - \alpha_d\ r_d\|w\|\ | \leq \frac{1}{2}$ for $d \neq 0$ and $|n_0 - \alpha_0\|w\|\ | \leq \frac{1}{2}$ for $d = 0$. Put $w' = \Sigma_{d \in S_0} n_d(d + v_d) \in V + D$. Then w' has the following properties:

$$\|w - w'\| \leq \frac{1}{2} \sum_{\substack{d \in S_0 \\ \alpha_d \neq 0}} n_d \|d + v_d\| \leq C_1$$

$$\text{with} \quad C_1 = \frac{1}{2}|S|\varepsilon^{-1} + 1,$$

$$\|d + v_d\| \leq \max\left(\frac{1}{\varepsilon}, 1\right) \quad \text{if } n_d > 0$$

and

$$\begin{aligned} \|w'\| &\leq \sum_{d \in S_0} n_d \|d + v_d\| \\ &\leq \sum_{d \in S_0} |n_d - \alpha_d\|w\|\ |\ \|d + v_d\| + \sum_{d \in S_0} \alpha_d\ r_d \|d + v_d\|\ \|w\| \\ &\leq C_1 + \|w\| \sum_{\alpha_d \neq 0} \alpha_d \|w_d\| = C_1 + \|w\|, \end{aligned}$$

which implies our claim. □

References

[1] A. Borel, Some metric properties of arithmetic quotients of symmetric spaces and an extension theorem, *J. Diff. Geom.* **6** (1972), 543–560 (reprinted in: A. Borel, Oeuvres, vol. III, 153–170).

[2] J. Ding, A proof of a conjecture of C. L. Siegel, *J. Number Theory* **46** (1994), 1–11.

[3] S. Grosser and M. Moskowitz, Compactness conditions in topological groups, *J. Reine u. Angew. Math.* **246** (1971), 1–40.

[4] L. Ji, Metric compactifications of locally symmetric spaces, *Int. J. Math.* **9** (1998), 465–491.

[5] L. Ji and R. MacPherson, Geometry of compactifications of locally symmetric spaces, preprint.

[6] E. Leuzinger, Tits geometry and arithmetic groups, preprint.

[7] G. D. Mostow, *Strong rigidity of locally symmetric spaces*, *Annals of Math. Studies* vol. 78, Princeton, 1973.

[8] C. L. Siegel, Zur Reduktionstheorie quadratischer Formen, *The Mathematical Society of Japan*, 1959 (reprinted without the introduction in: C. L. Siegel, Gesammelte Abhandlungen, vol. 3, 275–327).

2

Finiteness properties of groups acting on twin buildings

P. Abramenko[1]

1. Twin BN–pairs

In the theory of (reductive) algebraic groups, the notion of a BN–pair introduced by Jacques Tits plays an important role. Similarly important are twin BN–pairs in the theory of Kac–Moody groups over fields. This concept was already implicitly present in Tits' fundamental paper [5] on Kac–Moody groups. (By the way, I am using the notion Kac–Moody group here in the sense of Tits, i. e., we are talking about the "minimal version" of Kac–Moody groups here.) An explicit definition of a twin BN–pair was given by Tits in [6, Subsection 3.2]. To fix ideas, let us reproduce this definition here.

Definition 1. *Let G be a group with two BN–pairs (B_+, N), (B_-, N) with the same subgroup N and such that $B_+ \cap N = B_- \cap N$. Denote by W the common Weyl group $W = N/B_+ \cap N = N/B_- \cap N$, and let S be the distinguished set of generators of W. We denote the length function on W with respect to S by l. Then (B_+, B_-, N) is called a twin BN–pair in G if the following two conditions are satisfied.*

(TBN1) $B_\epsilon w B_{-\epsilon} s B_{-\epsilon} = B_\epsilon ws B_{-\epsilon}$ *for* $\epsilon \in \{+, -\}$, *all* $w \in W$, *and all* $s \in S$ *such that* $l(ws) < l(w)$,

(TBN2) $B_+ s \cap B_- = \emptyset$ *for all* $s \in S$.

The rank r of the twin BN–pair (B_+, B_-, N) is defined as the cardinality $r = |S|$.

Analogously as the usual BN–axioms imply the Bruhat decomposition, for a twin BN–pair one obtains the *Birkhoff decomposition* $G = \bigcup_{w \in W} B_+ w B_-$, which is also a disjoint union (see [6] or [1, Lemma 1]).

[1] Research supported by the Deutsche Forschungsgemeinschaft through a Heisenberg Fellowship.

A subgroup P of G is called *parabolic* (with respect to the twin BN–pair (B_+, B_-, N)) if it contains a conjugate of B_+ or of B_-. Any parabolic subgroup P of G is conjugate to a *standard parabolic* subgroup of the form

$$P_{\epsilon,J} = B_\epsilon W_J B_\epsilon = \bigcup_{w \in W_J} B_\epsilon w B_\epsilon$$

where $\epsilon \in \{+, -\}$, $J \subseteq S$, and where $W_J =< J >$ is the (special) subgroup of W generated by J. The rank of $P_{\epsilon,J}$ and of any of its conjugates is by definition the cardinality $|J|$. The parabolic subgroup $P_{\epsilon,J}$ and any of its conjugates is called *spherical* if W_J is finite.

The twin BN–pair (B_+, B_-, N) is called *n–spherical* (with $n \in \mathbb{N}$) if each of its parabolic subgroups of rank $\leq n$ is spherical, i. e. if and only if W_J is finite for any $J \subseteq S$ of cardinality $|J| \leq n$.

Examples. As indicated above, a Kac–Moody group over a field k always possesses a twin BN–pair (see [6, Subsection 3.3] or [1, Example 5]). As a special case, one obtains in a canonical way a twin BN-pair of rank $m + 1$ of affine type in $G = \mathcal{G}(k[t, t^{-1}])$ if $\mathcal{G}$ is a simple and simply connected Chevalley group (scheme) of rank m. (The twin BN–pair in G can also be described without any Kac–Moody theory, see [2, Section 3].) Note that this twin BN–pair is m–spherical and that $\mathcal{G}(k[t])$ is a spherical parabolic subgroup of $\mathcal{G}(k[t, t^{-1}])$ in the above sense. This example generalizes to (absolutely almost) simple and simply connected isotropic k–groups $\mathcal{G}$ of k–rank m and $G = \mathcal{G}(k[t, t^{-1}])$, see [6, Subsection 3.2] and [1, Section 3.1] for the explicit description of the twin BN–pair in $\mathcal{G}(k[t, t^{-1}])$ for some classical groups $\mathcal{G}$.

Having the above examples in mind, we are going to discuss some results about finiteness properties of groups with twin BN–pairs, respectively of their parabolic subgroups. By finiteness properties we here mean finite generation, finite presentation, as well as the higher (homological) finiteness properties FP_l, respectively F_l (which is FP_l plus finite presentability in case $l \geq 2$). In the examples it is clear that one even cannot expect finite generation if the field k is infinite. So one would deal with finite fields $k = \mathbb{F}_q$ in these examples, and we need to introduce the parameter q in the general context of twin BN–pairs in some appropriate way. We shall also have to take into account that in non-split situations different parameters (like in unitary groups the cardinality of a field with an involution and the cardinality of the fixed field under this involution) can play a role.

Definition 2. *For a group G with twin BN–pair (B_+, B_-, N), Weyl group W, and distinguished set of generators S of W, we define the parameters*

$q_{\min}, q_{\max} \in \mathbb{N} \cup \{\infty\}$ *as follows:*

$$q_{\min} = \min_{\epsilon \in \{+,-\}, s \in S} [P_{\epsilon,\{s\}} : B_\epsilon] - 1,$$
$$q_{\max} = \max_{\epsilon \in \{+,-\}, s \in S} [P_{\epsilon,\{s\}} : B_\epsilon] - 1.$$

2. Some results

In this section, we are working with the following

Standing Assumption: G is a group with a twin BN–pair (B_+, B_-, N) of finite rank r and with finite parameter $q_{\max}$. We also assume that the intersections $\bigcap_{g \in G} gB_+g^{-1}$ and $\bigcap_{g \in G} gB_-g^{-1}$ are finite.

It is clear that r and $q_{\max}$ ought to be finite if any finiteness results are to be expected. Also, we cannot deduce anything for the group G in this context if it has big normal subgroups $\bigcap_{g \in G} gB_+g^{-1}$ and $\bigcap_{g \in G} gB_-g^{-1}$. However, in the examples described above, these intersections will always be finite, provided that the field k is finite. Since finite generation and finite presentation are of interest in their own right and since the corresponding proofs work under less technical assumptions in these cases, I shall first state what can be proved here.

Theorem 1. *If the twin BN–pair (B_+, B_-, N) is 2-spherical and $4 \leq q_{\min}$, then any parabolic subgroup of G is finitely generated. If, in addition, the twin BN–pair is not 3–spherical, then any spherical parabolic subgroup of G is not of type FP_2 (and, hence, in particular not finitely presented).*

Counter–Example 1.

(1) It is neither surprising nor hard to show that the spherical parabolic subgroups of twin BN–pairs which are not 2-spherical are not finitely generated. The most prominent example in this context is the observation due to Nagao and reproved by Serre that $SL_2(\mathbb{F}_q[t])$ is not finitely generated. (However, G itself is always finitely generated under our standing assumption, which is easily seen.)
(2) It is much more surprising and considerably harder to show that the parameter $q_{\min}$ does matter in this context. In fact, it can be shown that certain Kac–Moody groups of compact hyperbolic type of rank 3 over the fields $\mathbb{F}_2$ and $\mathbb{F}_3$ have proper parabolic subgroups which are **not** finitely generated.

Theorem 2. *If the twin BN–pair (B_+, B_-, N) is 3-spherical and $7 \leq q_{\min}$, then any parabolic subgroup of G is finitely presented. If, in addition, the twin*

BN–pair is not 4–spherical, then any spherical parabolic subgroup of G is not of type FP_3.

For the full group G, one can derive finite presentability under less restrictive assumptions. This follows directly from the main result of [3].

Theorem 3. *If the twin BN–pair (B_+, B_-, N) is 2-spherical and $4 \leq q_{\min}$, then G is finitely presented.*

Counter-Example 2.

(1) G cannot be expected to be finitely presented if the twin BN–pair is not 2–spherical. In fact, if the Weyl group W of our twin BN–pair is the infinite dihedral group, then it can be proved that G is not finitely presented. This generalizes a result of Stuhler stating that $SL_2(\mathbb{F}_q[t, t^{-1}])$ is not finitely presented.

(2) It is again more surprising that the assumption on $q_{\min}$ in Theorem 3 is indeed necessary. Recent studies (which have to be elaborated) indicate that there exist Kac–Moody groups of compact hyperbolic type of rank 3 over $\mathbb{F}_2$ (and maybe also over $\mathbb{F}_3$) which are not finitely presented.

The theorem about higher finiteness properties is now in a similar spirit as the Theorems 1 and 2, but it requires (in view of the method of proof) additional assumptions.

Theorem 4. *Assume that the twin BN–pair (B_+, B_-, N) is n–spherical and that $2^{2n-1} \leq q_{\min}$. Assume further that the Coxeter diagram of the the Coxeter system (W, S) associated to the twin BN–pair does not contain any subdiagrams of type F_4, E_6, E_7, or E_8. Then any parabolic subgroup of G is of type F_{n-1}. If, in addition, the twin BN–pair is not $(n+1)$–spherical, then any spherical parabolic subgroup of G is not of type FP_n.*

Theorem 4 has the following corollary, which is also stated as Theorem C in Chapter III of [1]. (The special case $\mathcal{G} = SL_{n+1}$ was already treated before independently by Abels and the author.)

Corollary. *Let $\mathcal{G}$ be an absolutely almost simple classical group, defined over $\mathbb{F}_q$ and of $\mathbb{F}_q$–rank $n > 0$. Assume that $2^{2n-1} \leq q$. Then $\mathcal{G}(\mathbb{F}_q[t, t^{-1}])$ and $\mathcal{G}(\mathbb{F}_q[t])$ are of type F_{n-1}, and $\mathcal{G}(\mathbb{F}_q[t])$ is not of type FP_n.*

3. Brief comment on the proofs

The main idea is to exploit the action of the group G and of its parabolic subgroups on the twin building $(\Delta_+, \Delta_-, \delta^*)$, where δ^* is the W–valued co-distance on the chambers of this twin building (cf. [6] or [1, Section I.2] for the definition of the twin building associated to a twin BN–pair; one uses the Birkhoff decomposition for the twin BN–pair to define the co-distance in a similar way as the Bruhat decomposition for BN–pairs is used to define the usual W–valued distance in a building). This W–valued co-distance has a numerical variant d^* (if c_+, c_- are chambers in Δ_+, Δ_-, respectively, and if $\delta^*(c_+, c_-) = w \in W$, then $d^*(c_+, c_-) = l(w)$), and this numerical co-distance can be used in order to filter the twin building, respectively one of its "halves". For instance, if P is a spherical parabolic subgroup of G, then it is the stabilizer of some simplex, say $a_- \in \Delta_-$, so that $P = G_{a_-}$. Now we consider the action of G_{a_-} on Δ_+ and study the G_{a_-}–invariant filtration of Δ_+ given by $\Delta_{+,j} = \{x_+ \in \Delta_+ \mid d^*(x_+, a_-) \leq j\}$, where $d^*(x_+, a_-)$ is defined to be the minimum of all $d^*(c_+, d_-)$ running over all chambers c_+ containing x_+ and all chambers d_- containing a_-. In this situation we can apply the FP_n criterion derived by Brown (cf. [4]) provided that we have enough (or rather the "right") local information about the building Δ_+ in order to deduce the desired homotopy properties about the G_{a_-}–invariant filtration $(\Delta_{+,j})_{j\in\mathbb{N}}$. It is in connection with this local information that one has to make some assumptions concerning $q_{\min}$ and where one has to exclude the exceptional types, which are (still) technically too complicated.

On the other hand, if we consider G and not a spherical parabolic subgroup, then it is more effective to study the action of G on the (topological) product $\Delta_+ \times \Delta_-$ and to consider the G–invariant filtration given by the subcomplexes

$$\big\{(x_+, y_-) \in \Delta_+ \times \Delta_- : d^*(x_+, y_-) \leq j\big\}.$$

In order to determine the precise "finiteness length" of G as in the case of spherical parabolic subgroups (where the investigation of one half at a time was sufficient), one still needs some local information concerning this new filtration. The results in Section 2 show that G in general enjoys better finiteness properties than its spherical parabolic subgroups. If this last piece of local information alluded to above was available then I could give a complete proof of (for instance) the following result.

Claim. Let G be as in Theorem 4, and assume in addition that the twin BN–pair has rank $n+1$ and is not $(n+1)$–spherical (like, for instance, $SL_{n+1}(\mathbb{F}_q[t, t^{-1}])$). Then G is of type F_{2n-1}, but not of type FP_{2n}.

Detailed proofs of the theorems stated in Section 2, hopefully together with a completed proof of the above claim, will be published elsewhere.

References

[1] P. Abramenko, *Twin Buildings and Applications to S-Arithmethic Groups*, Lecture Notes in Math. vol. 1641, Springer, 1996.

[2] P. Abramenko, Group actions on twin buildings, *Bull. Belg. Math. Soc.* **3** (1996), 391–406.

[3] P. Abramenko und B. Mühlherr, Présentations de certaines BN-paires jumelées comme sommes amalgamées, *C. R. Acad. Sci. Paris* **325**, Série I (1997), 701–706.

[4] K. Brown, Finiteness properties of groups, *J. Pure Appl. Algebra* **44** (1987), 45–75.

[5] J. Tits, Uniqueness and presentation of Kac–Moody groups over fields, *J. Algebra* **105** (1987), 542–573.

[6] J. Tits, Twin buildings and groups of Kac–Moody type, in: Proceedings of a conference on groups, combinatorics and geometry, Durham 1990, LMS Lecture Notes Series vol. 165, Cambridge University Press, 1992, 249–286.

3

Higher finiteness properties of S-arithmetic groups in the function field case I

H. Behr

It is well known that S-arithmetic subgroups of reductive algebraic groups over number fields have "all" finiteness properties (see [BS 2]). On the other hand there exist many counterexamples in the function field case. Let F be a finite extension of $\mathbb{F}_q(t)$, G an almost simple algebraic group of F-rank r, 0_S an S-arithmetic subring of F with $|S| = s$, r_v the F_v-rank of G over the completion F_v of F for $v \in S$, and finally Γ an S-arithmetic subgroup of $G(F)$. We are interested in the following question:

Is it true that Γ is of type F_{n-1} but not F_n if and only if $r > 0$ and $\sum_{v \in S} r_v = n$?

(For the definition of finiteness properties, see the introduction of [Ab].) The answer is known to be 'yes' in the following cases:

(a) $G = SL_2$: see [St 2],
(b) $n = 1$ or 2 (finite generation or finite presentability): see [B 2],
(c) G classical, $0_S = \mathbb{F}_q[t]$ under the assumption that q is big enough compared with r: see [Ab] and [A] for SL_n.

In particular it is not known if the assumption on q in (c) is necessary. For $r = 0$, in the so-called cocompact case, Γ is of type F_∞ (cf. [BS 2]).

This paper is an attempt to attack the above question with some new methods — old in other contexts. First of all, inspired by the work of Serre, Quillen, Stuhler, and Grayson (cf. [G1,2]), we use *semi-stability for reduction theory*, and the idea to determine the homotopy type of the boundary of the unstable region by *retraction*. In this part we only deal with Chevalley groups G and arithmetic rings O_S for $|S| = 1$. The groups $G(F)$ and Γ act on the Bruhat–Tits building $X = X_v$, corresponding to G and F_v; Γ leaves the unstable region X' invariant. X' has a cover whose nerve is given by the spherical Tits building X_0, so it is $(r-1)$-spherical. The retraction to its boundary is not possible as in the number field case, since the geodesic lines are branching (discretely). Therefore we have to "split up" X' into apartments, thereby constructing a bigger complex $\widetilde{X}'$, which has a cover with nerve $\mathrm{Opp}X_0$, defined by an opposition relation in X_0. This complex was first considered by Charney for $G = GL_n$, by Lehrer and Rylands for classical groups who called it "split building", finally v. Heydebreck showed in the general case that this complex is also $(r-1)$-spherical — so is $\widetilde{X}'$. $\widetilde{X}'$ can be retracted to its boundary $\widetilde{Y}$, but $\widetilde{Y}$ is not finite mod Γ. Thus we have to consider a subcomplex $\widetilde{X}'_\Gamma$, where opposition is defined only with respect to Γ and to show that $\widetilde{X}'_\Gamma$ is a deformation retract of $\widetilde{X}'$. Now we obtain that $\widetilde{Y}_\Gamma$ is finite modulo Γ and can deduce the F_{n-1}-property of Γ. For the negative part, i.e., Γ is not of type F_n, one should come back to filtrations, the method used for the proofs of (a), (b) and (c) above, but for the moment I have no detailed argument. Thus we sketch the proof of the following.

Theorem. *For $s = 1$, the S-arithmetic subgroup $\Gamma = G(O_S)$ of a simply connected almost simple Chevalley group G of rank r is of type F_{r-1}.*

Conjecture. *Γ is not of type F_r.*

I hope that this program will turn out to be useful even in more general situations: for coefficient rings which are defined by more than one prime or, on the other side, for non-split groups.

I am grateful to Peter Abramenko for constructing a very instructive counter-example to an earlier version of this paper, and I would like to thank him, Kai-Uwe Bux, and Anja von Heydebreck for helpful discussions, and Mrs. Christa Belz for carefully typing several versions.

1. Notations

Let us denote by

F	a finite extension of the field of rational functions $\mathbb{F}_q(t)$ in t with coefficients in the finite field $\mathbb{F}_q$, $q = p^m$;
$\widehat{F} = F_v$	the completion of F with respect to the valuation v of F;
O and $\widehat{O}$	the valuation rings with respect to v in F or $\widehat{F}$;
G	a simply connected almost simple Chevalley group, defined over F;
r	the F-rank of G, $I = \{1, \dots, r\}$;
T	a maximal (split) F-torus of G;
$\Delta = \{\alpha_i\}_{i \in I}$	a set of simple roots of G with respect to T;

P_{Δ_0}	a parabolic subgroup of G of cotype Δ_0, $\Delta_0 \subseteq \Delta$, which means that $\Delta - \Delta_0$ is a set of simple roots for the semi-simple part of P_{Δ_0}, especially
$B = P_\Delta$	the Borel subgroup, defined by Δ, and
P_α	the maximal parabolic subgroup for $\Delta_0 = \{\alpha\}$.
X	the Bruhat–Tits–building, corresponding to G and v with its simplicial structure and its metric topology;
$A = X_T$	the apartment of X corresponding to T, thus $A \sim \mathbb{R}^r$;
$\{\alpha_i(x)\}_{i \in I}$	the coordinates of $x \in A$ which means by abuse of notation the following: If $x = t \cdot x_0$, x_0 the "origin" of A, $t \in T(\widehat{F})$, then $\alpha_i(x) := -v(\alpha_i(t))$;
X_0	the spherical Tits building of $G(F)$;
Γ	the S-arithmetic subgroup of $G(F)$ for $S = \{v\}$.

2. Reduction theory and the unstable region

We shall use reduction theory for arithmetic groups over function fields in the version described by Harder in [H2], 1.4. He defines

$$\pi(x, P) := \text{vol } (K_x \cap U(\widehat{F}))$$

for a special point $x \in X$, corresponding to a maximal compact subgroup K_x of $G(\widehat{F})$ and an F-parabolic group P and its unipotent radical U; the volume vol comes from the adelic Tamagawa measure. The function

$$d_P(x) := \log_q \pi(x, P)$$

can be extended by linear interpolation to all points x in an apartment $A = X_T$, defined by a maximal split $\widehat{F}$-torus T, contained in P and thereby uniquely for all $x \in X$. We may consider d_P as a *co-distance* with respect to the simplex σ_P, given by P in the spherical building X_∞ at infinity (cf. [Br2], VI.9).

For the action of $T(\widehat{F})$ on X_T via ad T we have the formula

$$d_P(t \cdot x) = d_P(x) + \log_q \; |\delta_P(t)|$$

where δ_P is the character "sum of roots in U", which is a multiple of the dominant weight ω_P and the q-logarithm is the negative additive valuation $-v(\delta_P(t))$. For each Borel group B over F and its set $\Delta = \{\alpha_1, \dots, \alpha_r\}$ of simple roots (with respect to an F-torus T), the maximal parabolic groups P_α $(\alpha \in \Delta)$

containing B and their fundamental weights ω_{P_α}, one has

$$\alpha = \sum_{\beta\in\Delta} c_{\alpha,\beta}\omega_{P_\beta} = \sum_{\beta\in\Delta} c'_{\alpha,\beta}\delta_{P_\beta},$$

where $c_{\alpha,\beta}$ are the integral coefficients of the Cartan-matrix, such that $c'_{\alpha,\beta} \in \mathbb{Q}$; in particular, $c'_{\alpha,\alpha}$ is positive and $c'_{\alpha,\beta}$ for $\beta \neq \alpha$ is zero or negative (for at most 3 β's). Using these coefficients, Harder defines *numerical invariants*

$$n_\alpha(x, B) := \prod_{\beta\in\Delta} \pi(x, P_\beta)^{c'_{\alpha,\beta}}.$$

Again we pass to the additive version, setting

$$c_{B,\alpha}(x) := \log_q\ [n_\alpha(x, B)]$$

and obtain for each $b \in B(F)$ the relation

$$\begin{aligned} c_{B,\alpha}(b \cdot x) &= c_{B,\alpha}(x) + \log_q\ |\alpha(b)| \\ &= c_{B,\alpha}(x) - v(\alpha(b)). \end{aligned}$$

$c_{B,\alpha}$ is an affine linear function on the apartment X_T; we define the origin O_B by $c_{B,\alpha}(O_B) = 0$ for all $\alpha \in \Delta$ and by abuse of notation $\alpha(t \cdot O_B) := -v(\alpha(t))$ for $t \in T(F)$, thus we get by linear interpolation a set of affine coordinates $\{\alpha_1(x), \dots, \alpha_r(x)\}$ for each point $x \in X_T$. We are now able to state the **main theorems of reduction theory** (for Chevalley groups).

(A) There exists a constant C_1 such that for all $x \in X$ there is an F-Borel group B with $c_{B,\alpha}(x) \geq C_1$ for all $\alpha \in \Delta$; then x is called *"reduced with respect to B"*.

(B) There exists a constant $C_2 \geq C_1$, such that for $x \in X$ reduced with respect to B and B', and $c_{B_\alpha}(x) \geq C_2$ for all $\alpha \in \Delta_0 \subseteq \Delta$, $P = P_{\Delta_0} \supseteq B$, it follows $P \supseteq B'$; then x is called *"close to P"*, P is uniquely determined.

(C) There exists a constant $C_3 \geq C_2$, depending on the arithmetic group Γ, such that for $x \in X$, reduced with respect to B and with $c_{B,\alpha}(x) \geq C_3$ for all $\alpha \in \Delta_0 \subseteq \Delta$, we have for the unipotent radical U of the parabolic group $P = P_{\Delta_0} \supseteq B$

$$U(\widehat{F}) = (U(\widehat{F}) \cap K_x)(U(F) \cap \Gamma);$$

x is then called *"very close to P"*.

(D) For each constant $C \geq C_1$ the set

$$X_C := \left\{ x \in X \;\middle|\; \begin{array}{l} c_{B,\alpha}(x) \leq C \text{ for all } \alpha \in \Delta \text{ and all } B \\ \text{for which } x \text{ is reduced with respect to } B \end{array} \right\}$$

is Γ-invariant and X_C/Γ is compact.

(E) Borel subgroups over F of G belong to finitely many classes under Γ-conjugation (see [B1], 8).

Remark. The constant C_1 can be chosen as $C_1 \leq -2g - 2(h-1)$ where g denotes the genus of F and h is a *"class-number"* (for the precise definition see [H1], 2.2.6). For example, if $\Gamma = SL_n(\mathbb{F}_q[t])$ we may use $C_1 = 0$, but in general C_1 is negative.

We define the cone or sector of points in X_T, reduced with respect to $B \supset T$ by

$$D_{B,T} := \{x \in X_T \mid \alpha_i(x) \geq C_1 \text{ for } i = 1, \ldots, r\}.$$

Warning: For different Borel groups B and B', containing the same torus T, the origins O_B und $O_{B'}$ need not coincide and therefore the sectors $D_{B,T},\ B \supset T$ do not cover in general thc apartment X_T: see the example below. For an F-parabolic group P of cotype $\Delta_0 \neq \emptyset$, we denote by X'_P the set of all points $x \in X$ which are close to P:

$$X'_P := \left\{ x \in X \;\middle|\; \begin{matrix} c_{B,\alpha}(x) \geq C_1 \text{ for all } \alpha \in \Delta \setminus \Delta_0 \\ c_{B,\alpha}(x) \geq C_2 \text{ for all } \alpha \in \Delta_0 \end{matrix} \quad \text{for some } B \subseteq P \right\}$$

or $X'_P = \bigcup_{B \subseteq P} D_B := \bigcup_{B \subseteq P} \left(\bigcup_{T \subseteq B} D_{B,T} \cap X'_P \right)$,

and call

$$X' := \bigcup_P X'_P = \bigcup_{P \max} X'_P$$

the **unstable region** of X; the name is given in analogy to the description with vector bundles for the group $G = SL_n$ (cf. [G1], 4). For an F-parabolic group Q let P run over all maximal F-parabolic groups which contain Q; then we have

$$X'_Q = \bigcap_{P \supseteq Q} X'_P .$$

We obtain a *polyhedral decomposition* of X', defining

$$X''_Q := \overline{X'_Q \setminus \bigcup_{Q_1 \subsetneq Q} (X'_Q \cap X'_{Q_1})} .$$

In the special case, where $C_1 = 0$, we have in a fixed sector $D_{B,T}$ the following descriptions:

$$\begin{aligned} X'_Q \cap D_{B,T} &= \{x \in D_{B,T} \mid \alpha(x) \geq 0 \text{ for all } \alpha \in \Delta, \\ &\qquad\qquad\qquad \alpha(x) \geq C_2 \text{ for all } \alpha \in \Delta_0\} \\ X''_Q \cap D_{B,T} &= \{x \in D_{B,T} \mid 0 \leq \alpha(x) \leq C_2 \text{ for all } \alpha \in \Delta \setminus \Delta_0, \\ &\qquad\qquad\qquad \alpha(x) \geq C_2 \text{ for all } \alpha \in \Delta_0\} \end{aligned}$$

where $Q = P_{\Delta_0}$. In particular for $Q = B$, which means $\Delta_0 = \Delta$, $X''_B \cap D_{B,T}$ is a cone inside $D_{B,T}$, for $Q = P$ maximal, i.e. $\Delta_0 = \{\alpha\}$, we get for $X''_P \cap D_{B,T}$ a cylindric convex set, furthermore infinite prisms etc. Finally we have $X'_P = \bigcup_{Q \subseteq P} X''_Q$.

Remark. Assume we have an enumeration of the set of simple roots, given by a type function on the vertices of the spherical building X_0, then for $x \in X''_Q$ the set of maximal parabolic subgroups P containing Q defines a chain which generalizes the "canonical filtration" of vector bundles for $G = SL_n$ (cf. [G1]) or respectively lattices in the number field case (cf. [St1] and [G2]).

Above all we are interested in the *boundary* $Y := \partial X'$ of the unstable region, which can be described for a parabolic group Q of cotype $\Delta_0 \neq \emptyset$ as follows:

$$Y_Q := \partial X''_Q := \left\{ x \in X''_Q \;\middle|\; \begin{array}{ll} c_{B,\alpha}(x) \geq C_1 & \text{for all } \alpha \in \Delta \setminus \Delta_0 \text{ and all} \\ & B \subseteq Q \\ c_{B,\alpha}(x) \geq C_2 & \text{for all } \alpha \in \Delta_0 \text{ and equality} \\ & \text{for at least one } B \subseteq Q \end{array} \right\}$$

$$Y = \partial X' := \bigcup_Q \partial X''_Q .$$

In the next step we distinguish *geodesic lines* in X''_Q: A point $x \in X''_Q$ with coordinates $\alpha(x)$ for an appropriate B determines uniquely a boundary point $y \in Y_Q$ by setting $\alpha(y) = \alpha(x)$ for all $\alpha \in \Delta - \Delta_0$ and $\alpha(y) = C_2$ for all $\alpha \in \Delta_0$, the segment $\overline{xy}$ lies on a geodesic. The "geodesic action" on this line in the apartment X_T is given by the torus $T_{\Delta_0} := \{t \in T \mid \alpha(t) = 0 \text{ for all } \alpha \in \Delta - \Delta_0\}$, contained in the radical of $Q = P_{\Delta_0}$, centralizing its semi-simple part. Along these geodesic lines we can define a retraction of X''_Q to its boundary Y_Q, for instance parametrized by the distance function d_Q. Therefore the local definitions fit together for X'_Q, but unfortunately they define no retraction from X'_Q to $\partial X'_Q$ since the geodesic lines are branching into different apartments. We shall need a further retraction from the sets X'_P to "infinity" along geodesics of "type P_{Δ_0}", given by the action of T_{Δ_0}, see next section.

Example. $G = SL_n,\ \Gamma = SL_n(\mathbb{F}_q[t])$

(1) In this case Γ admits a strict simplicial fundamental domain D which is a sector $D_{B,T}$ for a fixed pair $T \subset B$: see [Ab], I.3); this result can also be deduced from reduction theory with Siegel sets. This corresponds to the fact that we can choose $C_1 = 0,\ C_2 = 1$ in Harder's theory for this case. One may then define the polyhedral decomposition locally in D and extend it to X by the action of Γ.

(2) In order to show that origins O_B and $O_{B'}$ of different sectors in an apartment need not coincide, we use $n = 3$: denote by B^+ the upper triangular matrices in SL_3, by $B^- = w\, B^+ w^{-1}$ with $w = \begin{pmatrix} 0 & 0 & 1 \\ 0 & -1 & 0 \\ 1 & 0 & 0 \end{pmatrix}$ the lower triangular matrices, define $B' = g \cdot B^- := g\, B^- g^{-1} = g\, w\, B^+ w^{-1} g^{-1}$ with $g = \begin{pmatrix} 1 & 0 & t^{-n} \\ 0 & 1 & 0 \\ 0 & 0 & 1 \end{pmatrix}$, $n \in \mathbb{N}$, such that B^+ and B' are opposite Borel groups, defining an apartment A. We obtain an equation $gw = \gamma b$ with $\gamma \in \Gamma,\ b \in B\,(\mathbb{F}_q(t))$, explicitly

$$\begin{pmatrix} t^{-n} & 0 & 1 \\ 0 & -1 & 0 \\ 1 & 0 & 0 \end{pmatrix} = \begin{pmatrix} 1 & 0 & 0 \\ 0 & -1 & 0 \\ t^n & 0 & -1 \end{pmatrix} \begin{pmatrix} t^{-n} & 0 & 0 \\ 0 & 1 & 0 \\ 0 & 0 & t^n \end{pmatrix} \begin{pmatrix} 1 & 0 & t^n \\ 0 & 1 & 0 \\ 0 & 0 & 1 \end{pmatrix}.$$

Compute

$$\begin{aligned} c_{B',\alpha'}(O_B) &= c_{gw\cdot B^+,\, gw(\alpha)}(gw(O_B)) \\ &\qquad \text{(since } w(O_B) = O_B \text{ and } g \text{ fixes a half-plane} \\ &\qquad \text{containing } O_B) \\ &= c_{\gamma b\cdot B^+, \gamma b(\alpha)}(\gamma b(O_B)) \\ &= c_{B^+,\alpha}(b(O_B)) \ \text{(by left-invariance of the} \\ &\qquad \text{measure under } \Gamma) \\ &= c_{B^+,\alpha}(O_B) + v(\alpha(b^2)) = 0 - 2n, \end{aligned}$$

which is valid for $\alpha = \alpha_1$ and $\alpha = \alpha_2$, thus $O_B \neq O_{B'}$: to get $O_{B'}$, we have to shift O_B in "direction of B'", precisely: with the coordinates α_1, α_2 corresponding to B one has $O_{B'} = (-2n, -2n)$.

3. Compactification of the Bruhat–Tits building

For the boundary at infinity of X we do not use the topologization of the building at infinity due to Borel–Serre; it is more convenient to have the compactification, constructed by *Landvogt* in [L], but we restrict it to the part defined over F. For a local field $\widehat{F}$ and a reductive algebraic group H denote by $X(H)$ the Bruhat–Tits building for the pair $(H, \widehat{F})$, then define

$$\overline{X} := \overline{X}(G) := \bigcup_{P \in \mathcal{P}} X\left(P/R_u(P)\right),$$

where $\mathcal{P}$ is the set of all parabolic F-subgroups of G and $R_u(P)$ the unipotent radical of P (cf. [L], 14.21). $\overline{X}$ is equipped with a topology which comes from the $\widehat{F}$-analytic topology on $G(\widehat{F})$ and the compactification of apartments, described below, and it induces the metric topology on each of the buildings $X(P/R_u(P))$. Consequently we consider only the — incomplete, but good (cf. [Br2], VI.9) — apartment system $\mathcal{A}$, defined over F, which is in 1-1-correspondence with the apartment system $\mathcal{A}_0$ of the Tits building X_0 of $G(F)$.

For $A \in \mathcal{A}$ denote by V the underlying $\widehat{F}$-vectorspace, by Σ the Coxeter complex with respect to G in V, by C a chamber of Σ and by $\Delta(C)$ a set of simple roots, such that $C = \{x \in A \mid \alpha(x) \geq 0 \text{ for all } \alpha \in \Delta(C)\}$. For an open face C' of C, set $\Delta(C') := \{\alpha \in \Delta(C) \mid \alpha_{|C'} > 0\}$ and denote by $\langle C' \rangle$ the subspace of V, generated by C'. $V^C := \bigcup_{\substack{C' \in \Sigma \\ C' \subseteq C}} V/\langle C' \rangle$ is called the *corner* defined by C. Provide $\widetilde{\mathbb{R}} := \mathbb{R} \cup \{\infty\}$ with its natural topology and topologize V^C in such a way that the map $f : V^C \longrightarrow \widetilde{\mathbb{R}}^n$, given by

$$f(x + \langle C' \rangle) := \begin{cases} \infty & \text{for } \alpha \in \Delta(C') \\ \alpha(x) & \text{for } \alpha \in \Delta(C) \setminus \Delta(C') \end{cases}$$

is a homeomorphism. A set $U \subseteq \overline{V} := \bigcup_{C' \in \Sigma} V/\langle C' \rangle$ is called open if $U \cap V^C$ is open for all chambers $C \in \Sigma$; by that $\overline{V}$ becomes compact and is called the compactification of V. $\overline{A} := A \times^V \overline{V}^{\Sigma}$ is then the *compactification* of A with corners A^C (cf. [L], §2). We abbreviate in the following: $X(P) := X(P/R_u(P))$, and we define the **boundary** of $\overline{X}$ to be

$$\partial \overline{X} := \overline{X} \setminus X = \bigcup_{P \neq G} X(P).$$

The closure of $X(P)$ in $\overline{X}$ is given by $\bigcup_{Q \subseteq P} X(Q)$; we shall also need $X_P := X \cup X(P)$.

Our next aim is to determine the *homotopy type of the unstable region* X', using the cover with the sets X'_P, P a maximal parabolic F-group. The nerve of this cover is the spherical Tits building X_0 which is known to be $(r-1)$-spherical. For this purpose we have to show that the sets X'_P and their intersections X'_Q (Q an arbitrary F-parabolic group) are contractible, and to prove this we construct *retractions to infinity*, more precisely to $X(Q)$, defined by the geodesic action of the torus T_{Δ_0} for $Q = P_{\Delta_0}$. To describe it in a sector $D_{B,T}$, $T \supseteq T_{\Delta_0}$, it is helpful not to use all local coordinates α for $D_{B,T}$ ($\alpha \in \Delta$), but only those α, lying in $\Delta - \Delta_0$ and to complete them with the functions d_P for all $P = P_\alpha$, $\alpha \in \Delta_0$ (this is admissible since the roots in $\Delta - \Delta_0$ and the fundamental weights for Δ_0 are linearly independent). Then we can define the map

$$r_{Q,B,T} : D_{Q,B,T} \times [0,\infty] \longrightarrow D_{Q,B,T} \ (Q \supseteq B)$$

for $D_{Q,B,T} := \overline{D_{B,T} \cap X'_Q} \cap X_Q$ where the closure is meant in $\overline{X}$, given by $r_{Q,B,T}(x,t) = x_t$ with

$$\begin{aligned} \alpha(x_t) &= \alpha(x) && \text{for all } \alpha \in \Delta - \Delta_0 \text{ and } x \in X \\ d_P(x_t) &= d_P(x) + t && \text{for all } P = P_\alpha,\ \alpha \in \Delta_0 \text{ and } x \in X \\ \alpha(x) &= x && \text{for all } \alpha \in \Delta,\ x \in \overline{X} \setminus X. \end{aligned}$$

For different tori T and T', containing T_{Δ_0}, points $x \in D_{B,T}$ and $x' \in D_{B',T'}$ can have the same image for $t = \infty$ in $X(Q)$, described by different systems of coordinates α, coming from the apartments X_T and $X_{T'}$ respectively, but the coordinates d_P for $P \supseteq Q$ are defined independently from these apartments. Thus the maps $r_{Q,B,T}$ fit together, defining for $t = \infty$ a retraction

$$r_Q : \overline{X'_Q} \cap X_Q \longrightarrow X(Q) .$$

The map r_Q is continuous since its restrictions to the sectors $D_{B,T}$ are fibrations. Moreover, the map r_Q is surjective: for each point $x \in X(Q)$ we find a point x' projecting to x for sufficiently large values $d_P(x')$ for all $P \supseteq Q$ such that x' is close to Q, and therefore exists $B \subseteq Q$ for which x' is reduced, so $x' \in D_{B,T}$ for some $T \subseteq B$ and $x' \in D_{B,T} \cap X'_Q$. Finally the affine building $X(Q)$ is contractible, thus by the retraction r_Q the set $\overline{X'_Q} \cap X_Q$ is also contractible and as a metrizable manifold the same is true for its interior X'_Q (cf. [BS1], 8.3.1).

Proposition 1. *The unstable region X' is $(r-1)$-spherical.*

Proof. $X' = \bigcup_{P \in \mathcal{P}_{\max}} X'_P$ with $\mathcal{P}_{\max} := \{P \text{ maximal } F\text{-parabolic in } G\}$, the non-empty intersections of the covering sets X'_P are of type X'_Q, Q F-parabolic, and we have seen above that all these sets are contractible. The covering sets are closed and the cover is locally finite because X is a locally finite simplicial complex. Its nerve is given by the spherical Tits building X_0 as an abstract complex which is known to be $(r-1)$-spherical. Thus we obtain that X' is $(r-1)$-spherical, using the same theorem as Borel–Serre in [BS1], 8.2. □

Remark. For the group $G = SL_n$ (or $G = GL_n$), Proposition 1 was proved by Grayson with a similar argument using vector bundles (cf. [G1], thm. 4.1). The same idea can be used for $\partial\overline{X} := \overline{X} - X = \bigcup_{P \neq G} X(P)$. We have the natural cover $\partial\overline{X} = \bigcup_{P \neq G} \overline{X(P)}$ with $\overline{X(P)} = \bigcup_{Q \subseteq P} X(Q)$; all these sets are contractible as Bruhat–Tits buildings or closures of them and their intersection pattern is given again by X_0. So we get the following.

Corollary 1. *$\partial\overline{X}$ is $(r-1)$-spherical.*

4. Buildings with opposition

(a) In each apartment A_0 of a *spherical building* X_0 there exists a natural opposition involution. If A_0 is described as an abstract Coxeter complex $\Sigma = \Sigma(W, S)$ with group W and generating set S, $W_J = \langle J \rangle$ for $J \subseteq S$, i.e. $\Sigma = \{wW_J \mid w \in W,\ J \subseteq S\}$ and w_0 denotes the element of maximal length in W, then define

$$\mathrm{op}_\Sigma(w\, W_J) := w w_0 W_{w_o J w_0}\ ;$$

in particular if the Coxeter diagram has no non-trivial symmetry, then $w_0 J w_0 = J$ for all J. If X_0 is the spherical *Tits building* of a group $G(F)$ (G reductive, F a field), the simplices of X_0 may be identified with the proper F-parabolic subgroups of $G(F)$. Each such group has a Levi decomposition $P = L \ltimes R_u(P)$, and two parabolics are called *opposite* if they have a common Levi subgroup, more precisely,

$$P \text{ op } P' :\iff P \cap P' \text{ is a Levi subgroup of } P \text{ and } P'.$$

$[R_u(P)](F)$ acts simply transitive on the set of all parabolic subgroups opposite P (cf. [BT], § 4), thus we can identify them with the elements of this radical if we distinguish one opposite group.

(b) Pairs of opposite simplices of a spherical building with incidence in both components provide again a simplicial complex. It was introduced by

R. Charney (see [C]) for $G = GL_n$, even over Dedekind domains in the language of flags; she showed that it has the same homotopy type as the spherical building of GL_n itself. Lehrer and Rylands (see [LR]) defined such a complex for reductive groups G — they called it the *"split building"* of G — and proved the corresponding homological result for types A_n and C_n. A. von Heydebreck (see [vH]) considered this complex for arbitrary spherical buildings and showed that it is also $(n-1)$-spherical in dimension n. We use the definition

$$\operatorname{Opp} X_0 := \{(P, P') \mid P \operatorname{op} P'\}.$$

(c) Moreover, we need a subcomplex of $\operatorname{Opp} X_0$, where the opposition relation is defined with respect to Γ.

As a first step we distinguish an apartment $A_1 = X_{T_1}$ of X, T_1 a maximal split F-torus such that $N(T_1) \cap \Gamma$ contains (a copy of) the Weyl group W of X_0 (for instance, A_1 could contain a vertex with stabilizer $G(\widehat{O}) \supset G(\mathbb{F}_q) \supset W$). We fix a Borel group $B_1 \supset T_1$ and its opposite B_1' in A_1. The choice of B_1' defines an identification of $\operatorname{Opp} B_1 := \{B' \mid B' \operatorname{op} B_1\}$ with $U_{B_1}(F)$, and we can consider the subset $\operatorname{Opp}_\Gamma B_1$, corresponding to $U_{B_1}(F) \cap \Gamma =: U_1 \cap \Gamma$ such that

$$\operatorname{Opp}_\Gamma B_1 := \{B' = \gamma_1 B_1' \gamma_1^{-1} \mid \gamma_1 \in U_1 \cap \Gamma\}.$$

We extend this notion Γ-invariant: for $B = \gamma B_1 \gamma^{-1}$ with $\gamma \in \Gamma$, the element γ is determined up to $B_1(F) \cap \Gamma$, so we obtain different opposite Borel groups $B' = \delta B_1' \delta^{-1}$ with $\delta \in \gamma \cdot (U_1 \cap \Gamma)$ — neglecting the torus component in $T_1 \subset B_1$ since it fixes also B_1'. Consequently the identification of $\operatorname{Opp} B$ with $U_B(F)$ depends on the choice of δ, but this has no influence on the definition

$$\operatorname{Opp}_\Gamma B := \{B' = \gamma' B'(\gamma')^{-1} \mid \gamma' \in U_B(F) \cap \Gamma\}$$

because $U_B = \gamma U_1 \gamma^{-1}$, which implies with $u, u' \in U_1 \cap \Gamma$:

$$\gamma' B'(\gamma')^{-1} = \gamma u' \gamma^{-1} \gamma u B_1' (\gamma u)^{-1} (\gamma u' \gamma^{-1})^{-1} = \gamma u' u B_1' (\gamma u' u)^{-1}$$

thus $\operatorname{Opp}_\Gamma B = \gamma \cdot \operatorname{Opp}_\Gamma B_1 \gamma^{-1}$. In general, not all F-Borel groups are conjugate under Γ; there exist finitely many Γ-conjugacy classes (see part E of reduction theory). We fix a set

$$B_1, B_2 = g_2 B_1 g_2^{-1}, \ldots, B_h = g_h B_1 g_h^{-1} \; (g_i \in G(F))$$

of representatives and also of their opposite groups

$$B_1', B_2' = g_2 B_1' g_2^{-1}, \ldots, B_h = g_h B_1' g_h^{-1},$$

and define in the same way as above

$$\mathrm{Opp}_\Gamma B_i := \{B' = \gamma_i B_i' \gamma_i^{-1} \mid \gamma_i \in U_{B_i}(F) \cap \Gamma\},\ i = 1, \dots, h$$

and for $B = \gamma B_i \gamma^{-1},\ B' = \gamma B_i' \gamma^{-1}$

$$\mathrm{Opp}_\Gamma B := \{B' = \gamma' B (\gamma')^{-1} \mid \gamma' \in U_B(F) \cap \Gamma\},$$

which does not depend on the special choice of B' (but we do not have $g_i \,\mathrm{Opp}_\Gamma B_1 g_i^{-1} = \mathrm{Opp}_\Gamma B_i$ in general). Finally we can make the same procedure with parabolic groups, starting with the set of standard parabolic groups Q_1 containing B_1 and their opposites $Q_1' \supseteq B_1'$. Since Q_1 and Q_1' have a Levi subgroup in common, we obtain all Γ-opposites of Q_1 by conjugation of Q_1' with elements from $U_{Q_1}(F) \cap \Gamma$ and we have to restrict in all definitions above the groups $U_B(F) \cap \Gamma$ to their subgroups $U_Q(F) \cap \Gamma$ for $Q \supseteq B$. We denote this relation by Opp_Γ and define

$$\mathrm{Opp}_\Gamma X_0 := \{(Q, Q') \mid Q \,\mathrm{op}_\Gamma\, Q'\}.$$

5. Proof of the theorem (sketch)

In order to define a retraction from the unstable region to its inner boundary, we have to split it up into apartments, thereby constructing a bigger complex (part of an *"affine split building"*) as follows: denote by $\mathcal{T}$, $\mathcal{B}$, $\mathcal{Q}$ and $\mathcal{P}$ the sets of maximal tori, Borel groups, parabolic and maximal parabolic groups in G, all defined over F (for other notations cf. section 2)

$$Z := \{(x, T) \in X' \times \mathcal{T} \mid \exists B \in \mathcal{B} : x \in D_{B,T}\},$$
by definition $D_{B,T} \subset X_T$ and $T \subset B$.

Since a maximal torus T is uniquely determined by a pair of opposite Borel groups (B, B'), say $T = T_{B,B'}$, there exists an equivalent description

$$Z = \{(x, B') \in X' \times \mathcal{B} \mid \exists B \in \mathcal{B} : B \,\mathrm{op}\, B',\ x \in D_{B,T} \text{ for } T = T_{B,B'}\}.$$

In Z we need an equivalence relation, according to the structure of $\mathrm{Opp} X_0$, so we define

$$(x_1, T_1) \sim (x_2, T_2) \iff \begin{cases} x_1 = x_2 =: x \in D_{B_1,T_1} \cap D_{B_2,T_2} \\ \exists Q \in \mathcal{Q} : Q \supseteq B_1,\ Q \supseteq B_2,\ x \in X_Q''. \end{cases}$$

The group Q is uniquely determined by reduction theory and this fact implies the transitivity of the relation. We can define the equivalence also using the second description of Z:

$$
\begin{aligned}
&(x_1, B_1') \sim (x_2, B_2') \\
&\qquad \Longleftrightarrow \begin{cases} x_1 = x_2 =: x \\ \exists (Q, Q') \in \mathrm{Opp} X_0 : B_i \subseteq Q,\ B_i' \subseteq Q' \text{ for } i = 1, 2 \\ x \in X''_Q. \end{cases}
\end{aligned}
$$

In this situation the common Levi subgroup L of Q and Q' is the centralizer of a torus T_L (not necessarily maximal), contained in $T_1 \cap T_2$. Let us denote by

$$
\begin{aligned}
&[x, B'] \text{ the class of } (x, B') \quad \text{and by} \\
&\widetilde{X}' := Z/\sim = \{[x, B'] \mid (x, B') \in Z\} \quad \text{and} \\
&\widetilde{X}'_{Q,Q'} := \{[x, B'] \in \widetilde{X}' \mid x \in X'_Q, B' \subseteq Q'\} \text{ for } (Q, Q') \in \mathrm{Opp}\, X_0\,,
\end{aligned}
$$

and finally the analogous definition for $\widetilde{X}''_{Q,Q'}$ with $x \in X''_Q$. The *topology* of $\widetilde{X}'$ is given as follows: we choose for X' the metric topology as a subspace of the affine building X, for $\mathcal{T}$ and $\mathcal{B}$ the $\widehat{F}$-analytic topology induced from $G(\widehat{F})$, since all maximal tori in $\mathcal{T}$ or all Borel groups in $\mathcal{B}$ are conjugate under $G(F)$; finally we have the product topology on Z and the quotient topology on $\widetilde{X}'$. One should emphasize that every point (x, B') has an open neighbourhood in Z of the form $U \times V$, where U is the disjoint union of open sets U_T in X_T, because the complex X is locally finite, so we can avoid ramification inside U_T. For a point $[x, B']$ in $\widetilde{X}''_{Q,Q'} \subset \widetilde{X}'$ there exists a neighbourhood $U \times V$, where U is the union of segments of geodesic lines in $\widetilde{X}''_{Q,Q'}$, defined by the torus $T = T_{\Delta_0}$ if Q and Q' are both of cotype Δ_0. We want moreover to define a *boundary at infinity* for $\widetilde{X}'$, generalizing the construction of Landvogt. There the Bruhat–Tits buildings $X(Q) := X(Q_{/R_u(Q)})$, which contribute to the boundary $\partial\overline{X}$ are defined only by quotient groups. For a pair (Q, Q') of opposite parabolic groups, the common Levi group $L = Q \cap Q'$ is isomorphic to $Q_{/R_u(Q)}$, so we may consider $X(L)$ instead of $X(Q)$, defined by a subgroup of G. For $\widetilde{X}'$ it is more convenient to split up also $\partial\overline{X}$, using the different buildings $X(L)$ instead of a single $X(Q)$. Therefore we set

$$
\begin{aligned}
\partial_\infty \widetilde{X}' &:= \bigcup_L X(L)\,, \text{ where } L = Q \cap Q',\ (Q, Q') \in \mathrm{Opp}\, X_0. \\
\overline{X}' &:= \widetilde{X}' \cup \partial_\infty \widetilde{X}'\,.
\end{aligned}
$$

The details are the same as in Landvogt's construction, but let us remark that for a point of $\partial_\infty \widetilde{X}'$ each neighbourhood meets infinitely many "apartments"

$\widetilde{X}'_T := \{[x, T] \in \widetilde{X}' \mid x \in X_T\}$. Now we can imitate the proof of Proposition 1, in order to determine the *homotopy type* of $\widetilde{X}'$. We have a cover

$$\widetilde{X}' = \bigcup_{\mathrm{Opp}\, X_0} \widetilde{X}'_{Q,Q'} = \bigcup_{(P,P')} \widetilde{X}'_{P,P'} \text{ with } (P, P') \in \mathrm{Opp}\, X_0 \cap (\mathcal{P} \times \mathcal{P})$$

with closed sets; their intersections are given by

$$\widetilde{X}'_{Q,Q'} = \bigcap \left\{ \widetilde{X}'_{P,P'} \,\middle|\, (P, P') \supseteq (Q, Q') \right\},$$

thus this cover has the nerve Opp X_0. The covering sets and their intersections can be surjectively contracted to $X(L) \subset \partial_\infty \widetilde{X}$ along geodesic lines defined by the torus T_L in the center of $L = Q \cap Q'$ and $X(L)$ is a contractible space, so $\widetilde{X}'_{Q,Q'}$ is also contractible. Using the result of v. Heydebreck, cited in section 4, we know that Opp X_0 is $(r-1)$-spherical, and therefore we have the following.

Proposition 2. $\widetilde{X}'$ *is* $(r-1)$*-spherical.*

But in contrast to X' it is now possible to retract $\widetilde{X}'$ to its "inner boundary" (cf. section 2)

$$\widetilde{Y} := \partial_0 \widetilde{X}' := \{[x, B'] \in \widetilde{X}' \mid x \in Y\}$$

along geodesic lines in $\widetilde{X}''_{Q,Q'}$, which do not ramify in $\widetilde{X}'$, because we identified different apartments only in these sets $\widetilde{X}''_{Q,Q'}$, and the geodesics coincide in their intersections. Thus we obtain the following.

Corollary 2. $\widetilde{Y}$ *is* $(r-1)$*-spherical.*

We need the analogous results for a subcomplex $\widetilde{X}'_\Gamma$ of $\widetilde{X}'$, replacing in the definitions the relation "op" by "op_Γ", consequently we have to admit only pairs of Borel groups (B, B') with $B \,\mathrm{op}_\Gamma\, B'$ and tori $T_{B,B'}$ for $(B, B') \in \mathrm{Opp}_\Gamma X_0$. For this purpose we require that also $\mathrm{Opp}_\Gamma X_0$ is $(r-1)$-spherical which is true for $G = SL_n$ by the proof of Charney (see [C]), for the general case see the appendix. Then we obtain the following result.

Proposition 3. $\widetilde{X}'_\Gamma$ *and* $\widetilde{Y}_\Gamma := \widetilde{Y} \cap \widetilde{X}'_\Gamma$ *are* $(r-1)$*-spherical.*

The next step is to show that $\widetilde{Y}_\Gamma$ *is modulo* Γ *a finite complex* — this is the only point where we need $\widetilde{X}'_\Gamma$ instead of $\widetilde{X}'$. For the points of $\widetilde{Y}_\Gamma$ the numerical invariants of reduction theory are bounded from above (and below by definition), so part D of the "main theorem" says that $\widetilde{Y}_\Gamma / \Gamma$ is compact. Moreover, by part (E) there exist only finitely many conjugacy classes of Borel groups, therefore in a set of representatives $[y, B']$ for $\widetilde{Y}_\Gamma / \Gamma$ with $y \in D_{B,T}$ only finitely many

Borel groups B occur, and since B' op_Γ B, there is only one B' modulo Γ for each B : $\widetilde{Y}_\Gamma/\Gamma$ is a finite complex. Since all stabilizers in Γ are finite, we can apply the finiteness criterion of K. Brown (see [Br1], 1.1 and 3.1) to get the following result.

Proposition 4. Γ *is of type* F_{r-1}.

Remark (concerning the conjecture "Γ is not of type F_r"): Construct an infinite series of $(r-1)$-spheres S_k in $Y = \partial_0 X'$, which are contractible only in growing parts X_k, defined by a (rough) filtration of X; then $\{\pi_{r-1}(X_k)\}$ is not "essentially trivial" in the sense of K. Brown (see [Br1], 2).

6. Appendix

For the group $G = SL_n$ the complex $\mathrm{Opp}_\Gamma X_0$ is also $(r-1)$-spherical by [C] and so are $\widetilde{X}'_\Gamma$ and $\widetilde{Y}_\Gamma$. It is not true that $\mathrm{Opp}_\Gamma X_0$ is a deformation retract of $\mathrm{Opp}\, X_0$, as was shown by Abramenko, who constructed a counter-example. But we have the following.

Lemma. $\widetilde{X}'_\Gamma$ *is a deformation retract of* $\widetilde{X}'$.

Proof. We wish to map a point $[x, B']$ of $\widetilde{X}'$ with $x \in D_{B,T} \subset X_T$, B op B', $T = T_{B,B'}$ to $[x, B'_0]$ with the same $x \in X'$ and B'_0 op_Γ B, obtaining a new torus $T_0 := T_{B,B'_0}$. Identifying the Borel groups opposite to B with elements of $U(F)$ (the unipotent radical of B), for $[x, B']$ the group B' corresponds to an element of $U(F) \cap K_x$ with $K_x = \mathrm{Stab}_{G(F)}(x)$ since $x \in X_{T_{B,B'}}$. This compact group contains finitely many elements of the discrete group $U(F) \cap \Gamma$; we have to make a choice: there is one element, defining B'_0 and T_0, such that $X_{T_0} \cap X_T$ is maximal because the intersection is given as the intersection of half-apartments, defined by root groups, and for a Chevalley group, U is the semi-direct product of its root-groups. This definition is compatible with the equivalence relation in Z and the map induces the identity on $\widetilde{X}'_\Gamma$. Moreover, this map is also continuous: the topology in the second component is induced by the analytic topology of the group $G(\widehat{F})$; an element of $U(F) \cap K_x$ has a neighbourhood which contains only one element of $U(F) \cap \Gamma$, due to its discreteness. □

Remark. Since $\mathrm{Opp}_\Gamma X_0$ is the nerve of a cover of $\widetilde{X}'_\Gamma$, we proved indirectly that it is $(r-1)$-spherical. A direct proof for groups over Dedekind rings would be of interest.

References

[A] H. Abels, Finiteness properties of certain arithmetic groups in the function field case, *Israel J. Math.* **76** (1991), 113–128.

[Ab] P. Abramenko, *Twin buildings and Applications to S-arithmetic groups*, Springer Lecture Notes in Math. vol. 1641, 1996.

[B1] H. Behr, Endliche Erzeugbarkeit arithmetischer Gruppen über Funktionenkörpern, *Invent. Math.* **7** (1969), 1–32.

[B2] H. Behr, Arithmetic groups over function fields I, *J. Reine und Angewandte Math.* **495** (1998), 79–118.

[BS1] A. Borel and J.P. Serre, Corners and Arithmetic Groups, *Comm. Math. Helv.* **48** (1973), 436–491.

[BS2] A. Borel and J.P. Serre, Cohomologie d'immeubles et des groupes S-arithmetiques, *Topology* **15** (1976), 211–232.

[Br1] K. Brown, Finiteness properties of groups, *J. Pure and Applied Algebra* **44** (1987), 45–75.

[Br2] K. Brown, *Buildings*, Springer, 1989.

[C] R. Charney, Homology stability for GL_n of a Dedekind Domain, *Invent. Math.* **56** (1980), 1–17.

[G1] D. Grayson, Finite generation of K-groups of a curve over a finite field, in: Springer Lecture Notes in Math. vol. 966 (1980), 69–90.

[G2,3] D. Grayson, Reduction theory using semi-stability I,II, *Comm. Math. Helv.* **59** (1984), 600–634, and **61** (1986), 661–676.

[H1] G. Harder, Halbeinfache Gruppenschemata über vollständigen Kurven, *Invent. Math.* **6** (1968), 107–149.

[H2] G. Harder, Die Kohomologie S-arithmetischer Gruppen über Funktionenkörpern, *Invent. Math.* **42** (1977), 135–175.

[vH] A. von Heydebreck, Homotopy properties of certain complexes associated to spherical buildings, *Israel J. Math.*, **133** (2003), 369–379.

[L] E. Landvogt, *A compactification of the Bruhat–Tits Building*, Springer Lecture Notes in Math. vol. 1619, 1996.

[LR] G.I. Lehrer and L.J. Rylands, The split building of a reductive group, *Math. Ann.* **296** (1993), 607–624.

[R] M. Ronan, *Lectures on Buildings*, Academic Press, 1989.

[S] J.P. Serre, *Trees*, Springer, 1980.

[St1] U. Stuhler, Eine Bemerkung zur Reduktionstheorie quadratischer Formen, *Arch. Math.* **27** (1976), 604–610, und: Zur Reduktionstheorie der positiven quadratischen Formen, *Arch. Math.* **28** (1977), 611–619.

[St2] U. Stuhler, Homological properties of certain arithmetic groups in the function field case, *Invent. Math.* **57** (1980), 263–281.

4

Controlled topology and group actions

R. Bieri* and R. Geoghegan†

This is a report on our work during the last few years on extending the Bieri-Neumann-Strebel-Renz theory of "geometric invariants" of groups to a theory of group actions on non-positively curved (= CAT(0)) spaces. With the exception of Theorem 8, which is proved here, and the related material in §5.3, proofs of all our theorems can be found in our papers [BG_I] (controlled connectivity and openness results), [BG_{II}] (the geometric invariants) and [BG_{III}] (SL_2 actions on the hyperbolic plane). An earlier expository paper [BG 98] is also relevant.

1. The geometric invariants

Here we recall the "geometric" or "Σ-" invariants of groups developed during the 1980's by Bieri, Neumann, Strebel and Renz (abbrev. BNSR); see [BNS 87], [BR 88], [Re 88]. We set things out in a way which leads directly to generalizations which were not anticipated in the original literature. Let G be a group of type[1] F_n, $n \geq 1$. Let X be a contractible G-CW complex which is either (a) free with cocompact n-skeleton, or (b) properly discontinuous and cocompact. Case (a) exists by the definition of F_n; Case (b) is often useful but can only exist when G has finite virtual cohomological dimension.

1.1. Controlled connectivity

Let $\chi : G \to \mathbb{R}$ be a non-zero character, i.e., a homomorphism to the additive group of real numbers. Reinterpret $\mathbb{R}$ as the group of translations, Transl($\mathbb{E}^1$),

* Supported in part by a grant from the Deutsche Forschungsgemeinschaft.
† Supported in part by grants from the National Science Foundation.

[1] G is of type F_n if there is a $K(G, 1)$-complex with finite n-skeleton. All groups are of type F_0, F_1 is "finitely generated", F_2 is "finitely presented", etc. F_∞ means F_n for all n.

of the Euclidean line $\mathbb{E}^1$, and thus reinterpret χ as an action of G on $\mathbb{E}^1$ by translations. In an obvious way, χ defines a (unique up to bounded homotopy) G-equivariant "control function" $h : X \to \mathbb{E}^1$. We say that the action χ is *controlled* $(n-1)$*-connected* (abbrev. CC^{n-1}) *over* ∞ if for any $s \in \mathbb{R}$ and $p \leq n-1$ there exists[2] $\lambda(s) \geq 0$ such that every map $f : S^p \to h^{-1}([s, \infty))$ extends to a map $\tilde{f} : B^{p+1} \to h^{-1}([s - \lambda(s), \infty))$, and $s - \lambda(s) \to \infty$ as $s \to \infty$. Clearly, χ is CC^{n-1} over ∞ if and only if $r\chi$ is CC^{n-1} over ∞ for some (equivalently, any) $r > 0$.

1.2. The geometric invariants

Let $S(G)$ be the sphere of non-zero characters on G modulo positive scalar multiplication: when $\chi \in \mathrm{Hom}(G, \mathbb{R})$ is a non-zero character we denote by $[\chi]$ the point of $S(G)$ represented by χ. The preferred endpoint of $\mathbb{R}$ is ∞. Define

$$\Sigma^n(G) := \{[\chi] \in S(G) \mid \chi \text{ is } CC^{n-1} \text{ over } \infty\}.$$

This is the nth *(homotopical) geometric invariant* of [BR 88] and [Re 88], and coincides with the invariant $\Sigma_{G'}$ of [BNS 87] when $n = 1$.

1.3. A recasting in terms of a single translation action

Consider $G \to G/G'$, the abelianization homomorphism. Write $V := G/G' \otimes \mathbb{R}$. There is an induced homomorphism $G \to G/G' \to V$. We reinterpret V as $\mathrm{Transl}(V)$, the group of translations, and we rewrite this homomorphism as $\alpha : G \to \mathrm{Transl}(V)$, the *canonical translation action of G on V*. The vector space V has a natural base point 0. An *endpoint* of V is a ray starting at 0. The set, ∂V, of endpoints is a sphere of dimension $(\dim V) - 1$ with the obvious topology. We choose an inner product $(\cdot, \cdot)$ for V. Then each $e \in \partial V$ defines a "projection on the e-direction" functional $\pi_e \in V^*$ by $\pi_e(x) = (x, u_e)$, where u_e is the point on the ray e of unit distance from 0. Thus we get, for $e \in \partial V$, a character $\chi_e = \pi_e \circ \alpha : G \to \mathbb{R}$. The map $\partial V \to S(G)$, $e \mapsto [\chi_e]$ is a homeomorphism. Looking ahead, we say that the action α *is CC^{n-1} over* $e \in \partial V$ if and only if χ_e is CC^{n-1} over ∞. We define

$$\Sigma^n(\alpha) := \{e \in \partial V \mid \alpha \text{ is } CC^{n-1} \text{ over } e\}.$$

Then the above homeomorphism $e \mapsto [\chi_e]$ identifies $\Sigma^n(\alpha)$ with $\Sigma^n(G)$. It is from this point of view – $\Sigma^n(\alpha)$ as an invariant of an action α of G on V – that our generalization takes off.

[2] We will see in §7 that it would be equivalent to have λ independent of s in which case the convergence to ∞ would be automatic.

2. Isometric actions on CAT(0) spaces

In Section 1 we considered a canonically defined translation action of G on a certain finite-dimensional real vector space V. We chose an inner product for V and hence a metric. Translations are isometries with respect to any such metric, and V (with this metric) is an example of a proper CAT(0) space. In our generalization we consider actions of G by isometries on proper CAT(0) spaces. A general reference on CAT(0) spaces is [BrHa].

2.1. Proper CAT(0) spaces

A metric space (M, d) is a *proper* CAT(0) *space* if (i) it is a geodesic metric space: this means that an isometric copy of the closed interval $[0, d(a, b)]$ called a *geodesic segment* joins any two points $a, b \in M$; (ii) for any geodesic triangle Δ in M with vertices a, b, c let Δ' denote a triangle in the Euclidean plane with vertices a', b', c' and corresponding side lengths of Δ' and Δ equal; let ω and ω' be geodesic segments from b to c and from b' to c' respectively; then for any $0 \le t \le d(b, c)$, $d(a, \omega(t)) \le ||a' - \omega'(t)||$; and (iii) d is *proper*, i.e., the closed ball $B_r(a)$ around any $a \in M$ of any radius r is compact. In a CAT(0) space the geodesic segment from a to b is unique and varies continuously with a and b. This implies that CAT(0) spaces are contractible.

Examples of CAT(0)-spaces are Euclidean space $\mathbb{E}^m$, hyperbolic space $\mathbb{H}^m$, locally finite affine buildings, complete simply connected open Riemannian manifolds of non-positive sectional curvature, and any finite cartesian product $(\prod_i M_i, d)$ of CAT(0) spaces (M_i, d_i) with $d(a, b) := (\sum_i d_i(a_i, b_i)^2)^{\frac{1}{2}}$.

A *geodesic ray* in M is an isometric embedding $\gamma : [0, \infty) \to M$. Two geodesic rays γ, γ' are *asymptotic* if there is a constant $r \in \mathbb{R}$ such that $d(\gamma(t), \gamma'(t)) \le r$ for all t. The set of all geodesic rays asymptotic to γ is called[3] the *endpoint of* γ and denoted $e = \gamma(\infty)$. The collection of all endpoints of geodesic rays form the *boundary* ∂M of M. Since d is proper it is complete, so for every pair $(a, e) \in M \times \partial M$ there is a unique geodesic ray $\gamma : [0, \infty) \to M$ with $\gamma(0) = a$ and $\gamma(\infty) = e$. Hence there is a natural bijection between ∂M and the set of all geodesic rays emanating from a base point, and ∂M acquires the compact-open topology of the latter via this bijection. This topology, which is independent of the choice of a, is compact and metrizable. It is called the *cone topology*. An action on M by isometries induces a topological action on ∂M. Associated to each geodesic ray γ of M is its *Busemann*

[3] In Section 1, V has a natural basepoint. In general, we do not wish to prefer a basepoint in M so the definition of endpoint must be given in a basepoint-free manner.

function $\beta_\gamma : M \to \mathbb{R}$ given by $\beta_\gamma(b) = \lim_{t\to\infty}(d(\gamma(0), \gamma(t)) - d(b, \gamma(t))$. For each $s \in \mathbb{R}$ the associated *horoball* $HB_s(\gamma) := \beta_\gamma^{-1}([s, \infty))$. Horoballs "centered" at γ play a role analogous to that of balls centered at $a \in M$. Indeed,[4] $HB_s(\gamma) = \mathrm{cl}_M(\bigcup\{B_{t-s}(\gamma(t))|s < t\})$. Horoballs are contractible.

2.2. CC^{n-1} over end points

Let $n \geq 0$ be an integer, let M be a proper CAT(0) space, let X be an n-dimensional contractible free G-CW-complex such that $G\backslash X^n$ is finite or a properly discontinuous cocompact contractible G-CW complex, let $\rho : G \to \mathrm{Isom}(M)$ be an isometric action and let $h : X \to M$ be a control function (i.e., a G-map). We pick a geodesic ray $\gamma : [0, \infty) \to M$ and let $e = \gamma(\infty)$. We write $X_{(\gamma,s)}$ for the largest subcomplex of X lying in $h^{-1}(HB_s(\gamma))$. We say that X *is controlled* $(n-1)$*-connected* (CC^{n-1}) *in the direction* γ (with respect to ρ) if for any horoball $HB_s(\gamma)$ and $-1 \leq p \leq n-1$ there exists $\lambda(s) \geq 0$ such that every map $f : S^p \to X_{(\gamma,s)}$ extends to a map $\tilde{f} : B^{p+1} \to X_{(\gamma,s-\lambda(s))}$ and $s - \lambda(s) \to \infty$ as $s \to \infty$. The number λ depends on the horoball $HB_s(\gamma)$. When $p = -1$ this says that each $X_{(\gamma,s)}$ is non-empty. Equivalent forms of this definition are discussed in Section 7. The property "X is CC^{n-1} in the direction γ" is a property of the endpoint e rather than the ray γ. It is also independent of the choice of X and of h, i.e., it is a property of the action ρ. So, if X is CC^{n-1} in the direction γ we will say that ρ is CC^{n-1} *over* (or *in the direction*) $e = \gamma(\infty)$. Our generalization of $\Sigma^n(G) \cong \Sigma^n(\alpha)$ in Section 1 is

$$\Sigma^n(\rho) := \{e \in \partial M \mid \rho \text{ is } CC^{n-1} \text{ over } e\}.$$

This is the nth *(homotopical) geometric invariant* of the action ρ of G on M. A second generalization of $\Sigma^n(G)$, the nth "dynamical invariant" $\mathring{\Sigma}^n(\rho)$ will be introduced in Section 3.

2.3. The case $n = 0$

Clearly, $e \in \Sigma^0(\rho)$ if and only if $h(X)$ has non-empty intersection with every horoball $HB_s(\gamma)$ (where γ defines e). For example, if μ is the action of $SL_2(\mathbb{Z})$ on the hyperbolic plane $\mathbb{H}$ (upper half space model) by Möbius transformations, then $\partial\mathbb{H} = \mathbb{R} \cup \{\infty\}$ and $\Sigma^0(\mu) = \partial\mathbb{H} - (\mathbb{Q} \cup \{\infty\})$; see Section 6.1. As

[4] In Section 1, V has a natural base point 0, e is identified with a geodesic ray γ starting at 0, $\beta_\gamma = \pi_e$, and the horoballs $HB_s(\gamma)$ are the half spaces containing (most of) this ray whose boundaries are orthogonal to γ.

another example, consider the action α in Section 1. Then clearly[5] $\Sigma^0(\alpha) = \partial V$. It is no accident that one of these actions is cocompact and the other is not; in fact, we have the following.

Theorem 1. $\Sigma^0(\rho) = \partial M$ *if and only if the action* ρ *is cocompact.*[6]

3. The dynamical invariants

Most of our results on the invariant $\Sigma^n(\rho)$ depend on the fact that a certain subset $\mathring{\Sigma}^n(\rho)$ of $\Sigma^n(\rho)$, which often coincides with $\Sigma^n(\rho)$, has a dynamical description. This requires

3.1. A new tool

If X and Y are two CW complexes, we write $\hat{\mathcal{F}}(X, Y)$ for the set of all cellular maps $f : D(f) \to Y$, where $D(f)$ is a finite subcomplex of X. By a *sheaf of maps on X with values in Y* we mean any subset $\mathcal{F}$ of $\hat{\mathcal{F}}(X, Y)$ which is closed under restrictions and finite unions. The sheaf $\mathcal{F}$ is *complete* (resp. *locally finite*) if each finite subcomplex of X occurs as the domain of some member (resp. finitely many members) of $\mathcal{F}$. A *cross section* of the complete sheaf $\mathcal{F}$ is a map $X \to Y$ whose restrictions to all finite subcomplexes lie in $\mathcal{F}$. For example, every cellular map $\phi : X \to Y$ is a cross section of its "restriction", the sheaf $\text{Res}(\phi)$ consisting of all restrictions of ϕ to finite subcomplexes. These concepts become useful if X and Y are endowed with cell permuting actions of a group G. Then $\hat{\mathcal{F}}(X, Y)$ has a natural G-action: if $g \in G$ and $f \in \hat{\mathcal{F}}(X, Y)$ then the g-translate of f, which we write $gf \in \hat{\mathcal{F}}(X, Y)$, has domain $D(gf) = gD(f)$ and maps gx to $gf(x)$ for each $x \in D(f)$. A *G-sheaf* is a sheaf which is invariant under this action. If $\phi : X \to Y$ is a G-equivariant cellular map then $\text{Res}(\phi)$ is a G-sheaf and is, of course, locally finite. If ϕ is an arbitrary cellular map then the G-sheaf generated by $\text{Res}(\phi)$ will not, in general, be locally finite. But if it is so – and the important fact is that this happens far beyond the equivariant case – we call ϕ a *finitary* (more precisely: *G-finitary*) map. Thus a finitary map $\phi : X \to Y$ is just a cellular map which can be exhibited as a cross section of a locally finite G-sheaf.

In our situation, finitary maps will occur as cellular endomorphisms $\phi : X^n \to X^n$. Recall that X^n is endowed with a chosen G-equivariant control map

[5] We omitted $n = 0$ in Section 1 to be faithful to the history. It is a nice exercise to consider $n = 0$ in the context of Section 1.

[6] An action of G on M is *cocompact* if there is a compact subset $K \subset M$ with $GK = M$.

$h : X^n \to M$ into the CAT(0)-space M. We call a cellular map $\phi : X^n \to X^n$ a *contraction towards* $e \in \partial M$ if there is a number $\varepsilon > 0$ with

$$\beta_\gamma h\phi(x) \geq \beta_\gamma h(x) + \varepsilon, \text{ for all } x \in X^n,$$

where $\gamma : [0, \infty) \to M$ is a geodesic ray with $\gamma(\infty) = e$. The nth *dynamical invariant of* ρ is

$$\overset{\circ}{\Sigma}{}^n(\rho) := \{e \in \partial M \mid X^n \text{ admits a } G\text{-finitary contraction towards } e\}.$$

This turns out to be independent of the choices of h and X^n.

3.2. Relationship with $\Sigma^n(\rho)$

Now we consider the action of G on ∂M induced by ρ. It is clear that both $\overset{\circ}{\Sigma}{}^n(\rho)$ and $\Sigma^n(\rho)$ are G-invariant subsets of ∂M. The following result asserts that $\overset{\circ}{\Sigma}{}^n(\rho)$ is a "characteristic subset" of $\Sigma^n(\rho)$.

Theorem 2. $\overset{\circ}{\Sigma}{}^n(\rho) = \{e \in \partial M \mid \mathrm{cl}_{\partial M}(Ge) \subseteq \Sigma^n(\rho)\}$.

As an immediate consequence we see that if $E \subseteq \partial M$ is a closed G-invariant subset of ∂M then $E \subseteq \overset{\circ}{\Sigma}{}^n(\rho)$ if and only if $E \subseteq \Sigma^n(\rho)$. This applies, in particular, to the following two important special cases.

Corollary. a) $\overset{\circ}{\Sigma}{}^n(\rho) = \partial M$ *if and only if* $\Sigma^n(\rho) = \partial M$.
b) *Assume* $e \in \partial M$ *is fixed under* G. *Then* $e \in \Sigma^n(\rho)$ *if and only if* $e \in \overset{\circ}{\Sigma}{}^n(\rho)$.

In the "classical" context of Section 1 (BNSR-theory) G acts on a Euclidean space V by translations, $\alpha : G \to \mathrm{Transl}(V)$, and so the action induced on ∂V is trivial. Hence $\overset{\circ}{\Sigma}{}^n(\alpha) = \Sigma^n(\alpha)$; this special case of assertion b) of the corollary is the so-called "Σ^n-Criterion" of the older theory.

3.3. Openness results

The "angular distance" between two points e and e' of ∂M is the supremum over points $a \in M$ of the angle between the geodesic rays γ and γ' representing e and e' where $\gamma(0) = \gamma'(0) = a$. This is a metric on ∂M and the corresponding length metric is called the "Tits distance", denoted $Td(e, e')$. In general, the topology on ∂M given by Td is finer than the cone topology: i.e., id: $(\partial M, Td) \to (\partial M$, cone topology) is continuous. The space $(\partial M, Td)$ is a complete CAT(1) metric space. Two extremes are represented by $M = \mathbb{E}^k$ where the two topologies agree giving S^{k-1}, and $M = \mathbb{H}^k$, k-dimensional hyperbolic space, where $(\partial M, Td)$

is discrete while (∂M, cone topology) is S^{k-1}. If a cellular endomorphism $\phi : X^n \to X^n$ is a contraction towards some $e \in \partial M$, then ϕ is also a contraction towards all $e' \in \partial M$ sufficiently close to e in the Tits metric. Hence, $\mathring{\Sigma}^n(\rho)$ is open in the Tits metric topology of ∂M. In fact, when the function spaces Isom(M) and Hom(G,Isom(M)) are given the compact-open topology, one can show the following.

Theorem 3. $\{(\rho, e) \mid e \in \mathring{\Sigma}^n(\rho)\}$ *is open in* $\mathrm{Hom}(G, \mathrm{Isom}(M)) \times \partial M$ *when* ∂M *carries the Tits distance topology.*

For a different and deeper openness theorem we need ∂M compact, i.e., the cone topology on ∂M.

Theorem 4. *If E is a closed subset of (∂M, cone topology) then* $\{\rho \mid E \subseteq \Sigma^n(\rho)\}$ *is an open subset of* $\mathrm{Hom}(G,\mathrm{Isom}(M, E))$. *In particular,* $\{\rho \mid \Sigma^n(\rho) = \partial M\}$ *is open in* $\mathrm{Hom}(G,\mathrm{Isom}(M))$.

Here, $\mathrm{Isom}(M, E)$ is the space of isometries of M which leave E invariant. The core idea in the proof of these openness theorems is the following: the control function $h : X^n \to M$ can be chosen to vary continuously with the action $\rho \in \mathrm{Hom}(G, \mathrm{Isom}(M))$. Let $e \in \mathring{\Sigma}^n(\rho)$ for a given action ρ, so that we have a finitary contraction $\phi : X^n \to X^n$ towards e. If we could describe ϕ in terms of a finite number of equations we might expect that the very same ϕ would also be a contraction towards e if the action ρ were subjected to a small perturbation. However, a description of ϕ requires not only the finitary G-sheaf $\mathcal{F}(\phi)$ generated by $\mathrm{Res}(\phi)$ but also an infinite number of choices of members of $\mathcal{F}(\phi)$. Thus we cannot expect the same ϕ to work for all ρ' near ρ. But the sheaf $\mathcal{F}(\phi)$ itself can be described in terms of a finite number of equations and we can pin down a finite number of inequalities which are necessary and sufficient for $\mathcal{F}(\phi)$ to have a cross section which contracts towards e. Thus, even though perturbing the action ρ slightly to ρ' requires a new finitary contraction ϕ' towards e to establish $e \in \Sigma^n(\rho')$, we are able to guarantee that ϕ' does exist as a cross section of the old sheaf $\mathcal{F}(\phi)$. A similar reasoning applies to prove Theorem 4 once the following preliminary result is established.

Proposition. *If $E \subseteq \partial M$ is a G-invariant closed (and hence compact) subset, with $E \subseteq \mathring{\Sigma}^n(\rho)$, then there is a locally finite G-sheaf $\mathcal{F}$ with the property that for each $e \in E$ the sheaf*[7] *$\mathcal{F}$ admits a cross section $\phi_e : X^n \to X^n$ which is a contraction towards e.*

[7] Note that the same $\mathcal{F}$ works for all $e \in E$.

4. The case when $\overset{\circ}{\Sigma}{}^n(\rho) = \partial M$

4.1. Contractions towards a point $a \in M$

The case when $\overset{\circ}{\Sigma}{}^n(\rho) = \partial M$ deserves special attention. By definition we then have finitary contractions $\phi_e : X^n \to X^n$ towards each $e \in \partial M$, and we just mentioned in the Proposition that these can be obtained as cross sections of a single locally finite sheaf $\mathcal{F}$. It seems natural to expect that parallel to the construction of these ϕ_e one ought to be able to construct cross sections $\phi_a : X^n \to X^n$ contracting toward points $a \in M$ in the following obvious sense: a cellular map $\phi : X^n \to X^n$ is said to be a *contraction towards* $a \in M$ if there exists a radius $r > 0$ and a number $\epsilon > 0$ such that $d(a, h\phi(x)) \le d(a, h(x)) - \epsilon$ for every $x \in X^n$ with $d(a, h(x)) > r$ (this is independent of a). In order to deduce the existence of a contraction towards $a \in M$ from the hypothesis $\overset{\circ}{\Sigma}{}^n(\rho) = \partial M$ we will actually need the mild assumption that M be almost geodesically complete (see below). But quite generally it can be expressed in terms of

4.2. The CC^{n-1} property over $a \in M$

For $a \in M$ and $r > 0$ we denote by $X_{(a,r)}$ the largest subcomplex of X lying in $h^{-1}(B_r(a))$. We say X is *controlled* $(n-1)$*-connected* (CC^{n-1}) over a (with respect to h) if for all $r \ge 0$ and $-1 \le p \le n-1$ there exists $\lambda \ge 0$ such that every map $f : S^p \to X_{(a,r)}$ extends to a map $\tilde{f} : B^{p+1} \to X_{(a,r+\lambda)}$. If X is CC^{n-1} over some $a \in M$ it is easy to see that X is in fact CC^{n-1} over each point of M, so we can speak of X being CC^{n-1} over M without reference to a point $a \in M$. The number λ in the definition of "CC^{n-1} over a" depends on a and on r. See Section 7 for more on this definition. The number λ can be chosen to be a function $\lambda(r)$ independent of a if and only if the G-action on M is cocompact. The property of X being CC^{n-1} is independent of the choice of X and of h, i.e., is a property of the action ρ. So, if X is CC^{n-1} we will say that ρ is CC^{n-1}.

Theorem 5. *Assume the action $\rho : G \to$ Isom(M) is cocompact. Then the following two conditions are equivalent:*

(i) *ρ is CC^{n-1} over $a \in M$,*
(ii) *there exists a G-finitary contraction $\phi : X^n \to X^n$ towards $a \in M$.*

Moreover, (i) *or* (ii) *implies*

(iii) $\overset{\circ}{\Sigma}{}^n(\rho) = \partial M = \Sigma^n(\rho)$.

4.3. Almost geodesic completeness

The (proper) CAT(0) space (M, d) is *geodesically complete* if every geodesic segment $[0, t] \to M$ can be extended to a geodesic ray $[0, \infty) \to M$. We say the CAT(0) space (M, d) is *almost geodesically complete* if there is a number $\mu \geq 0$ such that for any two points $a, b \in M$ there is a geodesic ray γ with $\gamma(0) = a$ such that $\gamma([0, \infty))$ meets $B_\mu(b)$. A recent Theorem of P. Ontaneda shows that this property is often guaranteed in cases of interest.

Otaneda's Theorem. [On] *Let M be a non-compact proper $CAT(0)$-space such that* Isom(M) *acts cocompactly. A sufficient condition for almost geodesic completeness is that the cohomology with compact supports $H_c^*(M)$ be non-trivial. This condition is satisfied whenever some subgroup of* Isom(M) *acts cocompactly with discrete orbits.*[8]

Theorem 5′. (Boundary Criterion) *If $\rho : G \to Isom(M)$ is an isometric action on a proper and almost geodesically complete $CAT(0)$-space M then the conditions* (i), (ii) *and* (iii) *in Theorem* 5 *are equivalent.*[9]

As a consequence of Theorem 5′ the openness result of Section 3 implies that (i), (ii) and (iii) are "open conditions" for $\rho \in \mathrm{Hom}(G, \mathrm{Isom}(M))$ when G is of type F_n and M is almost geodesically complete. In [BG$_\mathrm{I}$] we give a direct proof of the openness of (i) and (ii) without reference to (iii). This proof is more technical but has the advantage that the assumption that M be almost geodesically complete is not needed.

Theorem 6. *If G is a group of type F_n then the set*

$$\{\rho \mid \rho \ \text{ is cocompact and } \ CC^{n-1} \ \text{ over } \ a \in M\}$$

is an open subset of $\mathrm{Hom}(G, \mathrm{Isom}(M))$.

Remark. We believe that the equivalent conditions (i) and (ii) in Theorem 5 define basic properties of an isometric action ρ, and that Theorem 5′ should be viewed as a tool to establish these properties, via the Boundary Criterion. In fact, this criterion is a local–global principle: it breaks up the problem of deciding whether ρ is CC^{n-1} into "local" questions over each $e \in \partial M$, so that different methods and viewpoints may be used for different endpoints. An example of

[8] D. Farley has shown that this condition is also satisfied if M (as above) is an M_κ-complex with finite shapes (see [BrHa] for the relevant definitions).

[9] We do not know whether the assumption that M be almost geodesically complete is necessary in Theorem 5′. It is not needed in the case $n = 0$, which is Theorem 1. It is an open problem as to whether there exists a non-compact CAT(0) space M which is not almost geodesically complete but admits a cocompact group action by isometries.

this occurs in Section 6.1 where we establish the precise CC^{n-1} properties of the Möbius action of various subgroups of $SL_2(\mathbb{R})$ on the hyperbolic plane.

5. Applications

5.1. Openness Results for the F_n-property

When the action $\rho : G \to \mathrm{Isom}(M)$ has discrete orbits[10] then the property that ρ be CC^n over $a \in M$ has an interpretation in terms of the finiteness property "type F_n" on the stabilizer G_a of a.

Theorem 7. *Let $\rho : G \to \mathrm{Isom}(M)$ be a cocompact action which has discrete orbits. Then ρ is CC^{n-1} if and only if the stabilizer G_a has type F_n.*

Theorem 7 is a consequence of the following homotopy version of K.S. Brown's finiteness criterion [Br 87, Theorem 2.2]:

F_n-Criterion. *Let H be a group, Y a contractible free H-CW complex and $(K_r)_{r\in\mathbb{R}}$ an increasing filtration of Y by H-subcomplexes so that $Y = \bigcup_r K_r$ and each K_r has cocompact n-skeleton. Then H is of type F_n if and only if Y is CC^{n-1} with respect to the filtration (K_r).*[11]

Indeed, Theorem 7 follows by setting $Y = X$, $K_r = X_{(a,r)}$ and $H = G_a$. Clearly $X_{(a,r)}$ is a G_a-subcomplex. The remaining part of the proof, that each $X^n_{(a,r)}$ is cocompact as a G_a-complex, is not hard.

A special case of Theorem 7 is worth noting. If $N = \ker \rho$ we have a short exact sequence $N \rightarrowtail G \twoheadrightarrow Q$ with $Q \le \mathrm{Isom}(M)$, and short exact sequences for the stabilizers $N \rightarrowtail G_a \twoheadrightarrow Q_a$. If we replace the assumption that ρ have discrete orbits by the stronger assumption that the induced action of Q on M be properly discontinuous[12] then Theorem 7 applies – but since all Q_a are finite the assertion that G_a be of type F_n is equivalent to N being of type F_n. Hence Theorem 7 becomes the following.

Theorem 7′. *Let Q act cocompactly and properly discontinuous on M. Then ρ is CC^{n-1} if and only if N has type F_n.*

From Theorem 6 we get:

[10] An action of G on M *has discrete orbits* if every orbit is a closed discrete subset of M.

[11] In Sections 2.2 and 3.1 we defined CC^{n-1} using filtrations which came from control functions, but the definition makes sense with respect to any filtration.

[12] An action of Q on M is *properly discontinuous* if every point $a \in M$ has a neighbourhood U such that $\{q \in Q | qU \cap U \neq \emptyset\}$ is finite (equivalently: if the action has discrete orbits and has finite point stabilizers).

Corollary. *Let $\mathcal{R}(G, M)$ denote the space of all isometric actions of G on M which have discrete orbits. Then the set of all isometric actions $\rho \in \mathcal{R}(G, M)$ which are cocompact and have point stabilizers of type F_n is open in $\mathcal{R}(G, M)$.*

Corollary. *Let $\mathcal{R}_0(G, M)$ denote the subspace of all $\rho \in \mathcal{R}(G, M)$ with the property that $\rho(G)$ acts properly discontinuously on M. Then the set of all $\rho \in \mathcal{R}_0(G, M)$ which are cocompact and have* $\ker \rho$ *of type F_n is open in $\mathcal{R}_0(G, M)$.*[13]

There is no hope of a general openness result in $\mathrm{Hom}(G, \mathrm{Isom}(M))$ for the finiteness properties "$\ker \rho$ is of type F_n" or "the point stabilizers of ρ are of type F_n". This indicates the advantage of the property CC^{n-1} over these traditional finiteness properties. To get a counterexample, consider a finitely generated group G whose abelianization G/G' is free of rank 2, and take M to be the Euclidean line $\mathbb{E}^1$. Then every non-discrete translation action of G on $\mathbb{E}^1$ has kernel the commutator subgroup G'. But the non-discrete translation actions are dense in the space of all translation actions. So if we had an openness result for the property "$\ker \rho$ is finitely generated", it would imply "G' is finitely generated if (and only if) some homomorphism $\chi : G \twoheadrightarrow \mathbb{Z}$ has finitely generated kernel". This is absurd as is shown by the direct product $G = \langle a, x | xax^{-1} = a^2 \rangle \times \mathbb{Z}$ which has commutator subgroup isomorphic to the dyadic rationals, i.e., $G' \cong \mathbb{Z}[\frac{1}{2}]$.

5.2. Connections with Lie groups and local rigidity

The following examples explain how our openness results are related to locally rigid isometric actions of discrete groups on classical symmetric spaces.

Example. Let M be a locally symmetric space of non-compact type (e.g. the quotient of a virtually connected non-compact linear semisimple Lie group by a maximal compact subgroup). The natural Riemannian metric makes M a proper CAT(0) space. The group $\mathrm{Isom}(M)$ is a Lie group. Call its Lie algebra $\mathfrak{g}$. Each representation $\rho \in \mathrm{Hom}(G, \mathrm{Isom}(M))$ makes $\mathfrak{g}$ into a $\mathbb{Z}G$-module which we denote by $\mathfrak{g}(\rho)$. A theorem of Weil [We 64] says that if G is finitely generated and if $H^1(G; \mathfrak{g}(\rho)) = 0$ then all nearby representations are conjugate to ρ in $\mathrm{Isom}(M)$, i.e., ρ has a neighbourhood N in $\mathrm{Hom}(G, \mathrm{Isom}(M))$ such that every $\rho' \in N$ is of the form $\rho'(g) = \gamma\rho(g)\gamma^{-1}$ where γ (dependent on ρ') is an isometry of M; then ρ is said to be *locally rigid* (see [Rag p. 90]). In that

[13] This Corollary has predecessors in the literature for the case of homomorphisms $\rho : G \to \mathbb{Z}$. Openness of the condition "$\ker \rho$ is finitely generated" was proved in [Ne 79], and of the condition "$\ker \rho$ is finitely presented" in [FrLe 85]. See also [BRe 88] and [Re 88].

case $\ker(\rho') = \ker(\rho)$ for all $\rho' \in N$ – a much stronger statement than the conclusion of the last Corollary. But this Corollary holds in situations where $H^1(G; \mathfrak{g}(\rho)) \neq 0$, so one may wish to think of it as a weak form of local rigidity: the kernels may not be locally constant, but their finiteness properties are locally constant. The next example illustrates this.

Example. Let G be the group presented by $\langle x, y | xy^2 = y^2 x \rangle$. For $n \geq 0$ define $\rho_n : G \to \mathbb{Z}$ by $\rho_0(x) = 0$, $\rho_0(y) = 1$, and when $n \geq 1$ $\rho_n(x) = n$, $\rho_n(y) = n^2$. It is shown in [BS] that $\ker(\rho_0)$ is a free group of rank 2 and when $n \geq 1$, $\ker(\rho_n)$ is a free group of rank $n^2 + 1$. For $n \geq 1$ define $\tilde{\rho}_n : G \to \mathbb{R}$ by $\tilde{\rho}_n(g) = \frac{1}{n^2}\rho_n(g)$. Identifying $\mathbb{R}$ with the translation subgroup of $\mathrm{Isom}(\mathbb{R})$, we see that $\{\tilde{\rho}_n\}$ converges to ρ_0 in $\mathrm{Hom}(G, \mathrm{Isom}(\mathbb{R}))$, and $\ker(\tilde{\rho}_n) = \ker(\rho_n)$ for all $n \geq 1$. Indeed, each $\tilde{\rho}_n$ is a cocompact action and $\tilde{\rho}_n(G)$ acts properly discontinuously on $\mathbb{R}$. This is a case where our Corollary applies but local rigidity fails.

Remark. The paper [Fa 99] contains results in a Lie group context which can be seen as analogous to Theorem 3 and the last Corollary.

5.3. The problem of fibering a manifold over a manifold

In this section we point out a connection between this work and the problem of deforming a map of closed connected manifolds to a fiber bundle map. The various definitions of CC^{n-1} in Sections 1-3 have homological analogues. We will only state one in detail. We say that X is *controlled* $(n-1)$*-acyclic* (CA^{n-1}) in the direction γ (with respect to the action ρ) if for any horoball $HB_s(\gamma)$ and $-1 \leq p \leq n-1$ there exists $\lambda(s) \geq 0$ such that every singular p-cycle (with integer coefficients) in $X_{(\gamma,s)}$ bounds a singular $(p+1)$-chain in $X_{(\gamma,s-\lambda(s))}$, and $s - \lambda(s) \to \infty$ as $s \to \infty$. By standard methods, one proves, for this situation (see, for example, [Ge]):

Hurewicz Theorem. *For $n \geq 2$, CC^1 in the direction γ and CA^{n-1} in the direction γ is equivalent to CC^{n-1} in the direction γ.*

A group G is of *type* F [resp. *type* FD] if there exists a finite [resp. finitely dominated[14]] $K(G, 1)$-complex. Let $N \rightarrowtail G \twoheadrightarrow Q$ be a short exact sequence of groups where G has type F and Q acts properly discontinuously and cocompactly on the proper CAT(0) space M. Assume that M is almost geodesically complete.

Question. When is N of type FD?

[14] Y is *finitely dominated* if there exist a finite complex W and maps $Y \to W \to Y$ so that the indicated composition is homotopic to id_Y.

We may choose our free G-complex X to be contractible, cocompact and of finite dimension, say dimension d. It follows that N has geometric dimension $\leq d$, since $N \backslash X$ is a $K(N, 1)$-complex of dimension d. It is well-known (see [Ge; Chapter 2] for example) that geometric dimension $\leq d$ together with type F_d implies type FD. By Theorem 2′ it follows that if the action ρ of G on M (induced by $G \to Q \hookrightarrow \text{Isom}(M)$) is CC^{d-1} then N has type F_d. If we combine the Boundary Criterion (Theorem 5′), Theorem 7′ and the Hurewicz Theorem stated above, we conclude the following.

Theorem 8. *If N is finitely presented and the action ρ is CA^{d-1} over every $e \in \partial M$ then N is of type FD.*

In [BG 98] we replace the condition "CA^{d-1} over every e" by a condition "$H_i(G; \mathcal{N}(\rho)) = 0 \; \forall i \leq d-1$" where $\mathcal{N}(\rho)$ is a G-module of "Novikov coefficients" analogous to the Novikov Ring in, for example, [No], [Pa], [Ra]. We will not repeat that discussion here. But it is worth noting the following connection with the fibering problem for manifolds.

Let $f : V \to W$ be a map from a closed aspherical triangulable d-manifold to a closed k-manifold of non-positive sectional curvature. Assume both manifolds are connected and that f induces on fundamental group an epimorphism $G \twoheadrightarrow Q$ with kernel N. Then the space $N \backslash \tilde{V}$ is the homotopy fiber[15] of f; it is aspherical, so it is finitely dominated if and only if N has type FD. Since $\tilde{W}$ is a proper almost geodesically complete CAT(0) space, we are in the situation of Theorem 8, and thus, once we know that N is finitely presented (something that can be checked fairly easily in practice) Theorem 8 reduces the problem of checking that the homotopy fiber of f is finitely dominated to a local homological condition over each $e \in \partial \tilde{W}$. The relevance of this to geometric topology is the following theorem of Farrell and Jones [FJ 89].

Theorem. *Let $d - k \geq 5$ and let W be hyperbolic (i.e., of constant negative sectional curvature). Then f is homotopic to a block bundle map if and only if the homotopy fiber of f is finitely dominated.*

We will not define "block bundle" here – see [FJ 89]; it is a weaker concept than "fiber bundle". Note that the K-theoretic torsions which are required to vanish in [FJ 89, Theorem 10.7] do vanish under these hypotheses.[16]

[15] Given a map $\phi : A \to B$ between topological spaces there is a space A' containing A as a strong deformation retract and a Hurewicz fibration $\phi' : A' \to B$ with $A \hookrightarrow A' \xrightarrow{\phi'} B$ homotopic to ϕ; see [Sp; Ch. 2]. The fiber of ϕ' (or any space of the same homotopy type) is the *homotopy fiber* of ϕ.

[16] We thank Tom Farrell for drawing our attention to this connection with [FJ 89].

6. Examples

6.1. SL_2-actions on the hyperbolic plane

Let m be a positive integer. The *Möbius action*, $\begin{bmatrix} a & b \\ c & d \end{bmatrix} z = \frac{az+b}{cz+d}$, of $SL_2(\mathbb{Z}[\frac{1}{m}])$ on the hyperbolic plane $\mathbb{H}$ (the upper half space model $\{z \in \mathbb{C} \mid \operatorname{Im} z > 0\}$ with metric ds/y and $\partial\mathbb{H} = \mathbb{R} \cup \{\infty\}$) is an action by isometries. Except for the case of $SL_2(\mathbb{Z})$, i.e., $m = 1$, it is an indiscrete action. We denote this action by $\mu_m : SL_2(\mathbb{Z}[\frac{1}{m}]) \to \operatorname{Isom}(\mathbb{H})$.

Theorem 9. *Let m be divisible by exactly s different primes. Then*

$$\Sigma^n(\mu_m) = \begin{cases} \partial\mathbb{H}, & \text{if } n < s \\ \mathbb{R} - \mathbb{Q}, & \text{if } n \geq s. \end{cases}$$

The lowest case, $m = 1$ ($s = 0$) arose in Section 2.3. Theorem 9 contains the well-known fact that under the Möbius action of $SL_2(\mathbb{Z})$ on $\mathbb{H}$, the irrational boundary points are precisely the "points of approximation"; see [Be] and [BG 98, §3.4]. By Theorem 1 it also includes the easily proved fact that the Möbius action μ_m is cocompact if and only if $m \geq 2$. In the cocompact ($m \geq 2$) case the Boundary Criterion (Theorem 5′) yields

Corollary. *For $m \geq 2$, μ_m is CC^{n-1} over $a \in M$ if and only if $n < s$.*

6.2. Generalization to other subgroups of $SL_2(\mathbb{R})$.

We briefly comment on the proof of Theorem 9 and we will observe that it goes through in more general circumstances. The full proof is in [BG_{III}]. The crucial facts about a group $G \leq SL_2(\mathbb{R})$ and its Möbius action $\mu : G \to \operatorname{Isom}(\mathbb{H})$ which we need are the following: G acts cocompactly on a contractible locally finite CW-complex Y so that the base point stabilizer $H \leq G$ is a Fuchsian group[17]. (In Section 6.1 we had $G = SL_2(\mathbb{Z}[\frac{1}{m}])$, $H = SL_2(\mathbb{Z})$, and Y the product of the Bruhat-Tits trees for the s different prime divisors of m.) The space $X = \mathbb{H} \times Y$ is contractible and the diagonal action of G on X is properly discontinuous. Hence X, endowed with projection $h : X \twoheadrightarrow \mathbb{H}$, is suitable for computing $\Sigma^n(\mu)$ provided G acts cocompactly, i.e., when H is a cocompact Fuchsian group. If so we find easily that $\Sigma^n(\mu) = \partial\mathbb{H}$ for all $n \geq 0$.

If H is a non-compact Fuchsian group (such as $SL_2(\mathbb{Z})$), X is not permitted in the limited definition of $\Sigma^n(\rho)$ given in Section 2. However, the more elaborate

[17] That is, a discrete subgroup of $SL_2(\mathbb{R})$ with finite covolume.

definitions given in Section 7.2 below, show us how to check the CC-properties using X and h. In the absence of cocompactness we need a filtering of X by cocompact subspaces, and these must be chosen with care. In the special case when H has finite index in G then G is itself Fuchsian and we can take $X = \mathbb{H}$ filtered by the truncated planes

$$C_t = \mathbb{H} - H\{z \in \mathbb{C} \mid \mathrm{Im}(z) < \log t\}, \quad t \geq 0.$$

One finds for all $n \geq 0$

$$\begin{aligned}\Sigma^n(\mu) &= \{e \in \partial\mathbb{H} \mid e \text{ not a parabolic fixed point of } H\} \\ &= \{e \in \partial\mathbb{H} \mid e \text{ not a cusp of any Dirichlet domain for } H\}.\end{aligned}$$

The assumption that the G-CW-complex Y be locally finite implies that its stabilizers are commensurable; hence G is in the commensurator of H. By a famous theorem of Margulis [Ma] it follows that if the index of H in G is not finite then H must be an arithmetic Fuchsian group. These groups are classified, and from this classification (e.g. [Ka]) we find that if H is not cocompact then, up to conjugacy, we have $G \leq SL_2(\mathbb{Q})$ and H commensurable with $SL_2(\mathbb{Z})$. This brings us close to the special case of Section 6.1 where $G = SL_2(\mathbb{Z}[\frac{1}{m}])$ but does leave room for more examples.

The proof of Theorem 9 (and its generalizations) proceeds then as follows. Once the filtration of the G-space X by cocompact subspaces is set up, the inclusion $\mathbb{R} - \mathbb{Q} \subseteq \Sigma^\infty(\mu)$ can be established. The discussion of the rational endpoints $e \in \mathbb{Q} \cup \{\infty\}$ is based on the following somewhat surprising result which - in the restricted situation of this subsection - relates the question "is $e \in \Sigma^n(\mu)$?" to the restricted action of the stabilizer G_e on $\mathbb{H}$.

Theorem 10. *For $e \in \mathbb{Q} \cup \{\infty\}$ and $n \geq 0$, we have $e \in \Sigma^n(\mu)$ if and only if $e \in \Sigma^n(\mu \mid G_e)$.*

The advantage of passing to G_e is twofold. On the one hand G_e fixes the endpoint e and hence permutes the horoballs at e. In fact the action μ induces a character $\chi_e : G_e \to \mathbb{R}$ with $gHB_s(\gamma) = HB_{s+\chi_e(g)}(\gamma)$, for all $g \in G_e$, where γ is a geodesic ray with $\gamma(\infty) = e$. In [BG$_{\text{II}}$] we prove that $e \in \Sigma^n(\mu \mid G_e)$ if and only if $[\chi_e] \in \Sigma^n(G_e)$. On the other hand the stabilizer G_e is a metabelian group and this allows us to extract what we need from the literature on the geometric invariants in the sense of Section 1: namely, a theorem of H. Meinert [Me 96] which reduces $\Sigma^n(G_e)$ to $\Sigma^1(G_e)$, and the computation of $\Sigma^1(G_e)$ by means of valuations on fields in [BS 81] and [BGr 84].

6.3. Tree actions

Let T be an infinite locally finite tree and let $\rho : G \to \mathrm{Isom}(T)$ be a cocompact action of G by simplicial automorphisms. Then, by Bass-Serre theory, G is the fundamental group of a finite graph of groups $(\Gamma, \mathcal{G})$, where $\Gamma = G\backslash T$ and $\mathcal{G}$ is the system of edge and vertex stabilizers along a fundamental transversal of T. The edge stabilizers are of finite index in the vertex stabilizers since T is locally finite. Following [Bi 98] we define the *finiteness length* of G [resp. $\mathcal{G}$] to be $\mathrm{fl}G := \sup\{k \mid G \text{ is of type } F_k\}$ [resp. $\mathrm{fl}\mathcal{G} := \inf\{\mathrm{fl}H \mid H \in \mathcal{G}\}$], and the *connectivity length* of a character $\chi : G \to \mathbb{R}$ to be $\mathrm{cl}(\chi) := \sup\{k \mid k \le \mathrm{fl}G$ and $[\chi] \in \Sigma^k(G)\}$. In this case, $\mathrm{fl}\mathcal{G} = \mathrm{fl}H$ for any $H \in \mathcal{G}$. We begin by noting three elementary facts. First, T is almost geodesically complete so we may apply Theorem 5′. Secondly, if the fixed point set $(\partial T)^G$ is a proper subset of ∂T then it is either empty or a singleton; for if there are two singleton orbits then ∂T consists of just those points[18]. Thirdly, any orbit consisting of more than one point is dense[19], so that its closure is ∂T. It follows that if S is the union of closures of orbits – and $\mathring{\Sigma}^n(\rho)$ is such a set – then $S = \partial T$ or is a singleton or is empty. It is well known that $\mathrm{fl}\mathcal{G} \le \mathrm{fl}G$, so *if* $(\partial T)^G = \emptyset$ *then*[20]

$$(*) \qquad \mathring{\Sigma}^n(\rho) = \begin{cases} \partial T & \text{if } 0 \le n \le \mathrm{fl}\mathcal{G} \\ \emptyset & \text{if } \mathrm{fl}\mathcal{G} < n \le \mathrm{fl}G. \end{cases}$$

There remains the case when the G-tree T has exactly one fixed end e. For such a tree we have the associated non-zero character $\chi_e : G \to \mathbb{R}$ of Section 6.2 measuring the shift towards e. Thus if $n \le \mathrm{fl}\mathcal{G}$, $\Sigma^n(\rho)$ is defined, and by Theorems 7 and 5′, $\Sigma^n(\rho) = \partial T$. In particular, $e \in \Sigma^n(\rho)$ and therefore $[\chi_e] \in \Sigma^n(G)$, implying $n \le \mathrm{cl}(\chi_{\rho,e})$. In summary: $\mathrm{fl}\mathcal{G} \le \mathrm{cl}(\chi_{\rho,e}) \le \mathrm{fl}G$ and

$$(**) \qquad \mathring{\Sigma}^n(\rho) = \begin{cases} \partial T & \text{if } 0 \le n \le \mathrm{fl}\mathcal{G} \\ \{e\} & \text{if } \mathrm{fl}\mathcal{G} < n \le \mathrm{cl}(\chi_{\rho,e}) \\ \emptyset & \text{if } \mathrm{cl}(\chi_{\rho,e}) < n \le \mathrm{fl}G. \end{cases}$$

In [BG$_{\mathrm{II}}$], we compute the parameters $\mathrm{fl}\mathcal{G}$, $\mathrm{cl}(\chi_{p,e})$ and $\mathrm{fl}G$ occurring in $(**)$ when G is metabelian of finite Prüfer rank[21]. Among such groups one can achieve (in $(**)$) $\mathring{\Sigma}^n(\rho) = \partial T$ and $\mathring{\Sigma}^n(\rho) = \{e\}$ in arbitrarily large ranges of n, while $\mathring{\Sigma}^n(\rho) = \emptyset$ only occurs when $n = \mathrm{fl}G < \infty$.

[18] And in that case T is essentially a line, a case already understood.

[19] Let e be in such an orbit. The union of lines from $g_1 e$ to $g_2 e$, for all $g_1, g_2 \in G$ is non-empty: it is a G-invariant subtree. This together with cocompactness implies Ge is dense.

[20] When $\mathrm{fl}G$ or $\mathrm{fl}\mathcal{G} = \infty$, this is to be understood as meaning "for all n".

[21] This means: the commutator subgroup G' is abelian with finite torsion and finite torsion free rank ($\dim_{\mathbb{Q}}(G' \otimes \mathbb{Q} < \infty)$).

7. Controlled connectivity

The usefulness of the theorems in this paper depends on one's ability to check the CC or CA properties. It turns out that there are various formulations of these conditions which are equivalent, but not obviously so. We discuss these for the CC case; the CA case is entirely analogous.

7.1. G of type F_n

Up to now we have only used the properties CC^{n-1} in contexts where G is of type F_n; specifically we required that $G\backslash X^n$ be a finite complex. In this case we can characterize CC^{n-1} in the direction γ (see Section 2) in various ways.

Theorem 11. *The following are equivalent:*

(i) *for any $m \in \mathbb{Z}$ and $-1 \leq p \leq n-1$, there exists an integer $\lambda(m) \geq 0$ such that $m - \lambda(m) \to \infty$ as $m \to \infty$ and every map $f : S^p \to X_{(\gamma,m)}$ extends to a map $\tilde{f} : B^{p+1} \to X_{(\gamma,m-\lambda(m))}$,*

(ii) *for any $m \in \mathbb{Z}$ and $-1 \leq p \leq n-1$, there exists an integer $\mu(m) \geq 0$ such that $m + \mu(m) \to -\infty$ as $m \to -\infty$ and every map $f : S^p \to X_{(\gamma,m+\mu(m))}$ extends to a map $\tilde{f} : B^{p+1} \to X_{(\gamma,m)}$,*

(iii) *for any $m \in \mathbb{Z}$ and $-1 \leq p \leq n-1$ there exist integers $\lambda(m), \mu(m) \geq 0$ such that every map $f_1 : S^p \to X_{(\gamma,m)}$ extends to a map $\tilde{f}_1 : B^{p+1} \to X_{(\gamma,m-\lambda(m))}$, and every map $f_2 : S^p \to X_{(\gamma,m+\mu(m))}$ extends to a map $\tilde{f}_2 : B^{p+1} \to X_{(\gamma,m)}$.*

Condition (i) is clearly equivalent to saying that ρ is CC^{n-1} over $e := \gamma(\infty)$. Hence (ii) and (iii) are equivalent formulations of that definition. If $[S^p, Z]$ denotes the set of homotopy classes of maps from S^p to the space Z, then (iii) can be rephrased as saying that the sequence

$$\cdots \leftarrow [S^p, X_{(\gamma,m)}] \leftarrow [S^p, X_{(\gamma,m+1)}] \leftarrow \cdots$$

of sets and functions (induced by inclusions) is "essentially trivial" both as an inverse sequence and as a direct sequence. The terms "pro-trivial" in the inverse case and "ind-trivial" in the direct case are also used. It often turns out that Condition (i) holds with constant $\lambda(m) =: \lambda \geq 0$. The proof of Theorem 11 then shows that $\mu(m)$ in Condition (ii) can also be taken to be constant; in fact $\mu(m) = \lambda$. Theorem 11 is proved in the Appendix to [BG$_{\text{II}}$].

7.2. More general G and X

In checking the CC conditions, there are occasions, such as the examples described in Sections 6.1 and 6.2, when the natural X is not cocompact. And we would wish for a definition of CC^{n-1} even when G does not have type F_n. These two issues are addressed in this section. For proofs see [BG$_{\text{III}}$].

Let G be a group, (M, d) a proper CAT(0), X a non-empty G-space (i.e. G acts on X by homeomorphisms) and $h : X \to M$ a G-map (a "control function"). We say that X is *controlled n-connected* (abbrev. CC^n) *over the point* $a \in M$ if each cocompact G-subspace K of X lies in a cocompact G-subspace K' such that for each $r \geq 0$ and each $-1 \leq p \leq n$ there exists $\lambda \geq 0$ satisfying:

($*_a$) every singular p-sphere in K over $B_r(a)$ bounds a singular $(p+1)$-ball in K' over $B_{r+\lambda}(a)$.

This property is independent of the choice of control function h, because if h' is another then for each cocompact set $K \subset X$ there is a number δ_K such that $d(h(x), h'(x)) \leq \delta_K$ for all $x \in K$. So only the existence of a control function $X \to M$ is needed for the definition to make sense. In general, λ depends on K, r and a. If X is CC^n over $a \in M$ then X is CC^n over every point of M, but with varying λ. If λ can be chosen independent of a (i.e., depending only on K and r), we say that X *is uniformly* CC^n *over* M. Again, this is independent of the control function. The case $p = -1$ is included in ($*_a$): it means that $h^{-1}(B_{r+\lambda}(a)) \cap K'$ is non-empty. Of course any non-empty X is CC^{-1} over each $a \in M$. But "uniformly CC^{-1}" has a useful interpretation:

Proposition 1. *X is uniformly CC^{-1} over M if and only if the G-metric space M is cocompact.* □

The proof is straightforward, as is the proof of:

Proposition 2. *Let M be cocompact. Then X is CC^n over some $a \in M$ if and only if X is uniformly CC^n over M.* □

These CC^n-properties of X over M are homotopy invariant in a strong sense:

Theorem 12. (Invariance Theorem) *Let $h : X \to M$ and $h' : X' \to M$ be G-maps and let X' be G-homotopically dominated by X.*[22] *If X is CC^n over a (resp. uniformly CC^n over M) then X' is CC^n over a (resp. uniformly CC^n over M).*

[22] That is, there are G-maps $\phi : X \to X'$ and $\psi : X' \to X$ such that $\phi \circ \psi$ is G-homotopic to $\mathrm{id}_{X'}$.

Proof. The proof is routine. □

There is a "best" X to choose over M, namely the universal cover of a $K(G, 1)$-complex, by definition a contractible free G-CW complex. Since the action on X is free and M is contractible, control functions $h : X \to M$ exist. This brings us to our main definition: we say that the G-action on M *is* CC^n *over* $a \in M$ [resp. *is uniformly* CC^n] if this X is CC^n over a [resp. is uniformly CC^n over M]. This definition is independent of X and h by Theorem 16. More precisely, we can check whether the G-action on M is CC^n over a, or uniformly CC^n, by using any contractible space X on which G acts freely and properly discontinuously, and any control map, provided X has the G-homotopy type of a G-CW complex. For example, we may use the universal cover of a $K(G, 1)$-space which is an ANR rather than a CW complex. By a *G-ANR* we simply mean an ANR equipped with an action of G by homeomorphisms.

Proposition 3. *Let X be a contractible G-ANR, let the action of G on X be properly discontinuous, let $h : X \to M$ be a control function, and let $a \in M$. If G has a torsion-free subgroup of finite index, then X is CC^n over $a \in M$ (resp. is uniformly CC^n over M) if and only if the G-action on M is CC^n over a (resp. is uniformly CC^n over M).*

In other words, in this case one can read off the CC-properties of the action on M using this X and h.

Let $e \in \partial M$ be represented by the geodesic ray γ. The pair (X, h) is CC^n *over* e if each cocompact G-subspace K of X lies in a cocompact subspace K' such that for each $t \in \mathbb{R}$ and each $-1 \leq p \leq n$ there exists $\lambda \geq 0$ satisfying

($*_e$) every singular p-sphere in K over $HB_t(\gamma)$ bounds a singular $(p+1)$-ball in K' over $HB_{t-\lambda}(\gamma)$

and

($**_e$) every singular p-sphere in K over $HB_{t+\lambda}(\gamma)$ bounds a singular $(p+1)$-ball in K' over $HB_t(\gamma)$.

Of course this[23] depends only on e, not on γ. Moreover, independence of h and the obvious analogue of Theorem 16 concerning G-homotopy invariance hold here too. We say that the G-action on M *is* CC^n *over* $e \in \partial M$ if some (equivalently, any) contractible free[24] G-CW complex X is CC^n over e. If G has

[23] The case $p = -1$ says, in ($*_e$), that X is non-empty, and, in ($**_e$), that $h^{-1}(HB_t(\gamma)) \cap K$ is non-empty.

[24] CAT(0) spaces are contractible so the necessary control function exists.

a torsion-free subgroup of finite index the obvious analogue of Proposition 7.3 holds and "the G-action on M being CC^n over e" can be measured using a properly discontinuous G-ANR.

It should be clear to the reader that the definitions of CC^{n-1} given here are consistent with those given in earlier sections.

8. A Non-finitely generated example

With this extended definition of the CC properties, we can state a natural extension of Theorem 9. Let S be an infinite set of primes and let $\mathbb{Z}_S$ denote the subring of $\mathbb{Q}$ whose denominators, in lowest terms, only involve primes in S. Let $\mu_S : SL_2(\mathbb{Z}_S) \to \text{Isom}(\mathbb{H})$ denote the Möbius action. Whereas the groups $SL_2(\mathbb{Z}[\frac{1}{m}])$ are of type F_∞, the group $SL_2(\mathbb{Z}_S)$ is not finitely generated. Nevertheless, using our extended definition we have the following.

Theorem 13. *The action μ_S is CC^∞ over all points of $\partial\mathbb{H}$.*[25]

Note that this applies in particular to the $SL_2(\mathbb{Q})$ action on $\mathbb{H}$.

References

[Be] A.F. Beardon, *The geometry of discrete groups*, Springer, New York, 1995.

[Bi 98] R. Bieri, Finiteness length and connectivity length for groups, in: *Geometric group theory down under*, (Canberra, 1996), de Gruyter, Berlin, 1999, 9–22.

[BG 98] R. Bieri and R. Geoghegan, Kernels of actions on non-positively curved spaces, in: *Geometry and cohomology in group theory* (P.H. Kropholler, G. Niblo and R. Stöhr, ed.) LMS Lecture Notes vol. 252, Cambridge University Press, Cambridge, 1998, 24–38.

[BG_I] R. Bieri and R. Geoghegan, Connectivity properties of group actions on non-positively curved spaces I: controlled connectivity and opennenss results, Memoirs of AMS vol. 161 (2003) no. 764.

[BG_{II}] R. Bieri and R. Geoghegan, Connectivity properties of group actions on non-positively curves spaces II: the geometric invariants, Memoirs of AMS vol. 161 (2003) no. 764.

[BG_{III}] R. Bieri and R. Geoghegan, Topological properties of SL_2 actions on the hyperbolic plane, *Geometriae Dedicata*, to appear.

[BGr 84] R. Bieri and J.R.J. Groves, The geometry of the set of characters induced by valuations, *J. Reine und Angew. Math.* **347** (1984), 168–195.

[BNS 87] R. Bieri, W. Neumann and R. Strebel, A geometric invariant of discrete groups, *Invent. Math.* **90** (1987), 451–477.

[25] CC^∞ means CC^n for all n.

[BR 88] R. Bieri and B. Renz, Valuations on free resolutions and higher geometric invariants of groups, *Comm. Math. Helvetici* **63** (1988), 464–497.

[BS] R. Bieri and R. Strebel, *Geometric invariants for discrete groups*, monograph in preparation.

[BS 81] R. Bieri and R. Strebel, A geometric invariant for modules over an Abelian group, *J. Reine und Angew. Math.* **322** (1981), 170–189.

[BrHa] M. Bridson and A. Haefliger, *Metric spaces of non-positive curvature*, Grundlehren der Mathematischen Wissenschaften 319, Springer, Berlin, 1999.

[Br 87] K.S. Brown, Finiteness properties of groups, *J. Pure and Applied Algebra* **44** (1987), 45–75.

[Fa 99] F.T. Farrell, Fibered representations, an open condition, *Topology Appl.* **96** (1999) 185–190.

[FJ 89] F.T. Farrell and L.E. Jones, A topological analogue of Mostow's rigidity theorem, *J. Amer. Math. Soc.* **2** (1989), 257–370.

[FrLe 85] D. Fried and R. Lee, Realizing group automorphisms, *Contemp. Math.* **36** (1985), 427–433.

[Ge] R. Geoghegan, *Topological methods in group theory*, monograph in preparation.

[Ka] S. Katok, Fuchsian groups, Chicago Lectures in Mathematics, *University of Chicago Press, Chicago, IL*, 1992.

[Ma] G.A. Margulis, Discrete groups of motions of manifolds of nonpositive curvature, *Proceedings of the International Congress of Mathematicians (Vancouver, B.C., 1974), vol. 2*, pp. 21–34. *Canad. Math. Congress, Montreal, Que.*, 1975.

[Me 96] H. Meinert, The homological invariants of metabelian groups of finite Prüfer rank: a proof of the Σ^m-conjecture, *Proc. London Math. Soc.* (3) **72** (1996), 385–424.

[Ne 79] W.D. Neumann, Normal subgroups with infinite cyclic quotient, *Math. Sci.* **4** (1979), 143–148.

[No 82] S.P. Novikov, The hamiltonian formalism and a multivalued analogue of Morse theory, *Uspekhi Mat. Nauk* **37**, 3–49 (1982) English: *Russ. Math. Surv.* **37**, 1–56 (1982).

[On] P. Ontaneda, Cocompact CAT(0) spaces are almost extendible, preprint.

[Pa 95] A.V. Pazhitnov, On the Novikov complex for rational Morse formes, *Annales de la Faculté de Sciences de Toulouse*, **4** (1995), 297–338.

[Rag] M.S. Ragunathan, *Discrete subgroups of Lie groups*, Ergebnisse der Math. vol. 68, Springer, Berlin, 1972.

[Ra 95] A. Ranicki, Finite domination and Novikov rings, *Topology* **34** (1995) 619–632.

[Re 88] B. Renz, *Geometrische Invarianten und Endlichkeitseigenschaften von Gruppen*, Dissertation, Frankfurt am Main, 1988.

[Se] J.-P. Serre, *Trees*, Springer, Berlin, 1980.

[Sp] E. Spanier, *Algebraic Topology*, McGraw-Hill, New York, 1966.

[We 64] A. Weil, Remarks on the cohomology of groups, *Ann. Math.* **80** (1964), 149–157.

5

Finiteness properties of soluble S-arithmetic groups: a survey

K.-U. Bux

1. Introduction: groups and geometries

Every group is supposed to act upon a certain set preserving some additional structure. This set, which is almost always a space, should be associated to the group in a natural way – in many cases, the set comes first and the group is associated to the set. Linear groups act on vector spaces. Fuchsian groups act on the hyperbolic plane. Symmetry groups of geometric configurations act upon these objects.

In some cases the right space to act on is not that easily found. The mapping class group of a surface, i.e., the group of homotopy classes of homeomorphisms of that surface does not act upon the surface since it is a proper quotient of its automorphism group. Nevertheless, it acts upon the Teichmüller space of the surface. The group of outer automorphisms of a free group of finite rank acts on the Culler-Vogtmann space (outer space). These spaces have been christened in honor of those who found them. This indicates that it is not at all easy to find the right space.

How can we distinguish the right action from other actions? Most groups admit actions on many spaces. What distinguishes a good choice? As a rule of thumb, we will aim at small stabilizers as well as a small quotient space. In the case of a free action, every orbit is isomorphic to the group, and if the quotient space is a point, then there is only one orbit to deal with. Hence, at least philosophically, small quotient and stabilizers will force the space acted upon to look roughly like the group itself. Hence one may be able to deduce some properties of the group by proving some analogues for the space.

1.1. Finiteness properties

To give a flavour of the theorems one obtains by employing geometrically arising group actions, we shall deal with finite generation and finite presentation first. Both of the following theorems are due to A. M. Macbeath [Macb64, Theorem 1]. J.-P. Serre also gave a proof [Serr77, Chapitre 1 § 3 Appendice, page 45].

Let X be topological space and let U be an open subset of X. We assume that X is locally connected and locally simply connected. Hence covering space theory applies. Suppose that the group G acts on X such that

$$X = GU := \bigcup_{g \in G} gU.$$

Put $\mathcal{H} := \{h \in G \mid hU \cap U \neq \emptyset\}$. Note that $\mathcal{H}$ is symmetric, i.e., $h^{-1} \in \mathcal{H}$ for every $h \in \mathcal{H}$.

Theorem 1.1. *If X is nonempty and path connected, $\mathcal{H}$ generates G. In particular, if $\mathcal{H}$ is finite, then G is finitely generated.*

We can refine this result to obtain presentations as follows. Let $\mathcal{X} := \{x_h \mid h \in \mathcal{H}\}$ be a set of letters, one for each element of $\mathcal{H}$. Fix the following set of relations $\mathcal{R} := \{x_{h_1} x_{h_2} x_{h_3}^{-1} \mid h_i \in \mathcal{H},\ h_1 h_2 = h_3\}$. The group

$$\tilde{G} := \langle \mathcal{X} \mid \mathcal{R} \rangle$$

admits an obvious homomorphism $\varphi : \tilde{G} \to G$ taking x_h to h.

Theorem 1.2. *Suppose U is path connected and X is 1-connected, i.e., nonempty, connected, and simply connected. Then φ is an isomorphism.*

Corollary 1.1. *If $\mathcal{H}$ is finite in addition to the assumptions of Theorem 1.2, then G has a finite presentation.*

So we see that finite generation or finite presentability of a group are somehow geometric or topological properties of the group.

Example 1. *The integers $\mathbb{Z}$ act on $\mathbb{R}$ by translations. The translates of the open interval $(-\frac{1}{2}, 1\frac{1}{2})$ cover $\mathbb{R}$. We find $\mathcal{H} = \{-1, 0, 1\}$ and obtain the finite presentation*

$$\mathbb{Z} = \Big\langle x_{-1}, x_0, x_1 \Big| x_{-1}x_0 = x_0x_{-1} = x_{-1},\ x_0x_0 = x_{-1}x_1 = x_1x_{-1} = x_0,\ x_1x_0 = x_0x_1 = x_1 \Big\rangle.$$

Example 2. *Let $X = \mathbb{E}^2$ be the Euclidean plane and consider the group G of those isometries of X that leave invariant the tiling by equilateral triangles shown below. G is a Euclidean Coxeter group.*

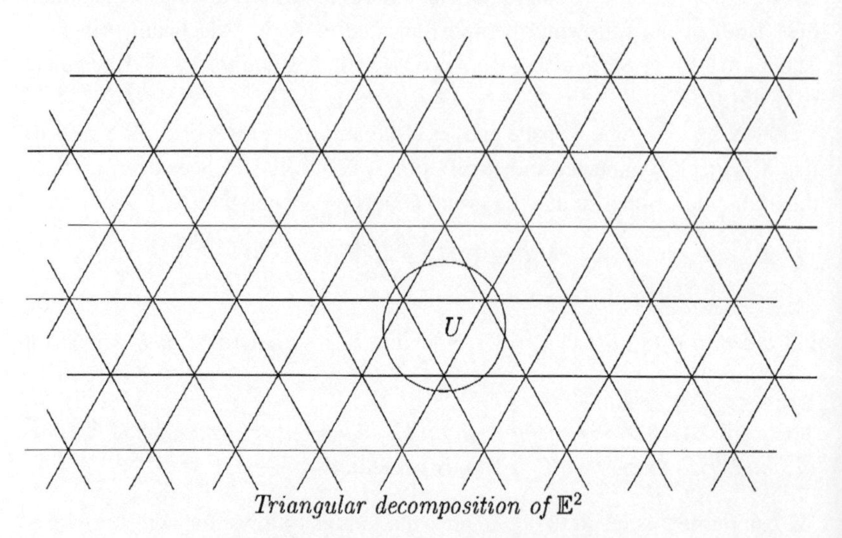

Triangular decomposition of $\mathbb{E}^2$

Let U be an open disc containing a fixed triangle Δ. Since an element of G being an isometry is uniquely determined by the images of the three vertices of Δ, only finitely many elements of G can take Δ to a nearby triangle. Hence the assumptions of Theorem 1.2 *are satisfied and we conclude that G is finitely presented.*

Example 3. *More interesting is that we can do the very same thing for $SL_2(\mathbb{Z})$. This group acts on the upper half plane $\mathbb{H}^2 = \{z \in \mathbb{C} \mid \Im(z) > 0\}$ via*

$$\begin{pmatrix} a & b \\ c & d \end{pmatrix} z = \frac{az+b}{cz+d}.$$

The upper half plane is a model for hyperbolic geometry and the action of $\mathrm{SL}_2(\mathbb{Z})$ *is by hyperbolic isometries. It leaves invariant a set of hyperbolic geodesics as shown in the figure – recall that, within the upper half plane model, geodesic lines are represented by vertical lines and half circles centered on the real line. Thus, we obtain a decomposition into* fundamental domains.

Taking an open neighbourhood $U \subset \mathbb{H}^2$ of one of these domains and applying Theorem 1.2 *shows that the group $SL_2(\mathbb{Z})$ is finitely presented. In fact even more is true. At every vertex of the decompositions there are three intersecting lines. So six geodesic rays issue from each vertex. Three of these run straight away to infinity (that is, they approach the ideal boundary of the hyperbolic*

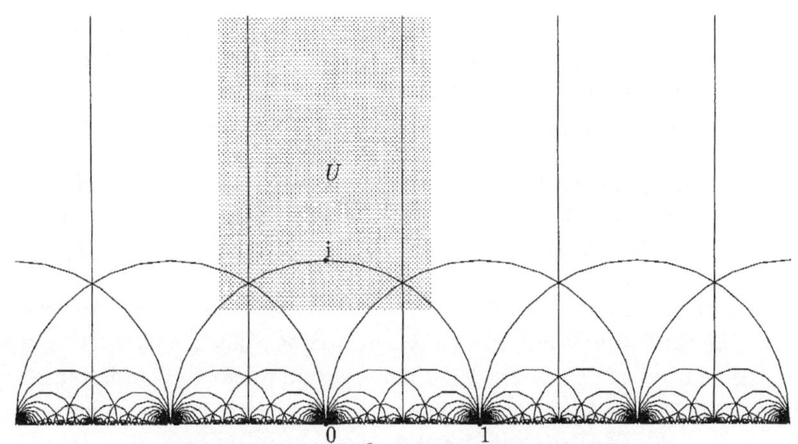

Decomposition of $\mathbb{H}^2$ in fundamental domains

plane by going vertically upwards or approaching the real line) whereas the other three join the vertex to its three neighbouring vertices. The edges (line segments of finite length joining neighbouring vertices) form a three-valent tree. The group $SL_2(\mathbb{Z})$ acts on this tree since an isometry cannot take an edge to an infinite geodesic ray. Furthermore, $SL_2(\mathbb{Z})$ has a subgroup of finite index that acts freely on this tree. Finally, the action of this subgroup has a compact quotient. This is one of the strongest finiteness conditions a group can satisfy. A group G is of finite type *if it acts cocompactly (i.e., with compact quotient) and freely on a contractible CW-complex. An equivalent condition is that G is the fundamental group of a finite complex whose homotopy groups in dimensions ≥ 2 vanish. G is* virtually of finite type *if it has a subgroup of finite index which is of finite type. As we have seen, $SL_2(\mathbb{Z})$ is virtually of finite type. We will see below that this statement generalises to all arithmetic groups for which $SL_2(\mathbb{Z})$ is the most simple non-trivial example.*

There are finiteness properties of intermediate strength between finite generation and finite presentability on the one hand and being of finite type on the other hand. We can weaken the condition of being of finite type by restricting the cocompactness condition to a low dimensional skeleton: a group G is *of type F_m* if it acts freely on a contractible CW complex whose m-skeleton has finite quotient mod G. The cellular chain complex of such a complex provides a free resolution of the trivial $\mathbb{Z}G$-module $\mathbb{Z}$ which is finitely generated up to dimension m. We can define a further weakening by saying that G is *of type FP_m* if $\mathbb{Z}$ considered as a trivial $\mathbb{Z}G$-module has a projective resolution that is finitely generated up to dimension m. Hence every group of type F_m is of type FP_m. The converse, however, does not hold [BeBr97, Examples 6.3(3)].

A group is finitely generated if and only if it is of type F_1 (or FP_1; the difference does not show up in this dimension), and it is finitely presented if and only if it is of type F_2. Recall that the m-skeleton of a contractible CW complex is $(m-1)$-*connected*, that is, its homotopy groups are trivial in dimensions less than m.

Theorem 1.3 ([Brow87, Proposition 1.1]). *Suppose G acts cocompactly on an $(m-1)$-connected CW complex X by cell permuting homeomorphisms such that the stabilizer of each cell c is of type $F_{m-\dim(c)}$. Then G is of type F_m.*

In particular, this theorem allows for finite cell stabilizers whereas the definition of higher finiteness properties uses a free action. Although the relationship of Theorem 1.3 and Corollary 1.1 is not apparent, the former generalises the latter. Let G, X, and U be as in Theorem 1.2. The *nerve* of the covering $X = \bigcup_{g\in G} gU$ is the simplicial complex N defined by the following two conditions.

(1) The vertex set of N is G.
(2) A finite subset $\sigma \subseteq G$ spans a simplex if and only if

$$U_\sigma := \bigcap_{g\in\sigma} gU \neq \emptyset.$$

The group G acts on N on the left by multiplication. The stabilizer of any simplex is finite. The action is cocompact if the set $\mathcal{H} = \{h \in G \mid U \cap hU \neq \emptyset\}$ is finite. Hence Corollary 1.1 follows from Theorem 1.3 and the following

Proposition 1.1. *If X is 1-connected and U is path connected then the geometric realization $|N|$ is 1-connected.*

Proof. Our proof is modeled on an argument of Quillen's [Quil78, Section 7].

We may assume that X is locally path connected. Otherwise we retopologize X by the topology generated by path components of open subsets of X. With this topology, X is locally path connected and still 1-connected [Span66, Chapter 2, Exercises A2-3]. The subset U remains open and path connected.

Let X be a topological space and $\mathcal{X}$ the category of open subsets in X with inclusions as morphisms. Given a partially ordered set D, a *good cover of X over D* is a functor $\mathcal{U} : D \to \mathcal{X}$ satisfying $X = \varinjlim_{\alpha\in D} \mathcal{U}(\alpha)$. A *$\mathcal{U}$-covering* of X is a covering space of X that covers every $\mathcal{U}(\alpha)$ evenly. Let $\mathcal{S}et^{\mathrm{bij}}$ be the category of sets with bijections as the only morphisms. There is a one-to-one correspondence between functors $\boldsymbol{E} : D \to \mathcal{S}et^{\mathrm{bij}}$ and $\mathcal{U}$-coverings $\pi : E \to X$ given by

$$E = \varinjlim_{\alpha\in D} \mathcal{U}(\alpha) \times \boldsymbol{E}(\alpha).$$

The covering $\pi : E \to X$ covers X evenly if and only if $\boldsymbol{E}$ is naturally equivalent to a constant functor.

The set of simplices in the nerve N, also denoted by N, is partially ordered by the face relation. The functor $\boldsymbol{U}$ assigning to every simplex $\sigma \in N$ the set

$$\boldsymbol{U}(\sigma) := U_\sigma = \bigcap_{g \in \sigma} gU$$

is a good cover of X over N. Since X is 1-connected, every covering of X covers X evenly and is a $\boldsymbol{U}$-covering for that reason. Hence every functor $N \to \mathcal{S}\mathrm{et}^{\mathrm{bij}}$ is naturally equivalent to a constant functor.

There also is a good cover $\boldsymbol{N}$ of the geometric realization $|N|$ over N given by

$$\boldsymbol{N}(\sigma) := \mathrm{St}_N^\circ(\sigma).$$

Since the open star of every simplex is contractible, any covering of $|N|$ covers every $\boldsymbol{N}(\sigma)$ evenly. Hence all coverings of $|N|$ are $\boldsymbol{N}$-coverings. These, corresponding to functors from N to $\mathcal{S}\mathrm{et}^{\mathrm{bij}}$ and therefore to coverings of X, cover $|N|$ evenly. Hence $|N|$ is 1-connected. □

In Theorem 3.1 we will quote a refinement of Theorem 1.3 that provides necessary and sufficient conditions for higher finiteness properties. There are many other finiteness properties. For instance, a group could be of finite cohomological or geometrical dimension. See [Brow82, Chapter VIII] for more about this.

Example 4. *Finally, let us consider a group that leads us directly to the heart of the matter. Fix a finite field k and let $K := k(t)$ be the field of rational functions over k. This field contains the ring $L := k[t, t^{-1}]$ as a subring. We will consider the group*

$$\mathrm{B}_2^0(L) := \left\{ \begin{pmatrix} u & p \\ 0 & u^{-1} \end{pmatrix} \middle| \, u \in L^*, \ p \in L \right\} \subseteq \mathrm{SL}_2(L).$$

It acts on L via

$$\begin{pmatrix} u & p \\ 0 & u^{-1} \end{pmatrix} q := \frac{uq + p}{u^{-1}} = u^2 q + up$$

as inspired by the action of $SL_2(\mathbb{Z})$ on $\mathbb{C}$ in the preceding example, and as above, we will turn this action into an action on a tree. For each $m \in \mathbb{Z}$ put

$$V_m := \left\{ \sum_{i \in \mathbb{Z}} \alpha_i t^i \in L \,\middle|\, \alpha_i = 0 \ \forall i \leq m \right\}.$$

We regard these sets as neighbourhoods of $0 \in L$. Translation yields *neighbourhoods for each Laurent polynomial* $p \in L$

$$V_m(p) := \{q \in L \mid p - q \in V_m\}.$$

We call m *the* radius *of* $V_m(p)$. *Let*

$$\mathcal{V} := \{V_m(p) \mid p \in L,\ m \in \mathbb{Z}\}$$

be the set of all these neighbourhoods partially ordered by inclusion. Observe that $\mathrm{B}_2^0(L)$ *acts on* $\mathcal{V}$ *preserving the ordering. The inclusion relation gives rise to a directed graph with vertex set* $\mathcal{V}$. *For two elements* $V,\ V' \in \mathcal{V}$, *insert an edge pointing from* V *to* V' *if there is no* $W \in \mathcal{V}$ *with* $V' \subsetneq W \subsetneq V$ *while* $V' \subsetneq V$. *So the edges point from* V *precisely to those elements of* $\mathcal{V}$ *that are maximal with respect to inclusion in* V. *Note that these are finite in number whence we obtain a locally finite graph* T_+, *on which* $\mathrm{B}_2^0(L)$ *acts.* T_+ *does not contain directed cycles, and given two neighbourhoods* $V,\ V' \in \mathcal{V}$ *with nonempty intersection* $V \cap V' \neq \emptyset$, *one of them contains the other. Hence each vertex of* T_+ *is the terminal vertex of at most one edge, though* $|k|$ *edges issue from there. It follows that* T_+ *does not contain undirected cycles. Furthermore, every neighbourhood* V *is contained in some neighbourhood* V_m *of* 0. *Hence every vertex in* T_+ *is joined by an edge path to the line*

$$\cdots \rightarrow V_{m-1} \rightarrow V_m \rightarrow V_{m+1} \rightarrow \cdots$$

whence T_+ *is a tree. Since the action of* $\mathrm{B}_2^0(L)$ *on* T_+ *does not have finite stabilizers, we cannot obtain finite generation of this group by applying Theorem* 1.2 *to* T_+. *Nevertheless, there is a remedy. The "co-neighbourhoods"*

$$W_m := \left\{ \sum_{i \in \mathbb{Z}} \alpha_i t^i \in L \,\middle|\, \alpha_i = 0\ \forall i \geq m \right\}$$

give rise to a second tree T_- *in a completely analogous manner. The diagonal action of* $\mathrm{B}_2^0(L)$ *on the product* $T_+ \times T_-$ *permutes the cells (vertices, edges, and squares) of this complex. Stabilizers of cells are now finite, however, the action has no compact quotient. So we pass to an appropriate subspace. As neighbourhoods have radii, co-neighbourhoods have "co-radii". The map that associates to every vertex* $(V_m(p), W_n(q))$ *in the Product* $T_+ \times T_-$ *the number* $m - n$ *extends linearly to edges and squares of* $T_+ \times T_-$. *We denote this map by*

$$\pi : T_+ \times T_- \to \mathbb{R}$$

and observe that it is invariant under the action of $\mathrm{B}_2^0(L)$. *The space*

$$X := \pi^{-1}(0)$$

is connected, and Theorem 1.3 *implies that* $\mathrm{B}_2^0(L)$ *is finitely generated. This construction will be considered again in Section* 4 *where we shall also prove that* $\mathrm{B}_2^0(L)$ *is not finitely presented.*

1.2. Arithmetic groups

We will think of a *linear algebraic group* as a group of matrices of determinant 1 which is defined by polynomial equations that the matrix coefficients are supposed to satisfy. These coefficients can be taken from any fixed ring containing the constants that occur in the defining equations. A typical example is SO_r. Note that a set of polynomial equations chosen at random will almost always fail to define a subgroup of the special linear group SL. Nevertheless, if it does, then the fact that the determinant is 1 shows that if the coefficients of a matrix belong to a certain ring, the coefficients of its inverse will do so, too. [Bore91] may serve as a reference on linear algebraic groups. Among the linear algebraic groups, there are two extreme types: soluble groups on the one hand and semi-simple groups on the other hand. A linear algebraic group is *semi-simple* if it does not contain a nontrivial connected soluble normal subgroup. Therefore, any linear algebraic group is an extension of a semi-simple group by a soluble group. Reductive groups are close to semi-simple groups. A linear algebraic group is *reductive* if it contains only "small" connected soluble normal subgroups, that is, these soluble normal subgroups do not contain unipotent elements. Maximal connected soluble subgroups of linear algebraic group are called *Borel subgroups*. Since they need not be normal, semi-simple and reductive groups may contain many Borel subgroups. They are a most important tool in studying reductive groups.

An *arithmetic group* or an *S-arithmetic* is obtained from a linear algebraic group when the coefficients of its matrices are chosen from an *arithmetic ring* or an *S-arithmetic ring*. We already have encountered such rings in the examples. $\mathbb{Z}$ is the most simple arithmetic ring. Generally, any ring of algebraic integers in a number field is called arithmetic. In the number field case, e.g., over the rationals $\mathbb{Q}$, S-arithmetic rings are obtained from arithmetic rings by localizing at a finite set S of prime elements. That is, one allows some primes to be inverted. All these rings live inside of *global number fields* which are by definition finite extensions of the rationals $\mathbb{Q}$. Number fields have a nice arithmetic structure, and it has been recognized long ago that there are fields of positive characteristic which exhibit a very similar behaviour. These are the so called *global function fields* which are finite extensions of fields $k(t)$ of rational functions over finite fields k. It does not make sense to look for algebraic integers in these fields – they form a finite subfield, the so called *constant functions*. Hence there are no interesting arithmetic rings. Nevertheless, S-arithmetic subrings can be defined even though the notion of primes has to be recasted in terms of valuations. The precise definitions, which apply to number fields as well, are given in the next section. At the moment it suffices to know that the polynomial ring $k[t]$ and the ring $k[t, t^{-1}]$ of Laurent polynomials are S-arithmetic rings over a set of one prime and two primes, respectively. $\mathrm{SL}_2(\mathbb{Z})$ and $\mathrm{B}_2^0(k[t, t^{-1}])$ are typical examples. $\mathrm{SL}_2(\mathbb{Z})$ is an arithmetic subgroup of a semi-simple linear algebraic group and $\mathrm{B}_2^0(k[t, t^{-1}])$ is an S-arithmetic subgroup of a Borel group.

As long as people have studied arithmetic groups, they have been investigating finiteness properties. This is not surprising since, as we have seen, finiteness properties are of geometric nature and the geometries associated to algebraic groups are a fundamental tool in studying them and their arithmetic subgroups. This goes for S-arithmetic subgroups as well. For an arithmetic group over a number field, the symmetric space of the real Lie group of real points of the algebraic group is a geometric model whereas, over function fields, Bruhat-Tits buildings provide good geometries for S-arithmetic groups. The hyperbolic plane of Example 3 is a symmetric space, and the trees T_+ and T_- of Example 4 are Bruhat-Tits buildings associated to the two primes defining the ring of Laurent polynomials.

Let us discuss the number field case first. Let $\mathcal{G}$ be a linear group defined over the global number field K with ring of integers $\mathcal{O}$. Beside finite generation and finite presentability (see [BoHa62] and [Behr62]), the first result on higher finiteness properties is a theorem of M.S. Raghunatan [Ragh68, Theorem 1, Corollaries 2 and 4] implying that an arithmetic subgroup of $\mathcal{G}(K)$ is of type F_∞ provided that $\mathcal{G}$ is semi-simple. He proved even more, namely that any arithmetic subgroup of a semi-simple group is virtually of finite type. This is a far-reaching generalisation of Example 3. A. Borel and J.-P. Serre reproved Raghunathan's result by different means [BoSe73, Theorem 9.3]. They also observed that this theorem implies that an arithmetic subgroup of $\mathcal{G}(K)$ is of type F_∞ regardless of whether the linear group $\mathcal{G}$ is semi-simple. Furthermore, they generalised the result to S-arithmetic groups [BoSe76, Proposition 6.10]. In this case however, they had to assume that the linear algebraic group is reductive. Then, as above, an S-arithmetic subgroup is virtually of finite type.

What about non-reductive linear algebraic groups? M. Kneser proved that $\mathcal{G}(\mathcal{O}_S)$ is finitely presented if and only if for each non-Archimedian prime $v \in S$ the locally compact group $\mathcal{G}(\overline{K_v})$ is compactly presented where $\overline{K_v}$ denotes the completion of K at v [Knes64]. Compact presentability is the analogue of finite presentability in the category of locally compact topological groups. Using Kneser's result, H. Abels characterized all finitely presented S-arithmetic groups over number fields [Abel87]. The main point is that $\mathcal{G}(K_v)$ is compactly presented if and only if a maximal $\overline{K_v}$-split soluble subgroup $\mathcal{B}(K_v)$ is compactly presented. This reduces the problem to soluble groups. For these, he proved:

Theorem 1.4. *Let Γ be a soluble S-arithmetic group. Then there is a short exact sequence*

$$N \hookrightarrow \Gamma \twoheadrightarrow Q$$

where N is nilpotent and Q contains a finitely generated Abelian subgroup of finite index. Γ is finitely presented if and only if the following two conditions hold

(1) *The abelianization $N/_{[N,N]}$ is a tame $\mathbb{Z}Q$-module,*
(2) *$H_2(N, \mathbb{Z})$ is finitely generated over $\mathbb{Z}\Gamma$.*

To explain tameness, we have to consider homomorphisms $\chi : Q \to \mathbb{R}$. Such a homomorphism defines a submonoid $Q_\chi := \{q \in Q \mid \chi(q) \geq 0\}$. A $\mathbb{Z}Q$-module is said to be *tame* if, for every homomorphism χ, the module is finitely generated either over $\mathbb{Z}Q_\chi$ or over $\mathbb{Z}Q_{-\chi}$ or over both rings. This concept was introduced by R. Bieri and R. Strebel in [BiSt80] to study finite presentability of metabelian groups. This line of thought has been generalised by H. Abels and A. Tiemeyer to deal with higher finiteness properties. It yields the following

Theorem A. *Let K be a global number field and S a finite nonempty set of primes containing all Archimedean primes. Let $\mathcal{O}_S$ be the corresponding S-arithmetic subring of K. Furthermore let $\mathcal{B}$ be a Borel subgroup scheme of a reductive group defined over K. Then the S-arithmetic group $\mathcal{B}(\mathcal{O}_S)$ is of type* F_∞.

P. Abramenko (unpublished) proved this result before. His proof establishes that $\mathcal{B}(\mathcal{O}_S)$ is virtually of finite type. In spite of these results, there does not yet exist a generalisation of Theorem 1.4, i.e., a list of necessary and sufficient conditions for a soluble S-arithmetic group to be of type F_m. Further examples of soluble arithmetic groups are due to H. Abels (see [Abel79] and [AbBr87]). We will discuss the proof of Theorem A in Section 3 following Abels and Tiemeyer.

Less is known in the function field case. Even for reductive groups the problem of determining the finiteness properties of S-arithmetic subgroups is unsolved. For these, H. Behr has given a complete solution for finite generation and finite presentability in [Behr98]. Concerning higher finiteness properties, there are mainly two series of examples. U. Stuhler proved that $\mathrm{SL}_2(\mathcal{O}_S)$ is of type $\mathrm{F}_{|S|-1}$ but not of type $\mathrm{FP}_{|S|}$ ([Stuh76] and [Stuh80]), where $\mathcal{O}_S$ is an S-arithmetic subring of any global function field. This series shows a positive influence of the number of primes. On the other hand, P. Abramenko [Abra87] and H. Abels [Abel91], have proved independently, that $\mathrm{SL}_n(\mathbb{F}_q[t])$ is of type F_{n-2} but not of type FP_{n-1} provided that q is big enough. P. Abramenko generalised this series to other classical groups [Abra96]. It is not yet known whether the assumption on q can be dropped. H. Behr, though, has proposed a strategy for proving a positive answer recently [Behr99].

For soluble groups, the author obtained the following result ([Bux97a] and [Bux97b], see [Bux99] for an English version).

Theorem B. *Let K be a global function field and S a finite nonempty set of primes. Let $\mathcal{O}_S$ be the corresponding S-arithmetic subring of K. Furthermore let $\mathcal{B}$ be a Borel subgroup scheme of a Chevalley group defined over $\mathbb{Z}$. Then the S-arithmetic group $\mathcal{B}(\mathcal{O}_S)$ is of type* $\mathrm{F}_{|S|-1}$ *but not of type* $\mathrm{FP}_{|S|}$.

Concerning other soluble groups, little is known in the function field case (see [Bux97b, Bemerkungen 8.10 and 8.11] or [Bux99, Remarks 8.6 and 8.7]). We will outline the proof of Theorem B in Section 4. Since the detailed proofs will be published elsewhere, we focus on the geometric aspects.

Acknowledgement. I would like to thank Prof. Heinz Helling and Prof. Thomas Müller for giving me the opportunity to present part of this work at the conference they organized at Bielefeld in August 1999. I am indebted to Peter Brinkmann and Blake Thornton for very helpful criticism on the manuscript.

2. Notation

Let K be a *global field*. Hence K is either a *global number field* (that is, a finite extension of the rationals $\mathbb{Q}$) or a *global function field* (that is a finite extension of the field of rational functions $k(t)$ over a finite field k). For detailed information about global fields and their arithmetic, see [O'Me73], [CaFr67], or [Weil73]. Let A be the finite set of non-Archimedian primes of K, which is empty if K is a function field. Recall that a *prime* or *place* v of K is an equivalence class of valuations. To this we can associate the corresponding completion K_v of K. This is a local field. Its additive group therefore has a Haar measure. For each $x \in K_v$, the multiplication by x induces an automorphism of the additive group, which rescales the Haar measure by a positive real number $|x|_v$. If v is non-Archimedian, the corresponding *valuation ring* $\mathcal{O}_v := \{x \in K_v \mid |x|_v \leq 1\}$ of v-adic integers is a local ring. For any finite, nonempty set S of primes containing A let $\mathcal{O}_S := \{f \in K \mid |f|_v \leq 1 \ \forall v \notin S\}$ denote the corresponding S-arithmetic ring and $\mathbb{A}_S := \times_{v \in S} K_v \times \times_{v \notin S} \mathcal{O}_v$ the ring of S-adeles. The adele ring is the direct limit $\mathbb{A} := \varinjlim_S \mathbb{A}_S$. This is a locally compact ring. It contains K as a discrete subring via the diagonal embedding because each element in K belongs to $\mathcal{O}_v$ for all but finitely many primes v – in the language of the function field case: a function has at most finitely many poles. Let $\mathcal{G}$ be a linear algebraic group defined over K represented as a group of matrices with determinant 1. Then for any subring R of any K-algebra, the group $\mathcal{G}(R)$ of R-points is defined. Hence we regard $\mathcal{G}$ as a functor taking (topological) subrings of K-algebras to (topological) groups. The group $\mathcal{G}(\mathcal{O}_S)$ is called an S-arithmetic subgroup of $\mathcal{G}$. It does depend on the matrix representation chosen for $\mathcal{G}$. Nevertheless, any two representations of $\mathcal{G}$ yield commensurable S-arithmetic groups. For this reason, any group commensurable with $\mathcal{G}(\mathcal{O}_S)$ is called *S-arithmetic*, as well. In the number field case, the set A of Archimedian primes is nonempty. Hence, over number fields, A-arithmetic groups exist. These are simply called *arithmetic*. Since commensurable groups have identical finiteness properties, we may confine our investigation to groups of the form $\mathcal{G}(\mathcal{O}_S)$.

3. The number field case

Throughout this section, K is a number field. H. Abels and A. Tiemeyer introduced higher compactness properties C_m and CP_m for locally compact groups which extend compact generation (equivalent to C_1 and to CP_1) and compact presentability (equivalent to C_2) [AbTi97]. For a discrete group, the properties C_m (respectively CP_m) and F_m (respectively FP_m) are equivalent. In [Tiem97], a Hasse Principle is derived relating finiteness properties of S-arithmetic groups to compactness properties of linear algebraic groups over local fields. We will discuss this in Section 3.2. Applying this, Tiemeyer settled Theorem A mainly using the technique of contracting automorphisms already present in [Abel87]. This approach yields a comparatively simple proof, also due to Tiemeyer, for the result of Borel's and Serre's on the finiteness properties of S-arithmetic groups. We will present this argument in Section 3.3. It does not yield, however, that these groups are virtually of finite type.

3.1. Finiteness properties and compactness properties

The starting point of the definition of compactness properties is the following celebrated criterion of K. Brown's.

Theorem 3.1 ([Brow87, Theorems 2.2 and 3.2]). *Let Γ be a discrete group and $(X_\alpha)_{\alpha\in D}$ a directed system of cocompact Γ-CW-complexes with equivariant connecting morphisms and $(m-1)$-connected limit. Suppose that, for each X_α, the stabilizers of all j-cells are of type F_{n-j} (respectively FP_{n-j}). Then Γ is of type F_m (respectively FP_m) if and only if the directed systems of homotopy groups $(\pi_n(X_\alpha))_{\alpha\in D}$ (respectively the directed systems of reduced homology groups $(H_n(X_\alpha))_{\alpha\in D}$) are essentially trivial for all $n < m$.*

Herein, a directed system $(G_\alpha)_{\alpha\in D}$ of groups is called *essentially trivial* if for each element $\alpha \in D$ there is a $\beta \geq \alpha$ such that the natural map $G_\alpha \to G_\beta$ is trivial. Defining

$$\mathrm{Mor}\big((G_\alpha)_{\alpha\in D}, (G'_\beta)_{\beta\in D'}\big) := \varprojlim_{\alpha\in D} \varinjlim_{\beta\in D'} \mathrm{Mor}\big(G_\alpha, G'_\beta\big)$$

we turn the class of all directed systems of groups into the category $\mathrm{ind}_{\mathrm{Gr}}$ [AbTi97, Section 1]. In $\mathrm{ind}_{\mathrm{Gr}}$, the essentially trivial directed systems of groups are precisely those that are isomorphic to the initial element of this category represented by the trivial system of trivial groups. Henceforth we consider directed systems of groups as elements of $\mathrm{ind}_{\mathrm{Gr}}$.

For every group Γ, there is a canonical directed system of cocompact simplicial Γ-complexes indexed by the finite subsets $F \subseteq \Gamma$, namely the Γ-orbit of F in the simplicial complex of all finite subsets of Γ, acted upon by left translation. Simplex stabilizers are finite and the limit is contractible. Hence finiteness properties can – in principle – be tested with this directed system in view of Theorem 3.1. We will somehow mimic this construction for locally compact groups. Let G be a locally compact group, X a locally compact space together with an action of G on X, and $\mathrm{E}X$ the free simplicial set over X whose m skeleton is just X^{m+1}. The face (respectively degeneration) maps are given by deleting (respectively doubling) a coordinate. Note that G acts on $\mathrm{E}X$ diagonally and therefore also on its geometric realization $|\mathrm{E}X|$. This is a contractible space. Furthermore, let $\mathcal{C}X$ denote the set of all compact subsets of X directed by inclusion. For each set $C \in \mathcal{C}X$, $G \cdot \mathrm{E}C := \{(gx_0, \ldots, gx_m) \mid g \in G,\ \{x_0, \ldots, x_m\} \subseteq C,\ m \in \mathbb{N}\}$ is a G-invariant simplicial subset of $\mathrm{E}X$. Hence we have a directed system $\mathrm{E}_G X := (G \cdot \mathrm{E}C)_{C\in\mathcal{C}\in X}$ of simplicial G-sets depending functorially on G and X, whose limit is $\mathrm{E}X$. Taking the geometric realization first and passing afterwards to either homotopy groups π_n or reduced homology H_n, induces the directed systems of groups $\mathrm{ind} - \pi_n(\mathrm{E}_G X) := (\pi_n(|G \cdot \mathrm{E}C|))_{C\in\mathcal{C}X} \in \mathrm{ind}_{\mathrm{Gr}}$ and $\mathrm{ind} - H_n(\mathrm{E}_G X := (\mathrm{H}_n(|G \cdot \mathrm{E}C|))_{C\in\mathcal{C}X} \in \mathrm{ind}_{\mathrm{Gr}}$, respectively.

Definition 1. *The pair (G, X) satisfies* condition P_m *(respectively* condition PP_m*) if* $\mathrm{ind} - \pi_n(\mathrm{E}_G X) \cong 0$ *(respectively* $\mathrm{ind} - \mathrm{H}_n(\mathrm{E}_G X) \cong 0$*) for all $n < m$. That is, these associated directed systems are essentially trivial. Now we can define compactness properties of locally compact groups. The group G is* of type C_m *(respectively* of type CP_m*) if the pair (G, G) satisfies condition P_m (respectively PP_m) where G is considered as a G-space via left multiplication.*

It is immediate from Brown's criterion that these notions agree for discrete groups with the properties F_m and FP_m, respectively.

Remark 1. Recall that an object R in a category is a retract of the object O if there are arrows $R \hookrightarrow O$ and $O \twoheadrightarrow R$ whose composition is the identity on R. Retract diagrams are preserved by covariant and contravariant functors. Hence a retract of the group G enjoys at least the same compactness and finiteness properties as G because of the categorial characterisation – vanishing of some functorially assigned objects – of compactness and finiteness properties given above.

We will use the following lemmata.

Lemma 1 ([AbTi97, Lemma 3.2.2]). *The P_m- and PP_m-conditions are independent of the choice of the G-space X provided the action of G on X is proper. Hence the conditions C_m and CP_m can be tested using any proper G-space.*

Corollary 3.1 ([AbTi97, Corollary 3.2.3]). *If G is a locally compact group and $B \leq G$ is a closed subgroup with compact quotient $G/_B$ then G and B have the same compactness properties.*

In the case of finiteness properties, one can use Brown's Criterion 3.1 to slightly improve the lemma.

Lemma 2. *Let X be a G-space such that, for all k-tuples of points in X, the intersection of their stabilizers is of type F_{m-k}. Then G is of type F_m (respectively FP_m) if and only if the pair (G, X) satisfies P_m (respectively PP_m).*

Lemma 3 ([AbTi97, Lemma 3.1.1]). *A finite direct product of locally compact groups is of type C_m (respectively CP_m) if each of its factors is of type C_m (respectively CP_m).*

3.2. The Hasse principle

In this section we shall outline the proof of

Theorem 3.2 (Hasse Principle [Tiem97, Theorem 3.1]). *Let $\mathcal{G}$ be a linear algebraic group. Then the S-arithmetic group $\mathcal{G}(\mathcal{O}_S)$ is of type F_m (respectively FP_m) if and only if for each non-Archimedian prime $v \in S$ the locally compact completion $\mathcal{G}(K_v)$ is of type C_m (respectively CP_m).*

Proof. For a non-Archimedian prime v, the group $\mathcal{G}(\mathcal{O}_v)$ is a compact open subgroup of $\mathcal{G}(K_v)$, which is a locally compact group. Let us put $\mathcal{G}_S := \times_{v \in S}\mathcal{G}(K_v)$. This is a locally compact group since S is finite. The subgroup

$K_S := \times_{v \in S-A} \mathcal{G}(\mathcal{O}_v)$ is compact and open in $\mathcal{G}_{S-A}$. Since we have

$$\mathcal{G}(\mathbb{A}_S) = \underset{v \in S}{\times} \mathcal{G}(K_v) \times \underset{v \notin S}{\times} \mathcal{G}(\mathcal{O}_v)$$

we can, on the one hand, consider K_S as a subgroup of $\mathcal{G}(\mathbb{A}_S)$; on the other hand, there is a natural projection map $\pi_S : \mathcal{G}(\mathbb{A}_S) \to \mathcal{G}_S$. Since K_S is open in $\mathcal{G}_{S-A}$, $X := \mathcal{G}_{S-A}/K_S$ is a discrete set upon which $\mathcal{G}_{S-A}$ acts properly. $\mathcal{G}(\mathcal{O}_S)$ also acts on X, via π_S. The stabilizers are of the form $\mathcal{G}(\mathcal{O}_S) \cap gK_S g^{-1}$ and can be shown to be commensurable with $\mathcal{G}(\mathcal{O}_S) \cap K_S = \mathcal{G}(A)$ whence they are arithmetic and therefore of type F_∞. Hence, by Lemma 2, $\mathcal{G}(\mathcal{O}_S)$ is of type F_m (respectively FP_m) if and only if the pair $(\mathcal{G}(\mathcal{O}_S), X)$ is of type P_m (respectively PP_m). On the other hand, by Lemma 3, $\mathcal{G}(K_v)$ is of type C_m (respectively CP_m) for each $v \in S - A$ if and only if the direct product $\mathcal{G}_{S-A} = \times_{v \in S-A} \mathcal{G}_{K_v}$ is of type C_m (respectively CP_m). Because of Lemma 1, this in turn holds if and only if the pair $(\mathcal{G}_{S-A}, X)$ is of type P_m (respectively PP_m). Hence we have to compare the actions of $\mathcal{G}(\mathcal{O}_S)$ and $\mathcal{G}_{S-A}$ on X. In particular it suffices to show that the filtrations $\mathrm{E}_{\mathcal{G}(\mathcal{O}_S)}X$ and $\mathrm{E}_{\mathcal{G}_{S-A}}X$ are cofinal. So let K be a compact (that is finite) subset of X. Obviously

$$\mathcal{G}(\mathcal{O}_S) \cdot \mathrm{E}K \le \mathcal{G}S - A \cdot \mathrm{E}K$$

hence we only have to prove that there is another finite subset $L \subset X$ such that

$$\mathcal{G}_{S-A} \cdot \mathrm{E}L \le \mathcal{G}(\mathcal{O}_S) \cdot \mathrm{E}K.$$

In [Tiem97, Corollary 2.3], it is deduced from Borel's Finiteness Theorem [Bore63, Theorem 5.1] that there are finitely many $g_1, \ldots, g_m$ such that $\mathcal{G}_{S-A} = \bigcup_{i=1}^m \mathcal{G}'(\mathcal{O}_S) g_i K_S$, where $\mathcal{G}'(\mathcal{O}_S) := \pi_S(\mathcal{G}(\mathcal{O}_S))$. Observe that for finite $K \subset X$ the set $L := \bigcup_{i=1} m g_i K_S K$ is finite since K_S acts with finite orbits on X as it is a compact group. Then, for every $g \in \mathcal{G}_{S-A}$, there is a g_i and an element $\gamma \in \mathcal{G}(\mathcal{O}_S)$, acting via its image in $\mathcal{G}'(\mathcal{O}_S)$, such that $g \in \gamma g_i K_S$, whence $gK = \gamma g_i K_S K \subseteq \gamma L$. From this,

$$\mathcal{G}_{S-A} \cdot \mathrm{E}L \le \mathcal{G}(\mathcal{O}_S) \cdot \mathrm{E}K$$

follows immediately. □

Corollary 3.2. *Let $\mathcal{P}$ be a parabolic subgroup of $\mathcal{G}$. Then $\mathcal{G}(\mathcal{O}_S)$ and $\mathcal{P}(\mathcal{O}_S)$ have the same finiteness properties.*

Proof. For each prime v, since $\mathcal{G}(K_v)/\mathcal{P}(K_v)$ is compact by [BoTi65, Proposition 9.3], the groups $\mathcal{G}(K_v)$ and $\mathcal{P}(K_v)$ have the same compactness properties by Corollary 3.1. Now the claim follows by the Hasse principle. □

Remark 2. *Corollary 3.2 yields a quick and dirty proof for Theorem A, since S-arithmetic subgroups of Chevalley groups are of type F_∞ by [BoSe76, Proposition 6.10]. Using Corollary 3.2 this way, however, is somehow approaching the problem from the wrong angle. The corollary should be considered the other way round, namely, as reducing the problem of determining the finiteness properties of general S-arithmetic groups to S-arithmetic subgroups of connected soluble linear algebraic groups.*

3.3. Stand alone proofs

In this section we shall indicate how S-arithmetic subgroups of soluble groups can be studied. Let $G = T \ltimes U$ be a locally compact semi-direct product of locally compact groups where T is Abelian. An element $t \in T$ is *contracting* or acts *by contraction* on U if its positive powers converge uniformly to the identity on compact subsets of U. The following proposition, which generalises [Abel87, Proposition 1.3.1], shows that one is fortunate having a contracting element at hand.

Proposition 3.1 ([Tiem97, Theorem 4.3]). *If T contains a contracting element, G and T have identical compactness properties.*

Proof. Since T is a retract of G, we only have to show, that compactness properties of T are inherited by G. So assume that T is of type C_n. Let K_G be a compact subset of G. Enlarging this subset if necessary, we might assume that it is of the form

$$K_G = K_T K_U$$

where K_T is compact in T and K_U is a compact neighbourhood of the identity in U satisfying $tK_Ut^{-1} \subseteq K_U$ after replacing t by a power if necessary. Since we can test compactness properties of T by means of its action on G, there is a compact subset $L \supseteq K_G \cup K_G t^{-1}$ such that the inclusion of $T \cdot \mathrm{E}K_G \hookrightarrow T \cdot \mathrm{E}L$ induces trivial maps in homotopy groups up to dimension n. We claim that this holds also for the inclusion of $G \cdot \mathrm{E}K_G \hookrightarrow G \cdot \mathrm{E}L$. Consider the map

$$\alpha_t : \mathrm{E}G \to \mathrm{E}G$$
$$(g_0, \ldots, g_k) \mapsto (g_0 t^{-1}, \ldots, g_k t^{-1})$$

and observe that it takes $G \cdot \mathrm{E}K_G$ to itself because any simplex therein is of the form

$$\sigma = (gx_0u_0, \ldots, gx_ku_k)$$

with $x_i \in K_T$ and $u_i \in K_U$, and is taken to

$$\begin{aligned}\alpha_t(\sigma) &= (gx_0u_0t^{-1}, \ldots, gx_ku_kt^{-1})\\ &= (gx_0t^{-1}tu_0t^{-1}, \ldots, gx_kt^{-1}tu_kt^{-1})\\ &= (gt^{-1}x_0tu_0t^{-1}, \ldots, gt^{-1}x_ktu_kt^{-1}),\end{aligned}$$

which belongs to $G \cdot \mathrm{E}K_G]$ because of $tu_it^{-1} \in K_U$. From the first line of these equations, moreover, it follows that σ and $\alpha_t(\sigma)$ are faces of a common simplex in $G \cdot \mathrm{E}L$ since $K_Gt^{-1} \subseteq L$. Hence a sphere $\mathbb{S}$ in $G \cdot \mathrm{E}K_G$ is homotopic to the sphere $\alpha_t(\mathbb{S})$ inside $G \cdot \mathrm{E}L$. The map α_t, however, has another remarkable property. It contracts $\mathrm{E}G$ towards $T \cdot \mathrm{E}K_G$: consider a vertex xu in $\mathrm{E}G$, $x \in T$, $u \in U$. We have

$$\alpha_t^n(xu) = t^{-n}xt^nut^{-n}$$

and eventually $t^nut^{-n} \in K_U$, since t is contracting. Hence, any compact subset of $\mathrm{E}G$ is taken to $G \cdot \mathrm{E}K_G$ by some positive power of α_t. Combining the properties of α_t, we can move any sphere $\mathbb{S}$ of dimension $\leq n$ in $G \cdot \mathrm{E}K_G$ inside $G \cdot \mathrm{E}L$ into $T \cdot \mathrm{E}K_G$. This sphere, however, is homotopically trivial inside $T \cdot \mathrm{E}L \subseteq G \cdot \mathrm{E}L$ by choice of L. □

Proposition 3.2. *Let $\mathcal{P}$ be a parabolic subgroup of the reductive group $\mathcal{G}$, both defined over a local field F of characteristic* 0 *whose associated valuation $v : F^* \to \mathbb{R}$ is non-Archimedian. Then $\mathcal{P}(F)$ is of type C_∞.*

Remark 3. *Combined with the Hasse principle Theorem* 3.2*, this yields a proof of Theorem A which does not depend on* [BoSe76]*. Moreover, Corollary* 3.2 *implies that $\mathcal{G}(F)$ is of type C_∞, too.*

Proof. In view of [BoTi65, Proposition 9.3] and Corollary 3.1, it suffices to construct a maximal connected F-split soluble subgroup $\mathcal{B} \leq \mathcal{G}$ such that $\mathcal{B}(F)$ is of type C_∞. We start with a maximal F-split torus $\mathcal{T}$ in $\mathcal{G}$. The set of all characters $\mathcal{T} \to \mathfrak{Mult}$ is a finitely generated Abelian group, denoted by $X^+(\mathcal{T})$. Therefore $X^+(\mathcal{T}) \otimes_{\mathbb{Z}} \mathbb{R}$ is a real vector space. This is where the *root system* $\Phi := \Phi(\mathcal{G}, \mathcal{T})$ lives. It is the set of weights of the adjoint representation of $\mathcal{T}$ over the Lie algebra $\mathfrak{g}$ of $\mathcal{G}$. Thus, for every root $\alpha \in \Phi$, we are given a corresponding weight space and a corresponding unipotent subgroup $\mathcal{U}_\alpha \leq \mathcal{G}$. Fix a base of Φ thereby determining the positive half Φ^+, and let be the group generated by all $\mathcal{U}_\alpha$ for $\alpha \in \Phi^+$. This is a unipotent group [BoTi65, 3.8 (iv)], upon which $\mathcal{T}$ acts by conjugation inside $\mathcal{G}$ because we started with the adjoint representation. Furthermore, $\mathcal{B} := \mathcal{T} \ltimes \mathcal{U}$ is a maximal connected F-split soluble subgroup of $\mathcal{G}$. Hence, it suffices to show, that $\mathcal{T}(F) \ltimes \mathcal{U}(F)$ is of type C_∞ for each prime v. To prove this, we will find a contracting element $t \in \mathcal{T}(F)$. Then the claim follows from Proposition 3.1 since $\mathcal{T}(F)$ is of type C_∞ for it contains a cocompact finitely generated Abelian group. Whether or not an element of $\mathcal{T}(F)$ is contracting can be checked via its action on the Lie algebra: an element t acts by contraction on $\mathcal{U}(F)$ if for every positive root $\alpha \in \Phi^+$, the inequality $v(\alpha(t)) > 0$ holds. Consider now the one parameter subgroups in $\mathcal{T}$. They form a lattice of maximum rank in the dual of $X^+(\mathcal{T}) \otimes_{\mathbb{Z}} \mathbb{R} \supset \Phi$. Hence there is a one parameter subgroup $\xi : \mathfrak{Mult} \to \mathcal{T}$ representing a point on which all positive roots are strictly positive. Finally, $t := \xi(x)$ fits our needs provided $v(x) > 0$. □

4. The function field case

From now on, K is a global function field. In this case even S-arithmetic subgroups of reductive groups have finite finiteness length, and no Hasse Principle holds – the different primes have to work together in order to ensure higher finiteness properties. The main tool for establishing finiteness properties in this setting is to study the action of the S-arithmetic group $\mathcal{G}(\mathcal{O}_S)$ on the product of Bruhat-Tits buildings associated to $\mathcal{G}$ and the primes in S. Such buildings always exist if $\mathcal{G}$ is reductive [BrTi72]. For general $\mathcal{G}$, no replacement is known. Nevertheless, since Theorem B only deals with Borel subgroups of Chevalley groups, we can use the buildings associated to the latter in order to investigate the S-arithmetic subgroups of the former. We shall freely use the terminology related to buildings and the reader should have a basic knowledge of these geometric objects and their relationship to algebraic groups. Standard references are [Brow89] and [Rona89]. Nevertheless, the main geometric argument in Section 4.2 does only depend on properties of buildings that are explicitly stated there.

4.1. Preliminaries on Chevalley groups and their associated Bruhat-Tits buildings

Let $\mathcal{G}$ be a *Chevalley group*, i.e., a semi-simple linear algebraic group scheme defined over $\mathbb{Z}$. They have been introduced by Chevalley in [Chev60]. A standard reference is [Stei68]. Fix a Borel subgroup $\mathcal{B}$ of $\mathcal{G}$ and a maximal torus $\mathcal{T}$ inside $\mathcal{B}$ such that both are also defined over $\mathbb{Z}$. Since a global function field K has only non-Archimedian primes, there is a Euclidean Bruhat-Tits building $X_v := X(\mathcal{G}, K_v)$ associated to each prime $v \in S$. This is a CAT(0)-space whose boundary at infinity is the spherical building $\tilde{X}_v$ canonically associated to $\mathcal{G}$ and K_v. The group $\mathcal{B}(K_v)$ is the stabilizer of the *fundamental chamber* $C_v \in \tilde{X}_v$ at infinity. Likewise, $\mathcal{T}(K_v)$ stabilizes the *standard apartment* Σ_v in X_v, which corresponds to an apartment in $\tilde{X}_v$ containing C_v. It is a Euclidean Space in which we choose a simplicial cone $S_v \subseteq \Sigma_v$ representing C_v. We take the cone point to be the origin of Σ_v, turning it into a Euclidean vector space.

Recall that the root system Φ of $\mathcal{G}$ with respect to $\mathcal{T}$ consists of morphisms from $\mathcal{T}$ to $\mathfrak{Mult}$, the group scheme associating to each ring its group of multiplicative units, e.g., represented as diagonal 2×2 matrixes of determinant 1. So we can represent the root system on Σ_v by a set of linear forms Φ_v. Furthermore, we are given a base and a system of positive roots $\Delta \subseteq \Phi^+ \subset \Phi$ corresponding to the Borel subgroup $\mathcal{B}$. These give rise to sets of linear forms $\Delta_v \subseteq \Phi_v^+ \subset \Phi_v$. The linear forms in Δ_v form a system of coordinates $\xi_v := \left(\alpha_v^{(1)}, \ldots, \alpha_v^{(r)}\right) : \Sigma_v \to \mathbb{R}^r$ on Σ_v such that

$$S_v = \{s_v \in \Sigma_v \mid \alpha_v^{(j)}(s_v) \geq 0 \; \forall j \in \{1, \ldots, r\}\}$$

holds. This description remains true if all positive roots are taken into account. We normalize the coordinates such that an element $t_v \in \mathcal{T}(K_v)$ acts on Σ_v as a translation with coordinates

$$\left(\log(|\alpha^{(1)}(t_v)|_v), \ldots, \log(|\alpha^{(r)}(t_v)|_v)\right).$$

The subgroup of elements in $\mathcal{B}(K_v)$ that fix at least one point in Σ_v is the group of K_v-points $\mathcal{U}(K_v)$ of the *unipotent radical* $\mathcal{U} := \ker \mathcal{B} \to \mathcal{T}$. Moreover, Σ_v is a strong fundamental domain for the action of $\mathcal{U}(K_v)$ on X_v, and there is a well defined map $\rho_v : X_v \to \Sigma_v$ called the *retraction centered at* C_v.

We will study the diagonal action of $\mathcal{B}(\mathcal{O}_S)$ on the product $\boldsymbol{X} := \times_{v \in S} X_v$. The retractions ρ_v induce a retraction $\boldsymbol{\rho} : \boldsymbol{X} \to \boldsymbol{\Sigma} := \times_{v \in S} \Sigma_v$ onto the product of standard apartments. The coordinates ξ_v defined above give rise to a system of coordinates $\boldsymbol{\xi} : \boldsymbol{s} = (s_v)_{v \in S} \mapsto \left(\alpha_v^{(1)}, \ldots, \alpha_v^{(r)}\right)_{v \in S} \in \mathbb{R}^{r|S|}$ on $\boldsymbol{\Sigma}$. Since we normalized the coordinates ξ_v, the map $\boldsymbol{\zeta} : \boldsymbol{\Sigma} \to \mathbb{R}^r$, $\boldsymbol{s} = (s_v)_{v \in S} \mapsto \left(\sum_{v \in S} \alpha_v^{(1)}(s_v), \ldots, \sum_{v \in S} \alpha_v^{(r)}(s_v)\right)_{v \in S}$ is invariant with respect to the action of $\mathcal{T}(\mathcal{O}_S)$ since the product formula [O'Me73, Theorem 33.1] implies that for every $x \in \mathcal{O}_S$,

$$\prod_{v \in S} |x|_v = 1.$$

Hence, the map $\boldsymbol{\pi} := \boldsymbol{\zeta} \circ \boldsymbol{\rho} : \boldsymbol{X} \to \mathbb{R}^r$ is invariant under the action of $\mathcal{B}(\mathcal{O}_S)$ on $\boldsymbol{X}$.

Lemma 4 ([Bux97b, Lemma 2.3]). *For every compact subset $C \subset \mathbb{R}^r$, the S-arithmetic group $\mathcal{B}(\mathcal{O}_S)$ acts cocompactly on $\pi^{-1}(C)$.*

Sketch of Proof. We use adele topology. The main ingredient is that $\mathcal{U}(K)$ is a discrete subgroup of $\mathcal{U}(\mathbb{A})$ with compact quotient [Bux97b, Lemma 1.1]. This generalises the well known result that the additive group of K is discrete in $\mathbb{A}$ and that $\mathbb{A}/_K$ is compact [Weil73, Theorem 2, page 64]. It follows that for every polysimplex $\boldsymbol{\sigma} \in \boldsymbol{X}$ the double quotient

$$\mathcal{U}(\mathbb{A}_S)\backslash \mathcal{U}(\mathcal{O}_S) / \mathrm{Stab}_{\mathcal{U}(\mathbb{A}_S)}(\boldsymbol{\sigma})$$

is discrete and compact; hence finite. This implies that the map

$$\mathcal{U}(\mathcal{O}_S)\backslash \boldsymbol{X} \to \Sigma$$

induced by ρ is proper. Now the claim follows from Dirichlet's Unit Theorem [CaFr67, page 72].

4.2. Positive results

We outline the proof of the first half of Theorem B, namely

Theorem 4.1. *The group $\mathcal{B}(\mathcal{O}_S)$ is of type $F_{|S|-1}$.*

We apply Theorem 1.3. Lemma 4 exhibits subspaces of $\boldsymbol{X}$ on which $\mathcal{B}(\mathcal{O}_S)$ acts co-compactly. Cell stabilizers are finite as they are intersections of the compact stabilizers in $\mathcal{B}(\mathbb{A}_S)$ the discrete group $\mathcal{B}(\mathcal{O}_S)$. Hence it suffices to show that the preimage $\pi^{-1}(C)$ is $(|S|-2)$-connected for some compact set $C \subset \mathbb{R}^r$, e.g., for a point. We put $\boldsymbol{Y} := \pi^{-1}(0) = \rho(H) \subseteq \boldsymbol{X}$ where $H = \ker \zeta : \Sigma \to \mathbb{R}^r$. We write $\boldsymbol{Y}$ as an ascending union

$$\boldsymbol{Y} = \bigcup_{j=0}^{\infty} \boldsymbol{Y}_j$$

where $\boldsymbol{Y}_0$ is contractible and $\boldsymbol{Y}_{j+1}$ is obtained from $\boldsymbol{Y}_j$ by glueing in a convex Euclidean set along its boundary such that, homotopically, this amounts to attaching a cell of dimension at least $|S|-1$. This way, no nontrivial homotopy elements are introduced up to dimension $|S|-2$.

4.2.1. Moufang buildings and Λ-complexes

The Moufang condition describes a certain interplay between the geometry of a building and its group of automorphisms. We shall not quote the precise definition here since Proposition 4.1 below, which being the technical core of the proof of Theorem 4.1, spells out all properties of Moufang buildings we use. It is known that the buildings X_v are locally finite Euclidean Moufang buildings [Bux97b, Fact 6.1]. Hence, the following proposition applies.

Proposition 4.1 ([Bux97b, Lemma 6.2]). *Let X be a locally finite Euclidean Moufang building with a distinguished chamber C at infinity and a* *distinguished apartment* Σ *containing* C. *Then there is a sequence* $\Sigma = \Sigma_0, \Sigma_1, \ldots$ *of apartments such that the following conditions are satisfied:*

(1) *Each* Σ_j *contains* C.
(2) *The* Σ_j *cover* X, *i.e.*,

$$X = \bigcup_{j=0}^{\infty} \Sigma_j.$$

(3) *For every* $j > 0$, *the new part*

$$N_j := \Sigma_j \setminus \bigcup_{i<j} \Sigma_i$$

is an intersection of open half apartments in Σ_j.

None of the N_j *contains* C *because this chamber is already present in* $\Sigma_0 = \Sigma$.

Remark 4. *The Moufang assumption is crucial to the proof of Proposition* 4.1 *given in [Bux97b], and I do not know whether the statement holds for non-Moufang locally finite Euclidean buildings. Since the buildings arising in the number field case are non-Moufang, a proof of Theorem A along the lines of the proof of Theorem B presented here does not yet exist. We mention that A. von Heydebreck [Heyd99, Lemma 3.3] proved an exact analogue of Proposition* 4.1 *for finite buildings by purely geometric means that do not make use of any Moufang type assumption. For finite buildings, the distinguished chamber at infinity is to be replaced by a chamber inside the building.*

Proposition 4.1 says that we can build X starting from Σ by attaching first the closure $\overline{N_1}$ along its boundary which lies in Σ. Then we glue in $\overline{N_2}, \overline{N_3}$, etc. Since $\overline{N_j}$ is an intersection of closed half apartments all of which avoid C, it can be retracted to its boundary. Hence the homotopy type does not change during this procedure. This recovers the important fact that the building X is contractible. Note that the retraction $\rho := \rho_{C,\Sigma} : X \to \Sigma$ restricts to an isomorphism of Coxeter complexes $\rho : \Sigma_j \to \Sigma$ for every j. Since these are Euclidean Coxeter complexes, this map is an isometry. Since Σ is a Euclidean Coxeter complex, the orthogonal part of its automorphism group is a finite reflection group associated to a root system $\Phi := \Phi(\Sigma)$, which we represent as a finite set of linear forms on Σ. The chamber C at infinity determines a base and a subset of positive roots Φ^+ such that C is represented by the cone

$$\{s \in \Sigma \mid \alpha(s) \geq 0 \ \forall \alpha \in \Phi^+\}.$$

We state some geometric axioms describing this setting.

Definition 2. *Let Y be a metric, piecewise Euclidean CW complex and $\pi : Y \to \mathbb{E}$ a projection from Y onto a Euclidean vector space $\mathbb{E}$. Furthermore, let Λ be a finite set of linear forms on $\mathbb{E}$. The pair (Y, π) is called a* Λ-complex *if there exists a sequence $E_0, E_1, \ldots$ of subcomplexes in Y satisfying the following conditions:*

(1) *The map π restricts to an isometry $\pi|_{E_j} : E_j \to \mathbb{E}$ for every j.*
(2) *The subcomplexes E_j cover Y, i.e.,*

$$Y = \bigcup_{j \geq 0} E_j.$$

(3) *For every $j > 0$, the new part*

$$N_j := E_j \setminus \bigcup_{i<j} E_i$$

is the interior of a subcomplex in E_j whose π-image in $\mathbb{E}$ is of the form

$$\{e \in \mathbb{E} \mid \lambda(e) \leq c_\lambda\},$$

where $c_\lambda \in \mathbb{R} \cup \{\infty\}$ are constants not all of which are ∞. If $c_\lambda = \infty$, it imposes no restriction on e.

A sequence of this form is called increasing.

Example 5. *Proposition* 4.1 *implies immediately that, keeping the notation from above, $(X, \rho : X \to \Sigma)$ is a Φ^+-complex.*

The following constructions provide many other examples. First, we consider direct products in order to deal with $\boldsymbol{X} = \times_{v \in S} X_v$.

Lemma 5. For $i \in \{1, 2\}$, let $(Y^i, \pi^i : Y^i \to \mathbb{E}^i)$ be a Λ^i-complex. Let $\mathrm{pr}_i : \mathbb{E}^1 \times \mathbb{E}^2 \to \mathbb{E}^i$ denote the canonical projection and put $\Lambda^1 \uplus \Lambda^2 := \{\lambda^i \circ \mathrm{pr}_i : \mathbb{E}^1 \times \mathbb{E}^2 \to \mathbb{R} \mid i \in \{1, 2\},\ \lambda^i \in \Lambda^i\}$. Then $(Y^1 \times Y^2, \pi^1 \times \pi^2 : Y^1 \times Y^2 \to \mathbb{E}^1 \times \mathbb{E}^2)$ is a $\Lambda^1 \uplus \Lambda^2$-complex.

Proof. For $i \in \{1, 2\}$, let $E_0^i, E_1^i, E_2^i, \ldots$ be an increasing sequences for Y^i. It is easy to check that $E_0^1 \times E_0^2,\ E_1^1 \times E_0^2,\ E_0^1 \times E_1^2,\ E_1^1 \times E_1^2,\ E_2^1 \times E_0^2,\ E_0^1 \times E_2^2,\ E_2^1 \times E_1^2,\ E_1^1 \times E_2^2,\ E_2^1 \times E_2^2, \ldots$ is an increasing sequence for $Y^1 \times Y^2$. □

Corollary 4.1. $(\boldsymbol{X} = \times_{v \in S} X_v, \boldsymbol{\rho} : \boldsymbol{X} \to \boldsymbol{\Sigma})$ *is a $\biguplus_{v \in S} \Phi_v^+$-complex.*

Proof. All X_v are locally finite Euclidean Moufang buildings [Bux97b, Fact 6.1]. □

Now we turn to subcomplexes because we are interested in $\boldsymbol{Y} \subseteq \boldsymbol{X}$.

Observation 4.1. *Let $(Y, \pi : Y \to \mathbb{E})$ be a Λ-complex and let $\mathbb{E}'$ be a linear subspace of $\mathbb{E}$. Put $\Lambda|_{\mathbb{E}'} := \{\lambda|_{\mathbb{E}'} \mid \lambda \in \Lambda\}$. Then $(\pi^{-1}(\mathbb{E}'), \pi_{\pi^{-1}(\mathbb{E}')} : \pi^{-1}(\mathbb{E}') \to \mathbb{E}')$ is a $\Lambda|_{\mathbb{E}'}$-complex. We see this by taking an increasing sequence for Y and intersecting it with the preimage of $\mathbb{E}'$.*

Corollary 4.2. *$(\boldsymbol{Y}, \rho|_{\boldsymbol{Y}} : \boldsymbol{Y} \to H)$ is a $\left(\biguplus_{v \in S} \Phi_v^+\right)|_H$-complex.*

4.2.2. Connectivity of Λ-complexes

We call a set Λ of linear forms on a real vector *m-tame* if no positive linear combination of up to m elements of Λ vanishes, i.e., if 0 is not contained in the convex hull of up to m elements of Λ.

Proposition 4.2 ([Bux97a, Lemma 7.3]). *Let $(Y, \pi : Y \to \mathbb{E})$ be a Λ-complex and suppose Λ is m-tame. Then Y is $(m-1)$-connected.*

Proof. Let $E_0, E_1, E_2, \ldots$ be an increasing sequence for Y. Put

$$Y_j := \bigcup_{i \leq j} E_j.$$

Since Y is covered by the E_j, it suffices to show that every Y_j is $(m-1)$-connected. Since $Y_0 = E_0$ is contractible, we can prove the claim by induction if we control the process of obtaining Y_{j+1} from Y_j. From

$$Y_{j+1} = Y_j \cup \left(E_{j+1} \setminus Y_j\right)$$

and the fact that $E_0, E_1, E_2, \ldots$ is increasing it follows that Y_{j+1} is obtained from Y_j by attaching, along its boundary, a convex set N_j' of the form

$$N_j' = \{e \in \mathbb{E} \mid \lambda(e) \leq c_\lambda \forall \lambda \in \Lambda'\}$$

where $\Lambda' \subseteq \Lambda$ is a nonempty subset and $c_\lambda \in \mathbb{R}$. We have to control the homotopy type of the pair $(N_j', \partial N_j')$. If $N_j' = \partial N_j'$ there is nothing to do. Hence we can assume that N_j' has nonempty interior. First suppose that Λ' spans the dual $\mathbb{E}^*$, that is,

$$0 = \bigcap_{\lambda \in \Lambda'} \ker \lambda.$$

There are two cases.

(i) N_j' is unbounded. The interior N_j contains an infinite ray r, but it does not contain a whole line, since we assumed $0 = \bigcap_{\lambda \in \Lambda'} \ker \lambda$. Hence we may assume that r starts at $x \in \partial N_j'$. If the boundary $\partial N_j'$ does not contain a ray parallel to

r we can argue that moving all points parallel to r towards the boundary $\partial N'_j$ defines a deformation retraction of N'_j onto its boundary.

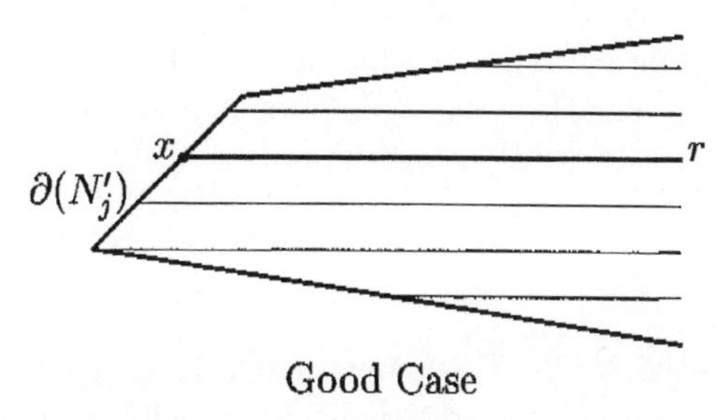

Good Case

Unfortunately, this does not work if the boundary does contain a ray parallel to r.

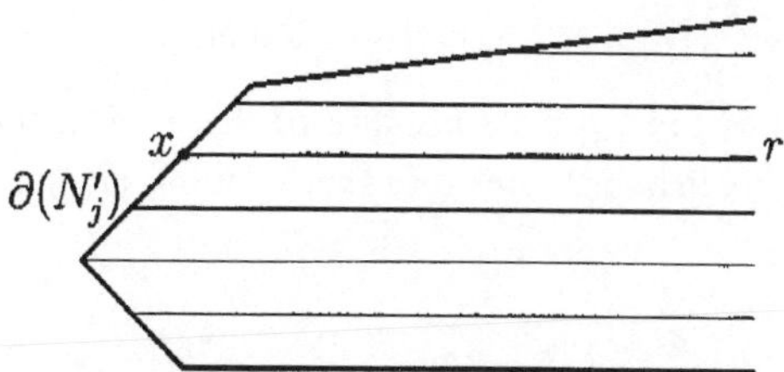

Bad case: no continuous retraction since points of $\partial(N'_j)$ are not to move

Nevertheless, modifying the lines of motion slightly such that points are departing from r when they are approaching $\partial N'_j$ works.

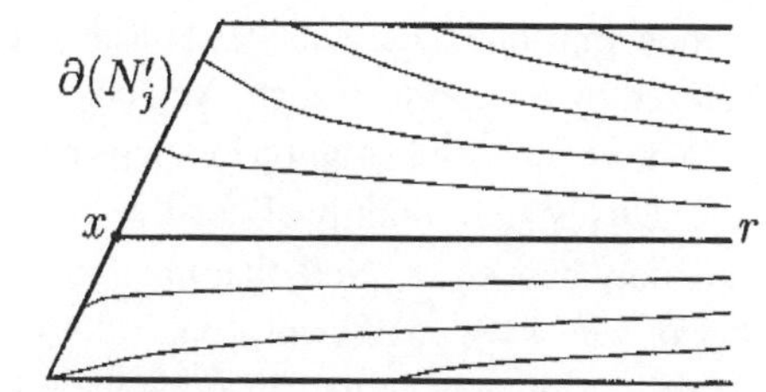

Modified lines of motion yielding a retraction

We just have to make sure that these lines intersect the boundary $\partial N'_j$. This is achieved most easily by ensuring that they even intersect a supporting hyperplane of N'_j at x which can be accomplished, e.g., by choosing the lines of motion to form a family of affine hyperbolas.

(ii) N'_j is bounded. Topologically N'_j is a disc. Since we are assuming that it has interior points, its dimension is the dimension of the surrounding space $\mathbb{E}$ which equals the dimension of its dual $\mathbb{E}^*$. In this case, furthermore, 0 lies in the convex hull of Λ'. By Caratheodory's Theorem [Eggl58, Remark (ii), page 38], there are $\dim \mathbb{E}^* + 1$ elements in Λ' whose convex hull contains 0. Since Λ and therefore Λ' are m-tame, $m < \dim \mathbb{E}^* + 1$ follows. Therefore N_j' is a disc of dimension at least m. In either case, attaching N_j' does not introduce nontrivial elements in the homotopy groups up to π_{m-1}. Suppose now Λ' does not span $\mathbb{E}^*$. Then we write N_j' as a direct product of

$$V := \bigcap_{\lambda \in \Lambda'} \ker \lambda$$

and a convex subset $C \subseteq \mathbb{E}/_V$. This quotient is dual to the span of Λ', and the considerations of the previous paragraph apply to C, yielding the same result. □

In order to apply this proposition to $\boldsymbol{Y}$ we have to prove

Lemma 6. The set $\left(\biguplus_{v \in S} \Phi_v^+\right)|_H$ is $(|S| - 1)$-tame.

This set is certainly not $|S|$-tame because of the product formula. Thus the lemma says roughly that there are no further relations among the places.

Proof. Let

$$\sum_{v \in S, \alpha \in \Phi^+} \mu_{v,\alpha} \alpha_v \tag{1}$$

be a convex combination, that is, all coefficients α_v are nonnegative and add up to 1. We have to show that this combination does not vanish on H unless at least $|S|$ coefficients are $\neq 0$. In fact, we will see that for each prime $v \in S$ there is at least one non-vanishing coefficient $\mu_{v,\alpha}$. Call a prime v *void* if all coefficients $\mu_{v,\alpha}$ indexed by that prime vanish. Any positive root is a positive integer combination of base roots. Substitution of all the positive roots by these combinations in (1) yields again a nontrivial positive combination vanishing on H, now involving only base roots. Note that this process may increase or decrease the number of vanishing coefficients, but it does not alter the set of void primes. Therefore, since we want to prove that the set of void primes is empty, we may assume without loss of generality that we are dealing with a convex combination

$$\sum_{v \in S, \alpha \in \Delta} \mu_{v,\alpha} \alpha_v \tag{2}$$

of base roots. If this combination vanishes on H we conclude that

$$\sum_{v \in S, \alpha \in \Delta} \mu_{v,\alpha} \alpha_v \in \operatorname{im}(\zeta^*).$$

Hence, for each base root α involved, we have

$$\mu_{v,\alpha} = \mu_{v',\alpha} \ \forall v, v' \in S$$

by definition of ζ. Hence none of these coefficients vanishes because we started with a nontrivial combination. $\square$

4.3. Negative results

In order to complete the proof of Theorem B we have to show that $\mathcal{B}(\mathcal{O}_S)$ is not of type $\mathrm{FP}_{|S|}$. We proof this by reduction to the rank-1-case, but we present only the algebraic reduction that applies to the Borel subgroups $\mathrm{B}_n^0 \leq \mathrm{SL}_n$ and $\mathrm{B}_n \leq \mathrm{GL}_n$. The general case is dealt with by an analogous geometric reduction to the rank-1-case [Bux97b, Section 5].

Theorem 4.2 ([Bux97a, Theorem 8.1]). *The group* $\mathrm{B}_2^0(\mathcal{O}_S)$ *is not of type* $FP_{|S|}$.

This can be proven by various means. Since $\mathrm{B}_2^0(\mathcal{O}_S)$ is metabelian, one can use the Σ-theory of geometric invariants introduced by R. Bieri and R. Strebel [BiSt80]. This line of reasoning was taken in [Bux97a].

Another way is via Bestvina-Brady-Morse theory [BeBr97]. For $\mathrm{B}_2^0(\mathcal{O}_S)$ the Euclidean buildings X_v are trees and the map

$$\pi : X \to \mathbb{R}^r = \mathbb{R}$$

takes values in the real line since SL_2 has rank $r = 1$. So we regard it as a *height* on X. To this map, Bestvina-Brady-Morse theory applies, which can be used to reprove that $\mathrm{B}_2^0(\mathcal{O}_S)$ is of type $\mathrm{F}_{|S|-1}$ as follows. Ascending and descending links are easily computed because of the product structure of X. They turn out to be joins of ascending, respectively descending, links in the factors. Hence they are points or joins of $|S|$-spheres, and in either case $(|S| - 1)$-connected. From this the claim follows by the Morse lemma of [BeBr97].

The method can be refined to yield as well a

Proof of Theorem 4.2. Consider the family of compact intervals of $\mathbb{R} = \mathbb{R}^r$ as a directed system via the relation of inclusion. We claim that the corresponding directed system

$$(\pi_{|S|-1}(\pi^{-1}(I)))_I \subset \mathbb{R}$$

of homotopy groups of preimages in X is not essentially trivial. To show that, for any $I \ni 0$, the map

$$\pi_{|S|-1}(\pi^{-1}(0)) \to \pi_{|S|-1}(\pi^{-1}(I))$$

is not trivial, we take a vertex $\boldsymbol{x} \in X$ above $\max(I)$. In its descending link there are spheres that are nontrivial in the whole link of $\boldsymbol{x}$. Pick one and project it along geodesics until it is on height 0. We see, that this height-0-sphere is nontrivial in $X \setminus \{\boldsymbol{x}\}$ because X is CAT(0) and there is a geodesic projection $(X \setminus \{\boldsymbol{x}\}) \to \mathrm{Lk}_X(\boldsymbol{x})$. This projection takes back the height-0-sphere to the nontrivial homotopy element in $\mathrm{Lk}_X(\boldsymbol{x})$ we started with. Hence, the height-0-sphere does not become trivial in $\pi^{-1}(I)$.

Lemma 7. $\mathrm{B}_n(\mathcal{O}_S)$ and $\mathrm{B}_n^0(\mathcal{O}_S)$ have identical finiteness properties.

Proof. Let D_n denote the scheme of diagonal matrices of rank n and let D_n^0 denote the subscheme of those matrices that have determinant 1. Note that $\mathrm{D}_n(\mathcal{O}_S)$ is finitely generated and contains a free Abelian subgroup A of finite index that decomposes as a product $A = A^0 \times A'$ where A^0 is a free Abelian subgroup of finite index in $\mathrm{D}_n^0(\mathcal{O}_S)$. Moreover, we can choose A' to be a group of multiples of the identity matrix. Hence A' acts trivially on the unipotent radical. Writing

$$\mathrm{B}_n(\mathcal{O}_S) = \mathrm{D}_n(\mathcal{O}_S) \ltimes \mathrm{U}_n(\mathcal{O}_S)$$

and

$$\mathrm{B}_n^0(\mathcal{O}_S) = \mathrm{D}_n^0(\mathcal{O}_S) \ltimes \mathrm{U}_n(\mathcal{O}_S),$$

we see that $\mathrm{B}_n(\mathcal{O}_S)$ is commensurable to $A \ltimes \mathrm{U}_n(\mathcal{O}_S) = A' \times \left(A^0 \ltimes \mathrm{U}_n(\mathcal{O}_S)\right)$, which has the same finiteness properties as $A^0 \ltimes \mathrm{U}_n(\mathcal{O}_S)$. This, in turn, is commensurable to $\mathrm{B}_n^0(\mathcal{O}_S)$. □

Hence there is no loss in confining ourselves to $\mathrm{B}_n(\mathcal{O}_S)$. For this group, the following observation easily accomplishes the reduction to $\mathrm{B}_2(\mathcal{O}_S)$.

Observation 4.2. *$\mathrm{B}_2(\mathcal{O}_S)$ is a retract of $\mathrm{B}_n(\mathcal{O}_S)$ in the following way:*

$$\begin{pmatrix} * & * & \cdots & * \\ 0 & * & \ddots & \vdots \\ \vdots & \ddots & \ddots & * \\ 0 & 0 & 0 & * \end{pmatrix}$$

with the obvious inclusion and projection map.

By Remark 1 it follows that $\mathrm{B}_2(\mathcal{O}_S)$ has all finiteness properties that $\mathrm{B}_n(\mathcal{O}_S)$ enjoys. Hence $\mathrm{B}_n(\mathcal{O}_S)$ is not of type $\mathrm{FP}_{|S|}$ in view of Theorem 4.2, whence $\mathrm{B}_n^0(\mathcal{O}_S)$ is neither.

References

[Abel79] H. Abels, An example of a finitely presented solvable group, in: *Homological Group Theory*, LMS Lecture Notes vol. 35 (C. T. C. Walls ed.), Cambridge University Press, Cambridge, 1979, 205–211.

[Abel87] H. Abels, *Finite presentability of S-arithmetic groups – Compact presentability of solvable groups*, Springer Lecture Notes in Mathematics vol. 1261, 1987.

[Abel91] H. Abels, Finiteness properties of certain arithmetic groups in the function field case, *Israel J. Math.* **76** (1991), 113–128.

[AbBr87] H. Abels, K. S. Brown, Finiteness properties of solvable S-arithmetic groups: an example, *J. Pure and Applied Algebra* **44** (1987), 77–83.

[AbTi97] H. Abels, A. Tiemeyer, Compactness properties of locally compact groups, *Transformation Groups* **2** (1997), 119–135.

[Abra87] P. Abramenko, *Endlichkeitseigenschaften der Gruppen* $\mathrm{SL}_n(\mathbb{F}_q[t])$, Dissertation, Frankfurt am Main, 1987.

[Abra96] P. Abramenko, *Twin Buildings and Applications to S-Arithmetic Groups*, Springer Lecture Notes in Mathematics vol. 1641, 1996.

[Behr62] H. Behr, Über die endliche Definierbarkeit von Gruppen, *J. Reine u. Angew. Math.* **211** (1962), 116–122.

[Behr98] H. Behr, Arithmetic groups over function fields I. A complete characterization of finitely generated and finitely presented arithmetic subgroups of reductive algebraic groups, *J. Reine u. Angew. Math.* **495** (1998), 79–118.

[Behr99] H. Behr, Higher finiteness properties of S-arithmetic groups in the function field case I: Chevalley groups with coefficients in $\mathbb{F}_q[t]$, preprint.

[BeBr97] M. Bestvina, N. Brady, Morse theory and finiteness properties of groups, *Invent. math.* **129** (1997), 445–470.

[BiSt80] R. Bieri, R. Strebel, Valuations and finitely presented metabelian groups, *Proc. London Math. Soc.* **41** (1980), 439–464.

[Bore63] A. Borel, Some finiteness properties of adele groups over number fields, *Publ. Math. IHES* **16** (1963), 5–30.

[Bore91] A. Borel, *Linear Algebraic Groups*, second edition, Springer, New York, 1991.

[BoHa62] A. Borel, Harish-Chandra, Arithmetic subgroups of algebraic groups, *Annals of Math.* **75** (1962), 485–535.

[BoSe73] A. Borel, J-P. Serre, Corners and arithmetic groups *Comm. Math. Helv.* **48** (1973), 436–491.

[BoSe76] A. Borel, J.-P. Serre, Cohomologie d'immeubles et de groupes S-arithmétiques, *Topology* **15** (1976), 211–232.

[BoTi65] A. Borel, J. Tits, Groupes réductifs, *Publ. Math. IHES* **27** (1965), 55–150.

[Brow82] K.S. Brown, *Cohomology of Groups*, Springer, New York, 1982.

[Brow87] K. S. Brown, Finiteness properties of groups, *J. Pure and Applied Algebra* **44** (1987), 45–75.

[Brow89] K.S. Brown, *Buildings*, Springer, New York, 1989.

[BrTi72] F. Bruhat, J. Tits, Groupes Réductives sur un Corps Local I, *Publ. Math. IHES* **41** (1972), 5–252.

[Bux97a] K.-U. Bux, Finiteness properties of some metabelian S-arithmetic groups, *Proc. London Math. Soc.* **75** (1997), 308–322.

[Bux97b] K.-U. Bux, Endlichkeitseigenschaften auflösbarer arithmetischer Gruppen über Funktionenkörpern, Dissertation, Frankfurt am Main, 1997. (http://math.kubux.net/math/bux98.ps.gz)

[Bux99] K-U. Bux, Finiteness properties of soluble arithmetic groups over global function fields, preprint. (http://math.kubux.net/math/bux99b.ps.gz)

[CaFr67] *Algebraic Number Theory. Proceedings of an Instructional Conference organized by the London Mathematical Society* (J.W.S. Cassels, A. Fröhlich eds.), Academic Press, London, 1967.

[Chev60] C. Chevalley, Certain Schémas de Groupes Semi-Simples, *Séminaire Bourbaki 13e année (1960/61) no. 219*, 1–16.

[Eggl58] H.G. Eggleston, *Convexity*, Cambridge Tracts in Mathematics and Mathematical Physics, Cambridge, 1958.

[Heyd99] A. von Heydebreck, Homotopy properties of certain complexes associated to spherical buildings, *Israel J. Math.* **133** (2003), 369–379.

[Knes64] M. Kneser, Erzeugende und Relationen verallgemeinerter Einheitengruppen, *J. Reine u. Angew. Math.* **214/215** (1964), 345–349.

[Macb64] A.M. Macbeath, Groups of homeomorphisms of a simply connected space, *Annals of Math.* **79** (1964), 473–488.

[O'Me73] O.T. O'Meara, *Introduction to Quadratic Forms*, Springer, New York, 1973.

[Quil78] D. Quillen, Homotopy properties of the poset of nontrivial p-subgroups of a group, *Adv. in Math.* **28** (1978), 101–128.

[Ragh68] M.S. Raghunatan, A note on quotients of real algebraic groups by arithmetic subgroups, *Invent. Math.* **4** (1968), 318–335.

[Rona89] M. Ronan, *Lectures on Buildings*, Academic Press, San Diego, 1989.

[Serr77] J.-P. Serre, *Arbres, amalgames, SL_2, Asterique* **46**, 1977.

[Span66] E. H. Spanier, *Algebraic Topology*, Springer, New York, 1966.

[Stei68] R. Steinberg, *Lectures on Chevalley Groups, Notes prepared by John Faulkner and Robert Wilson*, Yale University, New Haven, Connecticut, 1968.

[Stuh76] U. Stuhler, Zur Frage der endlichen Präsentiertheit gewisser arithmetischer Gruppen im Funktionenkörperfall, *Mathematische Annalen* **224** (1976), 217–232.

[Stuh80] U. Stuhler, Homological properties of certain arithmetic groups in the function field case, *Invent. Math.* **57** (1980), 263–281.

[Tiem97] A. Tiemeyer, A Local-Global Principle for Finiteness Properties of S-Arithmetic Groups over Number Fields, *Transformation Groups* **2** (1997), 215–223.

[Wall79] *Homological Group Theory* (C.T.C. Wall ed.), LMS Lecture Notes vol. 35, Cambridge University Press, Cambridge, 1979.

[Weil73] A. Weil, *Basic Number Theory*, reprint of the second edition (1973), Springer, Heidelberg, 1995.

6

Topology in permutation groups

P. J. Cameron

1. Introduction

In a lecture at Oberwolfach in the 1980s with the title "Topology in permutation groups", Helmut Wielandt defended the proposition that topology in permutation groups is of no use, and came to the conclusion that (with the addition of the word "almost") this was indeed the case. Wielandt made it clear that he was not speaking about the topology *of* permutation groups. (There is a natural topology on the symmetric group, namely the topology of pointwise convergence. In the case of countable degree, the open subgroups are those lying between the pointwise and setwise stabiliser of a finite set, and the closed subgroups are the automorphism groups of first-order structures.) Nevertheless, many in the audience felt that Wielandt's conclusion was too pessimistic. Since then, further results have supported this view. The present paper is a survey of some of these results. I begin with an account of Wielandt's own work on the subject, the connection of non-Hausdorff topologies preserved by G and the notions of primitivity and strong primitivity. The next topic is a theorem of Macpherson and Praeger, according to which a primitive group which preserves no non-trivial topology is highly transitive. Their proof uses some deep results from model theory. I give an elementary argument to replace part of the proof.

Some topologies low in the separation hierarchy can be interpreted as relational structures (specifically, preorders), so that their homeomorphism groups are closed subgroups of the symmetric group. This observation is a bridge to the next topic, the topology of permutation groups. This topic is too large for a survey, but some of the most important results relating to the earlier material are discussed. By convention, a structure on a set Ω is *trivial* if its automorphism group is the full symmetric group on Ω. Clearly, knowing only that a group preserves a trivial structure tells us nothing about the group.

2. Primitivity and strong primitivity

A transitive permutation group G on a set Ω is *primitive* if G preserves no equivalence relation on Ω apart from the trivial ones (equality and the relation Ω^2). In his pioneering study of infinite permutation groups, Wielandt [20] observed that this condition is not strong enough to derive analogues of results on finite primitive groups. Specifically, let G be a primitive group on a finite set Ω, and let Δ be a non-empty proper subset of Ω. Then the following separation property holds: for any distinct $\alpha, \beta \in \Omega$, there exists $g \in G$ with $\alpha g \in \Delta$ and $\beta g \notin \Delta$. This fails for infinite primitive groups: if G is the group of all order-preserving permutations of the real numbers, Δ the set of positive real numbers, and $\alpha < \beta$, then no such g can exist. Wielandt proposed the property of *strong primitivity* to remedy this defect. What follows is a slightly more general approach.

A *preorder* on a set Ω is a reflexive and transitive relation on Ω. A transitive permutation group G on Ω is called *strong* if every G-invariant preorder is symmetric (and so is an equivalence relation); and G is *strongly primitive* if it is strong and primitive, that is, the only G-invariant preorders on Ω are the trivial equivalence relations. For any transitive permutation group G, there is a natural bijection between G-congruences and subgroups of G containing the point stabiliser G_α. Hence G is primitive if and only if G_α is a maximal proper subgroup of G. The analogue for strong primitivity is as follows:

Proposition 2.1. *Let G be a transitive permutation group on Ω. There is a natural bijection between G-invariant preorders on Ω and submonoids of G containing G_α. Hence G is strong if and only if any submonoid of G properly containing the point stabiliser G_α is a subgroup of G; and G is strongly primitive if and only if G_α is a maximal proper submonoid of G.*

Proof. Suppose that $M(\alpha)$ is a submonoid of G containing G_α. By conjugation, we can define M_β for all $\beta \in \Omega$. Now define $\alpha \to \beta$ if there is an element of $M(\alpha)$ mapping α to β (equivalently, every element mapping α to β lies in $M(\alpha)$. This relation is a preorder. For suppose that $\alpha \to \beta$ and $\beta \to \gamma$. Choose $h \in M(\beta)$ with $\beta h = \gamma$. Then $h = g^{-1}kg$ for some $g, k \in M(\alpha)$ (where $\alpha g = \beta$); then $\alpha gh = \gamma$ and $gh = kg \in M(\alpha)$, so $\alpha \to \gamma$. Conversely, if $\to$ is a G-invariant preorder, and

$$M(\alpha) = \{g \in G : \alpha \to \alpha g\},$$

then $M(\alpha)$ is a submonoid of G containing G_α; and the two constructions are mutually inverse. □

For example, if G is strong, then $N_G(G_\alpha)/G_\alpha$ is a torsion group. In particular, the regular representation of G is strong if and only if G is a torsion group.

It is now a simple exercise to show that, if G is strongly primitive, then the separation property mentioned above holds.

3. Separation, preorders, and primitivity

If $\rightarrow$ is a preorder, then the relation $\equiv$ defined by

$$x \equiv y \text{ if and only if } x \rightarrow y \text{ and } y \rightarrow x$$

is an equivalence relation; and $\rightarrow$ induces a (partial) order on the set of equivalence classes of $\equiv$. Given a topology $\mathcal{T}$ on Ω, construct a preorder on Ω by the rule that $x \rightarrow y$ if and only if every open set containing x contains y. Conversely, given a preorder $\rightarrow$, take the neighbourhoods

$$U_x = \{y \in \Omega : x \rightarrow y\} \qquad \text{for } x \in \Omega$$

as a base for a topology on Ω. It is easy to see that the map

$$\text{preorder} \rightarrow \text{topology} \rightarrow \text{preorder}$$

is always the identity; the map

$$\text{topology} \rightarrow \text{preorder} \rightarrow \text{topology}$$

gives a stronger topology which is in general strictly stronger, though if Ω is finite, this map is also the identity. Note that the latter map is idempotent. We call a topology *relational* if it is fixed by this map (that is, if it is derived from a preorder). By definition, G is strongly primitive if and only if it preserves no non-trivial preorder.

Theorem 3.1. *Let G be a transitive permutation group on Ω.*

(a) *G is primitive if and only if every non-trivial topology invariant under G is $T0$.*
(b) *G is strongly primitive if and only if every non-trivial topology invariant under G is $T1$.*

Proof. We observe that the topology derived from a preorder $\rightarrow$ satisfies T0 if and only if the equivalence relation $\equiv$ is equality; and it satisfies T1 if and only if $\rightarrow$ has no arcs except loops. Hence, if G preserves a non-T0 topology, then it preserves a non-trivial equivalence, and if G preserves a non-T1 topology, then it preserves a preorder which is not symmetric. The converse result in (a) is clear (if G is imprimitive, take open sets to be all unions of blocks). For (b), if G is not strongly primitive, then by definition it preserves a non-trivial preorder. □

Remarks. 1. If G preserves a non-T0 (resp. non-T1) topology, then it preserves a relational topology with the same property.

2. There is a unique countable homogeneous preorder which is universal (in the sense that it contains all finite or countable preorders) and homogeneous (in the sense that any isomorphism between finite sub-preorders extends to an automorphism). The corresponding relational topology is non-T0 and admits a transitive group of homeomorphisms. Similarly, there is a unique countable homogeneous partial order containing all finite or countable orders. These follow immediately from Fraïssé's construction method [5] and the fact that the classes of finite preorders and finite partial orders both have the amalgamation property.

4. Topologies and filters

Macpherson and Praeger [8] proved that a primitive permutation group on a countable set Ω which is not highly transitive (that is, is not n-transitive for some n) is contained in a maximal subgroup of $\mathrm{Sym}(\Omega)$. The main part of their argument was to show that such a group preserves a non-trivial filter on Ω. (A *filter* is a non-empty family $\mathcal{F}$ of subsets of Ω which is closed under finite intersections and closed under taking supersets, but does not contain the empty set. As usual, it is non-trivial if it is not invariant under $\mathrm{Sym}(\Omega)$. If Ω is countable, the only trivial filters are the *indiscrete filter* $\{\Omega\}$ and the *cofinite filter* consisting of all cofinite subsets.) Their proof comes in two parts: first, prove that such a group preserves a non-trivial topology; and then show that any primitive group preserving a non-trivial topology must preserve a non-trivial filter. Their argument requires substantial machinery from model theory (the theorems of Ehrenfeucht–Mostowski, Engeler–Ryll-Nardzewski–Svenonius, and Cherlin–Harrington–Lachlan), and the connection between the filter and the topology is not clear. By contrast, the proof of the first step given below is more elementary, and only two constructions of filters from topologies are used: sets containing a finite intersection of open dense sets; and complements of finite unions of discrete sets. The condition that G preserves a filter is a "strong condition", see Dixon, Neumann and Thomas [3], Macpherson and Neumann [7]. On the other hand, McDermott [6] has proved strong restrictions on permutation groups which preserve no non-trivial topology.

Remarks. 1. I make essential use of Neumann's Separation Lemma [10, Lemma 2.3]: if G is a permutation group on Ω with no finite orbits, and Γ and Δ are finite subsets of Ω, then some translate of Γ by G is disjoint from Δ.

2. Primitivity is necessary. The group of all permutations preserving a partition of Ω into two parts of the same cardinality preserves a non-trivial topology but no non-trivial filter. Indeed, any imprimitive permutation group preserves a non-trivial topology, namely, the set of all unions of blocks in a given system of imprimitivity.

3. In some cases, one can see directly that a permutation group preserves a non-trivial topology. For example, from the countable random graph (see [2]) we derive a topology by taking the closed vertex neighbourhoods (consisting of a vertex v and all its neighbours) as a basis for the closed sets. In other cases, the construction of a topology is much less obvious.

Before stating the theorem, I give a simple characterisation of primitive automorphism groups of non-trivial topologies or filters. A *moiety* of the countable set Ω is an infinite subset of Ω whose complement is also infinite.

Lemma 4.1. *Let G be primitive on the countable set Ω.*

(a) *G preserves a non-trivial topology if and only if there exists a moiety Δ of Ω such that, for all $g_1, \ldots, g_n \in G$, the set $\Delta g_1 \cap \ldots \cap \Delta g_n$ is empty or infinite.*
(b) *G preserves a non-trivial filter if and only if there exists a moiety Δ of Ω such that, for all $g_1, \ldots, g_n \in G$, the set $\Delta g_1 \cap \ldots \cap \Delta g_n$ is infinite.*

Proof. (a) Let G be primitive. Assume that G preserves a non-trivial topology. If there is a non-empty finite open set, take a minimal such set U; then U is a block of imprimitivity, necessarily a singleton, and the topology is discrete. So take U to be a non-empty non-cofinite open set; by what has just been proved, it is a moiety, and the condition of the lemma obviously holds. Conversely, let Δ be such a set. Then the class of all non-empty sets of the form $\Delta g_1 \cap \ldots \cap \Delta g_n$, for $g_1, \ldots, g_n \in G$, is a basis for a non-trivial G-invariant topology on Ω.
(b) Neumann's Lemma implies that no filter invariant under an infinite transitive group can contain a finite set. So if G preserves a non-trivial filter, take U to be any non-cofinite set in the filter. Conversely, if Δ is a moiety with the property of the lemma, then the class of all sets containing $\Delta g_1 \cap \ldots \cap \Delta g_n$ for some $g_1, \ldots, g_n \in G$ is a non-trivial filter admitting G. □

Theorem 4.1. *Let Ω be an infinite set, and G a primitive permutation group on Ω. Then G preserves a non-trivial topology on Ω if and only if G preserves a non-trivial filter on Ω.*

Proof. I begin with some remarks on topologies admitting primitive groups. If such a topology contains a non-empty finite open set, then it is discrete: for the minimal open sets form a system of imprimitivity, and so all singletons are open. Similarly, if the topology contains a cofinite open set, then every cofinite set is open (but we cannot conclude that the topology is trivial in this case).

A set is dense if it meets every non-empty open set. Any finite intersection of dense open sets is non-empty. (The proof is by induction on the number of sets. Given dense open sets $U_1, \ldots, U_n$, the set $U_1 \cap \ldots \cap U_{n-1}$ is non-empty and open, by the inductive hypothesis; so it meets the dense open set U_n.) So the family of sets containing finite intersections of dense open sets is a filter.

If a topology admits a primitive group, and is the union of finitely many discrete subsets, then it is itself discrete. (To show this, it suffices to find a finite open set. Let $\Omega = X_1 \cup \ldots \cup X_n$, where $X_1, \ldots, X_n$ are pairwise disjoint discrete sets. Choose $x_1 \in X_1$. There exists an open set U_1 with $U_1 \cap X_1 = \{x_1\}$. If $U_1 = \{x_1\}$, we are done; so suppose not. Then, without loss, $U_1 \cap X_2$ is non-empty; choose $x_2 \in U_1 \cap X_2$. Then there is an open set U_2 with $U_2 \cap X_2 = \{x_2\}$, so that

$$\{x_2\} \subseteq (U_1 \cap U_2) \cap (X_1 \cup X_2) \subseteq \{x_1, x_2\}.$$

Continuing in this way, we end up with a finite open set.) It follows that, if the topology is not discrete, then the set of complements of finite unions of discrete sets is a filter.

Now we begin the proof of the theorem. One direction is elementary: if $\mathcal{F}$ is a filter, then $\{\emptyset\} \cup \mathcal{F}$ is a topology, which is non-trivial if $\mathcal{F}$ is. So we suppose that G is primitive and preserves a non-trivial topology $\mathcal{T}$, and (for a contradiction) that G preserves no non-trivial filter. By our two constructions above, it follows that every open dense set is cofinite, and every discrete set is finite. Form a graph Γ on Ω by joining x and y whenever there exist disjoint open sets containing x and y.

Case 1. Γ contains no infinite clique. In this case, every finite clique is contained in a finite maximal clique. Let $\{x_1, \ldots, x_n\}$ be a maximal clique. We can find pairwise disjoint open sets $U_1, \ldots, U_n$ with $x_i \in U_i$ for $i = 1, \ldots, n$. (For let W_{ij} and W_{ji} be disjoint open sets containing x_i and x_j respectively, for $i \neq j$; then put $U_i = \bigcap_{j \neq i} W_{ij}$.) Now $U = U_1 \cup \ldots \cup U_n$ is open and dense (by the maximality of the clique $\{x_1, \ldots, x_n\}$); so $F = \Omega \setminus U$ is finite. Let $\{x'_1, \ldots, x'_m\}$ be another maximal clique, and define $U'_1, \ldots, U'_m, F'$ analogously. By Neumann's Separation Lemma, we may assume that the finite sets $\{x_1, \ldots, x_n\} \cup F$ and $\{x'_1, \ldots, x'_m\} \cup F'$ are disjoint. Suppose that $m > n$. Then $x_1, \ldots, x_n$ lie in the union of the pairwise disjoint sets $U'_1, \ldots, U'_m$; say $x_i \in U'_{k_i}$ for $i + 1, \ldots, n$. Then one of the sets, say U'_l, does not contain any of $x_1, \ldots, x_n$. But then $U_1 \cap U'_{k_1}, \ldots, U_n \cap U'_{k_n}$ are pairwise disjoint open sets containing $x_1, \ldots, x_n$, and U_l is disjoint from all of them, a contradiction. So $m > n$ is impossible, and dually $n > m$ is impossible. Thus, we have $m = n$.

We conclude that all maximal cliques have the same size n. Now the induced subgraph of Γ on $U_1 \cup \ldots \cup U_n$ is complete n-partite. (It clearly contains the complete multipartite graph; but any additional edge would produce a clique of size greater than n.) We claim that Γ is complete n-partite. For suppose not. Since complete n-partite graphs are characterised by their finite subgraphs, there exists a finite subset Z of Ω on which the induced subgraph is not complete n-partite. By Neumann's Separation Lemma, we may assume that $Z \cap F = \emptyset$. Then $Z \subseteq U_1 \cup \ldots \cup U_n$, and we have a contradiction. So the claim is proved. Now, if $n > 1$, we have a contradiction to the primitivity of G, since G preserves the complete n-partite graph Γ. On the other hand, if $n = 1$, then Γ is a null graph, and so any two non-empty open sets intersect. So the class of sets which contain a non-empty open set is a filter, and is non-trivial since the topology is.

Case 2. Γ contains an infinite clique Y. Then the induced topology on Y is Hausdorff, and so Y contains an infinite discrete subset (see below). So the filter of complements of discrete sets is non-trivial. □

The fact that an infinite Hausdorff space contains an infinite discrete subset is an exercise in the book by Sierpiński [17]. Since this is not very helpful, I include the proof. Let X be an infinite Hausdorff space. Call a point $x \in X$ *bad* if every open set containing x is cofinite. Clearly there is at most one bad point. (Bad points can exist; for example, the added point in the one-point compactification of an infinite discrete space.) Now define a sequence $x_1, x_2, \ldots$ of points of X and a decreasing sequence $Y_0, Y_1, Y_2, \ldots$ of infinite sets as follows:

- $Y_0 = X$.
- Given Y_{n-1}, let x_n be any non-bad point of Y_{n-1}, U_n an open set containing X_n such that $Y_n = Y_{n-1} \setminus U_n$ is infinite.

Now we claim that $Z = \{x_1, x_2, \ldots\}$ is discrete. For any x_n, we have an open set U_n containing x_n and no x_m for $m > n$. Also, by the Hausdorff property, there is an open set V_n containing x_n and no x_m for $m < n$. Then $Z \cap (U_n \cap V_n) = \{x_n\}$, and so Z is discrete.

Problem. Can this result be quantified? That is, if X is a sufficiently vast Hausdorff space, must it contain an uncountable discrete set? The space $\mathbb{R}$ shows that cardinality $2^{\aleph_0}$ is not enough.

The companion theorem of Macpherson and Praeger [8] is:

Theorem 4.2. *Let Ω be countable, and assume that the permutation group G on Ω preserves no non-trivial filter on Ω. Then G is highly transitive (that is, G is n-transitive for all natural numbers n).*

I have been unable to produce an elementary proof of this theorem. In a special case, this can be done. Let G be any permutation group on Ω. The *algebraic closure* of a finite subset X of Ω is the union of the finite orbits of the pointwise stabiliser of X in G. We say that *algebraic closure is trivial* if the algebraic closure of X is precisely X, for all finite sets X.

Proposition 4.1. *Let G be a permutation group on Ω, for which algebraic closure is trivial. Then either G preserves a non-trivial topology on Ω, or G is highly transitive.*

Proof. Suppose that algebraic closure is trivial, and that G is not highly transitive. Then there exists a finite set X whose pointwise stabiliser has more than one orbit outside X, and all its orbits are infinite. Let Y be one of these orbits. Then, for any elements $g_1, \dots, g_n$ of G, the set $Yg_1 \cap \dots \cap Yg_n$ is fixed by the pointwise stabiliser of the finite set $F = Xg_1 \cup \dots \cup Xg_n$, and contains no point of F. So $Yg_1 \cap \dots \cap Yg_n$ is empty or infinite. Now the result follows from Lemma 4.1. □

One would expect that, if algebraic closure were non-trivial, one could use it to construct a non-trivial topological closure operation. But I have not succeeded in doing this. In particular, defining a set to be closed if it contains the algebraic closure of each of its finite subsets does not work.

5. Automorphism groups of filters

In this section we assume that the set Ω is countable. We show that any filter has a large automorphism group, and in particular, the automorphism group of an ultrafilter is a maximal subgroup of $\mathrm{Sym}(\Omega)$. This section is based on the work of Neumann [12], see also [7].

A filter $\mathcal{F}$ on Ω is *principal* if there is a subset X of Ω (necessarily non-empty) such that $\mathcal{F}$ consists of all subsets of Ω containing X. An *ultrafilter* is a maximal filter. Thus, a principal ultrafilter consists of all sets which contain a specified point $x \in \Omega$. Given a filter $\mathcal{F}$, we let $A(\mathcal{F})$ denote the automorphism group of $\mathcal{F}$, and $B(\mathcal{F})$ the set of all permutations $g \in \mathrm{Sym}(\Omega)$ with the property that $\mathrm{Fix}(g) \in \mathcal{F}$, where $\mathrm{Fix}(g)$ is the set of fixed points of g. If $\mathcal{F}$ is the principal filter defined by X, then $A(\mathcal{F})$ and $B(\mathcal{F})$ are the setwise and pointwise stabiliser of X respectively. We will see that they behave in a similar way in general.

Proposition 5.1. (a) *For any filter $\mathcal{F}$, $B(\mathcal{F})$ is a normal subgroup of $A(\mathcal{F})$.*
(b) *If $\mathcal{F}_1$ and $\mathcal{F}_2$ are filters with $\mathcal{F}_1 \subseteq \mathcal{F}_2$, then $B(\mathcal{F}_1) \leq B(\mathcal{F}_2)$.*

(c) *If $\mathcal{F}$ is a non-principal ultrafilter then $A(\mathcal{F}) = B(\mathcal{F})$ is a highly transitive maximal subgroup of* Sym(Ω) *which acts transitively on the sets of $\mathcal{F}$ which are moieties (that is, are not cofinite).*

Proof. All is straightforward. For (a), if $X \in \mathcal{F}$ and $g \in B(\mathcal{F})$, then $Xg \supseteq X \cap \text{Fix}(g)$, so $Xg \in \mathcal{F}$. So $B(\mathcal{F}) \subseteq A(\mathcal{F})$. If $g, h \in B(\mathcal{F})$, then $\text{Fix}(gh^{-1}) \supseteq \text{Fix}(g) \cap \text{Fix}(h)$, so $gh^{-1} \in B(\mathcal{F})$; so this set is a group. Normality is clear. Part (b) is trivial. For (c), we show first that $A(\mathcal{F}) = B(\mathcal{F})$. So let g be any permutation in $A(\mathcal{F})$. Assume that $g \notin B(\mathcal{F})$, so that $\text{Fix}(g) \notin \mathcal{F}$. Since $\mathcal{F}$ is an ultrafilter, $\Omega \setminus \text{Fix}(g) \in \mathcal{F}$, so there exists $h \in B(\mathcal{F})$ with $\text{Fix}(h) = \Omega \setminus \text{Fix}(g)$. Then $gh \in A(\mathcal{F})$ has no fixed points. Now let X consist of alternate points from the cycles of gh (where, if a cycle has finite odd length, we leave one gap of length 3), $Y = Xgh$, and $Z = \Omega \setminus (X \cup Y)$. The three sets X, Y, Z comprise a partition of Ω, so one of them lies in the ultrafilter $\mathcal{F}$. But each is disjoint from its image under gh, so $\emptyset \in \mathcal{F}$, a contradiction. To prove the high transitivity, we use the fact that a non-principal ultrafilter $\mathcal{F}$ contains the filter $\mathcal{F}_c$ of cofinite sets; and $B(\mathcal{F}_c)$ is the finitary symmetric group. (Indeed, if $\mathcal{F}$ is any filter containing all the cofinite sets, then $B(\mathcal{F})$ is highly transitive.) Let X and Y be moieties belonging to $\mathcal{F}$. Then $X \cap Y$ is an infinite set in $\mathcal{F}$. Partition this set into infinite subsets A and B. Since $\mathcal{F}$ is an ultrafilter, one of A and B is in $\mathcal{F}$, say A; then the pointwise stabiliser of A contains a permutation mapping X to Y. Finally, let G be a subgroup of Sym(Ω) which properly contains $A(\mathcal{F})$. Then G must move some moiety in $\mathcal{F}$ to one not in $\mathcal{F}$; hence G acts transitively on moieties. Since the pointwise stabiliser of any moiety in G is the symmetric group on its complement, it follows easily that $G = \text{Sym}(\Omega)$. □

6. Maximal subgroups

If M is a maximal proper subgroup of Sym(Ω), where Ω is countable, then clearly either M contains the finitary symmetric group F, or $MF = \text{Sym}(\Omega)$. In the latter case, M has countable index in Sym(Ω) (since F is countable), and so by Theorem 8.1 below, M is the stabiliser of a finite set. Conversely, the stabiliser of a finite set is a maximal subgroup. In the former case, M is highly transitive. More generally, Macpherson and Neumann [7] proved that any supplement for F in Sym(Ω) lies between the pointwise and setwise stabiliser of a finite set. As noted earlier, Macpherson and Praeger [8] proved:

Theorem 6.1. *A primitive subgroup of* Sym(Ω) *which is not highly transitive is contained in a maximal subgroup of* Sym(Ω).

Not every proper subgroup of $\mathrm{Sym}(\Omega)$ is contained in a maximal subgroup. Like every uncountable group, $\mathrm{Sym}(\Omega)$ is the union of a chain of subgroups, none of which are maximal. Macpherson and Neumann proved that any chain of subgroups with union $\mathrm{Sym}(\Omega)$ has length strictly greater than $|\Omega|$. It follows that some subgroup in the chain necessarily contains the finitary symmetric group.

7. The rational world

Among countable topologies, there is one which is particularly interesting: the set $\mathbb{Q}$ of rational numbers, with the usual topology. Unlike the case of $\mathbb{R}$, the topology of $\mathbb{Q}$ does not determine the order up to reversal; indeed, $\mathbb{Q}$ is homeomorphic to $\mathbb{Q}^n$ for any positive integer n, and its homeomorphism group is highly transitive. More generally, Sierpiński [16] showed that a countable topological space is homeomorphic to $\mathbb{Q}$ if and only if it is totally disconnected and $T1$ and without isolated points. Neumann [11] calls such a space a *rational world*. He has investigated the rational world and its homeomorphism group, and among other things has determined all the possible cycle structures of homeomorphisms. Strikingly, there are $2^{\aleph_0}$ non-conjugate homeomorphisms which permute all the points in a single cycle. The most dramatic theorem on the rational world is due to Mekler [9], with an alternative proof by Truss [19]. It states that, for countable groups, our necessary and sufficient conditions of Lemma 4.1 for a group to act on a topology are actually necessary and sufficient for it to act on the rational world:

Theorem 7.1. *Let G be a countable permutation group on a countable set Ω. Then the following are equivalent:*

(a) *G is permutation-isomorphic to a subgroup of the group of homeomorphisms of $\mathbb{Q}$;*
(b) *There is a subset Δ of Ω such that, for any elements $g_1, \ldots, g_n \in G$, the set $\Delta g_1 \cap \cdots \cap \Delta g_n$ is empty or infinite.*

8. Topology of permutation groups

The symmetric group itself carries a natural topology, the topology of pointwise convergence. A basis of neighbourhoods of the identity is formed by the pointwise stabilisers of finite sets of points. The basic facts about this topology are well known; this section contains only a few brief remarks. For the remainder of this section, we assume that Ω is countable. In this case, $\mathrm{Sym}(\Omega)$ is metrisable:

taking $\Omega = \mathbb{N}$, we may set $d(g, h) = 1/n$ if $ig = ih$ and $ig^{-1} = ih^{-1}$ for all $i < n$ but one of these equations fails for $i = n$, where $g, h \in \mathrm{Sym}(\Omega)$, $g \neq h$. The topology of $\mathrm{Sym}(\Omega)$ can be recovered from the group structure. This is a consequence of the following theorem due to Semmes [13] and Dixon, Neumann and Thomas [3]:

Theorem 8.1. *Let Ω be countable. Then a subgroup of index smaller than $2^{\aleph_0}$ in* $\mathrm{Sym}(\Omega)$ *contains the pointwise stabiliser of a finite subset X of Ω, and is contained in the setwise stabiliser of X.*

The first assertion means that the open subgroups of $\mathrm{Sym}(\Omega)$ are precisely those of countable index. We say that $\mathrm{Sym}(\Omega)$ has the *small index property*. The two assertions together are called the *strong small index property*.

Closed subgroups of $\mathrm{Sym}(\Omega)$ are characterised by the next result. Recall that a structure is *homogeneous* if every isomorphism between finite substructures can be extended to an automorphism.

Proposition 8.1. *The following conditions are equivalent for a subgroup G of* $\mathrm{Sym}(\Omega)$, *where Ω is countable:*

(a) *G is closed in* $\mathrm{Sym}(\Omega)$;
(b) *G is the automorphism group of a first-order structure on Ω*;
(c) *G is the automorphism group of a homogeneous relational structure on Ω.*

Now we can ask whether the topology of a closed subgroup of $\mathrm{Sym}(\Omega)$ can be recovered from its group structure, that is, whether the group has the small index property. This has been shown in many cases (for example, projective and affine spaces over finite fields, the countable random graph, the countable atomless Boolean algebra), but is not true for all closed subgroups (see Evans and Hewitt [4] for a counterexample).

9. Higher cardinalities

The questions considered here can of course be generalised to larger infinite sets. Typically, positive theorems tend to take on the more uncertain status of conjectures or independence results. I refer interested readers to the papers [1, 14, 15, 18].

Added in proof: I am grateful to Peter Biryukov for a number of comments on this paper. I would like to summarise some of the information here.

The spread $s(X)$ of a topological space X is defined as the least upper bound of the cardinalities of discrete subspaces. You ask about the relation between the cardinality and the spread of a Hausdorff space X. The best result provable in ZFC is $|X| \leq 2^{2^{s(X)}}$ (A. Hajnal and I. Juhasz, Discrete subspaces of topological spaces, *Indag. Math.* **29** (1967), 343–356). Additional set-theoretic assumptions lead to contrasting results in the case of countable spread (S. Todorcevic, Partition problems in topology, *Contemp. Math.* **84**, Amer. Math. Soc. 1989; V. Fedorchuk, On the cardinality of hereditarily separable compact Hausdorff spaces, *Soviet Math. Dokl.* **16** (1975), 651–655).

P. Biryukov (Spaces with maximal homeomorphism group, *Proc. Kemerovo St. Univ.* **4** (2000)) shows that if a non-discrete topology on an infinite set X is invariant under the alternating group on X, then its non-empty open sets form a filter F, and either $F = \{X\}$, or F contains the cofinite filter. Hence a non-discrete topology is trivial if and only if its non-empty open sets form a trivial filter.

References

[1] J. E. Baumgartner, S. Shelah, and S. Thomas, Maximal subgroups of infinite symmetric groups, *Notre Dame J. Formal Logic* **34** (1993), 1–11.

[2] P. J. Cameron, The random graph, in: *The Mathematics of Paul Erdős* (J. Nešetřil and R. L. Graham eds), Springer, Berlin, 1996, 331–351.

[3] J. D. Dixon, P. M. Neumann, and S. Thomas, Subgroups of small index in infinite symmetric groups, *Bull. London Math. Soc.* **18** (1986), 580–586.

[4] D. M. Evans and P. R. Hewitt, Counterexamples to a conjecture on relative categoricity, *Annals of Pure and Applied Logic* **46** (1990), 201–209.

[5] R. Fraïssé, Sur certains relations qui généralisent l'ordre des nombres rationnels, *C. R. Acad. Sci. Paris* **237** (1953), 540–542.

[6] J. P. J. McDermott, personal communication.

[7] H. D. Macpherson and P. M. Neumann, Subgroups of infinite symmetric groups, *J. London Math. Soc.* (2) **42** (1990), 64–84.

[8] H. D. Macpherson and C. E. Praeger, Maximal subgroups of infinite symmetric groups, *J. London Math. Soc.* (2) **42** (1990), 85–92.

[9] A. H. Mekler, Groups embeddable in the autohomeomorphisms of $\mathbb{Q}$, *J. London Math. Soc.* (2) **33** (1986), 49–58.

[10] P. M. Neumann, The lawlessness of finitary permutation groups, *Arch. Math.* **26** (1975), 561–566.

[11] P. M. Neumann, Automorphisms of the rational world, *J. London Math. Soc.* (2) **32** (1985), 439–448.

[12] P. M. Neumann, personal communication.

[13] S. Semmes, Endomorphisms of infinite symmetric groups, *Abstracts Amer. Math. Soc.* **2** (1981), 426.

[14] S. Shelah and S. Thomas, Implausible subgroups of infinite symmetric groups, *Bull. London Math. Soc.* **20** (1988), 313–318.

[15] S. Shelah and S. Thomas, Subgroups of small index in infinite symmetric groups, II, *J. Symbolic Logic* **54** (1989), 95–99.

[16] W. Sierpiński, Une propriété topologique des ensembles dénombrables denses en soi, *Fund. Math.* **1** (1920), 11–16.

[17] W. Sierpiński, *General Topology* (translated by C. C. Krieger), University of Toronto Press, Toronto, 1956.

[18] S. Thomas, Aspects of infinite symmetric groups, in: *Infinite groups and group rings* (J. M. Corson et al. eds), World Scientific, Singapore, 1993, 139–145.

[19] J. K. Truss, Embeddings of infinite permutation groups, in: *Proceedings of Groups – St. Andrews 1985* (E. F. Robertson and C. M. Campbell eds), LMS Lecture Notes vol. 121, Cambridge University Press, Cambridge, 1986, 335–351.

[20] H. Wielandt, *Unendliche Permutationsgruppen* (notes prepared by A. Mader), Math. Inst. Univ. Tübingen 1960; English translation *Infinite Permutation Groups* by P. V. Bruyns, pp. 199–235 in: H. Wielandt, *Mathematische Werke*, Volume 1: Group Theory (B. Huppert and H. Schneider eds), de Gruyter, Berlin, 1994.

7

Euler characteristics of discrete groups

I. M. Chiswell

1. Introduction

This is a write-up of the notes of a course given by the author in 1983/84. Apart from the addition of several references, only minor changes have been made. To compensate, a final section on progress since then has been added. The course originally was to have been accessible to Master's students taking basic algebra courses, although none actually attended the course. Much of the material is therefore quite elementary. Thus the necessary material on homological algebra and (co)homology of groups is developed. At the same time, it is not claimed to be an introduction to (co)homology of groups. In particular, the reader unfamiliar with this should be aware of various interpretations of homology and cohomology modules in dimension at most 2. (See Ch.II, §§3, 5, Ch.III §1 and Ch.IV, §§1–3 in [17], or Ch.10 in [100].) It is hoped the notes will be accessible to beginning research students, but also lead to renewed interest in the subject.

Throughout, by a ring we mean an associative ring with an identity element, which is assumed to be non-zero. By a module we mean a left module unless otherwise indicated. An Euler characteristic on a class of groups $\mathfrak{X}$ is a mapping $\chi : \mathfrak{X} \to R$, where R is a commutative ring, such that if $G \in \mathfrak{X}$ and $(G : H) < \infty$ then $H \in \mathfrak{X}$ and $\chi(H) = (G : H)\chi(G)$, and $\chi(G) = \chi(H)$ if G is isomorphic to H.

A method of defining such a mapping was given by Wall [116]. If G is a a group having a finite complex as classifying space, say X, then define $\chi(G)$ to be the Euler characteristic of X. Wall observed that this can be extended as follows. Let $\mathrm{v}\mathfrak{X}$ denote the class of groups which have a subgroup of finite index in $\mathfrak{X}$. If χ is an Euler characteristic on $\mathfrak{X}$ and R is a field of characteristic 0, then χ can be extended to $\mathrm{v}\mathfrak{X}$ by defining $\chi(G) = \dfrac{1}{(G : H)}\chi(H)$, where $H \in \mathfrak{X}$ and $(G : H) < \infty$. Using Wall's definition (with $R = \mathbb{Q}$), $\chi(G) = 1/|G|$ for finite groups G, and for finitely generated free groups F, $\chi(F) = 1 - \text{rank(F)}$, the rank being the number of elements in a basis. Wall also noted the following properties of χ:

$$\chi(G \times H) = \chi(G)\chi(H);$$
$$\chi(G * H) = \chi(G) + \chi(H) - 1.$$

assuming the right-hand side is defined. Generalising this suggests other desirable properties of an Euler characteristic on a class $\mathfrak{X}$:

(1) if $Q = G/N$ and $N, Q \in \mathfrak{X}$, then $G \in \mathfrak{X}$ and $\chi(G) = \chi(N)\chi(Q)$;
(2) if $G = A *_C B$ is a free product with amalgamation and $A, B, C \in \mathfrak{X}$, then $G \in \mathfrak{X}$ and $\chi(G) = \chi(A) + \chi(B) - \chi(C)$.

Another desirable property is the analogue of (2) for HNN-extensions:

(2)$'$ if $G = \langle A, t \mid tBt^{-1} = C \rangle$ is an HNN-extension and $A, B \in \mathfrak{X}$ (so $C \in \mathfrak{X}$), then $G \in \mathfrak{X}$ and $\chi(G) = \chi(A) - \chi(C)$.

Major progress was made by Serre [107], who defined an Euler characteristic on a class of groups called FL, by means of an alternating sum of ranks of free $\mathbb{Z}G$-modules. This is an algebraic version of Wall's definition. The work of Serre has been generalised in two ways. Free modules can be replaced by projective modules over the group ring RG, where R is any commutative ring. A suitable notion of rank for projective modules was defined by Hattori [60] and by Stallings [109]. Stallings defined a function taking values in the free R-module on the set of conjugacy classes of G. One obtains an Euler characteristic μ by taking the coefficient of the trivial conjugacy class ([10], [26], [40]). Taking $R = \mathbb{Q}$ generalises the definition on FL. The function μ satisfies (1), (2) and (2)$'$ above.

A different generalisation was given by Brown [15] using analogues of Betti numbers. The class of groups on which it is defined is different from that on which μ (with $R = \mathbb{Q}$) is defined, and neither is contained in the other. (Examples are given at the end of § 10.) It is unknown whether or not the two characteristics are the same for groups on which both are defined. The Brown characteristic satisfies (1), (2) and (2)$'$ only with the extra assumption that the relevant group G has a torsion-free subgroup of finite index. It has some interesting arithmetic properties which have not been proved for μ.

Although these definitions have a natural geometric origin, it should not be assumed they are the only possibilities. For example, given an Euler characteristic on a class $\mathfrak{X}$ with values in a field of characteristic 0, one can change its value by a scalar multiple on a commensurability class in $\mathfrak{X}$ and obtain a new Euler characteristic. Other definitions of an Euler characteristic have been given by Bailey [9], by Paschke [92] and by Reznikov [97]. The Euler characteristic

ρ of Paschke is defined for all finitely generated groups and is zero on the class of finite groups. However, if G, H are infinite, finitely generated groups having a common finite subgroup F, then $\rho(G *_F H) = \rho(G) + \rho(H) + 1/|F|$, so ρ does not satisfy (2). Finally the "virtual signature" of Roy [101] is defined for virtual Poincaré duality groups, and fails to satisfy some arithmetic properties of the Brown characteristic.

The reader is assumed to be familiar with basic module theory, and the functors Hom and tensor product, including their exactness properties, as well as the use of the universal property of tensor products to define maps on them. The ideas of projective and flat module, and their basic properties, are assumed known (but projectives are briefly discussed in §2, reflecting the expected audience for the course). This material is covered, for example, in [100, Chapters 2 and 3]. A short exact sequence of groups will usually be called a group extension. Familiarity is also assumed with the various ways of viewing an associative algebra over a commutative ring (see [8, Ch.4, §5, Exerc.2], [32, §3.2], [64, §3.9] and [73, Ch.V, §1]). When discussing Properties (1) and (2) above, a knowledge of free products with amalgamation and HNN extensions will be assumed. See, for example, [79, Ch.4].

Further, in places a knowledge of commutative algebra, elementary algebraic number theory or singular homology theory is needed. Also, it is assumed the reader has seen a version of the Artin-Wedderburn theory of semisimple rings. However, it is difficult to find a reference giving all aspects of the theory needed, so we comment on it.

First, if D is a division ring, the ring $M_n(D)$ of $n \times n$ matrices over D decomposes as a direct sum of simple left ideals, $M_n(D) = L_1 \oplus \ldots \oplus L_n$, where L_i is the set of matrices having all columns except column i equal to zero. This is an easy exercise. Clearly the L_i are all isomorphic as $M_n(D)$-modules to the module L of all column vectors of length n with entries in D. Note that D embeds as a subring in $M_n(D)$ ($d \mapsto d1_{M_n(D)}$, where $1_{M_n(D)}$ is the $n \times n$ identity matrix). This gives the usual action of D on L of matrix multiplication, by restriction, so each L_i has dimension n over D. Further, $L_i = M_n(D)e_{ii}$, where e_{ii} is the matrix with 1 in the (i, i) position and zeros elsewhere, and the identity matrix is the sum $e_{11} + \cdots + e_{nn}$ of orthogonal idempotents. It is another easy exercise to show $M_n(D)$ is a simple ring.

Let S be a simple $M_n(D)$-module (simple modules are discussed in [8, Ch.6, §8]). Choose $x \in S$, $x \neq 0$, so $M_n(D)x$ is not the zero submodule of S, hence $S = M_n(D)x$. The map $M_n(D) \to S$, $a \mapsto ax$, restricted to L_i, is a module homomorphism, so is either the zero map or an isomorphism (its kernel is a submodule of L_i and its image a submodule of S). It is not the zero map for all i, otherwise $x = \sum_{i=1}^n e_{ii}x = 0$, a contradiction. It follows that $S \cong L$. Thus $M_n(D)$ has exactly one simple module, up to isomorphism.

If A is a semisimple ring, then A is a direct product of finitely many matrix rings over division rings, say $A = A_1 \times \ldots \times A_r$ (see [32, §4.6]). If L is a simple A-module, then $A_i L$ is a submodule, so is 0 or L, and is equal to L for at least one i, otherwise $L = AL$ would be zero. If $A_i L = A_j L = L$ with $i \neq j$, then $L = A_j(A_i L) = 0$ since $A_i A_j = 0$, a contradiction. Hence $A_i L = L$ for exactly one value of i, and $A_j L = 0$ for $j \neq i$. It follows that L is a simple module for A_i, and any simple A_i-module L is a simple A-module (defining $A_j L = 0$ for $j \neq i$). From the previous paragraph, A has exactly r isomorphism classes of simple modules, represented by $S_1, \ldots, S_r$, where S_i is a simple module for A_i. (They are pairwise non-isomorphic as A-modules since $A_i S_i = S_i$ and $A_j S_i = 0$ for $j \neq i$.)

Also, we can write the identity element as a sum of orthogonal idempotents, $1 = f_1 + \ldots + f_m$, where f_i is the identity element of A_i. Decomposing f_i as a sum of orthogonal idempotents as above, we can write $1 = e_1 + \cdots + e_s$, where the e_i are orthogonal idempotents, and number so that every simple A-module is isomorphic to exactly one of $Ae_1, \ldots, Ae_r$.

Since A-modules are semisimple ([73, Ch.XVII, §4, Prop.4]), a finitely generated A-module M is isomorphic to $(Ae_1)^{m_1} \oplus \cdots \oplus (Ae_r)^{m_r}$ for some integers $m_1, \ldots, m_r$, which are uniquely determined by M. For $A_i M \cong (Ae_i)^{m_i}$ and if $A_i \cong M_{n_i}(D_i)$, $(Ae_i)^{m_i}$ has dimension $m_i n_i$ over D_i.

Recall that an idempotent e in a ring A is called *primitive* if it cannot be written as $e = f + g$, where f, g are orthogonal idempotents. If A is semisimple, e is primitive if and only if Ae is a simple left ideal. This is left as an exercise.

Concerning notation, 1_G denotes the identity element of the group G, although frequently just 1 is used. Similar comments apply to the multiplicative identity element of a ring. Usually id_A is used to mean the identity map on the set A, but sometimes it is just written as 1.

2. Trace functions and projective modules

Let A be a ring. A trace function on A is a map $t : A \to B$, where B is an (additively written) abelian group, such that for all $x, y \in A$,

$$t(x + y) = t(x) + t(y)$$
$$\text{and} \qquad t(xy) = t(yx).$$

Let $[A, A]$ be the additive subgroup of A generated by $\{xy - yx \mid x, y \in A\}$ and let $T_A : A \to A/[A, A]$ be the canonical map. Then T_A is a trace function on A and $A/[A, A] = T_A(A)$; we shall drop the subscript A and just write $T(A)$ for $A/[A, A]$. If $t : A \to B$ is any trace function, then $[A, A] \subseteq \mathrm{Ker}(t)$, and our first result is an easy consequence.

Lemma 2.1. *If $t : A \to B$ is any trace function, there is a unique homomorphism $s : T(A) \to B$ such that the diagram*

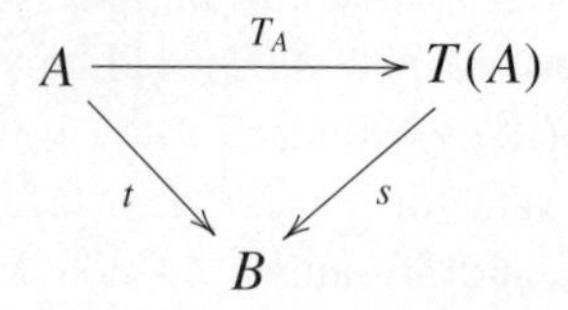

is commutative.

For this reason, T_A is called the *universal trace function* on A. If $f : A \to A'$ is a ring homomorphism, then $f([A, A]) \subseteq [A', A']$, so f induces a homomorphism $f_* : T(A) \to T(A')$, and this gives a functor from the category of rings to the category of abelian groups, sending A to $T(A)$.

Examples.

(1) If A is commutative, $T(A) = A$ and $T_A = \mathrm{id}_A$.

(2) If $A = A_1 \times A_2$ (product of rings), then $[A, A] = [A_1, A_1] \oplus [A_2, A_2]$, so $T(A) \cong T(A_1) \oplus T(A_2)$.

(3) Let R be a ring, $A = M_n(R)$ (the ring of $n \times n$ matrices over R). Define $t : A \to T(R)$ by $t(r_{ij}) = T_R\left(\sum_{i=1}^n r_{ii}\right)$. Then t is a trace function, so by Lemma 2.1 there is a homomorphism $s : T(A) \to T(R)$, given by $s(T_A(r_{ij})) = T_R((\sum_{i=1}^n r_{ii})$. We claim that s is an isomorphism. In fact, the map $R \to T(A)$, $r \mapsto T_A \begin{pmatrix} r & \dots & 0 \\ \vdots & & \vdots \\ 0 & \dots & 0 \end{pmatrix}$ is a trace function on R, and by Lemma 2.1 there is an induced map $u : T(R) \to T(A)$ given by $uT_R(r) = T_A \begin{pmatrix} r & \dots & 0 \\ \vdots & & \vdots \\ 0 & \dots & 0 \end{pmatrix}$, and s and u are inverse maps. For clearly $su = 1$, and to see $us = 1$, we need to show

$$(r_{ij}) \equiv \begin{pmatrix} \sum_{i=1}^n r_{ii} & \dots & 0 \\ \vdots & & \vdots \\ 0 & \dots & 0 \end{pmatrix} \quad (\mathrm{mod}\ [A, A])$$

that is, $\sum_{1 \le i,j \le n} r_{ij} e_{ij} \equiv \left(\sum_{i=1}^n r_{ii}\right) e_{11} \pmod{[A, A]}$, where e_{ij} is the matrix with 1 in the (i, j) position and 0 elsewhere. Using $e_{ij} e_{kl} = \delta_{jk} e_{il}$, if $i \neq j$ then $e_{ij} = e_{ii} e_{ij} - e_{ij} e_{ii} \equiv 0 \pmod{[A, A]}$ and $e_{ii} = (e_{i1} e_{1i} - e_{1i} e_{i1}) + e_{11} \equiv e_{11} \pmod{[A, A]}$ and the result follows.

If A is a ring, then $\mathrm{End}_A(A^n) \cong M_n(A)$ and Example 3 can be used to define a trace function on $\mathrm{End}_A(A^n)$ with values in $T(A)$. Stallings [109] went on to define a trace function on $\mathrm{End}_A(P)$, where P is a finitely generated projective module, using the fact that P is a summand of A^n for some n. Instead, we shall follow Hattori's basis-free approach [60], which is a little more complicated, but useful. To begin, we recall that a module P is projective if and only if the functor $\mathrm{Hom}_A(P, -)$ is exact, and this property is characterised as follows.

Lemma 2.2. *Let P be an A-module. The following are equivalent.*

(1) *P is projective;*
(2) *every short exact sequence $0 \to K \to M \to P \to 0$ of A-modules splits;*
(3) *P is a summand of a free A-module.*

Moreover, P is finitely generated projective if and only if it is a summand of a finitely generated free A-module.

Proof. This is proved in many texts, for example [100]. □

We also recall that it follows easily that a summand of a projective module is projective, and a direct sum of projective modules is projective. We shall need another characterisation of projective modules. If P is an A-module, P^* denotes $\mathrm{Hom}_A(P, A)$.

Definition. A coordinate system *for an A-module P is a family $\{x_i\}$ of elements of P together with a family $\{f_i\}$ of elements of P^* (where i runs through some index set) such that, for all $x \in P$, $f_i(x) = 0$ for all but finitely many i, and $x = \sum_i f_i(x)x_i$.*

Example. If P is free, let $\{x_i\}$ be a basis and let $\{f_i\}$ be the dual basis.

Lemma 2.3. *An A-module P is projective if and only if it has a coordinate system, and is finitely generated projective if and only if it has a finite coordinate system.*

Proof. Suppose P is projective. There are a free module F and a surjective A-homomorphism $\pi : F \to P$, giving a short exact sequence $0 \to \mathrm{Ker}\,\pi \to F \xrightarrow{\pi} P \to 0$, which splits by Lemma 2.2, that is, there is a homomorphism $\varphi : P \to F$ such that $\pi\varphi = \mathrm{id}_P$. Let $\{y_i\}$ be a basis for F and define $x_i = \pi(y_i)$. Let $\{g_i\}$ be the dual basis and define $f_i(x) = g_i(\varphi(x))$ for $x \in P$. Then $f_i \in P^*$ and for $x \in P$,

$$\begin{aligned} x = \pi(\varphi(x)) &= \pi\Big(\sum_i g_i(\varphi(x))y_i\Big) \\ &= \sum_i g_i(\varphi(x))\pi(y_i) = \sum_i f_i(x)x_i. \end{aligned}$$

showing $\{x_i\}$, $\{f_i\}$ is a coordinate system. If P is finitely generated, F can be chosen with $\{y_i\}$ finite.

Conversely, assume $\{x_i\}$, $\{f_i\}$ is a coordinate system for P. Let F be free with basis $\{y_i\}$ indexed by the same set, and let $\pi : F \to P$ be the A-homomorphism with $\pi(y_i) = x_i$. Define $\varphi : P \to F$ by $\varphi(x) = \sum_i f_i(x)y_i$. Then $\pi\varphi = \mathrm{id}_P$ and there is a split exact sequence

$$0 \longrightarrow \operatorname{Ker}\pi \longrightarrow F \underset{\varphi}{\overset{\pi}{\rightleftarrows}} P \longrightarrow 0$$

so $F = \varphi(P) \oplus \operatorname{Ker}\pi \cong P \oplus \operatorname{Ker}\pi$ and P is projective by Lemma 2.2. If $\{x_i\}$, $\{f_i\}$ is finite, the x_i generate P, so P is finitely generated. □

Now let P be an A-module. Then P^* is a right A-module (if $f \in P^*$ and $a \in A$, fa is defined by $(fa)(x) = f(x)a$, for $x \in P$). The mapping

$$P^* \times P \to \operatorname{End}_A(P)$$
$$(f, x) \mapsto g$$

where $g(y) = f(y)x$, induces a homomorphism of abelian groups $\varphi : P^* \otimes_A P \to \operatorname{End}_A(P)$.

Lemma 2.4. *The map φ is an isomorphism if and only if P is finitely generated projective.*

Proof. If φ is an isomorphism, it is surjective, so we can write $\mathrm{id}_P = \varphi(\sum_{i=1}^n (f_i \otimes x_i))$ for some n, where $f_i \in P^*$, $x_i \in P$ for $1 \leq i \leq n$. Then for $x \in P$,

$$x = \sum_{i=1}^n \varphi(f_i \otimes x_i)(x) = \sum_{i=1}^n f_i(x)x_i$$

so $\{x_i\}$, $\{f_i\}$ is a finite coordinate system for P.

Conversely, assume $\{x_i\}$, $\{f_i\}$ is a finite coordinate system. Define $\psi : \operatorname{End}_A(P) \to P^* \otimes_A P$ by $\psi(f) = \sum_i f_i \otimes f(x_i)$. We show that φ and ψ are inverse maps. First, for $x \in P$ and $f \in \operatorname{End}_A(P)$,

$$(\varphi\psi(f))(x) = \sum_i \varphi(f_i \otimes f(x_i))(x) = \sum_i f_i(x)f(x_i)$$
$$= f\left(\sum_i f_i(x)x_i\right) = f(x)$$

showing $\varphi\psi = 1$. Also, for $f \in P^*, x \in P$, $\psi\varphi(f \otimes x) = \psi(g)$, where $g(y) =$

$f(y)x$, and

$$\begin{aligned}\psi(g) &= \sum_i f_i \otimes g(x_i)\\ &= \sum_i f_i \otimes f(x_i)x\\ &= \sum_i f_i f(x_i) \otimes x\\ &= \left(\sum_i f_i f(x_i)\right) \otimes x = f \otimes x\end{aligned}$$

because $(\sum_i f_i f(x_i))(y) = \sum_i f_i(y) f(x_i) = f(y)$. Since φ, ψ are additive homomorphisms and the tensors $f \otimes x$ $(x \in P,\ f \in P^*)$ $\mathbb{Z}$-generate $P^* \otimes_A P$, $\psi\varphi = 1$. □

For any A-module P, the map $\begin{cases} P^* \times P \to T(A) \\ (f, x) \mapsto T_A(f(x)) \end{cases}$ induces a homomorphism of abelian groups $\theta : P^* \otimes P \to T(A)$. If P is finitely generated projective, using Lemma 2.4, we obtain an additive homomorphism

$$T_{P/A} : \mathrm{End}_A(P) \to T(A).$$

This is given explicitly by the next lemma. When reference to the ring A is unnecessary, $T_{P/A}$ will be abbreviated to T_P.

Lemma 2.5. *If $\{x_i\}$, $\{f_i\}$ is a finite coordinate system for P and $f \in \mathrm{End}_A(P)$, then*

$$T_P(f) = \sum_i T_A(f_i(f(x_i))).$$

Proof. We have

$$\begin{aligned}T_{P/A}(f) &= \theta(\psi(f)) \quad (\psi \text{ being the map in the proof of Lemma 2.4})\\ &= \theta\left(\sum_i f_i \otimes f(x_i)\right)\\ &= \sum_i \theta(f_i \otimes f(x_i)) = \sum_i T_A(f_i(f(x_i))).\end{aligned}$$

□

In the case of a free module, T_P can be described even more explicitly.

Lemma 2.6. *If P is finitely generated free and a dual pair of bases $\{x_i\}$, $\{f_i\}$ $(1 \leq i \leq n)$ is taken as coordinate system, then for $f \in \mathrm{End}_A(P)$,*

$$T_P(f) = \sum_{i=1}^{n} T_A(a_{ii})$$

where (a_{ij}) is the matrix of f with respect to $\{x_i\}$.

Proof. Just observe that $f_i(f(x_i)) = a_{ii}$. □

We are now ready to define a notion of *rank* for finitely generated projectives.

Definition. *Let P be a finitely generated projective A-module. The rank element $r_{P/A}$ of P is the element $T_P(\mathrm{id}_P) \in T(A)$.*

Again this will be abbreviated to r_P when reference to A is unnecessary.

Examples.

(1) From Lemma 2.6, $r_{A^n} = T_A(n.1_A) = nT_A(1_A)$.
(2) Let e be an idempotent in A, so $A = Ae \oplus A(1-e)$ and Ae is a projective A-module with coordinate system e, f, where $f(ae) = ae$. Hence $r_{Ae} = T_A(e)$.
(3) If $P \cong Q$ then $r_P = r_Q$. For if $f : P \to Q$ is an isomorphism and $\{x_i\}$, $\{f_i\}$ is a coordinate system for P, then $\{f(x_i)\}$, $\{f_i f^{-1}\}$ is a coordinate system for Q. Now apply Lemma 2.5.
(4) Let $A = M_n(D)$ where D is a division ring. If e is a primitive idempotent in A, then $Ae \cong Ae_{11}$, where $e_{11} = \begin{pmatrix} 1 & 0 & \dots & 0 \\ 0 & & & 0 \\ \vdots & & & \vdots \\ 0 & 0 & \dots & 0 \end{pmatrix}$ (see §1). Identifying $T(A)$ and $T(D)$ (Example 3 after Lemma 2.1), from the previous examples $T_A(e) = T_A(e_{11}) = T_D(1)$. In particular, if D is a field, $T_A(e) \neq 0$. Note, however, that there are division rings D with $T(D) = 0$. See [59] for an example.

Lemma 2.7. *Let P, Q be finitely generated projectives over A. Then*

(1) *$P \oplus Q$ is finitely generated projective and if $f \in \mathrm{End}_A(P)$, $g \in \mathrm{End}_A(Q)$ then $T_{P\oplus Q}(f \oplus g) = T_P(f) + T_Q(g)$. In particular $r_{P\oplus Q} = r_P + r_Q$;*
(2) *if $f, g \in \mathrm{End}_A(P)$ then $T_P(f+g) = T_P(f) + T_P(g)$;*
(3) *if $f \in \mathrm{Hom}_A(P, Q)$ and $g \in \mathrm{Hom}_A(Q, P)$, then $T_P(gf) = T_Q(fg)$.*

Proof. (1) Let $\{x_i\}$, $\{f_i\}$ $(i \in I)$ be a coordinate system for P and let $\{y_j\}$, $\{g_j\}$ $(j \in J)$ be a coordinate system for Q, where I and J are disjoint. Then $\{(x_i, 0), (0, y_j)\}$, $\{f_i \oplus 0, 0 \oplus g_j\}$ is a coordinate system for $P \oplus Q$ (indexed by $I \cup J$). Now apply Lemma 2.5.

(2) Immediate from Lemma 2.5.

(3) Define $\tilde{f}, \tilde{g} : P \oplus Q \to P \oplus Q$ by $\tilde{f}(p, q) = (0, f(p))$ and $\tilde{g}(p, q) = (g(q), 0)$, where $p \in P$, $q \in Q$. Then $\tilde{f}\tilde{g} = 0 \oplus fg$ and $\tilde{g}\tilde{f} = gf \oplus 0$, so by Part 1 we can assume $P = Q$. Let $\{x_i\}, \{f_i\}$ $(i \in I)$ be a coordinate system for P. Then

$$g(x_i) = \sum_{j \in I} f_j(g(x_i))x_j$$

$$\text{so} \quad fg(x_i) = \sum_{j \in I} f_j(g(x_i))f(x_j)$$

$$\text{and by Lemma 2.5} \quad T_P(fg) = T_A\left(\sum_{i,j} f_j(g(x_i))f_i(f(x_j))\right)$$

$$= T_P(gf) \quad \text{by symmetry.} \qquad \square$$

Remarks.

(1) If P is finitely generated projective, we can write $F = P \oplus Q$ where F is finitely generated free. Then if $f \in \mathrm{End}_A(P)$, $T_P(f) = T_F(f \oplus 0) = T_A(\sum_i a_{ii})$, where (a_{ij}) is the matrix of $f \oplus 0$ with respect to a basis of F, by Lemmas 2.5 and 2.7. This shows the equivalence with Stallings' definition of rank ([109]).

(2) By Lemma 2.7(1), the mapping $P \mapsto r_P$ induces a homomorphism of abelian groups $K_0(A) \to T(A)$, where $K_0(A)$ is the Grothendieck group of A.

(3) By Lemma 2.7(2) and (3), $T_{P/A} : \mathrm{End}_A(P) \to T(A)$ is a trace function. If F is finitely generated free, then choosing a basis gives an isomorphism $\mathrm{End}_A(F) \xrightarrow{\cong} M_n(A)$. By Lemma 2.6, there is a commutative diagram

$$\begin{array}{ccc} \mathrm{End}_A(F) & \xrightarrow{\cong} & M_n(A) \\ \downarrow{\scriptstyle T_{F/A}} & & \downarrow{\scriptstyle T_{M_n(A)}} \\ T(A) & \xleftarrow[s]{\cong} & T(M_n(A)) \end{array}$$

where s is the isomorphism constructed in Example (3) after Lemma 2.1. Hence $T_{F/A}$ can be identified with the universal trace function on $\mathrm{End}_A(F)$.

Now suppose $f : A \to B$ is a ring homomorphism. Then B, and indeed every B-module becomes an A-module. Suppose B is a finitely generated projective A-module. For $b \in B$, define $\rho_b \in \mathrm{End}_A(B)$ by $\rho_b(x) = xb$ for $x \in B$. This gives a map $B \to T(A)$, $b \mapsto T_{B/A}(\rho_b)$, which is a trace function on B by

Lemma 2.7, so by Lemma 2.1, there is a homomorphism $\mathrm{tr}_{B/A} : T(B) \to T(A)$ given by:

$$\boxed{\mathrm{tr}_{B/A}(T_B(b)) = T_{B/A}(\rho_b)}$$

Example. If B/A is a finite field extension (with f the inclusion map), then

$$\mathrm{tr}_{B/A}(b) = [B : A]_{\mathrm{ins}} \sum_{\tau} \tau(b)$$

where τ runs through the A-embeddings of B into a normal closure (or algebraic closure), and $[B : A]_{\mathrm{ins}}$ means inseparable degree. This is left as an exercise. (Hint: for $b \in B$, $\mathrm{tr}_{B/A}(b)$ is the sum of the roots of the characteristic polynomial of ρ_b over A computed with multiplicities, which has degree $[B : A]$ and whose roots are the roots of the minimum polynomial of ρ_b, which is the minimum polynomial of b over A.) In particular, if B/A is Galois, then

$$\mathrm{tr}_{B/A}(b) = \sum_{\tau \in \mathrm{Gal}(B/A)} \tau(b),$$

the familiar trace in Galois theory.

Lemma 2.8. *Let $f : A \to B$ be a ring homomorphism and assume B is finitely generated A-projective via f. Then if P is finitely generated B-projective, it is finitely generated A-projective, and if $g \in \mathrm{End}_B(P)$ then $g \in \mathrm{End}_A(P)$ and*

$$T_{P/A}(g) = \mathrm{tr}_{B/A}(T_{P/B}(g)).$$

Proof. For some n and Q, we can write $B^n \cong P \oplus Q$, and this is a decomposition as A-modules. Since B^n is a finitely generated projective A-module (direct sum of finitely many projectives) and P is a summand, P is finitely generated A-projective. Clearly $\mathrm{End}_B(P) \subseteq \mathrm{End}_A(P)$. By Lemma 2.7(1),

$$T_{P/B}(g) = T_{B^n/B}(g \oplus 0)$$
$$T_{P/A}(g) = T_{B^n/A}(g \oplus 0)$$

so it is enough to prove the formula when $P = B^n$. By Remark (3) above, there is a homomorphism $t : T(B) \to T(A)$ making the following diagram commutative

$$\begin{array}{ccc} \mathrm{End}_B(B^n) & \xrightarrow{T_{B^n/B}} & T(B) \\ \downarrow{\scriptstyle T_{B^n/A}} & \swarrow{\scriptstyle t} & \\ T(A) & & \end{array}$$

and it is enough to show $t = \mathrm{tr}_{B/A}$. Let $b \in B$ and let $g = \rho_b \oplus \underbrace{0 \oplus \cdots \oplus 0}_{(n-1)} \in \mathrm{End}_B(B^n)$. By Lemma 2.7(1) and Lemma 2.6,

$$T_{B^n/B}(g) = T_{B/B}(\rho_b) = T_B(b)$$

$$\text{and} \quad T_{B^n/A}(g) = T_{B/A}(\rho_b).$$

$$\text{Hence} \quad t(T_B(b)) = T_{B/A}(\rho_b), \quad \text{as required.}$$

□

Corollary 2.9. *In the situation of Lemma* 2.8, $r_{P/A} = \mathrm{tr}_{B/A}(r_{P/B})$.

If $f : A \to B$ is a ring homomorphism, B is also a right A-module, and if M is an A-module, $B \otimes_A M$ is a B-module (via $b(x \otimes m) = (bx) \otimes m$ for $b, x \in B$, $m \in M$). We call $B \otimes_A M$ the B-module *induced* from M. Also, f induces a homomorphism of abelian groups $f_* : T(A) \to T(B)$ ($f_*(T_A(a)) = T_B(f(a))$).

Remark 2.10. If additionally f is surjective (so $B \cong A/I$, where $I = \mathrm{Ker}(f)$) then M/IM is a B-module, defining $b(m + IM) = am + IM$, for any $a \in A$ such that $b = f(a)$ (for $b \in B$, $m \in M$). The map $\varphi : M \to B \otimes_A M$, $m \mapsto 1 \otimes m$ induces a map $M/IM \to B \otimes_A M$ which is a B-homomorphism. The map $B \times M \to M/IM$, $(b, m) \mapsto am + IM$, where $f(a) = b$, is well-defined and induces a homomorphism $B \otimes_A M \to M/IM$ which is an inverse to φ. Thus $M/IM \cong B \otimes_A M$. Taking $f = \mathrm{id}_A$ gives $A \otimes_A M \cong M$.

Lemma 2.11. *Let $f : A \to B$ be a ring homomorphism and let P be finitely generated A-projective. Then $B \otimes_A P$ is finitely generated B-projective, and if $g \in \mathrm{End}_A(P)$, then $1 \otimes g \in \mathrm{End}_B(B \otimes_A P)$ and $T_{B \otimes_A P/B}(1 \otimes g) = f_*(T_{P/A}(g))$. In particular, $r_{B \otimes_A P/B} = f_*(r_{P/A})$.*

Proof. Write $F = P \oplus Q$ where F is a finitely generated free A-module. Then

$$B \otimes_A F = (B \otimes_A P) \oplus (B \otimes_A Q)$$

and if $e_1, \ldots, e_n$ is a basis for F, then $1 \otimes e_1, \ldots, 1 \otimes e_n$ is a B-basis for $B \otimes_A F$, hence $B \otimes_A P$ is finitely generated B-projective. It is easily seen that $1 \otimes g \in \mathrm{End}_B(B \otimes_A P)$. Let $h = g \oplus 0 \in \mathrm{End}_A(F)$, so $1 \otimes h = (1 \otimes g) \oplus 0$. By Lemma 2.7(1),

$$T_{B \otimes_A P/B}(1 \otimes g) = T_{B \otimes_A F/B}(1 \otimes h), \ \ T_{P/A}(g) = T_{F/A}(h).$$

If (a_{ij}) is the matrix of h with respect to $e_1, \ldots, e_n$, then $(f(a_{ij}))$ is the matrix of $1 \otimes h$ with respect to $1 \otimes e_1, \ldots, 1 \otimes e_n$. By Lemma 2.6,

$$T_{B\otimes_A F/B}(1 \otimes h) = T_B\left(\sum_{i=1}^{n} f(a_{ii})\right) = f_*\left(T_A\left(\sum_{i=1}^{n} a_{ii}\right)\right)$$
$$= f_*(T_{F/A}(h))$$

proving the lemma. □

Algebras

Let R be a commutative ring, A an (associative) R-algebra. Then $[A, A]$ is an R-submodule of A, hence $T(A)$ is an R-module and and T_A is an R-linear map. Also, if X generates A as R-module, then $\{xy - yx \mid x, y \in X\}$ R-generates $[A, A]$.

Lemma 2.12. *Let A, B be R-algebras, where R is a commutative ring. Then there is an isomorphism of R-modules $T(A \otimes_R B) \to T(A) \otimes_R T(B)$, sending $T_{A\otimes_R B}(a \otimes b)$ to $T_A(a) \otimes T_B(b)$ (for $a \in A$, $b \in B$).*

Proof. Let $C = A \otimes_R B$. Then $T_A \otimes T_B : C \to T(A) \otimes_R T(B)$ is a trace function, because the elements $a \otimes b$ generate C as R-module, so $[C, C]$ is R-generated by the elements

$$[a \otimes b, a' \otimes b'] = [a, a'] \otimes bb' + a'a \otimes [b, b']$$

where a, a' run through A and b, b' through B, and $T_A \otimes T_B$ clearly maps these elements to 0. By Lemma 2.1, there is a map $\varphi : T(C) \to T(A) \otimes_R T(B)$ sending $T_C(a \otimes b)$ to $T_A(a) \otimes T_B(b)$, and it is R-linear. Now $T_C([a, a'] \otimes b) = T_C([a \otimes b, a' \otimes 1]) = 0$ and similarly $T_C(a \otimes [b, b']) = 0$. Hence there is a well-defined map

$$T(A) \times T(B) \to T(C)$$
$$(T_A(a), T_B(b)) \mapsto T_C(a \otimes b)$$

and this induces an R-linear map

$$T(A) \otimes T(B) \to T(C)$$
$$T_A(a) \otimes T_B(b) \mapsto T_C(a \otimes b)$$

and this is an inverse for φ. □

Lemma 2.13. *Assume the hypotheses of Lemma 2.12. If P is finitely generated A-projective and Q is finitely generated B-projective, $u \in \operatorname{End}_A(P)$ and $v \in$*

$\text{End}_B(Q)$, *then* $P \otimes_R Q$ *is finitely generated* $A \otimes_R B$*-projective and*

$$T_{P\otimes_R Q}(u \otimes v) = T_P(u) \otimes T_Q(v)$$

(identifying $T(A \otimes_R B)$ *and* $T(A) \otimes_R T(B)$ *via the isomorphism of Lemma* 2.12).

Proof. Let $\{x_i\}$, $\{f_i\}$ be a finite coordinate system for P and let $\{y_j\}$, $\{g_j\}$ be one for Q. Then $\{x_i \otimes y_j\}$, $\{f_i \otimes g_j\}$ is a finite coordinate system for $P \otimes_R Q$, so by Lemma 2.5,

$$\begin{aligned} T_{P\otimes_R Q}(u \otimes v) &= T\left(\sum_{i,j}(f_i \otimes g_j)(u(x_i) \otimes v(y_j))\right) \\ &= T\left(\sum_i f_i(u(x_i)) \otimes \sum_j g_j(v(y_j))\right) \\ &= T_P(u) \otimes T_Q(v), \end{aligned}$$

(where T means $T_{A\otimes_R B}$). □

Corollary 2.14. *Let* R' *be a commutative* R*-algebra,* A *an* R*-algebra. Then there is an isomorphism of* R'*-modules* $R' \otimes_R T(A) \to T(R' \otimes_R A)$ *sending* $r' \otimes T_A(a)$ *to* $T_{R'\otimes_R A}(r' \otimes a)$ *for* $r' \in R'$, $a \in A$. *If* P *is finitely generated* A*-projective and* $u \in \text{End}_A(P)$, *then* $R' \otimes_R P$ *is finitely generated* $(R' \otimes_R A)$*-projective and*

$$T_{R'\otimes_R P}(1 \otimes u) = 1 \otimes T_P(u).$$

In particular, $r_{R'\otimes_R P} = 1 \otimes r_P$.

Proof. Immediate from Lemmas 2.12 and 2.13. □

3. Complexes and homology

Let A be a ring and let Z be an abelian group (written additively).

Definition. *A* Z*-graded* A*-module is a family* $\{M_z \mid z \in Z\}$ *of* A*-modules. If* $M = \{M_z\}$ *and* $N = \{N_z\}$ *are* Z*-graded modules, and* $a \in Z$, *a homomorphism* $f : M \to N$ *of degree* a *is a family* $\{f_z \mid z \in Z\}$, *where* $f_z : M_z \to N_{z+a}$ *is an* A*-homomorphism.*

Note that, for any Z, there is a zero Z-graded module with $0_z = 0$ for all $z \in Z$. Also, given a Z-graded module M, there is a zero homomorphism $0 : M \to$

M of each degree with $0_z = 0$ for all $z \in Z$ and an identity homomorphism $\mathrm{id}_M : M \to M$ with $(\mathrm{id}_M)_z = \mathrm{id}_{M_z}$ for all $z \in Z$. If $f : M \to N$ is of degree a and $g : N \to P$ is of degree b, then we can define $gf : M \to P$ of degree $a + b$ by $(gf)_z = g_{z+a} f_z$:

$$M_z \xrightarrow{f_z} N_{z+a} \xrightarrow{g_{z+a}} P_{z+a+b}.$$

If $f, g : M \to N$ are of the same degree, we can define $f + g$ and af (for $a \in A$) by

$$(f+g)_z = f_z + g_z$$
$$(af)_z = af_z$$

(these are homomorphisms $M \to N$ of the same degree as f and g).

Definition. *A Z-graded module N is a* submodule *of a Z-graded module M if N_z is a submodule of M_z for all $z \in Z$. If N is a submodule of M, the* quotient module *M/N is defined by: $(M/N)_z = M_z/N_z$ for all $z \in Z$.*

If N is a submodule of M, there is a quotient map $\pi : M \to M/N$, where π_z is the quotient map $M_z \to M_z/N_z$ for all $z \in Z$. Suppose $f : M \to N$ is of degree a. If P is a submodule of M, define $f(P)$ by $f(P)_z = f_z(P_{z-a})$, so $f(P)$ is a submodule of N. If Q is a submodule of N, define $f^{-1}(Q)$ by $f^{-1}(Q)_z = f_z^{-1}(Q_{z+a})$ for $z \in Z$. In particular, define $\mathrm{Im}(f) = f(M)$ and $\mathrm{Ker}(f) = f^{-1}(0)$, that is, $\mathrm{Ker}(f)_z = \mathrm{Ker}(f_z)$. It is left to the reader to formulate the First Isomorphism Theorem for graded modules.

Direct sum

If M, N are Z-graded modules, define $M \oplus N$ by $(M \oplus N)_z = M_z \oplus N_z$.

We shall be interested in Z-graded modules in three cases:

(1) $Z = \{0\}$ (ordinary modules)
(2) $Z = \mathbb{Z}$
(3) $Z = \mathbb{Z} \oplus \mathbb{Z}$ (here a Z-graded module is called a $\mathbb{Z}$-bigraded module, and if a map has degree $a = (\alpha, \beta)$, it is said to have bidegree α, β).

In this section we shall consider $\mathbb{Z}$-graded modules, but the next definition applies to any abelian group Z.

Definition. *A* differential *on a Z-graded module M is a homomorphism $d : M \to M$ such that $dd = 0$.*

If d has degree a, this means that $d_{z+a}d_z$ is the zero map for all $z \in Z$. Further,

$$\begin{aligned} d_{z+a}d_z = 0 \quad \forall z \in Z &\iff d_z(M_z) \subseteq \mathrm{Ker}(d_{z+a}) \qquad \forall z \in Z \\ &\iff d_{z-a}(M_{z-a}) \subseteq \mathrm{Ker}(d_z) \quad \forall z \in Z \\ &\iff d(M) = \mathrm{Im}(d) \text{ is a submodule of } \mathrm{Ker}(d). \end{aligned}$$

This leads to the following important idea.

Definition. *If d is a differential on a Z-graded module M, the Z-graded module* Ker(d)/*Im*(d) *is called the* homology module *of the pair (M, d), written $H(M, d)$.*

It is customary to write $H_z(M, d)$ rather than $H(M, d)_z$; thus, if d has degree a,

$$H_z(M, d) = \mathrm{Ker}(d_z)/\mathrm{Im}(d_{z-a}).$$

Now we come to the main object of study in this section.

Definition. *A chain complex of A-modules is a $\mathbb{Z}$-graded module C together with a differential d of degree -1 :*

$$\dots C_{n+1} \xrightarrow{d_{n+1}} C_n \xrightarrow{d_n} C_{n-1} \xrightarrow{d_{n-1}} C_{n-2} \longrightarrow \dots \qquad (*)$$

The homology of (C, d) is thus given by $H_n(C, d) = \mathrm{Ker}(d_n)/\mathrm{Im}(d_{n+1})$. Often, d is suppressed and we just write C for the complex and $H_n(C)$ for the homology module.

Note

The sequence $(*)$ is exact at C_n if and only if $H_n(C) = 0$.

Definition. *A subcomplex of a chain complex (C, d) is a $\mathbb{Z}$-graded submodule D of C such that $d(D)$ is a submodule of D. Then D is also a chain complex with d restricted to D (in the obvious sense) as differential.*

If D is a subcomplex of C, the quotient graded module C/D can be made into a chain complex, defining $\bar{d}_n : (C/D)_n \to (C/D)_{n-1}$ by

$$\bar{d}_n(c + D_n) = d_n(c) + D_{n-1} \quad \text{for} \quad c \in C_n.$$

Definition. *The $\mathbb{Z}$-graded module $H(C/D, \bar{d})$ is the* relative homology *of the pair (C, D), denoted by $H(C, D)$.*

Chain maps

Let (C, d), (D, d') be chain complexes of A-modules. A *chain map* $f : C \to D$ is a homomorphism of $\mathbb{Z}$-graded modules of degree 0 such that $fd = d'f$. This means $f_{n-1}d_n = d'_n f_n$ for all $n \in \mathbb{Z}$, that is, all squares in the following diagram are commutative.

$$\begin{array}{ccccccccc} \cdots & \longrightarrow & C_{n+1} & \longrightarrow & C_n & \xrightarrow{d_n} & C_{n-1} & \longrightarrow & \cdots \\ & & \downarrow{\scriptstyle f_{n+1}} & & \downarrow{\scriptstyle f_n} & & \downarrow{\scriptstyle f_{n-1}} & & \\ \cdots & \longrightarrow & D_{n+1} & \longrightarrow & D_n & \xrightarrow{d'_n} & D_{n-1} & \longrightarrow & \cdots \end{array}$$

If $f : C \to D$ is a chain map, there is an induced homomorphism of degree 0, $H(f) : H(C) \to H(D)$, defined by

$$H_n(f)(c + \mathrm{Im}(d_{n+1})) = f(c) + \mathrm{Im}(d'_{n+1}) \qquad (\text{for } c \in \mathrm{Ker}(d_n)).$$

Further, if $f : C \to D$ and $g : D \to E$ are chain maps then $H(gf) = H(g)H(f)$ and $H(\mathrm{id}_C) = \mathrm{id}_{H(C)}$. In fact, there is a category $\mathcal{A}$ whose objects are chain complexes of A-modules and whose morphisms are chain maps, and a category $\mathcal{B}$ whose objects are graded A-modules and whose morphisms are homomorphisms, and we have defined a functor $H : \mathcal{A} \to \mathcal{B}$.

The connecting homomorphism

Let $0 \longrightarrow C \xrightarrow{i} D \xrightarrow{\pi} E \longrightarrow 0$ be a short exact sequence of chain complexes of A-modules (which means i and π are chain maps and $0 \longrightarrow C_n \xrightarrow{i_n} D_n \xrightarrow{\pi_n} E_n \longrightarrow 0$ is exact for every $n \in \mathbb{Z}$). Let d_C, d_D, d_E be the differentials on C, D, E. There is a commutative diagram with exact rows (ignoring the subscripts on the differentials):

$$\begin{array}{ccccccccc} 0 & \longrightarrow & C_n & \xrightarrow{i_n} & D_n & \xrightarrow{\pi_n} & E_n & \longrightarrow & 0 \\ & & \downarrow{\scriptstyle d_C} & & \downarrow{\scriptstyle d_D} & ① & \downarrow{\scriptstyle d_E} & & \\ 0 & \longrightarrow & C_{n-1} & \xrightarrow{i_{n-1}} & D_{n-1} & \xrightarrow{\pi_{n-1}} & E_{n-1} & \longrightarrow & 0 \\ & & \downarrow{\scriptstyle d_C} & ② & \downarrow{\scriptstyle d_D} & & \downarrow{\scriptstyle d_E} & & \\ 0 & \longrightarrow & C_{n-2} & \xrightarrow{i_{n-2}} & D_{n-2} & \xrightarrow{\pi_{n-2}} & E_{n-2} & \longrightarrow & 0 \end{array}$$

Suppose $e \in E_n$ and $d_E(e) = 0$. Then $e = \pi_n(d)$ for some $d \in D_n$, so $\pi_{n-1}(d_D(d)) = 0$, by commutativity of square ①. Hence $d_D(d) = i_{n-1}(c)$ for some $c \in C_{n-1}$ (uniquely determined by d since i_{n-1} is injective). Then

$$i_{n-2}(d_C(c)) = d_D(d_D(d)) = 0$$

by commutativity of square (2), and since i_{n-2} is injective, $d_C(c) = 0$. Using commutativity of squares and exactness of rows in a similar manner, we find that $c + \mathrm{Im}(d_C)$ does not depend on the choice of d, so there is a map $(\mathrm{Ker}(d_E))_n \to H_{n-1}(C)$ sending e to $c + \mathrm{Im}(d_C)$ which is clearly an A-homomorphism. Further, if $e \in \mathrm{Im}(d_E)$, we may take $d \in \mathrm{Im}(d_D)$ (using surjectivity of π_{n+1} and commutativity: $d_E\pi_{n+1} = \pi_n d_D$). Hence $c = 0$, so there is an induced A-homomorphism $\partial_n : H_n(E) \to H_{n-1}(C)$ given by:

$$\partial_n(e + \mathrm{Im}(d_E)) = c + \mathrm{Im}(d_C) \quad (e \in \mathrm{Ker}(d_E))$$

for every $n \in \mathbb{Z}$. This gives a homomorphism $\partial : H(E) \to H(C)$ of degree -1.

Definition. *The homomorphism ∂ is called the* connecting homomorphism *of the short exact sequence* $0 \longrightarrow C \xrightarrow{i} D \xrightarrow{\pi} E \longrightarrow 0$.

Theorem 3.1. *In this situation there is an exact triangle*

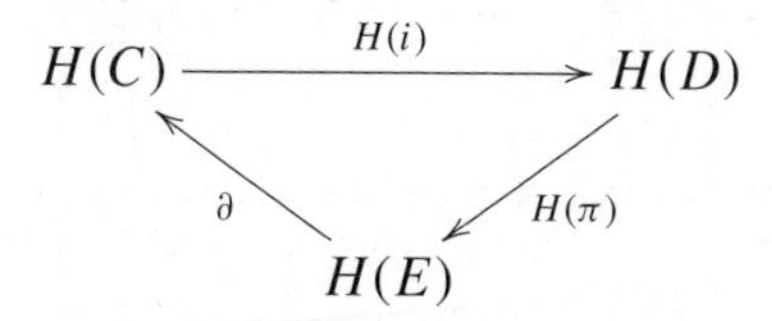

of $\mathbb{Z}$-graded modules, that is, there is an exact sequence of A-modules:

$$\dots H_n(C) \xrightarrow{H_n(i)} H_n(D) \xrightarrow{H_n(\pi)} H_n(E) \xrightarrow{\partial_n} H_{n-1}(C) \xrightarrow{H_{n-1}(i)} H_{n-1}(D) \xrightarrow{H_{n-1}(\pi)} \dots$$

Proof. Given n, there are six inclusions to establish, and we shall prove one of them, $\mathrm{Im}(H_n(\pi)) \subseteq \mathrm{Ker}(\partial_n)$. Now an element of $\mathrm{Im}(H_n(\pi))$ looks like $\pi_n(d) + \mathrm{Im}(d_E)$, for some $d \in \mathrm{Ker}(d_D)$. By definition of ∂, $\partial_n(\pi_n(d) + \mathrm{Im}(d_E)) = c + \mathrm{Im}(d_C)$, where $i_{n-1}(c) = d_D(d)$. But $d \in \mathrm{Ker}(d_D)$ and i is injective, so $c = 0$, giving the desired inclusion. The other five inclusions involve similar arguments and are left to the reader. □

Corollary 3.2. *If D is a subcomplex of a chain complex C, there is an exact sequence of A-modules:*

$$\dots H_n(D) \longrightarrow H_n(C) \longrightarrow H_n(C, D) \longrightarrow H_{n-1}(D) \longrightarrow H_{n-1}(C) \longrightarrow \dots$$

Proof. There is a short exact sequence of chain complexes

$$0 \longrightarrow D \xrightarrow{i} C \xrightarrow{\pi} C/D \longrightarrow 0,$$

where i_n is the inclusion map and π_n is the quotient map. □

Corollary 3.3. *More generally, if E is a subcomplex of D and D is a subcomplex of C, there is an exact sequence of A-modules*

$$\dots H_n(D,E) \longrightarrow H_n(C,E) \longrightarrow H_n(C,D) \longrightarrow H_{n-1}(D,E) \longrightarrow H_{n-1}(C,E) \longrightarrow \dots$$

Proof. The exact sequence

$$0 \longrightarrow D_n \longrightarrow C_n \longrightarrow C_n/D_n \longrightarrow 0$$

corresponding to the inclusion and quotient maps induces (by the isomorphism theorems for modules) an exact sequence

$$0 \longrightarrow D_n/E_n \longrightarrow C_n/E_n \longrightarrow C_n/D_n \longrightarrow 0$$

for all n, giving a short exact sequence of chain complexes

$$0 \longrightarrow D/E \longrightarrow C/E \longrightarrow C/D \longrightarrow 0.$$

□

Theorem 3.4 (Naturality of ∂). *Given short exact sequences of chain complexes*

$$0 \longrightarrow D \longrightarrow C \longrightarrow E \longrightarrow 0$$
$$0 \longrightarrow D' \longrightarrow C' \longrightarrow E' \longrightarrow 0$$

with connecting homomorphisms ∂, ∂' respectively, and chain maps f_C, f_D, f_E such that

$$\begin{array}{ccccc} D & \longrightarrow & C & \longrightarrow & E \\ \downarrow f_D & & \downarrow f_C & & \downarrow f_E \\ D' & \longrightarrow & C' & \longrightarrow & E' \end{array}$$

is commutative (in an obvious sense), there is a commutative diagram

$$\begin{array}{ccccccccccc} \cdots \longrightarrow & H_n(D) & \longrightarrow & H_n(C) & \longrightarrow & H_n(E) & \xrightarrow{\partial} & H_{n-1}(D) & \longrightarrow & H_{n-1}(C) & \longrightarrow \cdots \\ & \downarrow H_n(f_D) & & \downarrow H_n(f_C) & & \downarrow H_n(f_E) & & \downarrow H_{n-1}(f_D) & & \downarrow & \\ \cdots \longrightarrow & H_n(D') & \longrightarrow & H_n(C') & \longrightarrow & H_n(E') & \xrightarrow{\partial'} & H_{n-1}(D') & \longrightarrow & H_{n-1}(C') & \longrightarrow \cdots \end{array}$$

whose rows are given by Theorem 3.1.

Proof. This is left as an exercise, but note that commutativity of squares not involving ∂ follows from the fact, already noted, that H is a functor. □

Direct sums

If $(C, d), (C', d')$ are chain complexes of A-modules, $C \oplus C'$ is a chain complex with differential d'' given by $d''_n(c, c') = (d_n(c), d'_n(c'))$ (for $c \in C_n$, $c' \in C'_n$). Further, $H_n(C \oplus C') \cong H_n(C) \oplus H_n(C')$. This will usually be all that is needed, but clearly one can take the direct sum of any collection of chain complexes of A-modules, and the analogous result for homology is true.

The mapping cone

This is an idea from algebraic topology, where chain complexes and so homology modules are associated to topological spaces, and corresponds to the mapping cone of a map of topological spaces.

Definition. *Let $f : (C, d) \to (C', d')$ be a chain map. The* mapping cone *of f is the chain complex (M, d'') defined by:*

$$M_n = C_{n-1} \oplus C'_n$$

$$d''(c, c') = (-d(c), d'(c') + f(c))$$

omitting the subscripts on the differentials.

There is a chain map $i : C' \to M$, where $i_n : C'_n \to C_{n-1} \oplus C'_n$ is the canonical embedding $c' \mapsto (0, c')$. Define a chain complex $(\tilde{C}, \tilde{d})$ by $\tilde{C}_n = C_{n-1}$ and $\tilde{d}_n = -d_{n-1}$. There is also a chain map $\pi : M \to \tilde{C}$, where $\pi_n : C_{n-1} \oplus C'_n \to \tilde{C}_n = C_{n-1}$ is the canonical projection. Then

$$0 \longrightarrow C' \xrightarrow{\ i\ } M \xrightarrow{\ \pi\ } \tilde{C} \longrightarrow 0$$

is a short exact sequence of chain complexes.

Proposition 3.5. *In these circumstances there is an exact sequence*

$$\dots H_n(C') \xrightarrow{\ i_*\ } H_n(M) \xrightarrow{\ \pi_*\ } H_{n-1}(C) \xrightarrow{\ f_*\ } H_{n-1}(C') \xrightarrow{\ i_*\ } H_{n-1}(M) \longrightarrow \dots$$

(where subscripts have been omitted and $i_* = H(i)$, etc).

Proof. By Theorem 3.1 there is an exact sequence

$$\dots H_n(C') \xrightarrow{\ i_*\ } H_n(M) \xrightarrow{\ \pi_*\ } H_n(\tilde{C}) \xrightarrow{\ \partial\ } H_{n-1}(C') \longrightarrow \dots$$

and clearly $H_n(\tilde{C}) = H_{n-1}(C)$. It is also routine to check that $\partial = f_*$. □

Chain homotopy

This is another idea from algebraic topology, where homotopies between maps of topological spaces give rise to chain homotopies between chain maps.

Definition. *Let $f, g : (C, d) \to (D, d')$ be chain maps. A chain homotopy from f to g is a homomorphism $s : C \to D$ of $\mathbb{Z}$-graded modules of degree $+1$ such that $sd + d's = f - g$. That is,*

$$s_{n-1}d_n + d'_{n+1}s_n = f_n - g_n \qquad (\textit{for all } n \in \mathbb{Z})$$

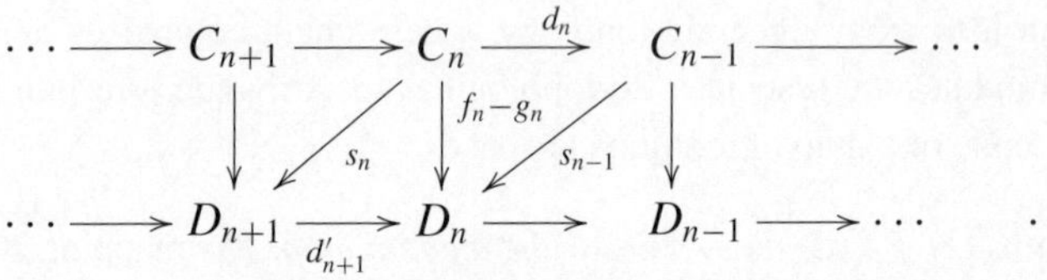

$$\begin{array}{ccccccccc}
\cdots & \longrightarrow & C_{n+1} & \longrightarrow & C_n & \xrightarrow{d_n} & C_{n-1} & \longrightarrow & \cdots \\
 & & \downarrow & \swarrow_{s_n} & \downarrow{\scriptstyle f_n - g_n} & \swarrow_{s_{n-1}} & \downarrow & & \\
\cdots & \longrightarrow & D_{n+1} & \xrightarrow[d'_{n+1}]{} & D_n & \longrightarrow & D_{n-1} & \longrightarrow & \cdots
\end{array}$$

It is easy to see that there is an equivalence relation $\simeq$ on the set of chain maps from C to D, where $f \simeq g$ if there is a chain homotopy from f to g. (This is read as "f is chain homotopic to g".)

Lemma 3.6. *If f is chain homotopic to g, then $H(f) = H(g)$.*

Proof. Ignoring the grading (i.e., omitting subscripts),

$$H(f)(c + \mathrm{Im}(d)) = f(c) + \mathrm{Im}(d')$$
$$H(g)(c + \mathrm{Im}(d)) = g(c) + \mathrm{Im}(d')$$

for $c \in \mathrm{Ker}(d)$, and we need to show that $f(c) - g(c) \in \mathrm{Im}(d')$. But if s is a chain homotopy from f to g, then

$$f(c) - g(c) = sd(c) + d's(c) = d'(s(c)) \in \mathrm{Im}(d'). \qquad \square$$

Projective resolutions

These will be used in the next section to generalise the trace of an endomorphism and rank element of a projective module to a wider class of modules. We begin with the definitions.

Definition. *A chain complex C is* positive *if $C_n = 0$ for $n < 0$. A chain complex C is* projective *(resp.* free*) if C_n is a projective (resp. free) module for all $n \in \mathbb{Z}$.*

This seems to be the usual terminology, although "non-negative" would be more accurate than "positive".

Definition. *A complex* over *an A-module M is a positive complex (C, d) together with an A-homomorphism $\varepsilon : C_0 \to M$ such that $\varepsilon d_1 = 0$.*

In this situation there is a chain complex $\dots C_1 \xrightarrow{d_1} C_0 \xrightarrow{\varepsilon} M \to 0 \dots$, called the *augmented complex*, written $C \xrightarrow{\varepsilon} M$. The map ε is called the *augmentation map.*

Definition. *A* resolution *of an A-module M is a complex (C, d) over M whose augmented complex $C \xrightarrow{\varepsilon} M$ is exact.*

This means $H_n(C) = 0$ for $n \neq 0$, $\mathrm{Im}(d_1) = \mathrm{Ker}(\varepsilon)$ and ε is onto. Since $H_0(C) = C_0/\mathrm{Im}(d_1)$, there is an isomorphism $H_0(C) \xrightarrow{\cong} M$. Conversely, given a positive complex C with $H_n(C) = 0$ for $n \neq 0$ and an isomorphism $H_0(C) \cong M$, we obtain a resolution of M.

Note. Any module M has a projective (indeed free) resolution. Choose a free module F_0 mapping onto M, to get a short exact sequence

$$0 \to K_0 \xrightarrow{i} F_0 \to M \to 0.$$

Now choose a free module F_1 mapping onto K_0, to get a short exact sequence

$$0 \to K_1 \to F_1 \xrightarrow{\pi} K_0 \to 0.$$

These sequences can be "spliced" to obtain an exact sequence

$$0 \to K_1 \to F_1 \xrightarrow{i\pi} F_0 \to M \to 0.$$

This can clearly be continued to construct a free resolution of M (putting $F_n = 0$ for $n < 0$).

Theorem 3.7 (The Comparison Theorem)**.** *Let*

$\gamma : M \to M'$ be a homomorphism of A-modules;
$C \xrightarrow{\varepsilon} M$ be an augmented projective complex over M;
$C' \xrightarrow{\varepsilon'} M'$ be an augmented resolution of M'.

Then there is a chain map $f : C \to C'$ such that $\varepsilon' f_0 = \gamma\varepsilon$, and any two such chain maps are chain homotopic.

Proof. We have to construct f_n for $n \geq 0$ such that all squares in the following diagram are commutative.

$$\begin{array}{ccccccccccccc} \cdots & \longrightarrow & C_3 & \longrightarrow & C_2 & \xrightarrow{d_2} & C_1 & \xrightarrow{d_1} & C_0 & \xrightarrow{\varepsilon} & M & \longrightarrow & 0 \\ & & \downarrow{\scriptstyle f_3} & & \downarrow{\scriptstyle f_2} & & \downarrow{\scriptstyle f_1} & & \downarrow{\scriptstyle f_0} & & \downarrow{\scriptstyle \gamma} & & \\ \cdots & \longrightarrow & C'_3 & \longrightarrow & C'_2 & \xrightarrow[d'_2]{} & C'_1 & \xrightarrow[d'_1]{} & C'_0 & \xrightarrow[\varepsilon']{} & M' & \longrightarrow & 0. \end{array}$$

To construct f_0, there is a map f_0 making the following diagram commutative:

$$\begin{array}{ccccc} & & C_0 & & \\ & \swarrow{\scriptstyle f_0} & \downarrow{\scriptstyle \gamma\varepsilon} & & \\ C'_0 & \xrightarrow[\varepsilon']{} & M' & \longrightarrow & 0. \end{array}$$

since C_0 is projective. We can now repeat this procedure on

$$\begin{array}{ccccccc} \cdots & \longrightarrow & C_2 & \xrightarrow{d_2} & C_1 & \xrightarrow{d_1} & \mathrm{Im}(d_1) & \longrightarrow & 0 \\ & & & & \downarrow{\scriptstyle f_1} & & \downarrow{\scriptstyle f_0} & & \\ \cdots & \longrightarrow & C'_2 & \xrightarrow[d'_2]{} & C'_1 & \xrightarrow[d'_1]{} & \mathrm{Im}(d'_1) & \longrightarrow & 0. \end{array}$$

to construct f_1, on checking that $f_0(\mathrm{Im}(d_1)) \subseteq \mathrm{Im}(d'_1)$. But this follows since $\varepsilon' f_0 d_1 = \gamma\varepsilon d_1 = 0$ and $\mathrm{Im}(d'_1) = \mathrm{Ker}(\varepsilon')$. Clearly this procedure can be continued to recursively construct the map f.

Suppose that $g : C \to C'$ is another chain map with $\varepsilon' g_0 = \gamma\varepsilon$. Let $h = f - g$. There is a diagram with commutative squares:

$$\begin{array}{ccccccccc} \cdots & \longrightarrow & C_1 & \longrightarrow & C_0 & \xrightarrow{\varepsilon} & M & \longrightarrow & 0 \\ & & \downarrow{\scriptstyle h_1} & \swarrow{\scriptstyle s_0} & \downarrow{\scriptstyle h_0} & & \downarrow{\scriptstyle 0} & & \\ \cdots & \longrightarrow & C'_1 & \xrightarrow[d'_1]{} & C'_0 & \xrightarrow[\varepsilon']{} & M' & \longrightarrow & 0. \end{array}$$

and to construct a chain homotopy s from f to g we first want to construct s_0 such that $d'_1 s_0 = h_0$ ($s_n = 0$ for $n < 0$). Now $\varepsilon' h_0 = 0$ implies $\mathrm{Im}(h_0) \subseteq \mathrm{Ker}(\varepsilon') = \mathrm{Im}(d'_1)$, and since C_0 is projective, there is a map s_0 making the diagram

$$\begin{array}{ccccc} & & C_0 & & \\ & \swarrow{\scriptstyle s_0} & \downarrow{\scriptstyle h_0} & & \\ C'_1 & \xrightarrow[d'_1]{} & \mathrm{Im}(d'_1) & \longrightarrow & 0. \end{array}$$

commutative, as required. Next, we want s_1 such that $d'_2 s_1 = h_1 - s_0 d_1$

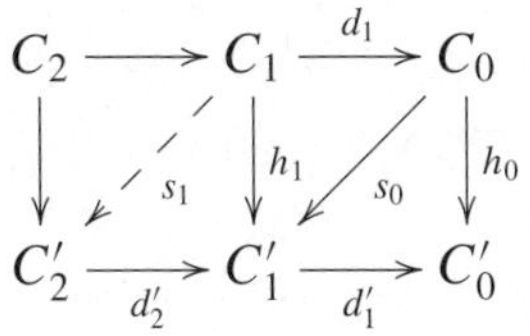

$$\begin{array}{ccccc} C_2 & \longrightarrow & C_1 & \xrightarrow{d_1} & C_0 \\ \downarrow & \swarrow s_1 & \downarrow h_1 & \swarrow s_0 & \downarrow h_0 \\ C'_2 & \xrightarrow[d'_2]{} & C'_1 & \xrightarrow[d'_1]{} & C'_0 \end{array}$$

and to repeat the construction for s_0, it is enough to show

$$\operatorname{Im}(h_1 - s_0 d_1) \subseteq \operatorname{Ker}(d'_1) = \operatorname{Im}(d'_2).$$

But $d'_1(h_1 - s_0 d_1) = d'_1 h_1 - d'_1 s_0 d_1 = d'_1 h_1 - h_0 d_1 = 0$ (h is a chain map), as required. We can now continue to construct $s_2, s_3 \ldots$. At the next stage, the calculation is a little more elaborate: $d'_2(h_2 - s_1 d_2) = d'_2 h_2 - d'_2 s_1 d_2 = d'_2 h_2 - (h_1 - s_0 d_1) d_2 = d'_2 h_2 - h_1 d_2 = 0$, since $d_1 d_2 = 0$. After that, the calculations are similar, successively increasing all subscripts by 1. □

Note. The map f in Theorem 3.7 is called a *lift* of γ.

Homotopy equivalence

Definition. *A chain map $f : C \to D$ is a* homotopy equivalence *if there is a chain map $g : D \to C$ such that $fg \simeq \mathrm{id}_D$ and $gf \simeq \mathrm{id}_C$. A chain complex C is* homotopy equivalent *to a chain complex D if there is a homotopy equivalence $f : C \to D$.*

Clearly homotopy equivalence is an equivalence relation on chain complexes.

Lemma 3.8. *Let P, Q be projective resolutions of the same A-module M. Then P and Q are homotopy equivalent.*

Proof. By Theorem 3.7, there are lifts f, g of id_M as indicated:

$$\begin{array}{ccc} P & \longrightarrow & M \\ f\downarrow & & \downarrow = \\ Q & \longrightarrow & M \end{array} \qquad \begin{array}{ccc} Q & \longrightarrow & M \\ g\downarrow & & \downarrow = \\ P & \longrightarrow & M. \end{array}$$

Then gf and id_P are lifts of id_M to P, so by Theorem 3.7, $gf \simeq \mathrm{id}_P$. Similarly, $fg \simeq \mathrm{id}_Q$. □

Split complexes

A chain complex is *split* (or *contractible*) if $\mathrm{id}_C \simeq 0_C$. Then by Lemma 3.6, $H(\mathrm{id}_C) = H(0_C)$, which implies $H(C) = 0$, so a split complex is exact. Suppose (C, d) is an exact complex and let $K_n = \mathrm{Ker}(d_n) = \mathrm{Im}(d_{n+1})$. For each $n \in \mathbb{Z}$, there is a short exact sequence $0 \to K_n \xrightarrow{i_n} C_n \xrightarrow{\pi_n} K_{n-1} \to 0$, where i_n is the inclusion map and $\pi_n(c) = d_n(c)$. Then $d_n = i_{n-1}\pi_n$ and C is obtained by "splicing" these short exact sequences:

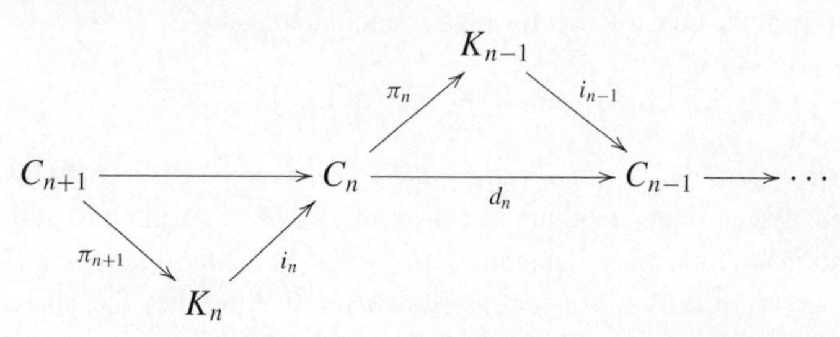

Suppose additionally that C is split, and let s be a chain homotopy from id_C to 0_C, so $sd + ds = \mathrm{id}_C$:

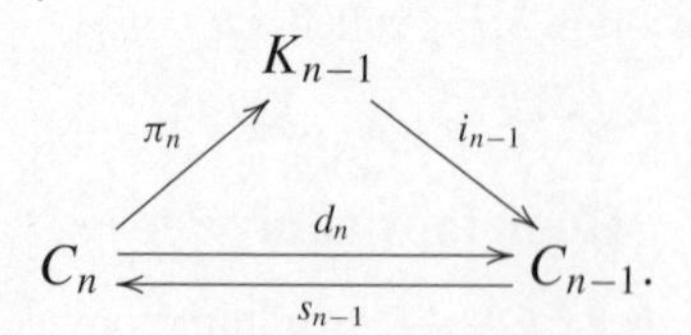

Let $\varphi_n = s_{n-1}i_{n-1}$. Then:

$$0 \longrightarrow K_n \xrightarrow{i_n} C_n \underset{\varphi_n}{\overset{\pi_n}{\rightleftarrows}} K_{n-1} \longrightarrow 0$$

is a split short exact sequence. For

$$\begin{aligned} \pi_n = \pi_n(s_{n-1}d_n + d_{n+1}s_n) &= \pi_n s_{n-1} d_n \quad (\pi_n d_{n+1} = 0) \\ &= \pi_n s_{n-1} i_{n-1} \pi_n \\ &= (\pi_n \varphi_n)\pi_n \end{aligned}$$

and since π_n is surjective, $\pi_n\varphi_n = \mathrm{id}_{K_{n-1}}$.

Lemma 3.9. *Let (P, d) be an exact projective complex and suppose there exists an integer k such that $P_n = 0$ for $n < k$. Then P is split.*

Proof. Replacing (P, d) by $(\tilde{P}, \tilde{d})$, where $\tilde{P}_n = P_{n+k}$ and $\tilde{d}_n = d_{n+k}$, we can assume $k = 0$. Then $P \xrightarrow{d_{-1}} 0$ is a projective resolution of 0, and id_P, 0_P are both lifts of the zero map $0 \to 0$. By Theorem 3.7, $\mathrm{id}_P \simeq 0_P$. □

Despite their simplicity, the next result and its corollary (the "long" version) are very useful in dealing with projective complexes.

Lemma 3.10 (Schanuel's Lemma). *Let*

$$0 \to K \xrightarrow{i} P \xrightarrow{\varepsilon} M \to 0$$

$$0 \to K' \xrightarrow{i'} P' \xrightarrow{\varepsilon'} M \to 0$$

be short exact sequences of A-modules, with P, P' projective. Then $K \oplus P' \cong K' \oplus P$.

Proof. Let Q be the pullback of the diagram

$$\begin{array}{ccc} & & P \\ & & \downarrow{\scriptstyle \varepsilon} \\ P' & \xrightarrow[\varepsilon']{} & M \end{array}$$

that is, let $Q = \{(p, p') \in P \oplus P' \mid \varepsilon(p) = \varepsilon'(p')\}$, a submodule of $P \oplus P'$. There is a short exact sequence

$$0 \to K' \xrightarrow{j'} Q \xrightarrow{\pi} P \to 0,$$

where $\pi(p, p') = p$, $j'(k') = (0, i'(k'))$ $(k' \in K')$, and a similar short exact sequence

$$0 \to K \xrightarrow{j} Q \xrightarrow{\pi'} P' \to 0.$$

Since P, P' are projective, these sequences split, so $K \oplus P' \cong Q \cong K' \oplus P$. $\square$

Corollary 3.11. *Let*

$$0 \to K \to P_n \to \ldots \to P_1 \to P_0 \to M \to 0$$

$$0 \to K' \to P'_n \to \ldots \to P'_1 \to P'_0 \to M' \to 0$$

be exact sequences of A-modules with all P_i, P'_i projective and $M \cong M'$. Then

$$K \oplus P'_n \oplus P_{n-1} \oplus P'_{n-2} \oplus \ldots \cong K' \oplus P_n \oplus P'_{n-1} \oplus P_{n-2} \oplus \ldots$$

(the sums end with P_0, P'_0).

Proof. There are exact sequences

$$0 \to K \to P_n \to \ldots P_2 \to P_1 \xrightarrow{d_1} L \to 0$$

$$0 \to K' \to P'_n \to \ldots P'_2 \to P'_1 \xrightarrow{d'_1} L' \to 0$$

and

$$0 \to L \to P_0 \to M \to 0$$
$$0 \to L' \to P'_0 \to M' \to 0$$

(where $L = \mathrm{Ker}(P_1 \to P_0)$, etc). By Lemma 3.10, $L \oplus P'_0 \cong L' \oplus P_0$, and there are exact sequences:

$$0 \longrightarrow K \longrightarrow P_n \longrightarrow \ldots P_2 \longrightarrow P_1 \oplus P'_0 \xrightarrow{d_1 \oplus \mathrm{id}_{P'_0}} L \oplus P'_0 \longrightarrow 0$$

$$0 \longrightarrow K' \longrightarrow P'_n \longrightarrow \ldots P'_2 \longrightarrow P'_1 \oplus P_0 \xrightarrow{d'_1 \oplus \mathrm{id}_{P_0}} L' \oplus P_0 \longrightarrow 0.$$

The result now follows by induction on n. □

Cochain complexes

A *cochain complex* is a $\mathbb{Z}$-graded module C with a differential d of degree $+1$

$$\ldots \longleftarrow C^{n+1} \overset{d^n}{\longleftarrow} C^n \overset{d^{n-1}}{\longleftarrow} C^{n-1} \ldots$$

The *cohomology* of (C, d) is the $\mathbb{Z}$-graded module $\mathrm{Ker}(d)/\,\mathrm{Im}(d)$. Note that it is customary to write the grading as a superscript; thus we write

$$H^n(C, d) = \mathrm{Ker}(d^n)/\,\mathrm{Im}(d^{n-1}).$$

A *cochain map* $f : (C, d) \to (\bar{C}, \bar{d})$, where (C, d), $\bar{C}, \bar{d})$ are cochain complexes, is a map $f : C \to \bar{C}$ of $\mathbb{Z}$-graded modules of degree 0 such that $fd = \bar{d}f$ ($f^{n+1}d^n = \bar{d}^n f^n$ for all n). Such a map induces a homomorphism $H(f) : H(C, d) \to H(C, \bar{d})$ and there are analogues of the results 3.1–3.4 for cochain complexes. Given a short exact sequence of cochain complexes $0 \to C \to D \to E \to 0$, there is an exact sequence

$$\ldots \to H^n(C) \to H^n(D) \to H^n(E) \overset{\partial}{\to} H^{n+1}(C) \to \ldots$$

etc. This can be deduced from the corresponding results for chain complexes, by noting that if (C, d) is a cochain complex, then defining $\tilde{C}_n = C^{-n}$, $\tilde{d}_n = d^{-n}$ gives a chain complex with $H^n(C) = H_{-n}(\tilde{C})$. Despite this, it is useful to have the notion of cochain complex, since there are situations where a natural grading leads to a cochain complex.

A *cochain homotopy* from f to g, where $f, g : (C, d) \to (\bar{C}, \bar{d})$ are cochain maps, is a map of $\mathbb{Z}$-graded modules $s : C \to \bar{C}$ of degree -1 such that $sd + \bar{d}s = f - g$ (in full, $s^{n+1}d^n + \bar{d}^{n-1}s^n = f^n - g^n$). There is an analogue of Lemma 3.6 and a notion of *split* cochain complex, and split implies exact.

4. Modules of type FP

In this section we shall generalise the trace of an endomorphism and rank element of a projective module to a wider class of modules. Once more fix a ring A.

Definition. *A finite projective complex of A-modules is a chain complex P such that, for all $n \in \mathbb{Z}$, P_n is finitely generated projective, and $P_n = 0$ for all but finitely many values of n. Given such a complex, if $f : P \to P$ is a chain map, we define*

$$L(f) = \sum_i (-1)^i T_{P_i}(f_i).$$

Here T_{P_i} is the function defined in Section 2, so $L(f) \in A/[A, A]$. Note that "finite" refers to two things: the modules are finitely generated, and the complex is of finite length (after truncating infinite strings of zero modules on the left and right).

Lemma 4.1. (1) *If P, Q are finite projective chain complexes and $f : P \to Q$, $g : Q \to P$ are chain maps, then $L(fg) = L(gf)$;*
(2) *If $f, g : P \to P$ are chain maps (where P is a finite projective complex) and $f \simeq g$, then $L(f) = L(g)$.*

Proof. (1) By Lemma 2.7, $T_{P_i}(g_i f_i) = T_{Q_i}(f_i g_i)$.

(2) Let s be a chain homotopy from f to g, so $f - g = sd + ds$, where d is the differential on P. Then

$$\begin{aligned} L(f) - L(g) &= \sum_i (-1)^i T_{P_i}(f_i - g_i) \\ &= \sum_i (-1)^i T_{P_i}(s_{i-1} d_i + d_{i+1} s_i) \\ &= \sum_i (-1)^i (T_{P_i}(s_{i-1} d_i) - T_{P_{i-1}}(d_i s_{i-1})) \\ &= 0 \qquad \text{by Lemma 2.7.} \end{aligned}$$

□

Definition. *An A-module M is of type FP over A if M has a finite projective resolution, that is, if there exists an exact sequence*

$$0 \to P_n \to P_{n-1} \to \ldots \to P_1 \to P_0 \xrightarrow{\varepsilon} M \to 0.$$

Let $P \xrightarrow{\varepsilon} M$ be a finite augmented projective resolution, $\varphi \in \mathrm{End}_A(M)$. By Theorem 3.7, there is a lift of φ to a chain map $f : P \to P$.

Lemma 4.2. *In these circumstances, $L(f)$ depends only on φ.*

Proof. Suppose $Q \stackrel{\varepsilon'}{\to} M$ is also a finite augmented projective resolution and $g : Q \to Q$ is a lift of φ. By Theorem 3.7, there exist lifts h, k of id_M as indicated in the diagrams:

$$\begin{array}{ccccccc} P & \xrightarrow{\varepsilon} & M & \quad & Q & \xrightarrow{\varepsilon'} & M \\ {\scriptstyle h}\downarrow & & \downarrow{\scriptstyle =} & & {\scriptstyle k}\downarrow & & \downarrow{\scriptstyle =} \\ Q & \xrightarrow[\varepsilon']{} & M & & P & \xrightarrow[\varepsilon]{} & M \end{array}$$

giving a commutative diagram:

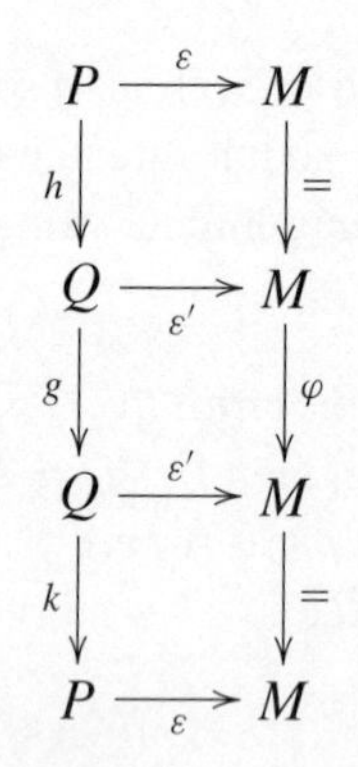

so kgh is a lift of φ to P. By Theorem 3.7, $kgh \simeq f$, so by Lemma 4.1,

$$L(f) = L(kgh) = L(gh.k).$$

But similarly, ghk is a lift of φ to Q, so $ghk \simeq g$ and $L(ghk) = L(g)$, hence $L(f) = L(g)$. □

Hence if M is of type FP and $\varphi \in \mathrm{End}_A(M)$, we can define $T_M(\varphi) = L(f)$, for any lift $f : P \to P$ of φ to a finite projective resolution P of M. (Thus $T_M(\varphi) = \sum(-1)^i T_{P_i}(f_i)$.) Note that $T_M(\varphi) \in T(A) = A/[A, A]$. We shall denote T_M by $T_{M/A}$ where necessary. This generalises the function T_P for finitely generated projectives P ($\ldots 0 \to P \stackrel{\mathrm{id}}{\to} P \to 0$ is a finite augmented projective resolution of P, having an obvious lift of id_P). Also, we can define the rank element

$$r_M = T_M(\mathrm{id}_M) \qquad \text{(for } M \text{ of type FP)}$$

which will be denoted by $r_{M/A}$ if necessary. Thus $r_M = \sum(-1)^i r_{P_i}$, where $P \to M$ is a finite augmented projective resolution (since id_P is a lift of id_M). We continue to establish some properties of the function T_M.

Lemma 4.3. *Suppose M, N are modules of type FP over A. Then*

(1) *$M \oplus N$ is of type FP, and if $\varphi \in \mathrm{End}_A(M)$, $\psi \in \mathrm{End}_A(N)$, then*

$$T_{M\oplus N}(\varphi \oplus \psi) = T_M(\varphi) + T_N(\psi)$$

(in particular, $r_{M\oplus N} = r_M + r_N$);

(2) *if φ, $\psi \in \mathrm{End}_A(M)$, then $T_M(\varphi + \psi) = T_M(\varphi) + T_M(\psi)$;*

(3) *if $\varphi \in \mathrm{Hom}_A(M, N)$ and $\psi \in \mathrm{Hom}_A(N, M)$, then $T_M(\psi\varphi) = T_N(\varphi\psi)$.*

Proof. Let $P \overset{\varepsilon}{\to} M$ and $Q \overset{\varepsilon'}{\to} N$ be finite augmented projective resolutions.

(1) There is a finite augmented projective resolution $P \oplus Q \overset{\varepsilon\oplus\varepsilon'}{\longrightarrow} M \oplus N$, so (1) follows from Lemma 2.7(1).
(2) If $f, g : P \to P$ are lifts of φ, ψ respectively, then $f + g$ is a lift of $\varphi + \psi$, and (2) follows from Lemma 2.7(2).
(3) Let $f : P \to Q$, $g : Q \to P$ be lifts of φ, ψ respectively. Then $gf : P \to P$ is a lift of $\psi\varphi$ and $fg : Q \to Q$ is a lift of $\varphi\psi$, so (3) follows from Lemma 4.1. □

There is a converse to Part (1) of the previous lemma.

Proposition 4.4. *If $M \oplus N$ is of type FP, then so are M and N.*

Proof. Let

$$0 \to P_n \overset{d_n}{\longrightarrow} P_{n-1} \to \ldots \to P_1 \overset{d_1}{\longrightarrow} P_0 \overset{\varepsilon}{\longrightarrow} M \oplus N \to 0$$

be a finite augmented projective resolution of $M \oplus N$ and let $S_i = \mathrm{Im}(d_{i+1})$, so S_i is a finitely generated submodule of P_i. Since ε is onto, and M, N are homomorphic images of $M \oplus N$ under the canonical projections, M and N are finitely generated, so we can find short exact sequences

$$0 \to K_0 \to Q_0 \to M \to 0$$
$$0 \to L_0 \to R_0 \to N \to 0$$

where Q_0, R_0 are finitely generated projective, giving a short exact sequence:

$$0 \to K_0 \oplus L_0 \to Q_0 \oplus R_0 \to M \oplus N \to 0. \qquad (*)$$

Also, there is a short exact sequence

$$0 \to S_0 \to P_0 \overset{\varepsilon}{\longrightarrow} M \oplus N \to 0.$$

By Schanuel's Lemma (3.10), $S_0 \oplus Q_0 \oplus R_0 \cong K_0 \oplus L_0 \oplus P_0$, hence K_0 and L_0 are finitely generated. Therefore we can find short exact sequences

$$0 \to K_1 \to Q_1 \to K_0 \to 0$$
$$0 \to L_1 \to R_1 \to L_0 \to 0$$

with Q_1, R_1 finitely generated projective. On taking direct sum and splicing with the sequence $(*)$, we obtain an exact sequence

$$0 \to K_1 \oplus L_1 \to Q_1 \oplus R_1 \to Q_0 \oplus R_0 \to M \oplus N \to 0.$$

There is also an exact sequence

$$0 \to S_1 \to P_1 \to P_0 \to M \oplus N \to 0$$

and by Cor. 3.11, $(K_1 \oplus L_1) \oplus P_1 \oplus (Q_0 \oplus R_0) \cong S_1 \oplus (Q_1 \oplus R_1) \oplus P_0$, hence K_1 and L_1 are finitely generated. Continuing in this way, we eventually construct exact sequences

$$0 \to K_{n-1} \to Q_{n-1} \to \ldots \to Q_1 \to Q_0 \to M \to 0$$
$$0 \to L_{n-1} \to R_{n-1} \to \ldots \to R_1 \to R_0 \to N \to 0$$

with all Q_i, R_i finitely generated projective. Now using Cor. 3.11 in the same way again shows that K_{n-1}, L_{n-1} are finitely generated projective, proving the result. □

We consider how the property FP and the function T_M behave with respect to short exact sequences.

Lemma 4.5. *Suppose* $0 \to C \xrightarrow{i} D \xrightarrow{\pi} E \to 0$ *is a short exact sequence of* A*-modules. Let* $P \xrightarrow{\varepsilon} C$ *and* $Q \xrightarrow{\varepsilon'} E$ *be augmented projective resolutions. Then there exists an augmented projective resolution* $R \xrightarrow{\theta} D$ *such that* $R_n = P_n \oplus Q_n$ *for all* $n \in \mathbb{Z}$,

$$0 \to P \xrightarrow{\tilde{i}} R \xrightarrow{\tilde{\pi}} Q \to 0$$

is a short exact sequence of chain complexes (where $\tilde{i}_n(p) = (p, 0)$ *and* $\tilde{\pi}_n(p, q) = q$*), and such that* $\tilde{i}$ *is a lift of* i *and* $\tilde{\pi}$ *is a lift of* π.

Proof. Let $K = \mathrm{Ker}(\varepsilon)$ and let $L = \mathrm{Ker}(\varepsilon')$. Since Q_0 is projective, there is an A-homomorphism $\varphi : Q_0 \to D$ such that

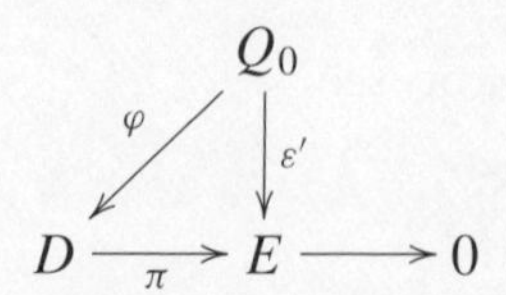

is commutative. Also, define ψ to be $i\varepsilon : P_0 \to D$ and let $\theta = \psi \oplus \varphi : P_0 \oplus Q_0 \to D$. There is a commutative diagram

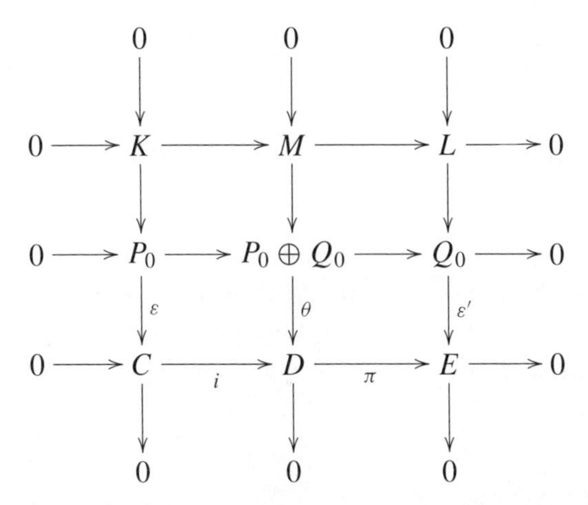

with columns and bottom two rows exact ($M = \mathrm{Ker}(\theta)$ and the maps in the top row are obtained by restriction from those in the middle row, which are the obvious ones). It follows that the top row is exact. (To avoid a direct argument, view the rows as chain complexes by adding zero modules, so the columns become a short exact sequence of complexes, and apply Theorem 3.1, recalling that a complex is exact if and only if its homology is zero.) Now repeat this procedure with $0 \to K \to M \to L \to 0$ in place of $0 \to P \to R \to Q \to 0$ and splice columns. Continue, to recursively construct the desired resolution. $\square$

Lemma 4.6. *Suppose*

$$\begin{array}{ccccccccc} 0 & \longrightarrow & C & \xrightarrow{i} & D & \xrightarrow{\pi} & E & \longrightarrow & 0 \\ & & \downarrow{\scriptstyle f} & & \downarrow{\scriptstyle g} & & \downarrow{\scriptstyle h} & & \\ 0 & \longrightarrow & C & \xrightarrow{i} & D & \xrightarrow{\pi} & E & \longrightarrow & 0 \end{array}$$

is a commutative diagram of A-modules with exact rows. Let $P \xrightarrow{\varepsilon} C$ and $Q \xrightarrow{\varepsilon'} E$ be augmented projective resolutions and let $R \xrightarrow{\theta} D$ be as in Lemma 4.5. *Let $\tilde{f} : P \to P$ and $\tilde{h} : Q \to Q$ be lifts of f, h respectively. Then there is a lift $\tilde{g} : R \to R$ of g such that*

$$\begin{array}{ccccccccc} 0 & \longrightarrow & P & \xrightarrow{\tilde{i}} & R & \xrightarrow{\tilde{\pi}} & Q & \longrightarrow & 0 \\ & & \downarrow{\scriptstyle \tilde{f}} & & \downarrow{\scriptstyle \tilde{g}} & & \downarrow{\scriptstyle \tilde{h}} & & \\ 0 & \longrightarrow & P & \xrightarrow{\tilde{i}} & R & \xrightarrow{\tilde{\pi}} & Q & \longrightarrow & 0 \end{array}$$

is commutative (where $\tilde{i}$ and $\tilde{\pi}$ are as in 4.5*).*

Proof. To recursively construct $\tilde{g}$ as in 4.5, it is enough to define $\tilde{g}_0 : R_0 \to R_0$ such that $\theta\tilde{g}_0 = g\theta$ and

$$\begin{array}{ccccccccc} 0 & \longrightarrow & P_0 & \longrightarrow & P_0 \oplus Q_0 & \longrightarrow & Q & \longrightarrow & 0 \\ & & \Big\downarrow{\scriptstyle \tilde{f}_0} & & \Big\downarrow{\scriptstyle \tilde{g}_0} & & \Big\downarrow{\scriptstyle \tilde{h}_0} & & \\ 0 & \longrightarrow & P_0 & \longrightarrow & P_0 \oplus Q_0 & \longrightarrow & Q_0 & \longrightarrow & 0 \end{array}$$

is commutative. (For then $\tilde{f}_0, \tilde{g}_0, \tilde{h}_0$ induce maps $K \to K, L \to L, M \to M$ in the proof of 4.5, making the appropriate diagram commutative in order to continue.) To make the diagram commutative, we need to find $\tilde{g}_0$ of the form $\tilde{g}_0(p,q) = (\tilde{f}_0(p) + \lambda(q), \tilde{h}_0(q))$ $(p \in P_0,\ q \in Q_0)$, where $\lambda \in \mathrm{Hom}_A(Q_0, P_0)$. If $\tilde{g}_0$ has this form, then by definition of θ in 4.5,

$$\theta\tilde{g}_0(p,q) = i\varepsilon\tilde{f}_0(p) + i\varepsilon\lambda(q) + \varphi\tilde{h}_0(q)$$

for $p \in P_0, q \in Q_0$, where $\pi\varphi = \varepsilon'$. Also,

$$\begin{aligned} g\theta(p,q) &= g(i\varepsilon(p) + \varphi(q)) \\ &= gi\varepsilon(p) + g\varphi(q). \end{aligned}$$

But $gi\varepsilon(p) = if\varepsilon(p) = i\varepsilon\tilde{f}_0(p)$, so we must find λ such that

$$i\varepsilon\lambda(q) = (g\varphi - \varphi\tilde{h}_0)(q)$$

for $q \in Q_0$. Now

$$\begin{aligned} \pi(g\varphi - \varphi\tilde{h}_0)(q) &= \pi g\varphi(q) - \pi\varphi\tilde{h}_0(q) \\ &= h\pi\varphi(q) - \pi\varphi\tilde{h}_0(q) \\ &= h\varepsilon'(q) - \varepsilon'\tilde{h}_0(q) = 0. \end{aligned}$$

Hence $(g\varphi - \varphi\tilde{h}_0)(q) = i(c)$ for some unique $c \in C$, so there is a map $\mu = i^{-1}(g\varphi - \varphi\tilde{h}_0) : Q_0 \to C$, and we need to choose λ so that $\varepsilon\lambda = \mu$. But this can be done since Q_0 is projective:

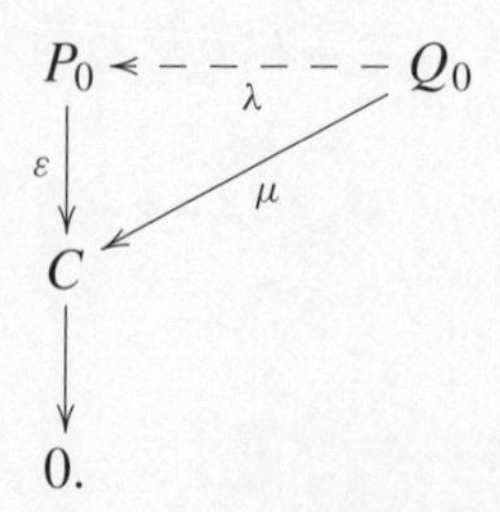

□

Theorem 4.7. *Suppose* $0 \to C \xrightarrow{i} D \xrightarrow{\pi} E \to 0$ *is a short exact sequence of A-modules. If two of* C, D, E *are of type* FP, *then so is the third.*

Proof. We distinguish three cases.

Case 1. C, E of type FP. Then D is of type FP by Lemma 4.5.
Case 2. C, D of type FP. Let $P \to C$, $Q \to D$ be finite augmented projective resolutions. Let $f : P \to Q$ be a lift of i (using Theorem 3.7). Let M be the mapping cone of f. Recall that $M_n = P_{n-1} \oplus Q_n$, so M is a positive, finite projective complex. By Prop. 3.5, there is an exact sequence

$$\ldots H_n(Q) \longrightarrow H_n(M) \longrightarrow H_{n-1}(P) \xrightarrow{f_*} H_{n-1}(Q) \longrightarrow \ldots$$

Since $H_n(P) = H_n(Q) = 0$ for $n \neq 0$, $H_n(M) = 0$ for $n \neq 0, 1$. For $n = 0, 1$, there is a commutative diagram with exact rows:

$$\begin{array}{ccccccccc} 0 \longrightarrow H_1(M) & \longrightarrow & H_0(P) & \xrightarrow{f_*} & H_0(Q) & \longrightarrow & H_0(M) & \longrightarrow & 0 \\ & & \downarrow \cong & & \downarrow \cong & & & & \\ 0 & \longrightarrow & C & \xrightarrow[i]{} & D & \longrightarrow & E & \longrightarrow & 0 \end{array}$$

hence f_* is injective, so $H_1(M) = 0$, and $H_0(M) \cong H_0(Q)/f_*(H_0(P)) \cong D/i(C) \cong E$. Thus M is a finite projective resolution of E, as required.

Case 3. D, E of type FP. Let $P \to D$, $Q \to E$ be finite augmented projective resolutions. Let $f : P \to Q$ be a lift of π. The mapping cone, M, of f is again a positive, finite projective complex. By a similar argument to that in Case 2, $H_n(M) = 0$ for $n \neq 0, 1$, $H_1(M) \cong \mathrm{Ker}(f_*) \cong C$ and f_* is onto, so $H_0(M) = 0$. Let d be the differential on M. Then there are short exact sequences

$$0 \to \mathrm{Ker}(d_1) \to M_1 \xrightarrow{d_1} M_0 \to 0$$
$$0 \to \mathrm{Im}(d_2) \to \mathrm{Ker}(d_1) \to C \to 0.$$

Since M_0 is projective, the first sequence splits, so $\mathrm{Ker}(d_1)$ is finitely generated projective, being a summand of M_1. Then the exact sequence $\ldots \to M_3 \to M_2 \to \mathrm{Ker}(d_1) \to C \to 0$ (obtained by splicing the second short exact sequence with $\ldots M_3 \to M_2 \to \mathrm{Im}(d_2) \to 0$) shows C is of type FP. □

Theorem 4.8. *Suppose*

$$\begin{array}{ccccccccc} 0 & \longrightarrow & C & \xrightarrow{i} & D & \xrightarrow{\pi} & E & \longrightarrow & 0 \\ & & \downarrow f & & \downarrow g & & \downarrow h & & \\ 0 & \longrightarrow & C & \xrightarrow{i} & D & \xrightarrow{\pi} & E & \longrightarrow & 0 \end{array}$$

is a commutative diagram of A-modules with exact rows, and C, D, E are of type FP. *Then $T_D(g) = T_C(f) + T_E(h)$. In particular, $r_D = r_C + r_E$.*

Proof. Choose finite projective resolutions P, Q of C, E respectively, and let $\tilde{f}$, $\tilde{g}$, $\tilde{h}$ be as in Lemma 4.6. We need to show $L(\tilde{g}) = L(\tilde{f}) + L(\tilde{h})$, and it is enough to show that, for all $n \in \mathbb{Z}$, $T_{R_n}(\tilde{g}_n) = T_{P_n}(\tilde{f}_n) + T_{Q_n}(\tilde{h}_n)$. Thus (see the proof of Lemma 4.5) we have reduced to the case C, E finitely generated projective, $D = C \oplus E$ and i, π are the obvious maps. Then for $c \in C, e \in E$, $g(c, e) = (f(c) + \lambda(e), h(e))$ for some A-homomorphism $\lambda : E \to C$. That is, $g = (f \oplus h) + \mu$, where $\mu(c, e) = (\lambda(e), 0)$. By Lemma 2.7, it is enough to show $T_{C\oplus E}(\mu) = 0$. Let $\rho : C \oplus E \to C \oplus E$ be the map $(c, e) \mapsto (0, e)$. Then $\mu = \mu\rho$, and by Lemma 2.7, $T(\mu) = T(\mu\rho) = T(\rho\mu) = T(0) = 0$, where T means $T_{C\oplus E}$. The last part follows on taking f, g, h to be the respective identity maps. □

Corollary 4.9. *Suppose $0 \to E_n \to E_{n-1} \to \ldots \to E_0 \to E \to 0$ is an exact sequence of A-modules with E_i of type* FP *for $0 \le i \le n$. Then E is of type* FP. *Also, if*

$$\begin{array}{ccccccccc} 0 & \longrightarrow & E_n & \longrightarrow \cdots \longrightarrow & E_0 & \longrightarrow & E & \longrightarrow & 0 \\ & & \downarrow{\scriptstyle f_n} & & \downarrow{\scriptstyle f_0} & & \downarrow{\scriptstyle f} & & \\ 0 & \longrightarrow & E_n & \longrightarrow \cdots \longrightarrow & E_0 & \longrightarrow & E & \longrightarrow & 0 \end{array}$$

is commutative, then $T_E(f) = \sum_{i=0}^{n}(-1)^i T_{E_i}(f_i)$. In particular, $r_E = \sum_{i=0}^{n}(-1)^i r_{E_i}$.

Proof. Decompose the sequence into exact sequences $0 \to E_n \to E_{n-1} \to \ldots \to E_1 \to F \to 0$ and $0 \to F \to E_0 \to E \to 0$, where $F = \operatorname{Ker}(E_0 \to E)$. Now use Theorems 4.7 and 4.8 and induction on n (in the second part, note that f_0 induces a map $g : F \to F$ by restriction). □

Exercise. Suppose $0 \to E_n \to E_{n-1} \to \ldots \to E_0 \to E \to 0$ is an exact sequence of A-modules and P_i is a projective resolution of E_i for $0 \le i \le n$. Writing P_{ij} for $(P_i)_j$, show that E has a projective resolution Q with $Q_k = \bigoplus_{i+j=k} P_{ij}$ for $k \in \mathbb{Z}$. (Use induction as in Cor. 4.9 and the mapping cone for the case $n = 1$, as in Case 2 of Theorem 4.7.)

Modules of type FL

A module M is of type FL if it has a finite free resolution, that is, there is an exact sequence

$$0 \to F_n \to F_{n-1} \to \ldots F_1 \to F_0 \xrightarrow{\varepsilon} M \to 0.$$

for some $n \in \mathbb{N}$, with each F_i finitely generated free. Clearly FL implies FP, and if M is of type FL, then $r_M = mT_A(1)$ for some $m \in \mathbb{Z}$. Also, if M, N are of type FL, then $M \oplus N$ is of type FL, and there are analogues of Theorem 4.7 and Cor. 4.9 for modules of type FL.

Now suppose $f : A \to B$ is a ring homomorphism and B is an A-module of type FP via f. As in Section 2, we can define

$$\mathrm{tr}_{B/A} : T(B) \to T(A)$$

by

$$\mathrm{tr}_{B/A}(T(b)) = T_{B/A}(\rho_b),$$

where $\rho_b : B \to B$ is defined by $x \mapsto xb$.

Lemma 4.10. *Let $f : A \to B$ be a ring homomorphism and assume B is an A-module of type FP via f. If M is a B-module of type FP, then M is an A-module of type FP, and if $g \in \mathrm{End}_B(M)$, then $g \in \mathrm{End}_A(M)$ and*

$$T_{M/A}(g) = \mathrm{tr}_{B/A}(T_{M/B}(g)).$$

Proof. A finitely generated free B-module is isomorphic to B^n for some n, so by Lemma 4.3(1) and induction on n, any finitely generated free B-module is of type FP over A. Hence any finitely generated projective B-module is of type FP over A by Prop. 4.4. Let

$$0 \to P_n \to \ldots \to P_0 \to M \to 0.$$

be a finite augmented projective resolution of M over B. Applying Cor. 4.9 to this sequence, M is of type FP over A. Let $g \in \mathrm{End}_B M$, so obviously $g \in \mathrm{End}_A(M)$. Let $\tilde{g}$ be a lift of g to the resolution.

$$\text{By definition,} \quad T_{M/B}(g) = \sum_i (-1)^i T_{P_i/B}(\tilde{g}_i)$$

$$\text{and by Cor. 4.9,} \quad T_{M/A}(g) = \sum_i (-1)^i T_{P_i/A}(\tilde{g}_i),$$

and since $\mathrm{tr}_{B/A}$ is $\mathbb{Z}$-linear, it is enough to show $T_{P_i/A}(\tilde{g}_i) = \mathrm{tr}_{B/A}(T_{P_i/B}(\tilde{g}_i))$, i.e., we can assume M is finitely generated B-projective. Now argue as in Lemma 2.8, using Lemma 4.3 (1) in place of Lemma 2.7(1). □

Automorphisms

Let $\alpha : A \to A$ be a ring homomorphism, M an A-module. Let ${}^{\alpha}M$ be the A-module with the same addition as M, but scalar multiplication defined by

$a \cdot m = \alpha(a)m$ ($a \in A$, $m \in M$), where on the right-hand side the scalar multiplication of M is used. Note that $\mathrm{End}_A(M) = \mathrm{End}_A({}^{\alpha}M)$. Also, there is an A-isomorphism

$$\tilde{A} \otimes_A M \xrightarrow{\cong} {}^{\alpha}M$$

sending $1 \otimes m$ to m, for $m \in M$, where $\tilde{A}$ is A made into a right A-module via α^{-1} ($a \cdot b = a\alpha^{-1}(b)$ for $a, b \in A$, where the multiplication of A is used on the right hand side) and a left A-module as usual (so $a(b \otimes m) = (ab) \otimes m$ for $m \in M$).

Lemma 4.11. *If M is of type FP, then so is ${}^{\alpha}M$ and $T_{{}^{\alpha}M}(f) = \alpha_*^{-1}(T_M(f))$ for $f \in \mathrm{End}_A(M)$, where $\alpha_*^{-1} : T(A) \to T(A)$ is the homomorphism induced by α^{-1}. In particular, $r_{{}^{\alpha}M} = \alpha_*^{-1}(r_M)$.*

Proof. If $\ldots P_1 \xrightarrow{d_1} P_0 \xrightarrow{\varepsilon} M \to 0$ is a finite augmented projective resolution of M, then

$$\ldots \longrightarrow {}^{\alpha}P_1 \xrightarrow{d_1} {}^{\alpha}P_0 \xrightarrow{\varepsilon} {}^{\alpha}M \to 0$$

is one of ${}^{\alpha}M$ (since ${}^{\alpha}P_i$ is projective, using ${}^{\alpha}P_i \cong \tilde{A} \otimes_A P_i$ and applying Lemma 2.11 to $\alpha^{-1} : A \to A$). If $\tilde{f}$ is a lift of f (to both resolutions), it also follows from Lemma 2.11 that $T_{{}^{\alpha}P_i}(\tilde{f}_i) = \alpha_*^{-1}(T_{P_i}(\tilde{f}_i))$ and the result follows since α_*^{-1} is $\mathbb{Z}$-linear. □

Algebras

We note just one result concerning them. First, recall that if M is an A-module, $m \in M$ and $a \in A$, then *a annihilates m* means that $am = 0$, and we put $\mathrm{Ann}_A(m) = \{a \in A \mid am = 0\}$, and $\mathrm{Ann}_A(M) = \bigcap_{m \in M} \mathrm{Ann}_A(m)$ (an ideal of A).

Lemma 4.12. *Let A be an R-algebra, where R is a commutative ring and let M be an A-module of type* FP. *Then $T_M : \mathrm{End}_A(M) \to T(M)$ is R-linear, and r_M is annihilated by* $\mathrm{Ann}_R(M)$.

Proof. Let $P \to M$ be a finite augmented A-projective resolution of M. If $g \in \mathrm{End}_A(M)$, take a lift $\tilde{g} : P \to P$ of g, so $T_M(g) = \sum(-1)^i T_{P_i}(\tilde{g}_i)$. If $r \in R$, $r\tilde{g}$ is a lift of rg, so $T_M(rg) = \sum(-1)^i T_{P_i}(r\tilde{g}_i)$. Therefore to prove T_M is R-linear, we can assume M is finitely generated A-projective. Let $\{x_j\}$, $\{f_j\}$ be a finite coordinate system for M. By Lemma 2.5,

$$T_M(rg) = T_A\left(\sum_j f_j(rg(x_j))\right) = T_A\left(r\sum_j f_j(g(x_j))\right) = rT_M(g)$$

(since f_j is A-linear, so R-linear, and T_A is R-linear). Thus T_M is R-linear, and if $r \in \mathrm{Ann}_R(M)$, then $rg = 0$, and taking $g = \mathrm{id}_M$, the last part of the lemma follows. □

Remark 4.13. *If $f : A \to B$ is a homomorphism of R-algebras and B is an A-module of type FP via f, it follows that $\mathrm{tr}_{B/A}$ is R-linear (because $\rho_{rb} = r\rho_b$ for $r \in R$, $b \in B$, and $T_{B/A}$ is R-linear by Lemma 4.12.)*

5. Cohomological dimension

Here we introduce the homology and cohomology modules of a group, and an associated invariant of the group, its cohomological dimension.

Group rings

Let R be a ring and G a group, and denote by RG the free R-module with basis G. Thus RG is an additive group, and an element of RG has the form $\sum_{g\in G} r_g g$, where $r_g \in R$ and $r_g = 0$ for all but finitely many $g \in G$, and $\sum_{g\in G} r_g g = \sum_{g\in G} s_g g$ if and only if $r_g = s_g$ for all $g \in G$. We can make RG into a ring, called the group ring of G over R. Multiplication is defined using the multiplication in G and R; thus $(r_1g_1)(r_2g_2) = (r_1r_2)(g_1g_2)$ for $r_i \in R$, $g_i \in G$. This is extended to RG in the only possible way so that the distributive laws will hold. We define

$$\left(\sum_{g\in G} r_g g\right)\left(\sum_{g\in G} s_g g\right) = \sum_{g\in G} t_g g$$

where $t_g = \sum_{xy=g} r_x s_y$. It can be checked that RG becomes a ring, with $1_{RG} = 1_R.1_G = 1_G$, and G is a subgroup of the group of units of R.

There is a short exact sequence $0 \to I_G \to RG \xrightarrow{\varepsilon} R \to 0$, where $\varepsilon(\sum_{g\in G} r_g g) = \sum_{g\in G} r_g$ and $I_G = \mathrm{Ker}(\varepsilon)$, which is called the *augmentation sequence*. The map ε is an R-module homomorphism and a ring homomorphism, so I_G is an ideal in RG, called the *augmentation ideal*, and ε is called the *augmentation map*. Also, it is easy to see that I_G is a free R-module on

$\{g - 1 \mid g \in G,\ g \neq 1\}$. There is an R-homomorphism $i : R \to RG, r \mapsto r1_G$, which is a ring embedding, and $\varepsilon i = \mathrm{id}_R$, so as a sequence of R-modules, the augmentation sequence splits. If M is an RG-module, it is an R-module (via i) and $(g, m) \mapsto gm$ ($g \in G$, $m \in M$) gives an action of G on M as R-automorphisms. Conversely, an R-module M with an action of G on M as R-automorphisms determines an RG-module; scalar multiplication is defined by: $(\sum_{g \in G} r_g g)m = \sum_{g \in G} r_g(gm)$. Similar comments apply to right RG-modules.

Definition. *An RG-module M is* trivial *if the G-action is* trivial *($gm = m$ for all $g \in G$, $m \in M$).*

Thus M is trivial if and only if $I_G M = 0$. Since G has a trivial action on any module, trivial RG-modules are in one to one correspondence with R-modules. We shall always view R as a trivial RG-module unless stated otherwise. Then ε is an RG-homomorphism, and the augmentation sequence can be made into the start of an augmented RG-projective resolution of R, with ε as augmentation map, whence the name.

Remark 5.1. *If $f : R \to R'$ is a ring homomorphism, there is a corresponding ring homomorphism $\bar{f} : RG \to R'G$, sending $\sum\limits_{g \in G} r_g g$ to $\sum\limits_{g \in G} f(r_g)g$.*

Suppose H is a subgroup of G. Denote by $R(G/H)$ the free R-module on $G/H = \{gH \mid g \in G\}$. Since G acts on G/H by left multiplication, $R(G/H)$ is an RG-module. Also, RH is a subring of RG, and viewing R as a trivial RH-module, $RG \otimes_{RH} R$ becomes an RG-module via the inclusion map $RH \to RG$ (see the remarks preceding Lemma 2.11). Thus $b(a \otimes r) = (ba) \otimes r$ (for $a, b \in RG, r \in R$).

Lemma 5.2. *There is an RG-isomorphism $RG \otimes_{RH} R \to R(G/H)$, given by $g \otimes 1 \mapsto gH$, for $g \in G$.*

Proof. This is left as an exercise. □

Lemma 5.3. *Let X be a subset of G. Then $G = \langle X \rangle$ if and only if $X - 1$ generates I_G as RG-module, where $X - 1 = \{x - 1 \mid x \in X\}$.*

Proof. Suppose X generates G. Write $g \in G$ as $g = y_1 \dots y_n$, where each $y_i \in X^{\pm 1}$. Then $g - 1 = \sum\limits_{i=1}^{n} (y_1 \dots y_{i-1})(y_i - 1)$ and $x^{-1} - 1 = -x^{-1}(x - 1)$ for $x \in X$, hence $X - 1$ generates I_G. Conversely, assume $X - 1$ generates I_G and let $H = \langle X \rangle$. Then $I_G = RGI_H$. In the G-action on $R(G/H)$, H fixes the coset $m = H$, so $I_H m = 0$. But then $I_G m = 0$ since $I_G = RGI_H$, so G fixes the coset H, hence $H = G$. □

Definition. *Let A be any ring, (P, d) a chain complex over A, M a right A-module. Then $M \otimes_A P$ is the chain complex*

$$\ldots \longrightarrow M \otimes_A P_{n+1} \xrightarrow{(d_{n+1})_*} M \otimes_A P_n \xrightarrow{(d_n)_*} M \otimes_A P_{n-1} \longrightarrow \ldots$$

where, if U, V are A-modules and $f \in \mathrm{Hom}_A(U, V)$, $f_* = 1 \otimes f : M \otimes_A U \to M \otimes_A V$. If N is an A-module, $\mathrm{Hom}_A(P, N)$ is the cochain complex

$$\ldots \longleftarrow \mathrm{Hom}_A(P_{n+1}, N) \xleftarrow{d^*_{n+1}} \mathrm{Hom}_A(P_n, N) \xleftarrow{d^*_n} \mathrm{Hom}_A(P_{n-1}, N) \longleftarrow \ldots$$

where, if $f \in \mathrm{Hom}_A(U, V)$,

$$f^* : \mathrm{Hom}_A(V, N) \to \mathrm{Hom}_A(U, N) \text{ is given by } f^*(g)(u) = g(f(u))$$

(for $g \in \mathrm{Hom}_A(V, N)$, $u \in U$).

Assume from now on that R is a commutative ring, so RG is an R-algebra. If M is an RG-module, then M is a right R-module, and there is a right G-action on M ($mg = g^{-1}m$), so M becomes a right RG-module.

Definition. *Let P be an RG-projective resolution of the trivial RG-module R. Let M be an RG-module. The* homology *of G with coefficients in M, denoted by $H_*(G, M)$, is the $\mathbb{Z}$-graded R-module $H(M \otimes_{RG} P)$.*

Thus $H_n(G, M) = H_n(M \otimes_{RG} P)$. We have to check this does not depend on the choice of P. If Q is another RG-projective resolution of R, then by Lemma 3.8, there are chain maps $f : P \to Q$ and $g : Q \to P$ such that $fg \simeq \mathrm{id}_Q$, $gf \simeq \mathrm{id}_P$. If s is a chain homotopy from gf to id_P, then $g_* f_* \simeq \mathrm{id}_{M \otimes_A P}$ via s_*, so by Lemma 3.6 and the fact that homology is a functor on chain complexes, $H_n(g_*)H_n(f_*) = \mathrm{id} H_n(M \otimes_A P)$. Similarly, $H_n(f_*)H_n(g_*) = \mathrm{id} H_n(M \otimes_A Q)$, so $H_n(f_*) : H_n(M \otimes_A P) \to H_n(M \otimes_A Q)$ is an isomorphism.

Definition. *Let P be an RG-projective resolution of R and let M be an RG-module. The* cohomology *of G with coefficients in M, denoted by $H^*(G, M)$, is the cohomology of the cochain complex* $\mathrm{Hom}_{RG}(P, M)$.

Thus $H^n(G, M) = H^n(\mathrm{Hom}_{RG}(P, M))$. By an argument similar to the one just given, this is independent of the choice of P.

Example. This example is intended for those familiar with elementary algebraic topology. Suppose G acts without fixed points on an acyclic topological space X. Then the singular chain complex $S(X)$ is a $\mathbb{Z}G$-free resolution of $\mathbb{Z}$. (See [100], Theorem 10.20.) If G acts properly on X (i.e., every $x \in X$ has an open neighbourhood U such that $gU \cap U = \emptyset$ for $1 \neq g \in G$), and M is a trivial $\mathbb{Z}G$-module, there is an isomorphism of complexes $\mathrm{Hom}_{\mathbb{Z}}(S(X/G), M) \cong$

$\mathrm{Hom}_{\mathbb{Z}G}(S(X), M)$, hence $H^n(G, M) \cong H^n(X/G, M)$ (see [80], pp 135–136). Similar comments apply to the homology of G.

Note. *Let G be a group, let R be a commutative ring and let $f : M \to N$ be an RG-homomorphism. Let $P \to R$ be an augmented projective resolution. There is an induced cochain map* $\mathrm{Hom}(P, f) : \mathrm{Hom}_{RG}(P, M) \to \mathrm{Hom}_{RG}(P, N)$, *sending $g \in \mathrm{Hom}_{RG}(P_n, M)$ to fg. There is also an induced chain map $f \otimes 1 : M \otimes_{RG} P \to N \otimes_{RG} P$ sending $m \otimes p$ to $f(m) \otimes p$ for $m \in M$, $p \in P_n$. These induce, for each n, homomorphisms $H^n(G, M) \to H^n(G, N)$ and $H_n(G, M) \to H_n(G, N)$, both of which will be denoted by $\bar{f}$. (This should not cause confusion).*

Lemma 5.4. *Let G be a group, R a commutative ring and $0 \to L \xrightarrow{i} M \xrightarrow{\pi} N \to 0$ a short exact sequence of RG-modules. Then there are exact sequences*

$$\ldots H_n(G, L) \xrightarrow{\bar{i}} H_n(G, M) \xrightarrow{\bar{\pi}} H_n(G, N) \to H_{n-1}(G, L) \to \ldots$$

$$\ldots H^n(G, L) \xrightarrow{\bar{i}} H^n(G, M) \xrightarrow{\bar{\pi}} H^n(G, N) \to H^{n+1}(G, L) \to \ldots$$

Proof. Let $P \to R$ be an augmented RG-projective resolution. Since P is projective, using the preceding note, there is an exact sequence of cochain complexes

$$0 \to \mathrm{Hom}_{RG}(P, L) \to \mathrm{Hom}_{RG}(P, M) \to \mathrm{Hom}_{RG}(P, N) \to 0. \quad (1)$$

Since projectives are flat ([100, Cor. 3.46]), there is also an exact sequence of chain complexes

$$0 \to L \otimes_{RG} P \to M \otimes_{RG} P \to N \otimes_{RG} P \to 0. \quad (2)$$

The lemma follows on applying Theorem 3.1 and its analogue for cochain complexes to (1) and (2). □

Exercise

If A, B are RG-modules, show that for all n,

$$H_n(G, A \oplus B) \cong H_n(G, A) \oplus H_n(G, B)$$

$$\text{and} \quad H^n(G, A \oplus B) \cong H^n(G, A) \oplus H^n(G, B).$$

We shall now prove two useful technical results, which are needed to define the important idea of cohomological dimension.

Lemma 5.5. *Let A be any ring, let (C, d) be a chain complex over A and let M be an A-module. Let $Y_n = \operatorname{Coker}(d_{n+1}) = C_n/\operatorname{Im}(d_{n+1})$ and let $f_n : Y_n \to C_{n-1}$ be the map induced by $d_n : C_n \to C_{n-1}$. Then there is an exact sequence:*

$$\operatorname{Hom}_A(C_{n-1}, M) \xrightarrow{f_n^*} \operatorname{Hom}_A(Y_n, M) \longrightarrow H^n(\operatorname{Hom}_A(C, M)) \longrightarrow 0.$$

Proof. Let $\pi_n : C_n \to Y_n$ be the quotient map, so

$$\begin{array}{ccc} C_n & \xrightarrow{d_n} & C_{n-1} \\ & \searrow^{\pi_n} \quad \nearrow_{f_n} & \\ & Y_n & \end{array}$$

is commutative and $C_{n+1} \xrightarrow{d_{n+1}} C_n \xrightarrow{\pi_n} Y_n \longrightarrow 0$ is exact. Since $\operatorname{Hom}_A(-, M)$ is left exact,

$$0 \longrightarrow \operatorname{Hom}_A(Y_n, M) \xrightarrow{\pi_n^*} \operatorname{Hom}_A(C_n, M) \xrightarrow{d_{n+1}^*} \operatorname{Hom}_A(C_{n+1}, M)$$

is exact. Thus $\operatorname{Ker}(d_{n+1}^*) = \pi_n^*(\operatorname{Hom}_A(Y_n, M))$ and since $d_n = f_n\pi_n$, $d_n^* = \pi_n^* f_n^*$, so $\operatorname{Im}(d_n^*) = \pi_n^*(\operatorname{Im}(f_n^*))$. Since π_n^* is injective,

$$H^n(\operatorname{Hom}_A(C, M)) = \operatorname{Ker}(d_{n+1}^*)/\operatorname{Im}(d_n^*) \cong \operatorname{Hom}_A(Y_n, M)/\operatorname{Im}(f_n^*)$$

and the lemma follows. □

Corollary 5.6. *If, in Lemma* 5.5, $H^n(\operatorname{Hom}_A(C, Y_n)) = 0$, *then* $H_n(C) = 0$ *and there is a split exact sequence*

$$0 \longrightarrow Y_n \xrightarrow{f_n} C_{n-1} \xrightarrow{\pi_{n-1}} Y_{n-1} \longrightarrow 0.$$

Proof. Clearly $\operatorname{Im}(f_n) = \operatorname{Im}(d_n) = \operatorname{Ker}(\pi_{n-1})$, and π_{n-1} is surjective. By Lemma 5.5, $f_n^* : \operatorname{Hom}_A(C_{n-1}, Y_n) \to \operatorname{Hom}_A(Y_n, Y_n)$ is surjective, in particular there exists $\varphi \in \operatorname{Hom}_A(C_{n-1}, Y_n)$ such that $\mathrm{id}_{Y_n} = f_n^*(\varphi) = \varphi f_n$. Hence f_n is injective and the sequence splits, as claimed. From the commutative triangle in the proof of Lemma 5.5, $\operatorname{Im}(d_{n+1}) = \operatorname{Ker}(\pi_n) = \operatorname{Ker}(d_n)$, since f_n is injective, so $H_n(C) = 0$. □

Theorem 5.7. *Let R be a commutative ring, G a group and $n \geq 0$ an integer. The following are equivalent:*

(i) *for any exact sequence $0 \to K_n \to P_{n-1} \to \ldots \to P_1 \to P_0 \to R \to 0$ of RG-modules with all P_i RG-projective, K_n is RG-projective;*
(ii) *there exists an augmented RG-projective resolution $\ldots 0 \to P_n \to P_{n-1} \to \ldots \to P_1 \to P_0 \to R \to 0$;*

(iii) *for all RG-modules M and all integers $k > n$, $H^k(G, M) = 0$;*
(iv) *for all RG-modules M, $H^{n+1}(G, M) = 0$.*

Proof. Clearly (i)⇒(ii)⇒(iii)⇒(iv), and to show (iv)⇒(i), assume (iv) and take an exact sequence $0 \to K_n \to P_{n-1} \to \ldots \to P_1 \to P_0 \to R \to 0$ as in (i). Let P_n be an RG-projective module mapping onto K_n, call the kernel of the mapping K_{n+1}, and let P_{n+1} be an RG-projective module mapping onto K_{n+1}. Continue, to construct a projective resolution of R, (P, d). By (iv) and Cor. 5.6, there is a split exact sequence

$$0 \to Y_{n+1} \to P_n \to Y_n \to 0$$

where $Y_i = \operatorname{Coker}(d_{i+1})$. Hence Y_n is projective, and $K_n = \operatorname{Im}(d_n) \cong P_n / \operatorname{Ker}(d_n) = P_n / \operatorname{Im}(d_{n+1}) = Y_n$, so K_n is projective. □

We now come to the topic of this section.

Definition. *Let G be a group, R a commutative ring. The* cohomological dimension *of G over R, denoted* $\operatorname{cd}_R(G)$, *is*

$$\inf\left\{n \in \mathbb{N} \mid n \text{ satisfies conditions (i)–(iv) of Theorem 5.7}\right\}$$

where $\mathbb{N}$ is the set of natural numbers, including 0.

If the set is empty, the infimum is ∞, otherwise it is the least element of the set, so $\operatorname{cd}_R(G) \in \mathbb{N} \cup \{\infty\}$.

Example. Once again this example is for those with a knowledge of algebraic topology, although we shall discuss CW-complexes later. Let X be a G-complex (a CW-complex with an action of the group G which permutes the cells). If X is a free G-complex (i.e. the stabilizer of each cell is trivial) and acyclic, then its cellular chain complex C is a $\mathbb{Z}G$-free resolution of $\mathbb{Z}$. If X is of dimension at most n, then $C_{n+1} = C_{n+2} = \ldots 0$, hence $\operatorname{cd}_{\mathbb{Z}}(G) \leq n$. Given a group G, there is a connected CW-complex Y with fundamental group $\pi_1(Y) \cong G$ which is aspherical ($\pi_n(Y) = 0$ for $n > 1$). See [17, Theorem 7.1, Ch.VIII]. Such a space is called a $K(G:1)$ space. The universal covering X of Y is a free acyclic G-complex, in fact it is contractible (see [17, Ch.I, §4]). Define the *geometric dimension* of G, $\operatorname{geomdim}(G)$, to be

$$\inf\left\{n \in \mathbb{N} \mid \text{there exists a } K(G:1) \text{ of dimension } n\right\}.$$

From above, $\operatorname{cd}_{\mathbb{Z}}(G) \leq \operatorname{geomdim}(G)$. Equality holds except for the possibility that $\operatorname{cd}_{\mathbb{Z}}(G) = 2$, $\operatorname{geomdim}(G) = 3$ (see Theorem 11.6 below and the comments preceding it, which refer to Brown's book [17]).

Remark 5.8. *Suppose H is a subgroup of a group G, R a commutative ring. Then RH is a subring of RG. If T is a transversal of $H\backslash G = \{Hg \mid g \in G\}$, then RG is a free RH-module, with basis T. ($RG = \bigoplus_{t\in T} RHt$, and $RH \to RHt$, $x \mapsto xt$ is an RG-isomorphism). Similarly, RG is a free right RH-module, with basis T^{-1} (or any transversal for G/H). Since RG is RH-free, any RG-free module is RH-free, hence any projective RG-module is RH-projective.*

Lemma 5.9. *If H is a subgroup of G, then $\mathrm{cd}_R(H) \le \mathrm{cd}_R(G)$.*

Proof. By the remark, if

$$0 \to P_n \to \ldots \to P_0 \to R \to 0$$

is exact and all P_i are RG-projective, then the P_i are RH-projective, and the lemma follows. □

Lemma 5.10. *Let G be a group, R a commutative ring. Then $\mathrm{cd}_R(G) = 0$ if and only if $|G|$ is finite and a unit in R, in which case R is isomorphic to RGe, where $e = \dfrac{1}{|G|} \sum_{g\in G} g$, an idempotent in RG.*

Proof. Suppose $\mathrm{cd}_R(G) = 0$. This means R is a projective RG-module, which implies the augmentation sequence $0 \longrightarrow I_G \longrightarrow RG \underset{\varphi}{\overset{\varepsilon}{\rightleftarrows}} R \longrightarrow 0$ splits as a sequence of RG-modules, via φ, say. Write $\varphi(1) = \sum_{g\in G} r_g g$; then for $x \in G$, $x\varphi(1) = \varphi(x1) = \varphi(1)$, hence $r_{xg} = r_g$ for all $x, g \in G$, so $r_g = r$ for all $g \in G$, where r is some fixed element of R. Since $r_g = 0$ for all but finitely many g, G is finite ($\varphi \neq 0$ since $\varepsilon\varphi = \mathrm{id}_R$). Thus $\varphi(1) = r\left(\sum_{g\in G} g\right)$, and since $\varepsilon\varphi(1) = 1$, $1 = r|G|$, so $|G|$ is a unit in R. Conversely suppose $|G|$ is finite and a unit in R. Define $\varphi : R \to RG$ by $\varphi(r) = re$, where $e = \dfrac{1}{|G|} \sum_{g\in G} g$. Then φ is an RG-homomorphism, since $ge = e$ for $g \in G$, and $\varepsilon\varphi = \mathrm{id}_R$, so the augmentation sequence splits as a sequence of RG-modules, hence R is RG-projective. Since $\mathrm{Im}(\varphi) = Re = RGe$ and φ is injective, $R \cong RGe$. □

Lemma 5.11. *Let G be a finite group. Then $\mathrm{cd}_R(G) < \infty$ if and only if $|G|$ is a unit in R.*

Proof. If $|G|$ is a unit in R, $\mathrm{cd}_R(G) = 0$ by Lemma 5.10. Conversely, suppose $\mathrm{cd}_R(G) < \infty$. It is enough to show every prime divisor p of $|G|$ is invertible in R. For every such p, G has an element of order p. Therefore, it is enough to show $|H|$ is a unit in R for every cyclic subgroup H of G of prime order. By Lemma 5.9, $\mathrm{cd}_R(H) < \infty$ for all subgroups H of G. Therefore we can assume

G is cyclic, say $G = \langle x \rangle$, where x has order m, say. There are exact sequences

$$0 \longrightarrow I_G \longrightarrow RG \stackrel{\varepsilon}{\longrightarrow} R \longrightarrow 0 \qquad ①$$

$$0 \longrightarrow R \stackrel{\nu}{\longrightarrow} RG \stackrel{\tau}{\longrightarrow} I_G \longrightarrow 0 \qquad ②$$

where $\nu(r) = r(1 + x + \ldots + x^{m-1})$ and $\tau(u) = u(x - 1)$. By successive splicing of ① and ②, we obtain an exact sequence

$$0 \to R \to RG \to RG \to \ldots \to RG \to R \to 0$$

where the number of terms RG is at least $\mathrm{cd}_R(G) - 1$. Then by Theorem 5.7(i), R is RG-projective, so $\mathrm{cd}_R(G) = 0$ and the lemma follows from Lemma 5.10. □

Corollary 5.12. *If* $\mathrm{cd}_R(G) < \infty$, *then* $|H|$ *is a unit in* R *for all finite subgroups* H *of* G. *In particular,* $\mathrm{cd}_{\mathbb{Z}}(G) < \infty$ *implies* G *is torsion-free.*

Proof. This is immediate from Lemmas 5.9 and 5.11. □

Lemma 5.13. *If* $G = \langle x \rangle$ *is infinite cyclic, then* $\mathrm{cd}_R(G) = 1$.

Proof. By Lemma 5.3, I_G is a cyclic RG-module generated by $(x - 1)$, and in fact I_G is free cyclic with basis $\{x - 1\}$. For an element $u \in RG$ can be written as $u = \sum_{i=-\infty}^{\infty} r_i x^i$ where $r_i \in R$, and $r_i = 0$ for all but finitely many i. Then $u(x - 1) = 0$ implies $r_i = r_{i-1}$ for all i, hence $r_i = 0$ for all i, so $u = 0$. Since the augmentation sequence $0 \to I_G \to RG \to R \to 0$ is exact and I_G, RG are RG-projective, $\mathrm{cd}_R(G) \leq 1$, and $cd_R(G) > 0$ by Lemma 5.10. □

Note. If $f : G \to H$ is a group homomorphism, there is an induced homomorphism of R-algebras $\overline{f} : RG \to RH$ (for any commutative ring R), where $\overline{f}\left(\sum_{g \in G} r_g g\right) = \sum_{g \in G} r_g f(g)$. (In fact, the assignment $G \to RG$ becomes a functor from the category of groups to the category of R-algebras.) This makes RH into an RG-module via $\overline{f}$. If $N \trianglelefteq G$ and $f : G \to G/N$ is the quotient map, the RG-module structure induced by f on $R(G/N)$ is the same as that induced by the isomorphism $R(G/N) \cong RG \otimes_{RN} R$ of Lemma 5.2.

Definition. *A projective resolution* P *of* M *is of length* n *if* $P_k = 0$ *for* $k > n$.

Proposition 5.14. *Let* $N \trianglelefteq G$. *Then* $cd_R(G) \leq \mathrm{cd}_R(N) + \mathrm{cd}_R(G/N)$.

Proof. Suppose $\mathrm{cd}_R(N) = n < \infty$. There is an exact sequence $0 \to P_n \to \ldots \to P_0 \to R \to 0$ with all P_i RN-projective. Since RG is a free right RN-module, it is flat, so we can apply the functor $RG \otimes_{RN} -$ to this sequence to obtain an exact sequence:

$$0 \to P'_n \to \ldots \to P'_0 \to R(G/N) \to 0 \qquad (*)$$

where $P'_i = RG \otimes_{RN} P_i$, a projective RG-module (arguing as in Lemma 2.11), and using Lemma 5.2. Hence any free $R(G/N)$-module has an RG-projective resolution of length n (take a direct sum of copies of $(*)$). It follows that any projective $R(G/N)$-module, being a summand of a free module, has an RG-projective resolution of length n, using Cor. 3.11. (The argument is a simplified version of that for Prop. 4.4, without the need to show the modules are finitely generated.) Now suppose also that $\mathrm{cd}_R(G/N) = m < \infty$. Let $0 \to Q_m \to \ldots \to Q_0 \to R \to 0$ be exact with all Q_i $R(G/N)$-projective. As just noted, each Q_i has an RG-projective resolution of length n, so R has an RG-projective resolution of length $m + n$, by the exercise after Cor. 4.9. Therefore $\mathrm{cd}_R(G) \le m + n$. The result is obvious if $\mathrm{cd}_R(N)$ or $\mathrm{cd}_R(G/N)$ is infinite. □

6. The Stallings characteristic

Here the first definition of an Euler characteristic is given, and it will be denoted by μ. It can take values in a commutative ring R and is defined on a class of groups denoted by $\mathrm{FP}(R)$. It is based on the "Stallings total characteristic", using the notion of rank of a projective module given in Section 2. The idea of algebraically defining an Euler characteristic for discrete groups in this way is due to Serre [107], who defined it for a smaller class FL, using the more usual definition of rank for a free module. (This rank will be discussed in detail in Section 8.)

Let R be a commutative ring, G a group and let $A = RG$ (an R-algebra since R embeds in the centre $Z(RG)$ by $r \mapsto r1_G$). Then $[A, A]$ is the R-submodule generated by

$$X = \{gh - hg \mid g, h \in G\} = \{gxg^{-1} - x \mid g, x \in G\}$$

(see the remarks preceding Lemma 2.12). It follows that $T(A) = A/[A, A]$ is isomorphic to $R[G]$, where $R[G]$ denotes the free R-module on the set of conjugacy classes of G. For the R-homomorphism $t : RG \to R[G]$ defined by $g \mapsto [g]$, where $[g]$ denotes the conjugacy class of g in G, clearly satisfies $[A, A] \subseteq \mathrm{Ker}(t)$. By Lemma 2.1, t induces a map $s : T(A) \to R[G]$. But the map $[g] \mapsto g + [A, A]$ ($g \in G$) is well-defined since $X \subseteq [A, A]$, so extends to an R-homomorphism $u : R[G] \to T(A)$, and s, u are inverse maps. We shall identify $T(A)$ with $R[G]$ via s. Thus $T_{RG}(g) = [g]$ for $g \in G$, and this specifies T_{RG} since it is R-linear. If $f : G \to H$ is a group homomorphism, then f extends to an algebra homomorphism $RG \to RH$ (see the note after Lemma 5.13) which we also denote by f. The induced homomorphism $f_* : T(RG) \to T(RH)$ is given by $f_*([g]_G) = [f(g)]_H$ for $g \in G$, where subscripts denote conjugacy classes in the relevant group. Suppose H is a subgroup of G and the index $(G : H) < \infty$. Then $RG = \bigoplus_{v \in V} RHv$, where V is a transversal for H/G (Remark 5.8). Thus RG is a finitely generated free RH-module, and there is an R-homomorphism $\mathrm{tr}_{RG/RH} : T(RG) \to T(RH)$ given by

$$\mathrm{tr}_{RG/RH}(T(b)) = T_{RG/RH}(\rho_b)$$

for $b \in RG$. (This definition is given after Lemma 2.7 in Section 2, and $\mathrm{tr}_{RG/RH}$ is R-linear by Remark 4.13.)

Lemma 6.1. *In this situation, suppose $\tau = T_{RG}(g) = [g]_G$, where $g \in H$. Then $\mathrm{tr}_{RG/RH}(\tau) = \sum_{\sigma \in T(H)} z_\sigma \sigma$, where*

$$z_\sigma = \begin{cases} 0 & \text{if } \sigma \cap \tau = \emptyset, \\ (C_G(h) : C_H(h)) & \text{for any } h \in \sigma, \text{ if } \sigma \subseteq \tau. \end{cases}$$

($C_G(h)$ means the centraliser of h in G, and $T(H) = T_{RH}(H)$ is the set of conjugacy classes of H).

Proof. The map ρ_g, $x \mapsto xg$, for $g \in G$, permutes the summands RHv of RG, and stabilizes RHv if and only if $Hvg = Hv$, that is, $vgv^{-1} \in H$, in which case $\rho_g(v) = vg = (vgv^{-1})v$. By Lemma 2.6,

$$\begin{aligned} \mathrm{tr}_{RG/RH}(\tau) &= \sum_v T_{RH}(vgv^{-1}) \quad (\text{sum over } \{v \in V \mid vgv^{-1} \in H\}) \\ &= \sum_{\sigma \in T(H)} z_\sigma \sigma, \end{aligned}$$

where z_σ is the number of $v \in V$ such that $vgv^{-1} \in \sigma$. If $\sigma \cap \tau = \emptyset$, this is zero. Suppose $vgv^{-1} \in \sigma$ and take $s \in V$. Then

$$\begin{aligned} sgs^{-1} \in \sigma &\Leftrightarrow sgs^{-1} = hvgv^{-1}h^{-1} \quad \text{for some } h \in H \\ &\Leftrightarrow s^{-1}hv \in C_G(g) \quad \text{for some } h \in H \\ &\Leftrightarrow s \in HvC_G(g). \end{aligned}$$

$$\begin{aligned} \text{Thus} \quad z_\sigma &= \text{the number of } s \in V \text{ such that } Hs \subseteq HvC_G(g) \\ &= (C_G(g) : C_G(g) \cap v^{-1}Hv) \\ &= (C_G(g) : C_{v^{-1}Hv}(g)) \\ &= (C_G(vgv^{-1}) : C_H(vgv^{-1})) \\ &= (C_G(h) : C_H(h)) \quad \text{for any } h \in \sigma \\ &\qquad (\text{this is independent of the choice of } h, \text{ and } vgv^{-1} \in \sigma). \end{aligned}$$

(Independence of the index on h comes from the fact that conjugation by an element of H is an automorphism of G mapping H onto H.) □

Definition. *Let G be a group, R a commutative ring. Then $G \in \mathrm{FP}(R)$ means that the trivial RG-module R is of type FP over RG, and $G \in \mathrm{FL}(R)$ means that R is of type FL over RG. If $G \in \mathrm{FP}(R)$, we define*

$$\chi_G = r_{R/RG}.$$

Thus $\chi_G = \sum_i (-1)^i r_{P_i}$, where $0 \to P_n \to \ldots \to P_0 \to R \to 0$ is a finite augmented RG-projective resolution of R. We call χ_G the *Stallings total characteristic* of G. Thus χ_G is an R-linear combination of conjugacy classes of G.

Definition. *If $G \in \mathrm{FP}(R)$, define $\mu(G) =$ the coefficient of $[1]$ in χ_G (denoted by $\mu(G : R)$ if necessary).*

Thus $\mu(G) \in R$, and if $G \in \mathrm{FL}(R)$, then $G \in \mathrm{FP}(R)$ and $\chi_G = \mu(G)[1]$. Also, $G \in \mathrm{FP}(R)$ obviously implies $\mathrm{cd}_R(G) < \infty$.

Notation. If $g \in G$, $\tau = [g]$ and $r \in T(RG)$, denote by $r(g)$ or $r(\tau)$ the coefficient of τ in r. Thus $\mu(G) = \chi_G(1)$.

Example. This is again for readers familiar with algebraic topology. Suppose $G = \pi_1(X)$, where X is a finite aspherical CW-complex. Then the cellular chain complex C of $\tilde{X}$, the universal covering space of X, is a finite $\mathbb{Z}G$-free resolution of $\mathbb{Z}$, and C_n has a $\mathbb{Z}G$-basis in one-to-one correspondence with the n-cells of X. Hence $G \in \mathrm{FL}(\mathbb{Z})$, $\chi_G = \chi(X)[1]$ and $\mu(G) = \chi(X)$ (the Euler characteristic of X). In fact, it suffices to assume the universal covering space $\tilde{X}$ is acyclic (see [17, Ch.1, §4]). Cellular homology is discussed in Section 11 below.

Proposition 6.2. *Suppose $G \in \mathrm{FP}(R)$ and H is a subgroup of finite index in G. Then $H \in \mathrm{FP}(R)$ and $\mu(H) = (G : H)\mu(G)$. If $G \in \mathrm{FL}(R)$, then $H \in \mathrm{FL}(R)$.*

Proof. By Lemma 4.10, $H \in \mathrm{FP}(R)$ and

$$\chi_H = r_{R/RH} = \mathrm{tr}_{RG/RH}(r_{R/RG}) = \mathrm{tr}_{RG/RH}(\chi_G).$$

By Lemma 6.1 and R-linearity of $\mathrm{tr}_{RG/RH}$ (Remark 4.13),

$$\chi_H(\sigma) = (C_G(h) : C_H(h))\chi_G(\tau),$$

for $\sigma \in T(H)$, where $h \in \sigma$ and τ is the G-conjugacy class containing σ. Now take $\sigma = [1]$ to obtain the formula for μ. Since RG is a finitely generated free RH-module, finltely generated free RG-modules, being direct sums of finitely many copies of RG, are finitely generated free RH-modules, and the last part follows. □

Remark 6.3. *Let $f : R \to R'$ be a homomorphism of commutative rings, G a group. Then there is an isomorphism of R'-algebras $R' \otimes_R RG \xrightarrow{\cong} R'G$,* *given by $r' \otimes rg \mapsto r'f(r)g$, for $r' \in R'$, $r \in R$ and $g \in G$. (The inverse map is given by $g \mapsto 1 \otimes g$; note that $r' \otimes rg = r'f(r) \otimes g$.)*

Hence, if M is an RG-module, $R' \otimes_R M$ is an $R'G$-module, and the G-action is given by $g(r' \otimes m) = r' \otimes gm$, for $m \in M$. According to Cor. 2.14, $R' \otimes_R T(RG) \cong T(R' \otimes_R RG)$, and $T(R' \otimes_R RG) \cong T(R'G) = R'[G]$, while $T(RG) = R[G]$. The resulting isomorphism $R' \otimes_R R[G] \to R'[G]$ is given by $r' \otimes r[g] \mapsto r'f(r)[g]$, for $r' \in R'$, etc. By Cor. 2.14 (with $u = \mathrm{id}_P$), if P is a finitely generated projective RG-module and $r_P = \sum_{\tau \in T(G)} r_P(\tau)\tau$, then $R' \otimes_R P$ is finitely generated $R'G$-projective, and

$$r_{R' \otimes_R P} = \text{image of } (1 \otimes r_P) \text{ under this isomorphism} = \sum_{\tau \in T(G)} f(r_P(\tau))\tau.$$

Lemma 6.4. *If $f : R \to R'$ is a homomorphism of commutative rings and $G \in \mathrm{FP}(R)$, then $G \in \mathrm{FP}(R')$ and $\mu(G : R') = f(\mu(G : R))$.*

Proof. If

$$\ldots 0 \to P_n \to \ldots P_0 \to R \to 0 \to \ldots \qquad (*)$$

is a finite augmented RG-projective resolution, it splits as an R-complex by Lemma 3.9. (Each P_i is R-projective since RG is R-free, applying Lemma 2.8 to the inclusion map $R \to RG$.) Applying the functor $R' \otimes_R -$ gives

$$\ldots 0 \to P'_n \to \ldots P'_0 \to R' \to 0 \to \ldots \qquad (**)$$

(where $P'_i = R' \otimes_R P_i$), which is R'-split, so exact by Lemma 3.6. (If s is a chain homotopy from the identity map to the zero map on $(*)$, $1 \otimes s$ is a chain homotopy from the identity map to the zero map on $(**)$.) By the comments preceding the lemma, each P'_i is finitely generated $R'G$-projective and $r_{P'_i}(1) = f(r_{P_i}(1))$. Thus $(**)$ is a finite augmented $R'G$-projective resolution of R', and the formula for μ follows. □

Remark 6.5. *If $0 \to K \to P_n \to \ldots \to P_0 \to L \to 0$ is an exact sequence of A-modules (for any ring A), with $L, P_0, \ldots, P_n$ all projective, then K is projective. (This follows easily by induction on n.)*

Theorem 6.6. *If $f : R \to R'$ is a homomorphism of commutative rings and R' is a free R-module via f, then $G \in \mathrm{FP}(R')$ implies $G \in \mathrm{FP}(R)$.*

Proof. (Stallings [110]) Let

$$\ldots 0 \to Q_m \to \ldots \to Q_0 \to R' \to 0 \qquad (1)$$

be a finite augmented $R'G$-projective resolution. We shall construct a finite RG-projective resolution of R inductively. Assume $n \leq m$ and finitely generated RG-projectives P_i have been defined for $0 \leq i \leq n-1$ so that there is an exact sequence of RG-modules

$$\ldots 0 \to K_n \to P_{n-1} \to \ldots P_0 \to R \to 0. \tag{2}$$

Apply $R' \otimes_R -$ to (2) to obtain

$$\ldots 0 \to K'_n \to P'_{n-1} \to \ldots P'_0 \to R' \to 0 \tag{3}$$

where $K'_n = R' \otimes_R K_n$, etc). This is an exact sequence of $R'G$-modules with P'_i finitely generated $R'G$-projective. (Argue as in Lemma 6.4, using Remark 6.5; exactness also follows since R' is free, so flat, as R-module.) Let $L = \mathrm{Ker}(Q_n \to Q_{n-1})$, so L is a finitely generated $R'G$-module, and compare (3) and the exact sequence

$$0 \to L \to Q_{n-1} \to \ldots \to Q_0 \to R' \to 0$$

using Cor. 3.11, to see that K'_n is a finitely generated $R'G$-module. Thus K'_n is $R'G$-generated by a set $\{r'_i \otimes k_i \mid i \in I\}$, where $r'_i \in R'$, $k_i \in K_n$ and I is finite. Let M be the RG-submodule of K_n generated by $\{k_i \mid i \in I\}$. Applying $R' \otimes_R -$ to the short exact sequence $0 \to M \to K_n \to K_n/M \to 0$ gives an exact sequence $M' \xrightarrow{\alpha} K'_n \to (K_n/M)' \to 0$ (only right exactness of $R' \otimes_R -$ is needed here). Clearly α is surjective, so $(K_n/M)' = R' \otimes_R (K_n/M) = 0$. Let $\{x_j\}$ be a basis for R' as R-module. Then

$$R' \otimes_R (K_n/M) = \bigoplus_j Rx_j \otimes_R (K_n/M) \cong \bigoplus_j (K_n/M).$$

Thus $K_n/M = 0$ and $K_n = M$ is a finitely generated RG-module. We can therefore take a finitely generated RG-projective module P_n mapping onto K_n, with kernel K_{n+1}, say, and by splicing we obtain an exact sequence

$$\ldots 0 \to K_{n+1} \to P_n \to \ldots P_0 \to R \to 0. \tag{4}$$

Thus inductively we construct a sequence (4) for $0 \leq n < m$ (for $n = 0$ we take the augmentation sequence, so $P_0 = RG$ and $K_1 = I_G$). When $n = m-1$, applying Cor. 3.11 as above shows that K'_m is finitely generated $R'G$-projective and K_m is a finitely generated RG-module. Since R' is R-free, K'_m is a projective RG-module ($R'G = R' \otimes_R RG \cong \bigoplus_j RG$ as RG-module, so $R'G$-free modules are RG-free, and an $R'G$-summand is an RG-summand). But as RG-module, $K'_m = R' \otimes_R K_m \cong \bigoplus_j K_m$, hence K_m is RG-projective, and when $n = m$, (4) is an augmented finite RG-projective resolution. □

Corollary 6.7. (1) *If $G \in \mathrm{FP}(R)$, then $\mu(G : R)$ is fixed by every ring endomorphism of R.*
(2) *If $G \in \mathrm{FP}(R)$, then $G \in \mathrm{FP}(K)$ for some prime field K.*

Proof. (1) is immediate from Lemma 6.4. If $G \in \mathrm{FP}(R)$, let I be a maximal ideal in R; by Lemma 6.4 $G \in \mathrm{FP}(R/I)$, and by Theorem 6.6, $G \in \mathrm{FP}(K)$, where K is the prime field of R/I. □

Lemma 6.8. *If $G \in \mathrm{FP}(R)$, then G is finitely generated.*

Proof. If $\ldots \to P_1 \to P_0 \to R \to 0$ is an augmented finite RG-projective resolution, let $K = \mathrm{Ker}(P_1 \to P_0)$, so $K = \mathrm{Im}(P_2 \to P_1)$ is finitely generated. There are exact sequences

$$0 \to K \to P_0 \to R \to 0$$

$$0 \to I_G \to RG \to R \to 0 \quad \text{(the augmentation sequence).}$$

By Lemma 3.10, $I_G \oplus P_0 \cong K \oplus RG$, so I_G is finitely generated as RG-module, by $y_1, \ldots, y_k$, say. Since I_G is (freely) generated as R-module by $\{g - 1 \mid g \in G, g \neq 1\}$, each y_i can be written as a finite R-linear combination of the elements of $G - 1$. Let X be the set of elements $g \in G$ such that $g - 1$ occurs in the expression of some y_i, so X is finite, and $X - 1$ generates I_G as RG-module. By Lemma 5.3, X generates G. □

Lemma 6.9. (1) *If G is a finite group, then $G \in \mathrm{FP}(R)$ if and only if $|G|$ is a unit in R, in which case $\chi_G = \dfrac{1}{|G|} \sum_{g \in G} [g]$, so $\mu(G) = 1/|G|$.*
(2) *If G is infinite cyclic, then $G \in \mathrm{FL}(R)$ for any R and $\chi_G = 0$ (so $\mu(G) = 0$).*

Proof. (1) follows from Lemmas 5.10 and 5.11 (see Example (2) after Lemma 2.6), and (2) follows from the proof of Lemma 5.13. □

We now turn our attention to the extension formula for the function μ. First some terminology is needed. If M is a free A-module with basis X (A being any ring) and $m \in M$, we can write $m = \sum_{x \in X} a_x x$, for unique $a_x \in A$. The *support* of m, denoted supp(m), is the finite set $\{x \in X \mid a_x \neq 0\}$. This applies to the R-free modules RG (basis G), $R[G]$ (basis $T(G)$) and $R(G/H)$ (basis G/H), where G is a group and H is a subgroup.

Lemma 6.10. *Let $1 \to N \xrightarrow{i} G \xrightarrow{\pi} Q \to 1$ be a group extension, R a commutative ring. Let $i_* : T(RN) \to T(RG)$, $\pi_* : T(RG) \to T(RQ)$ be the induced maps. Assume $N \in \mathrm{FP}(R)$. Then RQ is an RG-module of type FP, and is of type FL if $N \in \mathrm{FL}(R)$. Let $\pi^* = \mathrm{tr}_{RQ/RG}$. Then*

$$\pi^*(T_{RQ}(1)) = i_*(\chi_N)$$

and for $\tau \in T(Q)$,

$$\operatorname{supp}(\pi^*(\tau)) \subseteq \{\sigma \in T(G) \mid \pi_*(\sigma) = \tau\}.$$

Proof. Let $0 \to P_n \to \ldots \to P_0 \to R \to 0$ be an augmented finite RN-projective resolution. As in Prop. 5.14, we can apply $RG \otimes_{RN} -$ to obtain a finite augmented RG-projective resolution of RQ:

$$0 \to P'_n \to \ldots \to P'_0 \to RQ \to 0$$

where $P'_i = RG \otimes_{RN} P_i$. (As in Lemma 5.2, there is an isomorphism $RG \otimes_{RN} R \to RQ$ given by $g \otimes 1 \mapsto \pi(g)$.) If all P_i are RN-free, then all P'_i are RG-free. By definition,

$$\begin{aligned}
\pi^*(T_{RQ}(1)) &= T_{RQ/RG}(\mathrm{id}_{RQ}) = r_{RQ/RG} \\
&= \sum_{k=0}^{n} (-1)^k r_{P'_k/RG} \\
&= \sum_{k=0}^{n} (-1)^k i_*(r_{P_k/RN}) \quad \text{(by Lemma 2.11)} \\
&= i_*(\chi_N).
\end{aligned}$$

For the last part, we can assume for convenience that i is an inclusion map. Fix $x \in G$. For any RN-module M, let ${}^xM$ denote the RN-module with the same R-module structure as M, but with N-action $n.m = (x^{-1}nx)m$, for $n \in N, m \in M$, where on the right the original N-action on M is used. Then ${}^xR = R$, and

$$0 \to {}^xP_n \to \ldots \to {}^xP_0 \to R \to 0$$

is another augmented finite RN-projective resolution, with the same maps as the original one. (See Lemma 4.11-the conjugation $N \to N$, $n \mapsto x^{-1}nx$ induces an automorphism $\alpha : RN \to RN$, and ${}^xM$ is ${}^\alpha M$.) By Theorem 3.7, there is a lift of id_R to a chain map $h : P \to {}^xP$:

$$\begin{array}{ccccccccccc}
0 & \longrightarrow & P_n & \longrightarrow \cdots \longrightarrow & P_1 & \longrightarrow & P_0 & \longrightarrow & R & \longrightarrow & 0 \\
 & & \downarrow{\scriptstyle h_n} & & \downarrow{\scriptstyle h_1} & & \downarrow{\scriptstyle h_0} & & \| & & \\
0 & \longrightarrow & {}^xP_n & \longrightarrow \cdots \longrightarrow & {}^xP_1 & \longrightarrow & {}^xP_0 & \longrightarrow & R & \longrightarrow & 0
\end{array}$$

Define $h'_k : P'_k \to P'_k$ by $h'_k(g \otimes p) = gx \otimes h_k(p)$ (for $p \in P_k$, $g \in G$). It is easily checked that h'_k is well-defined and an RG-homomorphism. By definition, $\pi^*(T_{RQ}(\pi(x))) = T_{RQ/RG}(\rho_{\pi(x)})$, where $\rho_{\pi(x)} : RQ \to RQ$ is the map $u \mapsto u\pi(x)$. Under the isomorphism $RG \otimes_{RN} R \to RQ$ above, $\rho_{\pi(x)}$ corresponds to the map given by $g \otimes 1 \mapsto gx \otimes 1$ $(g \in G)$. It follows that

$h' : P' \to P'$ is a lift of $\rho_{\pi(x)}$. Hence

$$\pi^*(T_{RQ}(\pi(x))) = \sum_k (-1)^k T_{P'_k}(h'_k).$$

Fix k. Let $\{x_j\}$, $\{f_j\}$ be a finite coordinate system for the RN-module P_k. Then $\{1 \otimes x_j\}$, $\{1 \otimes f_j\}$ is a coordinate system for the RG-module P'_k. (Here $(1 \otimes f_j)(g \otimes p) = gf_j(p)$ for $g \in G$, $p \in P_k$.) By Lemma 2.5,

$$\begin{aligned} T_{P'_k}(h'_k) &= \sum_j T_{RG}((1 \otimes f_j)(h'_k(1 \otimes x_j))) \\ &= \sum_j T_{RG}((1 \otimes f_j)(x \otimes h_k(x_j))) \\ &= \sum_j T_{RG}(xf_j(h_k(x_j))) \\ &= T_{RG}(xa_k), \text{ where } a_k = \sum_j f_j(h_k(x_j)) \in RN. \end{aligned}$$

Therefore $\pi^*(T_{RQ}(\pi(x))) = T_{RG}(xa)$, where $a = \sum_k (-1)^k a_k \in RN$, and the lemma follows. □

Theorem 6.11 (Bass [10], Stallings [110]). *Let $1 \to N \xrightarrow{i} G \xrightarrow{\pi} Q \to 1$ be a group extension, R a commutative ring. If N, $Q \in \mathrm{FP}(R)$, then $G \in \mathrm{FP}(R)$ and $\mu(G) = \mu(N)\mu(Q)$. If N, $Q \in \mathrm{FL}(R)$, then $G \in \mathrm{FL}(R)$.*

Proof. By Lemma 6.10, RQ is of type FP over RG, and if R is of type FP over RQ, it is of type FP over RG by Lemma 4.10. Similarly N, $Q \in \mathrm{FL}(R)$ implies $G \in \mathrm{FL}(R)$ using the analogue of Cor. 4.9 for modules of type FL. Also by Lemma 4.10,

$$\begin{aligned} \chi_G = r_{R/RG} &= \pi^*(r_{R/RQ}) = \pi^*(\chi_Q) \\ &= \sum_{\tau \in T(Q)} \chi_Q(\tau)\pi^*(\tau) \end{aligned}$$

where π^* is as in Lemma 6.10 (and is R-linear by Remark 4.13). By Lemma 6.10, $[1]_G \in \mathrm{supp}(\pi^*(\tau))$ implies $\tau = [1]_Q$. Hence,

$$\begin{aligned} \mu(G) &= \text{the coefficient of } [1]_G \text{ in } \chi_Q(1)\pi^*([1]_Q) \\ &= \text{the coefficient of} [1]_G \text{ in } \chi_Q(1)i_*(\chi_N) \quad \text{(by Lemma 6.10)} \\ &= \chi_Q(1)\chi_N(1) = \mu(Q)\mu(N) \end{aligned}$$

(because i_* sends $[n]_N$ to $[n]_G$). □

To make some simple applications of our results, some definitions are needed.

Definition. *Let $\mathfrak{X}$ be a class of groups (customarily, closed under isomorphism, and containing the class $\mathfrak{I}$ of all trivial groups). The class* v$\mathfrak{X}$ *of groups which are virtually in $\mathfrak{X}$ is defined by:*

$$G \in \mathrm{v}\mathfrak{X} \textit{ if and only if } G \textit{ has a subgroup of finite index } H \textit{ such that } H \in \mathfrak{X}.$$

Also, the class P$\mathfrak{X}$ *of poly-$\mathfrak{X}$ groups is the class of all groups G having a series $1 = G_0 \leq G_1 \leq \ldots \leq G_n = G$, for some integer $n \geq 0$, such that $G_{i-1} \trianglelefteq G_i$ and $G_i/G_{i-1} \in \mathfrak{X}$ for $1 \leq i \leq n$.*

Examples of the use of this terminology are given in the corollaries below. A class that will occur later is the class of virtually torsion-free groups. Note that v is not an operator in the usual sense, since it fails to satisfy $\mathrm{v}\mathfrak{I} = \mathfrak{I}$. In fact, $\mathrm{v}\mathfrak{I} = \mathfrak{F}$, the class of all finite groups. Thus finite groups are virtually trivial, an observation which should not be made within earshot of a finite group theorist.

Note. *As in Section* 1, *the function $\mu(- : R)$ can be extended to* $\mathrm{vFP}(R)$ *by defining*

$$\mu(G : R) = \frac{\mu(H : R)}{(G : H)},$$

where $H \leq G$, $H \in \mathrm{FP}(R)$ and $(G : H) < \infty$. Further, the analogue of Prop. 6.2 *holds: $G \in \mathrm{vFP}(R)$ and $(G : H) < \infty$ implies $H \in \mathrm{vFP}(R)$ and $\mu(H : R) = (G : H)\mu(G : R)$, and similarly with $\mathrm{vFP}(R)$ replaced by $\mathrm{vFL}(R)$. However, this is unnecessary when $R = \mathbb{Q}$, by the next result.*

Corollary 6.12. *(a) For any commutative ring R, $\mathrm{FP}(R) = \mathrm{P}\,\mathrm{FP}(R)$ and $\mathrm{FL}(R) = \mathrm{P}\,\mathrm{FL}(R)$;*
(b) $\mathrm{FP}(\mathbb{Q}) = \mathrm{vFP}(\mathbb{Q})$ and $\mathrm{vFP}(\mathbb{Z}) \subseteq \mathrm{FP}(\mathbb{Q})$.

Proof. (a) This is immediate using Theorem 6.11 and induction.
(b) Let $(G : H) < \infty$, $H \in \mathrm{FP}(\mathbb{Q})$. Then $N = \bigcap_{g \in G} gHg^{-1}$ is normal in G and has finite index in H (and G), so $N \in \mathrm{FP}(\mathbb{Q})$ by Prop. 6.2. Also, $G/N \in \mathrm{FP}(\mathbb{Q})$ by Lemma 6.9, so $G \in \mathrm{FP}(\mathbb{Q})$ by Theorem 6.11. Since $\mathrm{FP}(\mathbb{Z}) \subseteq \mathrm{FP}(\mathbb{Q})$ by Lemma 6.4, (b) follows. □

Corollary 6.13. (a) *poly(infinite cyclic) groups are in* $\mathrm{FL}(R)$ *for any commutative ring R (in particular, finitely generated torsion-free nilpotent groups are in* $\mathrm{FL}(R)$*).*
(b) *virtually polycyclic groups are in* $\mathrm{vFL}(R)$ *for any commutative ring R (in particular, finitely generated virtually nilpotent groups are in* $\mathrm{vFL}(R)$*).*

Proof. (a) follows from Lemma 6.9 and Cor. 6.12, and (b) follows since polycyclic groups are virtually poly(infinite cyclic). See [98, 5.4.15]. □

Free groups

To investigate the Stallings characteristic of a free group, we shall use the idea of a derivation.

Definition. *let G be a group, R a commutative ring, M an RG-module. The set* $\mathrm{Der}(G, M)$ *of derivations from G to M is the set of all maps $d : G \to M$ such that*

$$d(xy) = d(x) + xd(y) \qquad (\text{for all } x, y \in G).$$

Note that $\mathrm{Der}(G, M)$ is an R-module, with addition and scalar multiplication defined pointwise. Any map $d : G \to M$ determines an R-linear map $\delta : I_G \to M$ by $\delta(g - 1) = d(g)$ for $g \in G, g \neq 1$, and δ is an RG-homomorphism if and only if d is a derivation. This gives a bijection

$$\mathrm{Hom}_{RG}(I_G, M) \longleftrightarrow \mathrm{Der}(G, M)$$

(which is an R-isomorphism). Since G acts on the abelian group M, we can form the split extension $M \rtimes G$ (the underlying set is $M \times G$, with multiplication $(m, g)(m', g') = (m + gm', gg')$). Let $s : G \to M \rtimes G$ be a map of the form $s(g) = (d(g), g)$. Then s is a group homomorphism if and only if d is a derivation.

Lemma 6.14. *If G is a free group with basis X, then I_G is a free RG-module with basis $X - 1$.*

Proof. The set $X - 1$ generates I_G by Lemma 5.3. Let M be an RG-module and let $f_0 : X - 1 \to M$ be a map. We have to show that f_0 extends to an RG-homomorphism $f : I_G \to M$. Define $d_0 : X \to M$ by $d_0(x) = f_0(x - 1)$, then $s_0 : X \to M \rtimes G$ by $s_0(x) = (d_0(x), x)$. Then s_0 has an extension to a group homomorphism $s : G \to M \rtimes G$, and s has the form $s(g) = (d(g), g)$. (The composite map $G \to M \rtimes G \xrightarrow{\pi} G$, where $\pi : M \times G \to G$ is the projection map, is id_G, since it is the identity map on X.) Hence $d \in \mathrm{Der}(G, M)$, and d corresponds to $f \in \mathrm{Hom}_{RG}(I_G, M)$ given by $f(g - 1) = d(g)$ for $g \in G$. Clearly f extends f_0. □

Corollary 6.15. *If G is a finitely generated free group, then $G \in \mathrm{FL}(R)$ for any commutative ring R, and $\chi_G = (1 - \mathrm{rk}(G))[1]_G$, so $\mu(G) = 1 - \mathrm{rk}(G)$, where $\mathrm{rk}(G)$ means the rank of G, i.e., the number of elements in a basis. If $G \neq 1$, then $\mathrm{cd}_R(G) = 1$.*

Proof. By Lemma 6.14, the augmentation sequence $0 \to I_G \to RG \to R \to 0$ is a finite augmented RG-free resolution of R, and the formula for χ_G follows. The last part follows by Lemma 5.10. □

Note. *There is converse to the last part of Cor. 6.15. If* $\mathrm{cd}_R(G) = 1$ *and G is torsion-free then G is a free group. This is a profound theorem of Stallings and Swan (see [30], and for a more recent account,* [37, Ch. IV, §3]).

Free products with amalgamation

To study these, we shall again use derivations. Let F be a free group with basis X. Then I_G is RG-free on $X - 1$ (Lemma 6.14), and we let $\{\delta_x \mid x \in X\}$ be the dual basis, with δ_x corresponding to $x - 1$. Thus, for $u \in I_F$,

$$u = \sum_{x \in X} \delta_x(u)(x - 1)$$

and $\delta_x \in \mathrm{Hom}_{RF}(I_F, RF)$. The corresponding element of $\mathrm{Der}(F, RF)$ is denoted by $\frac{\partial}{\partial x}$; thus $\frac{\partial g}{\partial x} = \delta_x(g - 1)$ for $g \in F$. Hence, for $g \in F$,

$$(g - 1) = \sum_{x \in X} \frac{\partial g}{\partial x}(x - 1).$$

Extending $\frac{\partial}{\partial x}$ from F to an R-linear map $RF \to RF$, we have

$$u - \varepsilon(u) = \sum_{x \in X} \frac{\partial u}{\partial x}(x - 1) \quad \text{for } u \in RF,$$

where $\varepsilon : RF \to R$ is the augmentation map. Then if $\pi : F \to G$ is an epimorphism of groups, we can extend π to an R-algebra homomorphism $RF \to RG$, to obtain

$$(*) \quad \boxed{v - \varepsilon(v) = \textstyle\sum_{x \in X} \pi\left(\dfrac{\partial u}{\partial x}\right)(\pi(x) - 1)}$$

where $u \in RF, \pi(u) = v$ and $\varepsilon : RG \to R$ is the augmentation map.

The maps $\partial/\partial x$ are called Fox derivatives.

Lemma 6.16. *Suppose* $G = A *_C B$*, a free product with amalgamation of groups. Then there is an exact sequence of* RG*-modules*

$$0 \to R(G/C) \xrightarrow{\alpha} R(G/A) \oplus R(G/B) \xrightarrow{\beta} R \to 0$$

where $\alpha(xC) = (xA, -xB)$ *and* $\beta(xA, 0) = \beta(0, xB) = 1$ *for* $x \in G$.

Proof. Clearly $\beta\alpha = 0$ and β is onto. We show that $\mathrm{Ker}(\beta) \subseteq \mathrm{Im}(\alpha)$. Let $\{a_i\}$, $\{b_j\}$ be generators for A, B respectively, let F be the free group on the disjoint union $X = \{a_i\} \cup \{b_j\}$ and let $\pi : F \to G$ be the epimorphism induced by the obvious map $X \to G$. Let $v \in RG$. Then by $(*)$ preceding the lemma,

$$v - \varepsilon(v) = \sum_i \pi\left(\frac{\partial \tilde{v}}{\partial a_i}\right)(a_i - 1) + \sum_j \pi\left(\frac{\partial \tilde{v}}{\partial b_j}\right)(b_j - 1)$$

where $\tilde{v}$ is chosen with $\pi(\tilde{v}) = v$. Hence

$$vA - \varepsilon(v)A = \sum_j \pi\left(\frac{\partial \tilde{v}}{\partial b_j}\right)(b_j - 1)A$$

$$vB - \varepsilon(v)B = \sum_i \pi\left(\frac{\partial \tilde{v}}{\partial a_i}\right)(a_i - 1)B.$$

Suppose $(vA, wB) \in \mathrm{Ker}(\beta)$, that is, $\varepsilon(v) + \varepsilon(w) = 0$. Choose $\tilde{v}$, $\tilde{w}$ with $\pi(\tilde{v}) = v$, $\pi(\tilde{w}) = w$. Then (using these formulas)

$$\alpha\left(\sum_j \pi\left(\frac{\partial \tilde{v}}{\partial b_j}\right)(b_j - 1)C + \varepsilon(v)C - \sum_i \pi\left(\frac{\partial \tilde{w}}{\partial a_i}\right)(a_i - 1)C\right)$$
$$= (vA, wB)$$

so $(vA, wB) \in \mathrm{Im}(\alpha)$ as required. It remains to show that α is injective. Suppose $\alpha(vC) = 0$, where $v \in RG$, that is, $vA = vB = 0$. Assume $\mathrm{supp}(vC) \neq \emptyset$ and choose $g \in G$ such that $gC \in \mathrm{supp}(vC)$ and g is of maximal length. (Here length means length of a reduced word representing g relative to the decomposition $G = A *_C B$.) If $g \in C$ (length 0) then $vC = rC$ for some $r \in R$, so $rA = rB = 0$, hence $r = 0$ and $vC = 0$, contrary to $\mathrm{supp}(vC) \neq \emptyset$. Hence $g \notin C$.

Case 1. A reduced word for g ends in a letter from $A \setminus C$. Since $vB = 0$, there must be some $g' \in G$ with $g'C \in \mathrm{supp}(vC)$ and $g'C \neq gC$, but $gB = g'B$. Then $g' = gb$ for some $b \in B \setminus C$, so g' has greater length than g, a contradiction.

Case 2. A reduced word for g ends in a letter from $B \setminus C$. This similarly leads to a contradiction since $vA = 0$.

It follows that $\mathrm{supp}(vC) = \emptyset$, that is, $vC = 0$, as required. □

Theorem 6.17. *Let $G = A *_C B$ and let R be a commutative ring. If A, B, $C \in \mathrm{FP}(R)$, then $G \in \mathrm{FP}(R)$ and*

$$\mu(G) = \mu(A) + \mu(B) - \mu(C).$$

If A, B, $C \in \mathrm{FL}(R)$, then $G \in \mathrm{FL}(R)$.

Proof. Let $0 \to P_n \to \ldots \to P_0 \to R \to 0$ be a finite augmented RC-projective resolution. Since RG is a free right RC-module, it is flat, so applying the functor $RG \otimes_{RC} -$ and using Lemma 5.2 gives an exact sequence

$$0 \to P'_n \to \ldots \to P'_0 \to R(G/C) \to 0$$

where $P'_i = RG \otimes_{RC} P_i$. Hence $R(G/C)$ is of type FP as RG-module, and by definition,

$$\begin{aligned} r_{R(G/C)} &= \sum_i (-1)^i r_{P'_i/RG} \\ &= \sum_i (-1)^i \gamma_*(r_{P_i/RC}) \qquad \text{by Lemma 2.11} \end{aligned}$$

(where $\gamma : RC \to RG$ is the map induced by the inclusion $C \hookrightarrow G$)

$$= \gamma_*(\chi_C).$$

Similarly there are maps $\alpha : RA \to RG$ and $\beta : RB \to RG$ induced by inclusion maps and $r_{R(G/A)} = \alpha_*(\chi_A)$, $r_{R(G/B)} = \beta_*(\chi_B)$. By Lemma 4.3, $R(G/A) \oplus R(G/B)$ is of type FP over RG and $r_{R(G/A)\oplus R(G/B)} = \alpha_*(\chi_A) + \beta_*(\chi_B)$. By Theorem 4.7 and Lemma 6.16, R is of type FP over RG (i.e. $G \in \mathrm{FP}(R)$). Moreover, by Theorem 4.8, $\chi_G = \alpha_*(\chi_A) + \beta_*(\chi_B) - \gamma_*(\chi_C)$, and comparing coefficients of $[1]_G$, $\mu(G) = \mu(A) + \mu(B) - \mu(C)$. The last part follows using the FL analogue of Theorem 4.7. □

Note. There is a similar result for HNN-extensions which will not be proved. If $G = \langle t, A \mid tBt^{-1} = C\rangle$ is an HNN-extension, and $A, B \in \mathrm{FP}(R)$ (so $C \in \mathrm{FP}(R)$ since $C \cong B$), then $G \in \mathrm{FP}(R)$ and $\mu(G) = \mu(A) - \mu(B)$.

The relation sequence

We shall next consider the Stallings characteristic for one-relator groups, and to do this we need an exact sequence called the relation sequence. We begin with two lemmas. If H is a subgroup of a group G and R is a commutative ring, then there is an RG-homomorphism $\pi : RG \to R(G/H)$, given by $g \mapsto gH$. We define $J_H = \mathrm{Ker}(\pi)$.

Lemma 6.18. *In these circumstances,* $J_H = RGI_H \cong RG \otimes_{RH} I_H$.

Proof. Apply $RG \otimes_{RH} -$ to the augmentation sequence for H. This gives the top row of a commutative diagram with exact rows:

$$\begin{array}{ccccccccc} 0 & \longrightarrow & RG \otimes_{RH} I_H & \longrightarrow & RG \otimes_{RH} RH & \longrightarrow & RG \otimes_{RH} R & \longrightarrow & 0 \\ & & \downarrow{\scriptstyle\theta} & & \downarrow{\scriptstyle\psi}\,\cong & & \downarrow{\scriptstyle\varphi}\,\cong & & \\ 0 & \longrightarrow & J_H & \xrightarrow{\text{inclusion}} & RG & \xrightarrow{\pi} & R(G/H) & \longrightarrow & 0 \end{array}$$

Here φ is given by Lemma 5.2, ψ is the usual isomorphism $u \otimes v \mapsto uv$ and θ is induced by ψ. Hence θ is an isomorphism and $J_H = \mathrm{Im}(\theta) = RGI_H$. □

If $1 \to K \to E \xrightarrow{\pi} G \to 1$ is a group extension, then the commutator quotient $\overline{K} = K/[K, K]$ is an abelian group, and G acts on it using conjugation by elements of E and lifting along π. That is, if $g \in G$, choose $e \in E$ such that $g = \pi(e)$ and define $g\overline{k} = \overline{eke^{-1}}$, where $k \in K$ and $\overline{k} = k[K, K]$. (For convenience we assume $K \to E$ is an inclusion map.) Thus $\overline{K}$ becomes a $\mathbb{Z}G$-module, with addition $\overline{k} + \overline{k'} = \overline{kk'}$ for $k, k' \in K$. There is an exact sequence

$$0 \to J_K \to RE \xrightarrow{\pi} RG \to 0 \tag{1}$$

which by restricting π gives an exact sequence

$$0 \to J_K \to I_E \to I_G \to 0$$

giving by the isomorphism theorems a short exact sequence

$$0 \to \frac{J_K}{J_K I_E} \to \frac{I_E}{J_K I_E} \to I_G \to 0. \tag{2}$$

From (1), $RG \cong RE/J_K$, so $I_E/J_K I_E$ is an RG-module, with scalar multiplication given by $g(u + J_K I_E) = eu + J_K I_E$, for any $e \in E$ such that $\pi(e) = g$, where $g \in G$, $u \in I_E$. (See Remark 2.10.) The map $\frac{I_E}{J_K I_E} \to I_G$ in (2) is an RG-homomorphism, so $\frac{J_K}{J_K I_E}$ is an RG-submodule of $\frac{I_E}{J_K I_E}$.

Lemma 6.19. *If $1 \to K \to E \xrightarrow{\pi} G \to 1$ is a group extension and R is a commutative ring, then there is an RG-isomorphism*

$$\psi : R \otimes_{\mathbb{Z}} \overline{K} \longrightarrow \frac{J_K}{J_K I_E}$$

sending $(1 \otimes \overline{k})$ to $(k - 1) + J_K I_E$, for $k \in K$.

Proof. Again we assume $K = \mathrm{Ker}(\pi)$. There is a homomorphism of groups $K \to \frac{J_K}{J_K I_E}$ given by $k \mapsto (k - 1) + J_K I_E$, because if $k, k' \in K$,

$$(kk' - 1) = (k - 1)(k' - 1) + (k - 1) + (k' - 1)$$

and $(k - 1)(k' - 1) \in J_K I_E$. This induces a homomorphism $\overline{K} \to \frac{J_K}{J_K I_E}$.

Therefore, the map $R \times \overline{K} \to \dfrac{J_K}{J_K I_E}$ given by $(r, \overline{k}) \mapsto r(k-1) + J_K I_E$ induces an R-homomorphism $\psi : R \otimes_{\mathbb{Z}} \overline{K} \to \dfrac{J_K}{J_K I_E}$. Thus $\psi(1 \otimes \overline{k}) = (k - 1) + J_K I_E$. Further, ψ is an RG-homomorphism. For if $g \in G$,

$$\psi(g(1 \otimes \overline{k})) = \psi(1 \otimes g\overline{k}) = \psi(1 \otimes \overline{eke^{-1}}) = (eke^{-1} - 1) + J_K I_E$$

where $\pi(e) = g$, while

$$g\psi(1 \otimes \overline{k}) = e(k-1) + J_K I_E.$$

But $e(k-1) - (eke^{-1} - 1) = (eke^{-1} - 1)(e - 1) \in J_K I_E$, so $g\psi(1 \otimes \overline{k}) = \psi(g(1 \otimes \overline{k}))$ and it follows that ψ is an RG-homomorphism. We construct an inverse map to ψ. Let T be a transversal for K/E with $1 \in T$. Then

$$RE = \bigoplus_{t \in T} RKt.$$

Since $K \trianglelefteq E$, using $t(k-1) = (tkt^{-1} - 1)t$ $(k \in K)$, we see that $J_K = REI_K = \bigoplus_{t \in T} I_K t$. Also, I_K is R-free on $\{k - 1 \mid k \in K, k \neq 1\}$, so there is an R-homomorphism $\theta : J_K \to R \otimes_{\mathbb{Z}} \overline{K}$ given by $(k-1)t \mapsto 1 \otimes \overline{k}$. Now take $k \in K$, $e \in E$ and write $e = k't$ with $k' \in K$, $t \in T$. Then $(k-1)(e-1)$ $= ((kk' - 1) - (k' - 1))t - (k - 1)$, and it follows that θ is zero on $J_K I_E$. Hence there is an induced map $\varphi : \dfrac{J_K}{J_K I_E} \to R \otimes_{\mathbb{Z}} \overline{K}$. For $k \in K$, $t \in T$, $(k-1)t - (k-1) = (k-1)(t-1) \in J_K I_E$, and it follows easily that φ and ψ are inverse maps. □

Theorem 6.20. *Let* $1 \to N \to F \xrightarrow{\pi} G \to 1$ *be an extension of groups, where* F *is a free group with basis* X *and* $N = \mathrm{Ker}(\pi)$. *Then there is an exact sequence of* RG*-modules*

$$0 \to R \otimes_{\mathbb{Z}} \overline{N} \xrightarrow{\theta} Z \xrightarrow{\zeta} RG \to R \to 0,$$

where $\overline{N} = N/[N, N]$ *with* G*-action* $\pi(f)\overline{n} = \overline{fnf^{-1}}$, Z *is* RG*-free with a basis* $\{e_x \mid x \in X\}$, *and*

$$\theta(1 \otimes \overline{n}) = \sum_{x \in X} \pi\left(\frac{\partial n}{\partial x}\right) e_x, \quad \zeta(e_x) = \pi(x) - 1.$$

Proof. From (2) preceding Lemma 6.19, there is an exact sequence of RG-modules

$$0 \to \frac{J_N}{J_N I_F} \to \frac{I_F}{J_N I_F} \to I_G \to 0 \tag{3}$$

By Lemma 6.14 I_F is free on $X - 1$, so $\frac{I_F}{J_N I_F}$ is RG-free on $\{(x - 1) + J_N I_F \mid x \in X\}$, so letting Z be a free RG-module with basis $\{e_x \mid x \in X\}$ in one-to-one correspondence with X, there is an isomorphism

$$\begin{cases} \dfrac{I_F}{J_N I_F} \to Z \\ u + J_N I_F \mapsto \sum_{x \in X} \pi\left(\dfrac{\partial u}{\partial x}\right) e_x \end{cases}$$

(using $(*)$ preceding 6.16). Also, by Lemma 6.19 there is an isomorphism $\psi : R \otimes_{\mathbb{Z}} \overline{N} \to \dfrac{J_N}{J_N I_F}$ which sends $1 \otimes \overline{n}$ to $(n - 1) + J_N I_F$. On splicing (3) with the augmentation sequence for G and using these isomorphisms, we obtain the desired sequence. □

An extension of groups as in Theorem 6.20 corresponds to a presentation of the group G, where N is the normal closure in F of the relators. The $\mathbb{Z}G$-module $\overline{N}$ is called the *relation module* of the presentation, and the exact sequence of RG-modules in Theorem 6.20 is called the *relation sequence* (over R) of the presentation.

One-relator groups

The treatment of these is based on using the relation sequence together with the Lyndon Identity Theorem, a result which we shall just quote. The other ingredient needed is the following lemma.

Lemma 6.21. *Let G be a group, R a commutative ring. Let C be a finite subgroup of G with $|C|$ invertible in R. Then $R(G/C)$ is RG-projective, and* $r_{R(G/C)} = \dfrac{1}{|C|} \sum_{c \in C} [c]_G.$

Proof. By Lemma 5.10, R is RC-projective and by Lemma 6.9, $r_{R/RC} = \chi_C = \dfrac{1}{|C|} \sum_{c \in C} [c]_C$. By Lemma 5.2, $R(G/C) \cong RG \otimes_{RC} R$, and the result follows by Lemma 2.11 applied to the inclusion map $RC \to RG$. □

Lyndon Identity Theorem. *If, in Theorem 6.20, $N = \langle r^n \rangle^F$, where $r \in F$, r is not a proper power, $n > 0$ and $C = \langle \pi(r) \rangle$, the map $R(G/C) \to R \otimes_{\mathbb{Z}} \overline{N}$ given by $gC \mapsto 1 \otimes g\,\overline{\pi(r)}$ $(for\ g \in G)$ is an RG-isomorphism.*

Proof. This was proved in [78]. □

Note. *It is also true that $\pi(r)$ has order n in G, so $|C| = n$. See* [81, Cor. 4.11, p. 266].

Theorem 6.22. *Let $G = \langle x_1, \ldots, x_m \mid r^n \rangle$ be a finitely generated one-relator group, where r is not a proper power and $n > 0$. Then*

(1) $G \in \mathrm{vFL}(\mathbb{Z})$;

(2) $\mu(G : \mathbb{Q}) = 1 - m + \dfrac{1}{n}$;

(3) *if* $n = 1$, *then* $G \in \mathrm{FL}(\mathbb{Z})$.

Proof. By Theorem 6.20 and the Lyndon Identity Theorem, there is an exact sequence of RG-modules

$$0 \to R(G/C) \to RG^m \to RG \to R \to 0 \qquad (*)$$

where $C = \langle \overline{r} \rangle$ and $\overline{r}$ is the image of r in G. Take $R = \mathbb{Z}$, and let H be a torsion-free subgroup of finite index in G (the existence of H is proved in [54]). Then H acts freely on G/C, so if T is a transversal for the double cosets HgC ($g \in G$), $\mathbb{Z}(G/C)$ is $\mathbb{Z}H$-free, with basis $\{tC \mid t \in T\}$. Thus $(*)$ is a finite augmented $\mathbb{Z}H$-free resolution of $\mathbb{Z}$, and (1) follows. Now take $R = \mathbb{Q}$. By Lemma 6.21, $(*)$ is an augmented $\mathbb{Q}G$-projective resolution of $\mathbb{Q}$, and

$$\begin{aligned} \chi_G &= (1-m)[1] + \frac{1}{n}\sum_{i=0}^{n-1}[\overline{r}^i] \\ &= \left(1 - m + \frac{1}{n}\right)[1] + \frac{1}{n}\sum_{i=1}^{n-1}[\overline{r}^i]. \end{aligned}$$

Since $\overline{r}$ has order n, $\mu(G : \mathbb{Q}) = 1 - m + \dfrac{1}{n}$. Finally, if $n = 1$, G is torsion-free and we may take $H = G$ (see [81, Theorem 4.12]). □

Centres of groups of type FP

Here we give the application of χ_G originally made by Stallings [109] (Gottlieb's Theorem). If G is a group and R is a commutative ring, let C be the centre of RG. Then RG is a C-algebra and $T(RG)$ is a C-module. In particular the center $Z(G)$ of G acts on $T(RG)$, and $\gamma[g] = [\gamma g]$ (for $\gamma \in Z(G)$, $g \in G$).

Lemma 6.23. *Let* $G \in \mathrm{FP}(R)$.

(i) $\gamma\chi_G = \chi_G$ *for all* $\gamma \in Z(G)$;

(ii) *if* $\mu(G) \neq 0$, *then* $Z(G)$ *is finite;*

(iii) *if* $\chi_G = \mu(G)[1]$ *(e.g., if* $G \in \mathrm{FL}(R)$*) and* $\mu(G) \neq 0$, *then* $Z(G) = 1$.

Proof. Let C be the centre of RG. By Lemma 4.12, $\chi_G = r_{R/RG}$ is annihilated by $\mathrm{Ann}_C(R) = \mathrm{Ann}_{RG}(R) \cap C = I_G \cap C$, in particular by $\gamma - 1$ for $\gamma \in Z(G)$, and (i) follows. Since χ_G has finite support, (ii) and (iii) follow from (i). □

Applications

(1) Let $G = \langle x_1, \ldots, x_m \mid r^n \rangle$, where r is not a proper power and $n > 0$. Then from the proof of Theorem 6.22, $G \in \mathrm{FP}(\mathbb{Q})$ and

$$\chi_G = \left(1 - m + \frac{1}{n}\right)[1] + \frac{1}{n}\sum_{i=1}^{n-1}[\bar{r}^i].$$

Also, no two of $\bar{r}, \bar{r}^2, \ldots, \bar{r}^{n-1}$ are conjugate in G (see [54]). Thus if $Z(G) \neq 1$, then by Lemma 6.23, either $1 - m + \frac{1}{n} = 0$ or $1 - m + \frac{1}{n} = \frac{1}{n}$. That is, either $m = 2$ and $n = 1$, or $m = 1$ (a result of Murasugi [91]).

(2) If X is a finite aspherical CW-complex and $G = \pi_1(X)$, then $\chi_G = \chi(X)[1]$ (example before Prop. 6.2). Thus by Lemma 6.23, if $\chi(X) \neq 0$ then $Z(G) = 1$ (Gottlieb's Theorem).

The homological characteristic

The function μ is obtained by taking the coefficient of [1] in χ_G. Another way of obtaining an element of R, which we now investigate, is to take the sum of all coefficients in χ_G.

Definition. *Let $G \in \mathrm{FP}(R)$. Then $\tilde{\chi}(G) = \sum_{\tau \in T(G)} \chi_G(\tau)$.*

Lemma 6.24. *If $G \in \mathrm{FP}(R)$ and $H_i(G, R)$ is of type FP over R for all i, then*

$$\tilde{\chi}(G) = \sum_{i \geq 0}(-1)^i h_i$$

where $h_i = r_{H_i(G,R)/R}$.

Proof. Let $0 \to P_n \to \ldots \to P_0 \to R \to 0$ be an augmented finite RG-projective resolution. Apply $R \otimes_{RG} -$ to obtain a complex

$$\ldots 0 \to \tilde{P}_n \xrightarrow{d_n} \ldots \to \tilde{P}_1 \xrightarrow{d_1} \tilde{P}_0 \xrightarrow{d_0} R \to 0 \ldots$$

(where $\tilde{P}_i = R \otimes_{RG} P_i$), with $H_*(G, R) = H(\tilde{P}, d)$. By Lemma 2.11 applied to the augmentation map $\varepsilon : RG \to R$, $\tilde{P}_i$ is R-projective and

$$\tilde{\chi}(G) = \varepsilon_*(\chi_G) = \sum_{i \geq 0}(-1)^i \varepsilon_*(r_{P_i/RG}) = \sum_{i \geq 0}(-1)^i r_{\tilde{P}_i/R}.$$

There are short exact sequences

$$0 \to \mathrm{Ker}(d_i) \to \tilde{P}_i \to \mathrm{Im}(d_i) \to 0 \tag{1}$$

$$0 \to \mathrm{Im}(d_{i+1}) \to \mathrm{Ker}(d_i) \to H_i(G, R) \to 0 \tag{2}$$

and it follows by induction from Theorem 4.7 that $\mathrm{Im}(d_i)$ and $\mathrm{Ker}(d_i)$ are of type FP over R for all $i \geq 0$. Let $z_i = r_{\mathrm{Ker}(d_i)/R}$, $p_i = r_{\tilde{P}_i/R}$ and $b_i = r_{\mathrm{Im}(d_i)/R}$. By Theorem 4.8 applied to these short exact sequences,

$$\left.\begin{aligned} p_i &= z_i + b_i \\ z_i &= b_{i+1} + h_i \end{aligned}\right\} \quad \text{which implies } p_i = b_i + b_{i+1} + h_i$$

hence $\tilde{\chi}(G) = \sum (-1)^i p_i = \sum (-1)^i h_i$. □

Corollary 6.25. *If R is a principal ideal domain then*

$$\tilde{\chi}(G) = \sum_{i \geq 0} (-1)^i (\text{free rank of } H_i(G, R)) 1_R.$$

Proof. The exact sequences (1) and (2) in the previous proof show that $H_i(G, R)$ is a finitely generated R-module. Also, any finitely generated R-module M is of type FL over R. For, by the well-known theory of such modules, we can find a short exact sequence

$$0 \to N \to F \to M \to 0$$

where F is free, with basis $e_1, \ldots, e_k$ and N is free with basis $\theta_1 e_1, \ldots, \theta_r e_r$ for some $r \leq k$, where $\theta_1 | \theta_2 | \ldots | \theta_r$ (and "|" means "divides"). This shows M is of type FL, and $r_{M/R} = (k - r)1_R =$ (free rank of M)1_R. The result now follows from Lemma 6.24. □

In general $\tilde{\chi}$ does not satisfy the index formula $\tilde{\chi}(H) = (G : H)\tilde{\chi}(G)$ (where $(G : H) < \infty$). For example, if $R = \mathbb{Q}$ and G is finite, then $\tilde{\chi}(G) = \tilde{\chi}(H) = 1$ by Lemma 6.9. However, this formula holds when $R = \mathbb{Z}$, and is the basis for our second definition of an Euler characteristic. The proof uses a theorem of Swan:

> *If G is finite and P is finitely generated $\mathbb{Z}G$-projective, then $\mathbb{Q} \otimes_{\mathbb{Z}} P$ is finitely generated $\mathbb{Q}G$-free.*

This will be generalised in the next section. A lemma of Hattori on characters is needed, and we finish this section by proving it.

Definition. *Let R be a commutative ring, A an R-algebra, M an A-module which is finitely generated projective as an R-module. The* character afforded *by M is the map $\varphi_M : A \to R$, $a \mapsto T_{M/R}(\lambda_a)$, where $\lambda_a \in \mathrm{End}_R(M)$ is defined by $\lambda_a(m) = am$ for $m \in M$.*

Lemma 6.26. *In the definition just given, if A is finitely generated projective as R-module, then so is every finitely generated projective A-module, and if P is such a module, then*

$$\varphi_P(a) = T_{A/R}(\lambda_a \rho_b), \text{ for any } b \in A \text{ such that } T_A(b) = r_{P/A}.$$

Proof. Let $\begin{cases} \{a_i\},\ \{f_i\} & \text{be a finite } R\text{-coordinate system for } A \\ \{x_j\},\ \{g_j\} & \text{be a finite } A\text{-coordinate system for } P. \end{cases}$

Then $\{a_i x_j\}$, $\{f_i g_j\}$ is a finite R-coordinate system for P. Hence

$$\varphi_P(a) = T_{P/R}(\lambda_a) = \sum_{i,j} f_i g_j(\lambda_a(a_i x_j)) \quad \text{by Lemma 2.5}$$

$$= \sum_{i,j} f_i(g_j(aa_i x_j)) = \sum_i f_i(aa_i b), \text{ where } b = \sum_j g_j(x_j)$$

$$(\text{so } r_{P/A} = T_A(b) \text{ by Lemma 2.5})$$

$$= \sum_i f_i(\lambda_a \rho_b(a_i)) = T_{A/R}(\lambda_a \rho_b), \quad \text{again by Lemma 2.5.}$$

If $T_A(b') = T_A(b)$, then $b - b' = b''$, where $b'' \in [A, A]$, and it is enough by Lemma 2.7(2) to show that $T_{A/R}(\lambda_a \rho_{b''}) = 0$, since $\lambda_a \rho_b - \lambda_a \rho_{b'} = \lambda_a \rho_{b''}$. Again by Lemma 2.7(2) we can assume $b'' = b_1 b_2 - b_2 b_1$, where $b_1, b_2 \in A$. Then

$$\lambda_a \rho_{b''} = \lambda_a \rho_{b_1} \rho_{b_2} - \lambda_a \rho_{b_2} \rho_{b_1} = (\lambda_a \rho_{b_1})\rho_{b_2} - \rho_{b_2}(\lambda_a \rho_{b_1}),$$

and the result follows by Lemma 2.7(2) and (3). □

Lemma 6.27 (Hattori)**.** *Let G be a finite group, R a commutative ring. Let P be a finitely generated projective RG-module and let φ_P be the character afforded by P. Then for $g \in G$,*

$$\varphi_P(g) = |C_G(g)| r_P(g^{-1}).$$

Proof. Take $b = \sum_{x \in G} b_x x \in RG$ such that $r_P = T_{RG}(b)$. By Lemma 6.26, if $\kappa_{g,v}(u) = guv$ for $g \in G$, $u, v \in RG$, then

$$\varphi_P(g) = T_{RG/R}(\kappa_{g,b}) = \sum_{x \in G} b_x T_{RG/R}(\kappa_{g,x}).$$

Now the map $G \to G$, $u \mapsto gux$ is a permutation of G, so by Lemma 2.6, using the R-basis G for RG,

$$T_{RG/R}(\kappa_{g,x}) = |\{u \in G \mid gux = u\}| = |\{u \in G \mid x = ug^{-1}u^{-1}\}|$$

$$= \begin{cases} 0 & \text{if } x \notin [g^{-1}] \\ |C_G(g)| & \text{if } x \in [g^{-1}]. \end{cases}$$

Therefore $\varphi_P(g) = \sum_{x \in [g^{-1}]} b_x |C_G(g)| = |C_G(g)|\ r_P(g^{-1})$. □

Corollary 6.28 (taking $g = 1$). *In Lemma* 6.27, $r_{P/R} = |G| r_P(1)$. □

7. Swan's Theorem

First, we prove a more general version of this theorem than that stated after Cor. 6.25, and this is the version which will be used. Then we prove Hattori's generalisation, which is an interesting use of the rank element defined in Section 1. The basic idea is that under certain circumstances, the rank element determines a finitely generated projective module up to isomorphism. An account of the material needed on linear topologies and completions is given, since it is difficult to extract all relevant results from a single source.

Denote by $\mathrm{Jac}(M)$ the Jacobson radical of an A-module M. We begin by quoting a well-known result.

Nakayama's Lemma. *Let M be a finitely generated module over a ring A, N a submodule of M, I an ideal in A. If $I \subseteq \mathrm{Jac}(A)$ and $M = N + IM$, then $M = N$.*

Proof. See [32, 10.3] □

Lemma 7.1. *Let A be a ring, I an ideal with $I \subseteq \mathrm{Jac}(A)$ and let P, Q be finitely generated projective A-modules. Then*

(i) *if $f : P \to Q$ is an A-homomorphism, the induced map $\overline{f} : P/IP \to Q/IQ$ is onto (resp. an isomorphism) if and only if f is onto (resp. an isomorphism);*
(ii) *$P \cong Q$ if and only if $P/IP \cong Q/IQ$.*

Proof. (i) If $\overline{f}$ is onto then $Q = f(P) + IQ$, so $Q = f(P)$ by Nakayama's Lemma. If $\overline{f}$ is an isomorphism then it is onto, so there there is a short exact sequence

$$0 \longrightarrow K \longrightarrow P \underset{\varphi}{\overset{f}{\rightleftarrows}} Q \longrightarrow 0$$

where $K = \mathrm{Ker}(f)$, which splits (because Q is projective), via φ, say. Thus $P = \varphi(Q) \oplus K$, hence $IP = \varphi(IQ) \oplus IK$ (φ is one-to one). Also, $\overline{f}$ is one-to-one, which means that $f(p) \in IQ$ implies $p \in IP$, hence $IP = \varphi(IQ) \oplus K$. Hence $K = IK$, so $K = 0$ by Nakayama's Lemma.

(ii) Let $g : P/IP \xrightarrow{\cong} Q/IQ$. Since P is projective, there is a map f making the following diagram commutative:

$$\begin{array}{ccc} P & \overset{f}{\dashrightarrow} & Q \\ \downarrow & & \downarrow \\ P/IP & \underset{g}{\longrightarrow} & Q/IQ. \end{array}$$

But then $g = \overline{f}$ and (ii) follows from (i). □

Definition. *A ring A is* local *if $A/\operatorname{Jac}(A)$ is a division ring.*

If A is commutative, A is local if and only if it has a unique maximal ideal (the maximal ideal being $\operatorname{Jac}(A)$).

Lemma 7.2. *Over a local ring, all finitely generated projective modules are free.*

Proof. Let P be finitely generated projective over a local ring A, $J = \operatorname{Jac}(A)$. Then P/JP is a vector space over the division ring A/J, of finite dimension, say n. Hence $P/JP \cong A^n/JA^n$, so $P \cong A^n$ by Lemma 7.1. □

Lemma 7.3. *Let K be a field of characteristic $p > 0$ and let G be a finite p-group. Then KG is a local ring.*

Proof. Because of the augmentation sequence $0 \to I_G \to KG \to K \to 0$ and the fact that K is a simple KG-module (being a field), I_G is a maximal left ideal, so $\operatorname{Jac}(KG) \subseteq I_G$. It is enough to show I_G is nilpotent, for then $I_G = \operatorname{Jac}(KG)$ (see [8, Ch.8, Prop. 3.3]). Choose an element $c \in Z(G)$ (the centre of G) of order p and let $C = \langle c \rangle$, $\overline{G} = G/C$. By Lemma 6.18 and the observations preceding Lemma 6.19, there is a short exact sequence $0 \to KGI_C \to I_G \to I_{\overline{G}} \to 0$. If $u \in KG$ then $(u(c-1))^p = u^p(c-1)^p = u^p(c^p - 1) = 0$, since $c - 1$ is central and K has characteristic p, hence KGI_C is nilpotent. Consequently, if $I_{\overline{G}}$ is nilpotent then so is I_G. The result now follows by induction on n, where $|G| = p^n$. □

Lemma 7.4. *Let A be a ring, suppose $e_1 \in A$ and let $f_1 = 1 - e_1$. Let R be the subring of A generated by e_1. Then for all $n > 0$, there exist $e_n, f_n \in A$ such that:*

(i) $e_n \in Re_1^n$ *and* $f_n \in Rf_1^n$;
(ii) $1 = e_n + f_n$.

Proof. Since e_1, f_1 commute, the Binomial Theorem applies:

$$1 = (e_1 + f_1)^{2n-1} = \underbrace{\sum_{r=0}^{n-1} \binom{2n-1}{r} e_1^r f_1^{2n-1-r}}_{f_n} + \underbrace{\sum_{r=n}^{2n-1} \binom{2n-1}{r} e_1^r f_1^{2n-1-r}}_{e_n}$$

and e_n, f_n can be taken as indicated. □

Now let R be a commutative ring. Recall that the *nilradical* of R, denoted $\mathrm{rad}(R)$, is the set of all nilpotent elements of R, and is an ideal. It is equal to the intersection of all prime ideals of R ([64, Theorem 7.1]). Also, $\mathrm{Jac}(R)$ is the intersection of all maximal ideals of R, and since maximal ideals are prime, $\mathrm{rad}(R) \subseteq \mathrm{Jac}(R)$. Let $X = \mathrm{Spec}(R)$ (the set of all prime ideals of R), and for $U \subseteq R$, let $\mathcal{V}(U) = \{\mathfrak{p} \in X \mid \mathfrak{p} \supseteq U\}$. Note that $\mathcal{V}(U) = \mathcal{V}(I)$, where I is the ideal of R generated by U. We endow $\mathrm{Spec}(R)$ with the Zariski topology, where the closed sets are the sets $\mathcal{V}(U)$ for $U \subseteq R$. Also, for $r \in R$, let $X_r = X \setminus \mathcal{V}(r) = \{\mathfrak{p} \in X \mid r \notin \mathfrak{p}\}$.

Note. *If* $\mathfrak{p} \in X$ *and* $e \in R$ *is an idempotent, then* $e(1 - e) = 0$, *so either* $e \in \mathfrak{p}$ *or* $1 - e \in \mathfrak{p}$, *but not both (since* $1 \notin \mathfrak{p}$*).*

The term "clopen" is used to mean "closed and open".

Lemma 7.5. *A clopen set in* X *has the form* X_e *for some idempotent* $e \in R$.

Proof. Let U be clopen; then $U = \mathcal{V}(I)$, $X \setminus U = \mathcal{V}(J)$, where I, J are ideals. Clearly

$$\mathcal{V}(I + J) = \mathcal{V}(I \cup J) = \mathcal{V}(I) \cap \mathcal{V}(J) = \emptyset$$

which implies $I + J = R$, so $1 = e_1 + f_1$ for some $e_1 \in J$, $f_1 \in I$. Also (see [64, §7.5])

$$\mathcal{V}(IJ) = \mathcal{V}(I) \cup \mathcal{V}(J) = X$$

which implies $IJ \subseteq \mathrm{rad}(R)$, hence $(e_1 f_1)^n = 0$ for some n. Let e_n, f_n be as in Lemma 7.4. Then $e_n f_n = 0$ and $1 = e_n + f_n$, so e_n, f_n are orthogonal idempotents, and $e_n \in J$, $f_n \in I$. By the note preceding the lemma, $\mathcal{V}(I) = \mathcal{V}(1 - e_n) = X_{e_n}$. □

Again let $X = \mathrm{Spec}(R)$ and fix a finitely generated projective R-module P. For $\mathfrak{p} \in X$, $P_\mathfrak{p} \cong R_\mathfrak{p} \otimes_R P$ (see [64, §7.3]). Hence $P_\mathfrak{p}$ is a finitely generated projective $R_\mathfrak{p}$-module by Lemma 2.11, so is free by Lemma 7.2. Since R is commutative, the number of elements in a basis for $P_\mathfrak{p}$ is uniquely determined, and is called its rank, denoted $\mathrm{rk}_R(P_\mathfrak{p})$. (See the discussion at the start of Section 8.) Thus there is a map

$$d : X \to \mathbb{Z}$$
$$d(\mathfrak{p}) = \mathrm{rk}_R(P_\mathfrak{p}).$$

Lemma 7.6. *The map d is continuous, where $\mathbb{Z}$ is given the discrete topology.*

Proof. Suppose $d(\mathfrak{p}) = n$, so $P_\mathfrak{p}$ has a basis of the form $\frac{p_1}{s_1}, \dots, \frac{p_n}{s_n}$, with $p_i \in P$ and $s_i \in R \setminus \mathfrak{p}$. Since the s_i are units in $R_\mathfrak{p}$, $\frac{p_1}{1}, \dots, \frac{p_n}{1}$ is a basis for $P_\mathfrak{p}$. If F is the free R-module on $p_1, \dots, p_n$, the obvious map $f : F \to P$ induces an isomorphism $f_\mathfrak{p} : F_\mathfrak{p} \to P_\mathfrak{p}$. Hence $\mathrm{Coker}(f)_\mathfrak{p} = 0$, and $\mathrm{Coker}(f)$ is finitely generated, so there is a single element $s \in R \setminus \mathfrak{p}$ such that $s\,\mathrm{Coker}(f) = 0$. Therefore, there is an epimorphism

$$f[s^{-1}] : F[s^{-1}] \to P[s^{-1}]$$

which becomes an isomorphism when localised at $\mathfrak{p}$. ($R_\mathfrak{p}$ is the localisation of $R[s^{-1}]$; see [64, Prop. 7.4].) Since $P[s^{-1}]$ is projective over $R[s^{-1}]$, this epimorphism splits, so $\mathrm{Ker}(f[s^{-1}])$ is finitely generated. It becomes zero when localised at $\mathfrak{p}$, so is annihilated by an element of $R[s^{-1}] \setminus \mathfrak{p}[s^{-1}]$, hence by some $t \in R \setminus \mathfrak{p}$. Let $u = st$, so there is an isomorphism

$$f[u^{-1}] : F[u^{-1}] \xrightarrow{\cong} P[u^{-1}].$$

But then $F_\mathfrak{q} \cong P_\mathfrak{q}$ for all $\mathfrak{q} \in X$ such that $u \notin \mathfrak{q}$, since $R_\mathfrak{q}$ is a localisation of $R[u^{-1}]$, for all such $\mathfrak{q}$. Hence $d(\mathfrak{q}) = n$ for all $\mathfrak{q} \in X \setminus \mathcal{V}(u)$. Thus d is continuous at $\mathfrak{p}$, for all $\mathfrak{p} \in X$. □

Lemma 7.7. *Let P be a finitely generated projective module over a commutative ring R. Then r_P belongs to the subring of R generated by all idempotents.*

Proof. Let $d : X \to \mathbb{Z}$ be the continuous map in Lemma 7.6, where $X = \mathrm{Spec}(R)$. Since X is compact ([64, Prop. 7.12]),

$$X = d^{-1}(n_1) \cup \ldots \cup d^{-1}(n_k)$$

for some distinct integers $n_1, \dots, n_k$. By Lemma 7.5, $d^{-1}(n_i) = X_{e_i}$ for some idempotent $e_i \in R$. Now

$$X_{e_ie_j} = X_{e_i} \cap X_{e_j} = \emptyset \text{ implies } e_ie_j \in \mathrm{rad}(R), \text{ hence } e_ie_j = 0\ (i \neq j).$$

Note that if $\mathfrak{p} \in X$, then $\mathfrak{p} \in X_{e_i}$ if and only if $\mathfrak{p} \in \mathcal{V}(1 - e_i)$, if and only if $\mathfrak{p} \supseteq R(1 - e_i)$, if and only if $\mathfrak{p} \notin X_{1-e_i}$ (see the note preceding Lemma 7.5). Since $X = X_{e_1} \cup \ldots \cup X_{e_k}$, taking complements gives $X_{1-e_1} \cap \ldots \cap X_{1-e_k} = \emptyset$, which similarly implies $(1 - e_1)\ldots(1 - e_k) = 0$. Hence $1 = e_1 + \ldots + e_k$. Let $\mathfrak{p} \in X_{e_i}$; under the canonical map $R \to R_{\mathfrak{p}}$, $r_{P/R} \mapsto d(\mathfrak{p}).1 = n_i.1$ by Lemma 2.11. Therefore there exists $s \notin \mathfrak{p}$ such that $s(r_P - n_i) = 0$.

Denote the image of $a \in R$ in the quotient $\overline{R} = R/R(1 - e_i)$ by $\overline{a}$. It follows that $\overline{r}_P - \overline{n}_i = 0$. (Otherwise $\mathrm{Ann}_{\overline{R}}(\overline{r}_P - \overline{n}_i) \subseteq \overline{\mathfrak{m}}$, where $\mathfrak{m}$ is a maximal ideal of R containing $R(1 - e_i)$ and $\overline{\mathfrak{m}}$ is its image in $\overline{R}$, so $\mathfrak{m} \in X_{e_i}$ and $\mathfrak{m} \supseteq \mathrm{Ann}_R(r_P - n_i)$, contradicting the previous paragraph.) Since $R = Re_i \oplus R(1 - e_i)$, $(r_P - n_i)e_i = 0$. Thus $r_P = \sum_{i=1}^{k} r_P e_i = \sum_{i=1}^{k} n_i e_i$. □

Theorem 7.8. *Let R be an integral domain, G a finite group such that no prime divisor of $|G|$ is invertible in R. Then RG has no idempotents except* 0 *and* 1.

Proof. Let F be the field of fractions of R. If R has characteristic $p > 0$ then G is a p-group, FG is local by Lemma 7.3, in fact we showed $I_G = \mathrm{Jac}(FG)$ is nilpotent. If e is an idempotent in FG, $e \neq 1$, then $\epsilon(e) = 0$, where ϵ is the augmentation map, so $e \in \mathrm{Jac}(FG)$, therefore $e = 0$.

Suppose R has characteristic 0. Let e be an idempotent in RG, $P = RGe$. By Cor. 6.28, $r_{P/R} = |G|\, r_{P/RG}(1)$, and by Lemma 2.11, $r_{P/R} = r_{F\otimes_R P/F}$, which is a non-negative integer, and is positive if $P \neq 0$ ($F \otimes_R P$ is the localisation of P at 0, so $F \otimes_R P = 0$ implies $P = 0$ since R-projective modules are torsion-free). Thus $\dfrac{r_{P/R}}{|G|} \in \mathbb{Q} \cap R$, and in fact is an integer by hypothesis. (Write $\dfrac{r_{P/R}}{|G|} = \dfrac{a}{b}$ with a, $b \in \mathbb{Z}$ and coprime. Suppose a prime p divides b. Then there are integers α, β such that $1 = \alpha a + \beta b$, so $\dfrac{1}{p} = \alpha\left(\dfrac{b}{p}\right)\left(\dfrac{a}{b}\right) + \beta\left(\dfrac{b}{p}\right) \in R$, a contradiction.) But $RG = P \oplus Q$, where $Q = RG(1 - e)$, so by Lemma 2.7, $r_{P/R} + r_{Q/R} = |G|$. By the above (applied to $1 - e$ as well as e), $r_{P/R}$, $r_{Q/R}$ are non-negative integers divisible by $|G|$, so one is zero, hence either P or Q is zero; that is, $e = 0$ or 1. □

In the situation of Theorem 7.8, let P be a finitely generated projective RG-module, F the field of fractions of R, $P_G = R \otimes_{RG} P$. By Lemma 2.11, P_G is finitely generated R-projective, and

$$r_{P_G/R} = \varepsilon_*(r_{P/RG}) = \sum_{\tau \in T(G)} r_P(\tau)$$

where $\varepsilon : RG \to R$ is the augmentation map. Let $n = \mathrm{rank}(P_G)$, that is, $n = \dim_F(F \otimes_R P_G)$. By Lemma 2.11 applied to the inclusion map $R \hookrightarrow F$,

$$n.1_R = \sum_{\tau \in T(G)} r_P(\tau). \qquad (*)$$

Lemma 7.9. *In these circumstances, $r_{P/RG} = r_{RG^n/RG} = T_{RG}(n)$.*

Proof. By $(*)$, it is enough to show $r_P(g) = 0$ for $1 \neq g \in G$. If R has characteristic $p > 0$ then by Lemmas 7.2 and 7.3, $F \otimes_R P$ is FG-free and the result follows from Cor. 2.14. Assume R has characteristic 0 and let $H = \langle g \rangle$ (where $1 \neq g \in G$). By Cor. 2.9 and Lemma 6.1, $r_{P/RH}(g) = r_{P/RG}(g)(C_G(g) : H)$, so it is enough to prove the result for the abelian group H. But then $R[H]$ is identified with RH, and by Lemma 7.7 and Theorem 7.8, $r_{P/RH}$ belongs to the prime subring of RH, which is contained in R. □

Theorem 7.10 (Swan). *Let R be an integral domain, G a finite group such that no prime divisor of $|G|$ is invertible in R. Let F be the field of fractions of R and let P be a finitely generated projective RG-module. Then $F \otimes_R P \cong FG^n$, where $n = \dim_F(F \otimes_R P_G)$.*

Proof. As in Lemma 7.9, we need only consider the case that R has characteristic 0. By Lemma 7.9, $r_{P/RG} = r_{RG^n/RG}$, where $n = \mathrm{rank}(P_G)$. Hence by Cor. 2.14, $r_{F \otimes_R P/FG} = r_{FG^n/FG}$. By Lemma 6.27, $F \otimes_R P$ and FG^n afford the same character, so are isomorphic. (See [73, Ch.XVIII, §2, Theorem 3].) □

Linear topologies and completions

Let A be a ring, M an A-module. A *filtration* of M is a sequence of submodules:

$$M = M_0 \supseteq M_1 \supseteq M_2 \supseteq \dots \qquad (*)$$

Given $(*)$, we can define a topology on M by taking the sets $x + M_n$ $(m \geq 0)$ as the open sets containing x, for each $x \in M$, and M becomes a topological group. Such a topology is called a *linear* topology on M. Note that the topology is Hausdorff if and only if $\bigcap_{i=0}^{\infty} M_i = 0$.

Definition. *A sequence $(x_n)_{n \geq 0}$ in M is* Cauchy *if for all $k \geq 0$, there exists an integer U such that $m, n > U$ implies $x_n - x_m \in M_k$. The sequence* converges *to $x \in M$ (written $x_n \longrightarrow x$) if for all k there is an integer U such that $n >$

U implies $x_n - x \in M_k$. The module M is complete *in this topology if every Cauchy sequence converges.*

Note. *A sequence $(x_n)_{n\geq 0}$ is Cauchy if and only if $x_{n+1} - x_n \longrightarrow 0$.*

In future we shall just write sequences as (x_n), etc. Given M and the filtration $(*)$, let C_M be the set of all Cauchy sequences in M. Then C_M is an A-module, with operations

$$(x_n) + (y_n) = (z_n), \quad \text{where } z_n = x_n + y_n \text{ for all } n$$

$$a(x_n) = (w_n), \quad \text{where } w_n = ax_n \text{ for all } n,$$

(for $a \in A$). Let $Z_M = \{(x_n) \in C_M \mid x_n \longrightarrow 0\}$, an A-submodule of C_M.

Definition. *The quotient module $\widehat{M} = C_M/Z_M$ is called the* completion *of M.*

There is an A-homomorphism $\varphi_M : M \to \widehat{M}$, $x \mapsto \widehat{x} + Z_M$, where $\widehat{x}$ is the constant sequence (x_n), with $x_n = x$ for all n. We leave it as an exercise to check that:

- $\mathrm{Ker}(\varphi_M) = \bigcap_{i=0}^{\infty} M_i$, so φ_M is one-to-one if and only if M is Hausdorff;
- φ_M is onto if and only if M is complete.

If N is a submodule of M, define

$$\begin{aligned}\widetilde{N} &= \{(x_n) + Z_M \mid \text{there exists } U \text{ such that } n > U \text{ implies } x_n \in N\} \\ &= \{(x_n) + Z_M \mid x_n \in N \text{ for all } n\}.\end{aligned}$$

Then $\widehat{M}$ has a linear topology defined by the filtration $\{\widetilde{M}_i\}$.

Lemma 7.11. *If $u_n = \varphi_M(x_n)$ for $n \geq 0$, then (u_n) is a Cauchy sequence in $\widehat{M}$ if and only if (x_n) is a Cauchy sequence in M, in which case $u_n \longrightarrow (x_n) + Z_M$ in $\widehat{M}$.*

Proof. If $(y_n) \in Z_M$, then for any k, $y_n \in M_k$ for sufficiently large n. Hence, for $m, n \geq 0$, $u_m - u_n = \varphi_M(x_m - x_n) \in \widetilde{M}_k$ if and only if $x_m - x_n \in M_k$, so (u_n) is Cauchy if and only if (x_n) is. If (x_n) is Cauchy, given k, choose an integer V such that $m, n > V$ implies $x_m - x_n \in M_k$. Let $x = (x_n) + Z_M$. Then $n > V$ implies $x - u_n \in \widetilde{M}_k$, so $u_n \longrightarrow x$. □

Lemma 7.12. *In the circumstances above,*

(1) *if N is a submodule of M then the closure $\overline{\varphi_M(N)}$ equals $\widetilde{N}$. In particular, $\varphi_M(M)$ is dense in $\widehat{M}$;*
(2) *the completion $\widehat{M}$ is complete and Hausdorff.*

Proof. Clearly $\bigcap_k \widetilde{M}_k = 0$, so $\widehat{M}$ is Hausdorff. To prove (1), take $x = (x_n) + Z_M \in \widetilde{N}$, where $x_n \in N$ for all n. Then $u_n = \varphi_M(x_n) \in \varphi_M(N)$ and $u_n \longrightarrow x$ by Lemma 7.11, so $x \in \overline{\varphi_M(N)}$. Conversely, let $u_n = \varphi(x_n)$, where $x_n \in N$ and suppose $u_n \longrightarrow u \in \widehat{M}$. By Lemma 7.11, $u_n \longrightarrow (x_n) + Z_M$, and since $\widehat{M}$ is Hausdorff, $u = (x_n) + Z_M \in \widetilde{N}$. It is clear from the definition that a linear topology is first countable, so this proves (1). Finally, take any Cauchy sequence (u_n) in $\widehat{M}$. By (1), there exists $x_n \in M$ such that $\varphi_M(x_n) - u_n \in \widetilde{M}_n$. Then $(\varphi_M(x_n))$ is Cauchy and converges to x, where $x = (x_n) + Z_M$, by Lemma 7.11, hence $u_n \longrightarrow x$, showing $\widehat{M}$ is complete. □

Let M, N be A-modules with linear topologies defined by filtrations (M_i), (N_i) respectively. Let $f : M \to N$ be a homomorphism. Then f is continuous if and only if it is continuous at 0, which is equivalent to: for all k, there exists l such that $f(M_l) \subseteq N_k$. Suppose f is continuous. Then there is an A-homomorphism $\widehat{f} : \widehat{M} \to \widehat{N}$ given by

$$\widehat{f}((x_n) + Z_M) = (f(x_n)) + Z_N.$$

Moreover, $\widehat{f}$ is continuous since $f(M_l) \subseteq N_k$ implies $\widehat{f}(\widetilde{M}_l) \subseteq \widetilde{N}_k$. By Lemma 7.12(1), $\widehat{f}$ is the unique continuous homomorphism such that

$$\begin{array}{ccc} \widehat{M} & \xrightarrow{\widehat{f}} & \widehat{N} \\ {\scriptstyle \varphi_M}\uparrow & & \uparrow{\scriptstyle \varphi_N} \\ M & \xrightarrow[f]{} & N. \end{array}$$

is commutative. (This makes $M \mapsto \widehat{M}$ into a functor, between categories we leave the reader to describe.) If $f : M \to N$ is an A-homomorphism and (M_i) is a filtration on M, then $(f(M_i))$ is a filtration on N, adding N as the first term of the filtration if necessary, and f becomes continuous. If L is a submodule of M, $(L \cap M_i)$ is a filtration on L, and gives L the relative (subspace) topology.

Note. *The inclusion map $i : L \to M$ induces $\widehat{i} : \widehat{L} \to \widehat{M}$ and if $(x_n) \in C_L \cap Z_M$, clearly $(x_n) \in Z_L$, so $\widehat{i}$ is injective, and its image is $\widetilde{L}$. By Lemma* 7.12, *$\widehat{L} \cong \overline{\varphi_M(L)}$.*

Lemma 7.13. *Suppose $0 \longrightarrow L \xrightarrow{i} M \xrightarrow{\pi} N \longrightarrow 0$ is an exact sequence of A-modules, and M has a linear topology induced by (M_i). Let L, N have the topologies indicated above (viewing i as an inclusion map). Then*

$$0 \longrightarrow \widehat{L} \xrightarrow{\widehat{i}} \widehat{M} \xrightarrow{\widehat{\pi}} \widehat{N} \longrightarrow 0$$

is exact.

Proof. Clearly $\widehat{\pi}\,\widehat{i} = \widehat{\pi i} = 0$ and $\widehat{i}$ is injective (note above). We show that $\mathrm{Ker}(\widehat{\pi}) \subseteq \mathrm{Im}(\widehat{i}\,)$. Let $(x_n) \in C_M$ be such that $(\pi(x_n)) \in Z_N$. Then $\pi(x_n) \in \pi(M_{\alpha(n)})$, where $\alpha(n) \longrightarrow \infty$ as $n \longrightarrow \infty$. Therefore we can write $x_n = u_n + i(z_n)$, where $u_n \in M_{\alpha(n)}, z_n \in L$. Then $(u_n) \in Z_M$ and $i(z_n) = x_n - u_n$, so $(i(z_n)) \in C_M$, hence $(z_n) \in C_L$. Thus $(x_n) + Z_M = (i(z_n)) + Z_M = \widehat{i}((z_n)) \in \mathrm{Im}(\widehat{i}\,)$. It remains to show that $\widehat{\pi}$ is onto. Let $(y_n) \in C_N$ and choose $x_n \in M$ such that $\pi(x_n) = y_n$. then $(\pi(x_{n+1} - x_n)) \in Z_N$, so as above we can write $x_{n+1} - x_n = u_n + i(z_n)$, where $(u_n) \in Z_M, z_n \in L$. Let $s_n = x_1 + \sum_{j=1}^{n-1} u_j$. Then $(s_{n+1} - s_n) \longrightarrow 0$, so $(s_n) \in C_M$, and $\pi(s_n) = y_n$, so $\widehat{\pi}((s_n)) = (y_n) + Z_N$. □

For an elegant proof of Lemma 7.13, and interpretation of $\widehat{M}$ as $\varprojlim M/M_i$, see [6, Ch.10]

Topologies determined by ideals

Suppose $A = I_0 \supseteq I_1 \supseteq I_2 \supseteq \ldots$ is a filtration of the ring A, where all I_j are ideals of A. Then, with the topology determined by this filtration, A is a topological ring (this is left as an exercise). Also, C_A is a ring, with multiplication

$$(a_n)(b_n) = (c_n), \qquad \text{where } c_n = a_n\, b_n \text{ for all } n.$$

Further, Z_A is an ideal in C_A, so $\widehat{A}$ is a ring, and φ_A is a ring homomorphism. Obviously, if A is commutative, so is $\widehat{A}$. Let M be an A-module, (M_i) a filtration on M which is compatible in the sense that $I_n M \subseteq M_n$ for all n. Then C_M is a C_A-module, with scalar multiplication $(a_n)(m_n) = (a_n m_n)$, Z_M is a C_A submodule and $Z_A C_M \subseteq Z_M$. Hence $\widehat{M}$ is an $\widehat{A}$-module. If N is another A-module with a compatible filtration and $f : M \to N$ is a continuous A-homomorphism, then $\widehat{f}$ is an $\widehat{A}$-homomorphism. If B is a ring with a filtration $B = J_0 \supseteq J_1 \supseteq J_2 \supseteq \ldots$ by ideals, and $f : A \to B$ is a continuous ring homomorphism, then $\widehat{f} : \widehat{A} \to \widehat{B}$ is a ring homomorphism.

Definition. *Let I be an ideal in the ring A. The I-*adic topology *on A is that defined by the filtration $(I^n)_{n\geq 0}$. The I-*adic topology *on an A-module M is that defined by $(I^n M)_{n\geq 0}$. A filtration (M_n) of M is I-*admissible *if $I M_n \subseteq M_{n+1}$ for all n. An I-admissible filtration is called* essentially I-adic *(or I-*stable*) if there exists r such that $I M_n = M_{n+1}$ for $n > r$.*

Exercise.

(1) An I-admissible filtration is compatible with the I-adic filtration on A.
(2) An essentially I-adic filtration induces the I-adic topology on M.

Example. Let R be a ring, $A = R[x_1, \ldots, x_n]$ the polynomial ring in n variables. Let I be the ideal generated by $\{x_1, \ldots, x_n\}$, so $I = \{f \in A \mid f(\underline{0}) = 0\}$. Then the I-adic completion $\widehat{A}$ is $R[[x_1, \ldots, x_n]]$, the formal power series ring in n variables. To see this, if $(a_m) \in C_A$ and $x_1^{k_1} \ldots x_n^{k_n}$ is a monomial, the coefficient of this monomial in a_m is independent of m for sufficiently large m. Call it $\alpha_{\underline{k}}$ (where $\underline{k} = (k_1, \ldots, k_n)$). Then the map

$$C_A \to R[[x_1, \ldots, x_n]]$$
$$(a_m) \mapsto \sum_{\underline{k}} \alpha_{\underline{k}} x_1^{k_1} \ldots x_n^{k_n}$$

is a ring homomorphism inducing an isomorphism $\widehat{A} \cong R[[x_1, \ldots, x_n]]$.

If $f : M \to N$ is an A-homomorphism, and M, N have the I-adic topologies, then f is continuous. Hence I-adic completion is a functor from the category of A-modules to the category of $\widehat{A}$-modules. Also, if f is an epimorphism, the I-adic filtration on M induces the I-adic topology on N via f (because $f(I^n M) = I^n f(M)$).

Artin-Rees Lemma. *Let A be a commutative Noetherian ring, let I be an ideal in A, let M be a finitely generated A-module and let N be a submodule of M. Then the filtration on N induced by the I-adic filtration on M is essentially I-adic. That is, there exists r such that for $n > r$, $I^n M \cap N = I^{n-r}(I^r M \cap N)$.*

Proof. See [6, Cor. 10.10] □

Lemma 7.14. *Let A be a commutative Noetherian ring and let I be an ideal in A. If $0 \longrightarrow L \xrightarrow{i} M \xrightarrow{\pi} N \longrightarrow 0$ is an exact sequence of finitely generated A-modules then*

$$0 \longrightarrow \widehat{L} \xrightarrow{\widehat{i}} \widehat{M} \xrightarrow{\widehat{\pi}} \widehat{N} \longrightarrow 0$$

is exact (where ^ denotes I-adic completion).

Proof. This is immediate from Lemma 7.13 and the Artin-Rees Lemma. □

Note. *If $0 \longrightarrow L \longrightarrow M \longrightarrow N \longrightarrow 0$ is a split exact sequence of A-modules (where A is any ring), L is a summand of M and the I-adic filtration on M induces the I-adic filtration on L, so $0 \longrightarrow \widehat{L} \xrightarrow{\widehat{i}} \widehat{M} \xrightarrow{\widehat{\pi}} \widehat{N} \longrightarrow 0$ is exact by Lemma 7.13. Also, it is split since I-adic completion is a functor. Hence I-adic completion commutes with finite direct sums.*

Let A be any ring, I an ideal in A, M an A-module. There is an $\widehat{A}$-homomorphism

$$\alpha_M : \widehat{A} \otimes_A M \to \widehat{M}, \quad \alpha_M(\widehat{a} \otimes m) = \widehat{a}\varphi_M(m),$$

where $\hat{}$ denotes I-adic completion and $\widehat{A}$ is a right A-module via φ_A. This gives a natural transformation α from $\widehat{A} \otimes_A -$ to the functor $M \mapsto \widehat{M}$.

Lemma 7.15. *Let A be a ring, I an ideal in A, M a finitely generated A-module. Then the map α_M just defined is onto. If A is commutative Noetherian, α_M is an isomorphism.*

Proof. There is a short exact sequence $0 \to N \to F \to M \to 0$ where $F \cong A^n$ for some integer n, giving a commutative diagram with top row exact:

$$\begin{array}{ccccccccc} & & \widehat{A} \otimes_A N & \longrightarrow & \widehat{A} \otimes_A F & \longrightarrow & \widehat{A} \otimes_A M & \longrightarrow & 0 \\ & & \downarrow{\scriptstyle \alpha_N} & & \downarrow{\scriptstyle \alpha_F} & & \downarrow{\scriptstyle \alpha_M} & & \\ 0 & \longrightarrow & \widehat{N} & \longrightarrow & \widehat{F} & \longrightarrow & \widehat{M} & \longrightarrow & 0 \end{array}$$

(since tensor product is right exact). Also, the map $\widehat{F} \to \widehat{M}$ is surjective by Lemma 7.13. Since α_A is an isomorphism, α_F is, by the note after Lemma 7.14, therefore α_M is an epimorphism by commutativity of the right-hand square. If A is commutative Noetherian, N is finitely generated, so α_N is an epimorphism by the first part, and the bottom row is exact by Lemma 7.14. The lemma follows by "diagram chasing" (see [100, Lemma 3.32]). □

Corollary 7.16. *Let R be a commutative Noetherian ring, I an ideal in R, M a finitely generated R-module and let $\hat{}$ denote I-adic completion. Then*

(1) *$\widehat{R}$ is a flat R-algebra;*
(2) *if R is complete Hausdorff in the I-adic topology, then so is M;*
(3) *if N is a submodule of M, then $\widehat{N} \cong \overline{\varphi_M(N)} = \widehat{R}\varphi_M(N)$;*
(4) *the topology on $\widehat{R}$ is the $\widehat{R}I$-adic topology.*

Proof. (1) By Lemmas 7.14 and 7.15, $\widehat{R} \otimes_R -$ is exact on the category of finitely generated R-modules, and (1) follows (see [100, Theorem 3.53]).

(2) If R is complete Hausdorff then φ_R is an isomorphism (earlier exercise), so $\beta : M \to \widehat{R} \otimes_R M$ defined by $m \mapsto 1 \otimes m$ is an isomorphism. It follows that φ_M is an isomorphism, since there is a commutative diagram

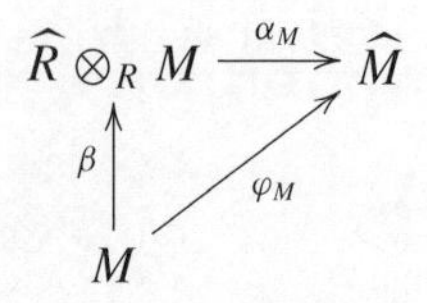

hence M is complete Hausdorff.

(3) Let $i : N \to M$ be the inclusion map. Since R is Noetherian, N is finitely generated. There is a commutative diagram

$$\begin{array}{ccc} \widehat{R} \otimes_R N & \xrightarrow{1 \otimes i} & \widehat{R} \otimes_R M \\ {\scriptstyle \alpha_N} \downarrow & & \downarrow {\scriptstyle \alpha_M} \\ \widehat{N} & \xrightarrow{\;\widehat{i}\;} & \widehat{M}. \end{array}$$

By Lemma 7.14, $\widehat{i}$ is injective, so $\widehat{i} : \widehat{N} \to \overline{\varphi_M(N)}$ is an isomorphism by the note preceding Lemma 7.13. Since α_N is an isomorphism (by Lemma 7.15), $\widehat{i}(\widehat{N}) = \alpha_M(1 \otimes i)(\widehat{R} \otimes_R N) = \widehat{R}\varphi_M(N)$.

(4) By definition, the topology on $\widehat{R}$ is that defined by the filtration $(\widetilde{I^k})_{k \geq 0}$. By Lemma 7.12,

$$\begin{aligned} \widetilde{I^k} &= \overline{\varphi_R(I^k)} \\ &= \widehat{R}\varphi_R(I^k) \quad \text{by (3)} \\ &= \widehat{R} I^k \\ &= (\widehat{R} I)^k. \end{aligned}$$

□

Corollary 7.17. *Again let R be commutative Noetherian, I an ideal, M a finitely generated R-module and let $\widehat{}$ denote I-adic completion. Then $\widehat{IM} \cong I\widehat{M}$ and $M/IM \cong \widehat{M}/I\widehat{M}$.*

Proof. By Lemma 7.14 there is a short exact sequence:

$$0 \longrightarrow \widehat{IM} \xrightarrow{\;\widehat{i}\;} \widehat{M} \longrightarrow \widehat{M/IM} \longrightarrow 0.$$

But M/IM is discrete in the I-adic topology, so complete and Hausdorff, hence $\varphi_{M/IM} : M/IM \to \widehat{M/IM}$ is an isomorphism. By Cor. 7.16(3), $\widehat{i}$ maps $\widehat{IM}$ isomorphically to $\widehat{R}\varphi_M(IM) = I\widehat{R}\varphi_M(M) = I\widehat{M}$. □

Remark 7.18. *If, in Cor. 7.17, A is an R-algebra, finitely generated as R-module, and M is a finitely generated A-module, then $\widehat{M}$ is the AI-adic* *completion of M, all homomorphisms in the proof are A-homomorphisms and so M/IM is A-isomorphic to $\widehat{M}/I\widehat{M}$ (because $(AI)^k M = I^k AM = I^k M$, so the I-adic topology on M coincides with the AI-adic topology).*

If R is a commutative local ring, $R/\operatorname{Jac}(R)$ is called the *residue field* of R.

Lemma 7.19. *Let R be a commutative Noetherian ring, I an ideal in R, $\widehat{R}$ the I-adic completion. Then:*

(1) *$\widehat{R}$ is commutative Noetherian;*
(2) *if R is local and $I = \operatorname{Jac}(R)$, then $\widehat{R}$ is local, $\operatorname{Jac}(\widehat{R}) = \widehat{R}I$ and R, $\widehat{R}$ have the same residue field.*

Proof. (1) First, I is finitely generated, by $y_1, \dots, y_n$ say. Let $B = R[x_1, \dots, x_n]$ be the polynomial ring, with the J-adic topology, where J is the ideal generated by $x_1, \dots, x_n$, so $\widehat{B} = R[[x_1, \dots, x_n]]$ (see the example preceding the Artin-Rees Lemma. Let $f : B \to R$ be the epimorphism given by $f|_R = \operatorname{id}_R$, $f(x_i) = y_i$ for $1 \leq i \leq n$. Then $f(J^n) = I^n$, so the I-adic topology is induced by the J-adic topology on B via f. By Lemma 7.13, the resulting ring homomorphism $\widehat{f} : \widehat{B} \to \widehat{R}$ is surjective. Since $\widehat{B}$ is Noetherian (see [73, Ch.VI,§3]), so is $\widehat{R}$.

(2) By Cor. 7.17, $\widehat{R}/\widehat{R}I \cong R/I$ (as rings), so $\widehat{R}/\widehat{R}I$ is a field, hence $\operatorname{Jac}(\widehat{R}) \subseteq \widehat{R}I$. Thus it is enough to show that $\widehat{R}I \subseteq \operatorname{Jac}(\widehat{R})$ ($\widehat{R}/\operatorname{Jac}(\widehat{R})$ is then a field, so $\widehat{R}$ is local). Let $x \in \widehat{R}I$. By Cor. 7.16(4), $x^n \longrightarrow 0$ as $n \longrightarrow \infty$ in $\widehat{R}$. Also,

$$(1 + (-1)^n x^{n+1}) = (1 + x)(1 - x + x^2 + \dots + (-1)^n x^n) = (1 + x)s_n$$

say. Now $\widehat{R}$ is Hausdorff by Lemma 7.12, so we can take limits, to see that $(1 + x)$ is invertible with inverse $1 - x + x^2 - x^3 + \dots$ ($s_{n+1} - s_n \to 0$, so (s_n) is Cauchy, hence the series converges since $\widehat{R}$ is complete by Lemma 7.12). It follows that $\widehat{R}I \subseteq \operatorname{Jac}(\widehat{R})$ (see [8, Ch.8, Prop. 3.1]). □

Lemma 7.20 (Krull Intersection Theorem). *Let R be a commutative Noetherian ring, I an ideal contained in $\operatorname{Jac}(R)$, M a finitely generated R-module. Then the I-adic topology on M is Hausdorff.*

Proof. Let $N = \bigcap_{n=0}^{\infty} I^n M$. By the Artin-Rees Lemma, there exists r such that, for $n > r$,

$$N = I^n M \cap N = I^{n-r}(I^r M \cap N) = I^{n-r} N \subseteq IN \subseteq N$$

so $N = IN$. By Nakayama's Lemma, $N = 0$, as required. □

Terminology. We shall use "complete local ring" to mean a commutative Noetherian local ring R, complete with respect to the I-adic topology, where $I = \mathrm{Jac}(R)$. This topology is Hausdorff by Lemma 7.20.

Proposition 7.21. *Let R be a complete local ring with $I = \mathrm{Jac}(R)$ its maximal ideal and let $F = R/I$. Suppose F'/F is a finite field extension. Then there is a complete local ring R', containing R as a subring, such that R' is R-free of rank $[F' : F]$, $\mathrm{Jac}(R') = IR'$ and R'/I' is F-isomorphic to F', where $I' = \mathrm{Jac}(R')$.*

Proof. We use induction on $[F' : F]$. The canonical map $R \to F$ induces a map $\varphi : R[x] \to F[x], g \mapsto \bar{g}$, on the polynomial rings. First suppose that $F' = F(a)$ for some $a \notin F$, and let the minimum polynomial of a over F be $\bar{f}$, where $f \in R[x]$ and f is monic. Then $F' \cong F[x]/F[x]\bar{f}$. Let $R' = R[x]/R[x]f$. Since f is monic, R embeds in R', and R' is R-free on $1, b, \ldots, b^{n-1}$, where $b = x + R[x]f$ and $n = \deg(\mathrm{f}) = \deg(\bar{\mathrm{f}}) = [\mathrm{F}' : \mathrm{F}]$. It follows that R' is Noetherian. Also, φ has kernel $IR[x]$, so induces a ring epimorphism $\psi : R' \to F'$ with kernel $\dfrac{IR[x] + R[x]f}{R[x]f}$, which we denote by I'. Thus $I' = IR'$and $R'/I' \cong F'$ (the isomorphism obtained is an F-isomorphism). Let J be an ideal of R'. Then $\psi(J) = 0$ or F'. Now $\psi(J) = F$ implies $R' = J + IR'$ (as R-modules), which implies $R' = J$ by Nakayama's Lemma. On the other hand, if $\psi(J) = 0$ then $J \subseteq I'$. Hence R' is a local ring, with $\mathrm{Jac}(R') = I' = IR'$. Also, the I'-adic topology on R' coincides with the I-adic topology on R' (by Remark 7.18 after Cor. 7.17), so R' is a complete local ring by Cor. 7.16(2). In general, if $F \neq F'$, we can find an intermediate field K such that $F' = K(a)$ for some a and $[K : F] < [F' : F]$. Inductively there is a complete local ring S satisfying the conclusions of Proposition 7.21 with K in place of F', and applying the argument already given to K, S in place of F, R gives the desired ring R'. (If $s_1, \ldots, s_m$ is an R-basis for S and $r_1, \ldots, r_n$ is an S-basis for R', then $\{s_i r_j\}$ is an R-basis for R'.) □

Before proving Hattori's theorem, some results on lifting idempotents are needed.

Lemma 7.22. *Let A be a ring, I an ideal such that A is complete and Hausdorff with respect to the I-adic topology. Then*

(a) *If ε is an idempotent in A/I, there exists an idempotent $e \in A$ such that $\bar{e} = \varepsilon$ (where $\bar{e} = e + I$).*
(b) *If e is an idempotent in A and $\bar{e} = \zeta + \eta$, where ζ and η are orthogonal idempotents in A/I, there exist orthogonal idempotents $f, g \in A$ such that $e = f + g$ and $\bar{f} = \zeta$, $\bar{g} = \eta$.*

Proof. (a) Take $e_1 \in A$ such that $\bar{e}_1 = \varepsilon$ and put $f_1 = 1 - e_1$. By Lemma 7.4, there are $e_n, f_n \in A$ for $n \geq 1$ such that

(1) $e_n \in Re_1^n$ and $f_n \in Rf_1^n$;
(2) $1 = e_n + f_n$.

where R is the subring of A generated by e_1. Now $e_1 f_1 \in I$, so by (1) $e_n f_n \in I^n$. By (2), we have

(3) $e_n^2 \equiv e_n \mod I^n$.

By (1),

$$e_n + e_{n+1} \equiv 0 \mod Re_1^n$$

$$e_n - e_{n+1} = f_{n+1} - f_n \equiv 0 \mod Rf_1^n$$

$$\text{Hence} \quad e_n^2 - e_{n+1}^2 \equiv 0 \mod I^n \quad (\text{since } Re_1^n f_1^n \subseteq I^n).$$

By (3),

(4) $e_n \equiv e_{n+1} \mod I^n$.

From (4), (e_n) is a Cauchy sequence. Since A is complete and Hausdorff, we can let $e = \lim_{n\to\infty} e_n$. By (3), $e_n^2 - e_n \longrightarrow 0$ and since $e_n^2 \longrightarrow e^2$ (A is a topological ring), $e^2 = e$. Since $e_n \longrightarrow e$, there exists n such that $e - e_n \in I$, and by (4), $e - e_1 \in I$. Therefore $\bar{e} = \varepsilon$.

(b) Choose $a \in A$ such that $\bar{a} = \zeta$. Let $e_1 = eae$, so $\bar{e}_1 = \zeta$. Use the argument of (a) to find a sequence (e_n) such that $e_n \longrightarrow f$, where $f^2 = f$ and $\bar{f} = \zeta$, and $e_n \in Re_1^n$ (R being the subring of A generated by e_1). Since $e_1 e = e_1$ and $e_n \in Re_1^n$, $e_n e = e_n$ for all n, hence $fe = f$ (since A is Hausdorff). Similarly $ef = f$. Now let $g = e - f$ to obtain the desired decomposition of e. □

Corollary 7.23. *In Lemma 7.22, if $\varepsilon_1, \ldots, \varepsilon_n$ are orthogonal idempotents in A/I and $e \in A$ satisfies $e^2 = e$ and $\bar{e} = \varepsilon_1 + \ldots + \varepsilon_n$, then there are orthogonal idempotents $e_i \in A$ such that $e = e_1 + \ldots + e_n$ and $\bar{e}_i = \varepsilon_i$ for $1 \leq i \leq n$.*

Proof. We use induction on n. For $n = 2$ it follows from Lemma 7.22(b). If $n > 2$, apply Lemma 7.22(b) to $e, \varepsilon_1 + \ldots + \varepsilon_{n-1}$ and ε_n. There exist orthogonal idempotents $f, e_n \in A$ such that $e = f + e_n$, $\bar{f} = \varepsilon_1 + \ldots + \varepsilon_{n-1}$ and $\bar{e}_n = \varepsilon_n$. Inductively we can write $f = e_1 + \ldots + e_{n-1}$, where the e_i (for $1 \leq i \leq (n-1)$) are orthogonal idempotents in A with $\bar{e}_i = \varepsilon_i$. Then $fe_n = 0$ implies $0 = (e_i f)e_n = e_i e_n$, and similarly $e_n e_i = 0$ for $1 \leq i \leq (n-1)$. □

Hattori's proof of Swan's theorem

Lemma 7.24. *Let R be a commutative local ring with maximal ideal M, and put $\overline{R} = R/M$. Let A be an R-algebra, finitely generated as an R-module, let $\overline{A} = A/MA$ and let $\varphi : A \to \overline{A}$ be the natural map. Then*

(1) $\mathrm{Jac}(A) = \varphi^{-1}(\mathrm{Jac}(\overline{A})) \supseteq MA$;
(2) *φ induces an isomorphism $A/\mathrm{Jac}(A) \cong \overline{A}/\mathrm{Jac}(\overline{A})$ (as R-algebras);*
(3) *$A/\mathrm{Jac}(A)$ is semisimple.*

Proof. Let S be a simple A-module. Since S is cyclic, it is a finitely generated R-module. Also, MS is an A-submodule of S, so is either 0 or S. By Nakayama's Lemma, $MS = S$ implies $S = 0$, hence $MS = (MA)S = 0$. It follows that $MA \subseteq \bigcap_S \mathrm{Ann}_A(S) = \mathrm{Jac}(A)$ (see the proof of Proposition 4.3, Chapter 4 in [64]). Hence all maximal left ideals of A contain MA, so $\mathrm{Jac}(A) = \varphi^{-1}(\mathrm{Jac}(\overline{A}))$ by the Correspondence Theorem. Now (2) follows by the Isomorphism Theorems, and (3) follows from (2) since $\mathrm{Jac}\left(\overline{A}/\mathrm{Jac}(\overline{A})\right) = 0$ and $\overline{A}/\mathrm{Jac}(\overline{A})$ is finite dimensional over $\overline{R}$, so has the descending chain condition on left ideals. (See [67, Part II, §3, Theorem 20] and [32, §4.6, Theorem 6]; note however, we are using the definition of semisimplicity in [32], which differs from that in [67].) □

Proposition 7.25. *Let R, M and $\overline{R}$ be as in Lemma 7.24. If $\overline{R}$ has characteristic $p > 0$ and G is a finite p-group, then RG is a local ring.*

Proof. Let $A = RG$ in Lemma 7.24. Then $\overline{A} = \overline{R}G$ and the result follows by Lemma 7.24(2) and Lemma 7.3. □

Now let R be a complete local ring with maximal ideal M and again let A be an R-algebra, finitely generated as an R-module. Put $\overline{R} = R/M$, $\overline{A} = A/MA$. The M-adic topology on A is the same as the AM-adic topology on A (see Remark 7.18), so A is complete and Hausdorff in the AM-adic topology by Cor. 7.16(2). By Lemma 7.24(1), $\mathrm{Jac}(A) \supseteq AM$ $(= MA)$ and $\mathrm{Jac}(\overline{A})$ is nilpotent since $\overline{A}$ is a finite dimensional $\overline{R}$-algebra (see [8, Ch.8, Prop.2.5]; $\overline{A}$ is left Artinian since its left ideals are $\overline{R}$-subspaces). Hence $\mathrm{Jac}(A)^k \subseteq AM$ for some k, so the AM-adic topology on A coincides with the J-adic topology, where $J = \mathrm{Jac}(A)$. By Lemma 7.24, A/J is semisimple, so there are primitive orthogonal idempotents $\varepsilon_1, \ldots, \varepsilon_s \in A/J$ such that $1 = \varepsilon_1 + \ldots + \varepsilon_s$ and for some $r \leq s$, any finitely generated (A/J)-module is isomorphic to $(\frac{A}{J}\varepsilon_1)^{m_1} \oplus \ldots \oplus (\frac{A}{J}\varepsilon_r)^{m_r}$ for unique integers $m_1, \ldots, m_r$ (see §1). By Cor. 7.23, there are orthogonal idempotents $e_1, \ldots, e_s \in A$ such that $e_i + J = \varepsilon_i$ for $1 \leq i \leq s$ and $1 = e_1 + \ldots + e_s$.

Lemma 7.26. *In this situation, suppose P is a finitely generated projective A-module. Then $P \cong (Ae_1)^{m_1} \oplus \ldots \oplus (Ae_r)^{m_r}$ for unique integers $m_1, \ldots, m_r$. Moreover, if Q is a finitely generated projective A-module and $P^m \cong Q^m$, then $P \cong Q$.*

Proof. The quotient P/JP is a finitely generated (A/J)-module, so is isomorphic to $(\frac{A}{J}\varepsilon_1)^{m_1} \oplus \ldots \oplus (\frac{A}{J}\varepsilon_r)^{m_r}$ for some integers $m_1, \ldots, m_r$. Let $Q = (Ae_1)^{m_1} \oplus \ldots \oplus (Ae_r)^{m_r}$. Then Q is A-projective (because $A = Ae_1 \oplus \ldots \oplus Ae_s$) and $Q/JQ \cong P/JP$. By Lemma 7.1, $Q \cong P$. Uniqueness is similarly proved and the last part follows easily. □

Lemma 7.27. *In Lemma 7.26, assume $A/\operatorname{Jac}(A)$ is a direct sum of matrix algebras over fields, and if $\overline{R}$ has characteristic $p > 0$, assume $T(A) = A/[A, A]$ has no p-torsion. Then $T_A(e_1), \ldots, T_A(e_r)$ are $\mathbb{Z}$-linearly independent.*

Proof. Suppose on the contrary that $\sum_{i=1}^{r} n_i T_A(e_i) = 0$, where the $n_i \in \mathbb{Z}$, not all zero. We can assume $n_1 \neq 0$ in $\overline{R}$. Applying the homomorphism $\varphi_* : T(A) \to T(A/J)$, where $\varphi : A \to A/J$ is the natural map, gives $\sum_{i=1}^{r} n_i T_{A/J}(\varepsilon_i) = 0$. Let $A/J = A_1 \times \ldots \times A_r$, where each A_i is a matrix algebra over a field F_i, with $\varepsilon_i \in A_i$ (see §1). Then

$$T(A/J) = T(A_1) \oplus \ldots \oplus T(A_r) \quad \text{(see Example (2) after Lemma 2.1).}$$

and $T(A_i) \cong F_i$ (Example (3) after Lemma 2.1). Also, F_i is a field extension of $\overline{R}$, so they have the same characteristic. To see this, A/J is an $\overline{R}$-algebra, since $MA \subseteq J$, so there is a ring homomorphism $\overline{R} \to Z(A/J)$ (the centre of A/J). Obviously $Z(A/J) = Z(A_1) \times \ldots \times Z(A_r)$ and by projection we obtain a homomorphism $\overline{R} \to Z(A_i)$ which is an embedding, $\overline{R}$ being a field. But $Z(A_i) = F_i 1_{A_i}$ (the set of scalar matrices) which is isomorphic to F_i, as required. Therefore all $n_i T_{A_i}(\varepsilon_i) = 0$, hence $T_{A_1}(\varepsilon_1) = 0$, a contradiction (see Example (4) after Lemma 2.6). □

We shall now quote another result without proof.

Theorem 7.28. *Let B be a finite-dimensional algebra over a field F. Then there is a finite field extension F' of F such that $F' \otimes_F B/\operatorname{Jac}(F' \otimes_F B)$ is split semisimple over F' (i.e., is a direct sum of matrix algebras over F').*

Proof. See [35, 7.12 and 7.13]. □

Theorem 7.29. *Let R be a complete local ring with maximal ideal M, let A be an R-algebra, finitely generated as R-module, and let $F = R/M$. If F*

has characteristic $p > 0$, assume $T(A)$ has no p-torsion. Let P, Q be finitely generated projective A-modules. Then $P \cong Q$ if and only if $r_{P/A} = r_{Q/A}$.

Proof. We show that $r_{P/A} = r_{Q/A}$ implies $P \cong Q$, the other implication being obvious. We first reduce to the case that $A/\operatorname{Jac}(A)$ is a direct sum of matrix rings over fields. By Theorem 7.28, there is a finite field extension F'/F such that $F' \otimes_F \overline{A}/\operatorname{Jac}(F' \otimes_F \overline{A})$ is split semisimple, where $\overline{A} = A/MA$. By Prop. 7.21 there is a complete local ring R', extending R, such that R' is R-free of rank $[F' : F]$, F' is F-isomorphic to R'/M' (where M' is the maximal ideal of R'), and $M' = MR'$. Let $A' = R' \otimes_R A$, an R'-algebra finitely generated as R'-module. Since $R' \otimes_R -$ is exact (R' is a free R-module), there is a short exact sequence

$$0 \to R' \otimes_R MA \xrightarrow{i} A' \to R' \otimes_R (A/MA) \to 0$$

and $\operatorname{Im}(i) = M'A'$, so $A'/M'A' \cong R' \otimes_R (A/MA)$. Also, there is a short exact sequence

$$0 \to MR' \to R' \to F' \to 0$$

giving an exact sequence:

$$0 = MR' \otimes_R (A/MA) \to R' \otimes_R (A/MA) \to F' \otimes_R (A/MA) \to 0.$$

Further, $F' \otimes_R (A/MA) = F' \otimes_F (A/MA)$ (literally equal, with the usual construction of tensor products [100, Theorem 1.4]). Hence $A'/M'A' \cong F' \otimes_F (A/MA)$ (as F-algebras), so $A'/M'A'$ becomes a product of matrix algebras over F' on factoring $\operatorname{Jac}(A'/M'A')$. Hence $A'/\operatorname{Jac}(A')$ is a product of matrix algebras over F' by Lemma 7.24(2). Moreover, if F has characteristic $p > 0$, the map $\lambda_p : T(A) \to T(A), x \mapsto px$ is one-to-one by assumption. Since $R' \otimes_R -$ is exact, $1 \otimes \lambda_p : R' \otimes_R T(A) \to R' \otimes_R T(A)$ is one-to-one. By Cor. 2.14, $R' \otimes_R T(A) \cong T(A')$, and $1 \otimes \lambda_p$ corresponds to multiplication by p on $T(A')$, hence $T(A')$ has no p-torsion. Let $P' = R' \otimes_R P$, $Q' = R' \otimes_R Q$. Then by Cor. 2.14, $r_{P/A} = r_{Q/A}$ implies $r_{P'/A'} = r_{Q'/A'}$. Also, $P' \cong P^n$ as A-module, where $n = [F' : F]$. Hence

$$P' \cong Q' \text{ implies } P^n \cong Q^n, \text{ which implies } P \cong Q \text{ by Lemma 7.26.}$$

Therefore, replacing R, A by R', A', we can assume that $A/\operatorname{Jac}(A)$ is a direct sum of matrix rings over fields. Using Lemma 7.26 we can write

$$P \cong (Ae_1)^{m_1} \oplus \ldots \oplus (Ae_r)^{m_r}$$

$$Q \cong (Ae_1)^{n_1} \oplus \ldots \oplus (Ae_r)^{n_r}$$

and by Lemma 2.7,

$$r_P = \sum_{i=1}^{r} m_i T_A(e_i), \quad r_Q = \sum_{i=1}^{r} n_i T_A(e_i)$$

(see Example (2) after Lemma 2.6).

Thus, if $r_P = r_Q$ then $\sum_{i=1}^{r}(m_i - n_i)T_A(e_i) = 0$. By Lemma 7.27, $m_i = n_i$ for all i, so $P \cong Q$. □

Corollary 7.30. *Let R be a commutative Noetherian local ring with maximal ideal M and let A be an R-algebra, finitely generated as R-module. If R/M has characteristic $p > 0$, assume $T(M)$ has no p-torsion. If P, Q are finitely generated projective A-modules, then $P \cong Q$ if and only if $r_{P/A} = r_{Q/A}$.*

Proof. Again we need only show that $r_{P/A} = r_{Q/A}$ implies $P \cong Q$. The idea is to take completions and apply Theorem 7.29. Let $\widehat{R}$ be the M-adic completion of R, a complete local ring by Cor. 7.16(4) and Lemma 7.19, with maximal ideal $\widehat{R}M$. Let $\widehat{A}$ be the M-adic completion of A, which is also the AM-adic completion of A (Remark 7.18 after Cor. 7.17). As M-adic completion, $\widehat{A}$ is an $\widehat{R}$-module, and as AM-adic completion, $\widehat{A}$ is a ring, and $\widehat{A}$ becomes an $\widehat{R}$-algebra. By Lemma 7.15, there is an isomorphism $\alpha_A : \widehat{R} \otimes_R A \to \widehat{A}$, and it is an $\widehat{R}$-algebra isomorphism. By Lemma 7.19, the residue fields R/M and $\widehat{R}/\widehat{R}M$ are isomorphic, so have the same characteristic. By Cor. 7.16, $\widehat{R} \otimes_R -$ is exact, and arguing as in Theorem 7.29, if this characteristic is $p > 0$, then $T(\widehat{A})$ has no p-torsion. By Lemma 7.15, there are isomorphisms $\widehat{R} \otimes_R P \cong \widehat{P}$, $\widehat{R} \otimes_R Q \cong \widehat{Q}$, where $\widehat{P}$, $\widehat{Q}$ are AM-adic completions. Hence,

$$\begin{aligned} r_P = r_Q &\Rightarrow r_{\widehat{P}} = r_{\widehat{Q}} \quad \text{by Cor. 2.14} \\ &\Rightarrow \widehat{P} \cong \widehat{Q} \quad \text{by Theorem 7.29} \\ &\Rightarrow \frac{\widehat{P}}{M\widehat{P}} \cong \frac{\widehat{Q}}{M\widehat{Q}} \\ &\Rightarrow \frac{P}{MP} \cong \frac{Q}{MQ} \quad \text{as } A\text{-modules} \end{aligned}$$

by Cor. 7.17 and Remark 7.18 following it. By Lemma 7.24(1), $AM = MA \subseteq \text{Jac}(A)$, and $MP = (AM)P$, so by Lemma 7.1, $P \cong Q$. □

Theorem 7.31. *Let R be an integral domain and let $\mathfrak{p} \in \text{Spec}(R)$ be such that $R_{\mathfrak{p}}$ is Noetherian. Let G be a finite group such that no prime divisor of $|G|$*

is invertible in R. Let P be a finitely generated projective RG-module. Then $P_{\mathfrak{p}} = R_{\mathfrak{p}} \otimes_R P$ is a free $R_{\mathfrak{p}}$-module.

Proof. If R has characteristic $p > 0$, the residue field of $R_{\mathfrak{p}}$ has characteristic p and G is a finite p-group. Hence $R_{\mathfrak{p}}G$ is local (Prop. 7.25) and $P_{\mathfrak{p}}$ is free by Lemma 7.2. If R has characteristic 0, so does $R_{\mathfrak{p}}$, and $T(R_{\mathfrak{p}}G)$ is a free $R_{\mathfrak{p}}$-module (on the conjugacy classes of G). Hence $T(R_{\mathfrak{p}}G)$ has no p-torsion for all primes p.

By Lemma 7.9, $r_{P/RG} = n.T_{RG}(1)$ for some n.

By Cor. 2.14, $r_{P_{\mathfrak{p}}} = n.T_{R_{\mathfrak{p}}G}(1) = r_{(R_{\mathfrak{p}}G)^n}$.

By Cor. 7.30, $P_{\mathfrak{p}} \cong (R_{\mathfrak{p}}G)^n$. □

8. Consequences of Swan's theorem

This section is preparatory to our second definition of an Euler characteristic. Before coming to the applications of Swan's Theorem, another notion of rank for modules of type FP is discussed, based on the rank of a free module in the usual sense. There is also a homological lemma whose elegant proof is due to K.S. Brown.

If A is a ring and F is a finitely generated free A-module, having a basis with n elements, then $r_F = nT_A(1)$ (Example (1) after Lemma 2.6). However, the integer n need not be uniquely determined by F, in general (it is only unique modulo $[A, A]$). For example, let V be a free module with a countably infinite basis over a ring R and let $U = V \oplus V$. Then it is easy to construct an isomorphism $U \cong V$. Let $A = \mathrm{End}_R(U)$. Then $A \cong \mathrm{Hom}_R(V, U) \oplus \mathrm{Hom}_R(V, U) \cong A \oplus A$, and the isomorphism obtained is an A-isomorphism. It follows that $A \cong A^n$, so A has a basis with n elements, for all $n \geq 1$. For a ring A, the following are equivalent:

(a) if M is a finitely generated free A-module, any two bases of M have the same number of elements;
(b) if M is a finitely generated free right A-module, any two bases of M have the same number of elements;

(because if $e_1, \ldots, e_n$ is a basis for M, the dual basis for $M^* = \mathrm{Hom}_A(M, A)$ has n elements). When these conditions are satisfied, we say that A has IBN (short for invariant basis number).

Examples.

(1) If $f : A \to B$ is a ring homomorphism and B has IBN, then A has IBN.

For if F is a free A-module with basis $e_1, \ldots, e_n$, then $B \otimes_A F$ is free with basis $1 \otimes e_1, \ldots, 1 \otimes e_n$.

(2) Left Noetherian rings have IBN, since they satisfy (a) (see [8, Ch.6, Prop.9.3] or [100, Theorem 4.9]), so (applying this to the opposite ring) right Noetherian rings satisfy (b) and have IBN. In particular, division rings have IBN (this can also be seen using the argument from elementary linear algebra, for vector spaces over a field).

(3) Consequently any ring admitting a homomorphism to a division ring has IBN. This includes commutative rings (they map onto the quotient by a maximal ideal, which is a field) and local rings.

(4) If R is a ring with IBN and G is a group, then RG has IBN, applying (1) to the augmentation map $RG \to R$.

Assume now that A is a ring with IBN. If M is a finitely generated free A module, we define the *rank* $\mathrm{rk}_A(M)$ to be the number of elements in a basis for M. (If A is a division ring this is the same as the dimension of M, denoted by $\dim_A(M)$.) More generally, if M is of type FL over A, we can define $\mathrm{rk}_A(M) = \sum_{i \geq 0}(-1)^i \mathrm{rk}_A(L_i)$, where $L \to M$ is a finite augmented projective resolution of M. (This is independent of the choice of L by Cor. 3.11.)

Remark 8.1. *If* $0 \to L_n \to \ldots \to L_0 \to 0$ *is exact and each* L_i *is of type* FL*, then*

$$\sum_{i=0}^{n}(-1)^i \mathrm{rk}_A(L_i) = 0.$$

The proof is similar to the proof of Cor. 4.9.

In particular, if $0 \to N \to K \to M \to 0$ is a short exact sequence of modules of type FL, then $\mathrm{rk}_A(K) = \mathrm{rk}_A(M) + \mathrm{rk}_A(N)$. Applying this to a split short exact sequence, if M, N are of type FL, $\mathrm{rk}_A(M \oplus N) = \mathrm{rk}_A(M) + \mathrm{rk}_A(N)$. If C is a positive chain complex of A-modules such that $H_n(C)$ is of type FL for all n, and zero for all but finitely many values of n, we define

$$\mathrm{rk}_A(C) = \sum_{i \geq 0}(-1)^i \mathrm{rk}_A(H_i(C)).$$

Lemma 8.2. *In this situation, if* C_n *is of type* FL *for all* n*, and zero for all but finitely many values of* n*, then*

$$\mathrm{rk}_A(C) = \sum_{i \geq 0}(-1)^i \mathrm{rk}_A(C_i).$$

Proof. This is similar to part of the proof of Lemma 6.24 and is left to the reader. □

If R is a principal ideal domain, then an R-module M is of type FL if and only if it is finitely generated, in which case $\mathrm{rk}_R(M)$ is the torsion-free rank of M (see the proof of Cor. 6.25). Let R be a principal ideal domain, let $p \in R$ be prime, M a finitely generated R-module and let $F = R/pR$. Define

$$M_p = M/pM = F \otimes_R M \quad \text{(see Remark 2.10)}$$

$$_pM = \{m \in M \mid pm = 0\}$$

Lemma. *With the definitions just given, let $r = \dim_F(M_p)$, $s = \dim_F({}_pM)$. Then $r = \mathrm{rk}_R(M) + s$.*

Proof. Since $(M \oplus N)_p = M_p \oplus N_p$ and ${}_p(M \oplus N) = {}_pM \oplus {}_pN$, we can assume (by the structure theory for R-modules) that $M \cong R/aR$, where a is 0 or a prime power. Let $m = \mathrm{rk}_A(M)$. There are three cases.

(1) If $a = 0$, then $m = 1, r = 1, s = 0$.
(2) If $a = q^n$ where q is a prime, not an associate of p, then $m = r = s = 0$.
(3) If $a = q^n$ where q is an associate of p, then $M \cong R/p^n R$ and $m = 0, r = 1$, $s = 1$. □

If M is a $\mathbb{Z}$-graded module, we say that M is finitely generated if $\bigoplus_{i \in \mathbb{Z}} M_i$ is finitely generated. (That is, M_i is finitely generated for all i, and is zero for all but finitely many i.)

Lemma 8.3. *Let C be a positive complex of finite length over a principal ideal domain R, with $H(C)$ finitely generated. Let p be a prime element of R and put $F = R/pR$. Then*

$$\mathrm{rk}_R(C) = \sum_i (-1)^i \dim_F(H_i(F \otimes_R C)).$$

Proof. Let $r_i = \dim_F(H_i(C)_p)$ and $s_i = \dim_F({}_pH_i(C))$. There is a short exact sequence of chain complexes $0 \to C \xrightarrow{p} C \xrightarrow{\pi} F \otimes_R C \to 0$, where p is short for λ_p (multiplication by p). By Theorem 3.1, there is an exact sequence

$$\ldots \to H_i(C) \xrightarrow{p} H_i(C) \xrightarrow{\pi_*} H_i(F \otimes_R C) \xrightarrow{\partial} H_{i-1}(C) \xrightarrow{p} H_{i-1}(C) \ldots$$

Thus $\mathrm{Ker}(\partial) = \mathrm{Im}(\pi_*) \cong H_i(C)/\mathrm{Ker}(\pi_*) = H_i(C)/\mathrm{Im}(p) = H_i(C)_p$ and $\mathrm{Im}(\partial) = \mathrm{Ker}(p) = {}_pH_{i-1}(C)$. Therefore, using the exact sequence

$$0 \to \mathrm{Ker}(\partial) \to H_i(F \otimes_R C) \to \mathrm{Im}(\partial) \to 0,$$

$\dim_F(H_i(F \otimes_R C)) = s_{i-1} + r_i$. By the lemma, $r_i = \mathrm{rk}_R(H_i(C)) + s_i$, and the result follows on taking alternating sums. □

We now introduce a new notion of equivalence for chain complexes, weaker than homotopy equivalence.

Definition. *A chain map $f : C \to C'$ is a* weak equivalence *if $H(f)$ is an isomorphism.*

Note. *If $f : A \oplus B \to C \oplus D$ is a homomorphism of direct sums, then we can write $f(a, b) = (\alpha(a) + \beta(b), \gamma(a) + \delta(b))$ for $a \in A$, $b \in B$, where $\alpha : A \to C$, $\beta : B \to C$ etc. Thus f can be defined by a formal matrix $\begin{pmatrix} \alpha & \beta \\ \gamma & \delta \end{pmatrix}$, and composition of such homomorphisms corresponds to matrix multiplication. This also applies when the modules are graded, with α etc having the same degree as f.*

Lemma 8.4. *Let $f : (C, d) \to (C', d')$ be a chain map, M the mapping cone of f. Then*

(1) *the map f is a weak equivalence if and only if $H(M) = 0$;*
(2) *the map f is a homotopy equivalence if and only if M is contractible.*

Proof. (1) This follows immediately from the exact sequence of Prop. 3.5:

$$\dots H_{n+1}(M) \longrightarrow H_n(C) \xrightarrow{H_n(f)} H_n(C') \longrightarrow H_n(M) \longrightarrow H_{n-1}(C) \xrightarrow{H_{n-1}(f)} H_{n-1}(C') \dots$$

(2) Recall $M_n = C_{n-1} \oplus C'_n$ with differential $d''(c, c') = (-d(c), d'(c') + f(c))$, which is represented by the matrix $\begin{pmatrix} -d & 0 \\ f & d' \end{pmatrix}$. Suppose r is a chain homotopy from id_M to 0 and r has matrix $\begin{pmatrix} u & g \\ w & v \end{pmatrix}$. Then g is a chain map, u is a chain homotopy from gf to id_C and v is a chain homotopy from $\mathrm{id}_{C'}$ to fg, hence f is a homotopy equivalence. Conversely, given g, u and v satisfying these conditions, define r by the matrix

$$\begin{pmatrix} u - g(vf + fu) & g \\ -v(vf + fu) & v \end{pmatrix}.$$

Then r is a chain homotopy from id_M to 0. □

Corollary 8.5. *If $f : C \to C'$ is a weak equivalence and C, C' are positive projective complexes, then f is a homotopy equivalence.*

Proof. This follows from Lemma 8.4 and Lemma 3.9 applied to the mapping cone M of f. □

Proposition 8.6. *Let A be a left Noetherian ring and let (C, d) be a positive chain complex of projective A-modules with $H_n(C)$ finitely generated for all n, and such that, for some integer k,*

$$H_n(\mathrm{Hom}_A(C, M)) = 0$$

for all $n > k$ and all finitely generated A-modules M. Then C is homotopy equivalent to a finite positive projective complex.

Proof. First we construct a positive complex (F, d') with F_n finitely generated free for all n and a weak equivalence $\tau : F \to C$. Assume F_i, τ_i, d'_i have been constructed for $i \leq n$ so that $H_i(\tau)$ is an isomorphism for $i < n$ and $H_n(\tau)$ is onto. (Here $H_n(\tau)$ refers to the complex $\ldots 0 \to F_n \to \ldots \to F_0 \to 0 \ldots$). Since A is left Noetherian, $\mathrm{Ker}(H_n(\tau))$ is finitely generated. Let $x_1, \ldots, x_r$ be elements of $\mathrm{Ker}(d'_n)$ representing a set of generators for $\mathrm{Ker}(H_n(\tau))$. Choose $y_1, \ldots, y_r$ in C_{n+1} such that $d_{n+1}(y_i) = \tau_n(x_i)$ for $1 \leq i \leq r$. Also, choose $z_1, \ldots, z_s$ in C_{n+1} representing generators for $H_{n+1}(C)$. Let F_{n+1} be a free A-module with basis $e_1, \ldots, e_r, f_1, \ldots, f_s$. Define $\tau_{n+1} : F_{n+1} \to C_{n+1}$ by $e_i \mapsto y_i$, $f_j \mapsto z_j$, and define $d'_{n+1} : F_{n+1} \to F_n$ by $e_i \mapsto x_i$, $f_j \mapsto 0$ $(1 \leq i \leq r,\ 1 \leq j \leq s)$. This constructs (F, d') and τ recursively. By Cor. 8.5, τ is a homotopy equivalence, hence $\mathrm{Hom}_A(F, M)$ is homotopy equivalent to $\mathrm{Hom}_A(C, M)$ via τ^*, for any A-module M (see the definition after Lemma 5.3). This is left as an exercise, using the fact that $\mathrm{Hom}_A(-, M)$ is an additive functor. Hence $H^n(\mathrm{Hom}_A(F, M)) = 0$ for all finitely generated A-modules M and all $n > k$. By Cor. 5.6 applied to F, $H_n(C) = H_n(F) = 0$ for $n > k$. Also by Cor. 5.6, putting $B_k = \mathrm{Im}(d'_{k+1})$, F_k/B_k is a summand of F_k, so is projective. Define

$$\overline{F}_n = \begin{cases} F_n & \text{if } n < k \\ F_k/B_k & \text{if } n = k \\ 0 & \text{if } n > k. \end{cases}$$

Then d'_k induces $\bar{d}_k : F_k/B_k \to F_{k-1}$. Putting $\bar{d}_n = d'_n$ for $n < d$ and $\bar{d}_n = 0$ for $n > k$ gives a finite positive projective complex $(\overline{F}, \bar{d})$. Define $\bar{\tau}_k : \overline{F}_k \to C_k$ to be the map induced by τ_k, $\bar{\tau}_n = \tau_n$ for $n < k$ and $\bar{\tau}_n = 0$ for $n > k$. Then $\bar{\tau}$ is a weak equivalence, so a homotopy equivalence by Cor. 8.5. □

We now come to the applications of Swan's Theorem.

Lemma 8.7. *Let R be a principal ideal domain, G a finite group such that no prime divisor of $|G|$ is invertible in R. Let C be a positive projective complex of finite length, with $H(C)$ finitely generated over R. Put $C_G = R \otimes_{RG} C$. Then*

$H(C_G)$ is finitely generated over R and

$$\mathrm{rk}_R(C) = |G| \ \mathrm{rk}_R(C_G).$$

Proof. The ring RG is a finitely generated R-module and its left ideals are R-submodules, so RG is left Noetherian (see [100, Theorem 4.1]). By Prop. 8.6, C is homotopy equivalent to a finite positive projective complex, say D. Since $R \otimes_{RG} -$ is a functor, C_G is homotopy equivalent to D_G (see the argument after the definition of homology of a group in §5), so we can assume C is a finite positive projective complex. By Lemma 8.2, it is enough to show that if P is a finitely generated projective RG-module, then $\mathrm{rk}_R(P) = |G| \, \mathrm{rk}_R(P_G)$, where $P_G = R \otimes_{RG} P$. Now $\mathrm{rk}_R(P)$ is the free rank of P, which is equal to $\dim_F(F \otimes_R P)$, where F is the field of fractions of R. By Theorem 7.10, $F \otimes_R P \cong FG^n$, where $n = \mathrm{rk}_R(P_G)$. Since FG is an F-vector space of dimension $|G|$, $\dim_F(FG^n) = n|G|$. □

Note. *If H is a subgroup of a group G and R is a ring, there is an isomorphism $R \otimes_{RH} RG \to R(H\backslash G)$ (the free R-module on $\{Hg \mid g \in G\}$), given by $1 \otimes g \mapsto Hg$ for $g \in G$. The proof is analogous to that of Lemma 5.2. If M is an RG-module, define $M_G = R \otimes_{RG} M$. Then $M_G \cong M/I_G M$ (apply Remark* 2.10 *to the augmentation map).*

Lemma 8.8. *Let $1 \to N \to G \xrightarrow{\pi} Q \to 1$ be a group extension, R a ring. Let M be an RG-module. Then the action of G induces an action of Q on M_N, and as RQ-modules,*

$$M_N \cong RQ \otimes_{RG} M.$$

Further, $M_G \cong (M_N)_Q$.

Proof. The Q-action is given by $q(r \otimes m) = r \otimes gm$, $(q \in Q, r \in R, m \in M)$ for any $g \in G$ such that $\pi(g) = q$. Now

$$\begin{aligned} RQ \otimes_{RG} M &\cong (R \otimes_{RN} RG) \otimes_{RG} M && \text{by the preceding note} \\ &\cong R \otimes_{RN} M = M_N && \text{by associativity of tensor product} \end{aligned}$$

and this gives an RQ-isomorphism. Also,

$$M_G \cong M/I_G M \cong \frac{M/I_N M}{I_G M/I_N M} \cong M_N/I_G M_N = M_N/I_Q M_N \cong (M_N)_Q$$

(or use associativity of tensor product and the first part). □

Note. *The lemma remains true if M is replaced by a chain complex of RG-modules.*

Definition. *Let G be a group such that $H(G,\mathbb{Z})$ is finitely generated. Then we define $\tilde{\chi}(G) = \sum_{i\geq 0}(-1)^i \operatorname{rk}_{\mathbb{Z}}(H_i(G,\mathbb{Z}))$.*

This is consistent with the earlier definition of $\tilde{\chi}$ on FP($\mathbb{Z}$), by Cor. 6.25. We now prove a limited version of the index formula for $\tilde{\chi}$.

Lemma 8.9. *Let G be a group with $\operatorname{cd}_{\mathbb{Z}}(G) < \infty$. Suppose N is a normal subgroup of finite index in G. If $H(N,\mathbb{Z})$ is finitely generated, then so is $H(G,\mathbb{Z})$, and*

$$\tilde{\chi}(N) = (G : N)\tilde{\chi}(G).$$

Proof. By Theorem 5.7, there is an augmented $\mathbb{Z}G$-projective resolution $P \to \mathbb{Z}$ of finite length. This is also a $\mathbb{Z}N$-projective resolution (see the argument of Lemma 5.9). Let $Q = G/N$ and let $C = P_N$, so $C \cong \mathbb{Z}Q \otimes_{\mathbb{Z}G} P$, by Lemma 8.8. Thus C is a projective $\mathbb{Z}Q$-complex of finite length, and $H(C) = H_*(N,\mathbb{Z})$. By Lemma 8.8, $C_Q = P_G$, so $H(C_Q) = H_*(G,\mathbb{Z})$. The result follows by Lemma 8.7. □

We are not quite ready to discuss the Brown definition of an Euler characteristic. In order to establish the extension formula, given a group extension $1 \to N \to G \to Q \to 1$, we need to relate $H(G,\mathbb{Z})$ with $H(N,\mathbb{Z})$ and $H(Q,\mathbb{Z})$. This is not done directly, but by means of a complex piece of machinery called a spectral sequence, which is the subject of the next section.

9. Spectral sequences

Exact couples

Spectral sequences are needed to discuss the behaviour of the Brown characteristic relative to group extensions. One of the best ways of introducing them is by means of exact couples.

Definition. *An exact couple is an exact triangle*

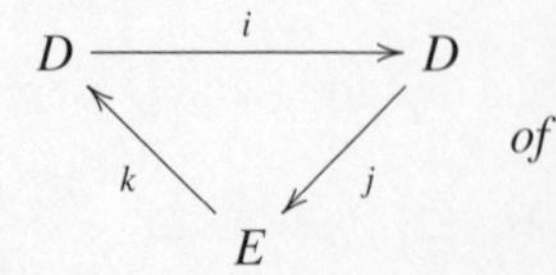

of A-modules (which may be graded by some abelian group), where A is a ring.

Given such an exact couple, $d = jk$ is a differential on E ($d^2 = j(kj)k = 0$).

We can form the *derived couple*:

$$\begin{array}{ccc} D' & \xrightarrow{i'} & D' \\ & {}_{k'}\nwarrow \quad \swarrow_{j'} & \\ & E' & \end{array}$$

where $D' = i(D)$, $E' = H(E, d) = \dfrac{\mathrm{Ker}(d)}{\mathrm{Im}(d)} = \dfrac{k^{-1}(\mathrm{Ker}(j))}{jk(E)} = \dfrac{k^{-1}i(D)}{ji^{-1}(0)}$, $i' = i|_{D'}$, k' is induced by k and j' by ji^{-1}. This is also an exact couple, and iteration leads to the following: given an exact couple as in the definition, define

$$D^r = i^{r-1}(D)$$
$$Z^r = k^{-1}i^{r-1}(D)$$
$$B^r = ji^{-r+1}(0)$$

and $E^r = Z^r/B^r$. Also, let
$i_r : D^r \to D^r$ be $i|_{D^r}$
$j_r : D^r \to E^r$ be the map induced by ji^{-r+1}
$k_r : E^r \to D^r$ be the map induced by k.

Lemma 9.1. *For $r \geq 1$,*

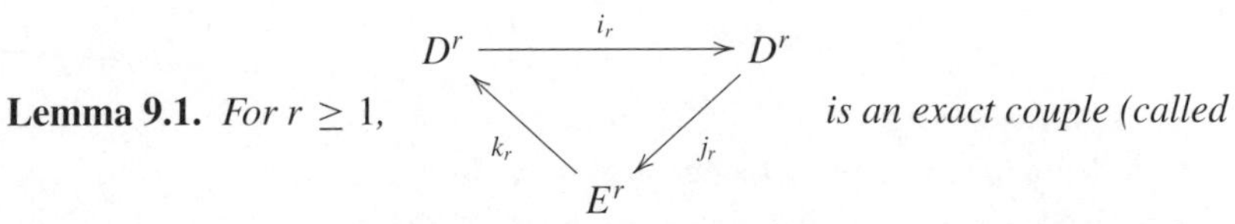

is an exact couple (called the rth derived couple of the original one) and if $d_r = j_r k_r$ then $E^{r+1} \cong H(E^r, d_r)$.

Proof. It is routine calculation to verify exactness. Calculation also shows

$$\mathrm{Ker}(d_r) = k^{-1}i^r(D)/ji^{-r+1}(0) = Z^{r+1}/B^r$$
$$\mathrm{Im}(d_r) = ji^{-r}(0)/ji^{-r+1}(0) = B^{r+1}/B^r$$

and the lemma follows. □

If the modules are graded, $\deg(i_r) = \deg(i)$, $\deg(k_r) = \deg(k)$ and $\deg(j_r) = \deg(j) - (r-1)\deg(i)$, where deg means degree of the map, hence

$$\boxed{\deg(d_r) = \deg(k) + \deg(j) - (r-1)\deg(i).}$$

Filtrations

Definition. *A filtration of a $\mathbb{Z}$-graded module H is a sequence $(F_pH)_{p\in\mathbb{Z}}$ of graded submodules F_pH of H, with F_pH a submodule of $F_{p+1}H$ for all $p \in \mathbb{Z}$.*

A filtration of a chain complex C is a filtration F_pC of C such that F_pC is a subcomplex of C for all $p \in \mathbb{Z}$.

Given a filtration (F_pC) of a chain complex C, there are exact homology sequences (see Cor. 3.2)

$$(*) \quad \ldots \longrightarrow H_n(F_{p-1}C) \xrightarrow{i} H_n(F_pC) \xrightarrow{j} H_n(F_pC, F_{p-1}C) \xrightarrow{k} H_{n-1}(F_{p-1}C) \ldots$$

(one for each $p \in \mathbb{Z}$). Define $\mathbb{Z}$-bigraded modules D, E by

$$D_{pq} = H_{p+q}(F_pC)$$
$$E_{pq} = H_{p+q}(F_pC, F_{p-1}C).$$

(In this context, p is called the filtration degree, q the complementary degree and $n = p + q$ the total degree.) The exact sequences $(*)$ then form an exact couple

$$(**) \quad \begin{array}{ccc} D & \xrightarrow{i} & D \\ {}_{k}\nwarrow & & \swarrow_{j} \\ & E & \end{array}$$

with bidegree$(i) = (1, -1)$, bidegree$(j) = (0, 0)$ and bidegree$(k) = (-1, 0)$. The derived couples give bigraded modules and differentials $(E^r, d_r)_{r\geq 1}$ with $E^{r+1} \cong H(E^r, d_r)$ and the differential d_r has bidegree

$$(-1, 0) + (0, 0) - (r - 1)(1, -1) = (-r, r - 1).$$

Definition. *A k-homology spectral sequence over a ring A is a sequence $(E^r, d^r)_{r\geq k}$ of $\mathbb{Z}$-bigraded A-modules E^r and differentials $d^r : E^r \to E^r$ of bidegree $(-r, r - 1)$ such that $E^{r+1} \cong H(E^r, d^r)$.*

(We are now writing d^r rather than d_r since superscripts will no longer be needed for products of maps, and subscripts will be used for the grading.) Thus a filtration of a complex C determines a 1-spectral sequence with

$$E^1_{pq} = H_{p+q}(F_pC, F_{p-1}C).$$

(We shall only consider homology spectral sequences and so just write k-spectral sequence.) Given a k-spectral sequence (E^r, d^r), we can define submodules B^r, Z^r of E^r with $B^r \subseteq B^{r+1} \subseteq Z^{r+1} \subseteq Z^r$, for all $r \geq k$, together with isomorphisms $\lambda_r : Z^r/B^r \to E^r$, as follows: $Z^k = E^k$, $B^k = 0$, $\lambda_k = \mathrm{id}_{E^k}$. Assuming Z^r, B^r, λ_r defined, let

$$Z^{r+1}/B^r = \lambda_r^{-1}(\mathrm{Ker}(d^r)), \quad B^{r+1}/B^r = \lambda_r^{-1}(\mathrm{Im}(d^r))$$

and let λ_{r+1} be the composite map: $\dfrac{Z^{r+1}}{B^{r+1}} \xrightarrow{\cong} \dfrac{\mathrm{Ker}(d^r)}{\mathrm{Im}(d^r)} \underset{\text{given isom.}}{\xrightarrow{\cong}} E^{r+1}$. (This defines Z^{r+1}, B^{r+1} using the Correspondence Theorem.) We obtain

$$0 = B^k \subseteq \ldots \subseteq B^r \subseteq \ldots \subseteq B^\infty \subseteq Z^\infty \subseteq \ldots \subseteq Z^r \subseteq \ldots \subseteq Z^k = E^k$$

where

$$B^\infty_{pq} = \bigcup_{r=k}^{\infty} B^r_{pq}, \quad Z^\infty_{pq} = \bigcap_{r=k}^{\infty} Z^r_{pq}.$$

Definition. *The* limit term *of the spectral sequence is the $\mathbb{Z}$-bigraded module $E^\infty = Z^\infty / B^\infty$.*

Lemma 9.2. *If (E^r, d^r) is the 1-spectral sequence derived from a filtration (F_pC) of a complex C as above, then*

$$Z^r_{pq} = \mathrm{Ker}(H_{p+q}(F_pC, F_{p-1}C) \longrightarrow H_{p+q-1}(F_{p-1}C, F_{p-r}C))$$

$$B^r_{pq} = \mathrm{Im}(H_{p+q+1}(F_{p+r-1}C, F_pC) \longrightarrow H_{p+q}(F_pC, F_{p-1}C))$$

(the maps come from the exact homology sequences of the triples $(F_pC, F_{p-1}C, F_{p-r}C)$ and $(F_{p+r-1}C, F_pC, F_{p-1}C)$. (See Cor. 3.3.)

Proof. The spectral sequence is derived from the exact couple (∗∗), and from the proof of Lemma 9.1, our notation is consistent with that for exact couples and

$$Z^r = k^{-1} i^{r-1}(D)$$

$$B^r = j i^{-r+1}(0)$$

(and the λ_r are identity maps). There is a commutative diagram with exact bottom row:

$$\begin{array}{ccccc}
 & & H_{p+q}(F_pC, F_{p-1}C) & \xrightarrow{=} & H_{p+q}(F_pC, F_{p-1}C) \\
(1) & & \downarrow k & & \downarrow \partial \\
H_{p+q-1}(F_{p-r}C) & \xrightarrow{i^{r-1}} & H_{p+q-1}(F_{p-1}C) & \longrightarrow & H_{p+q-1}(F_{p-1}C, F_{p-r}C)
\end{array}$$

The bottom row is part of the exact homology sequence of the pair $(F_{p-1}C, F_{p-r}C)$ (see Cor. 3.2), ∂ is the connecting homomorphism for the triple $(F_pC, F_{p-1}C, F_{p-r}C)$ (see Cor. 3.3). The indicated mapping is i^{r-1} by functoriality of H, and the square is commutative by naturality of ∂, that is,

Theorem 3.4 applied to

$$\begin{array}{ccccccccc}
0 & \longrightarrow & F_{p-1}C & \longrightarrow & F_pC & \longrightarrow & F_pC/F_{p-1}C & \longrightarrow & 0 \\
& & \downarrow & & \downarrow & & \downarrow = & & \\
0 & \longrightarrow & F_{p-1}C/F_{p-r}C & \longrightarrow & F_pC/F_{p-r}C & \longrightarrow & F_pC/F_{p-1}C & \longrightarrow & 0.
\end{array}$$

Hence $Z^r_{pq} = \mathrm{Ker}(\partial)$. The claim about B^r_{pq} comes from a similar diagram:

$$\begin{array}{ccccc}
H_{p+q+1}(F_{p+r-1}C, F_pC) & \longrightarrow & H_{p+q}(F_pC) & \xrightarrow{i^{r-1}} & H_{p+q}(F_{p+r-1}C) \\
\downarrow = & & \downarrow j & & \\
H_{p+q+1}(F_{p+r-1}C, F_pC) & \longrightarrow & H_{p+q}(F_pC, F_{p-1}C). & &
\end{array} \tag{2}$$

□

Lemma 9.3. *In Lemma* 9.2, $d^1_{pq} : H_{p+q}(F_pC, F_{p-1}C) \to H_{p+q-1}(F_{p-1}C, F_{p-2}C)$ *is the connecting homomorphism of the triple* $(F_pC, F_{p-1}C, F_{p-2}C)$.

Proof. In the case $r = 2$, the commutative square in (1) above gives a commutative triangle

$$\begin{array}{ccc}
H_{p+q}(F_pC, F_{p-1}C) & & \\
\downarrow k & \searrow \partial & \\
H_{p+q-1}(F_{p-1}C) & \xrightarrow{j} & H_{p+q-1}(F_{p-1}C, F_{p-2}C)
\end{array}$$

and $d^1 = jk$. □

Definition. *A filtration* (Φ_pH) *of a* $\mathbb{Z}$*-graded module* H *is* bounded *if, for each* n, *there exist integers* s *and* t *such that*

$$\Phi_pH_n = 0 \quad \textit{for } p < s$$

$$\Phi_pH_n = H_n \quad \textit{for } p > t.$$

Definition. *A* k*-spectral sequence* (E^r) converges *to a* $\mathbb{Z}$*-graded module* H, *written*

$$E^k_{pq} \underset{p}{\Longrightarrow} H_n,$$

if there are a bounded filtration (Φ_pH) *of* H *and isomorphisms* $E^\infty_{pq} \cong \Phi_pH_n/\Phi_{p-1}H_n$ *(where* $n = p + q$*).*

Theorem 9.4. *Let (F_pC) be a bounded filtration of a complex C and let (E^r) be the corresponding spectral sequence, viewed as a 2-spectral sequence by ignoring E^1. Then:*

(1) *the sequence $(\Phi_p H)$, where $\Phi_p H_n = \operatorname{Im}(H_n(F_pC) \to H_n(C))$ (image of the map induced by inclusion $F_pC \hookrightarrow C$) is a bounded filtration of $H(C)$;*
(2) *for each pair p, q, $E^\infty_{pq} = E^r_{pq}$ for all sufficiently large r;*
(3) *further, $E^2_{pq} \underset{p}{\Longrightarrow} H_n(C)$.*

Proof. By Lemma 9.2, for fixed p, q and all sufficiently large r,

$$Z^r_{pq} = Z^\infty_{pq} = \operatorname{Ker}(H_{p+q}(F_pC, F_{p-1}C) \xrightarrow{\alpha} H_{p+q-1}(F_{p-1}C))$$

$$B^r_{pq} = B^\infty_{pq} = \operatorname{Im}(H_{p+q+1}(C, F_pC) \longrightarrow H_{p+q}(F_pC, F_{p-1}C))$$

$$= \operatorname{Ker}(H_{p+q}(F_pC, F_{p-1}C) \xrightarrow{\beta} H_{p+q}(C, F_{p-1}C))$$

(from the exact homology sequence of $(C, F_pC, F_{p-1}C)$) and (2) follows. Using the exact homology sequences of (C, F_pC) and $(C, F_{p-1}C)$,

$$\Phi_p H_n = \operatorname{Ker}(H_{p+q}(C) \xrightarrow{\gamma} H_{p+q}(C, F_pC))$$

$$\Phi_{p-1} H_n = \operatorname{Ker}(H_{p+q}(C) \xrightarrow{\delta} H_{p+q}(C, F_{p-1}C)) \quad (n = p+q)$$

and boundedness of $(\Phi_p H)$ follows. With the maps α, etc. as indicated above, there is a commutative diagram with exact rows and columns:

$$
\begin{array}{ccccc}
H_{p+q}(C) & \xrightarrow{\gamma} & H_{p+q}(C, F_pC) & \longrightarrow & H_{p+q-1}(F_pC) \\
\| & & \uparrow \zeta & & \uparrow \\
H_{p+q}(C) & \xrightarrow{\delta} & H_{p+q}(C, F_{p-1}C) & \xrightarrow{\varepsilon} & H_{p+q-1}(F_{p-1}C) \\
 & & \beta \uparrow & & \uparrow \alpha \\
 & & H_{p+q}(F_pC, F_{p-1}C) & = & H_{p+q}(F_pC, F_{p-1}C).
\end{array}
$$

(The bottom square is commutative by applying Theorem 3.2 to the short exact sequences obtained from the pairs $(C, F_{p-1}C)$ and $(F_pC, F_{p-1}C)$ as in Corollary 3.1.) Now $\dfrac{\Phi_p H_n}{\Phi_{p-1} H_n} = \dfrac{\operatorname{Ker}(\gamma)}{\operatorname{Ker}(\delta)} = \dfrac{\delta^{-1}(\operatorname{Ker}(\zeta))}{\delta^{-1}(0)} \cong \operatorname{Im}(\delta) \cap \operatorname{Ker}(\zeta)$ and symmetrically $E^\infty_{pq} = \dfrac{\operatorname{Ker}(\alpha)}{\operatorname{Ker}(\beta)} \cong \operatorname{Im}(\beta) \cap \operatorname{Ker}(\varepsilon)$. By exactness, $\operatorname{Im}(\beta) = \operatorname{Ker}(\zeta)$ and $\operatorname{Ker}(\varepsilon) = \operatorname{Im}(\delta)$, and (3) follows. □

Bicomplexes

A *bicomplex* is a $\mathbb{Z}$-bigraded module $C = (C_{pq})$ with two differentials

$$d' \quad \text{of bidegree } (-1, 0)$$
$$d'' \quad \text{of bidegree } (0, -1)$$

such that $d'd'' + d''d' = 0$.

$$\begin{array}{ccc} C_{p-1,q} & \xleftarrow{d'} & C_{pq} \\ \downarrow{\scriptstyle d''} & & \downarrow{\scriptstyle d''} \\ C_{p-1,q-1} & \xleftarrow[d']{} & C_{p,q-1} \end{array}$$

The *total complex* $(T(C), d)$ is the chain complex with $T(C)_n = \bigoplus_{p+q=n} C_{pq}$ and $d|_{C_{pq}} = d'_{pq} + d''_{pq}$.

Note. *If instead of $d'd'' + d''d' = 0$ one has $d'd'' = d''d'$, one can obtain a bicomplex by replacing d'' by e'', where $e''_{pq} = (-1)^p d''_{pq}$.*

Example. Let (U, u), (V, v) be chain complexes over a ring A ((U, u) of right A-modules). Define $C_{pq} = U_p \otimes_A V_q$, $d' = u \otimes 1$, $d'' = 1 \otimes v$, so d', d'' are differentials with $d'd'' = d''d'$. Changing d'' as in the note gives a bicomplex C. The tensor product $U \otimes_A V$ is defined to be the total complex of C.

Given a bicomplex (C, d', d''), fix p, to get a chain complex

$$\ldots C_{p,q+1} \xrightarrow{d''} C_{pq} \xrightarrow{d''} C_{p,q-1} \to \ldots \quad .$$

Denote its homology by $H''_{p,*}(C)$, that is, $H''_{pq}(C) = \mathrm{Ker}(d''_{pq})/\mathrm{Im}(d''_{p,q+1})$. The map $d'_{pq} : C_{pq} \to C_{p-1,q}$ induces a differential

$$\delta'_{pq} : H''_{pq}(C) \to H''_{p-1,q}(C) \quad \text{by}$$
$$\delta'_{pq}(c + \mathrm{Im}(d''_{p,q+1})) = d'_{pq}(c) + \mathrm{Im}(d''_{p-1,q+1})$$

so for fixed q, $(H''_{*,q}(C), \delta')$ is a chain complex; denote its homology by $H'_*(H''_{*,q}(C))$, that is,

$$H'_p(H''_{pq}(C)) = \mathrm{Ker}(\delta'_{pq})/\mathrm{Im}(\delta'_{p+1,q}).$$

Similarly define $H'_{pq}(C) = \mathrm{Ker}(d'_{pq})/\mathrm{Im}(d'_{p+1,q})$. We obtain an induced differential $\delta''_{pq} : H'_{pq}(C) \to H'_{p,q-1}(C)$ and homology modules $H''_q(H'_{pq}(C)) = \mathrm{Ker}(\delta''_{pq})/\mathrm{Im}(\delta''_{p,q+1})$.

First filtration

With (C, d', d'') as above, let $T = T(C)$. We can define a filtration on T by

$$(F_pT)_n = \bigoplus_{i \le p} C_{i,n-i}.$$

Viewing the modules C_{pq} as situated at the lattice points (p, q) in the plane, (F_pT) is obtained by replacing C_{pq} by 0 to the right of the line $x = p$. (T_n is

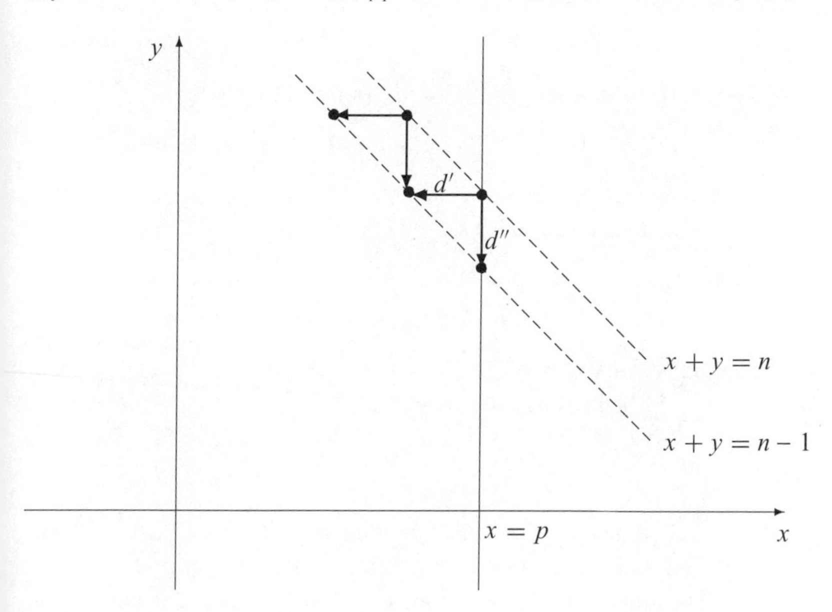

the direct sum of all modules on the line $x + y = n$.)

Lemma 9.5. *Let (E^r) be the* 1*-spectral sequence obtained from the filtration (F_pT) just defined. Then*

$$E^2_{pq} = H'_p(H''_{pq}(C)).$$

Proof. The quotient complex $F_pT/F_{p-1}T$ has $(F_pT/F_{p-1}T)_n = C_{p,n-p}$ and differential d'' (column $x = p$ in the picture), hence

$$E^1_{pq} = H_{p+q}(F_pT, F_{p-1}T) = H''_{pq}(C).$$

By Lemma 9.3, the spectral sequence differential d^1 is the connecting homomorphism, ∂, arising from the short exact sequence of chain complexes

$$0 \to \frac{F_{p-1}T}{F_{p-2}T} \xrightarrow{i} \frac{F_pT}{F_{p-2}T} \xrightarrow{\pi} \frac{F_pT}{F_{p-1}T} \to 0.$$

In total degrees n and $n-1$ (with $p+q=n$), this looks like:

$$\begin{array}{ccccccccc} 0 & \longrightarrow & C_{p-1,q+1} & \longrightarrow & C_{p-1,q+1} \oplus C_{pq} & \xrightarrow{\pi} & C_{pq} & \longrightarrow & 0 \\ & & \downarrow{\scriptstyle d''} & & \downarrow{\scriptstyle (d'',d'+d'')} & & \downarrow{\scriptstyle d''} & & \\ 0 & \longrightarrow & C_{p-1,q} & \xrightarrow{i} & C_{p-1,q} \oplus C_{p,q-1} & \longrightarrow & C_{p,q-1} & \longrightarrow & 0. \end{array}$$

Recall (Section 2) that, to define ∂, given $z \in C_{pq}$ with $d''(z)=0$, we first choose $y \in C_{p-1,q+1} \oplus C_{pq}$ such that $\pi(y)=z$. We take $y=(0,z)$. Then

$$\begin{aligned} \partial(z+\mathrm{Im}(d'')) &= i^{-1}(d'', d'+d'')(y) + \mathrm{Im}(d'') \\ &= i^{-1}(d'(z),0) + \mathrm{Im}(d'') = d'(z) + \mathrm{Im}(d'') \end{aligned}$$

(the choice of inverse image under i is immaterial, and we take it to be $d'(z)$). Hence $\partial = \delta'$ and the result follows. □

Second filtration

This filtration is denoted by (G_pT). To obtain G_pT, we replace all C_{pq} above the line $y=p$ by 0. Thus $(G_pT)_n = \bigoplus_{j \le p} C_{n-j,j}$.

Note.

(1) A $\mathbb{Z}$-bigraded module C is called *first-quadrant* if $C_{pq}=0$ for $p<0$ or $q<0$, and *third-quadrant* if $C_{pq}=0$ for $p>0$ or $q>0$.
(2) If C is a bicomplex, then both filtrations of T are bounded if and only if, for all n, there are only finitely many pairs (p,q) with $p+q=n$ and $C_{pq} \neq 0$.
(3) Hence, if C is first or third quadrant, both filtrations are bounded.

Theorem 9.6. *Let C be a first or third quadrant bicomplex. Then there are two convergent 2-spectral sequences arising from the first and second filtrations of $T(C)$:*

(1) $E^2_{pq} = H'_p(H''_{pq}(C)) \underset{p}{\Longrightarrow} H_n(T(C))$;
(2) $E^2_{pq} = H''_p(H'_{qp}(C)) \underset{p}{\Longrightarrow} H_n(T(C))$.

Proof. (1) This is immediate from Theorem 9.4 and Lemma 9.5.
(2) This follows from (1) applied to the *transpose* of C, that is, the bicomplex (D, e', e'') given by $D_{pq} = C_{qp}$, $e'_{pq} = d''_{qp}$ and $e''_{pq} = d'_{qp}$, since $T(D) = T(C)$ and the second filtration for $T(C)$ is the first filtration for $T(D)$. □

Definition. *Let M be a $\mathbb{Z}$-bigraded module over a ring A with IBN. Suppose M_{pq} is of type FL over A for all pairs p, q, and $M_{pq} = 0$ for all but finitely many pairs p, q. Then put*

$$\mathrm{rk}_A(M) = \sum_{p,q} (-1)^{p+q} \,\mathrm{rk}_A(M_{pq}).$$

Proposition 9.7. *Suppose A is a principal ideal domain, (E^r, d^r) is a k-spectral sequence over A and*

$$E^k_{pq} \underset{p}{\Longrightarrow} H_n.$$

If E^k_{pq} is finitely generated for all p, q, then H_n is finitely generated for all n. If also $E^k_{pq} = 0$ for all but finitely many p, q, then $H_n = 0$ for all but finitely many n, and

$$\mathrm{rk}_A(E^k) = \sum_{n \in \mathbb{Z}} (-1)^n \,\mathrm{rk}_A(H_n).$$

Proof. Suppose E^k_{pq} is finitely generated for all p, q. Since E^∞_{pq} is isomorphic to a section (subquotient) of E^k_{pq}, E^∞_{pq} is finitely generated for all p, q. There is a bounded filtration $(\Phi_p H)$ of H such that $(\Phi_p H)_n/(\Phi_{p-1} H)_n \cong E^\infty_{p,n-p}$ and it follows that H_n is finitely generated for all n. Now suppose also $E^k_{pq} = 0$ for all pairs $(p, q) \notin I$, where I is some finite subset of $\mathbb{Z} \times \mathbb{Z}$. Then for $(p, q) \notin I$ and all $r \geq k$, $E^r_{pq} = 0 = E^\infty_{pq}$ (again because they are sections of E^k_{pq}). Consequently, for each pair p, q, there exists l such that, for $r \geq l$, $E^r_{p-r,q+r-1} = E^r_{p+r,q-r+1} = 0$, so $\mathrm{Ker}(d^r_{pq}) = E^r_{pq}$ and $\mathrm{Im}(d^r_{p+r,q-r+1}) = 0$. Hence $E^l_{pq} \cong E^{l+1}_{pq} \cong \ldots \cong E^\infty_{pq}$. It follows that, for some r, $E^r \cong E^\infty$ as $\mathbb{Z}$-bigraded module. Using induction on r and the exact sequences

$$0 \to \mathrm{Ker}(d^r_{pq}) \to E^r_{pq} \to \xrightarrow{d^r_{pq}} \mathrm{Im}(d^r_{pq}) \to 0$$

$$0 \to \mathrm{Im}(d^r_{p+r,q-r+1}) \to \mathrm{Ker}(d^r_{pq}) \to E^{r+1}_{pq} \to 0$$

we see that $\mathrm{rk}_A(E^r) = \mathrm{rk}_A(E^k)$ for all integers $r \geq k$ (see Lemma 6.24 and Cor. 6.25 for a similar argument). Hence $\mathrm{rk}_A(E^k) = \mathrm{rk}_A(E^\infty)$. Also, we can find s, t such that $E^\infty_{pq} = 0$ for $|p| > s$ or $|q| > t$. Then for $|n| > s + t$, $E^\infty_{p,n-p} = 0$ for all p, so $H_n = 0$. Thus $H_n = 0$ for all but finitely many n. Since the filtration $(\Phi_p H)$ is bounded,

$$\mathrm{rk}_A(H_n) = \sum_{p \in \mathbb{Z}} \mathrm{rk}_A \left(\frac{\Phi_p H_n}{\Phi_{p-1} H_n} \right) = \sum_{p \in \mathbb{Z}} \mathrm{rk}_A(E^\infty_{p,n-p}).$$

Hence,

$$\sum_{n\in\mathbb{Z}}(-1)^n \operatorname{rk}_A(H_n) = \sum_{p,n}(-1)^n \operatorname{rk}_A(E^{\infty}_{p,n-p}) = \operatorname{rk}_A(E^{\infty}) = \operatorname{rk}_A(E^k).$$

□

Remark 9.8. *Let G be a group, R a commutative ring and M an RG-module. Then as noted in Section 5, M is a right RG-module, via $mg = g^{-1}m$ ($g \in G$, $m \in M$). If N is an RG-module, $M \otimes_R N$ is an RG-module with "diagonal" G-action: $g(m \otimes n) = gm \otimes gn$. From the construction of tensor products, $M \otimes_{RG} N$ is obtained from $M \otimes_R N$ by factoring out the RG-submodule generated by $X = \{m \otimes gn - mg \otimes n \mid g \in G, m \in M, n \in N\}$. Now*

$$\begin{aligned} X &= \{gm \otimes gn - m \otimes n \mid g \in G, m \in M, n \in N\} \text{ (replacing } m \text{ by } gm) \\ &= \{(g-1)(m \otimes n) \mid g \in G, m \in M, n \in N\}, \end{aligned}$$

so this submodule is $I_G(M \otimes_R N)$, hence $M \otimes_{RG} N \cong \dfrac{M \otimes_R N}{I_G(M \otimes_R N)} \cong (M \otimes_R N)_G$.
(Recall: if U is an RG-module, $U_G = R \otimes_{RG} U \cong U/I_G U$-see the note preceding Lemma 8.8.) The usual isomorphism $M \otimes_R N \cong N \otimes_R M$ induces an isomorphism $M \otimes_{RG} N \cong N \otimes_{RG} M$. Similar remarks apply when M, N are chain complexes.

Definition. *Let R be a commutative ring, G a group. An RG-module U is called H_*-acyclic if $H_n(G, U) = 0$ for all $n > 0$.*

Theorem 9.9. *Let R be a commutative ring, G a group, U a positive complex of RG-modules with every U_n H_*-acyclic. Then there is a convergent 2-spectral sequence*

$$E^2_{pq} = H_p(G, H_q(U)) \underset{p}{\Longrightarrow} H_n(U_G).$$

Proof. Let $V \to R$ be an augmented RG-projective resolution. View U as a complex of right RG-modules (see Remark 9.8: $ug = g^{-1}u$ for $g \in G, u \in U_n$). Then (example above) $U \otimes_{RG} V$ is the total complex of a bicomplex C with

$$C_{pq} = U_p \otimes_{RG} V_q.$$

Now C is first quadrant, so by Theorem 9.6, there are two spectral sequences converging to $H(U \otimes_{RG} V)$. Consider the first spectral sequence, with $E^2_{pq} =$

$H'_p(H''_{pq}(C))$. Now $(H''_{pq}(C))_{q\in\mathbb{Z}}$ is the homology of the pth column of C:

$$\begin{array}{c} U_p \otimes V_q \\ \downarrow \\ U_p \otimes V_{q-1} \\ \downarrow \\ \vdots \end{array}$$

which, up to sign of the differential, is $U_p \otimes_{RG} V$. Hence

$$H''_{pq}(C) = H_q(G, U_p) = 0 \text{ if } q \neq 0.$$

For $q = 0$, the exact sequence $V_1 \to V_0 \to R \to 0$ gives an exact sequence of chain complexes

$$U \otimes_{RG} V_1 \to U \otimes_{RG} V_0 \to U_G \to 0$$

($U_G \cong U \otimes_{RG} R$-see Remark 9.8 above). It follows that there is an isomorphism of chain complexes

$$(H''_{p,0}(C), \delta'_{p,0}) \cong U_G.$$

Thus

$$E^2_{pq} = \begin{cases} 0 & \text{if } q \neq 0 \\ H_p(U_G) & \text{if } q = 0. \end{cases}$$

Therefore all differentials d^r are 0 for $r \geq 2$ and $E^\infty_{pq} = E^2_{pq}$ for all p, q. Also, putting $H = H(U \otimes_{RG} V)$, there is a bounded filtration $(\Phi_p H)$ of H such that $(\Phi_p H)_n/(\Phi_{p-1} H)_n \cong E^\infty_{p,n-p}$. Hence

$$\begin{aligned} (\Phi_n H)_n/(\Phi_{n-1} H)_n &= H_n(U_G) \\ (\Phi_{n-1} H)_n = (\Phi_{n-2} H)_n &= \ldots = 0 \\ (\Phi_n H)_n = (\Phi_{n+1} H)_n &= \ldots = H_n \end{aligned}$$

and so $H_n = H_n(U_G)$. Now consider the second spectral sequence. First,

$$H'_{pq}(C) = H_p(U \otimes_{RG} V_q) \cong H_p(U) \otimes_{RG} V_q$$

since $- \otimes_{RG} V_q$ is exact (details are left as an exercise). Moreover, this gives an isomorphism of chain complexes

$$(H'_{p,*}(C), \delta''_{p,*}) \cong H_p(U) \otimes_{RG} V$$

up to sign of the differential. Hence $H''_q(H'_{pq}(C)) = H_q(G, H_p(U))$. Thus the second spectral sequence has $E^2_{pq} = H_p(G, H_q(U))$, so this is the required spectral sequence. □

Recall that, if $f : A \to B$ is a ring homomorphism and M is an A-module, $B \otimes_A M$ is called the B-module *induced* from M. The next result is about induced modules, where H is a subgroup of a group G and f is the corresponding inclusion map of group rings.

9.10 (Shapiro's Lemma)**.** *Let G be a group, H a subgroup of G, R a commutative ring, M an RH-module. Then for all n,*

$$H_n(H, M) \cong H_n(G, RG \otimes_{RH} M).$$

Proof. Let $P \to R$ be an augmented RG-projective resolution, so an RH-projective resolution (Remark 5.8). Then

$$\begin{aligned} H_*(H, M) &= H(M \otimes_{RH} P) \\ &\cong H(M \otimes_{RH} (RG \otimes_{RG} P)) \quad \text{(see Remark 2.10)} \\ &\cong H((M \otimes_{RH} RG) \otimes_{RG} P) \\ &= H_*(G, RG \otimes_{RH} M). \end{aligned}$$

(Some care is needed in using associativity of tensor product. We view $M \otimes_{RH} RG$ as a right RG-module via $(m \otimes g)g_1 = m \otimes (gg_1) (g, g_1 \in G, m \in M)$. But $RG \otimes_{RH} M$ is viewed as a right RG-module via $(g \otimes m)g_1 = g_1^{-1}(g \otimes m) = (g_1^{-1}g) \otimes m$. The isomorphism $M \otimes_{RH} RG \to RG \otimes_{RH} M$ used in the last step is given by $m \otimes g \mapsto g^{-1} \otimes m$.) □

Taking $H = 1$ gives the following

Corollary 9.11. *Induced modules $RG \otimes_R M$ are H_*-acyclic.*

Lemma 9.12. *Let G be a group, R a commutative ring, M an RG-module. Let M_0 be the underlying R-module corresponding to M. Then $RG \otimes_R M$ (with diagonal G-action) is RG-isomorphic to $RG \otimes_R M_0$ (as induced module).*

Proof. There are well-defined maps

$$\begin{aligned} \varphi : RG \otimes_R M_0 &\to RG \otimes_R M \\ g \otimes m &\mapsto g \otimes gm \\ \psi : RG \otimes_R M &\to RG \otimes_R M_0 \\ g \otimes m &\mapsto g \otimes g^{-1}m \end{aligned}$$

which are clearly inverse maps and RG-homomorphisms. □

Theorem 9.13 (Hochschild-Serre)**.** *Let $1 \to N \to G \to Q \to 1$ be a group extension, R a commutative ring, M an RG-module. Then there is a convergent*

2-*spectral sequence*

$$E^2_{pq} = H_p(Q, H_q(N, M)) \underset{p}{\Longrightarrow} H_n(G, M).$$

Proof. Let $P \to R$ be an augmented RG-projective resolution. Using Remark 9.8 preceding Theorem 9.9:

$$\begin{aligned} M \otimes_{RG} P &\cong (M \otimes_R P)_G \\ &\cong ((M \otimes_R P)_N)_Q \quad \text{by Lemma 8.8} \\ &\cong (M \otimes_{RN} P)_Q. \end{aligned}$$

Thus, putting $U = M \otimes_{RN} P$,

$$\begin{aligned} H_*(G, M) &\cong H(U_Q) \\ H_*(N, M) &\cong H(U). \end{aligned}$$

The theorem will follow from Theorem 9.9 provided each U_p is H_*-acyclic (as RQ-module). Summands and direct sums of H_*-acyclic modules are H_*-acyclic, since for any collection $\{M_i\}$ of RQ-modules, $H_*(Q, \bigoplus_i M_i) \cong \bigoplus_i H_*(Q, M_i)$ (because tensor product distributes over direct sum). Hence it is enough to show that $M \otimes_{RN} RG$ is H_*-acyclic. Now

$$\begin{aligned} M \otimes_{RN} RG &\cong (M \otimes_R RG)_N \cong (RG \otimes_R M)_N \\ &\cong (RG \otimes_R M_0)_N \quad \text{(see Lemma 9.12)} \\ &\cong RQ \otimes_{RG} (RG \otimes_R M_0) \quad \text{by Lemma 8.8} \\ &\cong RQ \otimes_R M_0 \quad \text{(by associativity of tensor product)} \end{aligned}$$

which is H_*-acyclic by Cor. 9.11. □

10. The Brown characteristic

We are now ready to give the second definition of an Euler characteristic and establish its properties.

Definition. *A group G is* of finite homological type *(written $G \in$ FHT) if*

(i) *G has a subgroup H of finite index with $\mathrm{cd}_{\mathbb{Z}}(H) < \infty$;*
(ii) *for all $\mathbb{Z}G$-modules M which are $\mathbb{Z}$-finitely generated, $H_n(G, M)$ is finitely generated for all n.*

Proposition 10.1. *Let G be a group, H a subgroup of G of finite index. Then $G \in$ FHT if and only if $H \in$ FHT.*

Proof. Suppose $G \in \mathrm{FHT}$ and let K be a subgroup of finite index with $\mathrm{cd}_{\mathbb{Z}}(K) < \infty$. Then $(H : H \cap K) < \infty$ and $\mathrm{cd}_{\mathbb{Z}}(H \cap K) < \infty$ by Lemma 5.9. If M is a $\mathbb{Z}$-finitely generated $\mathbb{Z}H$-module, then by Lemma 9.10,

$$H_n(H, M) \cong H_n(G, \mathbb{Z}G \otimes_{\mathbb{Z}H} M)$$

and $\mathbb{Z}G \otimes_{\mathbb{Z}H} M$ is $\mathbb{Z}$-finitely generated. For M is a quotient of $\mathbb{Z}^m$ for some integer m, hence $\mathbb{Z}G \otimes_{\mathbb{Z}H} M$ is a quotient of $(\mathbb{Z}G \otimes_{\mathbb{Z}H} \mathbb{Z})^m$, so of $\mathbb{Z}(G/H)^m$ by Lemma 5.2, and G/H is finite. It follows that $H_n(H, M)$ is finitely generated, so $H \in \mathrm{FHT}$.

Suppose $H \in \mathrm{FHT}$. If K is a subgroup of H of finite index such that $\mathrm{cd}_{\mathbb{Z}}(K) < \infty$, then $(G : K) < \infty$, so (i) in the definition holds for G. Since $\bigcap_{g \in G} gHg^{-1}$ has finite index in H, it is in FHT by what has already been proved, and is normal in G. Thus we can assume $H \trianglelefteq G$. Let $Q = G/H$, and let M be a $\mathbb{Z}$-finitely generated $\mathbb{Z}G$-module. By Theorem 9.13, there is a spectral sequence

$$E^2_{pq} = H_p(Q, H_q(H, M)) \underset{p}{\Longrightarrow} H_n(G, M)$$

and $H_q(H, M)$ is finitely generated by assumption, hence so is E^2_{pq} since Q is finite (this is left as an exercise). By Prop. 9.7, $H_n(G, M)$ is finitely generated for all n. □

Let $\mathcal{C}$ be the class of all $G \in \mathrm{FHT}$ with $\mathrm{cd}_{\mathbb{Z}}(G) < \infty$. Then for $G \in \mathcal{C}$, $\tilde{\chi}(G)$ is defined (recall that $\tilde{\chi}(G) = \sum_i (-1)^i \mathrm{rk}_{\mathbb{Z}}(H_i(G, \mathbb{Z}))$ -see the definition preceding Lemma 8.9).

Proposition 10.2. *The map $\tilde{\chi}$ is an Euler characteristic on $\mathcal{C}$. That is, if $G \in \mathcal{C}$ and H is a subgroup of G of finite index, then $H \in \mathcal{C}$ and $\tilde{\chi}(H) = (G : H)\tilde{\chi}(G)$.*

Proof. The class $\mathcal{C}$ is closed under taking subgroups of finite index by Prop. 10.1 and Lemma 5.9. If $G \in \mathcal{C}$ and $(G : H) < \infty$, let $N = \bigcap_{g \in G} gHg^{-1}$. It is enough to show that $\tilde{\chi}(N) = (G : N)\tilde{\chi}(G)$ and $\tilde{\chi}(N) = (H : N)\tilde{\chi}(H)$, so we can assume $H \trianglelefteq G$. The result then follows by Lemma 8.9. □

By Prop. 10.1, the class $\mathrm{v}\mathcal{C}$ of finite extensions of groups in $\mathcal{C}$ is just FHT. We can extend $\tilde{\chi}$ to an Euler characteristic χ on FHT in the usual way. If $G \in \mathrm{FHT}$, let H be a subgroup of finite index in G with $\mathrm{cd}_{\mathbb{Z}}(H) < \infty$, and define

$$\chi(G) = \frac{1}{(G : H)} \tilde{\chi}(H).$$

Proposition 10.3. *The map χ is an Euler characteristic on* FHT. *That is, if $G \in \mathrm{FHT}$ and $(G : H) < \infty$, then $H \in \mathrm{FHT}$ and $\chi(H) = (G : H)\chi(G)$.*

Proof. This follows easily from Prop. 10.2. □

Note.

(1) *Clearly* FP($\mathbb{Z}$) $\subseteq \mathcal{C}$. *Recall that, for* $G \in$ FP($\mathbb{Z}$), $\mu(G) = \chi_G(1)$, *while by Cor.* 6.25, $\tilde{\chi}(G) = \chi(G) = \sum_{\tau \in T(G)} \chi_G(\tau)$. *It is not known whether* $\mu(G) = \chi(G)$ *for all* $G \in$ FP($\mathbb{Z}$). *It is true if* $G \in$ FL($\mathbb{Z}$) ($\chi_G = \mu(G)[1]$ *for such* G).

(2) *By Prop.* 10.1, vFP($\mathbb{Z}$) $\subseteq$ FHT *and* χ, μ *agree on* vFL($\mathbb{Z}$). *It is false that* FP($\mathbb{Q}$) $\subseteq$ FHT. *There are groups in* FP($\mathbb{Q}$) *which are not virtually torsion-free. An example will be given later in this section.*

The Serre extension theorem

This is an important result relating the cohomological dimension of a group with that of a subgroup of finite index. The next few lemmas are needed in its proof. If $f : A \to B$ is a ring homomorphism and M is an A-module, as well as the induced module we can form the B-module *coinduced* from M. This is $\mathrm{Hom}_A(B, M)$, with scalar multiplication $bg(x) = g(xb)$ for $b, x \in B$, $g \in \mathrm{Hom}_A(B, M)$. In Section 9 we proved Shapiro's Lemma on homology of induced modules. There is a version for cohomology of coinduced modules.

Shapiro's Lemma for Cohomology. *Let G be a group, H a subgroup, R a commutative ring, M an RH-module. Then*

$$H^n(H, M) \cong H^n(G, \mathrm{Hom}_{RH}(RG, M)).$$

Proof. Let $P \to R$ be an augmented RG-projective resolution (so an RH-projective resolution). The adjoint isomorphism ([100, Theorem 2.11]) gives an isomorphism of cochain complexes

$$\mathrm{Hom}_{RH}(P, M) \cong \mathrm{Hom}_{RG}(P, \mathrm{Hom}_{RH}(RG, M))$$

and the result follows. □

For subgroups of finite index, coinduced and induced modules coincide.

Lemma 10.4. *If H is a subgroup of finite index in a group G and R is a commutative ring, then for any RH-module M, there is an RG-module isomorphism $RG \otimes_{RH} M \cong \mathrm{Hom}_{RH}(RG, M)$.*

Proof. Using the fact that RG is R-free on G, define $\varphi_0 : M \to \mathrm{Hom}_{RH}(RG, M)$ by $\varphi_0(m)(g) = \begin{cases} gm & \text{if } g \in H \\ 0 & \text{if } g \notin H \end{cases}$ and $\varphi : RG \otimes_{RH} M \to$

$\mathrm{Hom}_{RH}(RG, M)$ by $\varphi(g \otimes m) = g\varphi_0(m)$. It is easy to check that φ is an RG-homomorphism. Take a transversal T so that $G = \coprod_{t \in T} tH$ (disjoint union) and define $\psi : \mathrm{Hom}_{RH}(RG, M) \to RG \otimes_{RH} M$ by $\psi(f) = \sum_{t \in T} t \otimes f(t^{-1})$, and check that φ, ψ are inverse maps. □

In the next two lemmas, R is a commutative ring.

Lemma 10.5. *If $n = \mathrm{cd}_R(G) < \infty$, then there is a free RG-module F such that $H^n(G, F) \neq 0$.*

Proof. There is an RG-module M such that $H^n(G, M) \neq 0$, and there is a free module F mapping onto M, hence a short exact sequence of RG-modules $0 \to K \to F \to M \to 0$. The corresponding exact cohomology sequence (see the end of Section 3) is, in part:

$$\ldots \to H^n(G, K) \to H^n(G, F) \to H^n(G, M) \to 0$$

since $H^{n+1}(G, K) = 0$. Therefore $H^n(G, F) \neq 0$. □

Lemma 10.6. *Suppose $H \leq G$, $(G : H) < \infty$ and $\mathrm{cd}_R(G) < \infty$. Then $\mathrm{cd}_R(G) = \mathrm{cd}_R(H)$.*

Proof. Let $n = \mathrm{cd}_R(G)$. By Lemma 10.5, there is a free RG-module F such that $H^n(G, F) \neq 0$. Let F' be a free RH-module of the same rank as F. By Lemma 10.4,

$$\mathrm{Hom}_{RH}(RG, F') \cong F$$

and by Shapiro's Lemma for Cohomology $H^n(H, F') \neq 0$. Hence $\mathrm{cd}_R(H) \geq n$. But by Lemma 5.9, $\mathrm{cd}_R(H) \leq n$. □

Definition. *Let G be a group, R commutative ring. We say that G* has no R-torsion *if the order of every finite subgroup of G is invertible in R.*

By Corollary 5.12, if $\mathrm{cd}_R(G) < \infty$ then G has no R-torsion. Before proving Serre's theorem, we need to discuss the tensor product of chain complexes, which was defined in the discussion of bicomplexes in Section 9. Let A be a ring, $f : C \to E, g : D \to F$ chain maps (where C, E are chain complexes of right A-modules, D, F of A-modules). It is easy to see that there is a chain map $f \otimes g : C \otimes_A D \to E \otimes_A F$, where $(f \otimes g)_n(c \otimes d) = f_p(c) \otimes g_q(d)$, for $c \in C_p$, $d \in D_q$, where $p + q = n$.

Lemma 10.7. *Suppose $f, f' : C \to E$, $g, g' : D \to F$ are chain maps. If $f \simeq f'$ and $g \simeq g'$, then $f \otimes g \simeq f' \otimes g'$.*

Proof. If $f \simeq f'$ via s and $g \simeq g'$ via t, then $f \otimes g \simeq f' \otimes g'$ via u, where

$$u(c \otimes d) = s(c) \otimes g(d) + (-1)^p (f'(c) \otimes t(d)) \qquad \text{(where } c \in C_p, d \in D_q). \qquad \square$$

Remark 10.8. *Let A be a ring and let X be a positive chain complex of A-modules, $X \xrightarrow{\varepsilon} M$ an augmented complex. Then the augmented complex splits if and only if there is an A-homomorphism $\eta : M \to X_0$ such that $\varepsilon\eta = \mathrm{id}_M$, and $\mathrm{id}_X \simeq f$, where $f_n = 0$ $(n \neq 0)$ and $f_0 = \eta\varepsilon$. (If s is a chain homotopy from* id *to* 0 *on $X \xrightarrow{\varepsilon} M$, let $\eta = s_{-1}$, etc.)*

Let X, X' be positive complexes of A-modules (X of right modules), $X \xrightarrow{\varepsilon} M$, $X' \xrightarrow{\varepsilon'} M'$ augmented complexes. Then $(X \otimes_A X')_0 = X_0 \otimes_A X'_0$, and there is an augmented complex

$$X \otimes_A X' \xrightarrow{\varepsilon \otimes \varepsilon'} M \otimes_A M'.$$

Corollary 10.9. *In these circumstances, if $X \xrightarrow{\varepsilon} M$, $X' \xrightarrow{\varepsilon'} M'$ are split, then so is $X \otimes_A X' \xrightarrow{\varepsilon \otimes \varepsilon'} M \otimes_A M'$.*

Proof. By Remark 10.8, there exist $\eta : M \to X_0$, $\eta' : M' \to X'_0$ with $\mathrm{id}_M = \varepsilon\eta$, $\mathrm{id}'_M = \varepsilon'\eta'$, $\mathrm{id}_X \simeq f$, $\mathrm{id}_{X'} \simeq f'$, where f' is defined in a similar way to f. By Lemma 10.7, $\mathrm{id}_{X \otimes X'} \simeq f \otimes f'$. Putting $g = f \otimes f'$, $g_n = 0$ for $n \neq 0$ and $g_0 = (\eta \otimes \eta')(\varepsilon \otimes \varepsilon')$. Also, $(\varepsilon \otimes \varepsilon')(\eta \otimes \eta') = \mathrm{id}_{M \otimes M'}$, and the result follows by Remark 10.8. $\square$

We are finally ready to prove the Serre Extension Theorem.

Theorem 10.10. *Let G be a group, H a subgroup of finite index, R a commutative ring. Suppose $\mathrm{cd}_R(H) < \infty$ and G has no R-torsion. Then $\mathrm{cd}_R(G) = \mathrm{cd}_R(H)$.*

Proof. By Lemma 10.6, it suffices to show $\mathrm{cd}_R(G) < \infty$. Let $P \to R$ be an augmented RH-projective complex of finite length. Since R is commutative, we can iterate the construction of tensor product of complexes (starting from the right), and define

$$Q = \underbrace{P \otimes_R P \otimes_R \ldots \otimes_R P}_{n \text{ copies}}$$

where $n = (G : H)$. Thus $Q_m = \bigoplus P_{r_1} \otimes \ldots \otimes P_{r_n}$ (direct sum over all sequences $(r_1, \ldots, r_n)$ of non-negative integers such that $\sum_{i=1}^n r_i = m$). If d is the differential on P, the differential d' on Q is given by

$$d'(x_1 \otimes \ldots \otimes x_n) = \sum_{k=1}^{n} (-1)^{c_k} (x_1 \otimes \ldots \otimes d(x_k) \otimes \ldots \otimes x_n)$$

where $c_k = r_1 + \ldots + r_{k-1}$ and $x_i \in P_{r_i}$. Note that Q is of finite length. By Lemma 3.9, $P \to R$ is R-split, so by repeated use of Cor. 10.9, there is an augmented resolution $Q \to R$ which is an RH-resolution (the action is "diagonal", obtained by repeated use of Remark 9.8). We shall define an action of G on Q so that this becomes an RG-projective resolution, which will prove the theorem.

Let $\{t_1, \ldots, t_n\}$ be a transversal for G/H, so $G = \bigcup_{i=1}^{n} t_i H$. This determines a homomorphism $G \to S_n$ (where S_n is the symmetric group of degree n), denoted by $g \mapsto \sigma_g$, corresponding to the permutation representation of G on G/H. Thus, for $g \in G$, there are elements $h_1, \ldots, h_n \in H$ such that, if $\sigma = \sigma_g$,

$$gt_i = t_{\sigma(i)} h_{\sigma(i)} \qquad (1 \leq i \leq n).$$

Define the action of G on Q by

$$g(x_1 \otimes \ldots \otimes x_n) = (-1)^a (h_1 x_{\sigma^{-1}(1)} \otimes \ldots \otimes h_n x_{\sigma^{-1}(n)})$$

where, if $x_i \in P_{r_i}$,

$$a = \sum r_i r_j \quad \text{(sum over all pairs } (i, j) \text{ with } i < j \text{ and } \sigma(i) > \sigma(j)\text{)}.$$

One can check this makes $Q \to R$ into an augmented RG-resolution. (For details see [30, §1].)

To show Q is RG-projective, it suffices to show that, for any collection of RH-projective modules $\{P_i\}$, the corresponding RG-module $\bigoplus_m Q_m$ is RG-projective. If $\{P'_i\}$ is another such collection and $P''_i = P_i \oplus P'_i$, then $\bigoplus_m Q_m$ is an RG-summand of the corresponding module $\bigoplus_m Q''_m$, so we may assume all P_i are RH-free. Let X be the union of RH-bases for the P_i. Then $\bigoplus_m Q_m$ is R-free with basis

$$W = \{h_1 x_1 \otimes \ldots \otimes h_n x_n \mid x_i \in X, h_i \in H, 1 \leq i \leq n\}.$$

Moreover, $W \cup (-W)$ is invariant under the action of G. Let $N = \mathrm{Ker}(G \to S_n)$ (the homomorphism $g \mapsto \sigma_g$ above). If $g \in N$, then $gt_i = h_i t_i$ for $1 \leq i \leq n$ (where $h_i \in H$). If $w = k_1 x_1 \otimes \ldots \otimes k_n x_n \in W$ (where $k_i \in H$), then

$$gw = h_1 k_1 x_1 \otimes \ldots \otimes h_n k_n x_n$$

so $g \notin G_w$ (the stabilizer of w in G) unless $g = 1$. Thus $N \cap G_w = 1$ and G_w is isomorphic to a subgroup of G/N, hence is finite. Let W_0 be a set containing one element from each $(G \times C_2)$-orbit of $W \cup (-W)$ (where C_2 is cyclic of order 2, with its non-trivial element acting as multiplication by -1). Then

$$\bigoplus_m Q_m = \bigoplus_{w \in W_0} RGw$$

so it is enough to show every RGw is RG-projective. Let $\overline{G}_w = \{g \in G \mid gw = \pm w\}$, so $(\overline{G}_w : G_w) \leq 2$ and $\overline{G}_w$ is finite. Let $\eta : G_w \to \{\pm 1\}$ be the obvious homomorphism with kernel G_w (so $gw = \eta(g)w$). Define $e = \dfrac{1}{|\overline{G}_w|} \sum_{g \in \overline{G}_w} \eta(g)g$, so $e \in RG$. Let $\pi : RG \to RGw$ be the map defined by $g \mapsto gw$. There is a split exact sequence of RG-modules

$$0 \longrightarrow \mathrm{Ker}(\pi) \longrightarrow RG \underset{\varphi}{\overset{\pi}{\rightleftarrows}} RGw \longrightarrow 0$$

where φ is given by $gw \mapsto ge$ for $g \in G$ (well-defined since $ge = \eta(g)e$ for $g \in \overline{G}_w$). Hence RGw is a summand of RG. □

It follows from Theorem 10.10 that the class $\mathfrak{C}$ of groups in FHT with $\mathrm{cd}_R(G) < \infty$ is precisely the class of torsion-free groups in FHT, hence the following corollary.

Corollary 10.11. *If $G \in$ FHT is torsion-free, then $\chi(G) = \tilde{\chi}(G)$, so $\chi(G) \in \mathbb{Z}$.*

Note.

(1) *Let R be a commutative ring, G a group, M a trivial RG-module, flat as an R-module. Let $P \to R$ be an augmented RG-projective resolution. Then by associativity of tensor product,*

$$M \otimes_R (R \otimes_{RG} P) \cong M \otimes_{RG} P$$

(the usual isomorphism $M \otimes_R R \cong M$ is an RG-isomorphism since M has trivial G-action). Hence

$$\begin{aligned} H_*(G, M) &\cong H(M \otimes_R (R \otimes_{RG} P)) \\ &\cong M \otimes_R H(R \otimes_{RG} P) \quad \textit{(since } M \textit{ is } R\textit{-flat)} \\ &\cong M \otimes_R H_*(G, R). \end{aligned}$$

(2) *If $G \in \mathfrak{C}$, then $\chi(G) = \sum_{i \geq 0} (-1)^i \dim_{\mathbb{F}_p}(H_i(G, \mathbb{F}_p))$, where $\mathbb{F}_p = \mathbb{Z}/p\mathbb{Z}$ is the field of p elements. For if $P \to \mathbb{Z}$ is an augmented $\mathbb{Z}G$-projective*

resolution of finite length, then

$$\begin{aligned}\chi(G) &= \sum_{i\geq 0}(-1)^i \operatorname{rk}_{\mathbb{Z}}(H_i(\mathbb{Z}\otimes_{\mathbb{Z}G} P))\\ &= \sum_{i\geq 0}(-1)^i \dim_{\mathbb{F}_p}(H_i(\mathbb{F}_p \otimes_{\mathbb{Z}} (\mathbb{Z}\otimes_{\mathbb{Z}G} P))) \quad \text{by Lemma 8.3}\\ &= \sum_{i\geq 0}(-1)^i \dim_{\mathbb{F}_p}(H_i(\mathbb{F}_p \otimes_{\mathbb{Z}G} P)) \quad \text{by associativity, as in (1)}\\ &= \sum_{i\geq 0}(-1)^i \dim_{\mathbb{F}_p}(H_i(G,\mathbb{F}_p)).\end{aligned}$$

Next we prove the Extension Theorem for the Brown characteristic.

Theorem 10.12. *If* $1 \to N \xrightarrow{i} G \xrightarrow{\pi} Q \to 1$ *is a group extension,* $N, Q \in \mathrm{FHT}$ *and* G *is virtually torsion-free, then* $G \in \mathrm{FHT}$, *and* $\chi(G) = \chi(N)\chi(Q)$.

Proof. We may assume i is an inclusion map. Let Q', G' be torsion-free subgroups of finite index in Q, G respectively. Let $G_0 = \pi^{-1}(Q') \cap G'$, $Q_0 = \pi(G_0)$, so Q_0, G_0 are torsion-free and of finite index in Q, G respectively. Let $N_0 = N \cap G_0$, so $(N : N_0) < \infty$ and N_0 is torsion-free. There is an exact sequence

$$1 \to N_0 \to G_0 \to Q_0 \to 1.$$

Further,

$$\begin{aligned}(G : NG_0) &= (Q : Q_0)\\ (NG_0 : G_0) &= (N : N_0)\end{aligned}$$

Hence $(G : G_0) = (Q : Q_0)(N : N_0)$. Therefore by Prop. 10.3 we can replace G, N, Q by G_0, N_0, Q_0, i.e. assume G, N, Q are torsion-free. By the observation preceding Cor. 10.11, $\mathrm{cd}_{\mathbb{Z}}(N) < \infty$ and $\mathrm{cd}_{\mathbb{Z}}(Q) < \infty$, so $\mathrm{cd}_{\mathbb{Z}}(G) < \infty$ by Prop. 5.14. By Theorem 9.13, for any RG-module M there is a spectral sequence

$$E^2_{pq} = H_p(Q, H_q(N, M)) \underset{p}{\Longrightarrow} H_n(G, M),$$

and if M is $\mathbb{Z}$-finitely generated, then since $N, Q \in \mathrm{FHT}$, E^2_{pq} is finitely generated for all p, q. Hence $H_n(G, M)$ is finitely generated for all n by Prop. 9.7. This shows $G \in \mathrm{FHT}$, and being torsion-free, $G \in \mathcal{C}$. Take $M = \mathbb{F}_2$. Then $H_q(N, \mathbb{F}_2)$ is finite, and zero for all but finitely many values of q, so there is a subgroup Q_1 of Q with $(Q : Q_1) < \infty$ such that Q_1 acts trivially on all $H_q(N, \mathbb{F}_2)$ (the intersection of the stabilisers of all elements of $H_q(N, \mathbb{F}_2)$, for all q). Replacing Q by Q_1 and G by $\pi^{-1}(Q_1)$, we can assume Q acts trivially on $H_*(N, \mathbb{F}_2)$. Using Note (1) above,

$$E^2_{pq} \cong H_q(N, \mathbb{F}_2) \otimes_{\mathbb{F}_2} H_p(Q, \mathbb{F}_2),$$

so by Note (2) above,

$$\begin{aligned}
\chi(G) &= \sum_n (-1)^n \dim_{\mathbb{F}_2}(H_n(G, \mathbb{F}_2)) \\
&= \dim_{\mathbb{F}_2}(E^2) \qquad \text{by Prop. 9.7} \\
&= \left(\sum_q (-1)^q \dim_{\mathbb{F}_2}(H_q(N, \mathbb{F}_2))\right)\left(\sum_p (-1)^p \dim_{\mathbb{F}_2}(H_p(Q, \mathbb{F}_2))\right) \\
&= \chi(N)\chi(Q).
\end{aligned}$$

□

Remark 10.13. *Let H be a subgroup of a group G, R a commutative ring, M an RG-module. Then*

$$\begin{array}{ccc} R(G/H) \otimes_R M & \textit{is } RG\textit{-isomorphic to} & RG \otimes_{RH} M \\ \textit{(diagonal } G\textit{-action)} & & \textit{(as induced module)} \end{array}$$

The proof is similar to that of Lemma 9.12 *(the special case $H = 1$). There is also an RG-isomorphism*

$$\begin{array}{ccc} \sigma : \mathrm{Hom}_R(R(G/H), M) & \xrightarrow{\cong} & \mathrm{Hom}_{RH}(RG, M) \\ (\textit{“diagonal” action } (gf)(\gamma) = g(f(g^{-1}\gamma))) & & \textit{(as coinduced module)} \end{array}$$

where $\sigma(f)(g) = g(f(g^{-1}H))$. The inverse τ is given by $\tau(u)(gH) = g(u(g^{-1}))$.

Now we consider free products with amalgamation and HNN-extensions.

Lemma 10.14. *Suppose $G = A *_C B$ is a free product with amalgamation, R is a commutative ring, M an RG-module. Then there are exact “Mayer-Vietoris sequences”*

(1) $\ldots H^n(G, M) \to H^n(A, M) \oplus H^n(B, M) \to H^n(C, M) \to H^{n+1}(G, M) \to \ldots$

(2) $\ldots H_n(C, M) \to H_n(A, M) \oplus H_n(B, M) \to H_n(G, M) \to H_{n-1}(C, M) \to \ldots$

Proof. By Lemma 6.16 there is a short exact sequence

$$0 \to R(G/C) \xrightarrow{\alpha} R(G/A) \oplus R(G/B) \xrightarrow{\beta} R \to 0.$$

This R-splits, so applying $\mathrm{Hom}_R(-, M)$ and $- \otimes_R M$ gives exact sequences.

Using Remark 10.13, these become exact sequences of RG-modules:

(3) $0 \longleftarrow \mathrm{Hom}_{RC}(RG, M) \longleftarrow \mathrm{Hom}_{RA}(RG, M) \oplus \mathrm{Hom}_{RB}(RG, M) \longleftarrow M \longleftarrow 0$

(4) $0 \longrightarrow RG \otimes_{RC} M \longrightarrow (RG \otimes_{RA} M) \oplus (RG \otimes_{RB} M) \longrightarrow M \longrightarrow 0.$

The Lemma follows by applying Lemma 5.4 and the exercise following it to (3) and (4), then using both versions of Shapiro's Lemma. □

Note. *For an HNN-extension* $G = \langle t, A \mid tCt^{-1} = D\rangle$, *there are corresponding sequences*

(1)′ $\ldots H^n(G, M) \to H^n(A, M) \to H^n(C, M) \to H^{n+1}(G, M) \to \ldots$

(2)′ $\ldots H_n(C, M) \to H_n(A, M) \to H_n(G, M) \to H_{n-1}(C, M) \to \ldots$

(see [14, Theorem 2.12]).

Lemma 10.15. *If* $A, B, C \in \mathcal{C}$ *and* $G = A *_C B$, *then* $G \in \mathcal{C}$ *and*

$$\tilde{\chi}(G) = \tilde{\chi}(A) + \tilde{\chi}(B) - \tilde{\chi}(C).$$

Proof. Using the exact sequence (1) in Lemma 10.14 with $R = \mathbb{Z}$, we see that $\mathrm{cd}_{\mathbb{Z}}(G) < \infty$, and using sequence (2) we see that $H_n(G, M)$ is finitely generated for all $\mathbb{Z}$-finitely generated $\mathbb{Z}G$-modules M. Taking $M = \mathbb{Z}$ in (2), the formula for $\tilde{\chi}(G)$ follows using Remark 8.1. □

Lemma 10.16. *If* $A, C \in \mathcal{C}$ *and* $G = \langle t, A \mid tCt^{-1} = D\rangle$ *is an HNN-extension, then* $G \in \mathcal{C}$ *and*

$$\tilde{\chi}(G) = \tilde{\chi}(A) - \tilde{\chi}(C).$$

Proof. This is similar to that of Lemma 10.15 using sequences (1)′ and (2)′. □

Unfortunately, Lemmas 10.15 and 10.16 are not sufficient to obtain the corresponding result for FHT, because subgroups of free products and HNN-extensions are not necessarily of the same form. They have a more complicated structure (as "fundamental groups of a graph of groups"). This is best seen using the Bass-Serre theory of groups acting on trees, and we shall have to assume the reader is familiar with this. It follows by induction on the number of edges of X that, if $(\mathcal{G}, X)$ is a graph of groups, where X is a finite graph, G is its fundamental group and all edge and vertex groups are in $\mathcal{C}$, then $G \in \mathcal{C}$ and

$$\tilde{\chi}(G) = \sum_{v \in V(X)} \tilde{\chi}(G_v) - \sum_{e \in E(X)} \tilde{\chi}(G_e).$$

Here $V(X)$ is the set of vertices and $E(X)$ the set of unoriented edges of X; G_v is the group associated to v and G_e that associated to e.

Theorem 10.17. *If* A, B, $C \in \mathrm{FHT}$, $G = A *_C B$ *and* G *is virtually torsion-free, then* $G \in \mathrm{FHT}$ *and*

$$\chi(G) = \chi(A) + \chi(B) - \chi(C).$$

Proof. Let H be a torsion-free subgroup of finite index in G. Then according to the Bass-Serre theory, there is a tree on which G acts (see Theorem 7 and its proof in [108, §4.1]). Then H acts by restriction, and H is the fundamental group of a graph of groups, say $(\mathcal{H}, X)$ (see [31, Ch.8, Theorem 27] or [108, §5.4]). There are one-to-one correspondences:

$$V(X) \longleftrightarrow (H\backslash G/A) \coprod (H\backslash G/B) \quad \text{(disjoint union)}$$
$$E(X) \longleftrightarrow H\backslash G/C$$

where $H\backslash G/A = \{HgA \mid g \in G\}$, etc.

The vertex groups are $gAg^{-1} \cap H$ for $g \in T_A$ and $gBg^{-1} \cap H$ for $g \in T_B$, where T_A is a transversal for $H\backslash G/A$ and T_B is a transversal for $H\backslash G/B$. The edge groups are $gCg^{-1} \cap H$ for $g \in T_C$, where T_C is a transversal for $H\backslash G/C$. (See [31, Ch.8, Theorem 27].) Hence all edge and vertex groups are in $\mathcal{C}$ ($gAg^{-1} \cap H$ is of finite index in gAg^{-1} and is torsion-free, etc). By the observations preceding the theorem, $H \in \mathcal{C}$ (so $G \in \mathrm{FHT}$) and

$$\begin{aligned}
\tilde{\chi}(H) &= \sum_{g \in T_A} \tilde{\chi}(gAg^{-1} \cap H) + \sum_{g \in T_B} \tilde{\chi}(gBg^{-1} \cap H) \\
&\quad - \sum_{g \in T_C} \tilde{\chi}(gCg^{-1} \cap H) \\
&= \sum_{g \in T_A} \chi(gAg^{-1})(gAg^{-1} : gAg^{-1} \cap H) \\
&\quad + \sum_{g \in T_B} \chi(gBg^{-1})(gBg^{-1} : gBg^{-1} \cap H) \\
&\quad - \sum_{g \in T_C} \chi(gCg^{-1})(gCg^{-1} : gCg^{-1} \cap H) \\
&= \chi(A) \sum_{g \in T_A} (A : A \cap g^{-1}Hg) + \chi(B) \sum_{g \in T_B} (B : B \cap g^{-1}Hg) \\
&\quad -\chi(C) \sum_{g \in T_C} (C : C \cap g^{-1}Hg) \\
&= \chi(A)(G : H) + \chi(B)(G : H) - \chi(C)(G : H).
\end{aligned}$$

Therefore $\chi(G) = \dfrac{1}{(G : H)} \tilde{\chi}(H) = \chi(A) + \chi(B) - \chi(C)$. □

Theorem 10.18. *If* A, $C \in \mathrm{FHT}$, $G = \langle t, A \mid tCt^{-1} = D\rangle$ *is an HNN-extension and* G *is virtually torsion-free, then* $G \in \mathrm{FHT}$ *and*

$$\chi(G) = \chi(A) - \chi(C).$$

Proof. This is similar to the proof of Theorem 10.17, using the tree on which G acts given by the Bass-Serre theory (see [108, Example 3, §5.1]). □

Example. We give an example, due to H. R. Schneebeli, [105] to resolve several questions arising from the results in this section. It is based on a famous group studied by G. Higman [63]. Let

$$H = \langle a, b, c, d \mid aba^{-1} = b^2, bcb^{-1} = c^2, cdc^{-1} = d^2, dad^{-1} = a^2\rangle.$$

We list some properties of H and related groups. The proofs need some results on homology and cohomology of groups which have not been discussed.

(1) H is infinite, torsion-free and has no proper subgroups of finite index. (See [63].)
(2) $\mathrm{cd}_{\mathbb{Z}}(H) \leq 2$. (This follows from the way Higman constructs H by free products with amalgamation and HNN-extensions, and the Mayer-Vietoris sequences in Lemma 10.14 and the note following it.) In fact, by the Stallings-Swan Theorem (note after Cor. 6.15), $\mathrm{cd}_{\mathbb{Z}}(H) = 2$.
(3) $H \in \mathrm{FP}(\mathbb{Z})$. (This also follows from the construction of H, using Theorem 6.17 and the note following it.)
(4) $H_1(H, \mathbb{Z}) = H_2(H, \mathbb{Z}) = 0$. For by [100, Cor. 10.3], $H_1(H, \mathbb{Z}) = H/H' = 0$ (any abelian group with generators a, b, c, d satisfying the relations of H is clearly trivial). Also, $0 \leq \mathrm{def}(H) \leq \mathrm{rk}_{\mathbb{Z}}(H/H') - s$, where s is the minimal number of generators for $H_2(H, \mathbb{Z})$, so $s = 0$. (See [98, 14.1.5].) The deficiency of a finite group presentation is the number of generators minus the number of relations, and $\mathrm{def}(H)$ is the maximum deficiency of a finite presentation of H.
(5) $H^2(H, \mathbb{Z}) = H^1(H, \mathbb{Z}) = 0$. For $H^2(H, \mathbb{Z}) = 0$ by (4) and the Universal Coefficient Theorem (see [100, Theorem 8.26]). Also, the Universal Coefficient Theorem gives $H^1(H, \mathbb{Z}) \cong \mathrm{Ext}^1_{\mathbb{Z}}(H_0(H, \mathbb{Z}), \mathbb{Z})$, and $H_0(H, \mathbb{Z}) = \mathbb{Z}_H = \mathbb{Z}$ (see [100, Theorem 5.17]). Finally $\mathrm{Ext}^1_{\mathbb{Z}}(\mathbb{Z}, \mathbb{Z}) = 0$ by [100, Theorem 7.7].

Let C be an infinite cyclic subgroup of H, and define $G = H *_C H$. Then G also has no proper subgroups of finite index, and $G \in \mathrm{FP}(\mathbb{Z})$ by Theorem 6.17, so $G \in \mathrm{FHT}$. The Mayer-Vietoris sequence (1) in Lemma 10.14 shows that $\mathrm{cd}_{\mathbb{Z}}(G) \leq 2$, and that $H^2(G, \mathbb{Z}) \cong H^1(C, \mathbb{Z})$. We calculate this from the Universal Coefficient Theorem, which gives

$$H^1(C,\mathbb{Z}) \cong \mathrm{Hom}_{\mathbb{Z}}(H_1(C,\mathbb{Z}),\mathbb{Z}) \oplus \mathrm{Ext}^1_{\mathbb{Z}}(H_0(C,\mathbb{Z}),\mathbb{Z}).$$

Now $H_1(C,\mathbb{Z}) = \mathbb{Z}$ by [100, Cor.10.3], $H_0(C,\mathbb{Z}) = \mathbb{Z}$ by [100, Theorem 5.17] and as noted above, $\mathrm{Ext}^1_{\mathbb{Z}}(\mathbb{Z},\mathbb{Z}) = 0$. Hence $H^1(C,\mathbb{Z}) \cong \mathrm{Hom}_{\mathbb{Z}}(\mathbb{Z},\mathbb{Z}) \cong \mathbb{Z}$. (The isomorphism $\mathbb{Z} \to \mathrm{Hom}_{\mathbb{Z}}(\mathbb{Z},\mathbb{Z})$ is given by $n \mapsto f_n$, where $f_n(m) = mn$.) Thus

(6) $H^2(G,\mathbb{Z}) \cong \mathbb{Z}$.

For any integer k, the short exact sequence (of trivial $\mathbb{Z}G$-modules)

$$0 \to \mathbb{Z} \xrightarrow{k} \mathbb{Z} \to \mathbb{Z}/k\mathbb{Z} \to 0$$

(where the label "k" is short for multiplication by k) gives by Lemma 5.4 an exact sequence

$$\ldots H^2(G,\mathbb{Z}) \xrightarrow{k} H^2(G,\mathbb{Z}) \to H^2(G,\mathbb{Z}/k\mathbb{Z}) \to 0 \ldots$$

so

(7) $H^2(G,\mathbb{Z}/k\mathbb{Z}) \cong \mathbb{Z}/k\mathbb{Z}$.

(8) Similarly, $H^2(H,\mathbb{Z}/k\mathbb{Z}) = 0$.

Fix an integer $k > 1$. Since $H^2(G,\mathbb{Z}/k\mathbb{Z}) \neq 0$ and $\mathbb{Z}/k\mathbb{Z}$ is a trivial $\mathbb{Z}G$-module, there is a non-split central extension

$$0 \to \mathbb{Z}/k\mathbb{Z} \to X \xrightarrow{\pi} G \to 1$$

(see [17, IV Theorem 3.12]). If U is a torsion-free subgroup of finite index in X, then $\pi(U) = G$ and $\pi|_U$ is injective. But this means the extension splits, a contradiction. Hence X is not virtually torsion-free, so $X \notin$ FHT. Let Y_1, Y_2 be the inverse images under π of the two copies of H and let $Z = \pi^{-1}(C)$. By restricting π we obtain central extensions

$$0 \to \mathbb{Z}/k\mathbb{Z} \to Y_i \to H \to 1$$
$$0 \to \mathbb{Z}/k\mathbb{Z} \to Z \to C \to 1$$

which split, by (8) and because C is a free group. Hence Y_1, Y_2 and Z are in vFP($\mathbb{Z}$), so in FHT. Also, $X = Y_1 *_Z Y_2$. Thus the virtually torsion-free hypothesis is needed in Theorem 10.12 and Theorem 10.17. Further, by Theorem 6.11, $X \in$ FP($\mathbb{Q}$).

We can also give an example of a group in FHT which is not in FP($\mathbb{Q}$). Let H be as above, but now let G be the free product $\ast_{i\in I} H_i$, where the index set I is infinite and $H_i = H$ for all $i \in I$. Then G inherits from H the property of having no proper subgroups of finite index, and $\mathrm{cd}_{\mathbb{Z}}(G) = 2$ (see the Corollary to Theorem 3, §8.6 in [57]). Further, $H_n(G, \mathbb{Z}) = 0$ for $n > 0$. If $n > 1$, this follows from the general Mayer-Vietoris sequence in [17, §9, Ch. VII], using the usual graph of groups for a free product. (See the examples after Prop.18, §8.3 in [31], or Example (c) in Ch.1, §4.4 (with $A = 1$) in [108].) For $n = 1$, $G' \supseteq H_i' = H_i$ for all i, so $G' = G$ and $H_1(G, \mathbb{Z}) = G/G' = 0$ (see (4) above). Also, $H_0(G, \mathbb{Z}) = \mathbb{Z}_G = \mathbb{Z}$ ([100, Theorem 5.17]). If M is a $\mathbb{Z}G$-module finitely generated as a $\mathbb{Z}$-module, then G acts on M as $\mathbb{Z}$-automorphisms, corresponding to a homomorphism $\varphi : G \to \mathrm{Aut}_{\mathbb{Z}}(M)$. But M is a direct sum of cyclic groups, so residually finite, hence $\mathrm{Aut}_{\mathbb{Z}}(M)$ is residually finite (see [79, Theorem 4.8, Ch.IV]). Since G has no non-trivial finite homomorphic images, $\varphi(G) = \{1\}$, that is, M is a trivial $\mathbb{Z}G$-module. Using the Universal Coefficient Theorem (remark after Theorem 10.22 in [100]), $H_n(G, M) \cong \mathrm{Tor}_1^{\mathbb{Z}}(H_{n-1}(G, \mathbb{Z}), M) = 0$ for $n > 0$. (For $n = 1$, $\mathrm{Tor}_1^{\mathbb{Z}}(\mathbb{Z}, M) = 0$ since $\mathbb{Z}$ is $\mathbb{Z}$-projective – see [100, Theorem 8.4].) Again by [100, Theorem 5.17], $H_0(G, M) = M_G = M$. Hence $G \in$ FHT, but $G \notin \mathrm{FP}(\mathbb{Q})$ since G is not finitely generated (see Lemma 6.8).

11. Actions on CW-complexes

In the example preceding Prop. 6.2, we saw that, under certain circumstances, $\mu(G) = \chi(X)$, where X is a finite CW-complex and G is its fundamental group. This is in part because G acts freely on the universal covering space of X. Here we use actions on CW-complexes (not necessarily free) to obtain some interesting results on the Brown characteristic. There is no analogue of these for the other Euler characteristic μ. We shall have to assume that the reader is familiar with singular homology theory.

Notation. We put

$$\begin{aligned} E^n &= \{x \in \mathbb{R}^n \mid \|x\| \le 1\} \\ U^n &= \{x \in \mathbb{R}^n \mid \|x\| < 1\} \\ S^{n-1} &= \{x \in \mathbb{R}^n \mid \|x\| = 1\} \end{aligned}$$

where $\|\cdot\|$ denotes the usual euclidean norm on $\mathbb{R}^n$. Also, R denotes a principal ideal domain. If X is a topological space and Y is a subspace, then $H_n(X, Y; R)$ denotes the nth singular relative homology module, and $H_n(X; R)$ means $H_n(X, \emptyset; R)$.

Definition. *Let X be a Hausdorff space and let A be a closed subspace of X. We say X is* obtained from A by adjoining n-cells *if*

(a) *the complement $X \setminus A$ is a disjoint union $\coprod_{\lambda\in\Lambda} e_\lambda$, where each e_λ is open in X;*

(b) *for each $\lambda \in \Lambda$, there is a continuous map $f_\lambda : E^n \to \bar{e}_\lambda$ such that $f|_{U^n}$ is* *a homeomorphism onto e_λ, and $f_\lambda(S^{n-1}) \subseteq A$ (the f_λ are called "characteristic maps");*

(c) *a subspace Y of X is closed if and only if $Y \cap A$ is closed and, for all $\lambda \in \Lambda$, $f_\lambda^{-1}(Y)$ is closed.*

Proposition 11.1. *In this situation, $H_q(X, A; R) = 0$ for $q \neq n$. The maps $H_n(f_\lambda) : H_n(E^n, S^{n-1}; R) \to H_n(X, A; R)$ are injective and $H_n(X, A; R)$ is isomorphic to the direct sum of the image subgroups, so is isomorphic to $\bigoplus_{\lambda \in \Lambda} R$.*

Proof. See [85, Theorem IV.2.1] (the case $R = \mathbb{Z}$ is easily generalised). □

Definition. *A CW-complex is a Hausdorff space X together with subspaces*

$$\{X_n \mid n \in \mathbb{Z}, n \geq 0\}$$

such that $X_0 \subseteq X_1 \subseteq X_2 \subseteq \dots$, $X = \bigcup_{n \geq 0} X_n$ and:

(1) *the subspace X_0 is discrete;*
(2) *for all $n \geq 1$, X_n is obtained from X_{n-1} by adjoining n-cells;*
(3) *a subspace Y of X is closed if and only if $Y \cap \bar{e}$ is closed for all n-cells e and $n = 0, 1, 2, \dots$.*

(In (3), $Y \cap \bar{e}$ is closed if and only if $f^{-1}(Y)$ is closed, where f is the characteristic map corresponding to e.) The subspace X_n is called the n-skeleton of X.

Definition. *A subcomplex of a CW-complex X is a subspace Y such that Y is a union of cells of X, and for any cell e, $e \subseteq Y$ implies $\bar{e} \subseteq Y$.*

Defining $Y_n = X_n \cap Y$ then gives Y the structure of a CW-complex.

Definition. *A continuous map of CW-complexes, $f : X \to Y$, is* cellular *if $f(X_n) \subseteq Y_n$ for all $n \geq 0$.*

Products

If X, Y are CW-complexes, $X \times Y$ can be given the structure of a CW-complex. Define the n-cells of $X \times Y$ to be the sets $e_1 \times e_2$, where e_1 is a p-cell of X and e_2 is a q-cell of Y, for all p, q such that $p + q = n$. (Characteristic maps can be defined using a homeomorphism $E^n \to E^p \times E^q$.) We give $X \times Y$ the *weak topology*, that is, $U \subseteq X \times Y$ is closed if and only if $f_\lambda^{-1}(U)$ is closed for each cell e_λ of $X \times Y$. This is finer that the product (Tychonoff) topology, and agrees with the product topology if one of X, Y is locally compact, or if X, Y both have countably many cells ([86, Theorem 7.3.16]).

Cellular homology

Suppose X is a CW-complex and Y is a subcomplex. Define $C_n(X, Y; R) = H_n(X_n \cup Y, X_{n-1} \cup Y; R)$, and define

$$d_n : C_n(X, Y; R) \to C_{n-1}(X, Y; R)$$

to be the connecting homomorphism of the triple $(X_n \cup Y, X_{n-1} \cup Y, X_{n-2} \cup Y)$.

Proposition 11.2. *The pair $(C(X, Y; R), d)$ is a chain complex over R, $C_n(X, Y; R)$ is R-free with basis in one-to-one correspondence with the n-cells of X not in Y, and*

$$H_n(C(X, Y; R)) \cong H_n(X, Y; R).$$

Proof. The exact homology sequence of the pair $(X_n \cup Y, X_{n-1} \cup Y)$ has the form (suppressing R)

$$H_n(X_n \cup Y) \xrightarrow{j} C_n(X, Y) \xrightarrow{\partial} H_{n-1}(X_{n-1} \cup Y) \to \ldots \quad .$$

For each n there is a commutative diagram

$$\begin{array}{ccc} C_n(X, Y) & \xrightarrow{\quad\quad\partial\quad\quad} & H_{n-1}(X_{n-1} \cup Y) \\ & {\scriptstyle d_n} \searrow \qquad \swarrow {\scriptstyle j} & \\ & C_{n-1}(X, Y) & \end{array}$$

by naturality of the connecting homomorphism. (This is the argument of Lemma 9.3 applied to $(F_p S(X))$, where $S(X)$ is the singular chain complex of X and $F_p S(X) = S(X_p \cup Y)$.) Thus $d^2 = j(\partial j)\partial = 0$. Now $X_n \cup Y$ is obtained from $X_{n-1} \cup Y$ by adjoining the n-cells of X not in Y, so $C_n(X, Y; R)$ is R-free as claimed, by Prop. 11.1. There are maps $H_n(X, Y) \xleftarrow{k} H_n(X_n \cup Y, Y) \xrightarrow{l} H_n(X_n \cup Y, X_{n-1} \cup Y) = C_n(X, Y)$, where k and l are induced by inclusion maps. Further, lk^{-1} induces an isomorphism $H_n(X, Y) \to H_n(C(X, Y))$ (see [85, Theorem IV.4.2]). □

Definition. *If X is a CW-complex, put $C(X; R) = C(X, \emptyset; R)$.*

Lemma 11.3. *Let X be a CW-complex, Y a subcomplex. Then there is a short exact sequence of chain complexes*

$$0 \longrightarrow C(Y;R) \xrightarrow{\alpha} C(X;R) \xrightarrow{\beta} C(X,Y;R) \longrightarrow 0.$$

Proof. Let $\alpha_n : C_n(Y) \to C_n(X)$ be the map induced by the inclusion of pairs $(Y_n, Y_{n-1}) \to (X_n, X_{n-1})$ and let $\beta_n : C_n(X) \to C_n(X,Y)$ be that induced by $(X_n, X_{n-1}) \to (X_n \cup Y, X_{n-1} \cup Y)$ (again suppressing R). Then $\beta_n \alpha_n = 0$ since $Y_n \subseteq X_{n-1} \cup Y$. Let Λ index the n-cells of X, and let $M \subseteq \Lambda$ index those of Y. For $\lambda \in \Lambda$, there is a commutative diagram

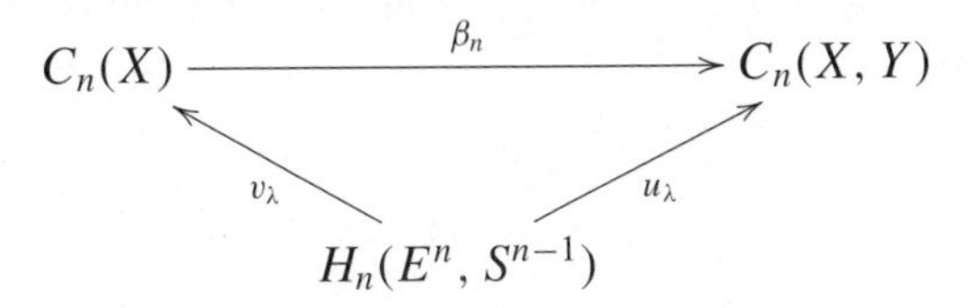

where u_λ, v_λ are induced by the corresponding characteristic map f_λ. If also $\lambda \in M$, there is a commutative diagram

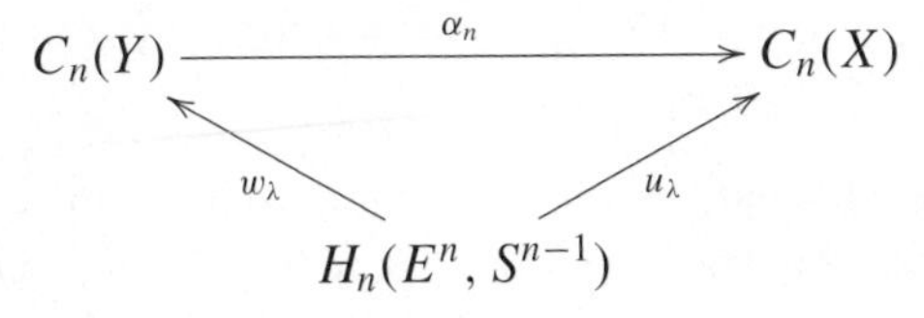

where w_λ is also induced by f_λ. It follows from Prop. 11.1 that $C_n(X) = \bigoplus_{\lambda\in\Lambda} R$, $C_n(Y) = \bigoplus_{\lambda\in M} R$, $C_n(X,Y) = \bigoplus_{\lambda\in\Lambda\setminus M} R$, and α_n, β_n are the obvious inclusion and projection maps, so

$$0 \longrightarrow C_n(Y;R) \xrightarrow{\alpha_n} C_n(X;R) \xrightarrow{\beta_n} C_n(X,Y;R) \longrightarrow 0.$$

is exact. Finally, α, β are chain maps by naturality of the connecting homomorphism. □

Proposition 11.4. *Suppose A, B are subcomplexes of a CW-complex X. Then (A, B) is an excisive couple, that is, the inclusion map $(A, A \cap B) \to (A \cup B, B)$ induces an isomorphism in singuar homology. Hence, there is an exact Mayer-Vietoris sequence*

$$\ldots H_n(A \cap B; R) \to H_n(A;R) \oplus H_n(B;R)$$
$$\to H_n(A \cup B; R) \to H_{n-1}(A \cap B; R) \to \ldots$$

Proof. See [77, Ch.V, Theorem 1.1]. □

Lemma 11.5. *Let $A_1, \ldots, A_m$ be a collection of subcomplexes of a CW-complex, closed under intersection. If $H_*(A_i; R)$ is R-finitely generated for $1 \leq i \leq m$, then $H_*(\bigcup_{i=1}^{k} A_i; R)$ is finitely generated for $1 \leq k \leq m$.*

Proof. The proof is by induction on k. Assume $B = A_1 \cup \ldots \cup A_k$ has finitely generated homology (where $k < m$). Then $A = A_{k+1}$ has finitely generated homology by assumption. Also,

$$A \cap B = (A_1 \cap A_{k+1}) \cup \ldots \cup (A_k \cap A_{k+1}) = A_{i_1} \cup \ldots \cup A_{i_k}$$

by assumption, for some indices $i_1, \ldots, i_k$. Applying the induction hypothesis to $A_{i_1}, A_{i_2} \ldots, A_{i_k}$, $A \cap B$ has finitely generated homology, hence so does $A \cup B$ by Prop. 11.4. □

Group actions

Definition. *Let G be a group. A G-*complex *is a CW-complex X with an action of G on X as cellular homeomorphisms (so G permutes the cells of X). A G-complex is* admissible *if $ge = e$, where $g \in G$ and e is a cell, implies $gx = x$ for all $x \in e$. A G-complex is* free *if $ge = e$, where $g \in G$ and e is a cell, implies $g = 1$.*

If X is a G-complex and Y is a G-invariant subcomplex, then $(C(X, Y; R), d)$ is an RG-complex. If G freely permutes the cells of $X \setminus Y$, then $C(X, Y; R)$ is RG-free, with a basis in one-to-one correspondence with the set of G-orbits of n-cells in $X \setminus Y$. See [17, I.3.1 and I.4].

Let K be a connected CW-complex, $p : \tilde{K} \to K$ a regular covering map, G the group of covering transformations. Then $\tilde{K}$ has the structure of a CW-complex, the cells being the components of $p^{-1}(e)$, where e is a cell of X. These cells are permuted freely and transitively by G, and map homeomorphically onto e via p (See [106, III.6.9].) This applies when $\tilde{K}$ is the universal covering space of K, so $G \cong \pi_1(K)$. In this case, the following are equivalent:

(1) $\pi_n(K) = 0$ for $n > 1$;
(2) $H_n(\tilde{K}) = 0$ for $n > 1$;
(3) $\tilde{K}$ is contractible.

See Ch.1, §4 (between 4.1 and 4.2) in [17].

Definition. *A CW-complex X is* n-dimensional *if $X = X_n$.*

Theorem 11.6. *Let G be a group, $n = \max\{\mathrm{cd}_{\mathbb{Z}}(G), 3\}$. Then there is an n-dimensional contractible free G-complex.*

Proof. This follows from [17, VIII 7.2] and the preceding remarks. □

The next result is a geometric version of the Serre Extension Theorem (Theorem 10.10). The proof is due to K. S. Brown.

Theorem 11.7. *Let G be a group, H a subgroup of finite index with* $\mathrm{cd}_{\mathbb{Z}}(H) < \infty$. *Then there is a finite dimensional contractible G-complex Y such that the stabilizer G_e is finite, for all cells e of Y.*

Proof. Using Theorem 11.6, let X be a finite dimensional contractible free H-complex. Define $Y = \mathrm{Hom}_H(G, X)$, the set of all mappings $f : G \to X$ of H-sets (where H acts on G by left multiplication). Then G acts on Y by $gf(g') = f(g'g)$ (for $g, g' \in G$, $f \in Y$). Choose a transversal $t_1, \ldots, t_n$ for $H\backslash G$. Then there is a bijection

$$\varphi : Y \to \prod_{i=1}^{n} X$$

given by $\varphi(f) = (f(t_1), \ldots, f(t_n))$, which gives Y the structure of a CW-complex. This is independent of the choice of transversal. For if $t_i' = h_i t_i$, where $h_i \in H$ for $1 \le i \le n$, and φ' is the corresponding map, there is a commutative diagram

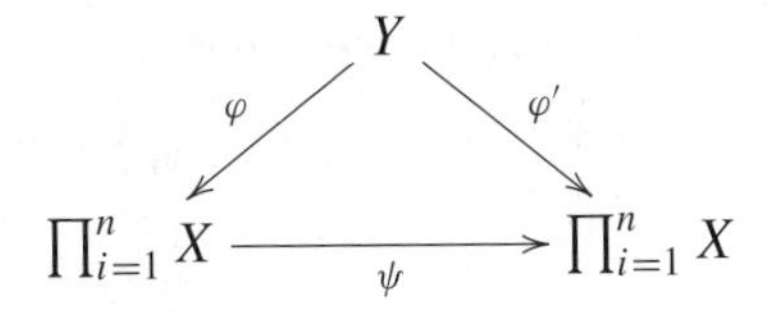

where ψ is the cellular homeomorphism $\prod_{i=1}^{n} h_i$. Also, if $t_1, \ldots, t_n$ is replaced by $t_{\sigma(1)}, \ldots, t_{\sigma(n)}$ (where $\sigma \in S_n$), this clearly does not change the CW structure on Y. If $g \in G$ and φ' is defined by $\varphi'(f) = (f(t_1 g), \ldots, f(t_n g))$, then

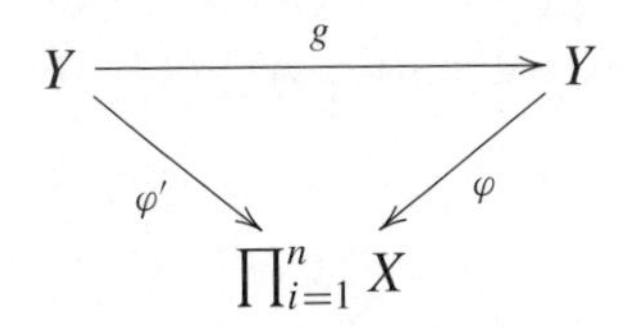

is commutative. Hence Y is a G-complex, and is clearly finite dimensional. The map $Y \to X$, $f \mapsto f(1)$ maps cells to cells and is an H-map. Since H acts freely on the cells of X, it acts freely on the cells of Y. Hence $G_e \cap H = 1$ for all cells e of Y, so G_e is finite. Finally, Y is contractible since $\prod_{i=1}^{n} X$ is. □

Note. *The complex X in Theorem* 11.7 *can be taken to be an ordered simplicial complex, and the resulting space Y is simplicial (see* [17, Ch.VIII, 7.3 and 11.2]). *Further, the G-action is simplicial. Taking barycentric subdivision if necessary, we can assume Y is ordered and G order-preserving. Then G acts admissibly on Y.*

Euler characteristics of complexes

The idea of Euler characteristic of a complex is well-known. We shall also use a relative version, for a CW-complex and a subcomplex.

Definition. *Let X be a CW-complex, Y a subcomplex. Put*

$$\chi(X, Y; R) = \mathrm{rk}_R(C(X, Y; R)) = \sum_{i \geq 0} (-1)^i \, \mathrm{rk}_R(H_i(X, Y; R)),$$

whenever this is defined, and put $\chi(X; R) = \chi(X, \emptyset; R)$.

Lemma 11.8. *The Euler characteristic satisfies:*

(1) *provided the right hand side is defined, $\chi(X, Y; R) = \chi(X; R) - \chi(Y; R)$;*
(2) *provided $\chi(X, Y; R)$ is defined, $\chi(X, Y; R) = \chi(X, Y; R/pR)$, for any prime element $p \in R$;*
(3) *if X is finite (i.e. has finitely many cells), then*

$$\chi(X; R) = \sum_{i \geq 0} (-1)^i (\text{number of } i\text{-cells of } X).$$

Proof. (1) follows from the exact sequence of Lemma 11.3 and the corresponding exact homology sequence (Theorem 3.1).

(2) For each n, the map $R \otimes_{\mathbb{Z}} H_n(X_n \cup Y, X_{n-1} \cup Y; \mathbb{Z}) \to H_n(X_n \cup Y, X_{n-1} \cup Y; R)$ in the Universal Coefficient Theorem is an isomorphism by Prop. 11.1, and this gives an isomorphism of chain complexes $R \otimes_{\mathbb{Z}} C(X, Y; \mathbb{Z}) \cong C(X, Y; R)$ (see [85, Ch.V, §§7,8]). Hence

$$\begin{aligned} C(X, Y; R/pR) &\cong R/pR \otimes_{\mathbb{Z}} C(X, Y; \mathbb{Z}) \\ &\cong R/pR \otimes_R (R \otimes_{\mathbb{Z}} C(X, Y; \mathbb{Z})) \\ &\cong R/pR \otimes_R C(X, Y; R) \end{aligned}$$

and the result follows by Lemma 8.3.

(3) This is immediate from Prop. 11.2 and Lemma 8.2. □

If Y is a G-complex and H is a subgroup of G, define

$$Y^H = \{y \in Y \mid hy = y \text{ for all } h \in H\}.$$

If Y is an admissible G-complex, Y^H is a subcomplex of Y.

Theorem 11.9. *Let G be a finite group (R a principal ideal domain), Y a finite dimensional admissible G-complex, with $H_*(Y^H; R)$ R-finitely generated for all subgroups H of G. If k is an integer which divides the length of every G-orbit, then $k|\chi(Y; R)$.*

Proof. For $H \leq G$, define $Y_H = \{y \in Y \mid G_y = H\}$, where G_y is the stabiliser of y in G. Then

$$Y_H = Y^H \setminus Y^{>H},$$

where $Y^{>H} = \bigcup_{G \geq H' \gneq H} Y^{H'}$ (a subcomplex of Y, so Y_H is a difference of subcomplexes).

Now $\{Y^{H'} \mid H' \gneq H\}$ is closed under intersection ($Y^{H_1} \cap Y^{H_2} = Y^{\langle H_1, H_2 \rangle}$), so by Lemma 11.5, $Y^{>H}$ has finitely generated homology. Hence $H_*(Y^H, Y^{>H}; R)$ is finitely generated (from the exact homology sequence of the pair $(Y^H, Y^{>H})$). Let $H_1, \ldots, H_n$ be a list of the subgroups of G, ordered so that $|H_i| \geq |H_{i+1}|$. Define subcomplexes $\emptyset = Y_0 \subseteq Y_1 \subseteq \ldots \subseteq Y_n = Y$ recursively by

$$Y_i = Y_{i-1} \cup Y^{H_i}.$$

Then $Y_{i-1} \cap Y^{H_i} = Y^{>H_i}$, and by Prop. 11.4, there is an isomorphism $H_*(Y_i, Y_{i-1}) \cong H_*(Y^{H_i}, Y^{>H_i})$ (suppressing R). Hence $H_*(Y_i, Y_{i-1})$ is finitely generated, and it follows by induction, using the exact homology sequence for (Y_i, Y_{i-1}), that $\chi(Y_i)$ is defined for $0 \leq i \leq n$. Therefore by Lemma 11.8(1),

$$\begin{aligned}
\chi(Y) &= \sum_{i=1}^{n} \chi(Y_i, Y_{i-1}) \\
&= \sum_{i=1}^{n} \chi(Y^{H_i}, Y^{>H_i}) \\
&= \sum_{H} \chi(Y^H, Y^{>H})
\end{aligned}$$

(sum over all subgroups H of G which are stabilisers of points in Y). If $H' = gHg^{-1}$ ($g \in G$), then $Y^{H'} = gY^H$ and $Y^{>H'} = gY^{>H}$, hence $\chi(Y^H, Y^{>H}) =$

$\chi(Y^{H'}, Y^{>H'})$, so

$$\chi(Y) = \sum_{H \in \mathfrak{C}} (G : N_G(H))\chi(Y^H, Y^{>H})$$

where $\mathfrak{C}$ is a set of representatives for the conjugacy classes of subgroups of G which occur as stabilisers. Now $N_G(H)/H$ acts freely on Y_H, so $C(Y^H, Y^{>H})$ is a chain complex of free $R(N_G(H)/H)$-modules. By Lemma 8.7, $(N_G(H) : H)$ divides $\mathrm{rk}_R(C(Y^H, Y^{>H})) = \chi(Y^H, Y^{>H})$. Thus $(G : N_G(H))\chi(Y^H, Y^{>H})$ is divisible by $(G : H)$, which is the length of an orbit (for $H \in \mathfrak{C}$), so is divisible by k. Hence $k|\chi(Y)$. □

Lemma 11.10. *Let $G = \langle g \rangle$ be cyclic of prime order p, $A = \mathbb{F}_p G$, $\gamma = g - 1 \in I_G$ (the augmentation ideal of A). For $1 \le i \le p-1$ there is a short exact sequence*

$$0 \to \gamma^{p-i} A \to A \xrightarrow{\gamma^i} \gamma^i A \to 0.$$

Proof. (We are using γ to denote both $g-1$ and multiplication by $g-1$; it will be clear from the context which is meant.) Recall that $\gamma A = I_G$ (by Lemma 5.3) and I_G is $\mathbb{F}_p$-free on $\{g^i - 1 \mid 1 \le i \le p-1\}$. Hence $\dim(\gamma A) = p-1$, and $\dim(A) = p$, so $\dim(\mathrm{Ker}(\gamma)) = 1$. Also,

$$\gamma^p = g^p - 1 = 0, \text{ so } \gamma^{p-1} g = \gamma^{p-1}, \text{ which implies } \gamma^{p-1} = \sum_{i=0}^{p-1} g^i.$$

There is an exact sequence

$$0 \to \gamma^{p-1} A \xrightarrow{j} A \xrightarrow{\gamma} \gamma A \to 0$$

where j is the inclusion map (γ^{p-1} corresponds to ν in the proof of Lemma 5.11). Hence $\mathrm{Ker}(\gamma) = \gamma^{p-1} A \subseteq \gamma^i A$ for $1 \le i \le p-1$. Therefore there is an exact sequence

$$0 \to \mathrm{Ker}(\gamma) \to \gamma^i A \xrightarrow{\gamma} \gamma^{i+1} A \to 0$$

which implies $\dim(\gamma^i A) = \dim(\gamma^{i+1} A) + \dim(\mathrm{Ker}(\gamma)) = \dim(\gamma^{i+1} A) + 1$, and so $\dim(\gamma^i A) = p - i$. Consequently, $\dim(\mathrm{Ker}(\gamma^i)) = \dim(A) - \dim(\gamma^i A) = i$, for $1 \le i \le p-1$. Finally, $\gamma^p = 0$ implies $\gamma^{p-i} A \subseteq \mathrm{Ker}(\gamma^i)$, and the lemma follows since they have the same dimension. □

Lemma 11.11. *Let G be cyclic of prime order p, X a finite-dimensional admissible G-complex, with $H_*(X; \mathbb{F}_p)$ finitely generated. Then $H_*(X^G; \mathbb{F}_p)$ is finitely generated.*

Proof. From the proof of Lemma 11.3, $C_n(X) = C_n(X^G) \oplus B_n$, where $B_n \cong C(X, X^G)$ as $\mathbb{F}_p G$-modules (suppressing $\mathbb{F}_p$). The stabiliser of a cell is either G or 1, so G freely permutes the cells of $X \setminus X^G$, hence B_n is a free $\mathbb{F}_p G$-module. Putting $\gamma = g - 1$, where g is a generator of G, there is a map $C_n(X) \to C_n(X)$, $x \mapsto \gamma^i x$, for $1 \le i \le p-1$, also denoted by γ^i. It follows from Lemma 11.10 that $\mathrm{Ker}(\gamma^i | B_n) = \gamma^{p-i} B_n$, hence

$$\mathrm{Ker}(\gamma^i) = C_n(X^G) \oplus \gamma^{p-i} B_n = C_n(X^G) \oplus \gamma^{p-i} C_n(X)$$

(because $\gamma^i C_n(X^G) = 0$ for $1 \le i \le p-1$). Thus there is a short exact sequence of chain complexes

$$0 \to C(X^G) \oplus \gamma^{p-i} C(X) \to C(X) \to \gamma^i C(X) \to 0.$$

The corresponding exact homology sequence (Theorem 3.1) is

$$\ldots H_{n+1}(\gamma^i C(X)) \to H_n(C(X^G)) \oplus H_n(\gamma^{p-i} C(X)) \\ \to H_n(C(X)) \to \ldots \quad . \qquad (*)$$

Let

$$\begin{aligned} a_{i,n} &= \dim(H_n(\gamma^i C(X))) \\ b_n &= \dim(H_n(C(X))) = \dim(H_n(X; \mathbb{F}_p)) \\ c_n &= \dim(H_n(C(X^G))) = \dim(H_n(X^G; \mathbb{F}_p)). \end{aligned}$$

From $(*)$,

$$c_n + a_{p-i,n} \le a_{i,n+1} + b_n \qquad (1 \le i \le p-1).$$

If X has dimension d, then $a_{i,n+1} = b_n = c_n = 0$ for $n > d$ (since $C_n(X) = 0$ by Prop. 11.2). Also, $a_{i,d+1} = 0$ and $b_d < \infty$, so by $(*)$, $c_d < \infty$ and $a_{p-i,d} < \infty$ for $1 \le i \le p-1$, that is, $a_{i,d} < \infty$ for $1 \le i \le p-1$. Now take $n = d-1$ to see $c_{d-1} < \infty$ and $a_{i,d-1} < \infty$ for $1 \le i \le p-1$. Continuing in this way, all $c_n < \infty$. □

Lemma 11.12. *Let P be a finite p-group, X a finite-dimensional admissible P-complex. If $H_*(X; \mathbb{F}_p)$ is finitely generated, then $H_*(X^H; \mathbb{F}_p)$ is finitely generated, for all subgroups H of P.*

Proof. It suffices to show $H_*(X^P; \mathbb{F}_p)$ is finitely generated. The proof is by induction on $|P|$, so assume $H_*(X^H; \mathbb{F}_p)$ is finitely generated for all $H \lneqq P$. Choose $G \le P$, central and of order p. By Lemma 11.11, $H_*(X^G; \mathbb{F}_p)$ is finitely generated. Also, the action of P induces an admissible action of P/G on X^G. By induction $H_*((X^G)^{P/G}; \mathbb{F}_p)$ is finitely generated. But $(X^G)^{P/G} = X^P$. □

Theorem 11.13. *Let G be a finite group, Y a finite-dimensional admissible G-complex with $H_*(Y;\mathbb{Z})$ finitely generated. If k is an integer which divides the length of every G-orbit, then $k|\chi(Y;\mathbb{Z})$.*

Proof. It suffices to show: if p^a is a prime power dividing k, then $p^a|\chi(Y;\mathbb{Z})$. Let P be a p-Sylow subgroup of G. Since $p^a|(G:G_y)$ for $y \in Y$, $p^a|(G:P_y)$ (where $P_y = G_y \cap P$), so $p^a|(P:P_y)$. That is, p^a divides the length of every P-orbit. Also, P clearly acts admissibly. By Lemma 11.8(2), $H_*(Y;\mathbb{F}_p)$ is finitely generated and $\chi(Y;\mathbb{Z}) = \chi(Y;\mathbb{F}_p)$. By Theorem 11.9 and Lemma 11.12, $p^a|\chi(Y;\mathbb{F}_p)$. □

Note. *Let X be a connected free G-complex. Any CW-complex is locally contractible (see* [86, Ch.7, Ex.14]*). A similar argument shows that, for every $x \in X$, there is a neighbourhood U of X such that $gU \cap U = \emptyset$ for all $g \in G$ with $g \neq 1$. It follows that the projection $p : X \to X/G$ is a regular covering map with G as group of covering transformations (see* [84, Prop. 5.8.2]*). Moreover, p induces the structure of a CW-complex on X/G.*

Theorem 11.14. *Suppose $G \in$ FHT and let m be the least common multiple of the orders of the finite subgroups of G. Then $m\chi(G) \in \mathbb{Z}$. (So by the Sylow Theorems, if a prime power p^a divides the denominator of $\chi(G)$, then G has a subgroup of order p^a).*

Proof. There is a normal subgroup of finite index H of G such that $\mathrm{cd}_{\mathbb{Z}}(H) < \infty$ (arguing as in Prop. 10.2), and H is torsion-free (Cor. 5.12). Thus any finite subgroup of G projects isomorphically onto the quotient G/H, so has order dividing $(G:H)$, hence m is a well-defined positive integer and $m|(G:H)$. By Theorem 11.7, there is a finite dimensional contractible G-complex X such that G_e is finite, for all cells e of X, and by the note following Theorem 11.7, we can assume X is admissible. Since H is torsion-free it acts freely on X, and since X is contractible it is acyclic, so $H_0(X,\mathbb{Z}) \cong \mathbb{Z}$ and there is an augmented $\mathbb{Z}H$-free resolution $C(X) \to \mathbb{Z}$. By the note above, $Y = X/H$ is a CW-complex. Moreover,

$$C(Y) \cong \mathbb{Z} \otimes_{\mathbb{Z}H} C(X) \qquad \text{(see [17, II.2.4])}.$$

Therefore $H_*(H,\mathbb{Z}) \cong H_*(C(Y)) \cong H_*(Y;\mathbb{Z})$, hence $H_*(Y;\mathbb{Z})$ is finitely generated and $\tilde{\chi}(H) = \mathrm{rk}_{\mathbb{Z}}(C(Y)) = \chi(Y;\mathbb{Z})$. Thus $\chi(G) = \dfrac{\chi(Y;\mathbb{Z})}{(G:H)}$ and we need to show that $\dfrac{m}{(G:H)}\chi(Y;\mathbb{Z}) \in \mathbb{Z}$, that is, $k|\chi(Y;\mathbb{Z})$, where $k = (G:H)/m \in \mathbb{Z}$. Let $Q = G/H$. The action of G on X induces a cellular action of Q on Y. Let $p : X \to Y$ be the quotient map. If e is a cell of Y and $\tilde{e}$ is a cell in $p^{-1}(e)$, then $Q_e = \pi(G_{\tilde{e}})$, where $\pi : G \to Q$ is the quotient map. (For if $\pi(g)e = e$, where $g \in G$, then $p(g\tilde{e}) = \pi(g)p(\tilde{e}) = \pi(g)e = e = p(\tilde{e})$, so $hg\tilde{e} = \tilde{e}$ for some $h \in H$, and $\pi(hg) = \pi(g)$.) Hence Y is an admissible Q-complex. Also, as already noted π maps $G_{\tilde{e}}$ isomorphically onto Q_e, so $|Q_e| = |G_{\tilde{e}}|$ divides m. Therefore $k = |Q|/m$ divides $(Q : Q_e)$, for all cells e of Y. The theorem now follows by Theorem 11.13. □

An application

Corollary 11.15. *Suppose $1 \to N \to G \to Q \to 1$ is a group extension, where $N \in \mathcal{C}$ (so torsion-free and in* FHT*) and Q is a finite p-group. If p does not divide $\chi(N)$, then the extension splits.*

Proof. By Prop. 10.3, $\chi(G) = \dfrac{\chi(N)}{|Q|}$, which is a fraction in lowest terms, so by Theorem 11.14, G has a subgroup H with $|H| = |Q|$, and H maps isomorphically onto Q since N is torsion-free. Hence the extension splits. □

The method of proof of Theorem 11.14 can be used to give formulas relating $\chi(G)$ and $\tilde{\chi}(G)$ in terms of torsion in G. See [15] and [17, IX, 11–14]. Concerning the Stallings characteristic, the following is proved in [26], and is a simple generalisation of the corresponding formula for groups of type FL($\mathbb{Z}$) in [107].

Theorem. *Let X be an acyclic G-complex such that there are only finitely many G-orbits of cells in X. Suppose $G_e \in$ FP(R) for all cells e, where R is a commutative ring. Then $G \in$ FP(R) and*

$$\mu(G : R) = \sum_{e \in \Sigma} (-1)^{\dim(e)} \mu(G_e : R),$$

where σ is a set of representatives for the G-orbits of cells.

For further work on actions on CW-complexes, see [18].

12. Projectives over group algebras

Here we study arithmetic properties of the rank element r_P of a finitely generated projective module P, following [10] and [11]. Let A be a ring of prime characteristic p, and let B be the polynomial ring over A in two non-commuting indeterminates x, y. (This can be constructed in a similar way to a group ring, replacing the group by the free monoid on $\{x, y\}$. See [32, §3.3].) Then

$$(x + y)^p = x^p + y^p + \sum z_1 z_2 \dots z_p$$

where the sum is over all sequences $(z_1, z_2, \dots, z_p)$ with $z_i \in \{x, y\}$ for $1 \le i \le p$, not all equal to x or all equal to y. The cyclic permutations

$$(z_1, z_2, \dots, z_p),\ (z_2, z_3, \dots, z_p, z_1),\ \dots, (z_p, z_1, z_2, \dots, z_{p-1})$$

give products that are all congruent mod $[B, B]$:

$$z_1 z_2 \dots z_p - z_2 z_3 \dots z_p z_1 = [z_1, z_2 z_3 \dots z_p], \text{ etc.}$$

Hence their sum is congruent to $p\, z_1 z_2 \dots z_p$, i.e. to 0, mod $[B, B]$. Since p is prime, these cyclic permutations are not proper powers (in the free monoid on $\{x, y\}$), so are pairwise distinct. (This is left as an exercise.) Hence $(x + y)^p = x^p + y^p + z \quad$ with $z \in [B, B]$. Given $a, b \in A$, there is a unique ring homomorphism $B \to A$ which sends x to a, y to b and is the identity map on A ([32, §3.3]). Hence

(1) For all $a, b \in A$, $(a + b)^p = a^p + b^p + c \quad$ with $c \in [A, A]$.

Since p is either 2 or odd, $(-a)^p = -a^p$ for $a \in A$, so by (1),

$$\begin{aligned}(ab - ba)^p &\equiv (ab)^p - (ba)^p \qquad \text{mod } [A, A]\\ &= [a, (ba)^{p-1} b]\\ &\equiv 0 \qquad \text{mod } [A, A]\end{aligned}$$

Therefore, using (1) again,

(2) For all $c \in A$, $c \in [A, A]$ implies $c^p \in [A, A]$.

It follows from (1) and (2) that there is a homomorphism $F = F_A : T(A) \to T(A)$ defined by $F(T_A(a)) = T_A(a^p)$. (For if $T_A(a) = T_A(b)$, then $a = b + c$ where $c \in [A, A]$, so $a^p \equiv b^p + c^p \equiv b^p \mod [A, A]$, hence F is well-defined, and is a homomorphism by (1)). We call F_A the *Frobenius homomorphism* of A. If A is an R-algebra, where R is a commutative ring, then T_A is R-linear, hence $F_A(ra) = T_A(r^p a^p) = r^p F_A(a)$ for $r \in R$, $a \in A$. Let L be a free A-module with basis $e_1, \dots, e_n$ and let $B = \mathrm{End}_A(L)$. Recall that, if $f \in B$, $T_{L/A}(f) = T_A\left(\sum_{i=1}^n a_{ii}\right)$, where (a_{ij}) is the matrix of f with respect to $e_1, \dots, e_n$ and $T_{L/A}$ can be identified with T_B (the universal trace function on B). (See Remark (3) after Lemma 2.7.) Hence $T_{L/A}(f^p) = F_B(T_{L/A}(f))$. Taking f to be the map defined by $\begin{cases} e_1 \mapsto a e_1 \\ e_i \mapsto 0 \quad (i > 1)\end{cases}$, where $a \in A$, we see that $F_B = F_A$. If P is a finitely generated projective A-module, write $P \oplus Q = L$, where L is finitely generated free. If $f \in \mathrm{End}_A(P)$, then $T_{P/A}(f) = T_{L/A}(f \oplus 0)$ by Lemma 2.7 and $(f \oplus 0)^p = f^p \oplus 0$, hence $F_A(T_P(f)) = T_P(f^p)$. Finally, if M is of type FP over A, $P \to M$ is a finite augmented projective resolution and $f \in \mathrm{End}_A(M)$, take a lift $\tilde{f} : P \to P$ of

f. Then $\tilde{f}^p$ is a lift of f^p, hence

$$F_A(T_M(f)) = F_A\left(\sum(-1)^i T_{P_i}(\tilde{f}_i)\right)$$
$$= \sum(-1)^i F_A(T_{P_i}(\tilde{f}_i)) = \sum(-1)^i T_{P_i}(\tilde{f}_i^p) = T_M(f^p).$$

We have proved the following.

Lemma 12.1. *If A is a ring of prime characteristic p, M is a module of type FP over A and $f \in \mathrm{End}_A(M)$, then $F_A(T_M(f)) = T_M(f^p)$.* □

Let R be a commutative ring of prime characteristic p, let G be a group, $A = RG$, $F = F_A$. If $r = \sum_{\tau \in T(G)} r(\tau)\tau \in T(A)$, then $F(r) = \sum_{\tau \in T(G)} r(\tau)^p F(\tau) = \sum_{\tau \in T(G)} r(\tau)^p \tau^p$, where if $\tau = [g]$, τ^p means $[g^p]$.

Lemma 12.2. *In this situation suppose $r = \sum_{\tau \in T(G)} r(\tau)\tau$ is fixed by some power F^m $(m > 0)$. If $S = \mathrm{supp}(r) = \{\tau \in T(G) \mid r(\tau) \neq 0\}$, then*

(a) *F^m permutes S, $r(\tau^{p^m}) = r(\tau)^{p^m}$ for $\tau \in S$ and $r(1)^{p^m} = r(1)$;*
(b) *if $s \in G$ and $r(s) \neq 0$, then s is conjugate in G to $s^{p^{mn}}$ for some $n \leq |S|$;*
(c) *for all τ, $r(\tau)$ is algebraic over $\mathbb{F}_p$.*

Proof. Since $r = F^m(r) = \sum_\tau r(\tau)^{p^m} F^m(\tau)$, if $\sigma \in S$, then $r(\sigma) = \sum_{\tau \in T(G), F^m(\tau)=\sigma} r(\tau)^{p^m}$, and $\sigma = F^m(\tau) = \tau^{p^m}$ for at least one conjugacy class $\tau \in S$. Hence $S \subseteq F^m(S)$ and since S is finite, F^m permutes S as claimed, and (a), (b) follow. If $\tau \in S$ and n is the length of the orbit of τ under F^m, then $r(\tau)^{p^{mn}} = r(\tau)$, so $r(\tau)$ is algebraic over $\mathbb{F}_p$. □

Corollary 12.3. *If M is an RG-module of type FP, the conclusions of Lemma 12.2, with $m = 1$, apply to $r = r_M$.*

Proof. By Lemma 12.1, $F_A(r_M) = r_M$, since $r_M = T_{M/A}(\mathrm{id}_M) = T_{M/A}(\mathrm{id}_M^p)$. □

We shall apply Corollary 12.3 to study projective modules over $\mathbb{C}G$. First, we quote some results from commutative algebra.

Noether Normalisation Lemma *Let $K[x_1, \ldots, x_n]$ be a finitely generated integral domain over a field K. Then there exist elements $y_1, \ldots, y_r$ $(r \geq 0)$ in $K[x_1, \ldots, x_n]$, algebraically independent over K, such that $K[x_1, \ldots, x_n]$ is integral over $K[y_1, \ldots, y_r]$.*

Proof. See [32, (11.10), Lemma 1]. □

Lemma 12.4. *Suppose B is an integral domain which is integral over a subdomain A and K is an algebraically closed field. Then any ring homomorphism $\varphi : A \to K$ can be extended to B.*

Proof. Let $\mathfrak{p} = \mathrm{Ker}(\varphi)$. Then $\mathfrak{p}$ is prime, so there is a prime ideal $\mathfrak{P}$ of B with $\mathfrak{P} \cap A = \mathfrak{p}$ (by the Cohen-Seidenberg Theorem; see [64, §7.6]). Then $B/\mathfrak{P}$ is integral over $A/\mathfrak{p}$ and it is enough to extend the injective map $A/\mathfrak{p} \to K$ induced by φ to $B/\mathfrak{P}$, so we can assume φ is injective. Then φ extends to the field of fractions A' of A, and it is enough to extend this map to the field of fractions B' of B. Thus we can assume B/A is an algebraic field extension, and the result is then well-known (see [73, Ch.VII, §2, Theorem 2]). □

Lemma 12.5. *Let $R = \mathbb{Z}[\alpha_1, \dots, \alpha_n]$ be a finitely generated integral domain and let $x \in R$ be transcendental over $\mathbb{Q}$. Then for all but finitely many prime numbers p, there are a field K of characteristic p and a homomorphism $\varphi : R \to K$ such that $\varphi(x)$ is transcendental over $\mathbb{F}_p$.*

Proof. Let L be the field $\mathbb{Q}(x)$. By the Noether Normalisation Lemma, there exist $y_1, \dots, y_r$ in $L[\alpha_1, \dots, \alpha_n]$, algebraically independent over L, such that $L[\alpha_1, \dots, \alpha_n]$ is integral over $L[y_1, \dots, y_r]$. Let $Y = \{y_1, \dots, y_r\}$ and let f_i be a monic polynomial satisfied by α_i over $L[Y]$. The coefficients of f_i are polynomials in Y over L, that is, polynomials whose coefficients are quotients of elements of $\mathbb{Z}[x]$. Let Π be the finite set of primes of $\mathbb{Z}[x]$ appearing as factors in the numerator or denominator of one of these quotients.

Let $S = \{a(x)/b(x) \mid a(x) \in \mathbb{Z}[x], b(x)$ is a product of primes in $\Pi\}$.

Then S is a subring of L and $f_i \in S[Y]$, so each α_i is integral over $S[Y]$. All but finitely many primes $p \in \mathbb{Z}$ do not divide the denominators of any of the polynomials in Π. For such primes p, extend the canonical map $\mathbb{Z} \to \mathbb{Z}/p\mathbb{Z} = \mathbb{F}_p$ to a homomorphism

$$\varphi : \mathbb{Z}[x] \to \mathbb{F}_p(t)$$

(where t is transcendental over $\mathbb{F}_p$) by letting $\varphi(x) = t$. We can further extend to $\varphi : S \to \mathbb{F}_p(t)$ and then to $\varphi : S[Y] \to \mathbb{F}_p(t)$ by sending each y_i to 0. By Lemma 12.4, φ then extends to a homomorphism from $S[Y][\alpha_1, \dots \alpha_n]$ to the algebraic closure of $\mathbb{F}_p(t)$, and restricting to R gives the desired homomorphism. □

Let P be a finitely generated projective module over a group ring RG, where R is a commutative ring. Then $P \oplus Q = L$ for some Q and finitely generated free RG-module L. If $p : L \to P$ is the projection map, then $p^2 = p$, so the matrix (e_{ij}) of P with respect to a basis of L is idempotent. Further, if

$e_{ij} = \sum_{g \in G} e_{ij}(g)g$, then $r_P(\tau) = \sum_{1 \le i \le n, g \in \tau} e_{ii}(g)$ for $\tau \in T(G)$, where (e_{ij}) is an $n \times n$ matrix. (See Remark (1) after Lemma 2.7.) Conversely, any idempotent in $M_n(RG)$ determines a finitely generated projective RG-module in this way.

Lemma 12.6. *Let M be a module of type FP over a group ring KG, where K is a field of characteristic 0. Then $r_M(\tau)$ is algebraic over $\mathbb{Q}$, for all $\tau \in T(G)$.*

Proof. $r_M = r_P - r_Q$ for some finitely generated projective RG-modules P, Q. (If $S \to M$ is a finite augmented projective resolution, let $P = S_0 \oplus S_2 \oplus \dots$, $Q = S_1 \oplus S_3 \oplus \dots$) Thus we can assume M is projective. Let $(e_{ij})_{1 \le i,j \le n}$ be an idempotent matrix over KG corresponding to M as above. Using the notation above, let

$$E = \{e_{ij}(g) \mid 1 \le i, j \le n, \ g \in G\}$$

(a finite set). Suppose $r_M(\sigma)$ is transcendental over $\mathbb{Q}$. By Lemma 12.5, there is a homomorphism

$$\varphi : \mathbb{Z}[E] \to F,$$

where F is a field of characteristic $p > 0$, such that $\varphi(r_M(\sigma))$ is transcendental over $\mathbb{F}_p$. We can extend φ to a homomorphism $\mathbb{Z}[E]G \to FG$ of group rings (Remark 5.1). Then $(\varphi(e_{ij}))_{1 \le i,j \le n}$ is an idempotent matrix over FG, giving a finitely generated FG-projective module N, with

$$r_N = \sum_{\tau \in T(G)} \varphi(r_M(\tau))\tau.$$

But then $\varphi(r_M(\sigma))$ is algebraic over $\mathbb{F}_p$ by Cor. 12.3, a contradiction. □

Lemma 12.7. *Let K be a field of characteristic 0, G a group, P a finitely generated projective KG-module. Then there exist a field K', such that $K'/\mathbb{Q}$ is a finite extension, and a finitely generated projective $K'G$-module Q such that $r_Q = r_P$.*

Proof. As in Lemma 12.6, P corresponds to some idempotent matrix $(e_{ij})_{1 \le i,j \le n}$ over KG. Again let

$$E = \{e_{ij}(g) \mid 1 \le i, j \le n, \ g \in G\}.$$

Let $L = \mathbb{Q}(r_P(\tau) \mid \tau \in T(G))$, so $L/\mathbb{Q}$ is finite by Lemma 12.6. Apply the Noether Normalisation Lemma to R, where $R = L[E] \subseteq K$: there is a set $Y = \{y_1, \dots, y_r\}$ such that $y_1, \dots, y_r$ are algebraically independent over L and R is integral over $L[Y]$. Let $\overline{L}$ be the algebraic closure of L and let $\lambda : L[Y] \to \overline{L}$ be the homomorphism which is the identity map on L and zero on Y. By Lemma 12.4, we can extend to a homomorphism $\lambda : R \to \overline{L}$, and $\lambda(L[Y]) = L$. Then, for $e \in E$, $\lambda(e)$ is algebraic over L, so over $\mathbb{Q}$. Let $K' = L(\lambda(e) \mid e \in E)$, so $K'/\mathbb{Q}$ is a finite extension and λ maps R to K'. Thus λ extends to a homomorphism $\lambda : RG \to K'G$. The matrix $(\lambda(e_{ij}))$ over $K'G$ is an idempotent. As in the proof of Lemma 12.6, there is a corresponding $K'G$-projective module Q, with $r_Q(\tau) = \lambda(r_P(\tau))$ for $\tau \in T(G)$. But λ restricted to L is the identity map, so $\lambda(r_P(\tau)) = r_P(\tau)$. □

We shall summarise some needed results from elementary algebraic number theory. Thus from now on, primes in $\mathbb{Z}$ will be called rational primes. More generally, if $K/\mathbb{Q}$ is a finite field extension, a *prime* of K means a prime ideal in $\mathcal{O}$, its ring of integers. If $\mathfrak{p}$ is a prime of K, then $\mathfrak{p} \cap \mathbb{Z} = \mathbb{Z}p$ for some rational prime p, and we say that $\mathfrak{p}$ *divides* p. We recall that $\mathcal{O}$ is a Dedekind domain, in particular its non-zero prime ideals are maximal (see [102, Theorem 1, §3.4]). Let $K/\mathbb{Q}$ be a finite Galois extension, $\mathfrak{p}$ a prime of K dividing the rational prime p. Let $\mathcal{O}$ be the ring of integers, $\mathcal{O}_\mathfrak{p}$ the localisation at $\mathfrak{p}$ and $\mathfrak{m}_\mathfrak{p}$ its maximal ideal. The residue field $F = \mathcal{O}_\mathfrak{p}/\mathfrak{m}_\mathfrak{p}$ is isomorphic to $\mathcal{O}/\mathfrak{p}$ (see [64, Prop.10.10]) and is a finite extension of $\mathbb{Z}/\mathbb{Z}p$ ([102, Prop.2, §6.2]). The group $\Gamma = \mathrm{Gal}(K/\mathbb{Q})$ permutes the primes of K dividing p (transitively-see [102, Prop.1, §6.2]), the stabiliser $\Gamma_\mathfrak{p}$ acts on F and there is a surjective map $\psi : \Gamma_\mathfrak{p} \to \mathrm{Gal}(F/\mathbb{F}_p)$ ([102, Prop.2, §6.2]). If p is unramified in K (i.e. ψ is one-to-one), $\Gamma_\mathfrak{p}$ is cyclic, generated by γ, where $\psi(\gamma)$ is the Frobenius map $x \mapsto x^p$ ([102, §6.3]). Denote γ by Frob($\mathfrak{p}$). Thus

$$\gamma(x) \equiv x^p \mod \mathfrak{m}_\mathfrak{p} \quad \text{for all } x \in \mathcal{O}_\mathfrak{p}.$$

Theorem 12.8. *For any $\gamma \in \mathrm{Gal}(K/\mathbb{Q})$, there are infinitely many rational primes p such that there exist primes $\mathfrak{p}$ of K dividing p such that Frob($\mathfrak{p}$) $= \gamma^m$ for some m with $\langle \gamma^m \rangle = \langle \gamma \rangle$.*

Proof. This follows easily from the Frobenius Density Theorem ([65, p.134]), using equation (1) in [102, §6.3]. □

Theorem 12.8 can be improved using the Tchebotarev Density Theorem ([74, 8.4, Theorem 10]) and this slightly simplifies the arguments which follow. This is left to the reader.

Theorem 12.9. *Let M be a module of type FP over a group ring KG, where K is a field of characteristic zero. Then*

(1) (Zalesskii) $r_M(1) \in \mathbb{Q}$;
(2) (Bass) $\mathbb{Q}(r_M(\tau) \mid \tau \in T(G))$ *is a finite Galois extension of* $\mathbb{Q}$ *with abelian Galois group.*

Proof. As in Lemma 12.6, write $r_M = r_{P_1} - r_{P_2}$ where P_1, P_2 are finitely generated projective KG-modules. By Lemma 12.7, there are finite field extensions $K_i/\mathbb{Q}$ and finitely generated projective K_iG-modules P_i' such that $r_{P_i} = r_{P_i'}$ (for $i = 1, 2$). Let $L = K_1K_2$ (we can view K_1, K_2 as subfields of $\mathbb{C}$, and L is the subfield of $\mathbb{C}$ they generate), so $L/\mathbb{Q}$ is finite. Let $P = L \otimes_{K_1} P_1'$, $Q = L \otimes_{K_2} P_2'$, so P, Q are finitely generated LG-modules, and $r_M = r_P - r_Q$. Let N be the normal closure of $L/\mathbb{Q}$, and let $\Gamma = \mathrm{Gal}(L/\mathbb{Q})$. As in Lemma 12.6, let $(e_{ij})_{1\le i,j\le n}$ be an idempotent matrix over LG corresponding to P, $(f_{ij})_{1\le i,j\le n}$ one corresponding to Q, and define

$$E = \{e_{ij}(g) \mid 1 \le i, j \le n,\ g \in G\} \cup \{f_{ij}(g) \mid 1 \le i, j \le n,\ g \in G\}.$$

If $0 \ne x \in N$, then $x \in \mathcal{O}_{\mathfrak{p}} \setminus \mathfrak{m}_{\mathfrak{p}}$ for all but finitely many primes $\mathfrak{p}$ of N ($\mathcal{O}$ is the ring of integers of N, $\mathfrak{m}_{\mathfrak{p}}$ is the maximal ideal of $\mathcal{O}_{\mathfrak{p}}$). Hence, for all but finitely many primes $\mathfrak{p}$ of N, the following two conditions are satisfied:

(a) $e \in \mathcal{O}_{\mathfrak{p}}$ for all $e \in E$, and $e \notin \mathfrak{m}_{\mathfrak{p}}$ if $e \ne 0$;
(b) for all $\gamma \in \Gamma$ and all $\sigma, \tau \in T(G)$,

$$\gamma r_M(\tau) - r_M(\tau) \ne 0 \quad \text{implies} \quad \gamma r_M(\tau) - r_M(\tau) \notin \mathfrak{m}_{\mathfrak{p}}.$$

We claim that:

$$(*) \text{ for all } \gamma \in \Gamma, \text{ there exists } n = n(\gamma) \text{ such that}$$
$$\gamma r_M(\tau) = r_M(\tau^n) \text{ for all } \tau \in T(G).$$

For let $\gamma \in \Gamma$. By Theorem 12.8, there is a prime $\mathfrak{p}$ of N satisfying (a) and (b) and such that $\mathrm{Frob}(\mathfrak{p}) = \gamma^m$, for some m with $\langle\gamma^m\rangle = \langle\gamma\rangle$. By (a), $(e_{ij}), (f_{ij})$ are idempotent matrices over $\mathcal{O}_{\mathfrak{p}}G$, so $(\bar{e}_{ij}), (\bar{f}_{ij})$ are idempotent matrices over FG, where $F = \mathcal{O}_{\mathfrak{p}}/\mathfrak{m}_{\mathfrak{p}}$ (and $e \mapsto \bar{e}$ denotes the canonical map $\mathcal{O}_{\mathfrak{p}}G \to FG$). Let $\bar{P}, \bar{Q}$ be the corresponding projective FG-modules. Note that $r_{\bar{P}}(\tau) = \overline{r_P(\tau)}$ for all $\tau \in T(G)$ (as in Lemma 12.6). Also, F has prime characteristic p, where $\mathfrak{p}$ divides p, so by Cor. 12.3, $r_{\bar{P}}(\tau^p) = r_{\bar{P}}(\tau)^p$ for all $\tau \in T(G)$. Hence:

$$\begin{aligned} & r_P(\tau^p) \equiv r_P(\tau)^p \mod \mathfrak{m}_{\mathfrak{p}} \\ \text{Also}\quad & \gamma^m r_P(\tau) \equiv r_P(\tau)^p \mod \mathfrak{m}_{\mathfrak{p}} \quad (\text{since } \gamma^m = \mathrm{Frob}(\mathfrak{p})) \\ \text{so}\quad & \gamma^m r_P(\tau) \equiv r_P(\tau^p) \mod \mathfrak{m}_{\mathfrak{p}}. \end{aligned}$$

Similarly $\gamma^m r_Q(\tau) \equiv r_Q(\tau^p) \mod \mathfrak{m}_{\mathfrak{p}}$, so $\gamma^m r_M(\tau) \equiv r_M(\tau^p) \mod \mathfrak{m}_{\mathfrak{p}}$, for all $\tau \in T(G)$. By (b), $\gamma^m r_M(\tau) = r_M(\tau^p)$ for all $\tau \in T(G)$. Since $\langle\gamma^m\rangle = \langle\gamma\rangle$, $\gamma = \gamma^{ml}$ for some $l > 0$, hence $\gamma r_M(\tau) = r_M(\tau^{p^l})$ for all $\tau \in T(G)$, proving $(*)$. Take $\tau = [\,1\,]$ to see that $\gamma r_M(1) = r_M(1)$ for all $\gamma \in \mathrm{Gal}(N/\mathbb{Q})$. Since $N/\mathbb{Q}$ is Galois, $r_M(1) \in \mathbb{Q}$. Let $A = \{r_M(\tau) \mid \tau \in T(G)\}$, $B = \mathbb{Q}(A)$. It follows from $(*)$

that $\gamma A \subseteq A$, hence $\gamma B \subseteq B$, for all $\gamma \in \Gamma$. It follows that $B/\mathbb{Q}$ is Galois (see the proof of Theorem 2(b), §6.1 in [102]). For $\gamma, \delta \in \Gamma$, there exist n, n' such that $\gamma r_M(\tau) = r_M(\tau^n)$, $\delta r_M(\tau) = r_M(\tau^{n'})$ for all $\tau \in T(G)$. Then $\gamma\delta r_M(\tau) = r_M(\tau^{nn'}) = \delta\gamma r_M(\tau)$. It follows that $\Gamma' = \mathrm{Gal}(B/\mathbb{Q})$ is abelian (the map $\Gamma \to \Gamma'$, $\gamma \mapsto \gamma|_B$ is onto-again see the proof of Theorem 2(b), §6.1 in [102]). □

In Theorem 12.9, if $B = \mathbb{Q}(r_M(\tau) \mid \tau \in T(G))$, the *Artin symbol* $\left(\frac{B/\mathbb{Q}}{p}\right)$ is defined, for any rational prime p which does not ramify in B. It equals $\mathrm{Frob}(\mathfrak{q})$, for any prime $\mathfrak{q}$ of B dividing p. See [102, §6.3].

Lemma 12.10. *In Theorem* 12.9, *let* $S_M = \mathrm{supp}(r_M) = \{\tau \in T(G) \mid r_M(\tau) \neq 0\}$. *Then there exists a finite set* Π *of rational primes, including those that ramify in* $B = \mathbb{Q}(r_M(\tau) \mid \tau \in T(G))$, *such that, for* $p \notin \Pi$

(1) $\delta r_M(\tau) = r_M(\tau^p)$ *for all* $\tau \in T(G)$, *where* $\delta = \left(\frac{B/\mathbb{Q}}{p}\right)$;
(2) *the map* $\tau \mapsto \tau^p$ *is a permutation of* S_M.

Proof. Let N be as in the proof of Theorem 12.9. Let Π be the set of rational primes which ramify in N or are such that, for some prime $\mathfrak{p}$ of N dividing p, condition (a) or (b) in the proof of Theorem 12.9 fails. Then Π is finite (see [102, Theorem 1, §5.3] and the proof of Theorem 12.9). The argument of Theorem 12.9 shows that, for $p \notin \Pi$, $\gamma r_M(\tau) = r_M(\tau^p)$ for all $\tau \in T(G)$, where $\mathfrak{p}$ is a prime of N dividing p and $\gamma = \mathrm{Frob}(\mathfrak{p})$, and (1) follows since $\gamma|_B = \left(\frac{B/\mathbb{Q}}{p}\right)$ (see [102, Prop.1(b), §6.3]). Further, $S_M \subseteq S_P \cup S_Q$ and by Condition (a), $S_P = S_{\bar{P}}$, $S_Q = S_{\bar{Q}}$ (with notation as in the proof of Theorem 12.9). By Cor. 12.3, the map $\tau \mapsto \tau^p$ is a permutation of $S_{\bar{P}}$ and of $S_{\bar{Q}}$, hence of $S_P \cup S_Q$, and so of S_M by Part (1). □

Note. *In Lemma* 12.10(2), *if* $r_M(g) \neq 0$, *it follows that* g *is conjugate to* g^{p^u} *for some* $u > 0$ *with* $u \leq |S_M|$.

Theorem 12.11. *Under the hypotheses of Theorem* 12.9,

(1) *if* $g \in G$ *has finite order* m, *and* $\omega = e^{2\pi i/m}$, *then* $r_M(g) \in \mathbb{Q}(\omega)$, *say* $r_M(g) = f(\omega)$ *(where* $f \in \mathbb{Q}[x]$, *the polynomial ring), and* $r_M(g^q) = f(\omega^q)$ *for all* q *prime to* m;
(2) *if* $g \in G$ *has infinite order and* $r_M(g) \neq 0$, *then* g *belongs to a subgroup* H *of* G *isomorphic to the additive group of* $\mathbb{Z}[1/p \mid p \text{ prime},\ p \notin \Pi]$ *(with* Π *as in Lemma* 12.10*).*

Proof. (1) By the Kronecker-Weber Theorem ([65, Ch.V, Theorem 5.9]), we can embed $B = \mathbb{Q}(r_M(\tau) \mid \tau \in T(G))$ in $\mathbb{Q}(z)$, where z is a primitive nth root of 1 for some n. Then both B and $\mathbb{Q}(\omega)$ embed in $\mathbb{Q}(\zeta)$, where ζ is a primitive Nth root of 1 and $N = mn$. By [102, Theorem 1, §6.4], there is an isomorphism $(\mathbb{Z}/N\mathbb{Z})^* \to \mathrm{Gal}(\mathbb{Q}(\zeta)/\mathbb{Q})$ induced by $q \mapsto \sigma_q$, where $(\mathbb{Z}/N\mathbb{Z})^*$ is the group of units of $\mathbb{Z}/N\mathbb{Z}$, and $\sigma_q(\zeta) = \zeta^q$ for integers q prime to N. In particular, if $q = q_1 q_2$ then $\sigma_q = \sigma_{q_1}\sigma_{q_2}$. Further, if q is prime, $\sigma_q = \left(\frac{\mathbb{Q}(\zeta)/\mathbb{Q}}{q}\right)$. Since $\mathbb{Q}(\zeta)/\mathbb{Q}$ is Galois (Example 2, §6.1 in [102]), there is (from the proof of Theorem 2(b), §6.1 in [102]) a short exact sequence

$$\mathrm{Gal}(\mathbb{Q}(\zeta)/\mathbb{Q}(\omega)) \rightarrowtail \mathrm{Gal}(\mathbb{Q}(\zeta)/\mathbb{Q}) \twoheadrightarrow \mathrm{Gal}(\mathbb{Q}(\omega)/\mathbb{Q}).$$

(The right-hand map is restriction and corresponds to the usual projection map $(\mathbb{Z}/N\mathbb{Z})^* \to (\mathbb{Z}/m\mathbb{Z})^*$.) Also, $\sigma_q(\omega) = \omega^q$ since ω is a power of ζ. Hence $\sigma_q \in \mathrm{Gal}(\mathbb{Q}(\zeta)/\mathbb{Q}(\omega))$ if and only if $\omega^q = \omega$, if and only if $\omega^{q-1} = 1$, i.e. $q \equiv 1 \mod m$. Any element of $(\mathbb{Z}/N\mathbb{Z})^*$ is represented by an integer q not divisible by any prime in Π. (For let $p_1, \dots, p_k$ be the elements of Π prime to N. Suppose $(z, N) = 1$. Then by the Chinese Remainder Theorem, there is an integer q such that

$$\begin{aligned} q &\equiv z \mod N \\ q &\equiv 1 \mod p_i \quad (1 \le i \le k) \end{aligned}$$

and q is not divisible by any prime in Π.) Writing such an integer q as a product of primes, $\sigma_q(r_M(g)) = r_M(g^q)$ by repeated use of Lemma 12.10. (For primes p not in Π, $\sigma_p = \left(\frac{\mathbb{Q}(\zeta)/\mathbb{Q}}{p}\right)$ restricted to B is $\left(\frac{B/\mathbb{Q}}{p}\right)$, by [102, Prop.1(b), §6.3].) Thus $r_M(g)$ is fixed by all σ_q with $q \equiv 1 \mod m$, i.e. by $\mathrm{Gal}(\mathbb{Q}(\zeta)/\mathbb{Q}(\omega))$. Hence $r_M(g) \in \mathbb{Q}(\omega)$, and (1) follows since $\sigma_q(\omega) = \omega^q$.

(2) By Lemma 12.10(2), the map $\varphi_q : \tau \mapsto \tau^q$ is a permutation of S_M, for any positive integer q not divisible by any prime in Π. Then $\varphi_q^N = 1$, where $N = |S_M|!$. Hence g is conjugate in G to g^{q^N} for all such q and g with $T(g) \in S_M$. Thus $g = tg^{q^N}t^{-1}$ for some $t \in G$, so $g = u^{q^N}$, where $u = tgt^{-1}$, and $T(u) = T(g) \in S_M$. Assume $g \in S_M$ has infinite order. Enumerate the primes not in Π, say $p_1, p_2, \dots$ and put $a_n = p_1^N \dots p_n^N$ for $n \ge 1$. Inductively we can find $g_n \in T(g)$ for $n \ge 0$ such that $g_0 = g$ and $g_n^{a_n} = g_{n-1}$ for $n > 0$. Let H be the subgroup of G generated by $\{g_0, g_1, \dots\}$. An element of H has the form g_n^k for some $n \ge 0$ and integer k. The map $H \to \mathbb{Q}$, $g_n^k \mapsto k/(a_1 \dots a_n)$ is easily seen to be well-defined and an injective homomorphism. Its image is clearly $\mathbb{Z}[1/p \mid p \text{ prime}, \ p \notin \Pi]$. □

Linnell [75, Lemma (4.1)], using results of Cliff [28], has proved the following.

Theorem. *Let G be a group.*

(1) *If p is a rational prime, Q is a finitely generated projective $\mathbb{Z}_pG$-module, and $g \in G$ is such that $r_Q(g) \neq 0$, then $[g] = [g^{p^n}]$ for some integer $n > 0$, and $r_Q(g) = r_Q(g^p)$;*
(2) *if P is a finitely generated $\mathbb{Z}G$-module, $g \in G$ and $r_P(g) \neq 0$, then there are subgroups C, H of G such that $g \in C \leq H \leq G$, C is isomorphic to the additive group of $\mathbb{Q}$, H is finitely generated, and the elements of C lie in finitely many H-conjugacy classes. (So if g has finite order, $r_P(g) = 0$.)*

Conjectures. Let R be a subring of $\mathbb{C}$ such that $R \cap \mathbb{Q} = \mathbb{Z}$, let G be a group and P a finitely generated projective RG-module.

(1) $r_P(1) = \sum_{\tau \in T(G)} r_P(\tau)$.
(2) $r_P(g) = 0$ for all $g \neq 1$ in G.
(3) P is stably free (i.e. there exist finitely generated free RG-modules F, L such that $P \oplus L = F$).

Conjectures (1) and (2) are from [11], where (3) was posed as a question. Clearly (3) $\Rightarrow$ (2) $\Rightarrow$ (1). Also, (1) (with $R = \mathbb{Z}$) implies the Brown characteristic χ and μ agree on FP($\mathbb{Z}$), so on vFP($\mathbb{Z}$) (see Cor. 6.25 and the note after Prop. 10.3). Using Lemma 12.10(2), Bass [10] shows that $r_P(g) = 0$ for all g of infinite order in G if G is linear, so by Linnell's Theorem, (2) is true for linear groups G (with $R = \mathbb{Z}$).

Lemma 12.12. *Let G be a group, M a $\mathbb{Z}G$-module of type FP. If $r_M(g) \neq 0$, then*

$$g \in \bigcap_{p \text{ prime}} G^{p^\infty},$$

where, for integers $n \geq 1$, $G^n = \{g^n \mid g \in G\}$ and $G^{p^\infty} = \bigcap_{j \geq 0} G^{p^j}$.

Proof. This follows easily from Part (1) of Linnell's Theorem. □

Corollary 12.13. *Conjecture 2, with $R = \mathbb{Z}$, is true for residually finite groups G.*

Proof. For residually finite groups G, $\bigcap_{p \text{ prime}} G^{p^\infty} = 1$, since this is true for finite groups G. □

Conjecture (2) is true for residually finite groups G and arbitrary R (this was proved by Moody[1]).

Further Conjectures.

(4) if $G \in \mathrm{FP}(\mathbb{Q})$, then $\chi_G(g) = 0$ for g of infinite order in G (recall $\chi_G = r_{\mathbb{Q}/\mathbb{Q}G}$).
(5) $\mathrm{FL}(\mathbb{Z}) = \mathrm{FP}(\mathbb{Z})$.

Conjecture (4) appears in [11], and (5) was asked as a question in [107]. Both (4) and (5) imply that μ and χ agree on $\mathrm{FP}(\mathbb{Z})$, because groups in $\mathrm{FP}(\mathbb{Z})$ have $\mathrm{cd}_{\mathbb{Z}}(G) < \infty$, so are torsion-free (Cor. 5.12). Conjecture (5) is related to (3). For if $G \in \mathrm{FP}(\mathbb{Z})$, there is an augmented projective resolution

$$0 \to P \to F_{n-1} \to \ldots \to F_0 \to \mathbb{Z} \to 0 \qquad (*)$$

with P finitely generated $\mathbb{Z}G$-projective and all F_i finitely generated $\mathbb{Z}G$-free. For if

$$0 \to P_n \to P_{n-1} \to \ldots \to P_0 \to \mathbb{Z} \to 0 \qquad (**)$$

is a finite augmented projective resolution, we can write $P_0 \oplus Q_0 = F_0$ where F_0 is finitely generated free. Put $P_1' = P_1 \oplus Q_0$. Then there is an augmented finite projective resolution

$$0 \to P_n \to P_{n-1} \to \ldots \to P_1' \xrightarrow{d_1'} F_0 \to \mathbb{Z} \to 0$$

where, if d is the differential on P, $d_1' = d_1 \oplus \mathrm{id}_{Q_0}$. Clearly repetition of this trick leads to a resolution of the form $(*)$. Further, $G \in \mathrm{FL}(\mathbb{Z})$ if and only if the projective module P in $(*)$ is stably free. For if it is stably free, we can repeat this trick one more time, writing $P \oplus F_{n+1} = F_n$, where F_n, F_{n+1} are finitely generated free, to obtain a finite augmented free resolution

$$0 \to F_{n+1} \to F_n \to F_{n-1} \to \ldots \to F_0 \to \mathbb{Z} \to 0.$$

Conversely, if $G \in \mathrm{FL}(\mathbb{Z})$, we may take the P_i in $(**)$ to be free, and comparing $(*)$ and $(**)$ using Cor. 3.11, P is stably free.

Further information on rank elements is provided by the following.

Theorem 12.14 (Kaplansky). *Let P be a finitely generated projective KG-module, where G is a group and K is a field of characteristic 0, corresponding to an $n \times n$ idempotent matrix over KG. Then $0 \le r_P(1) \le n$, and $r_P(1) = 0$ if and only if $P = 0$. (Recall that $r_P(1) \in \mathbb{Q}$ by Theorem 12.9.)*

[1] See Section 13.

Proof. See [88], also [93, §22]. □

13. Update

This is a survey of progress since the course was given. It is not claimed to be comprehensive.

Conjecture (2) at the end of Section 12 is known as the Bass conjecture (or strong Bass conjecture), and Conjecture (1) is called the weak Bass conjecture, often just for $R = \mathbb{Z}$. They are related to the idempotent conjecture, that if G is a torsion-free group, the only idempotents in $\mathbb{C}G$ are 0 and 1. (It is obviously necessary to assume torsion-free; see the idempotent in Lemma 5.10.) In fact, if e is an idempotent in $\mathbb{C}G$, let r_e denote $r_{\mathbb{C}Ge}$. Then the following are equivalent:

(1) G satisfies the idempotent conjecture;
(2) for all idempotents e in $\mathbb{C}G$, $r_e(g) = 0$ for all $g \neq 1$ in G;
(3) for all idempotents e in $\mathbb{C}G$, $\sum_{\tau \in T(G)} r_e(\tau) = r_e(1)$.

Proof. Clearly (1) ⇒ (2) ⇒ (3). Assume (3) and let e be an idempotent in $\mathbb{C}G$, say $e = \sum_{g \in G} e_g g$, with $e_g \in \mathbb{C}$. By Example (2) after Lemma 2.6,

$$r_e = T_{\mathbb{C}G}(e) = \sum_{g \in G} e_g[g] = \sum_{\tau \in T(G)} r_e(\tau)\tau$$

$$\text{so} \quad \varepsilon(e) = \sum_{g \in G} e_g = \sum_{\tau \in T(G)} r_e(\tau) = r_e(1)$$

where ε is the augmentation map. Since $\varepsilon(e)$ is an idempotent in $\mathbb{C}$, it is 0 or 1. Assume it is 0. By Theorem 12.14, $\mathbb{C}Ge = 0$, so $e = 0$. If $\varepsilon(e) = 1$, apply this argument to $1 - e$ to see $e = 1$. □

It was noted in [10] that the statement that finite groups satisfy the Bass Conjecture is equivalent to Swan's Theorem (Theorem 7.10). We survey progress on these conjectures. Formanek [55] showed that the idempotent conjecture is true for torsion-free Noetherian groups. In particular it is true for polycyclic by finite groups. (Another proof of this was given by Marciniak [83].) The Bass conjecture for polycyclic by finite groups was established by Weiss [118]. Results on the coefficients $r_P(g)$, where P is a finitely generated projective kG-module, k is a field and G is linear, were obtained by Cliff [29]. Moody's work (mentioned after Cor. 12.13) was not published. However, the following result appeared in [89]. Under suitable hypotheses on G and R, there are finite cyclic subgroups $C_1, \ldots, C_n$ and finitely generated RC_i-projectives P_i for $1 \leq i \leq n$ such that r_P is a $\mathbb{Q}$-linear combination of the rank elements $r_{RG \otimes_{RC_i} P_i}$. The hypotheses are that for $g \in G$ of infinite order, the set of g^m for which none of the prime divisors of m are invertible in R never fall into finitely many conjugacy classes. Schafer [103] proved a version of Linnell's Theorem with $\mathbb{Z}$ replaced by the ring of integers in an algebraic number field. Using this, he establishes the Bass conjecture for any subring of the algebraic integers in $\mathbb{C}$ and any group which is residually of bounded exponent. Schafer asked the following. If G is a finitely generated group having a subgroup A isomorphic to the additive group of $\mathbb{Q}$, such that A is contained in only finitely many conjugacy classes in G, must G be of infinite cohomological dimension? An affirmative answer would imply the Bass conjecture for all groups of finite cohomological dimension. However, a counterexample was given by Linnell [76].

Burghelea [20] gave a decomposition of the cyclic homology of a group algebra, with summands indexed by the conjugacy classes of G. Cyclic homology has proved useful in attacking the Bass conjecture. Let G be a group of finite homological dimension n over $\mathbb{Q}$. For $g \in G$, $C_G(x)$ denotes the centralizer of g in G. Eckmann [42] proved that, if g is an element of infinite order in G, then $H_i(C_G(g)/\langle g\rangle, \mathbb{Q}) = 0$ for $i \geq n$ in the following cases: (i) G is soluble of finite Hirsch rank, (ii) G is linear in characteristic 0 with finite homological dimension over $\mathbb{Q}$, (iii) $\mathrm{cd}_{\mathbb{Q}}(G) \leq 2$ (this includes finite groups, by Lemma 5.10). Using Burghelea's result, he shows that, for such groups G, if g is an element of infinite order then $r_P(g) = 0$ for finitely generated projective $\mathbb{C}G$-modules P. Eckmann's results were extended by Cornick [33]. Also, Schafer [104] generalised Burghelea's result to relative cyclic homology using Marciniak's algebraic proof of it [82]. This is then used to show the following. Let G be a group and suppose $g \in G$ has infinite order in G/G_n, where G_n is the nth term of the lower central series of G. Then for any finitely generated projective $\mathbb{Q}G$-module P, $r_P(g) = 0$.

More recent work has identified some classes of groups for which the Bass conjecture is true which have interesting closure properties. Let K be a field of characteristic zero. Following Eckmann's work, Chadha and Passi [23] defined the class $E(K)$ as follows. A group G is in class $E(K)$ if

(1) $\mathrm{hd}_K(G) < \infty$ and
(2) for all $g \in G$ of infinite order, $\mathrm{hd}_K(C_G(g)/\langle g\rangle) < \infty$,

where hd denotes homological dimension. Thus the groups in Eckmann's result above belong to $E(\mathbb{Q})$. Chadha and Passi [23] (see also Ji [66]) proved that $E(K)$ is closed under subgroups, extensions and free products, and under unions of bounded homological dimension over K. It follows that $E(\mathbb{Q})$ contains the class of elementary amenable groups of finite homological dimension over $\mathbb{Q}$, and so the class of polycyclic by finite groups. Chadha and Passi note that groups in $E(K)$ satisfy the Bass conjecture, and deduce that groups residually in $E(K)$ satisfy the weak Bass conjecture. In another paper [22], Chadha shows that, if $\mathrm{cd}_K(G) < \infty$ and G has a central free abelian subgroup of rank $\mathrm{cd}_K(G) - 1$, then $G \in E(K)$. Chadha and Passi [24] have also introduced the class of groups having what is called the group trace property in connection with the Bass conjecture.

More recently, Emmanouil [49] has introduced the class $\mathcal{C}(\mathbb{Q})$. To define this class, let g be an element of infinite order in a group G. There is a central extension

$$1 \to \mathbb{Z} \to C_G(g) \to N \to 1$$

where $N = C_G(g)/\langle g \rangle$, the mapping $\mathbb{Z} \to C_G(g)$ being $n \mapsto g^n$. Let $\alpha(g)$ be the element of $H^2(N, \mathbb{Z})$ classifying this extension.

Definition. *The group G is said to belong to $\mathcal{C}(\mathbb{Q})$ if, for all elements $g \in G$ of infinite order, the image of $\alpha(g)$ in the rational cohomology ring $H^*(N, \mathbb{Q})$ is nilpotent.*

Emmanouil shows that $\mathcal{C}(\mathbb{Q})$ is closed under subgroups, free products and finite direct products, and under extensions whose cokernels have finite homological dimension over $\mathbb{Q}$. Also, $\mathcal{C}(\mathbb{Q})$ contains the class of abelian groups and the class $E(\mathbb{Q})$. Consequently, the class of extensions of a group in $\mathcal{C}(\mathbb{Q})$ by one in $E(\mathbb{Q})$ is contained in $\mathcal{C}(\mathbb{Q})$. Further, the class $\mathrm{R}\mathcal{C}(\mathbb{Q})$ of groups residually in $\mathcal{C}(\mathbb{Q})$ satisfies Bass' Strong Conjecture. Denoting the class of torsion-free groups in $\mathcal{C}(\mathbb{Q})$ by $\mathcal{C}'(\mathbb{Q})$, it is further shown that a group residually in $\mathcal{C}'(\mathbb{Q})$ satisfies the idempotent conjecture. In another paper [50], Emmanouil considers the class $\mathcal{S}$ of all groups G, such that, if P is a projective $\mathbb{C}G$-module and $P_G = 0$, then $P = 0$. (Recall, from the Note preceding Lemma 8.8, that $P_G = \mathbb{C} \otimes_{\mathbb{C}G} P = P/I_G P$.) The class $\mathcal{S}$ is closed under subgroups, extensions, direct products and free products, and contains the class of torsion-free abelian groups. Emmanouil shows that, if N is a normal subgroup of a group G, $N \in \mathcal{S}$ and G/N satisfies the idempotent conjecture, then G satisfies the idempotent conjecture. The condition defining $\mathcal{S}$ is a relaxation of one used by Strebel [111], who considered groups satisfying the following. If $f : P \to Q$ is a homomorphism of projective $\mathbb{C}G$-modules and $1 \otimes f : \mathbb{C} \otimes_{\mathbb{C}G} P \to \mathbb{C} \otimes_{\mathbb{C}G} Q$ is injective, then f is injective. Any such group is in $\mathcal{S}$. (If $P_G = 0$, apply this condition to the zero map $f : P \to 0$ to see $P = 0$.) Strojnowski [112] showed that the class of groups satisfying Strebel's condition has the analogous closure properties to $\mathcal{S}$ and and contains the class of locally indicable groups. Indeed, one can replace $\mathbb{C}$ by an arbitrary ring R. Strojnowski also showed that all idempotents of RG are contained in R if and only if all idempotents in R are central.

Strojnowski [113] considered a class of groups G satisfying a condition called WD: if H is a finitely generated subgroup of G, $h \in H$, N is a positive integer and h is conjugate in H to h^{p^N} for all primes p, then $h = 1$. Strojnowski shows that groups with property WD satisfy the Bass conjecture, and gives several classes of groups satisfying WD.

Eckmann [44], [45] has considered various trace functions associated to a group, including $T_{\mathbb{C}G}$, and the corresponding ranks. In an appendix to [45], using results of Bass [10], he shows that torsion-free hyperbolic groups and CAT(0) groups G satisfy: if P is a finitely generated projective $\mathbb{C}G$-module, then $r_P(g) = 0$ for all $g \neq 1$ in G. In an addendum, he notes that this is true for the more general class of torsion-free semihyperbolic groups, by a simple argument of Bridson. The idempotent conjecture for torsion-free hyperbolic groups was proved by Ji [66].

There are several other papers on idempotents in group rings, and we shall not survey these, or papers on the existence of zero-divisors in group rings. We shall also not comment on other applications of Hattori-Stallings rank, nor on Euler characteristics for non-discrete groups.

There have been disappointingly few applications of the Euler characteristics we have studied. Stallings original application (Gottlieb's theorem) was generalised, first by Rosset [99], then by Dyer [41]. Dyer shows the following. Suppose $1 \to L \to G \to H \to 1$ is a group extension, H has a non-trivial torsion-free abelian normal subgroup and the homology groups $H_i(L, \mathbb{Z})$ are finitely generated. If $G \in \mathrm{FL}(\mathbb{Z})$ then $\mu(G) = 0$. Both Euler characteristics and the Hattori-Stallings rank are used in the proof that a group is a Poincaré duality group of dimension 2 if and only if it is a surface group ([46]). Euler characteristics have been used by Dicks and Leary in two papers [38], [39]. The second has an algebraic proof of some results of Bestvina and Brady [12], which give an example of a group of type $\mathrm{FP}(\mathbb{Z})$ which is not finitely presented. Other applications of Euler characteristics are given in [53], [34] and [47]. In [36], it is shown inter alia that if G is Poincaré duality group with $\chi(G) \neq 0$, then G is co-Hopfian and (using Gottlieb's Theorem) has trivial centre. Kulkarni [69] has used Euler characteristics in studying space forms, and mentions an application of Euler characteristics to free products in [71]. Euler characteristics of certain free products are used in [90], where a special case of Theorem 11.14 is encountered.

There have been some calculations of Euler characteristic. Serre [107] gave a recursive formula for the Euler characteristic of a finitely generated Coxeter group. An explicit formula was given by the author [27]. This has been used by Akita [5] and further calculations for Coxeter groups have been made by Maxwell [87]. For other work on Euler characteristics by Akita, see [3] and

[4]. There is also a formula for the Euler characteristic of a graph product (in a suitably defined sense) in [27], and results on the Euler characteristic of Artin groups appear in [25]. Computations of Euler characteristics for mapping class groups are made in [58] and [94]. Other calculations of Euler characteristic appear in [52]. Euler characteristics of amenable groups are studied in [43]; in particular, if G is an infinite amenable group admitting a finite $K(G, 1)$, then $\chi(G) = 0$.

There are some ideas in Brown's work that have been used by other authors. Firstly there is the notion of equivariant Euler characteristic $\chi_G(S)$ for a group G acting on a complex S and its relation to $\chi(G)$ has been studied. Indeed, if S is the complex obtained in a standard way from the poset of non-trivial finite subgroups of G, then $\chi(G) - \chi_G(S)$ is an integer (see [15], [17]). There is also a "local" version involving a prime p and the complex S_p obtained from the poset of non-trivial finite p-subgroups of G; $\chi(G) - \chi_G(S_p)$ is p-integral (see [16], [17]). These posets have remained objects of study to the present day. In particular, S_p, where G is finite and p divides the order of G has received considerable attention. Quillen [96] conjectured that S_p is contractible and Webb [117] conjectured that the quotient S_p/G for the action of G is contractible. Webb's conjecture has been established by Symonds [114]. Progress on Quillen's conjecture has been made in [61], [7] and [95]. For progress on the complex S obtained from the poset of non-trivial finite subgroups, we refer to [115] and [62]. These complexes were studied from a different point of view by Kurzweil [72]. Brown's work on equivariant Euler characteristics has also been used by Kulkarni [70]. We also mention the work of Brown and of Thevenaz on generalisations of one of the Sylow Theorems, and refer to [19] for details. Brown's result that, for a finite group G, $\chi(S_p) \equiv 1$ modulo the p-part of $|G|$, was also investigated by Gluck [56] and Yoshida [119]. The results of [16] and [19] have been applied by Yoshida [120]. There are other interesting results in [15], [17].There is a formula for the difference $\tilde{\chi}(G) - \chi(G)$, which has been used by Adem [1], [2]. Finally, there are some number-theoretic results on zeta functions which have been used by other authors; we mention only a recent application by Byeon [21].

In the discussion preceding Remark 5.8, the geometric dimension of a group was defined, and it was noted that this equals the cohomological dimension (over $\mathbb{Z}$) except for the possibility of a group with cohomological dimension 2 and geometric dimension 3. The statement that no such group exists, i.e. that cohomological dimension always equals geometric dimension, is known as the Eilenberg-Ganea conjecture ([48]). Bestvina and Brady [12] have an example which either gives a counterexample to this conjecture, or to the Whitehead asphericity conjecture. (This asserts that a subcomplex of an aspherical 2-complex is aspherical.)

There has been more recent progress on the Bass conjecture. Farrell and Linnell [51] have established it for elementary amenable groups (in particular, for soluble groups). Also, Berrick, Chatterji and Mislin [13] have proved it for, inter alia, amenable groups. They do this by establishing a link between the Bass conjecture and the "Bost assembly map", and using results of Lafforgue on this map.

References

[1] A. Adem, On the K-theory of the classifying space of a discrete group, *Math. Ann.* **292** (1992), 319–327; erratum ibid. **293** (1992), 385–386.

[2] A. Adem, Euler characteristics and cohomology of p-local discrete groups, *J. Algebra* **149** (1992), 183–196.

[3] T. Akita, Euler characteristics of groups and orbit spaces of free G-complexes, *Proc. Japan Acad. Ser. A Math. Sci.* **69** (1993), 389–391.

[4] T. Akita, On the Euler characteristic of the orbit space of a proper Γ-complex, *Osaka J. Math.* **36** (1999), 783–791.

[5] T. Akita, Euler characteristics of Coxeter groups, PL-triangulations of closed manifolds, and cohomology of subgroups of Artin groups, *J. London Math. Soc.* (2) **61** (2000), 721–736.

[6] M.F. Atiyah and I.G. Macdonald, *Introduction to Commutative Algebra*, Addison-Wesley, Reading, Mass., 1969.

[7] M. Aschbacher and S. D. Smith, On Quillen's conjecture for the p-groups complex, *Ann. of Math. (2)* **137** (1993), 473–529.

[8] M. Auslander and D. A. Buchsbaum, *Groups, Rings, Modules*, Harper& Row, New York, 1974.

[9] G.O. Bailey, *Uncharacteristically Euler!*, Ph.D. thesis, University of Birmingham, 1977.

[10] H. Bass, Euler characteristics and characters of discrete groups, *Invent. Math.* **35** (1976), 155–196.

[11] H. Bass, Traces and Euler characteristics, in: *Homological Group Theory*, Proc. Symp., Durham 1977, LMS Lecture Notes vol. 36, Cambridge University Press, 1979, 1–26.

[12] M. Bestvina and N. Brady, Morse theory and finiteness properties of groups, *Invent. Math.* **129** (1997), 445–470.

[13] A. J. Berrick, I. Chatterji and G. Mislin, From acyclic groups to the Bass conjecture for amenable groups, preprint.

[14] R. Bieri, *Homological Dimension of Discrete Groups*, Queen Mary College Mathematics Notes, London, Queen Mary College, University of London, 1976.

[15] K. S. Brown, Euler characteristics of discrete groups and G-spaces, *Invent. Math.* **27** (1974), 229–264.

[16] K. S. Brown, Euler characteristics of groups: The p-fractional part, *Invent. Math.* **29** (1975), 1–5.

[17] K. S. Brown, *Cohomology of Groups*, Graduate Texts in Mathematics vol. 87, Springer, New York, 1982.

[18] K. S. Brown, Complete Euler characteristics and fixed-point theory, *J. Pure Appl. Algebra* **24** (1982), 103–121 .

[19] K. S. Brown and J. Thevenaz, A generalization of Sylow's third theorem, *J. Algebra* **115** (1988), 414–430.

[20] D. Burghelea, The cyclic homology of the group rings, *Comm. Math. Helv.* **60** (1985), 354–365.

[21] D. Byeon, Class numbers and Iwasawa invariants of certain totally real number fields, *J. Number Theory* **79** (1999), 249–257.

[22] G. K. Chadha, Idempotents in non-archimedean group rings, *Comm. Algebra* **24** (1996), 1679–1693.

[23] G. K. Chadha and I. B. S. Passi, Centralizers and homological dimension, *Comm. Algebra* **22** (1994), 5703–5708.

[24] G. K. Chadha and I. B. S. Passi, Bass conjecture and the group trace property, *Comm. Algebra* **26** (1998), 627–639.

[25] R. Charney and M. Davis, Finite $K(\pi, 1)$s for Artin groups, in: *Prospects in Topology* (F. Quinn ed.), Ann. of Math. Stud. vol. 138, Princeton University Press, 1995, 110–124.

[26] I. M. Chiswell, Euler characteristics of groups, *Math. Z.* **147** (1976), 1–11.

[27] I. M. Chiswell, The Euler characteristic of graph products and of Coxeter groups, in: *Discrete Groups and Geometry*, LMS Lecture Notes vol. 173, Cambridge University Press, 1992, 36–46.

[28] G. H. Cliff, Zero divisors and idempotents in group rings, *Canad. J. Math.* **32** (1980), 596–602.

[29] G. H. Cliff, Ranks of projective modules of group rings, *Comm. Algebra* **13** (1985), 1115–1130.

[30] D. E. Cohen, *Groups of Cohomological Dimension One*, Lecture Notes in Mathematics vol. 245, Springer, Berlin-Heidelberg-New York, 1972.

[31] D. E. Cohen, *Combinatorial Group Theory: A Topological Approach*, LMS Student Texts vol. 14, Cambridge University Press, 1989.

[32] P. M. Cohn, *Algebra vol. 2*, Wiley, London, 1977.

[33] J. Cornick, On the homology of group graded algebras, *J. Algebra* **174** (1995), 999–1023.

[34] P.M. Curran, Subgroups of finite index in certain classes of finitely presented groups, *J. Algebra* **122** (1989), 118–129.

[35] C.W. Curtis and I. Reiner, *Methods of Representation Theory, with Applications to Finite Groups and Orders vol. I*, Wiley, New York, 1981.

[36] S. Deo and K. Varadarajan, Hopfian and co-Hopfian groups, *Bull. Austral. Math. Soc.* **56** (1997), 17–24.

[37] W. Dicks and M. J. Dunwoody, *Groups acting on Graphs*, Cambridge Stud. Adv. Math. vol. 17, Cambridge University Press, 1989.

[38] W. Dicks and I. J. Leary, On subgroups of Coxeter groups, in: *Geometry and Cohomology in Group Theory* (Kropholler et al. eds.), LMS Lect. Notes vol. 252, Cambridge University Press, 1998, 124–160.

[39] W. Dicks and I. J. Leary, Presentations for subgroups of Artin groups, *Proc. Amer. Math. Soc.* **127** (1999), 343–348.

[40] E. Dyer and A.T. Vasquez, An invariant for finitely generated projectives over Z(G), *J. Pure Appl. Algebra* **7**, (1976) 241–248.

[41] M. N. Dyer, Euler characteristics of groups, *Quart. J. Math. Oxford Ser. (2)* **38** (1987), 35–44.
[42] B. Eckmann, Cyclic homology of groups and the Bass conjecture, *Comm. Math. Helv.* **61** (1986), 193–202.
[43] B. Eckmann, Amenable groups and Euler characteristics, *Comm. Math. Helv.* **67** (1992), 383–393.
[44] B. Eckmann, Projective and Hilbert modules over group algebras, and finitely dominated spaces, *Comm. Math. Helv.* **71** (1997), 453–462. Addendum: ibid. **72** (1997), 329.
[45] B. Eckmann, Idempotents in a complex group algebra, projective modules and the von Neumann algebra, *Arch. Math.* **76** (2001), 241–249.
[46] B. Eckmann and P. Linnell, Poincaré duality groups of dimension two, II, *Comm. Math. Helv.* **58** (1983), 111–114.
[47] M. Edjvet and J. Howie, On the abstract groups $(3, n, p; 2)$, *J. London Math. Soc.* (2) **53** (1996), 271–288.
[48] S. Eilenberg and T. Ganea, On the Lusternik-Schnirelmann category of abstract groups, *Ann. of Math.* (2) **65**, (1957) 517–518.
[49] I. Emmanouil, On a class of groups satisfying Bass' conjecture, *Invent. Math.* **132** (1998), 307–330.
[50] I. Emmanouil, Projective modules, augmentation and idempotents in group algebras, *J. Pure Appl. Algebra* **158** (2001), 151–160.
[51] T. Farrell and P. Linnell, Whitehead groups and the Bass conjecture, preprint.
[52] B. Fine, G. Rosenberger, and M. Stille, Euler characteristic for one-relator products of cyclics, *Comm. Algebra* **21** (1993), 4353–4359.
[53] B. Fine and A. Peluso, Amalgam decompositions for one-relator groups, *J. Pure Appl. Algebra* **141** (1999), 1–11.
[54] J. Fischer, A. Karrass, and D. Solitar, On one-relator groups having elements of finite order, *Proc. Am. Math. Soc.* **33** (1972), 297–301.
[55] E. Formanek, Idempotents in Noetherian group rings, *Canad. J. Math.* **25** (1973), 366–369.
[56] D. Gluck, Idempotent formula for the Burnside algebra with applications to the p-subgroup simplicial complex, *Illinois J. Math.* **25** (1981), 63–67.
[57] K. W. Gruenberg, *Cohomology Topics in Group Theory*, Lecture Notes in Math. vol. 143, Springer, Berlin-Heidelberg-New York, 1970.
[58] J. Harer and D. Zagier, The Euler characteristic of the moduli space of curves, *Invent. Math.* **85** (1986), 457–485.
[59] B. Harris, Commutators in division rings, *Proc. Amer. Math Soc.* **9** (1958), 628–630.
[60] A. Hattori, Rank element of a projective module, *Nagoya Math. J.* **25** (1965), 113–120.
[61] T. Hawkes and I. M. Isaacs, On the poset of p-subgroups of a p-solvable group, *J. London Math. Soc.* (2) **38** (1988), 77–86.
[62] T. Hawkes, I. M. Isaacs, and M. Oezaydin, On the Moebius function of a finite group, *Rocky Mountain J. Math.* **19** (1989), 1003–1034.
[63] G. Higman, A finitely generated infinite simple group, *J. London Math. Soc.* **26** (1951), 61–64.
[64] N. Jacobson, *Basic Algebra II*, W. H. Freeman, San Francisco, 1980.

[65] G. J. Janusz, *Algebraic number fields*, Pure and Applied Mathematics. vol. 55, Academic Press, New York, 1973.

[66] R. Ji, Nilpotency of Connes' periodicity operator and the idempotent conjectures, *K-Theory* **9** (1995), 59–76.

[67] I. Kaplansky, *Fields and Rings*, Chicago Lectures in Math., University of Chicago Press, 1970.

[68] A. Karrass, W. Magnus and D. Solitar, Elements of finite order in groups with a single defining relation, *Commun. Pure Appl. Math.* **13** (1960), 57–66.

[69] R. S. Kulkarni, Proper actions and pseudo-Riemannian space forms, *Adv. in Math.* **40** (1981), 10–51.

[70] R. S. Kulkarni, Pseudofree actions and Hurwitz's $84(g-1)$ theorem, *Math. Ann.* **261** (1982), 209–226.

[71] R. S. Kulkarni, An extension of a theorem of Kurosh and applications to Fuchsian groups, *Michigan Math. J.* **30** (1983), 259–272.

[72] H. Kurzweil, A combinatorial technique for simplicial complexes and some applications to finite groups, *Discrete Math.* **82** (1990), 263–278.

[73] S. Lang, *Algebra*, Addison-Wesley, Reading, Mass., 1965.

[74] S. Lang, *Algebraic Number Theory*, Addison-Wesley, Reading, Mass., 1970.

[75] P. A. Linnell, Decomposition of augmentation ideals and relation modules, *Proc. London. Math. Soc.* **47** (1983), 83–127.

[76] P. A. Linnell, An example concerning the Bass conjecture, *Mich. Math. J.* **40** (1993), 197–199.

[77] A. T. Lundell and S. Weingram, *The Topology of CW Complexes*, Van Nostrand Reinhold Company, New York, 1969.

[78] R. C. Lyndon, Cohomology theory of groups with a single defining relation, *Ann. of Math* **52** (1950), 650–665.

[79] R. C. Lyndon and P. E. Schupp, *Combinatorial Group Theory*, Ergebnisse der Math. vol. 89, Springer, Berlin-Heidelberg-New York, 1977.

[80] S. MacLane, *Homology*, Springer, Berlin-Göttingen-Heidelberg, 1963.

[81] W. Magnus, A. Karrass and D. Solitar, *Combinatorial Group Theory: Presentations of Groups in Terms of Generators and Relations*, Wiley, New York-London-Sydney, 1966. (Reprinted and revised by Dover, 1976.)

[82] Z. Marciniak, Cyclic homology of group rings, in: *Geometric and Algebraic Topology*, Banach Center Publ. vol. 18, Polish Acad. Sci., Warsaw, 1986, 305–312.

[83] Z. Marciniak, Cyclic homology and idempotents in group rings, in: *Transformation Groups*, Proc. Symp., Poznan/Pol. 1985, Lecture Notes in Math. vol. 1217, Springer, New York, 1986, 253–257.

[84] W. S. Massey, *Algebraic Topology: An Introduction*, Harcourt Brace and World, New York, 1967.

[85] W. S. Massey, *Singular Homology Theory*, Graduate Texts in Mathematics vol. 70, Springer, New York, 1980.

[86] C. R. F. Maunder, *Algebraic Topology*, Cambridge University Press, 1980.

[87] G. Maxwell, Euler characteristics and imbeddings of hyperbolic Coxeter groups, *J. Austral. Math. Soc.* Ser. A **64** (1998), 149–161.

[88] M. S. Montgomery, Left and right inverses in group algebras, *Bull. Amer. Math. Soc.* **75** (1969), 539–540.

[89] J. A. Moody, Induction theorems for infinite groups, *Bull. Amer. Math. Soc.* **17** (1987), 113–116.
[90] T. Müller, Subgroup growth of free products, *Invent. Math.* **126** (1996), 111–131.
[91] K. Murasugi, The center of a group with a single defining relation, *Math. Ann.* **155** (1964), 246–251.
[92] W. L. Paschke, A numerical invariant for finitely generated groups via actions on graphs, *Math. Scand.* **72** (1993), 148–160.
[93] D. S. Passman, *Infinite Group Rings*, Marcel Dekker, New York, 1971.
[94] R. C. Penner, Perturbative series and the moduli space of Riemann surfaces, *J. Differential Geom.* **27** (1988), 35–53.
[95] J. Pulkus and V. Welker, On the homotopy type of the p-subgroup complex for finite solvable groups, *J. Austral. Math. Soc.* Ser. A **69** (2000), 212–228.
[96] D. Quillen, Homotopy properties of the poset of nontrivial p-subgroups of a group, *Adv. in Math.* **28** (1978), 101–128.
[97] A. Reznikov, Volumes of discrete groups and topological complexity of homology spheres, *Math. Ann.* **306** (1996), 547–554.
[98] D. J. S. Robinson, *A Course in the Theory of Groups*, Graduate Texts in Mathematics vol. 80, Springer, New York-Heidelberg-Berlin, 1982.
[99] S. Rosset, A vanishing theorem for Euler characteristics, *Math. Z.* **185** (1984), 211–215.
[100] J.J. Rotman, *An Introduction to Homological Algebra*, Academic Press, New York-San Francisco-London, 1979.
[101] R. Roy, A counterexample to questions on the integrality property of virtual signature, *Topology Appl.* **100** (2000), 177–185.
[102] P. Samuel, *Algebraic Theory of Numbers*, Hermann, Paris, 1970.
[103] J. A. Schafer, Traces and the Bass conjecture, *Michigan Math. J.* **38** (1991), 103–109.
[104] J. A. Schafer, Relative cyclic homology and the Bass conjecture, *Comm. Math. Helv.* **67** (1992), 214–225.
[105] H. R. Schneebeli, On virtual properties and group extensions, *Math. Z.* **159** (1978), 159–167.
[106] H. Schubert, *Topology*, Allyn and Bacon, Boston, 1968.
[107] J-P. Serre, Cohomologie des groupes discrets, in: *Prospects in Mathematics*, Ann. of Math. Stud. vol. 70, Princeton Univerity Press, 1971, 77–169.
[108] J.-P. Serre, *Trees*, Springer, New York, 1980.
[109] J. R. Stallings, Centerless groups–an algebraic formulation of Gottlieb's theorem, *Topology* **4** (1965), 129–134.
[110] J. R. Stallings, An extension theorem for Euler characteristics of groups, *Geom. Dedicata* **92** (2002), 3–39.
[111] R. Strebel, Homological methods applied to the derived series of groups, *Comm. Math. Helv.* **49** (1974), 302–332.
[112] A. Strojnowski, Idempotents and zero divisors in group rings, *Comm. Algebra* **14** (1986), 1171–1185.
[113] A. Strojnowski, On Bass' 'Strong conjecture' about projective modules, *J. Pure Appl. Algebra* **62** (1989), 195–198.
[114] P. Symonds, The orbit space of the p-subgroup complex is contractible, *Comm. Math. Helv.* **73** (1998), 400–405.

[115] A. Turull, Polynomials associated to characters, *Proc. Amer. Math. Soc.* **103** (1988), 463–467.
[116] C. T. C. Wall, Rational Euler characteristics, *Proc. Cambridge. Philos. Soc.* **57** (1961), 182–184.
[117] P. J. Webb, Subgroup complexes, in: *Representations of finite groups*, Proc. Conf., Arcata/Calif. 1986, Pt.1. *Proc. Sympos. Pure Math.* vol. 47, American Mathematical Society, Providence, 1987, 349–365.
[118] A. Weiss, Idempotents in group rings, *J. Pure Appl. Algebra* **16** (1980), 207–213.
[119] T. Yoshida, Idempotents of Burnside rings and Dress induction theorem, *J. Algebra* **80** (1983), 90–105.
[120] T. Yoshida, $|\mathrm{Hom}(A, G)|$, *J. Algebra* **156** (1993), 125–156.

8

Intersections of Magnus subgroups of one-relator groups

D. J. Collins

1. Introduction

The idea of specifying a group by *generators* and *relators* has a long history and goes back to around 1900 with the definition and study of the fundamental group of a topological space and the study of groups of transformations of geometric objects, notably tesselations of the hyperbolic plane. In this article we give an elementary description of the basics of the theory of a group given by a set of generators and a single relator and illustrate the use of this theory in proving a result - see Theorem 2 below - about intersections of what we call Magnus subgroups of such a group.

Following this introduction, in which we set out basic definitions and state our main theorem, in Section 2 we sketch the classical approach to one-relator groups due to Magnus [16, 17], which is the basis of our proof of Theorem 2. In Section 3 we give a detailed illustration of how examples of what we call 'exceptional' intersections of Magnus subgroups can arise. In Section 4 we state briefly some results which extend Theorem 2, and then, in Sections 5-7, we present the detailed proof.

To specify a group by *generators* and *relators*, one begins by defining the free group $F(X)$ on a basis X to consist of distinct strings of *letters* $x_1^{\varepsilon_1} x_2^{\varepsilon_2} \dots x_n^{\varepsilon_n}$ where $x_i \in X, \varepsilon_i = \pm 1$, which are *reduced* in the sense that $(x_i, \varepsilon_i) \neq (x_{i+1}, -\varepsilon_{i+1}), 1 \leq i \leq n-1$. Multiplication consists of concatenation of strings, followed by cancelling pairs of strings $x^{\varepsilon} x^{-\varepsilon}$ to obtain a reduced string. Checking the associativity axiom is a little awkward (in this version of the definition) but is not difficult and the result becomes a group when one allows the empty word to serve as identity. One then selects a set $\mathcal{R}$ of elements of $F(X)$ and defines a group G, with notation $G = \langle X \mid \mathcal{R} \rangle$, as the quotient of $F(X)$ defined by the normal closure N of $\mathcal{R}$ in $F(X)$ - this process defines a canonical epimorphism from $F(X)$ to G with kernel N. When a group G can be specified in this way, with $\mathcal{R}$ chosen to consist of a single element R, one calls G a one-relator group and writes $G = \langle X \mid R \rangle$ (or sometimes $G = \langle X \mid R = 1 \rangle$, indicating that the coset RN of R defines the identity element of G). Normally R is chosen to be cyclically reduced, that is, R is reduced but in addition the last and first letters of R are not inverse to one another. One may also, at will, replace R by any cyclic permutation of R or its inverse R^{-1}, which of course consists of the letters of R in reverse order and with the opposite signs.

The classical example of a one-relator group is the fundamental group

$$\pi_1(S_g) = \langle a_1, b_1, \dots, a_g, b_g \mid [a_1, b_1][a_2, b_2] \dots [a_g, b_g] \rangle$$

(writing $[a, b]$ for $aba^{-1}b^{-1}$) of a closed orientable surface of genus g, which appears in the work [9] of Dehn early in the 1900s, and the general theory begins with the work of Magnus [16] in 1930. While, overall, the theory of groups defined by generators and relators is less than wholly tractable, the restriction to a single relator is sufficiently restrictive that effort is rewarded and there is now a very substantial literature.

The most famous theorem concerning one-relator groups is the Freiheitssatz, first proved in [16] by Magnus, which asserts that if a proper subset of the generators omits a generator that appears in the (cyclically reduced) defining relation, then this subset is a free basis for the subgroup it generates. To make this statement precise, let $G = \langle X \mid R \rangle$ be a one-relator group, where R is a cyclically reduced (group) word on X; thus R is a finite string $x_1^{\varepsilon_1} x_2^{\varepsilon_2} \dots x_n^{\varepsilon_n}$ where $x_i \in X, \varepsilon_i = \pm 1$, and $(x_i, \varepsilon_i) \neq (x_{i+1}, -\varepsilon_{i+1})$, with subscripts taken modulo n. A subset $Y \subseteq X$ is a *Magnus subset* if Y omits an *essential* generator, that is one which appears in the relator R.

Theorem 1 (Freiheitssatz). *Let $G = \langle X \mid R \rangle$ be a one-relator group and let $\pi : F(X) \to G$ be the corresponding canonical epimorphism. Then the images $\{\pi(y), y \in Y\}$ of a Magnus subset Y of X are distinct elements of G and form a free basis of the subgroup of G that they generate.*

We call a subgroup M of G generated by the elements $\{\pi(y), y \in Y\}$ of a Magnus subset Y a *Magnus* subgroup. Since the elements $\{\pi(y), y \in Y\}$ are all distinct, the epimorphism π is usually suppressed and we shall follow this course henceforth. A naïve view might regard the theorem as unsurprising, since one might suspect that any possible relation among the elements of Y would have to be derived using the given relator R and would therefore require the use of a generator that does not appear in Y. While this idea can be verified in a number of varieties of algebras, non-trivial arguments are usually required and this is the case for groups. What is interesting about the classical argument used by Magnus, which is by induction on the length of the relator R, is the way in which the inductive hypothesis provides a structural understanding of the group G. The key technique is the use of amalgamated free products in which Magnus subgroups of a group with a shorter defining relator appear as the subgroups to be amalgamated - and since these are free all that is required to check that the construction is well-founded is to ensure that the two Magnus subsets which generate them have the same cardinality. Nowadays, following Moldavanskii [23], the use of amalgamated free products is usually replaced by the use of its sister construction, the still awkwardly named HNN-extension, and we shall describe this approach in greater detail below.

The specific question we shall address is the nature of the intersection of two Magnus subgroups of a given one-relator group $G = \langle X \mid R \rangle$. As a subgroup of a free group, the intersection is, by the Nielsen-Schreier theorem [15], a free group and therefore what is at issue is the *rank* of the intersection, that is the (uniquely defined) number of elements in any basis. Let Y and Z therefore be distinct Magnus subsets of X and let $F(Y)$ and $F(Z)$ be the corresponding Magnus subgroups they generate. It is clear that the intersection $F(Y) \cap F(Z)$ must contain the Magnus subgroup $F(Y \cap Z)$ and it is not difficult to show that there exists examples where $F(Y) \cap F(Z) \neq F(Y \cap Z)$. The simplest example of the latter occurs when Y and Z are disjoint and the relator R is of the form UV with $U \in F(Y)$ and $V \in F(Z)$ - for example the surface group displayed above with $Y = \{a_1, b_1\}$ and $Z = \{a_2, b_2, \ldots, a_g, b_g\}$ (see the brief argument after Theorem 2.1. We shall prove:

Theorem 2. *The intersection $F(Y) \cap F(Z)$ of two Magnus subgroups of the one-relator group G is either $F(Y \cap Z)$ or the free product of $F(Y \cap Z)$ with an infinite cyclic group and thus of rank $|Y \cap Z| + 1$.*

When the latter alternative holds we say that the two Magnus subgroups involved have *exceptional intersection*. A stronger form of this result has now been obtained by Howie[1] [13] and we shall say something about this below. In a forthcoming paper [7] we prove the following result concerning intersections of conjugates of Magnus subgroups of a one-relator groups. Much of the argument for this relies on the results developed in Section 6 of this article.

Theorem 3. *Let G be a one-relator group and let M and N be Magnus subgroups of G. For any $g \in G$, either $gMg^{-1} \cap N$ is cyclic (possibly trivial) or $g \in NM$.*

[1] It is a pleasure to ackowledge a stimulating exchange of ideas with Howie, including his suggestion that Theorem 3 might hold.

In this result we do not assume that M and N are distinct – however if $M = N$, the result is due to Bagherzadeh [1]. The two questions that we have addressed above are obvious and natural but have not, as far as we know, been fully answered before. However some special results have been obtained; thus Brodskii [Br] has proved that if S and T are disjoint Magnus subsets, then $F(S) \cap F(T)$ is cyclic and, as just noted, Bagherzadeh [Ba] has shown that if $g \notin F(S)$, then $gF(S)g^{-1} \cap F(S)$ is cyclic. Our results generalise these. Newman [25] also considers simple intersections of Magnus subgroups and obtains a number of results when r contains at least three distinct generators and one of S and T contains at least two elements not in $S \cap T$.

In the special case when the relator R is a proper power, much tighter conclusions can be obtained. In particular there are no exceptional intersections of Magnus subgroups and in Theorem 3 either $gMg^{-1} \cap N$ is trivial or $g \in NM$. A special case of the latter is the well-known result that in a one-relator group where the relator is a proper power, Magnus subgroups are *malnormal*, i.e., if $g \notin M$, then $gMg^{-1} \cap M$ is trivial. All of these results ultimately follow from the Spelling Lemma, first introduced in Newman [25, 26] – see Theorem 2.5 and the subsequent material.

2. Group theoretic approaches

We begin by describing the classical approach to one-relator groups and sketch the proofs of various standard results. Firstly, though, we need to describe some basic constructions in combinatorial group theory and state some standard results that we shall rely on. Necessarily our treatment is no more than a fairly detailed sketch and the reader is recommended to consult [8] and, especially, [15] – our debt to the latter is self-evident.

Much of combinatorial group theory is based on two 'sister' constructions, both of which have their origins in topology. These are *amalgamated free products* and *HNN-extensions*; both describe the form of the fundamental group arising from a topological construction, the former dealing with a space which is formed from two spaces by identifying homeomorphic subspaces, and the latter with a space which is formed by attaching a handle to a given space identifying the two ends of the handle with homeomorphic subspaces of the given space. Both group constructions are special cases of the more general concept of the fundamental group of a graph of groups (see [29]) but in fact constitute the basic building blocks thereof.

To construct an amalgamated free product, one must be given two groups H and K (two copies of the same group is not excluded), say via two disjoint presentations $H = \langle X \mid \mathcal{R}\rangle$ and $K = \langle Y \mid \mathcal{S}\rangle$, and an isomorphism φ between subgroups A and B of H and K respectively. The group G required, written either as $G = \langle H, K \mid A = B, \varphi\rangle$ or as $G = (H * K \mid A = B, \varphi)$, or variations and abbreviations thereof, is defined by the presentation $G = \langle X, Y|\mathcal{R}, \mathcal{S}, U_i = V_i, i \in I\rangle$ where $\{U_i, i \in I\}$, $\{V_i, i \in I\}$ are words which define (some) sets of generators of A and B which correspond under φ. Taking over the terminology of graphs of groups we call H and K the vertex groups and A and B the edge groups.

To construct an HNN-extension, one needs a single group H, say specified by a presentation $H = \langle X \mid \mathcal{R}\rangle$, and an isomorphism φ between subgroups A and B of H. The group G required, written as $G = \langle H, t \mid tAt^{-1} = B, \varphi\rangle$ or variations and abbreviations thereof, is defined by the presentation $G = \langle X, t \mid \mathcal{R}, tU_it^{-1} = V_i, i \in I\rangle$ where $\{U_i, i \in I\}$, $\{V_i, i \in I\}$ are words which define (some) sets of generators of A and B which correspond under φ. Again taking over the terminology of graphs of groups we call H the vertex group and A and B the edge groups. We note in passing that one can obviously generalise the concept of amalgamated free product to the case of arbitrarily many groups, all of which have a subgroup isomorphic to some fixed given group. In the special case where the common subgroup is trivial, the result is known as the free product (with 'trivial' amalgamation).

The main results that we shall use for these two constructions are known as the normal form theorems. We shall state these in the form we require, although this is not quite the strongest possible form. Details of the strong form are to be found in [8, 15, 19]. To begin with we observe that if $G = (H * K \mid A = B, \varphi)$, then elements of G can clearly be represented as strings $g = w_1w_2 \ldots w_n$ where the w_j are, alternately, non-empty words in the respective generating sets X and Y. [Strictly one need not assume that each w_i is non-empty but there is no loss of generality in so doing since otherwise one just concatenates the empty w_i and its two neighbours into a single word in one of the generating sets.]

Theorem 2.1. *Let $G = (H * K \mid A = B, \varphi)$ and let $g = w_1w_2 \ldots w_n$. If g represents the identity element of G then either*

(a) *$n = 1$ and w_1 represents the identity element of H or K as appropriate; or*
(b) *for some i, w_i represents an element of A or B as appropriate.*

Proof. What has now become accepted as the most satisfactory proof – known as the van der Waerden argument – of the strong normal form theorem can be found in either [8] or [15]. The result stated here follows easily from that. □

Corollary 2.1. *Each of the vertex groups H and K is embedded in the amalgamated product G via the natural identity map on words.*

Proof. This is immediate from part (a). □

We can use Theorem 2.1 to illustrate how Theorem 2 is satisfied in the case when

$$G = \pi_1(S_2) = \langle a_1, b_1, a_2, b_2 \mid [a_1, b_1][a_2, b_2] = 1 \rangle$$

and $Y = \{a_1, b_1\}$, $Z = \{a_2, b_2\}$. For suppose that an equality $u(a_1, b_1) = v(a_2, b_2)$ holds. Now G is in fact the amalgamated free product $\langle F(a_1, b_1) * F(a_2, b_2) \mid \langle [a_1, b_1] \rangle = \langle [a_2, b_2]^{-1} \rangle \rangle$ – the vertex groups are the two free groups on the generators indicated and the edge groups are the infinite cyclic subgroups $\langle [a_1, b_1] \rangle$ and $\langle [a_2, b_2]^{-1} \rangle$, where the identification $[a_1, b_1] = [a_2, b_2]^{-1}$ of the respective generators is of course just an alternative form of the single relation for G. Theorem 2.1 applied to uv^{-1} yields the conclusion that $u \in \langle [a_1, b_1] \rangle$ or $v^{-1} \in \langle [a_2, b_2]^{-1} \rangle$ (in fact both hold) and $F(Y) \cap F(Z)$ is cyclic. (Of course the same argument shows that in an amalgamated free product $G = (H * K \mid A = B, \varphi)$, the intersection $H \cap K$ is just $A = B$). When $Y = \{a_1, b_1, a_2\}$, $Z = \{b_1, a_2, b_2\}$, a similar argument can be applied to show that $F(Y) \cap F(Z)$ has rank three whereas $F(Y \cap Z)$ has rank two - but this time the amalgamated free product has the form $\langle F(a_1, b_1, a_2) * F(b'_1, a'_2, b_2) \mid \langle [a_1, b_1], b_1, a_2 \rangle = \langle [a_2, b_2]^{-1}, b'_1, a'_2 \rangle \rangle$ and one has to use the identification relations $b_1 = b'_1, a_2 = a'_2$ to eliminate the generators b'_1 and a'_2 and obtain the original presentation.

If $G = \langle H, t \mid tAt^{-1} = B, \varphi \rangle$, then its elements are expressible as strings $w_0 t^{\varepsilon_1} w_1 \dots t^{\varepsilon_n} w_n$, where the w_i are (possibly empty) words in the generating set X and $\varepsilon_i = \pm 1$. [This time it is more convenient to allow some of the w_i to be empty – an alternative version would allow the ε_i to be arbitrary non-zero integers and then one could require that every w_i be non-empty.]

Theorem 2.2. *Let $G = \langle H, t \mid tAt^{-1} = B, \varphi \rangle$, and let $g = w_0 t^{\varepsilon_1} w_1 \dots t^{\varepsilon_n} w_n$. If g represents the identity element of G then either*

(a) *$n = 0$ and w_0 represents the identity element of H; or*
(b) *for some i, g contains either tw_it^{-1} where w_i represents an element of A or $t^{-1}w_it$ where w_i represents an element of B.*

Proof. This too can be derived from a more precise version proved by the van der Waerden argument. □

Corollary 2.2. (a) *The vertex group H is embedded in the HNN-extension G via the natural identity map on words.*
(b) *If g represents an element of the vertex group H then either $n = 0$ or for some i, g contains tw_it^{-1} where w_i represents an element of A or $t^{-1}w_it$ where w_i represents an element of B.*

Proof. Part (a) is immediate from part (a) of the Theorem while (b) follows by applying (b) of the Theorem to the expression WV^{-1} where V is a word in the generators of H that represents the same element as W. □

In $\langle H, t \mid tAt^{-1} = B, \varphi\rangle$, we shall call an expression $g = w_0 t^{\varepsilon_1} w_1 \ldots t^{\varepsilon_n} w_n$, where $n \geq 1$, *reduced* if it does not satisfy the conclusion of (b) - so (b) asserts that if such an expression represents the identity element then it is not reduced. An easy inductive argument shows that if two reduced expressions represent the same element of G, then they contain the same number of occurrences of t and thus there is associated with an element of G a well-defined notion of *length* – indeed one can say more, namely, that the two sequences of exponents for the generator t must coincide, and we shall make extensive use of this.

Two further important results follow from these normal form theorems. We do not specifically require them for our problem concerning Magnus subgroups but it would be wrong to pass on without at least very brief mention. The first concerns what is known as *solving the word problem* and the second the determination of elements of finite order. To solve the word problem for a group presentation, one must prove the existence of an algorithm that will determine of any arbitrarily given word of the presentation whether or not it defines the identity. It follows from the normal form theorems that to solve the word problem for either an amalgamated free product or for an HNN-extension, one has to be able to solve the word problem for the vertex group(s) and one must be able to determine algorithmically whether or not an element of a vertex group lies in an edge group (as appropriate) (and, when the edge groups are not finitely generated, one must also know that the isomorphism φ is recursive, i.e., effectively calculable). The second consequence is that elements of finite order in an amalgamated free product or an HNN-extension are always just conjugates of elements of finite order of the (embedded copies of the) vertex groups and the parallel statement is true of finite subgroups.

We are now ready to turn our attention to the classical theory of one-relator groups. We begin, of course, with the Freiheitssatz.

Theorem 1 (Freiheitssatz)**.** *Let $G = \langle X \mid R\rangle$ be a one-relator group. Then a Magnus subset Y of X is a free basis of the Magnus subgroup of G that it generates.*

Proof. Let $G = \langle X \mid R\rangle$ be a one-relator group. If a generator $a \in X$ has zero exponent sum in R – and this is the key case – then G can be presented as an HNN-extension. By way of illustration, let $X = \{a, b, c\}$ and let $R = b^2ab^2ac^{-2}b^3a^{-3}c^2ac^2$. Now R is easily rewritten as a product of words of the form a^iba^{-i} and a^jca^{-j} with $0 \leq i \leq 2$ and $-1 \leq j \leq 2$. Using Tietze transformations – see [8, 15, 19] for a justification – one can therefore introduce generators $b_0 = b, b_1 = aba^{-1}, b_2 = a^2ba^{-2}, c_j = a^jca^{-j}$ where the range of values of j can be all integers or any interval of integers that contains $\{-1, 0, 1, 2\}$ One then obtains a presentation

$$G = \langle b_0, b_1, b_2, c_j, a \mid b_0^2b_1^2c_2^{-2}b_2^3c_{-1}^2c_0^2, ab_ia^{-1} = b_{i+1}, ac_ja^{-1} = c_{j+1} \rangle$$

where $i = 0, 1$ and j ranges over all integers in the chosen interval, except the largest if such exists. It is important to emphasize that while we can choose the range of either i or j to be infinite, we cannot do both and if we make the range of j infinite, then the range of i must be restricted to the interval (in this example), consisting of $\{0, 1, 2\}$. [When G has more than three generators, the subscript ranges can all be infinite save for one which must be restricted to the integer interval whose upper and lower boundaries are, respectively, the greatest and least subscripts occurring in the rewritten form R^* of R.]

By assuming, inductively on the length of the relator, that the Freiheitssatz holds for the group $G^* = \langle X^* \mid R^* \rangle = \langle b_0, b_1, b_2, c_j \mid b_0^2b_1^2c_2^2b_2^3c_{-1}^2c_0^2 \rangle$, and hence that the subgroups made conjugate by the element a are free of the same rank, we have described G as an HNN-extension. The normal form theorem for HNN-extensions ensures G^* is embedded in G by the natural map and hence that the two element set $\{b, c\} = \{b_0, c_0\}$ is a basis for a free subgroup. To see that $\{a, c\}$ is a basis of a free subgroup one need only observe that a potential relation between these two elements must have exponent sum zero in a and therefore lies in the free subgroup with basis c_j (provided that the range of j is chosen sufficiently large). The proof is completed in this case by exchanging the roles of b and c.

The case when no element of X has zero exponent sum in R is reduced to the above by a simple trick. Suppose that $G = \langle a, b, c \mid R(a, b, c) \rangle$ and that none of a, b, c has zero exponent sum in R. Write α and β for the exponent sums of a and b respectively. Define $\hat{G} = \langle x, y, c \mid R(x^\beta, yx^{-\alpha}, c) \rangle$ - it is easy to check that the cyclically reduced form of $R(x^\beta, yx^{-\alpha}, c)$ has exponent sum zero in x and can be rewritten as an HNN-extension over a group G^* whose relator is shorter than R. Inductively, therefore x and y form a basis of a free subgroup. However $\hat{G}$ is (isomorphic to) the amalgamated free product $\langle G * F(x) : a = x^\beta \rangle$. Thus G is embedded in $\hat{G}$ in such a way that a and b lie in the free group which has $\{x, y\}$ as basis and so $\{a, b\}$ is also a basis for a free group. This completes the argument for the Freiheitssatz, save that we have ignored the initial step when the length of R is small and the case when R contains only a single generator. Both are easily dealt with, usually by making use of the normal form theorem for free products. □

Before moving on to the next result relevant to our problem on Magnus subgroups we pause to state two further classical results.

Theorem 2.3. [18] *Let $G = \langle X \mid R \rangle$ be a one-relator group. Then the word problem for G is solvable.*

This is proved using the same inductive approach as the Freiheitssatz. From our remarks about solving the word problem for an HNN-extension, it should be clear to the reader that to push through the inductive argument, it is in fact necessary to prove a stronger result, namely that for any one-relator group and any of its Magnus subgroups, there is an algorithm to decide of an arbitrary word whether or not it represents an element of the given Magnus subgroup.

Theorem 2.4. [10] *Let $G = \langle X \mid R \rangle$ be a one-relator group. Then G has elements of finite order if and only if R is a proper power in the free group $F(X)$. In particular if $R = S^m$ in $F(X)$ and S is not a proper power, then*

(a) *the element of G represented by S has order m;*
(b) *every element of G of finite order is conjugate to an element represented by a power of S.*

Proof. The method of proof parallels that of the Freiheitsatz. The key observation is that when rewriting G as an HNN extension, the relator of the vertex group G^* is of the form S^{*m} when $R = S^m$, so that the induction hypothesis applies. By way of illustration, take $S = b^2ab^2ac^{-2}b^3a^{-3}c^2ac^2$, which was our actual relator in our illustration of the proof of the Freiheitssatz, and then take $R = S^3$, say. It is easy to check that when we pass to the vertex group $G^* = \langle X^* \mid R^* \rangle$, then R^* is indeed of the form $(S^*)^3$ where $S^* = b_0^2 b_1^2 c_2^{-2} b_2^3 c_{-1}^2 c_0^2$, and this holds generally (including the fact that if S is not a proper power then neither is S^*). The result then follows from the basic proporties of torsion in HNN-extensions referred to in the discussion following the statement of Theorem 2.2. □

Our next theorem is the key result in the theory of one-relator groups with torsion. Its original form is due to Newman [25, 26] but several authors have produced refinements, including a version for relators which are not proper powers, which we shall use later – see [11, 14, 21, 28]. Our account is based on [15] but obtains a slightly stronger conclusion than stated there.

Theorem 2.5 (Spelling Lemma). *Let $G = \langle X \mid R \rangle$ be a one-relator group and suppose that R is of the form S^m for some element S of $F(X)$, where $m \geq 2$ and S is not a proper power. Let M be a Magnus subgroup of G and let W be a freely reduced word which contains an essential generator of G that is not in the canonical basis of M. If W represents an element of M, then W contains a* *subword of the form $T^{m-1}T_1$ where T is a cyclic rearrangement of $S^{\pm 1}$ and T_1 is a proper initial segment of T containing all essential generators.*

Given a word W, we shall call a subword $T^{m-1}T_1$ satisfying the conditions in the theorem a *Gurevich* subword of W. The following, which is immediate, completely answers our question about intersections of Magnus subgroups when the single relator is a proper power.

Corollary 2.3. *Let $G = \langle X \mid R \rangle$ be a one-relator group and suppose that R is of the form S^m for some element S of $F(X)$ with $m \geq 2$. Then there are no exceptional intersections among the Magnus subgroups of G.*

Proof. If W and V are words in distinct Magnus subsets, and neither is a word in the intersection of the two Magnus subsets, any equality $W = V$ would immediately contradict Theorem 2.5. □

Proof of Theorem 2.5. Once again we can employ the technique of induction on the length of R, representing G as an HNN-extension when there exists $a \in X$ of zero exponent sum in R (and therefore also in S). As noted for Theorem 2.4, the form of the vertex group G^* when we express G as an HNN-extension permits the induction hypothesis to be applied to G^*. Suppose then that we have an equation $W = V$ where V is an element of the Magnus subgroup $F(Y)$, $Y \subseteq X$. The key case is that where the Magnus subgroup omits the generator a that has zero exponent sum in R. Let us continue with our illustrative example from Theorem 2.4:

$$G = \big\langle a, b, c \mid (b^2ab^2ac^{-2}b^3a^{-3}c^2ac^2)^3 \big\rangle$$

so that we have $W(a, b, c) = V(b, c)$, where W explicitly involves a. We can rewrite G as an HNN-extension

$$G = \big\langle G^*, a \mid aKa^{-1} = L \big\rangle \text{ where } G^*$$

$$= \big\langle b_0, b_1, b_2, c_j \mid (b_0^2 b_1^2 c_2^{-2} b_2^3 c_{-1}^2 c_0^2)^3 \big\rangle$$

with $K = F(b_0, b_1, c_j)$ and $L = F(b_1, b_2, c_j)$, choosing, for convenience, the range of j to be the whole of $\mathbb{Z}$. When we rewrite V in the generators of the HNN-extension, all that happens is that we add the subscript 0 to each generator and hence the resulting word V^* lies in G^*. By contrast, since W involves a, the process of rewriting is more complex. Since V omits a and WV^{-1} is a consequence of R, clearly W has zero exponent sum 0 in a. This means that W can be rewritten as a product of elements of the form a^iba^{-i} and a^jca^{-j}. We can immediately rewrite the latter as c_j but in general, however, the range of exponents of a that are involved will exceed the range of exponents that arise when R is rewritten as R^* and, in our example, we have no generator b_3 to represent the element a^3ba^{-3} should this expression arise. Instead we have to express a^3ba^{-3} as ab_2a^{-1} and leave it at that. More generally we express a^iba^{-i} as $a^{i-2}b_2a^{-(i-2)}$ when $i \geq 3$ and as $a^ib_0a^{-i}$ when $i < 0$, and reduce freely; let us call the result W^*. In the HNN-extension $G = \langle G^*, a \mid aKa^{-1} = L\rangle$ we then have an equation $W^* = V^*$. We want to apply the induction hypothesis and for this we need an equality of the appropriate type; if we are lucky enough to find that W^* is a word in the generators of G^* then this will be the desired equality. For $\{b_0, c_0\}$ is a Magnus subset in G^* and, since W was assumed to contain a, W^* will contain generators with non-zero subscript so that the induction hypothesis will apply to yield a Gurevich subword $T^{*2}T_1^*$ of W^*. Let us suppose that

$$T^{*2}T_1^* \equiv (c_2^{-2}b_2^3c_{-1}^2c_0^2b_0^2b_1^2)^2c_2^{-2}b_2^3c_{-1}^2c_0^2b_0^2b_1.$$

When we reverse the rewriting process to recover W from W^*, it is clear (after due consideration) that we must see

$$(c^{-2}b^3a^{-3}c^2ac^2b^2ab^2a)^2c^{-2}b^3a^{-3}c^2ac^2b^2ab$$

which is clearly a Gurevich subword of W – note that we must begin our Gurevich subword of W with c rather than with a^2c since if c^2 were to be proceeded in W^* by another occurrence of c_2, then we would not obtain a^2c when recovering W. Similarly we finish with b rather than ba^{-1}. We also note that since T_1^*, no matter what its form, must contain generators of G^* with different subscripts, we always obtain occurrences of a within the corresponding subword of W.

The alternative possibility is that we are in a situation where W^* contains occurrences of a but defines an element of the vertex group. It follows that W^* must contain, say, aW_ia^{-1} where W_i represents an element of $K = F(b_0, b_1, c_j)$ (continuing with our illustrative example to keep the notation simple). Two possibilities occur; it might happen that W_i already is a word in the generators $\{b_0, b_1, c_j\}$. In this event the relations defining the HNN-extension allow us to cancel the occurrences of a provided that we increase each subscript by 1. A similar observation applies if we have $a^{-1}W_ia$ with W_i a word in $\{(b_1, b_2, c_j)\}$ and we reduce, rather than increase, the subscripts by 1. In either case we make the change and then start over again. If we can eliminate all occurrences of a in this way then we are back to the situation where we can apply the induction hypothesis. So the awkward case is when we eventually get stuck and still have occurrences of a in what, by abuse of notation, we shall still call W^*. This time we have, say, a subword aW_ia^{-1} where W_i represents an element of $K = F(b_0, b_1, c_j)$ but W_i involves the generator b_2. But then in G^* we have a equation $W_i = Z$, where Z is a word of $F(b_0, b_1, c_j)$ and thus omits b_2, and we are back in a situation to which the induction hypothesis applies. As usual there are more cases to deal with – but these reduce to the above in the standard manner ([15, IV5.5] gives the full details, with only minor amendments needed to ensure that we actually verify that we have a Gurevich subword rather than just a word of the form $T^{m-1}T_1$ with T_1 non-empty). □

Theorem 2.5 also yields an easy proof of the following special case of Theorem 3.

Proposition 2.1. *Let $G = \langle X \mid R \rangle$ be a one-relator group and suppose that R is of the form S^m for some element S of $F(X)$ with $m \geq 2$. If M and N are Magnus subgroups of G and $g \in G$, then either $gMg^{-1} \cap N$ is trivial or $g \in NM$.*

Proof. Suppose firstly that $N = M$. If the conclusion is false then there exist $g \in G$ and non-trivial $u, v \in M$ such that $gug^{-1} = v$. Among all possible counterexamples, choose one in which the word representing g is of minimal length - observe that this ensures that gug^{-1} is freely reduced. In addition, by replacing u and v by large powers, we can assume that the words in the generators of M that represent u and v are of length very much greater than the length of R. Since $g \notin M$ and $v \in M$, Theorem 2.5 implies that gug^{-1} must contain a Gurevich subword, say $T^{m-1}T_1$. By choosing u and v to be very long we have ensured that this must occur either within gu or within ug^{-1}; suppose it lies within gu. Since u omits an essential generator, a non-trivial initial part of T_1 must lie within g and so g contains $T^{m-1}T_1'$ where T_1' is non-trivial. But there exists T_2' such that $T^{m-1}T_1'T_2'$ is a cyclic permutation of R or R^{-1} and, if we replace $T^{m-1}T_1'$ by $T_2'^{-1}$ we obtain a representation of g by a word of shorter length, contradicting the minimality of our choice. A similar argument applies when the Gurevich subword lies in ug^{-1}. The argument for the case when M and N are distinct is similar. We suppose that there exist $g \notin NM$ and non-trivial $u \in M, v \in N$ with $gug^{-1} = v$ and make the same assumptions about the lengths of the representatives of g, u and v. This time we apply Theorem 2.5 to the equality $gug^{-1}v^{-1} = 1$. Exactly the same considerations apply, save that we have also to exclude the possibility that $g^{-1}v^{-1}$ contains the Gurevich subword. □

The case when $M = N$ is established in [25] – where the trick of taking high powers of u and v is to be found – and is summarised by saying that Magnus subgroups are *malnormal*. The simplicity of the argument reveals the strength of the conclusion that a word representing the identity (or lying in a suitable

Magnus subgroup) must contain a Gurevich subword - as compared to merely containing a word of the form $T^{m-1}T_1$ where T_1 is a non-empty initial segment of T.

3. Two-generator groups with exceptional intersections

In this section we discuss two-generator one-relator groups, by which, leaving trivial exceptional cases aside, we mean groups with a presentation of the form $\langle a, b \mid R(a, b) = 1\rangle$, where both a and b are assumed to occur in R. Such groups have only two Magnus subgroups, name the two infinite cyclic groups $F(a)$ and $F(b)$ and their intersection is clearly either infinite cyclic or trivial. Thus Theorem 2 is trivial in this case. However, to demonstrate that even in this context there is rather more to the problem than immediately meets the eye, we discuss some examples.

There are obvious examples of two-generator one-relator groups that have exceptional intersections, namely the torus knot groups of the form $\langle a, b \mid a^m = b^n\rangle$, which are an amalgamated free product with the two Magnus subgroups as the vertex groups and infinite cyclic edge groups generated by a^m and b^n. Unfortunately there are also subtler examples. The group $G = \langle a, b \mid a^2b^{-3}a^2b^{-3}a^2 = 1\rangle$ has an exceptional intersection of its two Magnus subgroups; for it is easy to see that the given relation has as consequences $b^3 = (a^2b^{-3})^3$ - just multiply both sides of the defining equality by b^3 - and also $a^2 = (a^2b^{-3})^{-2}$. Then of course a^2 and b^3 commute since they are powers of the same element and it follows that $a^6 = b^6$ in G.

To understand this example more clearly and to set it in perspective, we introduce an additional generator $t = a^2b^{-3}$. Then, using Tietze transformations - see [8, 15, 19] - it is not hard to see that we can also present G in the form $G = \langle a, t, b \mid a^2 = t^2, t^3 = b^3\rangle$. In this form, the group is an example of what is usually known as a *stem product* of groups; to construct such a stem product one is given a linearly ordered sequence of groups H_i, $i = 0, 1, 2, \ldots n$ where H_i and H_{i+1} have a common isomorphic subgroup. One firstly forms G_1 as an amalgamated free product of H_0 and H_1. Proceeding iteratively, one forms G_{i+1} as an amalgamated free product of G_i and H_{i+1} with the stipulation that the edge subgroups are the two copies of the common isomorphic subgroup of H_i and H_{i+1}, regarding the former as embedded in G_i. Repeated application of Theorem 2.1 to our example shows that $F(a) \cap F(b) = \langle a^6\rangle = \langle b^6\rangle$.

A two-generator one-relator group with exceptional intersection of its two Magnus subgroups clearly has non-trivial centre and in our two examples it is not hard to see that the centre is precisely the intersection of the two Magnus subgroups. This is an instance of a more general phenomenon which is summarised by the following theorem of Pietrowski.

Theorem 3.1. [27] *If the two-generator one-relator group G, whose quotient by its derived group is not free abelian of rank* 2, *has a non-trivial centre, then G has a stem product presentation*

$$\left\langle x, t_1, t_2, \dots t_{n-1}, y \mid x^{p_1} = t_1^{q_1}, t_1^{p_2} = t_2^{q_2}, \dots, t_{n-2}^{p_{n-1}} = t_{n-1}^{q_{n-1}}, t_{n-1}^{p_n} = y^{q_n} \right\rangle$$

where p_i, q_j are integers such that $gcd(p_i, q_j) = 1$ when $i > j$. Moreover the centre of G is $F(x) \cap F(y) = \langle x^{p_1 \dots p_n} \rangle = \langle y^{q_1 \dots q_n} \rangle$.

It should, however, be noted that even if a two-generator one-relator group, whose quotient by its derived group is not free abelian of rank 2, has a non-trivial centre, not every two-generator presentation given by a single relator will have an exceptional intersection. For example, $\langle a, b \mid a^m = b^n \rangle$ also has presentation $\langle a, c \mid a^{m-1} = c(ac)^{n-1} \rangle$, obtained by writing $b = ac$ so that $c = a^{-1}b$ and it is easy to show that $F(a) \cap F(c)$ is trivial.

Pietrowski's theorem is oddly tantalising. It is clear that a group with such a presentation has non-trivial centre $F(x) \cap F(y)$ and the coprimeness conditions allow one to show that $\{x, y\}$ is a set of generators. However it is *not* always the case that the resulting group can be specified by a single relation in terms of x and y (although this can always be done with two relations). The simplest example where two relations are required seems to be $\langle x, t_1, t_2, y \mid x^2 = t_1^2, t_1^5 = t_2^5, t_2^3 = y^3 \rangle$ - see [5]. There is still no precise statement of the exact conditions under which a *Pietrowski* presentation will yield a two-generator one-relator group – see [22, 5, 20]. There are two earlier results concerning one-relator groups with centre that we must state for completeness.

Theorem 3.2. [24] *If a one relator group has non-trivial centre then it is a two generator group (except for the trivial case when it has only one generator).*

Theorem 3.3. [3] *There is an algorithm to decide of an arbitrary two generator one-relator group whether or not it has non-trivial centre.*

Example 3.1. *Let $G = \langle a, b, c \mid ba^2c^2b^{-1}a^2c^2 = 1 \rangle$. Then the intersection $F(a, b) \cap F(b, c)$ is exceptional.*

Proof. Since conjugation by b (on the left) inverts a^2c^2, it follows that b^2 commutes with a^2c^2. Thus we have a relation $b^2a^2c^2 = a^2c^2b^2$ which can be rewritten as $a^{-2}b^2a^2 = c^2b^2c^{-2}$ and the Magnus subgroups $F(a, b)$ and $F(b, c)$ have exceptional intersection. Since this group is actually an HNN-extension, it is not hard to show that the intersection is the free subgroup of $F(a, b)$ with basis $\{a^{-2}b^2a^2, b\}$. □

Our study of examples has so far been confined to illustrating that there are subtleties about the problems we are addressing that are not immediately apparent. To understand their implication further we return to the example $G = \langle a, b \mid a^2b^{-3}a^2b^{-3}a^2 = 1\rangle$. Suppose we have a group presentation $\hat{G}$ whose generating set is the union of three disjoint subsets A, B and C and which has a single relator of the form $u^2v^{-3}u^2v^{-3}u^2$, where u is a word of $F(A, B)$ and v is a word of $F(B, C)$, neither of which lies in $F(B)$. Then it follows that in $\hat{G}$ we have $u^6 = v^6$ (just copy the argument that shows that $a^6 = b^6$ in G) and therefore the two Magnus subgroups $F(A, B)$ and $F(B, C)$ have exceptional intersection since neither u^6 nor v^6 lies in $F(B)$. Can we determine exactly what the intersection is? The answer is that we can, based on the fact that we can copy our procedure for exhibiting G as a stem product to exhibit $\hat{G}$ as a stem product. If we introduce a generator $t = u^2v^{-3}$, then we can define $\hat{G}$ via the presentation $\langle A, B, C, t \mid u^2 = t^2, t^3 = v^3\rangle$. To see that this is a stem product we write $H_0 = F(A, B)$, $H_1 = F(t, B')$ and $H_2 = F(B'', C)$ where B' and B'' are sets in one-to-one correspondence with B. We then form, firstly, the amalgamated free product G_1 from H_0 and H_1 via the relations $u^2 = t^2$, $B = B'$ and then the amalgamated product G_2 from G_1 and H_2 via the relations $t^3 = v^3$, $B' = B''$ - it is routine to check that the various edge groups defined are isomorphic. The resulting presentation is obviously transformed into $\langle A, B, C, t|u^2 = t^2, t^3 = v^3\rangle$ by eliminating the sets B' and B" of generators. It is now a routine exercise in the normal form theorem for amalgamated free products to show that the intersection $F(A, B) \cap F(B, C)$ in $\hat{G}$ is just the free subgroup of $F(A, B)$ with basis $\{u^6\} \cup B$. In fact $\hat{G}$ can be written as a stem product in a different way - and this is more important to us as we shall see later. We again take $H_0 = F(A, B)$ and $H_2 = F(B'', C)$ but we take H_1 to be the the free product $G' = \langle a, b|a^2b^{-3}a^2b^{-3}a^2 = 1\rangle * F(B')$. The two amalgamations are given, respectively by the relations $u = a$, $B = B'$ and $b = v$, $B' = B''$. It is easy to check that this does define our group $\hat{G}$ and, via the normal form theorem, that $F(A, B) \cap F(B, C) = (F(a) \cap F(b)) * F(B)$, which is what is required. Moreover it is clear that if in any group given by $\langle x, y \mid R(x, y) = 1\rangle$ we obtain a non-trivial equality of the form $x^m = y^n$, then these latter remarks apply to any one-relator presentation of the form $\langle A, B, C|R(u, v) = 1\rangle$ where u lies in $F(A, B)$, v lies in $F(B, C)$, neither lies in $F(B)$ and $R(u, v)$ is the result of substituting u for x and v for y.

One might be tempted to conjecture that all one-relator presentations with an exceptional intersection of Magnus subgroups arise in this way from presentations

$$\langle x, y | R(x, y) = 1 \rangle$$

in which $F(x)$ and $F(y)$ have non-trivial intersection. However our example

$$G = \langle a, b, c | ba^2c^2b^{-1} = c^{-2}a^{-2} \rangle$$

shows that this is not the case – the reader may care to demonstrate that this group is not one that can be obtained in this way – and this example, too, lends itself to generalisation.

Suppose that in $G_0 = \langle x, y \mid R(x, y) = 1 \rangle$ a relation of the form $xy^m x^{-1} = y^n$ holds, with $m, n \neq 0$. (Of course the latter is the famous Baumslag-Solitar relation [2] which, when it is a defining relation, provides groups which have a fascinating range of properties – see [6].) Now consider a group $G = \langle A, B, C \mid R(vu, w) = 1 \rangle$ where u lies in $F(A, B)$, v lies in $F(B, C)$, with neither in $F(B)$, and w is a word of $F(B)$. Then it follows that in G the relation $vuw^m u^{-1} v^{-1} = w^n$ must hold and hence that $uw^m u^{-1} = v^{-1} w^n v$, which forces an exceptional intersection (since clearly $uw^m u^{-1} \notin F(B)$). Can we say anything about the intersection $F(A, B) \cap F(B, C)$? The answer is that we can because we can also express G as a stem product. We describe the precise form in the next section.

4. Further results

Here we summarise briefly some further results on intersections of Magnus subgroups that can be obtained when techniques that have a more geometric and topological flavour are brought to bear. In particular we describe results of Howie that will appear in [13].

Theorem 4. [13] *Let $G = \langle A, B, C \mid R = 1 \rangle$ be such that the Magnus subgroups $F(A, B)$ and $F(B, C)$ have exceptional intersection. Then there exists a two-generator one-relator group $G_0 = \langle x, y \mid R_0(x, y) = 1 \rangle$ such that one of the following hold.*

(a) *in G_0, $x^m = y^n$ and $R(A, B, C)$ is freely equal to $R_0(u, v)$, where u lies in $F(A, B)$, v lies in $F(B, C)$ and neither lies in $F(B)$;*
(b) *in G_0, $xy^m x^{-1} = y^n$ and $R(A, B, C)$ is freely equal to $R_0(vu, w)$ where u lies in $F(A, B)$, v lies in $F(B, C)$, neither lies in $F(B)$, and w is a word of $F(B)$.*

Theorem 2 can then be derived from this result, using Theorem 2.1, by observing that the group G can then be built from the group G_0 and free groups in the following ways.

(a) Write $\hat{G}_0 = \langle x, y \mid R_0(x, y) = 1 \rangle * F(B)$. Then G is expressed as a stem product:

$$F(A \cup B)) \underset{\langle u\rangle * F(B) = \langle x\rangle * F(B)}{*} \hat{G}_0 \underset{\langle y\rangle * F(B) = \langle v\rangle * F(B)}{*} F(B \cup C).$$

(b) Here G is expressible as the stem product

$$F(A \cup B)) \underset{\langle u\rangle * F(B) = \langle x\rangle * F(B)}{*} \hat{G}_0 \underset{\langle y\rangle * F(B) = \langle v\rangle * F(B)}{*} F(B \cup C),$$

but this time $\hat{G}_0$ is of the form

$$(\langle x, y \mid R_0(x, y) = 1 \rangle \underset{\langle y\rangle = \langle w\rangle}{*} F(B)) * F(t).$$

Finally Theorem 4, in combination with Theorem 3.3 and Theorem 3.1, yields:

Theorem 4.1. *Let $G = \langle A \cup B \cup C \mid R = 1 \rangle$ be a one-relator group. Then it is algorithmically decidable whether or not the intersection of $F(A \cup B)$ and $F(B \cup C)$ in G contains exceptional elements. If so, then the algorithm yields a word u such that $F(A \cup B) \cap F(B \cup C) = F(B) * \langle u \rangle$.*

We also describe briefly how Theorem 2 can be strengthened in a different way. A generalisation of the theory of one-relator groups has been developed, by various authors, to what are called one-relator products. Given a family $\{G_i, i \in I\}$ of groups their free product G – introduced in Section 2 as the amalgamated free product with trivial amalgamation – can be viewed as the set of all strings of the form $w_1 w_2 \ldots w_n$ where each w_j is a non-trivial element of some G_i and adjacent w_j lie in distinct G_i. Multiplication consists of concatenation (and then whatever cancellation and coalescence is possible within an individual G_i). A one-relator product is then the quotient of G by the normal closure of a single element, cyclically reduced in the sense that w_1 and w_n lie in different G_i, of length $n \geq 2$. By analogy a *Magnus* subset of the family $\{G_i, i \in I\}$ is a subset that omits a group which has an element explicitly appearing in the 'relator' which is factored out to define the one-relator product. The 'Freiheitssatz for a one-relator product' is then the statement that the subgroup generated by (the images of) the groups in the Magnus subset is the free product of these groups.

It is far from the case that the Freiheitssatz is valid in general – the simplest counterexample is when G is the free product of cyclic groups of orders two, three and infinity, and the relator makes the two generators of the finite cyclic groups conjugate by the generator of the infinite cyclic group. However there are situations where it does hold – the relevant sufficient condition may be a restriction on the *factors* G_i or on the relator. Among such situations we confine ourselves to the case when the factors are locally indicable, i.e., every finitely generated subgroup has the infinite cyclic group as a homomorphic image. This theory was introduced by Brodskii [4] and Howie [12], and Brodskii has obtained results which include special cases of Theorem 2 and Theorem 3. In particular, he showed that in a one-relator product $\langle A * B \mid r = 1\rangle$ of locally indicable groups, the intersections $gAg^{-1} \cap A$ and $gAg^{-1} \cap B$ are cyclic. Since A and B need not be freely indecomposable, the former corresponds to the results of Bagherzadeh [1] and the latter to the case in Theorem 2 when the two Magnus subsets are disjoint. The following result in [13] generalises Theorem 2 to one-relator products of locally indicable groups (in a similar way, the restriction in the statement to three factors is not a loss of generality).

Theorem 5. [13] *In a one-relator product $\langle A * B * C \mid r = 1\rangle$ of locally indicable groups, the intersection $(A * B) \cap (B * C)$ is either just B or the free product of B with an infinite cyclic group.*

5. Exceptional intersections, elements and equalities

We are now ready to begin the proof of Theorem 2. We shall employ the traditional approach to one-relator groups, using induction on the length of the relator, by expressing a one-relator group as an HNN-extension when there is a generator of exponent sum zero in the relator, using the standard trick when no generator has exponent sum zero in r. We note at the outset that, by the Corollary to Proposition 2.5, we only have to deal with the case when r is not a proper power. In one particular case we require a detailed understanding and analysis of the intersections of Magnus subgroups of the vertex group of the HNN-extension we construct and we turn first to an analysis of this situation before we embark upon our formal inductive arguments. Example 3.1 above is a paradigm for this case. The situation to be considered is as follows:

(a) $G = \langle X : r = 1\rangle$, where r is cyclically reduced;
(b) all generators in X appear in the relator r;
(c) X has been partitioned into three disjoint non-empty subsets A, B and C with $A = \{a\}$ and $C = \{c\}$ singletons;
(d) neither a nor c has exponent sum zero in r;
(e) there exists $b \in B$ which does have exponent sum zero in r.

The two Magnus subgroups that we wish to consider are $F(A, B)$ and $F(B, C)$ and of course we shall be interested in when $F(A, B) \cap F(B, C) \neq$

$F(B)$. The essence of the inductive method is that when there exists a generator with exponent sum zero in r, then G can be expressed as an HNN-extension over a vertex group G^* with a relator of shorter length. We have already described how there is a standard procedure for this but also noted that minor variations are sometimes useful, depending on the problem to hand. In our particular circumstances here we shall, in the usual manner, write $x_i = b^i x b^{-i}$, for any $x \in X, x \neq b$. Since r has exponent sum zero in b, it can be rewritten as a word r^* in certain x_i, with at least one such for every $x \in X, x \neq b$ (since we have assumed all elements of X appear in r) The element of choice lies in the range of subscripts for which we include x_i among the generators of G^*. We shall take X^* to consist of the generators $\{a_\kappa, \ldots, a_\lambda, c_\mu, \ldots, c_\nu\} \cup \{x_i, i \in \mathbb{Z}\}, x \neq a, c$, where $a_\kappa, a_\lambda, c_\mu, c_\nu$ are the respective minimal and maximal generators in r^* associated with a and c, and otherwise the subscript range is infinite. To express G as an HNN-extension over G^* we then add b as a generating letter together with the relations $bx_i b^{-1} = x_{i+1}$ whenever this makes sense (including when x is a or c). The introduction of some notation to make this precise is unavoidable.

Definition. *We write:*

$$A^* = \{a_\kappa, \ldots, a_\lambda\},\ A^*_+ = \{a_{\kappa+1}, \ldots, a_\lambda\},\ A^*_- = \{a_\kappa, \ldots, a_{\lambda-1}\},$$

$$C^* = \{c_\mu, \ldots, c_\nu\},\ C^*_+ = \{c_{\mu+1}, \ldots, c_\nu\},\ C^*_- = \{c_\mu, \ldots, c_{\nu-1}\},$$

and $B^* = \{x_i, x \in B', i \in \mathbb{Z}\}$ *where* $B' = B \setminus \{b\}$.

We are not excluding the possibility that $\lambda = \kappa$ so that A^*_+ and A^*_- are empty and similarly for C^*_+ and C^*_-. We shall refer to the four generators $a_\kappa, a_\lambda, c_\mu, c_\nu$ as *extremal* generators and will call a word of G^* *intermediate* if it omits all four extremal generators. With this notation, G^* is generated (disjointly) by $A^* \cup B^* \cup C^*$, and we define the edge group L to have free basis $A^*_- \cup B^* \cup C^*_-$ and the edge group U to have free basis $A^*_+ \cup B^* \cup C^*_+$. These are, of course, Magnus subgroups - we shall frequently write them as $L = F(A^*_-, B^*, C^*_-)$ and $U = F(A^*_+, B^*, C^*_+)$, and similarly for other Magnus subgroups such as $F(A^*, B^*)$. We can summarise our notation, with some loss of detail, by writing

$$G = \langle A^*, B^*, C^*, b \mid r^* = 1,\ \ bF(A^*_-, B^*, C^*_-)b^{-1} = F(A^*_+, B^*, C^*_+)\rangle.$$

We shall apply the results of this section during the inductive step of the proof of Theorem 2. Throughout this section, therefore we shall always assume:

Assumption 5.1. *Theorem* 2 *is valid for* G^* *in the situation described above.*

Our interest is in *exceptional* equalities, that is equalities of the form $h = k$ where $h \in F(A, B), k \in F(B, C)$, $h, k \notin F(B)$ and will sometimes refer to

h and k as *exceptional elements*. It will be helpful if we point out now that our terminology and notation will sometimes be ambiguous in that we shall not always distinguish between group elements and words which represent them. This is common enough practice but already above we have written $h \in F(A, B), h \notin F(B)$ to emphasise that h is a word in $A \cup B$ which explicitly involves a – however since (the element represented by) h lies in the intersection $F(A, B) \cap F(B, C)$ it would be legitimate to write $h \in F(B, C)$. In general, the reader must use the context to decide what is intended. We shall also, with apologies, be somewhat inconsistent in using both the expressions $v \equiv z$ and $v = z$ to mean that z and v are the same word - of course there is no ambiguity in the latter when z and v are words of the same free group and normally we shall use the former only for emphasis or to introduce notation. Again however the reader will need to refer to the context.

We firstly examine how elements $h \in F(A, B), k \in F(B, C)$ are represented as elements of the HNN-extension. To understand this we consider $h \in F(A, B)$; suppose that h has exponent sum σ in b so that h is freely equal to a word $h^* b^\sigma$ where h^* has exponent sum zero in b. Then we can write h^* as an expression in $b^i a b^{-i}$ and $b^i x b^{-i}, x \in B'$ where, at this point, there is no restriction on the values of i. We then replace $b^i a b^{-i}$ by a_i if $\kappa \leq i \leq \lambda$, by $b^{i-\kappa} a_\kappa b^{-(i-\kappa)}$ if $i < \kappa$ and by $b^{i-\lambda} a_\lambda b^{-(i-\lambda)}$ if $i > \lambda$. For the remaining terms $b^i x b^{-i}, x \in B'$, we simply write x_i. (The reader should recall the proof of Theorem 2.5.) By adding b^σ as a suffix to the expression just obtained we have now expressed h as an element of the HNN-extension. An identical procedure can be applied to an arbitrary k. However, we may not yet have obtained expressions for h and k which are in reduced form relative to the HNN-extension – and we need this if we are to use Theorem 2.2. To achieve this for h, we clearly must eliminate occurrences of subwords either of the form $bh_j(A^*_-, B^*)b^{-1}$ or $b^{-1}h_j(A^*_+, B^*)b$ which we can do by 'shifting subscripts' in h_j up or down one place as appropriate, and deleting the occurrences of b. In what we hope is a suggestive notation, we thus replace $bh_j b^{-1}$, where h_j is a word of $F(A^*_-, B^*)$, by $\overrightarrow{h_j}$ and $b^{-1}h_j b$, where h_j is a word of $F(A^*_+, B^*)$, by $\overleftarrow{h_j}$ and then freely reduce the whole result word. Let us call a word of $F(A^*, B^*)$ *shift-reduced* if no such 'shift' operation can be performed. There is an obvious parallel definition for $F(B^*, C^*)$.

Lemma 5.1. *A shift-reduced word of $F(A^*, B^*)$ or of $F(B^*, C^*)$ is in reduced form relative to the HNN-extension.*

Proof. Suppose not; then it must contain a subword, say $bh_j b^{-1}$, where h_j is equal in G^* to an element of the group L but must, as a word, contain occurrences of a_λ, since otherwise we could have applied a shift. We therefore have an equality of the form $h_j = w$ where w is a word of $F(A^*_-, B^*, C^*_-)$. However neither side of this equality involves the extremal generator c_ν and so the equality therefore must hold freely in the Magnus subgroup $F(A^*, B^*, C^*_-)$, which is impossible as h_j contains a_λ and w does not. □

Normal Form equalities. It follows from Lemma 5.1 that our analysis of an exceptional equality $h = k$ will involve a sequence of equalities derived from an application of Theorem 2.2 to an equality between reduced words of the form

$$h \equiv h_0 b^{\varepsilon_1} h_1 \ldots b^{\varepsilon_m} h_m = k_0 b^{\varepsilon_1} k_1 \ldots b^{\varepsilon_m} k_m \equiv k$$

where $h_0, \ldots h_m \in F(A^*, B^*)$ and $k_0, \ldots k_m \in F(B^*, C^*)$. Such an equality will yield a sequence of equalities

$$h_0 = k_0 z_0,\ \overline{z_0} h_1 = k_1 z_1,\ \overline{z_1} h_2 = k_2 z_2, \ldots, \overline{z_{m-1}} h_m = k_m$$

where the *auxiliary terms* $z_0, \ldots z_{m-1}$ lie variously in L or U according as each of $\varepsilon_1, \ldots, \varepsilon_m$ is ± 1 and $\overline{z_{i-1}}$ represents a 'downshift' or 'upshift' of subscripts according as $\varepsilon_i = \pm 1$ (and will usually be written as $\overleftarrow{z_{i-1}}$ or $\overrightarrow{z_{i-1}}$ accordingly). An elementary argument shows that we can always assume that the terms z_i in the Normal Form equalities are of *type* $(A^* : C^*)$, that is, when z_i is non-trivial, its initial generating letter lies in A^* and its terminal generating letter lies in C^*. (Strictly these initial letters will lie in either A^*_+ or A^*_- according to whether we are considering an element of U or L, but we shall usually leave the context to clarify which it is.) For clarity we shall, sometimes but not always, denote an auxiliary term by w_i when the element in question lies in L, leaving z_i for use when a term lies in U. We illustrate what is involved.

Example 5.1. *Suppose that we have*

$$h \equiv h_0 b h_1 b h_2 b^{-1} h_3 b h_4 = b k_1 b k_2 b^{-1} k_3 b k_4 \equiv k.$$

Then we can write

$$h = h_0^* b h_1^* b h_2^* b^{-1} h_3^* b h_4^* = b k_1^* b k_2^* b^{-1} k_3^* b k_4^* = k$$

so that the auxiliary terms in the normal form sequence for the latter are of type $(A^* : C^*)$.

Proof. By Theorem 2.2 applied to $h = k$, we obtain the sequence

$$h_0 = k_0 z_0,\ \overleftarrow{z_0} h_1 = k_1 z_1,\ \overleftarrow{z_1} h_2 = k_2 w_2,\ \overrightarrow{w_2} h_3 = k_3 z_3,\ \overleftarrow{z_3} h_4 = k_4.$$

Now we can write $z_0 \equiv y_0 z_0^* x_0$ where $y_0 \in F(B^*, C_+^*)$, $x_0 \in F(A_+^*, B^*)$ (either may be trivial) and z_0^* is either trivial or of type $(A^* : C^*)$; further we can obtain similar expressions for z_1, w_2, z_3. If we define $h_0^* \equiv h_0 x_0^{-1}$, $k_0^* = y_0 k_0$, $h_1^* \equiv \overleftarrow{x_0} h_1 x_1^{-1}$, $k_1^* \equiv \overleftarrow{y_0}^{-1} h_1 y_1$, $h_2^* \equiv \overleftarrow{x_1} h_1 x_2^{-1}$, $k_2^* \equiv \overleftarrow{y_1}^{-1} k_2 y_2$, $h_3^* \equiv \overrightarrow{x_2} h_3 x_3^{-1}$, $k_3^* \equiv \overrightarrow{y_2}^{-1} k_3 y_3$ and $h_4^* \equiv \overleftarrow{x_2} h_3 x_3^{-1}$, $k_4^* \equiv \overleftarrow{y_2}^{-1} k_3 y_3$ then it is easy to see that

$$h_0^* = k_0^* z_0^*,\ \overleftarrow{z_0^*} h_1^* = k_1^* z_1^*,\ \overleftarrow{z_1^*} h_2^* = k_2^* w_2^*,\ \overrightarrow{w_2^*} h_3^* = k_3^* z_3^*,\ \overleftarrow{z_3^*} h_4^* = k_4^*$$

and hence that

$$h = h_0^* b h_1^* b h_2^* b^{-1} h_3^* b h_4^* g^{-1} = b k_1^* b k_2^* b^{-1} k_3^* b k_4^* = k. \qquad \square$$

We shall always assume that we have made such a transformation and will refer to the latter expression and its associated equalities as being in *standardised form*. Example 5.1 reveals how certain intersections of Magnus subgroups arise in our analysis. Our first normal form equality is $h_0 = k_0 z_0$, which clearly defines an element $F(A^*, B^*) \cap F(A_+^*, B^*, C^*)$, and our last equality is $\overleftarrow{z_3} h_4 = k_4$ which defines an element $F(A^*, B^*, C_-^*) \cap F(B^*, C^*)$. Moreover the inequality $\overleftarrow{z_1} h_2 = k_2 w_2$ when rewritten as $\overleftarrow{z_1} h_2 w_2^{-1} = k_2$ also defines an element $F(A^*, B^*, C_-^*) \cap F(B^*, C^*)$ and, finally, $\overleftarrow{z_0} h_1 = k_1 z_1$ and $\overrightarrow{w_2} h_3 = k_3 z_3$ define elements of $F(A^*, B^*, C_-^*) \cap F(A_+^*, B^*, C^*)$. There are obvious connections among the intersections we have introduced since both $F(A^*, B^*) \cap F(A_+^*, B^*, C^*)$ and $F(A^*, B^*, C_-^*) \cap F(B^*, C^*)$ are subgroups of

$$F(A^*, B^*, C_-^*) \cap F(A_+^*, B^*, C^*),$$

and our first steps are to show how these intersections (and others) are intimately interlinked. As usual, let $G^* = \langle X^* : r^* = 1 \rangle$ and let $D \subset X^*$; we say a (reduced) word in X^* is *D-special* if both its initial and terminal generating letters belong to $D \cup D^{-1}$. If $D = \{d\}$ is a singleton we shall write d-special. Given a word t in X^* and subset D of X^* such that t contains occurrences of elements of D, we define the *D-core* of t to be the maximal subword of t that is D-special. The propositions following form the underlying basis of our argument.

Proposition 5.1. *Let $F(A^*, B^*) \cap F(A_+^*, B^*, C^*) = \langle v \rangle * F(A_+^*, B^*)$ be an exceptional intersection, where $v \in F(A_+^*, B^*, C^*)$ and v is C^*-special.*

(1) *Suppose that we are given elements $k \in F(B^*, C^*)$, $z \in U = F(A_+^*, B^*, C_+^*)$ such that*
  (i) *$kz \in \langle v \rangle * F(A_+^*, B^*)$;*
  (ii) *$k \notin F(B^*, C_+^*)$ and k has initial generator in C^*;*
  (iii) *if z is non-trivial, then z is of type $(A^* : C^*)$.*

If $v \in F(B^*, C^*)$, then z is trivial while if $v \notin F(B^*, C^*)$, then z is non-trivial and v (or v^{-1}) can be written as $v_0 v^*$ where $v_0 \equiv k \in F(B^*, C^*)$ and $v^* \equiv z \in U$.

(2) *Suppose that we are given elements $k \in F(B^*, C^*)$, $z_1, z_2 \in U = F(A^*_+, B^*, C^*_+)$ such that*
   (i) $z_1^{-1} k z_2 \in \langle v \rangle * F(A^*_+, B^*)$;
   (ii) $k \notin F(B^*, C^*_+)$;
   (iii) *z_1, z_2 are non-trivial of type $(A^* : C^*)$.*

   Then $v \notin F(B^*, C^*)$ and one of the following holds:
   (a) *v (or v^{-1}) can be written in the form $v_0 v^*$ with $z_1 = v^* = z_2$ and $k = v_0^{-1} \beta v_0$, where $\beta \in F(B^*)$, in particular $z_1^{-1} k z_2 = v^{*-1} v_0^{-1} \beta v_0 v^*$.*
   (b) *v (or v^{-1}) can be written in the form $v_1^{-1} v_0 v_2$ with $z_1 = v_1$, $z_2 = v_2$, $v_0 \in F(B^*, C^*)$ with $k = v_0^l$, where $l = \pm 1$ except perhaps when $z_1 = v_1 = v_2 = z_2$.*

Corollary 5.1. *Statements exactly parallel to Proposition* 5.1 *hold for each of the intersections $F(A^*, B^*) \cap F(A^*_-, B^*, C^*)$, $F(A^*, B^*, C^*_+) \cap F(B^*, C^*)$ and $F(A^*, B^*, C^*_-) \cap F(B^*, C^*)$.*

Proof of 5.1. (1) We have to consider an equality of the form

$$kz = \alpha_0 v^{i_1} \alpha_1 \dots v^{i_e} \alpha_e$$

where $\alpha_0, \alpha_1, \dots, \alpha_e \in F(A^*_+, B^*)$. The left hand side is clearly (freely) reduced as written. This is not quite true of the right hand side as there may be limited cancellations within powers of v if v is not cyclically reduced but these are the only possible cancellations. Since $k \notin U$, the left hand side must contain occurrences of c_μ and therefore c_μ occurs on the right hand side so that v and every reduced power of v must contain occurrences of c_μ. However, in the left hand side every occurrence of c_μ lies to the left of every possible occurrence of a generator from A^*_+ and we can use this to obtain information about the right hand side. Suppose that $v \in F(B^*, C^*)$. Then the only possible location within the right hand side for occurrences of generators from A^*_+ is within α_e. However, if z is non-trivial the left-hand side ends with a generator from C^* and this is contradictory. Hence $z = 1$. If, on the other hand, $v \notin F(B^*, C^*)$ then any reduced power of v must contain occurrences of generators from A^*_+ and hence the right hand side must contain occurrences of generators from A^*_+ - and hence z is non-trivial. Since reduced powers of v also contain c_μ, it follows that the equality must collapse to just $kz = \alpha_0 v^{i_1} \alpha_1$ with $\alpha_0, \alpha_1 \in F(B^*)$ - but our assumptions on k and z give $\alpha_0 = \alpha_1 = 1$. To complete the argument we have to show $|i_1| = 1$. When v is cyclically reduced one can see immediately that if $|i_1| \geq 2$, then there would be occurrences of c_μ in distinct powers of v

separated by occurrences of elements of A_+^* which contradicts the positioning implied by our analysis of kz and the desired conclusion follows. If $v = \hat{v}\bar{v}\hat{v}^{-1}$ with $\bar{v}$ cyclically reduced, then one first eliminates the possibility that c_μ lies in $\hat{v}$ and then argues as in the previous case.

(2) This time we analyse an equality $z_1^{-1}kz_2 = \alpha_0 v^{i_1}\alpha_1 \dots v^{i_e}\alpha_e$ where $\alpha_0, \alpha_1, \dots, \alpha_e \in F(A_+^*, B^*)$. The left hand side is reduced as written, must contain c_μ and no occurrences of c_μ (if there is more than one) can enclose between them any generators of A_+^*. The same therefore applies to the (reduced form of) the right hand side. Since z_1, z_2 are non-trivial, the left hand side does contain occurrences of generators from A_+^* and therefore so does the right hand side. These cannot occur in $\alpha_1, \dots, \alpha_{e-1}$ (if $e > 1$) since then they would be enclosed by occurrences of c_μ. Since the left hand side is C^*-special, we must have $\alpha_0 = 1 = \alpha_e$. The only place for occurrences of generators of A_+^* is therefore within v and so $v \notin F(B^*, C^*)$. In addition, if $e \geq 3$, then occurrences of c_μ in v^{i_1} and v^{i_3} enclose occurrences of generators from A_+^* in v^{i_2} which is impossible. Hence $e \leq 2$, and, if $e = 2$ then $\alpha_1 \in F(B^*)$. Arguing as in (i) we can show that $|i_1| = 1$ ($= |i_2|$ when $e = 2$) except when $e = 1$ and $z_1 = z_2$. The desired conclusion now follows by a simple inspection of the possibilities. □

Proposition 5.2. *Let $F(A^*, B^*) \cap F(B^*, C^*) = \langle u \rangle * F(B^*) = \langle v_0 \rangle * F(B^*)$ be an exceptional intersection, where $u \in F(A^*, B^*)$, $v_0 \in F(B^*, C^*)$ and $u = v_0$. Then u involves both extremal generators of A^*, v_0 involves both extremal generators of C^* and the four intersections*

$$F(A^*, B^*) \cap F(A_+^*, B^*, C^*),\ F(A^*, B^*) \cap F(A_-^*, B^*, C^*),$$
$$F(A^*, B^*, C_+^*) \cap F(B^*, C^*),\ F(A^*, B^*, C_-^*) \cap F(B^*, C^*),$$

are also exceptional. In particular

$$F(A^*, B^*) \cap F(A_+^*, B^*, C^*) = \langle u \rangle * F(A_+^*, B^*) = \langle v_0 \rangle * F(A_+^*, B^*)$$

and similar expressions hold in the remaining cases.

Proof. If any of the four extremal generators are omitted, the equality $u = v_0$ holds freely which is impossible. Then $u = v_0$ is an exceptional equality for each of the four intersections and, for example, we can write $F(A^*, B^*) \cap F(A_+^*, B^*, C^*) = \langle \hat{v} \rangle * F(A_+^*, B^*)$ for some C^*-special $\hat{v} \in F(A_+^*, B^*, C^*)$. Clearly $v_0 \in \langle \hat{v} \rangle * F(A_+^*, B^*)$ and, since this occurs within $F(A_+^*, B^*, C^*)$, it follows that $\hat{v} \in F(B^*, C^*)$ and $v_0 \in \langle \hat{v} \rangle * F(B^*)$. However we also know that $\hat{v}$ defines an element of $F(A^*, B^*)$ and so $\hat{v} \in \langle v_0 \rangle * F(B^*)$ and the desired conclusion follows. □

Proposition 5.3. *Let $F(A^*, B^*) \cap F(A^*_+, B^*, C^*) = \langle u \rangle * F(A^*_+, B^*) = \langle v \rangle * F(A^*_+, B^*)$ be an exceptional intersection, where $u = v$, $u \in F(A^*, B^*)$, $v \in F(A^*_+, B^*, C^*)$, $v \notin F(B^*, C^*)$ and v is C^*-special of the form $v_1^{-1} v_0 v_2$, with $v_0 \in F(B^*, C^*)$ and $v_1, v_2, \in F(A^*_+, B^*, C^*_+)$ of type $(A^* : C^*)$ (one but not both may be trivial). Then u and v together contain occurrences of the four extremal generators and, in particular, a_κ appears only in u and c_μ occurs only in v_0. Furthermore*

(a) *the intersection $F(A^*, B^*, C^*_+) \cap F(B^*, C^*)$ is also exceptional and equals*

$$\langle v_1 u v_2^{-1} \rangle * F(B^*, C^*_+);$$

(b) *both $F(A^*, B^*, C^*_-) \cap F(A^*_+, B^*, C^*)$ and $F(A^*, B^*, C^*_+) \cap F(A^*_-, B^*, C^*)$ are exceptional and*

$$F(A^*, B^*, C^*_-) \cap F(A^*_+, B^*, C^*) = \langle u \rangle * F(A^*_+, B^*, C^*_-)$$
$$= \langle v_1^{-1} v_0 v_2 \rangle * F(A^*_+, B^*, C^*_-)$$

and

$$F(A^*, B^*, C^*_+) \cap F(A^*_-, B^*, C^*) = \langle v_1 u v_2^{-1} \rangle * F(A^*_+, B^*, C^*_-)$$
$$= \langle v_0 \rangle * F(A^*_+, B^*, C^*_-).$$

Proof. If any extremal generator is omitted the equality would have to hold freely which is impossible. The positioning of the occurrences of a_κ and c_μ is immediate from the definitions of the sets of generators involved. (Without further information, we can say nothing about where the (necessary) occurrences of a_λ and c_ν are located – for example a_λ may occur in one or more of u, v_1^{-1}, v_2.)

To prove (a) we begin by noting that $F(A^*, B^*, C^*_+) \cap F(B^*, C^*)$ is exceptional since we have the equality $v_1 u v_2^{-1} = v_0$. We use Proposition 5.1 (in the form arising when A^* and C^* exchange roles). To do this we need to check that we have the appropriate hypotheses. We can write $F(A^*, B^*, C^*_+) \cap F(B^*, C^*) = \langle \hat{u} \rangle * F(B^*, C^*_+)$ where $\hat{u} \in F(A^*, B^*, C^*_+)$ and is A^*-special. Then, with minor adjustments such as inversion or transferring a word of $F(B^*)$ from one side to the other if either of v_1 or v_2 is trivial, the equality $v_1 u v_2^{-1} = v_0$ gives the necessary hypothesis to apply either the analogue of (2)(a) or 2(b) of Proposition 5.1. There is, however, one point that needs to be checked, namely that only the analogue of 2(b) can apply when both v_1 and v_2 are non-trivial. However if we had $\hat{u} = \hat{u}_0 \hat{u}_1$, with $\hat{u}_0 \in F(A^*, B^*)$ and $\hat{u}_1 \in F(A^*_+, B^*, C^*_+)$, then there would exist an equality $\hat{u}_0 \hat{u}_1 = \hat{v}_0$ yielding an exceptional equality $\hat{u}_0 = \hat{v}_0 \hat{u}_1^{-1}$. Possibly after transferring a word of $F(B^*)$ to the left hand side, we could then apply (1) of Proposition 5.1 to obtain a contradiction and thereby obtain (a).

For (b) it is immediate that $F(A^*, B^*, C^*_-) \cap F(A^*_+, B^*, C^*)$ is exceptional since $u = v$ defines an exceptional element. Hence the intersection can thus be written as $\langle \hat{u} \rangle * F(A^*_+, B^*, C^*_-)$ where $\hat{u}$ is an a_κ-special element of $F(A^*, B^*, C^*_-)$. Clearly $u \in \langle \hat{u} \rangle * F(A^*_+, B^*, C^*_-)$, say

$$u = \alpha_0 \hat{u}^{i_1} \alpha_1 \ldots \hat{u}^{i_e} \alpha_e,$$

where $\alpha_j \in F(A^*_+, B^*, C^*_-)$, and this holds in $F(A^*, B^*, C^*_-)$. Since $\hat{u}$ is a_κ-special, the right hand side is essentially reduced as written (only inessential cancellations within powers of $\hat{u}$ can occur) and therefore, since $u \in F(A^*, B^*)$, it follows that $\hat{u} \in F(A^*, B^*)$ and $u \in \langle \hat{u} \rangle * F(A^*_+, B^*)$. Then, however, $\hat{u} \in F(A^*, B^*) \cap F(A^*_+, B^*, C^*)$ and so $\hat{u} \in \langle u \rangle * F(A^*_+, B^*)$. This gives $\langle u \rangle * F(A^*_+, B^*) = \langle \hat{u} \rangle * F(A^*_+, B^*)$ and hence $\langle u \rangle * F(A^*_+, B^*, C^*_-) = \langle \hat{u} \rangle * F(A^*_+, B^*, C^*_-)$ as required. □

The next proposition parallels Proposition 5.3, but with the roles of A^* and C^* switched, and is proved in a parallel manner.

Proposition 5.4. *Let* $F(A^*, B^*, C^*_-) \cap F(B^*, C^*) = \langle p \rangle * F(B^*, C^*_-) = \langle q \rangle * F(B^*, C^*_-)$ *be an exceptional intersection, where* $p = q$, $p = p_1 p_0 p_2^{-1}$ *is* A^**-special, with* $p_0 \in F(A^*, B^*)$, $p_1, p_2 \in F(A^*_-, B^*, C^*_-)$ *of type* $(A^* : C^*)$ *(not both trivial) and* $q \in F(B^*, C^*)$. *Then* p *and* q *together contain occurrences of the four extremal generators and, in particular,* a_λ *appears only in* p_0 *and* c_ν *occurs only in* q. *Furthermore*

(a) *the intersection* $F(A^*, B^*) \cap F(A^*_-, B^*, C^*)$ *is also exceptional and equals*

$$\langle p_1^{-1} q p_2 \rangle * F(B^*, C^*_+);$$

(b) *both* $F(A^*, B^*, C^*_-) \cap F(A^*_+, B^*, C^*)$ *and* $F(A^*, B^*, C^*_+) \cap F(A^*_-, B^*, C^*)$ *are exceptional and*

$$F(A^*, B^*, C^*_-) \cap F(A^*_+, B^*, C^*) = \langle p_1 p_0 p_2^{-1} \rangle * F(A^*_+, B^*, C^*_-)$$
$$= \langle q \rangle * F(A^*_+, B^*, C^*_-)$$

and

$$F(A^*, B^*, C^*_+) \cap F(A^*_-, B^*, C^*) = \langle p_0 \rangle * F(A^*_+, B^*, C^*_-)$$
$$= \langle p_1^{-1} q p_2 \rangle * F(A^*_+, B^*, C^*_-).$$ □

These propositions indicate that one of three basic situations will occur:

(1) $F(A^*, B^*) \cap F(B^*, C^*)$ is exceptional;

(2) either $F(A^*, B^*) \cap F(A_+^*, B^*, C^*)$ or $F(A^*, B^*, C_-^*) \cap F(B^*, C^*)$ is exceptional;
(3) both $F(A^*, B^*) \cap F(A_+^*, B^*, C^*)$ and $F(A^*, B^*, C_-^*) \cap F(B^*, C^*)$ are exceptional.

We look particularly at the situation when (2) or (3) occurs. Before doing so, we step back briefly from the specific situation under consideration to introduce some further terminology. Suppose that we are given two Magnus subgroups $F(W, Y)$ and $F(Y, Z)$ of a one-relator group - assuming, as the notation implies, that W, Y, Z are disjoint. If $F(W, Y) \cap F(Y, Z) = \langle u \rangle * F(Y) = \langle v \rangle * F(Y)$, with $u = v$, there is not a unique choice for the generator of the exceptional infinite cyclic factor. However there are, up to inversion, two *canonical* choices defined by requiring that either u is chosen W-special or that v is chosen Z-special. We then have two equalities, say $\bar{u} = \bar{v}$ where $\bar{u}$ is W-special and $\hat{u} = \hat{v}$ where $\hat{v}$ is Z-special. When this happens then $\bar{u}$ is the W-core of $\hat{u}$ and $\hat{v}$ is the Z-core of $\bar{v}$.

We firstly examine Proposition 5.3 to see what the canonical choices are for the various intersections under consideration. When we have $F(A^*, B^*) \cap F(A_+^*, B^*, C^*) = \langle u \rangle * F(A_+^*, B^*) = \langle v \rangle * F(A_+^*, B^*)$, where $u = v$, then one canonical generator is a_κ-special and the other is C^*-special. For Proposition 5.3, our hypotheses stipulate that the given v is the C^*-special canonical generator. The other canonical generator will be a_κ-special and hence is the a_κ-core of u. We therefore write $u \equiv u_1 s u_2$ where $u_1, u_2 \in F(A_+^*, B^*)$ and s is the a_κ-core of u.

If we now turn to the intersection $F(A^*, B^*, C_-^*) \cap F(A_+^*, B^*, C^*)$, Proposition 5.3 tells us that s is the canonical a_κ-special generator of this intersection. The companion canonical generator is (represented by) a c_ν-special element of $F(A_+^*, B^*, C^*)$ which must be the c_ν-core of $t \equiv u_1^{-1} v u_2^{-1}$ - let us denote it by $\bar{t}$ and write $t \equiv t_1 \bar{t} t_2$. For the companion intersection $F(A^*, B^*, C_+^*) \cap F(A_-^*, B^*, C^*)$ we have the exceptional equality $v_1 u v_2^{-1} = v_0$. We therefore obtain a canonical c_μ-special generator as the c_μ-core τ of v_0 - if we write $v_0 \equiv v_{01} \tau v_{02}$ then we have an exceptional equality $\sigma = \tau$ where $\sigma \equiv v_{01}^{-1} v_1 u v_2^{-1} v_{02}^{-1}$ and the a_λ-core of σ is the companion a_λ-special canonical generator.

Our final result Proposition 5.5 in this section draws together all the threads of the analysis we have undertaken by determining what happens when both Proposition 5.3 and Proposition 5.4 can be applied. We shall see how this means that all the intersections that are discussed in these two propositions are exceptional and that there is a single 'common exceptional' equality which can expressed in the way appropriate to each intersection.

Proposition 5.5. *Let*

$$F(A^*, B^*) \cap F(A^*_+, B^*, C^*) = \langle u \rangle * F(A^*_+, B^*)$$
$$= \langle v_1^{-1} v_0 v_2 \rangle * F(A^*_+, B^*)$$

be an exceptional intersection, where $u, v_1^{-1} v_0 v_2$ *are as in 5.3 and*

$$F(A^*, B^*, C^*_-) \cap F(B^*, C^*) = \langle p_1 p_0 p_2^{-1} \rangle * F(B^*, C^*_-)$$
$$= \langle q \rangle * F(B^*, C^*_-)$$

where $p_1 p_0 p_2^{-1}, q$ *are as in 5.4. Then, in the notation introduced above,*

(a) *up to inversion, the* a_κ*-cores of* u *and* $p \equiv p_1 p_0 p_2^{-1}$ *coincide;*
(b) *up to inversion, the* c_μ*-cores of* v_0 *and* $p_1^{-1} q p_2$ *coincide;*
(c) *up to inversion, the* a_λ*-cores of* p_0, σ *and* $v_1 u v_2^{-1}$ *coincide;*
(d) *up to inversion, the* c_ν*-cores of* q, t *and* $v \equiv v_1^{-1} v_0 v_2$ *coincide.*

In particular

(a') *the* a_κ*-core of* $p_1 p_0 p_2^{-1}$ *lies in* $F(A^*, B^*)$*;*
(b') *the* c_μ*-core of* $p_1^{-1} q p_2$ *lies in* $F(B^*, C^*)$*;*
(c') *the* a_λ*-core of* $v_1 u v_2^{-1}$ *lies in* $F(A^*, B^*)$*;*
(d') *the* c_ν*-core of* $v \equiv v_1^{-1} v_0 v_2$ *lies in* $F(B^*, C^*)$.

Proof. To obtain (a) and (d), we simply have to observe that it follows from Propositions 5.3 and 5.4, and the subsequent discussion, that the equalities $u = v$, $s = t$ and $p = q$ can all be used to define the two canonical generators of $F(A^*, B^*, C^*_-) \cap F(A^*_+, B^*, C^*)$. Then, for example, if $\overline{p}$ is the a_κ-core of p, we deduce that $s \in \langle \overline{p} \rangle * F(A^*_+, B^*)$ and hence that $\overline{p} \in F(A^*, B^*)$, which yields $\overline{p} = s$. A similar obervation applies to (b) and (c). □

We need one further remark on notation. When we encounter $F(A^*, B^*) \cap F(A^*_+, B^*, C^*)$ as an exceptional intersection for which the hypotheses of 5.3 are satisfied, we shall have drawn this inference from a sequence of Normal Form equalities for one or more equalities which define elements of $F(A, B) \cap F(B, C)$. To maintain consistency with the notation for the Normal Form equalities, we have to make a choice to distinguish the canonical C^*-special generator from its inverse and therefore fix the notation $u = v_1^{-1} v_0 v_2$ relative to the notation for the Normal Form equalities. If, instead, we first encounter $F(A^*, B^*, C^*_-) \cap F(B^*, C^*)$, rather than $F(A^*, B^*) \cap F(A^*_+, B^*, C^*)$, and Proposition 5.4 applies, then we make a choice between inverses to define the notation $p_1 p_0 p_2^{-1} = q$. In many situations, however, we shall in fact encounter both $F(A^*, B^*, C^*_-) \cap F(B^*, C^*)$ and $F(A^*, B^*) \cap F(A^*_+, B^*, C^*)$

simultaneously, where both Proposition 5.3 and Proposition 5.4 apply and hence, by Proposition 5.5 the a_κ-core of u and the a_κ-core of $p_1 p_0 p_2^{-1}$ coincide up to inversion. In such circumstances it is useful to choose our notation so that they in fact coincide, although this does mean that if we specify $u = v_1^{-1} v_0 v_2$ from a sequence of Normal Form equalities, we no longer have a free choice for $p_1 p_0 p_2^{-1} = q$. Finally, before moving on to the next section we need one more definition which is conveniently located here. When we have a word of, for example $F(A_+^*, B^*, C^*)$, we then have a well-defined notion of *syllable length*, namely the standard length function associated to the representation of $F(A_+^*, B^*, C^*)$ as the amalgamated free product $F(A_+^*, B^*) *_{F(B^*)} F(B^*, C^*)$. The syllable length of a freely reduced word of $F(A_+^*, B^*, C^*)$ is also the number of *syllables*, that is, distinct subwords that

(i) lie in either $F(A_+^*, B^*)$ or $F(B^*, C^*)$;
(ii) do not lie in $F(B^*)$;
(iii) are maximal with respect to (i) and (ii).

Using this terminology, we shall refer to (a') – (d') of Proposition 5.5 as the *single syllable* criterion.

6. The equality $wh = kz$ in $F(A^*, B^*, C_-^*) \cap F(A_+^*, B^*, C^*)$

In the previous section we have examined the way in which the existence of 'small' exceptional intersections, i.e., those in which one of the two factors is either $F(A^*, B^*)$ or $F(B^*, C^*)$, determines the nature of 'large' intersections such as $F(A^*, B^*, C_-^*) \cap F(A_+^*, B^*, C^*)$. Small intersections arise from initial and final Normal Form equalities, so that the results of the previous section apply – they also arise from equalities where there is a change of sign in the occurrences of b and we shall need to use this. To exploit our Normal Form equalities to the full, however, it is also necessary to undertake an analysis of those equalities that are derived from successive occurences of b with the same sign. A standardised Normal Form equality that defines an exceptional element $F(A^*, B^*, C_-^*) \cap F(A_+^*, B^*, C^*)$ takes the form $wh = kz$ where w is an element of $F(A_-^*, B^*, C_-^*)$ of type $(A^* : C^*)$, z is an element of $F(A_+^*, B^*, C_+^*)$ of type $(A^* : C^*)$, $h \in F(A^*, B^*)$ and $k \in F(B^*, C^*)$. In the equalities we now examine, w and z will be assumed non-trivial (otherwise our equality defines a 'small' intersection) but we must explicitly allow h and k to be trivial.

We begin by returning to our analysis in Section 5, assuming that

$$F(A^*, B^*) \cap F(A_+^*, B^*, C^*) = \langle u \rangle * F(A_+^*, B^*) = \langle v \rangle * F(A_+^*, B^*),$$

where $u = v_1^{-1}v_0v_2$, so that in turn

$$F(A^*, B^*, C^*_-) \cap F(A^*_+, B^*, C^*) = \langle s \rangle * F(A^*_+, B^*, C^*_-)$$
$$= \langle t \rangle * F(A^*_+, B^*, C^*_-)$$

with $s \in F(A^*, B^*)$ a_κ-special, $t = t_1\bar{t}t_2$ where $\bar{t}$ is the c_ν-core of t and $s = t$ in G^*. There is a parallel situation, which we regard as being dealt with tacitly, if we know that

$$F(A^*, B^*, C^*_-) \cap F(B^*, C^*) = \langle p \rangle * F(B^*, C^*_-) = \langle q \rangle * F(B^*, C^*_-).$$

Now any exceptional element of

$$F(A^*, B^*, C^*_-) \cap F(A^*_+, B^*, C^*)$$

can be written, as a reduced word in $F(A^*, B^*, C^*_-)$, in the form $\alpha_0 s^{i_1}\alpha_1 \dots s^{i_e}\alpha_e$ where the words $\alpha_0, \alpha_1, \dots, \alpha_e \in F(A^*_+, B^*, C^*_-)$ and a power s^i denotes the reduced form obtained when s is not cyclically reduced. To express this as a reduced word in $F(A^*_+, B^*, C^*)$ one must substitute t for s in the given expression and then freely reduce the result which will, necessarily, be an expression of the form $\beta_0\bar{t}^{j_1}\beta_1 \dots \bar{t}^{i_e}\beta_e$, where $\beta_0, \beta_1 \dots, \beta_f \in F(A^*_+, B^*, C^*_-)$, since it is clear that any element of the intersection $F(A^*, B^*, C^*_-) \cap F(A^*_+, B^*, C^*)$ has such a representation as a word of $F(A^*_+, B^*, C^*)$. This means that any equality between a word $\tilde{g}$ of $F(A^*, B^*, C^*_-)$ and a word $\hat{g}$ of $F(A^*_+, B^*, C^*)$ which defines an element of their intersection is a consequence in the free group $F(A^*, B^*, C^*)$ of the relation $s = t$, that is $\tilde{g}^{-1}\hat{g}$ actually lies in the normal closure in $F(A^*, B^*, C^*)$ of $s^{-1}t$ (and not merely in the normal closure of the defining relation r^* of G^*.) It follows therefore that if we wish to analyse how an equality of the form $wh = kz$ can be obtained we can treat $s^{-1}t$ as if it were the defining relation of G^* and bring to bear the following result which is a strong form of Theorem 2.5. Our aim is to show, in all the cases we need consider, that the equality either holds freely, which means that $h = k = 1$ and $w \equiv z$ as words of $F(A^*_+, B^*, C^*_-)$, or that the possible forms of $wh = kz$ are restricted to simple explicit cases.

Proposition 6.1. [28] *Let $G = \langle X : r = 1 \rangle$, where r is cyclically reduced, be a one-relator group and let ω be a cyclically reduced consequence of r.*

(a) *If $r \equiv s^m$ where $m \geq 1$ and s is not a proper power, then either ω is simply a cyclic conjugate of $r^{\pm 1}$ or ω contains, cyclically, two disjoint subwords $p^{m-1}p_1$ and $q^{m-1}q_1$ where p and q are cyclic permutations of s or s^{-1} and p_1, q_1 are initial segments of p and q, respectively, each containing all the generating letters that appear in r.*

(b) *In particular if* $m = 1$ *so that* r *is not a proper power, then either* ω *is simply a cyclic conjugate of* $r^{\pm 1}$ *or* ω *contains, cyclically, two disjoint subwords which are, cyclically, subwords of* r *or* r^{-1}, *each containing all the generating letters that appear in* r.

We extend our definition of *Gurevich* subword to cover the case when r is not a proper power – (b) then asserts that ω contains two disjoint Gurevich subwords. We now apply Proposition 6.1 to the equality $wh = kz$. We need to do this only under additional assumptions which we set out below. There is a range of possibilities for $wh = kz$, depending on whether or not h or k is non-trivial and we examine these to see how, if at all, the cyclically reduced form of $h^{-1}w^{-1}kz$ can contain Gurevich subwords. We shall concentrate on what we shall refer to as *extremal Gurevich* subwords, that is words which are, cyclically, subwords of both $h^{-1}w^{-1}kz$ and $(u^{-1}v_1^{-1}v_0v_2)^{\pm 1}$ which contain occurrences of all four extremal generators $a_\kappa, a_\lambda, c_\mu, c_\nu$ – clearly any Gurevich subword must contain an extremal Gurevich subword. The additional assumptions we make are:

(i) Both $F(A^*, B^*) \cap F(A_+^*, B^*, C^*)$ and $F(A^*, B^*, C_-^*) \cap F(B^*, C^*)$ are exceptional with the common basic relator expressed as $u^{-1}v_1^{-1}v_0v_2$ or $p_2p_0^{-1}p_1^{-1}q$ as appropriate;
(ii) One of v_1, v_2 is intermediate (possibly trivial) but the other is not – and hence is non-trivial.

Suppose, for example, that v_1 is intermediate. Then, using the single syllable criterion, we see that the following possible forms for $u^{-1}v_1^{-1}v_0v_2$ occur:

(1) $u(a_\kappa, a_\lambda)^{-1}v_1^{-1}v_0(c_\mu)v_{21}q(c_\nu)v_{22}$;
(2) $u(a_\kappa)^{-1}v_1^{-1}v_0(c_\mu, c_\nu)v_{21}p_0(a_\lambda)^{-1}v_{22}$
(3) $u(a_\kappa)^{-1}v_1^{-1}v_0(c_\mu)v_{21}p_0(a_\lambda)^{-1}v_{22}q(c_\nu)v_{23}$;
(4) $u(a_\kappa)^{-1}v_1^{-1}v_0(c_\mu)v_{21}p_0(a_\lambda)^{-1}v_{22}q(c_\nu)v_{23}$.

In the above, the notation $u(a_\kappa, a_\lambda)$ indicates that u contains both a_κ and a_λ and the notation $u(a_\kappa)$ indicates that u contains a_κ but not a_λ, and so on. Also v_{21}, v_{22}, v_{23} are necessarily intermediate. By expressing the relator in the form $p_2p_0^{-1}p_1^{-1}q$, it follows routinely in each case that one of p_1, p_2 is intermediate (possibly trivial) and the other is not. In case (1), for example, $u(a_\kappa, a_\lambda)^{-1}v_1^{-1}v_0(c_\mu)v_{21}q(c_\nu)v_{22}$ becomes $v_{22}u(a_\kappa, a_\lambda)^{-1}v_1^{-1}v_0$ $(c_\mu)v_{21}q(c_\nu)$ with $p_0 \equiv u(a_\kappa, a_\lambda)$ so that $p_2 \equiv v_{22}$ is intermediate whereas $p_1^{-1} \equiv v_1^{-1}v_0(c_\mu)v_{21}$ is not. In this setting we define $d(a_\kappa, c_\mu)$ to be the syllable length of whichever of v_1, v_2 is intermediate and $d(a_\lambda, c_\nu)$ to be the syllable length of whichever of p_1, p_2 is intermediate.

Lemma 6.1. *Assume the hypotheses and notation described above.*

(a) *Let $h^{-1}w^{-1}kz$ be a cyclically reduced word, where $h \in F(A^*, B^*), k \in F(B^*, C^*)$ are non-trivial, $w \in L$, $z \in U$ are non-trivial and are of type $(A^* : C^*)$. If $L(w) \leq d(a_\kappa, c_\mu)$ or $L(z) \leq d(a_\lambda, c_\nu)$, then $h^{-1}w^{-1}kz$ does not contain two disjoint Gurevich subwords.*

(b) *Let $h^{-1}w^{-1}h'z$ and $k\prime^{-1}w^{-1}kz$ be cyclically reduced words, where $h, h' \in F(A^*, B^*)$ and $k, k' \in F(B^*, C^*)$ are all non-trivial, $w \in L$, $z \in U$ are non-trivial and are of type $(C^* : C^*)$ in $h^{-1}w^{-1}h'z$ and type $(A^* : A^*)$ in $k\prime^{-1}w^{-1}kz$. If $L(w) \leq d(a_\kappa, c_\mu)$ or $L(z) \leq d(a_\lambda, c_\nu)$, then neither $h^{-1}w^{-1}h'z$ nor $k\prime^{-1}w^{-1}kz$ contains two disjoint Gurevich subwords.*

Proof. Without loss of generality we can assume v_1 is intermediate and v_2 is not. Then, by inspection, $d(a_\lambda, c_\nu)$ is either $L(v_{21})$ or $L(v_{22})$ and hence $d(a_\lambda, c_\nu) < L(v_2)$.

To prove (a), we suppose we have two disjoint Gurevich subwords of $h^{-1}w^{-1}kz$; then there are two disjoint extremal Gurevich subwords. Now neither extremal Gurevich subword can be a subword of any of $h^{-1}w^{-1}$, $w^{-1}k$, kz, zh^{-1} - for each of these omits an essential generator. Moreover, neither extremal Gurevich subword can contain any of $h^{-1}w^{-1}$, $w^{-1}k$, kz, zh^{-1} for then one of $h^{-1}w^{-1}$, $w^{-1}k$, kz, zh^{-1} would contain its companion extremal Gurevich subword. It follows, therefore that an extremal Gurevich subword must take one of the four forms

$$h_1^{-1}w^{-1}k_1,\ w_1^{-1}kz_1,\ k_2zh_2^{-1},\ z_2h^{-1}w_2^{-1},$$

where w_1, w_2 denote proper, non-trivial, initial and terminal segments of w and similarly for h, k and z, and that a pair must be either $\{h_1^{-1}w^{-1}k_1, k_2zh_2^{-1}\}$ or $\{w_1^{-1}kz_1, z_2h^{-1}w_2^{-1}\}$. We consider a pair of the first type. Suppose that $L(w) \leq d(a_\kappa, c_\mu) = L(v_1)$; notice that then v_1 is non-trivial. In $h_1^{-1}w^{-1}k_1$ we can position a_κ, which must come from $u^{\pm 1}$, in h_1^{-1} or w^{-1} and c_μ, which must come from $v_0^{\pm 1}$, in w^{-1} or k_1. Between such occurrences we have to position either $v_1^{\pm 1}$ or $v_2^{\pm 1}$. Clearly it cannot be the latter since this is not intermediate and we deduce, using the length condition, that $w \equiv v_1^{\pm 1}$. But then both a_κ and a_λ would have to lie in h_1^{-1}, and thus in u, and both c_μ and c_ν would have to lie in k_1 and hence in v_0. This implies v_2 is intermediate, which is a contradiction. Alternatively, if $L(z) \leq d(a_\lambda, c_\nu)$, then in $k_2zh_2^{-1}$ we have to position a_λ from $p_0^{\pm 1}$ in h_2^{-1} or z and c_μ from $q^{\pm 1}$ in k_2 or z. Between these we have to position whichever of $p_1^{\pm 1}$ or $p_2^{\pm 1}$ is intermediate. However the length condition implies whichever it is constitutes the whole of z and in the same way as the previous case we deduce that both p_1 and p_2 are intermediate which is contradictory. Now we consider the second possible pair. Suppose that $L(w) \leq d(a_\kappa, c_\mu)$; in

$w_1^{-1}kz_1$ we have to place a_κ in w_1^{-1} and c_μ in w_1^{-1} or k. Again we have to fit either $v_1^{\pm 1}$ or $v_2^{\pm 1}$ between these. However the length condition rules out the former and the fact that it is not intermediate rules out the latter. So suppose that $L(z) \leq d(a_\lambda, c_\nu)$; in $z_2 h^{-1} w_2^{-1}$ we can place c_ν only in z_2 and a_λ in either h^{-1} or z_2. However the length condition means that we cannot place whichever of p_1,p_2 is intermediate between these; but the other is not intermediate and so this too is ruled out.

(b) It suffices to consider $h^{-1}w^{-1}h'z$. Inspection shows that the possible pairs must be of the form $\{h_1^{-1}w^{-1}h_1', h_2' z h_2^{-1}\}$ and $\{w_1^{-1}h'z_1, z_2 h^{-1} w_2^{-1}\}$. Suppose that $L(w) \leq d(a_\kappa, c_\mu) = L(v_1)$. In $h_1^{-1}w^{-1}h_1'$ we have to position c_μ in w^{-1}. While it appears that we can place a_κ anywhere, all the possibilities are promptly ruled out by the length condition and the fact that v_2 is not intermediate. If $L(z) \leq d(a_\lambda, c_\nu)$, then we can apply a similar argument to $h_2' z h_2^{-1}$ and so the first pair is ruled out. When we turn to the second pair, we see that in $w_1^{-1}h'z_1$ we can position c_μ only in w_1^{-1} and a_κ in $w_1^{-1}h'$. Similarly we can place c_ν only in z_1 and a_λ in h' or z_1. Using the appropriate length condition in combination with the non-intermediacy of the appropriate elements of $\{v_1, v_2, p_1, p_2\}$ yields the necessary contradiction. □

Proposition 6.2. . *Assume the hypotheses and notation prior to Lemma* 6.1. *Let the equality $wh = kz$, where $w \in L$ and $z \in U$ are both non-trivial of type $(A^* : C^*)$ and $h \in F(A^*, B^*)$, $k \in F(B^*, C^*)$, define an element of $F(A^*, B^*, C^*_-) \cap F(A^*_+, B^*, C^*)$. If $L(w) \leq d(a_\kappa, c_\mu)$ or $L(z) \leq d(a_\lambda, c_\nu)$, then one of the following holds:*

(a) *the element defined by $wh = hz$ is non-exceptional and the equality holds freely – in particular, $h = k = 1$ and $w \equiv z$ is intermediate.*
(b) *$h^{-1}w^{-1}kz$ is a cycle of $(u^{-1}v_1^{-1}v_0 v_2)^{\pm 1}$. In particular one of the following two possibilities occurs:*
   (i) *$L(w) \leq d(a_\kappa, c_\mu)$ and either v_1 is intermediate so that $d(a_\kappa, c_\mu) = L(v_1)$, in which case $wh = kz$ is $v_1 u = v_0 v_2$ (that is $w \equiv v_1, z \equiv v_2, h = u, k = v_0$), or v_2 is intermediate so that $d(a_\kappa, c_\mu) = L(v_2)$ and $wh = kz$ is $v_2 u^{-1} = v_0^{-1} v_1$.*
   (ii) *$L(z) \leq d(a_\lambda, c_\nu)$ and either p_1 is intermediate so that $d(a_\lambda, c_\nu) = L(p_1)$, in which case $wh = kz$ is $p_2 p_0^{-1} = q^{-1} p_1$, or p_2 is intermediate so that $d(a_\lambda, c_\nu) = L(p_2)$ and $wh = kz$ is $p_1 p_0 = q p_2$.*

Proof. Suppose the equality does not hold freely. The all four extremal generators must appear and, since a_κ can appear only in wh and c_ν only in kz, the equality must be exceptional for $F(A^*, B^*, C^*_-) \cap F(A^*_+, B^*, C^*)$.

(i) Suppose, firstly, that $h, k \neq 1$ so that $h^{-1}w^{-1}kz$ is cyclically reduced. Then by Lemma 6.1 $h^{-1}w^{-1}kz$ is a cycle of $(u^{-1}v_1^{-1}v_0v_2)^{\pm 1}$. Suppose, for example, that v_1 is intermediate and that $L(w) \leq L(v_1) = d(a_\kappa, c_\mu)$. We again look at the placing of a_κ and c_ν. The former can occur within h^{-1} or w^{-1} and the latter within w^{-1} or k and the intervening subword between the two syllables where a_κ and c_ν appear is either the whole of w^{-1} or a proper subword of w^{-1}. Now this intervening subword must be $v_1^{\pm 1}$ or $v_2^{\pm 1}$. The fact that v_2^{-1} is not intermediate rules it out and the only possibility consistent with the length inequality is that the intervening subword is the whole of w^{-1} and that it is $v_1^{\pm 1}$. It follows that $h^{-1}w^{-1}k \equiv u^{-1}v_1^{-1}v_0$ and hence that $wh = kz$ is $v_1u = v_0v_2$. A similar argument gives the desired conclusion if v_2 is intermediate. If on the other hand $L(z) \leq d(a_\lambda, c_\nu)$, then a similar argument – or a simple appeal to upper and lower symmetry – provides what is required.

(ii) Suppose that $h = 1$ and $k \neq 1$; again the equality cannot hold freely and we shall show that it cannot in fact occur. We have to examine the cyclically reduced form of kzw^{-1}. This is obtained by cancelling a common terminal segment of w and z that must be intermediate (in the sense that it contains no extremal generators). In particular, the occurrences of a_κ and a_λ, which necessarily appear in w and z respectively, will not be cancelled. Then, depending on the exact nature of the common terminal segment cancelled, the resulting cyclically reduced word will be either of the form $kz'h\prime^{-1}w\prime^{-1}$ with w', z' also both of type $(A^* : C^*)$, or $kz'k\prime^{-1}w\prime^{-1}$, with w', z' both of type $(A^* : A^*)$, and h', respectively k', non-trivial. Suppose that we get $kz'h\prime^{-1}w\prime^{-1}$; since $k, h' \neq 1$ this is cyclically reduced. Now $L(w') < L(w)$, since the final syllable of w must have been completely cancelled, so that if $L(w) \leq d(a_\kappa, c_\mu)$ then also $L(w') \leq d(a_\kappa, c_\mu)$. By repeating the argument for Case (i), we deduce that either v_1 is intermediate and $w' \equiv v_1$ or v_2 is intermediate and $w' \equiv v_2$. But now we have, for example,

$$L(v_1) = L(w') < L(w) = d(a_\kappa, c_\mu) = L(v_1)$$

which is impossible. Similar arguments apply when v_2 is intermediate and when $L(z) \leq d(a_\lambda, c_\nu)$. Suppose, alternatively, we obtain $kz'k\prime^{-1}w\prime^{-1}$. By way of illustration, assume that v_1 is intermediate and $L(w) \leq d(a_\kappa, c_\mu) = L(v_1)$. Since a_κ can occur only in $w\prime^{-1}$ and c_μ only in k or $k\prime^{-1}$, we are forced, by the kind of arguments previously used, to the conclusion that v_1 is a proper subword of $w\prime^{-1}$. However, we then have the contradictory inequalities $L(v_1) < L(w') \leq L(v_1)$. A similar argument works in the other possibilities. We are left to deal with the case when $h \neq 1$ and $k = 1$ and the case when $h = k = 1$. These are dealt with similarly. In the former we may have cancellation between intermediate initial segments of w and z and in the latter cancellation of both initial and terminal intermediate segments. However, the same basic approach always leads to a contradiction. □

7. The Intersection Theorem

Here we prove our main result.

Theorem 7.1. *Let $G = \langle X : r = 1\rangle$, where r is cyclically reduced, be a one-relator group and $F(S)$, $F(T)$ Magnus subgroups of G. Then either $F(S) \cap F(T) = F(S \cap T)$ or there are words $u = u(S)$ and $v = v(T)$ such that*

$$F(S) \cap F(T) = \langle u\rangle * F(S \cap T) = \langle v\rangle * F(S \cap T),$$

where we may choose $u = v$ in G. In particular,

$$\text{rank}(F(S \cap T)) \leq \text{rank}(F(S) \cap F(T)) \leq \text{rank}(F(S \cap T)) + 1.$$

The argument will proceed by induction on the length $|r|$ of the relator r, noting that for small values of $|r|$, the result is elementary by inspection. By use of the normal form theorem for free products, the general case follows from the case when $S \cup T = \text{Supp}(r)$ and so we can always assume the latter. Also when $|\text{Supp}(r)| = 2$, the conclusion is immediate, so that we can also assume that $|\text{Supp}(r)| \geq 3$. As already noted we write $B = S \cap T$ and then can choose A and C disjoint so that $S = A \cup B$ and $T = B \cup C$. We make one further mild simplification before proceeding – it is easy to check that the general case reduces to the case when A and C are singletons, say $A = \{a\}$ and $C = \{c\}$. To start the proof proper, let us suppose, then, that we have an equality $h(A, B) = k(B, C)$ defining an exceptional element of the intersection $F(A, B) \cap F(B, C)$. Clearly all elements of $X = \text{Supp}(r)$ must appear in h or k since otherwise the equality holds freely which is obviously contradictory.

Case Assumption 7.1. *Either a or c has exponent sum zero in r.*

Without loss of generality we may assume it is the former. We may further assume, by replacing r by a cyclic permutation if necessary, that $c^{\pm 1}$ is the initial letter of r. In the standard manner we can express G as an HNN-extension of the form $G = \langle G^*, a \mid aLa^{-1} = U\rangle$ where L and U are Magnus subgroups of the vertex group G^*. To do this we define $C^* = \{c_\mu, \ldots, c_\nu\}$ and $B^* = \{b_i, i \in \mathbb{Z}, b \in B\}$ where, as usual, b_i and c_i denote the conjugates $a^i ba^{-i}$ and $a^i ca^{-i}$ with μ and ν respectively the mimimal and maximal subscripts attached to c that appear when we rewrite r as a word r^* in $B^* \cup C^*$. With this notation $G^* = \langle B^*, C^* \mid r^* = 1\rangle$ and the two edge groups are $L = F(B^*, C^*_-)$,

$U = F(B^*, C_+^*)$, in the notation carried over from Section 5. We note that by requiring that r begins with $c^{\pm 1}$ we have ensured that $\mu \leq 0 \leq \nu$. Now we can transform our equality $h(A, B) = k(B, C)$ into one expressed in the generators of G as HNN-extension. Since k omits a, it follows that h has zero exponent sum in a and thus we obtain an equality $h^*(B^*) = k^*(B_0^*, c_0)$ where $B_0^* = \{b_0, b \in B\}$ (and k^* is literally the same word expression as k). This equality defines an element of the intersection of $F(B^*) \cap F(B_0^*, c_0)$ and since c_0 is explicitly involved, the element defined is exceptional. By the induction hypothesis, we obtain words $u^*(B^*)$ and $v^*(B_0, c_0)$ such that

$$F(B^*) \cap F(B_0^*, c_0) = \langle u^* \rangle * F(B_0^*) = \langle v^* \rangle * F(B_0^*).$$

Since our original equality defined an arbitrary exceptional element of $F(A, B) \cap F(B, C)$, clearly $F(A, B) \cap F(B, C) = F(B^*) \cap F(B_0^*, c_0)$ and we have completed the proof for Case 7.1 when we obtain the desired u and v by rewriting u^* and v^* in terms of the original generators.

Case Assumption 7.2. *There exists some generator $b \in B$ with exponent sum zero in r. (Of course we are also assuming that Case* 7.1 *does not apply).*

The simplicity of the argument in Case 7.1 is deceptive; the case to hand is distinctly harder to deal with as should already be apparent from our analysis in Section 5. However before dealing with it, we dispose of the remaining case when there are no generators that have exponent sum zero in r. This is achieved by reducing it to Case 7.2. Fortunately, the argument is as easy as that for Case 7.1.

So we are in a situation where no generator has exponent sum zero in r. Pick some generator $b \in B$ and $a \in A$ (or $c \in C$). Let a and b have exponent sums α and β in r. In the usual way, form the amalgamated free product $\hat{G} = \langle G * F(x) \mid b = y^{\alpha} \rangle$, where y is a new generator. Introduce a further new generator x and eliminate a via $a = xy^{-\beta}$. (The reader accustomed to this trick should note that the notation has been switched (deliberately) from the usual version in which one sets $a = x^{\beta}$ and then puts $b = yx^{-\alpha}$.) Now $\hat{G}$ is defined by a single relator $\hat{r}$ which has exponent sum zero in y – there is a contribution of $\alpha\beta$ derived from occurrences of b in r which is cancelled by a contribution of $-\alpha\beta$ from occurrences of a – and G is naturally embedded in $\hat{G}$. Moreover – and this explains our particular procedure – if $\hat{A}$ and $\hat{B}$ denote the sets of generators obtained by replacing a and b by x and y in A and B, then $F(A, B)$ is embedded in $F(\hat{A}, \hat{B})$ and $F(B, C)$ is embedded in $F(\hat{B}, C)$ The point of the trick is that the one-relator group $\hat{G}$ can be expressed as an HNN-extension in the usual way over a vertex group which has a relator that is of shorter length than the original relator r of G. Assuming that we have inductively established our theorem for both

Case 7.1 and Case 7.2, then we know it holds for $\hat{G}$. The exceptional intersection $F(A, B) \cap F(B, C)$ is embedded in the intersection $F(\hat{A}, \hat{B}) \cap F(\hat{B}, C)$, which is then necessarily exceptional and therefore expressible as $\langle \hat{v} \rangle * F(\hat{B})$ where $\hat{v}$ is a word of $F(\hat{B}, C)$ that can be assumed to be C-special. So if $h = k$ is an exceptional equality for $F(A, B) \cap F(B, C)$ it follows that k, expressed as a word in $\{y^\beta, B'\}$, is equal to a reduced word of the form $d_0 \hat{v}^{i_1} d_1 \ldots \hat{v}^{i_e} d_e$ where the $d_j \in F(\hat{B})$. It follows that all occurrences of y in this expression occur in powers of y^β and hence that, in $F(B, C)$, $k \in \langle v \rangle * F(B, C)$ where v is the word obtained from $\hat{v}$ by replacing each occurrence of y^β by b. This completes the reduction and we can now begin to deal with Case 7.2. We adopt all the notation set out in Sections 5 – 6.

Case Assumption 7.3. *G^* contains an exceptional element of $F(A, B) \cap F(B, C)$.*

It follows from Lemma 5.1 that $G^* \cap F(A, B) = F(A^*, B^*)$, $G^* \cap F(B, C) = F(B^*, C^*)$ and so an exceptional element of $F(A, B) \cap F(B, C)$ that lies in G^* is an exceptional element of $F(A^*, B^*) \cap F(B^*, C^*)$. By the induction hypothesis, there exist elements $u^* \in F(A^*, B^*)$ and $v^* \in F(B^*, C^*)$ such that

$$F(A^*, B^*) \cap F(B^*, C^*) = \langle u^* \rangle * F(B^*) = \langle v^* \rangle * F(B^*)$$

with $u^* = v^*$ in G^*.

Claim 7.1. *$F(A, B) \cap F(B, C) = \langle u \rangle * F(B) = \langle v \rangle * F(B)$ where u and v are simply u^* and v^* rewritten as words in $F(A, B)$ and $F(B, C)$.*

Proof. It suffices to show that an arbitrary exceptional element of $F(A, B) \cap F(B, C)$ lies in $\langle u \rangle * F(B) = \langle v \rangle * F(B)$. We consider such an element

$$h_0 b^{\varepsilon_1} h_1 \ldots b^{\varepsilon_n} h_n = k_0 b^{\varepsilon_1} k_1 \ldots b^{\varepsilon_n} k_n$$

of $F(A, B) \cap F(B, C)$, which we can take to be in standardised form as in Example 5.1, and argue by induction on n. If $n = 0$ we have just observed that our given element lies in $\langle u^* \rangle * F(B^*) = \langle v^* \rangle * F(B^*)$ and hence in $\langle u \rangle * F(B) = \langle v \rangle * F(B)$. So we may suppose $n > 0$. By 'upper and lower' symmetry we can assume $\varepsilon_1 = 1$. Thus we obtain an equality $h_0 = k_0 z_0$, where $z_0 \in U = F(A^*_+, B^*, C^*_+)$ and is of type $(A^* : C^*)$, which clearly defines an element of $F(A^*, B^*) \cap F(A^*_+, B^*, C^*)$. Now $h_0 \in U$ if and only if $k_0 \in U$ and if both lie in U, then $h \in F(A^*_+, B^*)$ and $k \in F(B^*, C^*_+)$. In this event, we can pass b from right to left across both and cancel; the resulting equality still defines an exceptional element and so by the induction on n, lies in $\langle u^* \rangle * F(B) = \langle v^* \rangle * F(B)$ whence of course the same is true of our original element. We can thus reduce to the case when $h_0 \notin F(A^*_+, B^*)$, $k_0 \notin F(B^*, C^*_+)$ and

$h_0 = k_0 z_0$ defines an exceptional element of $F(A^*, B^*) \cap F(A^*_+, B^*, C^*)$. By Proposition 5.2,

$$F(A^*, B^*) \cap F(A^*_+, B^*, C^*) = \langle u^* \rangle * F(A^*_+, B^*) = \langle v^* \rangle * F(A^*_+, B^*)$$

and hence $k_0 z_0 \in \langle v^* \rangle * F(A^*_+, B^*)$. Since v^* does not contain any occurrences of elements of A^*_+, it follows from Proposition 5.2 that z_0 is trivial and hence $h_0 = k_0$. Now $h_0 \notin F(B)$, since h has zero exponent sum in b and $h \notin F(A^*, B^*)$, and similarly $k_0 \notin F(B)$. Therefore $h_0 = k_0$ defines an exceptional element of $F(A, B) \cap F(B, C,)$ and, by induction on n, lies in $\langle u \rangle * F(B) = \langle v \rangle * F(B)$. Cancelling $h_0 b = k_0 b$ yields

$$h_1 b^{\varepsilon_2} \ldots b^{\varepsilon_n} h_n = k_1 b^{\varepsilon_2} \ldots b^{\varepsilon_n} k_n.$$

Either this defines an element of $F(B)$, in which case there is nothing further to prove, or we have an exceptional element of $F(A, B) \cap F(B, C)$. Again we can appeal to induction on n and the proof is complete in Case 7.3. □

Case Assumption 7.4. *There are no exceptional elements of $F(A, B) \cap F(B, C)$ in G^*.*

We consider equalities of the form

$$h_0 b^{\varepsilon_1} h_1 \ldots b^{\varepsilon_m} h_m = k_0 b^{\varepsilon_1} k_1 \ldots b^{\varepsilon_m} k_m,$$

which define exceptional elements of $F(A, B) \cap F(B, C)$. By our Case Assumption, $m > 0$ in all instances.

Lemma 7.1. *Any exceptional equality*

$$h_0 b^{\varepsilon_1} h_1 \ldots b^{\varepsilon_m} h_m = k_0 b^{\varepsilon_1} k_1 \ldots b^{\varepsilon_m} k_m,$$

of minimal length in b contains at most one sign change in its signature pattern $(\varepsilon_1, \varepsilon_2, \ldots, \varepsilon_m)$.

Proof. Let us suppose that there are least two such changes. As usual we can assume $\varepsilon_1 = 1$. The (standardised) system of normal form equalities must then contain $h_0 = k_0 z_0$, $\overleftarrow{z_{i-1}} h_i = k_i w_i$ and $\overrightarrow{w_{j-1}} h_j = k_j z_j$ where $i < j$ and $z_0, z_{i-1}, z_j \in U$, $w_i, w_{j-1} \in L$. Now, by an argument we have already used, $h_0 = k_0 z_0$ must define an exceptional element of $F(A^*, B^*) \cap F(A^*_+, B^*, C^*)$ since otherwise $h_0 = k_0$ and we can contradict minimality. Also $h_j = \overrightarrow{w_{j-1}}^{-1} k_j z_j$ must define an exceptional element of $F(A^*, B^*) \cap F(A^*_+, B^*, C^*)$ since otherwise we have not expressed h and k in reduced form. It then follows from Proposition 5.1 that $\overrightarrow{w_{j-1}} \equiv z_0 \equiv z_j$ and hence that

$z_j = k_0^{-1} h_0$. However we know that

$$h_0 b h_1 \ldots b h_i b^{-1} \ldots b^{-1} h_j = k_0 b k_1 \ldots b k_i b^{-1} \ldots b^{-1} k_j z_j$$

and hence that

$$h_0 b h_1 \ldots b h_i b^{-1} \ldots b^{-1} h_j h_0^{-1} = k_0 b k_1 \ldots b k_i b^{-1} \ldots b^{-1} k_j k_0^{-1}$$

and

$$h_0 b h_{j+1} b^{\varepsilon_{j+1}} \ldots b^{\varepsilon_m} h_m = k_0 b k_{j+1} b^{\varepsilon_{j+1}} \ldots b^{\varepsilon_m} k_m.$$

By the minimality of m, neither is exceptional and hence their product is not exceptional, which is a contradiction. □

We can strengthen Lemma 7.1.

Lemma 7.2. *Any exceptional equality*

$$h_0 b^{\varepsilon_1} h_1 \ldots b^{\varepsilon_m} h_m = k_0 b^{\varepsilon_1} k_1 \ldots b^{\varepsilon_m} k_m,$$

of minimal length in b contains no sign changes in its signature pattern.

Proof. Suppose, to the contrary, that there is one sign change. Making a choice of initial sign for b we can write this as $h_0 b h_1 \ldots b h_i b^{-1} \ldots b^{-1} h_m = k_0 b k_1 \ldots b k_i b^{-1} \ldots b^{-1} k_m$. This time we observe that the equalities $h_0 = k_0 z_0$ and $\overleftarrow{z_{i-1}} h_i w_i^{-1} = k_i$ in the Normal Form system define exceptional elements of $F(A^*, B^*) \cap F(A^*_+, B^*, C^*)$ and $F(A^*, B^*, C^*_-) \cap F(B^*, C^*)$ respectively. By Proposition 5.1, the first gives us a basic exceptional equality of the form $h_0 \equiv u = v_0 v_2 \equiv k_0 z_0 \in F(A^*, B^*) \cap F(A^*_+, B^*, C^*)$ (where we make a choice between inverses). For the second, the dual of Proposition 5.1 gives us a basic exceptional equality of the form $p_1 p_0 p_2^{-1} = q$, where one (but not both - otherwise we are back in Case 7.1) of p_1, p_2 may be trivial. If p_2 is trivial, then the basic exceptional relation is $p_1 p_0 = q$ and we have $\overleftarrow{z_{i-1}} h_i w_i^{-1} \equiv p_1 p_0 \beta p_0^{-1} p_1^{-1}$, where $\beta \in F(B^*)$. In particular $w_i \equiv p_1$ and the Normal Form equalities yield $h_0 b h_1 \ldots b h_i = k_0 b k_1 \ldots b k_i w_i$ and hence $h_0 b h_1 \ldots b h_i p_0 = k_0 b k_1 \ldots b k_i q$. As in the previous lemma, this decomposes the equality $h = k$ into two equalities which are of shorter b-length and yields a contradiction. A similar argument can be applied if p_1 is trivial.

The position is more complicated if p_1, p_2 are non-trivial. Since $u = v_0 v_2$ and $p_1 p_0 p_2^{-1} = q$ are versions of our basic exceptional equality, it is simple to check that one of p_1, p_2 is intermediate and the other is not. In particular, p_1, p_2 are distinct and $\overleftarrow{z_{i-1}} h_i w_i^{-1} \equiv (p_1 p_0 p_2^{-1})^{\pm 1}$. Therefore one of $\overleftarrow{z_{i-1}}, w_i$ is intermediate. Suppose that $\overleftarrow{z_{i-1}}$ is intermediate, and for definiteness, that it is p_2 which is intermediate so that $\overleftarrow{z_{i-1}} \equiv p_2$. Then, of course, $L(\overleftarrow{z_{i-1}}) = d(a_\lambda, c_\nu)$

and therefore $L(z_{i-1}) = d(a_\lambda, c_\nu)$. Applying this to the equality $\overleftarrow{z_{i-2}}h_{i-1} = k_{i-1}z_{i-1}$, we deduce from Proposition 6.2 that either this equality holds freely, in which case $\overleftarrow{z_{i-2}} \equiv z_{i-1}$ and hence $L(z_{i-2}) = L(\overleftarrow{z_{i-2}}) = L(z_{i-1}) = d(a_\lambda, c_\nu)$ or it is of the form $p_1p_0 = qp_2$. If the latter holds then we have $z_{i-1} \equiv p_2 \equiv \overleftarrow{z_{i-1}}$ which implies that p_2 is trivial. Therefore only the former holds. If $i > 1$ we can iterate this argument to obtain $L(z_0) = d(a_\lambda, c_\nu)$, which we would also obtain directly if $i = 1$. However it is straightforward to check that $d(a_\lambda, c_\nu) < L(v_2) = L(z_0)$ and hence we have a contradiction. A similar argument applies if $\overleftarrow{z_{i-1}} \equiv p_1$. If, on the other hand, it turns out that w_i is intermediate, then we argue in a similar manner but this time we work towards the equality $\overrightarrow{w}_{m-1}h_m = k_m$, which must be the same as $v_2u^{-1} = v_0^{-1}$, giving a similar contradiction. □

We are now at the final step of Case Assumption 7.4 and hence the completion of Case 7.2. As always we know we have the exceptional intersection $F(A, B) \cap F(B, C)$. Let $h = k$ be an exceptional equality of minimal b-length m; by our Case Assumption $m > 0$. We claim that

$$F(A, B) \cap F(B, C) = \langle h \rangle * F(B) = \langle k \rangle * F(B).$$

Let us suppose that the claim is false; then there must exist an exceptional equality $h' = k'$ of b-length $n \geq m$ such that $h' \notin \langle h \rangle * F(B)$. We choose $h' = k'$ so that n is minimal among all such possible equalities. We show firstly that $h' = k'$ has uniform signature pattern. Without loss of generalitry, we can assume, from Lemma 7.2, that $h = k$ has signature pattern $(1, 1, \ldots, 1)$. Suppose that the signature pattern for $h' = k'$ has a sign change of the form $(\ldots, -1, 1, \ldots)$. The argument of Lemma 7.1, using the equalities $h_0 = k_0z_0$ and $w'_{i-1}h'_i = k'_iz'_i$ from the two systems of Normal Form equalities, then shows that $h' = k'$ can be decomposed as $h'_1 = k'_1$, $h'_2 = k'_2$ where both are of b-length less than n. Then of course $h'_1, h'_2 \in \langle h \rangle * F(B)$ and hence $h' = h'_1h'_2 \in \langle h \rangle * F(B)$. If $h' = k'$ only contains a sign change $(\ldots, 1, -1, \ldots)$, then invert $h = k$ to $h^{-1} = k^{-1}$ and proceed in parallel fashion. We are therefore reduced to the case where we have

$$h_0bh_1 \ldots bh_m = k_0bk_1 \ldots bk_m \quad \text{and} \quad h'_0bh'_1 \ldots bh'_n = k'_0bk'_1 \ldots bk'_n.$$

From the respective systems of Normal Form equalities we obtain $h_0 = k_0z_0$ and $h'_0 = k'_0z'_0$; however these are both instances of the defining equality $u = v_0v_2$ for $F(A^*, B^*) \cap F(A^*_+, B^*, C^*)$ and, since the Normal Form equalities were standardised, we obtain $h'_0 = h_0$, $k'_0 = k_0$ and $z_0 \equiv z'_0$. If $m = 1$, then we form $h^{-1}h' = k^{-1}k'$ which, after cancellation, has b-length $n - 1$. Hence $h^{-1}h' \in \langle h \rangle * F(B)$ (if $n = 1$ we use Case Assumption 7.4) whence $h' \in \langle h \rangle * F(B)$. If $m > 1$, our second pair of Normal Form equalities is $\overleftarrow{z_0}h_1 = k_1z_1$

and $\overleftarrow{z_0} h'_1 = k'_1 z'_1$, since $z_0 \equiv z'_0$. Eliminating $\overleftarrow{z_0}$ yields $k_1 z_1 h_1^{-1} = k'_1 z'_1 h_1'^{-1}$ which we can rewrite as $z_1^{-1} k_1^{-1} k'_1 z'_1 = h_1^{-1} h'_1$. If the reduced form of $k_1^{-1} k'_1$ omits c_μ, then this holds freely yielding $h'_1 = h_1, k'_1 = k_1$ and $z'_1 \equiv z_1$. Otherwise Proposition 5.1(2) must apply and we obtain $z_1 \equiv v_2 = v_0^{-1} u$ - but then $h_0 b h_1 = k_0 b k_1 z_1 = k_0 b k_1 v_2 = k_0 b k_1 v_0^{-1} u$ and hence $h_0 b h_1 u^{-1} = k_0 b k_1 v_0^{-1}$. Again we have a decomposition argument giving us a contradiction and so only the case when we have $h'_1 = h_1, k'_1 = k_1$ and $z'_1 \equiv z_1$. occurs. We can iterate this argument to obtain $h'_0 = h_0, h'_1 = h_1, \ldots h'_{m-1} = h_{m-1}$ and $k'_0 = k_0, k'_1 = k_1, \ldots k'_{m-1} = k_{m-1}$. Then (after cancellation), either $h^{-1} h' = k^{-1} k'$ lies in $F(B)$ or is exceptional with b-length less than n. Either way, $h^{-1} h' \in \langle h \rangle * F(B)$ whence the same is true of h. This contradiction completes Case 7.4, and hence the whole proof of 7.1 is complete.

References

[1] G. H. Bagherzadeh, Commutativity in one-relator groups, *J. London Math. Soc.* **13** (1975), 459–471.

[2] G. Baumslag and D. Solitar, Some two-generator one-relator non-Hopfian groups, *Bull. Amer. Math. Soc.* **68** (1962), 199–201.

[3] G. Baumslag and T. Taylor, The centre of groups with one defining relator, *Math. Ann.* **175** (1968), 315–319.

[4] S. D. Brodskii, Equations over groups and groups with one defining relation, *Sib. Math. J.* **25** (1984), 235–251.

[5] D. J. Collins, Generation and presentation of one-relator groups with centre, *Math. Z.* **157** (1977), 63–77.

[6] D. J. Collins, Baumslag-Solitar group, in: *Encyclopaedia of Mathematics, Supplement III*, Kluwer, Dordrecht, 2001, 62–65.

[7] D. J. Collins, Intersections of conjugates of Magnus subgroups of one-relator groups, in preparation.

[8] D. J. Collins, R. I. Grigorchuk, P. F. Kurchanov, and H. Zieschang, *Combinatorial Group Theory and Applications to Geometry*, Springer, Berlin-Heidelberg-New York, 1998.

[9] M. Dehn, Über diskontinuierliche Gruppen, *Math. Ann.* **71** (1912), 116–144.

[10] J. Fischer, A. Karrass, and D. Solitar, On one-relator groups having elements of finite order, *Proc. Amer. Math. Soc.* **33** (1972), 297–301.

[11] G. A. Gurevich, On the conjugacy problem for groups with a single defining relation, *Proc. Steklov Inst. Math.* **133** (1977), 108–120.

[12] J. Howie, On pairs of 2-complexes and systems of equations over groups, *J. Reine u. Angew. Math.* **324** (1981), 165–174.

[13] J. Howie, Intersections of Magnus subgroups of one-relator products, in preparation.

[14] J. Howie and S. J. Pride, A spelling theorem for staggered generalized 2-complexes, with applications, *Invent. Math.* **76** (1984), 55–74.

[15] R. C. Lyndon and P. E. Schupp, *Combinatorial Group Theory*, Springer, Berlin-Heidelberg-New York, 1977.

[16] W. Magnus, Über diskontinuierliche Gruppen mit einer definierenden Relation, *J. Reine u. Angew. Math.* **163** (1930), 141–165.

[17] W. Magnus, Untersuchungen über einige unendliche diskontinuierliche Gruppen, *Math. Ann.* **105** (1931), 52–74.

[18] W. Magnus, Das Identätsproblem für Gruppen mit einer definierenden Relation, *Math. Ann.* **106** (1932), 295–307.

[19] W. Magnus, A. Karrass, and D. Solitar, *Combinatorial Group Theory*, Wiley, New York, 1966.

[20] J. McCool, A class of one-relator groups with centre, *Bull. Aust. Math. Soc.* **44** (1991), 245–252.

[21] J. McCool and P. E. Schupp, On one relator groups and HNN extensions, in: *Collection of articles dedicated to the memory of Hanna Neumann, II., J. Austral. Math. Soc.* **16** (1973), 249–256.

[22] S. Meskin, A. Pietrowski, and A. Steinberg, One-relator groups with center, *J. Aust. Math. Soc.* **16** (1973), 319–323.

[23] D. I. Moldavanskii, Certain subgroups of groups with one defining relation, *Sib. Math.* **8** (1967), 1039–1048.

[24] K. Murasugi, The center of a group with a single defining relation, *Math. Ann.* **155** (1964), 246–251.

[25] B. B. Newman, Ph.D Thesis, CUNY, New York, 1968.

[26] B. B. Newman, Some results on one-relator groups, *Bull. Amer. Math. Soc.* **74** (1968), 568–571.

[27] A. Pietrowski, The isomorphism problem for one-relator groups with non-trivial centre, *Math. Z.* **136** (1974), 95–106.

[28] S. J. Pride, One-relator quotients of free products, *Math. Proc. Camb. Philos. Soc.* **88** (1980), 233–243.

[29] J.-P. Serre, *Trees*, Springer, Berlin-Heidelberg-New York, 1980.

9

A minimality property of certain branch groups

R. I. Grigorchuk and J. S. Wilson

1. Introduction

In 1979, S. Pride [16] introduced a concept of 'largeness' for groups, which depends on a certain pre-order $\preceq$ on the class of groups (whose precise definition we recall below). The finitely generated groups which are 'largest' in Pride's sense are the ones having a subgroup of finite index which can be mapped onto the free group of rank 2, and so a number of important results can be reformulated as statements that certain groups are 'as large as' the free group of rank 2 (cf. for example [2], [10], [13], [14]). At the other extreme, the groups G satisfying $G \preceq 1$ are just the finite groups. Here we are concerned with the groups which are as small as possible with respect to being infinite, in the sense of Pride's pre-order: thus they are infinite groups G such that if H is infinite and $H \preceq G$ then $G \preceq H$. These groups are called atomic (or minimal) groups. There is a strong connection between atomic groups and just infinite groups. Question 5 of [16] was answered negatively, independently by the first author [6] and P. M. Neumann [15], by the construction of finitely generated just infinite groups with infinitely many commensurability classes of subnormal subgroups. We shall explain below how the theory of just infinite groups developed by the second author [19], [20] leads directly to the question of determining which finitely generated branch groups are atomic. We shall establish a sufficient condition for certain branch groups to be atomic, and using it we shall prove that some of the best known examples of branch groups are atomic.

2. Passage to branch groups

The pre-order on groups introduced by Pride [16] was modified slightly in [4], and it is the latter version of the definition that we adopt. Let G, H be groups.

Then we write $H \preceq G$ if there exist

(i) a subgroup G° of finite index in G;
(ii) a subgroup H° of finite index in H and a finite normal subgroup Y of H°;
(iii) an epimorphism from G° to H°/Y.

We write $H \sim G$ if $H \preceq G$ and $G \preceq H$ and we write $[G]$ for the equivalence class of G. The relation $\preceq$ induces a partial order (also denoted by $\preceq$) on the collection of equivalence classes. A group G is called atomic if $[G] \neq [1]$ and if there is no class lying strictly between $[G]$, $[1]$. Easy examples of atomic groups are provided by (a) direct products of countably many copies of a finite simple group and (b) groups in the class $[\mathbb{Z}]$ of virtually infinite cyclic groups. The atomic groups which are finitely generated are perhaps of greatest interest. Proposition 1 in Section 3 gives a criterion for certain finitely generated groups to be atomic.

Let G be a finitely generated atomic group. Then (as was known to Baer at least fifty years ago; see [1, §1, Lemma 1]) G has a normal subgroup N maximal with respect to having infinite index, and clearly $G/N \preceq G$, so that $[G]$ contains the just infinite group G/N. Thus each class $[G]$ of finitely generated atomic groups contains a representative which is just infinite. By the trichotomy for just infinite groups in [19], [20], each just infinite group G is of one of the following types:

(1) a finite extension of a free abelian group (of rank n, say);
(2) a finite extension of a direct product of n copies of an infinite group all of whose non-trivial subnormal subgroups have finite index, for some integer n;
(3) a just infinite group whose 'structure lattice' is infinite.

Now the groups of types (1), (2) are clearly atomic if and only if $n = 1$; thus the infinite cyclic and dihedral groups are two representatives of the class $[\mathbb{Z}]$ of atomic groups, and there is a class of atomic groups corresponding to each commensurability class of infinite groups all of whose non-trivial subnormal subgroups have finite index. It is known that not all just infinite groups of the type (3) are atomic; see [15, Comment 5.4].

It was shown in [20] that the groups in (3) have faithful actions as branch groups; we recall the definition. Let $(l_n)_{n\geqslant 0}$ be a sequence of integers with $l_n \geqslant 2$ for each n. The spherically homogeneous rooted tree of type (l_n) is a tree T with a vertex v_0 (called the root vertex) of valency l_0, such that every vertex at distance $n \geqslant 1$ from v_0 has valency $l_n + 1$. The distance from v_0 to a vertex v is called the level of v. We picture the tree with v_0 at the top, with l_n edges descending from each vertex of level n. Each vertex v of level m is the root of a spherically homogeneous subtree T_v of type $(l_n)_{n \geqslant m}$. Let G be a group which acts faithfully as a group of automorphisms of T. For each v, let $\mathrm{rst}_G(v)$ be the subgroup of G consisting of automorphisms which fix all vertices of T outside T_v, and for each n let $\mathrm{rst}_G(n)$ be the group generated by the subgroups $\mathrm{rst}_G(v)$ with v of level n. The group G is called a branch group if for each $n \geqslant 1$ the following two conditions hold:

(i) G acts transitively on the vertices of level n;
(ii) $\mathrm{rst}_G(n)$ has finite index in G.

For each choice of the sequence (l_n) there are finitely generated branch groups G on the tree T (see [7]). In the examples that we shall consider, the sequence (l_n) is constant, so that T is a regular rooted tree. We fix $d \geqslant 2$ and write T for the regular tree of degree d. Let K be a subgroup of the automorphism group $\mathrm{Aut}\, T$ of T and let v be a vertex of T. Then there is a canonical isomorphism from T to the subtree T_v with root v, and we write $v{*}K$ for the subgroup of $\mathrm{Aut}\, T$ which fixes all vertices outside T_v and whose action on T_v is induced by the isomorphism from T to T_v.

Definition 1. *A group G is said to be a d-regular branch group over the subgroup K if G acts faithfully on the d-regular tree T, if condition* (i) *above holds, if K has finite index, and if the product K_1 of the groups $v{*}K$ over all vertices v of level* 1 *is a subgroup of finite index in K.*

Theorem 1. *Let G be a d-regular branch group over the subgroup K, and for each $m \geqslant 1$ let K_m be the product of the groups $v{*}K$ over all vertices v of level m. Suppose that $K' \geqslant K_m$ for some m and that K is a subdirect product of finitely many just infinite groups $\Gamma_1, \ldots, \Gamma_k$, each of which is commensurable with G. Then G is atomic.*

The theorem will be proved in Section 3; it will be used in Section 4 to show that some well-known examples of branch groups are atomic and it will be clear that it can be applied to many related groups. However it does not apply to all of the p-groups G_ω introduced in [6], which are parametrized by certain sequences $\omega \in \{0, 1, 2, \ldots, p\}^{\mathbb{N}}$. It seems likely that if ω is not eventually periodic then G_ω has infinite height with respect to the pre-order $\preceq$ of Edjvet and Pride [4].

3. Proofs

We begin with a criterion for certain finitely generated groups to be atomic. We write $H \leqslant_f G$ to indicate that H is a subgroup of finite index in G.

Proposition 1. *Let G be a finitely generated group which has no non-trivial finite normal subgroups. Then the following are equivalent:*

(a) *G is atomic;*

(b) *G has a descending chain (G_n) of subgroups of finite index with the following properties:*

  (1) *every subgroup of G of finite index contains some G_n;*

  (2) *whenever Q is a just infinite image of one of the groups G_n, there is an epimorphism from a subgroup of finite index in Q to a subgroup of finite index in G.*

Proof. Suppose first that G is atomic. Choose any chain (G_n) satisfying (1); this is possible since G has only countably many subgroups of finite index. Suppose that G_n/L is just infinite; consideration of the quotient map $G_n \to G_n/L$ shows that $G_n/L \preceq G_n$, and so as G is atomic we must have $G_n \preceq G_n/L$. Therefore there are $H^\circ \leqslant_f G_n/L$, $G^\circ \leqslant_f G$, X finite with $X \lhd G^\circ$ and an epimorphism $H^\circ \to G^\circ/X$. The normal subgroup $\langle X^G \rangle$ is generated by finitely many torsion elements each having finitely many conjugates, and so it is finite by Dicman's Lemma (see [17, (14.5.7)]), and hence trivial. Thus X is trivial, and we obtain an epimorphism $H^\circ \to G^\circ$, and (2) holds.

Now suppose instead that G is a group having a chain (G_n) satisfying (1), (2). Suppose that H is an infinite group such that $H \preceq G$. Then there are subgroups $G^\circ \leqslant_f G, H^\circ \leqslant_f H$, and Y finite with $Y \lhd H^\circ$, and an epimorphism $\varphi : G^\circ \to H^\circ/Y$. We claim that there is an epimorphism ψ from a subgroup of finite index in H° to a subgroup of finite index in G°. From this it clearly follows that G is atomic. Because of condition (1), we can restrict attention to the case when $G^\circ = G_n$ for some n. Let $\omega : H^\circ/Y \to Q$ be an epimorphism to a just infinite group Q, and let M be the kernel of the composite of φ and ω. Thus G_n/M is isomorphic to Q, and so by (2) there is an epimorphism ε from a subgroup Q° of finite index in Q to a subgroup $G^{\circ\circ}$ of finite index in G. Let $H^{\circ\circ}$ be the preimage of Q° under ω; thus $H^{\circ\circ}$ has finite image in H, and we obtain an epimorphism from $H^{\circ\circ}$ to $G^{\circ\circ}$ by composing the the quotient map $H^{\circ\circ} \to H^{\circ\circ}/Y$, the restriction of ω to $H^{\circ\circ}/Y$, and ε. This completes the proof of Proposition 1. □

Proposition 2. (a) *Let G be a branch group. Then every non-trivial normal subgroup contains $\mathrm{rst}_G(n)'$ for some n.*

(b) *Let G be a regular branch group of degree d over the subgroup K, and for each $m \geqslant 1$ let K_m denote the product of the subgroups $v{*}K$ over all vertices v of level m. If the derived group K' of K contains K_m for some m, then every subgroup of finite index in G contains K_n for some n.*

Proof. (a) This is Theorem 4 in [7].
(b) Since every subgroup of finite index contains a normal subgroup of finite index, it suffices by (a) to prove that each subgroup $\mathrm{rst}_G(r)'$ contains K_n for some n. However since $K_m \leqslant K'$ we have

$$v{*}K_m \leqslant v{*}K' \leqslant (\mathrm{rst}_G(r))'$$

for each vertex v of level r, and hence $K_{m+r} \leqslant (\mathrm{rst}_G(r))'$. □

Proposition 3. *Let G be a group having normal subgroups $N_1, \ldots, N_r$ such that each quotient group G/N_i is just infinite and not virtually abelian and such that $\bigcap_{i=1}^{r} N_i = 1$. Let $M \lhd G$ and suppose that G/M is just infinite. Then $M = N_i$ for some i.*

Proof. We argue by induction on r. The result holds clearly for $r = 1$ and so we suppose that $r > 1$. Let $V = \bigcap_{i<r} N_i$. If $M \geqslant V$, the result follows by induction. Otherwise, VM/M is a non-trivial normal subgroup of G/M and so it has finite index and therefore $V/(V \cap M)$ is infinite. Now $V \cap N_r = 1$, and so (by the isomorphism theorem for G-operator groups [17, p. 29]), V is isomorphic as a G-operator group to VN_r/N_r, so that all non-trivial normal subgroups of G contained in V have finite index in V. Thus we conclude that $V \cap M = 1$. Hence $[V, M] = 1$, and $[VN_r, MN_r] \leqslant N_r$. Suppose that VN_r/N_r and MN_r/N_r are both non-trivial; then each has finite index in G/N_r and hence their intersection is an abelian normal subgroup of finite index in G/N_r, a contradiction. Therefore either $V \leqslant N_r$ or $M \leqslant N_r$; in the first case we have $\bigcap_{i=1}^{r} N_i = \bigcap_{i=1}^{r-1} N_i$ and the result follows by induction, and in the second case, since both G/M and G/N_r are just infinite, we have $M = N_r$. □

Proof of the Theorem. Suppose that G is as in the statement of Theorem 1. Since G is not virtually abelian (see for example [9, Lemma 2]), none of the groups Γ_i is virtually abelian. Now each term in the chain of subgroups (K_m) has finite index in G, and by Proposition 2, each subgroup of finite index in G contains some term of this chain. Therefore by Proposition 1 it suffices to verify that each just infinite image of each group K_m is commensurable with G. However each K_m is isomorphic to a direct power of K, and so the result follows from Proposition 3.

4. Examples

In this section we shall prove that certain d-regular branch groups (where $d \geqslant 2$) are atomic. Let T be the d-regular rooted tree. We regard the vertices of T as finite strings of elements from the set $\{0, \ldots, d-1\}$; the root vertex is $\emptyset$, the vertices of level n are the strings of length n and the edges descending from a vertex v of level n lead to the d vertices of level $(n+1)$ which begin with the string v. Let L be the set of vertices of level 1. We define a to be the automorphism of T which acts as the d-cycle $(0, 1, \ldots, d-1)$ on L and which leaves untouched the remaining entries of strings; thus a permutes cyclically the trees $T_0, \ldots, T_{d-1}$ with roots $0, \ldots, d-1$. Let G be a subgroup of Aut T. We write $\mathrm{st}_G(1)$ for the kernel of the action of G on L and ψ for the injective map from $\mathrm{st}_G(1)$ to $\prod_{v \in L} \mathrm{Aut}\, T$ induced by the restrictions of automorphisms to their actions on the trees with roots in L. As in Section 3, for $K \leqslant G$ we denote by K_m the product of the groups $v{*}H$ with v of level m. By Theorem 1, if G is a regular branch group over K, then G is atomic if

(1) $K_m \leqslant K'$ for some m, and
(2) K is a subdirect product of just infinite groups commensurable with G.

We write $\times^n G$ to denote the direct product of n copies of a group G. If L is a subgroup of $\times^n G$, we refer to the images of L under the n projection maps from $\times^n G$ to G as the projections of L.

Example 1. Let $d = 2$. Let b, c, d be the automorphisms of T which fix the vertices 0, 1 and are defined recursively by

$$\psi(b) = (a, c), \quad \psi(c) = (a, d), \quad \psi(d) = (1, b).$$

Let $G = \langle a, b, c, d \rangle$. This group G is the first of the two 2-groups introduced by the first author in [5]; for recent accounts of its properties, see [7], [8], [12]. Let $K = \langle (ab)^2 \rangle$. Then G is a 2-regular branch group over K and $K_m \leqslant K'$; see [12, Proposition 30, p. 230]. Therefore to prove that G is atomic it suffices to prove that K is a subdirect product of just infinite groups commensurable with G. Now $K = \langle (ab)^2, (abad)^2, (bada)^2 \rangle$ (see [12, Proposition 30, p. 230]), so that K fixes the vertices 0, 1, and ψ maps the three generators of K to $((ac)^{-1}, ac)$, $(1, (ab)^2)$, $((ab)^2)$. To show that G is atomic it will therefore now suffice to prove the following result.

Proposition 4. *The group $\Gamma = \langle ac, (ab)^2 \rangle$ is a just infinite branch group of finite index in G.*

Proof. The finiteness of index follows since

$$\psi(K) \leqslant G \times G$$

and since $|G : K|$ is finite and $|G \times G : \mathrm{im}\, \psi| = 8$; see [12, Theorem 28]. Assume for the moment that Γ is known to act transitively on the vertices of level m for each m. Since $\Gamma \leqslant_f G$ and $K' \leqslant_f G$ and G is a 2-regular branch group there are integers r, s such that $K_n \leqslant_f \Gamma$ for all $n \geqslant r$ and $K_s \leqslant_f K'$; clearly then Γ is a regular branch group over K_r and the derived group of K_r contains K_{r+s}, so that Γ is just infinite by Proposition 2 (a). (In fact, Γ is a normal subgroup of index 4 in G and it contains K, from [3].) It remains to prove that Γ acts transitively on the set of vertices of level m for each $m \geqslant 1$. Clearly it acts transitively on the set L of vertices of level 1. Consider

$$\mathrm{st}_\Gamma(1) = \langle (ab)^2, (ab)^{2ac}, (ac)^2 \rangle.$$

The map ψ sends the three generators to (ac, ca), (ac, dab) and (ad, da), and hence each projection of $\psi(\mathrm{st}_\Gamma(1))$ equals Ω where $\Omega = \langle b, ac \rangle$. It now suffices to prove that Ω acts transitively on the set of vertices of level m for each $m \geqslant 1$. Clearly it acts transitively on L, and we have $\mathrm{st}_\Omega(1) = \langle b, (ac)^2, b^{ac} \rangle$. The images of the three generators under ψ are (a, c), (da, ad) and (aca, dad), and so each projection of Ω is equal to G, which does act transitively on the set of vertices of level m for each m. This concludes the proof that Γ is a branch group. □

Example 2. Now let $d = 4$ and let G be the second 2-group introduced in [5]. This group is generated by a, b, where a is as described above and b is defined recursively by $\psi(b) = (a, 1, a, b)$. To show that G is atomic it suffices to prove the following result, in which we write $K = G'$ to simplify the notation.

Lemma 1. (a) *All of the projections of* $\mathrm{st}_G(1)$ *and of* $\psi(K)$ *are equal to* G.
(b) *G is a 4-regular branch group over K.*
(c) $K_2 \leqslant K'$.
(d) *G is just infinite.*

Proof. (a) The subgroup $\mathrm{st}_G(1)$ is generated by the conjugates of b under the elements of $\langle a \rangle$, and K contains the conjugates of $[a, b]$ under the elements of $\langle a \rangle$. Therefore the result follows since $\psi([a, b]) = (ba, a, a, b)$ and since the images under ψ of these conjugates of b, $[a, b]$ are obtained by permuting co-ordinates cyclically.
(b) We have

$$\psi(b^a) = (b, a, 1, a), \quad \psi([b, b^a]) = ([a, b], 1, 1, 1);$$

therefore $\psi(G')$ contains the latter element and its conjugates under the first projection of $\mathrm{st}_G(1)$, which is G by (a). Hence $K_2 \leqslant K = G'$, and the index in G of the latter is finite as G/G' is finite. It remains to prove that G acts transitively on the vertices of level m, for each m. We prove this by induction, the result being clear for $m = 1$. If G is transitive on the vertices of level m it

follows from (a) that $\mathrm{st}_G(1)$ has four orbits on the vertices of level $m+1$, and since these are permuted transitively by $\langle a \rangle$ we conclude that G has one orbit on the vertices of length $m+1$.
(c) The subgroup $K' = G''$ contains both

$$[[b, b^a], [a, b]^{a^2}] \quad \text{and} \quad [[b, b^a], [a, b]^a],$$

whose images under ψ are

$$([[a, b], a], 1, 1, 1) \quad \text{and} \quad ([[a, b], b], 1, 1, 1).$$

We conclude from (a) that $\times^4 \gamma_3(G) \leqslant \psi(K')$. Next, $\psi(\gamma_3(G))$ contains $\psi([[a, b]^a, b]) = ([b, a], 1, 1, 1)$, and hence contains $\times^4 K$. The result follows.
(d) This now follows from Proposition 2. □

Example 3. Let p be an odd prime. Define an automorphism b recursively by setting $\psi(b) = (a, a^{-1}, 1, \dots, 1, b)$. The group $G = \langle a, b \rangle$ is the Gupta–Sidki p-group, introduced in [11]. Note that $a^p = b^p = 1$, so that $|G : G'| = p^2$. Again, to simplify the notation, we write $K = G'$. It is known from [18] that G is a p-regular branch group over K and that each of the p projections of $\psi(K)$ is equal to G. Indeed, the inclusion $\times^p K \leqslant \psi(G)$ comes easily from the observation that

$$\psi(b^a) = (b, a, a^{-1}, 1, \dots, 1)$$

and

$$\psi([b^i, (b^j)^a]) = ([a^i, b^j], 1 \dots, 1). \qquad (*)$$

To show that G is atomic the following observation suffices.

Lemma 2. *The inclusion $K_2 \leqslant K'$ holds.*

Proof. Consider the commutator of the element in $(*)$ and the image under ψ of $([a, b]^k)^{a^s}$ for suitable values of s: we find that $\psi(G'')$ contains

$$([[a^i, b^j], a^k], 1 \dots, 1) \quad \text{and} \quad ([[a^i, b^j], b^k], 1 \dots, 1),$$

and we deduce easily that $\times^p \gamma_3(G) \leqslant \psi(G'')$. But ψ maps the element $[b, [a, b]^{a^2}]$ of $\gamma_3(G)$ to $(1, [a^{-1}, b], \dots)$, and hence $\times^p G' \leqslant \psi_3(G)$. The assertion follows. □

Acknowledgements. The authors acknowledge support from the Swiss National Science Foundation. They thank the University of Geneva for its hospitality while this work was being carried out.

References

[1] R. Baer, The hypercenter of a group, *Acta Math.* **89** (1953), 165–208.

[2] B. Baumslag and S. J. Pride, Groups with two more generators than relators, *J. London Math. Soc.* (2) **17** (1978), 425–426.

[3] T. Ceccherini-Silberstein, F. Scarabotti and F. Tolli, The top of the lattice of normal subgroups of the Grigorchuk group, *J. Algebra* **246** (2001), 292–310.

[4] M. Edjvet and S. J. Pride, The concept of "largeness" in group theory II, in: *Proceedings of Groups–Korea 1983*, Lecture Notes in Math. vol. 1098, Springer, 1985, pp. 29–54.

[5] R. I. Grigorchuk, On the Burnside problem for periodic groups, *Funktsional. Anal. i Prilozhen.* **14** (1980), 53–54. (English translation: *Functional Anal. Appl.* **14** (1980), 53–54.)

[6] R. I. Grigorchuk, The growth degrees of finitely generated groups and the theory of invariant means, *Izv. Akad. Nauk. SSSR. Ser. Mat.* **48** (1984), 939–985. (English translation: *Math. USSR. Izv.* **25** (1985), 259–300.)

[7] R. I. Grigorchuk, Just infinite branch groups, in: *New horizons in pro-p groups*, Birkhäuser, 2000, 121–179.

[8] R. I. Grigorchuk and J. S. Wilson, A structural property concerning abstract commensurability of subgroups, *J. Lond. Math. Soc.,* to appear.

[9] R. I. Grigorchuk and J. S. Wilson, The uniqueness of the actions of certain branch groups on rooted trees, *Geom. Dedicata*, to appear.

[10] F. J. Grunewald and J. Schwermer, Free nonabelian quotients of SL_2 over orders of imaginary quadratic numberfields, *J. Algebra* **69** (1981), 298–304.

[11] N. Gupta and S. Sidki, On the Burnside problem for periodic groups, *Math. Z.* **182** (1983), 385–388.

[12] P. de la Harpe, *Topics in geometric group theory*, University of Chicago Press, 2000.

[13] A. Lubotzky, Free quotients and the first Betti number of some hyperbolic manifolds, *Transform. Groups* **1** (1996), 71–82.

[14] G. A. Margulis and E. B. Vinberg, Some linear groups virtually having a free quotient, *J. Lie Theory* **10** (2000), 171–180.

[15] P. M. Neumann, Some questions of Edjvet and Pride about infinite groups, *Illinois J. Math.* **30** (1986), 301–316.

[16] S. J. Pride, The concept of largeness in group theory, in: *Word problems II*, North Holland Publishing Company, 1980, 299–335.

[17] D. J. S. Robinson, *A course in the theory of groups*, Springer, 1982.

[18] S. N. Sidki, On a 2-generated infinite 3-group: subgroups and automorphisms, *J. Algebra* **110** (1987), 24–55.

[19] J. S. Wilson, Groups with every proper quotient finite, *Proc. Cambridge Philos. Soc.* **69** (1971), 373–391.

[20] J. S. Wilson, On just infinite abstract and profinite groups, in: *New horizons in pro-p groups*, Birkhäuser, 2000, 181–203.

10

Lattices with non-integral character

H. Helling*

1. Introduction

Hyperbolic lattices in dimension three, that is, discrete cofinite subgroups of $SL(2, \mathbb{C})$, show a preference for having integrally valued character functions, see, e.g., [2], [3]. The first (and only) publicly known lattice with non-integral character seems to be the one presented by Vinberg at the very end of his fundamental paper [4] where it plays the rôle of an example for reflection groups. We in this paper first present a very geometric version of this example and then discuss a series of lattices which contains, most probably, infinitely many with no integer valued character. The main difference when compared to Vinberg's case is that the lattices exhibited here are cocompact. They appear as the result of Dehn surgery along the figure eight knot with parameter $(\pm 4n, n)$; all other Dehn surgery results are, as soon as they are hyperbolic, integrally valued on their character. We do not know of any geometric significance of this exceptional behaviour, yet. The construction hints where to look for more peculiarities of this type.

2. Vinberg's example

This is a lattice in three-dimensional hyperbolic space generated by reflections. Let P be the solid in $\mathbf{H}^3$ described combinatorially as a prism with two opposite triangular and three planar quadrangular faces as shown in figure 1.

* Partially supported by the Emmy Noether Institute for Mathematics and the Minverva Foundation.

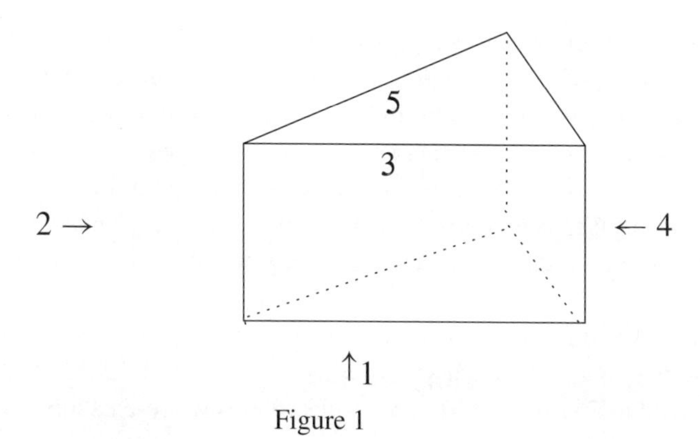

Figure 1

We number the faces in the following way:

1: the bottom triangle,
2: the left hidden quadrangle,
3: the front quadrangle,
4: the right hidden quadrangle,
5: the top triangle.

This labelling is also used as a labelling of the reflections performed on the faces. Compositions of reflections should be read from right to left, so 21 indicates the orientation preserving isometry which is created by first reflecting in face number 1 and then in face number 2. We use symbols as $1 \cap 2 = 2 \cap 1$ for the edge of P where 1 and 2 intersect, $1 \cap 2 \cap 3$ in any ordering for the vertex where 1, 2, and 3 intersect, and $\angle 12 = \angle 21$ for the spatial angle enclosed by the faces 1 and 2. Vinberg shows in [4] that the combinatorial object P may be given the following geometrical realization in $\mathbf{H}^3$:

(1) angular conditions:

$\angle 12 = \angle 13 = \angle 34 = \angle 25 = \angle 35 = \angle 45 = \pi/2$,
$\angle 14 = \pi/3$,
$\angle 23 = \angle 24 = \pi/6$;
(2) vertex locations:

All vertices which are visible in figure 1 are inside hyperbolic space; the hidden vertex $1 \cap 2 \cap 4$ is at infinity.

Theorem (Vinberg [4]). *The group Γ generated by reflections on the faces of P is a cofinite but not cocompact lattice in hyperbolic space $\mathbf{H}^3$. It is not arithmetic.*

The discreteness and cofiniteness of Γ comes from the very general discussion in [4] of reflection groups where Γ plays the rôle of an example. The non-arithmeticity comes from the observation that in order to describe Γ as a matrix group in $O(3, 1)$ matrices with no longer integral traces are needed: the denominators of the traces pick up powers of the prime 2 with exponent unbounded. We, using Poincaré's model of hyperbolic geometry in dimension 3, reprove non-arithmeticity and, as a complement to [4], determine the trace field of Γ.

Let Γ^+ be the index 2 subgroup of Γ consisting of all orientation preserving isometries of Γ. The elements $\sigma_1 = 21, \sigma_2 = 25, \tau_1 = 23, \tau_2 = 24$ of Γ are contained in Γ^+, already. They allow the following presentation of Γ^+:
Generators: $\sigma_1, \sigma_2, \tau_1, \tau_2$,
Relators:

(1) $\sigma_1^2 = \sigma_2^2 = (\sigma_1\tau_1)^2 = (\sigma_2\tau_1)^2 = (\sigma_2\tau_2)^2 = (\tau_2^{-1}\tau_1)^2 = \text{identity}$,
(2) $(\sigma_1\tau_2)^3 = \text{identity}$,
(3) $\tau_1^6 = \tau_2^6 = \text{identity}$.

In order to not overload the discussion with additional notation we interpret the letters $\sigma_1, \sigma_2, \tau_1, \tau_2$ as elements of $SL(2, \mathbb{C})$ instead of $PSL(2, \mathbb{C}) \cong \text{Iso}^+(\mathbf{H}^3)$; consequently Γ^+ is now a subgroup of $SL(2, \mathbb{C})$. In rows (1) and (2) we then have to interpret the identity as $\begin{pmatrix} -1 & 0 \\ 0 & -1 \end{pmatrix}$, and we may and shall do so in row (3), as well. Let now

$$s : \Gamma^+ \longrightarrow \mathbf{C}$$

be the character of the lattice Γ^+ as a subgroup of $SL(2, \mathbb{C})$. Then lines (1), (2) and (3) translate to

(1') $s(\sigma_1) = s(\sigma_2) = s(\sigma_1\tau_1) = s(\sigma_2\tau_1) = s(\sigma_2\tau_2) = s(\tau_2^{-1}\tau_1) = 0$,
(2') $s(\sigma_1\tau_2) = 1$,
(3') $s(\tau_1) = s(\tau_2) = \sqrt{3}$.

Here the value 1 for $s(\sigma_1\tau_2)$ comes from $(\sigma_1\tau_2)^3 = \begin{pmatrix} -1 & 0 \\ 0 & -1 \end{pmatrix}$; the choice of the value $\sqrt{3}$ for $s(\tau_1)$ and $s(\tau_2)$ instead of $-\sqrt{3}$ is compatible with lines (1') and (2'). We mention

$$s(\tau_2\tau_1) = 3,$$

which comes from

$$s(\tau_2\tau_1) + s(\tau_2^{-1}\tau_1) = s(\tau_2\tau_1) + 0 = s(\tau_2)s(\tau_1) = 3.$$

Secondly,

$$s(\sigma_1\tau_2^3) = (\sqrt{3}^2 - 1)s(\sigma_1\tau_2) - \sqrt{3}s(\sigma_1) = 2 = s(\tau_2\sigma_1\tau_2^2) = s(\tau_2^2\sigma_1\tau_2).$$

So the elements $\sigma_1\tau_2^3$, $\tau_2\sigma_1\tau_2^2$, and $\tau_2^2\sigma_1\tau_2$ are parabolic elements with fixed point $1 \cap 2 \cap 4$ from figure 1 which is a cusp. It is also easy to see that these three elements generate the torsion free part of the stabilizer of this cusp. Some general character formalism (see [2]) allows to calculate $s^2(\sigma_1\tau_1\tau_2)$ and $s^2(\sigma_2\tau_1\tau_2)$ and then also $s(\sigma_1\sigma_2)$:

$$4\Big(s(\sigma_1\tau_1\tau_2) - s(\sigma_1\tau_2\tau_1)\Big)^2 + \det\begin{pmatrix} -4 & 0 & 2 \\ 0 & -1 & 3 \\ 2 & 3 & -1 \end{pmatrix} = 0$$

yields

$$\Big(2s(\sigma_1\tau_1\tau_2) - \sqrt{3}\Big)^2 = -9,$$

so

$$s^2(\sigma_1\tau_1\tau_2) = 3\Big(\frac{1+\sqrt{-3}}{2}\Big)^2.$$

Similarly,

$$4\Big(s(\sigma_2\tau_1\tau_2) - s(\sigma_2\tau_2\tau_1)\Big)^2 + \det\begin{pmatrix} -4 & 0 & 0 \\ 0 & -1 & 3 \\ 0 & 3 & -1 \end{pmatrix} = 0$$

results in

$$s^2(\sigma_2\tau_1\tau_2) = -2.$$

Furthermore

$$4\Big(s(\sigma_1\tau_1\tau_2) - s(\sigma_1\tau_2\tau_1)\Big)\Big(s(\sigma_2\tau_1\tau_2) - s(\sigma_2\tau_2\tau_1)\Big)$$
$$+\det\begin{pmatrix} 2s(\sigma_1\sigma_2) & 0 & 2 \\ 0 & -1 & 3 \\ 0 & 3 & -1 \end{pmatrix} = 0$$

leads to

$$s^2(\sigma_1\sigma_2) = 9/2,$$

which is not integral any more. Some more computations along these lines result in explicit values of the character s, e.g., on $\sigma_1\sigma_2\tau_1$ and $\sigma_1\sigma_2\tau_2$. We collect everything in the following result.

Theorem. *The lattice* Γ^+ *has trace field equal to* $\mathbf{Q}(\sqrt{-3})$, *the field of cube roots of unity. Its character values (squared) are unbounded at the non-archimedian valuation at the prime 2 and integral at all other non-archimedian places. It is cofinite with exactly one cusp.*

We end this paragraph by drawing a fundamental domain for Γ^+ in hyperbolic space. The cusp we locate at infinity. It is then easy to see that the matrix solution with character values (1'), (2'), (3') is, up to conjugacy by the stabilizer of the cusp:

$$\sigma_1 = \begin{pmatrix} -i & \frac{3}{\sqrt{2}}i \\ 0 & i \end{pmatrix}, \quad \sigma_2 = \begin{pmatrix} 0 & i \\ i & 0 \end{pmatrix},$$

$$\tau_1 = \begin{pmatrix} \frac{\sqrt{3}-3i}{2} & i\sqrt{2} \\ -i\sqrt{2} & \frac{\sqrt{3}+3i}{2} \end{pmatrix}, \quad \tau_2 = \begin{pmatrix} \frac{\sqrt{3}+i}{2} & 0 \\ 0 & \frac{\sqrt{3}-i}{2} \end{pmatrix}.$$

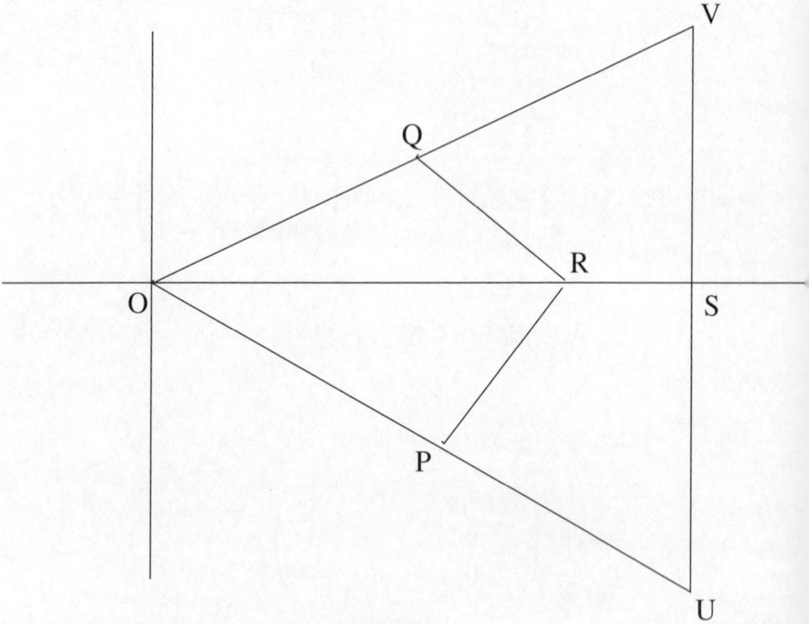

Figure 2

Figure 2 shows a Ford fundamental domain of Γ^+, viewed from the cusp at infinity. It is a triangular prism with three faces lying on the vertical Euclidian halfplains containing $\{\infty, O, V\}$, $\{\infty, O, U\}$, and$\{\infty$.U, V$\}$, respectively. The floor is composed by pieces of three isometric spheres:

$\{O, P, R, Q\}$ lies on the isometric sphere of σ_2,
$\{P, U, S, R\}$ lies on the isometric sphere of τ_1, and
$\{R, S, V, Q\}$ lies on the isometric sphere of τ_1^{-1}.

The coordinates of these points are:

$$\begin{aligned}
O &= (0, 1),\\
P &= \left(\tfrac{\sqrt{3}-i}{\sqrt{6}}, \tfrac{1}{\sqrt{3}}\right),\\
Q &= \left(\tfrac{\sqrt{3}+i}{\sqrt{6}}, \tfrac{1}{\sqrt{3}}\right),\\
U &= \left(\tfrac{3-i\sqrt{3}}{2\sqrt{2}}, \tfrac{1}{\sqrt{2}}\right),\\
V &= \left(\tfrac{3+i\sqrt{3}}{2\sqrt{2}}, \tfrac{1}{\sqrt{2}}\right),\\
S &= \left(\tfrac{3}{2\sqrt{2}}, \tfrac{1}{2\sqrt{2}}\right),\\
R &= \left(\tfrac{2\sqrt{2}}{3}, \tfrac{1}{3}\right).
\end{aligned}$$

The above generators identify the faces of the prism in the following way:

$$\sigma_1 : \begin{matrix} \infty \\ U \\ S \\ V \end{matrix} \longrightarrow \begin{matrix} \infty \\ V \\ S \\ U \end{matrix}, \quad \sigma_2 : \begin{matrix} O \\ P \\ R \\ Q \end{matrix} \longrightarrow \begin{matrix} O \\ Q \\ R \\ P \end{matrix},$$

$$\tau_1 : \begin{matrix} P \\ U \\ S \\ R \end{matrix} \longrightarrow \begin{matrix} Q \\ V \\ S \\ R \end{matrix}, \quad \tau_2 : \begin{matrix} \infty \\ O \\ P \\ U \end{matrix} \longrightarrow \begin{matrix} \infty \\ O \\ Q \\ V \end{matrix}.$$

The orbifold $\Gamma^+ \setminus \mathbf{H}^3$ obviously allows an orientation reversing isometry, realized, e.g., by the reflection of the fundamental domain in figure 2 through the hyperbolic plane above the real axis.

3. Figure eight knot Dehn surgery

We recall the presentation of the fundamental group $\pi_1(S^3 \setminus 4_1)$ of the figure eight knot complement in the 3-sphere in terms of an HNN-extension:

$$\pi_1(S^3 \setminus 4_1) \cong \langle \xi, \eta, \mu \mid \mu\xi\mu^{-1} = \eta\xi,\ \mu\eta\mu^{-1} = \eta\xi\eta \rangle. \tag{4}$$

This group has a faithful lattice representation in $PSL(2, \mathbb{C})$, which is the group $\mathrm{Iso}^+(\mathbf{H}^3)$ of orientation preserving isometries of hyperbolic space. As it is (two-) torsion free one can lift this representation to a representation in $SL(2, \mathbb{C})$, and it is easy to see that it is legitimate to interpret the above presentation as the presentation of a lattice Γ in $SL(2, \mathbb{C})$. The character variety of $\Gamma \cong \pi_1(S^3 \setminus 4_1)$, that is, the space of deformations of the character of Γ in $SL(2, \mathbb{C})$ is the affine algebraic curve

$$t^2 = \frac{x^2 + x - 1}{x - 1} \tag{5}$$

with the point $x = 1$ removed, see [2]. This means the following: let $s : \Gamma \longrightarrow \mathbf{C}$ be any character on Γ which results from a deformation of the lattice character of Γ. Then s is determined by its values $s(\xi) = x$ and $s(\mu) = t$, and these values are related by equation (5). In this paragraph we replace x by $x = q + 1$ which has the effect that formulae and notation become much more transparent respectively simplified. Equation (5) now reads

$$t^2 = \frac{q^2 + 3q + 1}{q} = q + q^{-1} + 3. \tag{5'}$$

It reveals the fact that the elliptic curve (5) allows two holomorphic involutions: $t \mapsto -t$ and $q \mapsto q^{-1}$. We may rewrite (5') as

$$q^2 - (t^2 - 3)q + 1 = 0. \tag{5''}$$

So if t is an algebraic integer, then so is q; it then is even a unit. We collect some information about s, see [2]:

$$s(\xi) = x = q + 1, \; s(\eta) = y = \frac{x}{x - 1} = 1 + q^{-1}, \; s(\eta\xi) = x = q + 1,$$

$$s(\mu) = s(\mu^{-1}\xi) = s(\mu^{-1}\eta) = s(\mu^{-1}\eta\xi) = t.$$

Note that the character s restricted to the rank 2 free subgroup of $\pi_1(S^3 \setminus 4_1)$ generated by ξ and η is determined by $s(\xi) = x$ and $s(\eta) = y$, which are related by $xy = x + y$, equivalently, $(x - 1)(y - 1) = 1$. We set $\lambda = \eta^{-1}\xi^{-1}\eta\xi$ and compute

$$s(\lambda) = x^2 + y^2 + z^2 - xyz - 2 = \frac{1}{q^2}(q^4 + q^3 - 2q^2 + q + 1)$$

$$= \left(q + \frac{1}{q}\right)^2 + \left(q + \frac{1}{q}\right) - 4 = t^4 - 5t^2 + 2. \tag{6}$$

We also need $s(\mu^{-1}\lambda)$ and $s(\mu\lambda)$ in terms of character coordinates:

$$s(\mu^{-1}\lambda) = s\Big((\xi\mu^{-1})(\eta^{-1}\xi^{-1})\eta\Big) = -s\Big((\eta^{-1}\xi^{-1})(\xi\mu^{-1}\eta)\Big)$$
$$-s(\xi\mu^{-1})s(\eta^{-1}\xi^{-1})s(\eta) + s(\xi\mu^{-1})s\Big((\eta^{-1}\xi^{-1})\eta\Big)$$
$$+s(\eta^{-1}\xi^{-1})s\Big((\xi\mu^{-1})\eta\Big) + s(\eta)s(\xi\mu^{-1}\eta^{-1}\xi^{-1}).$$

We use

$$s(\mu^{-1}\eta^{-1}) = -s(\mu^{-1}\eta) + s(\mu)s(\eta) = -t + t\Big(1 + \frac{1}{q}\Big) = \frac{t}{q}$$

and get

$$s(\mu^{-1}\lambda) = t\frac{q^3 - q^2 + 1}{q^2}. \tag{7}$$

From this follows

$$s(\mu\lambda) = t\frac{q^3 - q + 1}{q}. \tag{8}$$

We observe that $s(\mu^{-1}\lambda)$ and $s(\mu\lambda)$ are related to each other via the automorphism $q \mapsto q^{-1}$ of the curve (5'). To perform Dehn surgery at the knot 4_1 means to add a relator of the form $\lambda^m = \varepsilon\mu^n$ to the presentation (4) of $\Gamma < SL(2, \mathbb{C})$; here m and n are integers not simultaneously 0 and ε is central. In terms of character values we have, in this situation

$$s(\lambda^m) = \pm s(\mu^n). \tag{9}$$

Remember that $s(\gamma^k)$ is a monic polynomial with integer coefficients in $s(\gamma)$ of degree $|k|$, for $\gamma \in \Gamma$ and $k \in \mathbb{Z}$. We have seen in (6) that $s(\lambda)$ is a degree 4 polynomial in terms of $s(\mu) = t$, so if $|n| \neq 4|m|$, equation (7) defines $s(\mu) = t$ as an algebraic integer. In this case we see from (5") that q is an algebraic integer, even a unit, and so all values $s(\gamma)$, $\gamma \in \Gamma$ are algebraically integral. We shall deal with the case $|n| = 4|m|$ separately but first collect all information in the following result.

Theorem. *Let m, n be integers such that the orbifold created by Dehn surgery $\lambda^m = \varepsilon\mu^n$ along the figure eight knot in S^3 is hyperbolic, and let $\Gamma_{m,n} < SL(2, \mathbb{C})$ be the corresponding lattice. Then the character s on $\Gamma_{m,n}$ is algebraically integer valued if*

(1) $|n| \neq 4|m|$ *or*
(2) $n = \pm 4m$ and m not a power of a prime number.

In case $m = \pm p^k$, $n = \pm 4p^k$ with p a prime number and $k \geq 1$, the situation is as follows: define $c_m = 2 + 2\cos\frac{\pi}{m} = \left(2\cos\frac{\pi}{2m}\right)^2$, and $f_m(z)$ to be the degree 4 polynomial

$$f_m(z) = z^4 + 3z^3 + z^2 + c_m.$$

Then the character s on $\Gamma_{m,n}$ is algebraically integer valued if and only if f_m splits over the field $\mathbf{Q}(c_m)$ into two factors one of which has the form $z^2 + az + b$ with b a unit in that field and discriminant $a^2 - 4b$ negative.

What is left to prove is a detailed study of the situation $n = \pm 4m$. We assume m positive which does not mean any loss; requiring hyperbolicity means $m \geq 2$. First $n = -4m$. We have, from $\lambda^m = \varepsilon\mu^{-4m}$: $(\mu^4\lambda)^m = \varepsilon = \pm\begin{pmatrix}1&0\\0&1\end{pmatrix}$ as λ and μ commute. So $s(\mu^4\lambda) = -2\cos\frac{\pi}{m}$. Here the negative sign is mandatory for odd m: the group $\Gamma_{m,-4m}$, regarded as a group of isometries of hyperbolic space, has no 2-torsion and so may be lifted to a matrix group; omitting the negative sign would create $(\mu^4\lambda)^m = \begin{pmatrix}-1&0\\0&-1\end{pmatrix}$. For even m the negative sign is expected as, for $m \to \infty$, this value should converge to -2, which is the character value on λ in the complete hyperbolic case where λ and μ are parabolic. We derive from equations (6) and (7):

$$\begin{aligned} s(\mu^{-4}\lambda) &= (t^3 - 2t)s(\mu^{-1}\lambda) - (t^2 - 1)s(\lambda) \\ &= t^2(t^2 - 2)\frac{q^3 - q^2 + 1}{q^2} - (t^2 - 1)(t^4 - 5t^2 + 2). \end{aligned}$$

We express, using (5'), t^2 in terms of q and get

$$s(\mu^{-4}\lambda) = \frac{1}{q^4}(2q^4 + q^2 + 3q + 1).$$

So

$$s(\mu^4\lambda) = -s(\mu^{-4}\lambda) + s(\mu^4)s(\lambda) = q^4 + 3q^3 + q^2 + 2.$$

Thus, the algebraic equation for q is

$$\begin{aligned} q^4 + 3q^3 + q^2 + c_m &= 0 \text{ if}(\mu^4\lambda)^{\mathrm{m}} = \text{identity}, \\ c_m q^4 + q^2 + 3q + 1 &= 0 \text{ if}(\mu^{-4}\lambda)^{\mathrm{m}} = \text{identity}. \end{aligned}$$

Remember that both q and q^{-1} enter into character values. So if a lattice character is integer valued, it is necessary (and sufficient) that q be an algebraic unit which is not real (as otherwise the trace field would be real).

Proposition. *The number* $c_m = 2 + 2\cos\frac{\pi}{m} = (2\cos\frac{\pi}{2m})^2$ *is, for* $m \geq 2$, *an algebraic integer. It is a unit if and only if* m *is not a prime power; if* $m = p^k$ *with* p *a prime,* $k \geq 1$, *then it is a prime in the ring of integers in* $\mathbf{Q}(c_m)$ *with degree* 1. *Its norm as an element of this field is* p.

We sketch the proof which is presumably in the literature: let $g_m(z)$ be the minimal polynomial of the primitive $(2m)^{th}$ root of unity $e^{\frac{\pi i}{m}}$:

$$g_m(z) = \prod_{0<d|2m} \left(z^d - 1\right)^{\mu(\frac{2m}{d})}$$

with μ the Möbius function. Its degree is $\varphi(2m)$, which is an even natural number; φ is of course Euler's totient function. Its constant term is

$$g_m(0) = \left(-1\right)^{\sum_{0<d|2m} \mu(\frac{2m}{d})} = 1.$$

The rational function $g_m(z)/z^{\varphi(2m)/2}$ is invariant under change from z to z^{-1}. This means that it may be written as a monic polynomial $\tilde{g}_m$ with integer coefficients in terms of $z + 2 + z^{-1}$:

$$\frac{1}{z^{\varphi(2m)/2}} g_m(z) = \tilde{g}_m(z + 2 + z^{-1}).$$

$g_m(z)$ is, of course, the minimal polynomial of c_m. We have, in order to compute the norm of c_m to compute its constant coefficient:

$$\tilde{g}_m(0) = \frac{1}{z^{\varphi(2m)/2}} g_m(z)|_{z=-1} = (-1)^{\varphi(2m)/2} g_m(-1).$$

We have

$$g_m(-1) = \prod_{\substack{d|2m \\ d \text{ odd}}} \left(-2\right)^{\mu(\frac{2m}{d})} \prod_{\substack{d|2m \\ d \text{ even}}} \lim_{z\to 1} \left(\frac{z^d - 1}{z - 1}\right)^{\mu(\frac{2m}{d})}$$

$$\times \lim_{z\to 1} \prod_{\substack{d|2m \\ d \text{ even}}} \left(z - 1\right)^{\mu(\frac{2m}{d})}.$$

The first and the third product are easily seen to be 1, and the second is

$$\prod_{\substack{d|2m \\ d \text{ even}}} d^{\mu(\frac{2m}{d})} = 2^{\sum_{d|m} \mu(\frac{m}{d})} \prod_{d|m} d^{\mu(\frac{m}{d})} = \begin{cases} p & \mathrm{m} = \mathrm{p}^{\mathrm{k}} \\ 1 & \text{otherwise} \end{cases}.$$

This is the statement of the proposition.

The polynomial $f_m(z) = z^4 + 3z^3 + z^2 + c_m$ is easily seen to have 2 real and one pair of complex conjugate roots. Let m be a prime power, so c_m not a unit. If first f_m is irreducible over the field $\mathbf{Q}(c_m)$, then no root is an algebraic unit. If f_m splits over this field into a cubic and a linear factor one sees at once that the cubic factor cannot have a unit as its constant term (and, of course, the zero of the linear factor being real cannot lead to a lattice character). So in this case no character can be integer valued. There remains the case that f_m decomposes into two quadratic factors one having two real roots and constant term a unit times c_m and the other one having constant term a unit and two complex conjugates as the roots. It is undecided to which extent this situation has to be expected.

4. Examples

For $m = 2, 3, 4$, and 5 we have $c_m = 2, 3, 2 + \sqrt{2}$, and $\frac{5+\sqrt{5}}{2}$, respectively. The polynomial

$$f_m(z) = z^4 + 3z^3 + z^2 + c_m$$

is in all cases seen to be irreducible over $\mathbf{Q}(c_m)$. So the lattices $\Gamma_{m,\pm 4m}$, for these values of m, have trace fields which are degree 4 extensions of $\mathbf{Q}(c_m)$. Their character is not integer valued; its values have denominators which are powers of the prime number c_m. If m goes to infinity, the polynomial f_m becomes

$$f_\infty(z) = z^4 + 3z^3 + z^2 + 4 = (z + 2)^2(z^2 - z + 1).$$

The root of the quadratic factor defines the lattice representation of the fundamental group of $S^3 \setminus 4_1$, again.

We briefly sketch a situation where the figure eight knot complement is replaced by the manifold M with fundamental group

$$\pi_1(M) \cong \langle \xi, \eta, \mu \mid \mu\xi\mu^{-1} = \eta\xi, \mu\eta\mu^{-1} = (\eta\xi)^3\eta \rangle.$$

This is another standard hyperbolic manifold. It fibers over the circle with fibre the once punctured torus. Again, $\langle \xi, \eta \rangle$ is the (free) fundamental group of the fibre and μ represents the pseudo-Anosov on the level of $\langle \xi, \eta \rangle$. The character variety of $\pi_1(M)$ is a hyperelliptic curve:

$$t^2 = (z - 1)^2 \frac{z^3 + z^2 - 2z - 1}{z^2 - z - 1},$$

see [2]. Here the point (z, t) on this curve serves as a coordinate for the character

s on $\pi_1(M)$ determined by

$$s(\xi) = z,\ s(\eta) = z\frac{z-1}{z^2-z-1} = 1 + \frac{1}{z^2-z-1},\ s(\eta\xi) = z,$$

$$s(\mu) = t,\ s(\mu^{-1}\xi) = \frac{1}{z-1}t,\ s(\mu^{-1}\eta) = t,\ s(\mu^{-1}\eta\xi) = \frac{1}{z-1}t.$$

In this case, Dehn surgery leads to lattices with non-integral character in case $\mu^3 = \lambda^{\pm 4}$, and the algebraic equation for its character is, in analogy to the figure eight knot situation, $s(\mu^{\pm 3}\lambda^4) = -2\cos\frac{\pi}{m}$ for m sufficiently large in order to guarantee hyperbolicity. The resulting polynomial which replaces f_m has degree 16.

References

[1] P. Alestalo, H. Helling, On torus fibrations over the circle, SFB 343, Ergänzungsreihe no. 97-005, Bielefeld, 1977.

[2] H. Helling, The trace field of a series of hyperbolic manifolds, SFB 343, preprint no. 99-072, Bielefeld, 1999.

[3] T. Koch, Fordsche Fundamentalbereiche hyperbolischer einfach-punktierter Torus-Bündel, SFB 343, Ergänzungsreihe no. 99-009, Bielefeld, 1999.

[4] E. B. Vinberg, Discrete groups in Lobachecvskii spaces generated by reflections, *Mat. Sb.* **72** (1967), 471–488. Also: *Math. USSR-Sb.* **1** (1967), 429–444.

11

Some applications of probability in group theory

Avinoam Mann

My aim in this article is to show how simple probabilistic arguments can be applied to prove group theoretical results. While some of these results are formulated in probabilistic language, some have purely group theoretical formulations (see Theorems 2 and 3). I am going to give only a sample of such results, referring to [Sh] for a more exhaustive survey. Not all of the results here occur in [Sh], though, and Propositions 1 and 2 have not been published before.

Our subject apparently begins with E. Netto, who, more than a century ago, wrote: "If we arbitrarily select two or more substitutions of n elements, it is to be regarded as extremely probable that the group of lowest order which contains these is the symmetric group, or at least the alternating group" [N1, p.76]. Later, in the English version of his book, he added: "In the case of two substitutions the probability in favour of the symmetric group may be as about 3/4, and in favour of the alternating, but not symmetric, group as about 1/4" [N2, p.90]. These statements were made precise by J. D. Dixon [D], who proved that, as $n \to \infty$, the probability that two elements of S_n generate either S_n or A_n tends to 1. Dixon made then a generalized conjecture: let us write $P(G, k)$ for the probability that k random elements generate the group G, then Dixon conjectured that, letting S range over all finite simple groups, $P(S, 2) \to 1$, as $|S| \to \infty$. Applying the classification of the finite simple groups, that latter conjecture was proved by W. M. Kantor–A. Lubotzky [KL] for the classical and small exceptional groups, and by M. Liebeck–A. Shalev [LS] for the remaining ones.

Whereas for finite groups it is clear what we mean by probability, we simply count the number of elements, this is not so for infinite groups. Nevertheless, given any finitely generated group G, we would like to ask: what is the probability that a random finite subset generates G? Guided by Hilbert's dictum that one should start with the simplest examples, let us start with $G = Z$, the infinite cyclic group. This group can be generated by one element, but as only two of its infinitely many elements are such generators, the probability that one element generates Z seems to be 0. So let $p = P(Z, 2)$ be the probability that two elements generate Z. Choosing two integers at random, they generate some subgroup nZ. Again, the probabilty that $n = 0$ seems to be 0, so with probability 1 our pair of integers generates a non-trivial subgroup. They lie in nZ with probability $1/n^2$, and, once we know that they lie there, then, recalling that $nZ \cong Z$, they generate nZ with the same probability p. This yields

$$p(\sum_{n=1}^{\infty} 1/n^2) = 1,$$

and thus $p = 1/\zeta(2) = 6/\pi^2$. In the same way we see that $P(Z, k) = 1/\zeta(k)$. We can even apply this argument for $k = 1$, getting a "proof" that the harmonic series diverges.

Another route to the same result is the following: two integers generate Z if and only if they are relatively prime. Given a prime q, the probability that at least one of them is not divisible by q is $1 - 1/q^2$, and since this non-divisibility assumption has to be satisfied for all primes, we have $p = \Pi_q(1 - 1/q^2) = 1/\zeta(2)$. Again a similar consideration applies for each k. The case $k = 1$ yielding this time that the product $\Pi(1 - 1/q)$ diverges, hence so does the series $\sum 1/q$. We seem also to have derived the Euler factorization of the ζ-function. This, however, depends on our representing $P(Z, k)$ as a product over primes, and to justify this we have to assume that divisibilities by distinct primes are independent events. This hypothesis is equivalent to unique factorization, which, in turn, is equivalent to the Euler factorization.

I assume that by now the readers have both their eyebrows raised: just how are my probabilities defined? Before answering this question, let me work out a slightly more complicated example, still proceeding naively. Take $G = Z^2$, a free abelian group of rank 2. As this group is countable, and all its subgroups are finitely generated, it contains only countably many subgroups. The probability of a random k-tuple to lie in a subgroup of infinite index is then 0, so with probability 1 it generates a subgroup of finite index. Like Z, our group has the property of being isomorphic to all of its finite index subgroups, so a similar argument to the one above yields $P(Z^2, k) = (\sum a_n(Z^2)/n^k)^{-1}$, where $a_n(G)$ denotes the number of subgroups of G of index n. The problem is that we do not know what $a_n(Z^2)$ is. So let us try the other route. A k-tuple generates Z^2 if and only if it does not lie in any maximal subgroup. All maximal subgroups have prime indices, and for index q there are exactly $q + 1$ subgroups of that index, corresponding to subgroups of order q of Z^2/qZ^2, an elementary abelian group of order q^2. It is easy to count the number of generating k-tuples of the latter group, which shows that the probability of generating it is $(1 - 1/q^k)(1 - 1/q^{k-1})$. Now a k-tuple generates Z^2 if it generates it (mod qZ^2) for all q, so we obtain $P(Z^2, k) = \Pi(1 - 1/q^k)(1 - 1/q^{k-1}) = 1/\zeta(k)\zeta(k-1)$. Comparing the two expressions for our probability we have $\sum a_n(Z^2)/n^k = \zeta(k)\zeta(k-1)$. This is an equality between two Dirichlet series. Subgroups of index n correspond to subgroups of Z^2/nZ^2, and since the latter group has order n^2 and all its subgroups can be generated by two elements, the number of these subgroups is at most the number of pairs of elements, so $a_n(Z^2) \leq n^4$. Thus both the series $\sum a_n(Z^2)/n^k$, and the series for $\zeta(k)\zeta(k-1)$ converge for large enough k, and for these k's they have the same value. That means that the two Dirichlet series are identical: they have the same coefficients. Writing this in the form $\sum a_n(Z^2)/n^k = (\sum 1/n^k)(\sum n/n^k)$ and comparing coefficients, we obtain

$$a_n(Z^2) = \sum_{m|n} m.$$

Similarly we obtain for each integer d that $\sum a_n(Z^d)/n^k = 1/\zeta(k)\zeta(k-1)...\zeta(k-d+1)$, and we can derive the value of $a_n(Z^d)$ from that.

The readers may have been patient with me as long as all that I did was to offer a new way of looking at some well known facts such as that the sum of the prime reciprocals diverges, but now that I seem to be using these considerations in order to derive some less familiar results, they would, or at least should, ask for justification. Indeed, my arguments need a probability measure defined on Z that is both translation invariant and countably additive, and it is easy to see that such measures do not exist. But they exist on compact groups, known there as Haar measures. More precisely, I will be considering profinite groups, i.e., inverse limits of finite groups. The inverse limit structure endows such groups with a natural compact topology, and hence with a finite Haar measure, which we always normalize so that our group G has measure 1, and can be considered as a probability space. To justify considerations such as those that were applied above to Z, we pass from an arbitrary group G to its profinite completion, the inverse limit of the system $\{G/N\}$, where N ranges over all normal subgroups of G of finite index. If G is residually finite, i.e., the intersection of all the above N's is trivial, then G is embedded into its profinite completion. When talking about generators of a profinite group, we mean generators as a topological group, i.e., X generates G means that G is the minimal closed subgroup of G containing X. The closure of an arbitrary subset X is $\bar{X} = \cap XN$, with N ranging over all open subgroups of G (since G is compact, all open subgroups have finite index, and they constitute a basis for the neighbourhoods of the identity). It follows that X generates G if and only if each finite factor group G/N is generated by XN/N. Thus G is finitely generated, by d elements, say, if and only if each finite factor group G/N can be generated by d elements. Moreover, X generates G if and only if X is not contained in any proper open subgroup, so the set of k-tuples generating G is the complement in G^k of $\cup H^k$, where H ranges over all proper open subgroups. Therefore the set of k-tuples generating G is closed, and in particular measurable. Also, $P(G, k)$ is the infimum of $P(G/N, k)$, for N as above.

To make the considerations above about Z and Z^d precise, we first pass from these groups to their profinite completions, say $\hat{Z}^d$. Then we note that $a_n(\hat{Z}^d) = a_n(Z^d)$, and that, by an argument like the one for Z^2 above, these numbers grow polynomially. We quote the following well known probabilistic result [R, pp. 389-392].

Borell-Cantelli Lemma. *Let A_i be events in a probability space, with probabilities p_i.*

(1) *If $\sum p_i$ converges, then with probability* 1 *only finitely many of the events A_i happen.*
(2) *If $\sum p_i$ diverges, and the A_i are pairwise independent, then with probability* 1 *infinitely many of the A_i happen.*

Taking as our events "belonging to a given finite index subgroup", we see by part 1 that k elements in $\hat{Z}^d$, for large enough k, almost surely belong only to finitely many such subgroups. By the expression above for the closure of a subset, this means that these k elements generate a subgroup of finite index with probability 1. Given that fact, all the considerations above work. We now make the following definition.

Definition 1. *A profinite group G is* positively finitely generated (**PFG**) *if $P(G, k) > 0$, for some k.*

Thus our arguments above show that $\hat{Z}^d$ is **PFG**. Moreover, it can be shown that $P(\hat{Z}^d, k) > 0$ whenever $k > d$, but that $P(\hat{Z}^d, d) = 0$. Thus, even though d elements suffice for generating this group, they do so, as it were, by chance only, while if we pick out random $(d + 1)$-tuples one after the other, we are virtually certain to find sooner or later a set of generators. On the other hand, it is shown in [KL], that a free profinite group of rank not 1 is not **PFG**. If G is a finitely generated pro-p group (an inverse limit of finite p-groups) then its Frattini subgroup $\Phi(G)$ is of finite index, and a k-tuple of elements generates G if and only if their images generate the finite factor group $G/\Phi(G)$. From this it is obvious that G is a **PFG** group, and in this case the minimal number required to generate G and to generate it with positive probability coincide. It was also shown in [KL] that if G is pro-nilpotent and generated by d elements, then it is **PFG**, and $d + 1$ elements suffice to generate it with positive probability.

Now if some set of elements fails to generate G, they are contained in some open maximal subgroup of G (in a profinite group each proper closed subgroup is contained in a maximal one, and the latter is open, hence of finite index). So it seems that if G has only a few maximal subgroups, then a random subset stands a good chance not to belong to any one of these maximals, and thus generates G. We have learned in recent years from the computer science people that "a few" is to be interpreted as "at most polynomial". So let us write $m_n(G)$ for the number of maximal subgroups of G of index n, and let us make the following definition.

Definition 2. *A group G is termed a* polynomial maximal subgroup growth (**PMSG**) *group if there exist two constants C and s, such that, for all n,*

$$m_n(G) \leq Cn^s.$$

Here G is either an abstract or a profinite group, and in the latter case we consider, as usual, only closed subgroups. It may be somewhat surprising that our vague remarks above about "standing a good chance" and "few" are made precise by the following result.

Theorem 1 (A.Mann - A.Shalev [MS]). *A profinite group is* **PFG** *if and only if it is* **PMSG**.

The proof of one direction is easy, [M1]. Let G be a **PMSG** group, as defined above. If a k-tuple does not generate G, it lies in some open maximal subgroup M, and the probability of that happening is $1/|G:M|^k$. Thus the probability of a random k-tuple to lie in some maximal subgroup is at most $\sum_{n>1} m_n(G)/n^k = s(k)$, say, and by our assumption on $m_n(G)$ we see that $s(k) < 1$, if k be large enough. For such k we have $P(G,k) \geq 1 - s(k) > 0$, so G is **PFG**.

The proof of the reverse direction is much deeper. We first apply the classification of the finite simple groups, and the various results obtained following that classification about the subgroup structure of these groups, to show that there exists a constant c such that for any finite simple group S we have $m_n(S) \leq cn^2$. (Actually the exponent is a little better than 2, and has been improved since, but is still far from the conjectured $1 + o(1)$.) From this we deduce: *there exists a function $s(d)$, such that in any finite d-generated group, the number of maximal subgroups of index n and trivial core is at most $n^{s(d)}$*. Indeed $s(d) = \max(d, r)$, for some constant r. So far it is all finite group theory. But the final stage needs probability theory. Let G be a **PFG** group. Let H and K be two maximal subgroups with distinct cores, say $N = \mathrm{Core}(H) \nleq K$. Then $HK = NK = G$. This means that belonging to H and K are independent events in G. Say that two maximal subgroups are equivalent if they have the same core, and choose one representative from each equivalence class. If the number of equivalence classes of maximals of index n is not polynomial, then the second part of the Borel-Cantelli lemma shows that with probability 1 a random k-tuple (for any k) lies in infinitely many of these representatives. Such a k-tuple cannot generate G, contradicting G being **PFG**. Thus for some t the number of equivalence classes of maximals of index n is at most n^t, and then the italicised claim above shows that the number of all maximal subgroups of index n is at most $n^{t+s(d)}$.

This criterion certainly implies that finitely generated pro-nilpotent groups are **PFG**, because in a d-generated such group all maximal subgroups are normal of prime index, and the number of those of index p is less than p^d. In [M1] the criterion is used to show that a finitely generated prosoluble group is **PFG**. A much more general theorem was proved in [BPS]. It is shown there that if X is any finite group, and if G is a finitely generated profinite group that does not involve X as an upper section, i.e., it does not contain open subgroups H and K, with $K \lhd H$, such that $X \cong H/K$, then G is **PFG**. We term such a group X-*deficient*, and we term it *deficient*, if it is X-deficient for some X. That assumption on G is equivalent to requiring that the variety of profinite groups generated by it is not the variety of all profinite groups. Knowing that free profinite groups are not **PFG**, it was tempting to think that deficiency characterizes **PFG** groups, but this is not the case. Consider the Cartesian product of infinitely many non-isomorphic finite simple non-abelian groups. It can be shown that as a profinite group this product is generated by two elements with positive probability. Indeed we can allow also isomorphic copies of the same group, as long as for each n we have only polynomially many simple factors with subgroups of index n. That ensures that our group is a **PMSG** one. Obviously we can choose our set of simple groups so that they involve each finite group. Another construction was found independently in [B], where M. Bhattacharjee shows that the inverse limit of iterated wreath products of simple alternating groups can be positively generated by two elements. A structural characterization of **PFG** groups is still missing. I want to draw your attention also to the following property, which was proved in [M1] for prosoluble groups and was extended in [BPS] to all finitely generated deficient groups: *the subgroup growth is exponential*. This means that $a_n(G) \leq c^n$, for some c. This is a genuine restriction: in free groups the number of subgroups grows faster than exponential. The relationship between this **ESP** (*Exponential Subgroup Property*) and **PFG** is not known.

The proofs regarding Cartesian products above, as well as the proofs in [B], are by direct computation. An alternative approach, which was noted by A. Shalev, can be extended to the following result.

Proposition 1. *Let G be a finitely generated profinite group. Suppose that, for all numbers n, the number of open normal subgroups of G of index at most n is at most $(\log n)^s$, for some s. Then G is a* **PFG** *group.*

Proof. Let G be generated by d elements, and let M be a maximal subgroup of G of index n. Then $|G : \mathrm{Core(M)}| \leq n!$, so by assumption Core(M) is one of at most n^{2s} normal subgroups. By the italicised claim in the proof of the theorem above, for each possible core we have at most $n^{s(d)}$ maximal subgroups of index n with that core. It follows that G is a **PMSG** group. □

The growth assumption in this proposition is very strong. However, we cannot do much better. It is easy to construct 2-generated profinite groups in which the number of normal subgroups of index at most n is less than n^{ϵ}, for any ϵ, that are not **PFG** (see Example 2 on p. 457 of [M1]).

While both the Cartesian products above and Bhattacharjee's wreath products obviously satisfy the growth condition of the last proposition, for the latter it is not clear that proving that they are finitely generated is easier than proving directly that they are **PFG**. But D. Segal has recently proved that iterated wreath products of finite simple non-abelian groups, relative to any faithful primitive actions of these groups, are generated by three elements [Se]. Let us note also the following result.

Proposition 2. *Let $\{S_i\}$ be a sequence of finite simple non-abelian groups, let $G_1 = S_1$, let G_i be the standard wreath product of S_i by G_{i-1}, and let G be the inverse limit of G_i. Then G is a* **PFG** *group.*

Proof. Let N be the base group of G_i. Then N is a minimal normal subgroup of G_i. Let H be a complement of N in G_i. Then H permutes regularly the direct factors isomorphic to S_i of N. Therefore H normalizes some "diagonal" subgroup D of N, so $H < HD$, and H is not maximal in G_i. Now the proof in [MS] of the italicized claim above shows that the number of maximal subgroups of G of index n with a given core is at most n^c, for some constant c. Since the number of cores is restricted as in the previous proposition, our claim follows. □

We conjecture that a result similar to Proposition 2 holds also if the wreath product is not necessarily the standard one, but is taken with respect to any faithful permutation action of G_i. Note that the proposition implies that all the groups G_i can be generated by a bounded number of elements. A similar remark applies to the results of [B] and [Se].

We now come to the applications that do not mention probability in their formulation. These follow from essentially one simple computation. Let G be a **PFG** group. We ask what is the probability $Q(G, k)$ that k random elements of G generate a subgroup of finite index. We remark parenthetically that while it is easy to see that if H is open in G and H is **PFG**, then so is G, the reverse implication was not proved yet. Be that as it may, our question above makes sense, and since G has only countably many finite index subgroups, we have $Q(G, k) = \sum_{H \leq_f G} P(H, k)$. Now suppose that G is deficient. Let F be the free group of rank d in the profinite variety generated by G. Then we know that F is **PFG**, so $P(F, k) = C > 0$, for some k. It follows that all d-generated groups in the same variety are generated by k elements with probability at least C. In calculating $Q(G, k)$, let us consider only the contribution of subgroups that can be generated by d elements, and let us write $a_{n,d}(G)$ for the number of such subgroups of index n. We have then that k random elements of G generate a d-generated subgroup of finite index with probability at least $C(\sum a_{n,d}(G)/n^k)$. Being a probability, the last sum is at most 1, so we see that $\sum a_{n,d}(G)/n^k < 1/C$, and in particular, the sum converges. Thus we have the following.

Theorem 2. *Let G be a deficient profinite group. Then for each d there exists some k, depending only on d and the variety generated by G, such that*

$$a_{n,d}(G) = o(n^k).$$

E.g. for pro-p groups we obtain $a_{n,d}(G) = o(n^d)$ and for prosoluble groups $a_{n,d}(G) = o(n^{(13/4)d+4})$.

Next, we count normal subgroups in a similar manner. For this we have to restrict ourselves to pro-p groups. Let G a finitely generated pro-p group, and let N be a normal subgroup. We ask what is the probability that N is the normal closure in G of k elements. The subgroup $[N, G]N^p$ is a proper subgroup of N which is easily seen to be the intersection of all the maximal subgroups of N that are normal in G. Therefore a k-tuple generates N normally if and only if it generates $N/[N, G]N^p$ normally. But the latter group is central in $G/[N, G]N^p$, so our k-tuple generates it normally if and only if it generates it in the usual sense, so our probability is simply the probability that k elements of G generate $N/[N, G]N^p$. Again, this k-tuple has first to lie in N, the probability of this being $1/|G : N|^k$. If the elementary abelian group $N/[N, G]N^p$ has rank k, then the probability of a k-tuple to generate it is $(1 - 1/p)...(1 - 1/p^k) = C_k$, say, while if the rank is smaller than k the probability is even bigger, and for bigger rank the probability is 0. Let us write now $t_{n,r}(G)$ for the number of normal subgroups of G of index n that are the normal closure of r elements. Then the above remarks show that the probability that r elements of G generate normally a subgroup of finite index is at least $C_r(\sum t_{n,r}(G)/n^r)$. As above, we deduce that the sum converges, and that $t_{n,r}(G) = o(n^r)$. In particular, let F be a free pro-p group of finite rank $d \leq r$. Then each p-group P of order p^n that can be generated by d elements is a factor group of F, and if P can be defined by r relations, then the kernel N is generated normally by these relations. Recalling that if a finite group can be defined by r relations, then it can also be generated by the same number of elements, we obtain the following.

Theorem 3. *Let $h(n, r)$ be the number of groups of order n that can be defined by r relations. Then, as $k \to \infty$,*

$$h(p^k, r) = o(p^{kr}).$$

We conjecture that in general $h(n, r) = o(n^r)$. From the theorem above one can deduce this for nilpotent groups, but for finite groups in general we know only that $h(n, r) = O(n^{r \log n})$, [M2].

Added in proof: More details on our topic can be found in the recent book [LubSeg], especially Chapter 11.

References

[B] M. Bhattacharjee, The probability of generating certain profinite groups by two elements, *Israel J. Math.* **86** (1990), 311–320.

[BPS] A. V. Borovik, L. Pyber, and A. Shalev, Maximal subgroups in finite and profinite groups, *Trans. Amer. Math. Soc.* **348** (1996), 3745–3761.

[D] J. D. Dixon, The probability of generating the symmetric group, *Math. Z.* **110** (1969), 199–205.

[KL] W. M. Kantor and A. Lubotzky, The probability of generating a finite classical group, *Geom. Ded.* **36** (1990), 67–87.

[LS] M. Liebeck and A. Shalev, The probability of generating a finite simple group, *Geom. Ded.* **56** (1995), 103–113.

[LubSeg] A. Lubotzky and D. Segal, *Subgroup Growth*, Birkhäuser, Basel, 2003.

[M1] A. Mann, Positively finitely generated groups, *Forum Math.* **8** (1996), 429–459.

[M2] A. Mann, Enumerating finite groups and their defining relations II, submitted for publication.

[MS] A. Mann and A. Shalev, Simple groups, maximal subgroups, and probabilistic aspects of profinite groups, *Israel J. Math.* **96** (1996), 449–468.

[N1] E. Netto, *Substitutiontheorie und ihre Anwendungen auf die Algebra*, Teubner, Leipzig, 1882.

[N2] E. Netto, *The Theory of Substitutions and its Applications to Algebra*, George Weber, Ann Arbor, 1892.

[R] A. Renyi, *Probability Theory*, North Holland, Amsterdam, 1970.

[Se] D. Segal, The finite images of finitely generated groups, *Proc. London Math. Soc.* (3) **82** (2001), 597–613.

[Sh] A. Shalev, Probabilistic group theory, in: *Groups St Andrews 1997 in Bath, II*, LMS Lecture Notes in Mathematics vol. 261, Cambridge, 1999, 648–678.

12

Parity patterns in Hecke groups and Fermat primes

T. W. Müller

For Wilfried Imrich on the occasion of his sixtieth birthday

1. Introduction

For a group $\mathfrak{G}$ denote by $s_n(\mathfrak{G})$ the number of subgroups of index n in $\mathfrak{G}$. If $\mathfrak{G}$ is finitely generated or of finite (subgroup) rank, then $s_n(\mathfrak{G})$ is finite for all n. The present paper is concerned with the behaviour modulo 2 of the function $s_n(\mathfrak{G})$ and another arithmetic function counting free subgroups of finite index in the case when $\mathfrak{G}$ is a Hecke group. Rather surprisingly, it turns out that Fermat primes play an important special role in this context, a phenomenon hitherto unobserved in the arithmetic theory of Hecke groups, and, as a byproduct of our investigation, several new characterizations of Fermat primes are obtained.

The natural framework for our research is the theory of *subgroup growth* of finitely generated virtually free groups. The notion of subgroup growth, which has evolved over the last two decades in the work of Grunewald, Lubotzky, Mann, Segal, and others including the present author, brings together under a common conceptual roof investigations concerning arithmetic properties of the sequence $\{s_n(\mathfrak{G})\}_{n\geq 1}$ or related subgroup counting functions and their connection with the algebraic structure of the group $\mathfrak{G}$. The original motivation for these studies comes from three sources: the notion of word growth and, more specifically, Gromov's characterization in [12] of finitely generated groups with polynomial word growth, the theory of rings of algebraic integers and their zeta functions, and the work of M. Hall and T. Radó in the late 1940's on Schreier systems and their associated subgroups in free groups; cf. [13], [14], and [15]. Most of the major developments up to 1992 are described in Lubotzky's Galway notes [19], [20], and the literature cited therein. More recent contributions include [7], [22], [21], [26], [27], [28], [30], [31], and [8].

Whereas for instance the growth behaviour and the asymptotics of the function $s_n(\mathfrak{G})$ tend to react fairly smoothly to variation of $\mathfrak{G}$, say, over a commensurability class, divisibility properties of $s_n(\mathfrak{G})$ are usually severely deformed if not completely destroyed in this process. The latter type of arithmetic structure appears to be rather subtle and peculiar to the particular group under investigation. This tendency (apart from considerable technical difficulties) may account for the fact that very little is known concerning divisibility properties of subgroup counting functions. Indeed, the only published results of major interest in the present context seem to be Stothers' formulae for the modular group. Before stating his results let us introduce some general notation. For a finitely generated group $\mathfrak{G}$ define

$$\Pi(\mathfrak{G}) := \left\{ n \in \mathbb{N} : \; s_n(\mathfrak{G}) \equiv 1 \bmod 2 \right\}.$$

Moreover, for $\mathfrak{G}$ a finitely generated virtually free group, let

$$\Pi^*(\mathfrak{G}) := \left\{ \lambda \in \mathbb{N} : \; b_\lambda(\mathfrak{G}) \equiv 1 \bmod 2 \right\}.$$

Here, $b_\lambda(\mathfrak{G})$ is the number of free subgroups in $\mathfrak{G}$ of index $\lambda m_{\mathfrak{G}}$, and $m_{\mathfrak{G}}$ denotes the least common multiple of the orders of the finite subgroups in $\mathfrak{G}$. We call $\Pi(\mathfrak{G})$ and $\Pi^*(\mathfrak{G})$ the *parity pattern* respectively the *free parity pattern* of the group $\mathfrak{G}$. In this notation the main result of Stothers [39] is that

$$\Pi(PSL(2, \mathbb{Z})) = \left\{2^{\sigma+1} - 3\right\}_{\sigma \geq 1} \cup 2\left\{2^{\sigma+1} - 3\right\}_{\sigma \geq 1}. \tag{1}$$

This striking result, which had been conjectured for some time on the basis of numerical evidence, has, for more than 20 years, stood out as an indication that a fascinating chapter of subgroup arithmetic might still be awaiting its discovery. In the course of his proof Stothers also shows that

$$\Pi^*(PSL(2, \mathbb{Z})) = \left\{2^{\sigma} - 1\right\}_{\sigma \geq 1}. \tag{2}$$

The latter pattern has been shown to occur for a larger class of virtually free groups of free rank 2, including free products $\mathfrak{G} = G_1 *_S G_2$ of two finite groups G_i with an amalgamated subgroup S of odd cardinality, whose indices $(G_i : S)$ satisfy $\{(G_1 : S), (G_2 : S)\} = \{2, 3\}$ or $= \{2, 4\}$; cf. [25, Prop. 6].[1]

For an integer $q \geq 3$ let $\mathfrak{H}(q)$ be the Hecke group corresponding to q, i.e., the group of linear fractional transformations generated by the transformations $\tau' = -1/\tau$ and $\tau' = \tau + 2\cos(\pi/q)$. We have $\mathfrak{H}(3) = \mathrm{PSL}(2, \mathbb{Z})$, and in general $\mathfrak{H}(q)$ is a subgroup of $\mathrm{PSL}(2, R)$, where R is the ring of integers of the

[1] The free rank $\mu(\mathfrak{G})$ of a finitely generated virtually free group $\mathfrak{G}$ is defined as the rank of a free subgroup of index $m_{\mathfrak{G}}$ in $\mathfrak{G}$. See Section 2.1 below.

cyclotomic field $\mathbb{Q}(\zeta_q + \zeta_q^{-1})$ with $\zeta_q = e^{i\pi/q}$. The precise structure of the matrices entering into this representation of $\mathfrak{H}(q)$ is not in general known. As an abstract group, $\mathfrak{H}(q)$ is the free product of a cyclic group of order 2 and a cyclic group of order q; in particular, $\mathfrak{H}(q)$ is virtually free. The purpose of this paper is to investigate the parity patterns $\Pi(\mathfrak{H}(q))$ and $\Pi^*(\mathfrak{H}(q))$ for $q \geq 3$.

There are several a priori reasons to suspect that Hecke groups might prove to be an interesting and fruitful class of groups to study with regard to their parity patterns and similar number-theoretic properties: (i) The groups $\mathfrak{H}(q)$ form what is probably the most natural generalization of the modular group, sharing the latter's central position in large parts of mathematics, in particular its close ties with number theory. (ii) In 1998, building on results obtained in [23], an analogue of Stothers' formula (1) for $q = 5$ was discovered by the author, namely

$$\Pi(\mathfrak{H}(5)) = \left\{\frac{2^{2\sigma+1}-5}{3}\right\}_{\sigma\geq 1} \cup 2\left\{\frac{2^{2\sigma+1}-5}{3}\right\}_{\sigma\geq 1}; \tag{3}$$

cf. [32]. (iii) According to a result of Grady and Newman a free product $\mathfrak{G}$ of finitely many cyclic groups of prime order containing at least four copies of the cyclic group of order 2 in its free decompositon satisfies $\Pi(\mathfrak{G}) = \mathbb{N}$; cf. [11, Theorem 1]. In the same paper, Grady and Newman conjecture that two free factors C_2 should already suffice to ensure the same conclusion. This conjecture, if true, would place Hecke groups within a tight borderline region, which might well deserve further study. The more specific conjecture made in [10] to the effect that a free product of two or more copies of the cyclic group of order 2 should have all its subgroup numbers odd, follows immediately from one of our results (see Proposition G below). (iv) So far, parity patterns have only been studied for groups of free rank at most 3, and there are some indications that for fixed free rank only a very small number of non-trivial patterns arise. The fact that the free rank $\mu(\mathfrak{H}(q))$ becomes unbounded as $q \to \infty$, while accounting for technical difficulties well beyond the scope of any previously known method of investigation, adds considerably to the interest of Hecke groups as a class of "test samples" with regard to this type of arithmetic structure.

We now turn to the main results of this paper. Somewhat surprisingly, formulae (1) – (3) do not generalize to all Hecke groups. Instead, there exist canonical (and maximal) generalizations of Stothers' formulae (1) and (2), each characterizing a particular subclass of Hecke groups (see Theorems A and B below). Moreover, both these subclasses essentially correspond to certain "arithmetic singularities", outside of which the subgroup counting function under investigation displays a peculiar type of fractal behaviour, thus ruling out the possibility of describing the associated parity pattern by a closed formula à la Stothers.

For an integer $q \geq 3$ define

$$\Lambda_q^* := \left\{ \frac{\mu_q^\sigma - 1}{\mu_q - 1} : \sigma = 1, 2, \dots \right\},$$

i.e., Λ_q^* is the set of partial sums of the geometric series generated by the free rank $\mu_q := \mu(\mathfrak{H}(q))$ of $\mathfrak{H}(q)$. Our first main result characterizes those Hecke groups whose free parity pattern canonically generalizes the pattern found by Stothers for the modular group.

Theorem A. *Let $q \geq 3$ be an integer. Then the following assertions are equivalent:*

(i) $\Pi^*(\mathfrak{H}(q)) = \Lambda_q^*$.
(ii) $b_\lambda(\mathfrak{H}(q)) \equiv 0 \ (2)$ *for* $2 \leq \lambda \leq \mu_q$.
(iii) q *or* $q - 1$ *is a 2-power.*

As is well known, Fermat primes, i.e., prime numbers of the form $2^{2^\lambda} + 1$ with $\lambda \geq 0$, satisfy (or can even be characterized by) a number of curious regularity conditions; for instance, according to Gauss, [2] a regular p-gon ($p > 2$ a prime) can be constructed by compass and ruler if and only if p is a Fermat prime. By specializing Theorem A to the case when q is a prime number we obtain a new such characterization.

Corollary A′. *Let $q > 2$ be a prime. Then q is a Fermat prime if and only if* $\Pi^*(\mathfrak{H}(q)) = \Lambda_q^*$.

The connection with Fermat primes becomes even more striking when turning to the patterns $\Pi(\mathfrak{H}(q))$. For $q \geq 3$ define

$$\Lambda_q := \left\{ \frac{2(q-1)^\sigma - q}{q - 2} : \sigma = 1, 2, \dots \right\},$$

i.e., Λ_q is the set of partial sums of the series $1 + 2\sum_{\sigma \geq 1}(q-1)^\sigma$.

Theorem B. (a) *For even $q > 2$ we have* $\Pi(\mathfrak{H}(q)) = \mathbb{N}$.
(b) *Let $q \geq 3$ be an odd integer. Then the following assertions are equivalent:*

(i) $\Pi(\mathfrak{H}(q)) = \Lambda_q \cup 2\Lambda_q$.
(ii) $s_n(\mathfrak{H}(q)) \equiv 0 \ (2)$ *for* $n \in ([2q(q-1)] - \{1, 2q-1\}) \cap (\mathbb{N} - 2\mathbb{N})$.
(iii) q *is a Fermat prime.*

We also establish certain general properties of $\Pi^*(\mathfrak{H}(q))$ and $\Pi(\mathfrak{H}(q))$, a number of which are summarized in the next two results.

[2] Disquisitiones arithmeticae, § 366.

Theorem C. (a) *Let* $q_1, q_2 \geq 3$ *be integers. Then we have* $\Pi^*(\mathfrak{H}(q_1)) = \Pi^*(\mathfrak{H}(q_2))$ *if and only if* $\mu_{q_1} = \mu_{q_2}$.
(b) *Every entry of* $\Pi^*(\mathfrak{H}(q))$ *is congruent to* 1 *modulo* $2^{\nu_2(\mu_q)}$.[3]
(c) *The first two entries of* $\Pi^*(\mathfrak{H}(q))$ *are* 1 *and* $2^{\nu_2(\mu_q)} + 1$*; in particular, given an integer* $\alpha^* \geq 2$*, the set* $\{1, \alpha^*\}$ *can be extended to a free parity pattern of some Hecke group if and only if* $\alpha^* - 1$ *is a 2-power.*
(d) *The series* $1 + \sum_{\lambda \geq 1} b_\lambda(\mathfrak{H}(q))\, z^\lambda$ *is never rational over* $GF(2)$*; in particular, the set* $\Pi^*(\mathfrak{H}(q))$ *is always infinite.*

Theorem D. (a) *Let* $q_1, q_2 \geq 3$ *be odd integers, and suppose that* $\Pi(\mathfrak{H}(q_1)) = \Pi(\mathfrak{H}(q_2))$*. Then we have* $q_1 = q_2$.
(b) *If* $q \geq 3$ *is odd, then*

(i) $\Pi(\mathfrak{H}(q)) \subseteq (1 + 4\,\mathbb{N}_0) \cup (2 + 8\,\mathbb{N}_0)$,
(ii) *the first entries of* $\Pi(\mathfrak{H}(q))$ *are* $1, 2,$ *and* $2p(q) - 1$*, where* $p(q)$ *denotes the smallest prime divisor of* q,
(iii) *the series* $\sum_{n \geq 0} s_{n+1}(\mathfrak{H}(q))\, z^n$ *is not rational over* $GF(2)$*; in particular, the set* $\Pi(\mathfrak{H}(q))$ *is infinite.*

Furthermore, we obtain fairly explicit descriptions of the patterns $\Pi^*(\mathfrak{H}(q))$ and $\Pi(\mathfrak{H}(q))$ for arbitrary q.

Theorem E. *For every* $q \geq 3$,

$$\Pi^*(\mathfrak{H}(q)) = \Big\{\lambda \in \mathbb{N} : \ \mathfrak{s}(\lambda) + \mathfrak{s}((\mu_q - 1)\lambda + 1) - \mathfrak{s}(\mu_q \lambda) = 1\Big\}.$$

Here $\mathfrak{s}(x)$ denotes the sum of digits in the binary representation of x. The corresponding result for the patterns $\Pi(\mathfrak{H}(q))$ is somewhat more involved. For an odd integer $q \geq 3$ let $1 = d_0 < d_1 < \cdots < d_r = q$ be the set of divisors of q in increasing order, and let

$$\underline{d}_q := \left(\frac{d_1 - 1}{2}, \frac{d_2 - 1}{2}, \ldots, \frac{d_r - 1}{2}\right) \in \mathbb{N}_0^r.$$

As usual, define the norm of a vector $\underline{v} = (v_1, \ldots, v_r) \in \mathbb{N}_0^r$ as $||\underline{v}|| = \sum_{j=1}^r v_j$, and if $\underline{u} = (u_1, \ldots, u_r)$ and $\underline{v} = (v_1, \ldots, v_r)$ are two such vectors, then their scalar product is given by $\underline{u} \cdot \underline{v} = \sum_{j=1}^r u_j v_j$.

Theorem F. *Let* $q \geq 3$ *be an odd integer. Then we have*

$$\Pi_q = \Theta_q \cup 2\Theta_q,$$

[3] For a positive integer n, $\nu_2(n)$ denotes the 2-adic norm of n, i.e., the exponent of 2 in the prime decomposition of n.

where Θ_q *consists of all positive integers* $n \equiv 1$ (4) *such that the set*

$$\Big\{\underline{n} \in \mathbb{N}_0^r : \ \underline{d}_q \cdot \underline{n} = \frac{n-1}{4} \quad \text{and} \quad \sum_{j=1}^{r} \mathfrak{s}(n_j) + \mathfrak{s}(\frac{n+1}{2} - ||\underline{n}||) - \mathfrak{s}(\frac{n-1}{2}) = 1\Big\}$$

has odd cardinality.

The description of the parity pattern Π_q for q odd given in the last theorem simplifies considerably if q is a prime number.

Corollary F′. *Let* $q > 2$ *be a prime. Then* $\Pi_q = \Theta_q \cup 2\,\Theta_q$, *where*

$$\Theta_q = \Big\{2(q-1)n + 1 : \ n \in \mathbb{N}_0 \quad \text{and} \quad \mathfrak{s}(n) + \mathfrak{s}((q-2)n+1) - \mathfrak{s}((q-1)n) = 1\Big\}.$$

Sections 2 – 4 of the present paper are concerned with the free parity patterns $\Pi^*(\mathfrak{H}(q))$. In Section 2, building on results of [25], we derive a new recurrence relation for the function $b_\lambda(\mathfrak{G})$, which is not of Hall type, and is, at least in principle, fairly well adapted for studying divisibility properties, but rather complicated and awkward to handle. However, a careful analysis of this recursion, undertaken in Section 3, leads to the much simpler identity

$$X_q^*(z) = 1 + z\left(X_q^*(z)\right)^{\mu_q} \tag{4}$$

for the mod 2 projection $X_q^*(z) \in GF(2)[[z]]$ of the generating function $1 + \sum_{\lambda=1}^{\infty} b_\lambda(\mathfrak{H}(q))z^\lambda$, and our results for the patterns $\Pi^*(\mathfrak{H}(q))$ follow from a thorough discussion of (4). The analysis of the patterns $\Pi(\mathfrak{H}(q))$ is considerably more involved. Instead of working directly with $\mathfrak{H}(q)$ we first consider the group $\widetilde{\mathfrak{H}}(q) := C_q * C_q$ which embeds in the corresponding Hecke group as a subgroup of index 2. Indeed, if $\mathfrak{H}(q) = \langle \sigma, \tau | \sigma^2 = \tau^q = 1\rangle$, then the subgroup generated by τ and τ^σ is isomorphic to $\widetilde{\mathfrak{H}}(q)$ with transversal $\{1, \sigma\}$, so that $\mathfrak{H}(q)$ is in fact a split extension of the group $\widetilde{\mathfrak{H}}(q)$ by C_2. The connection between the parity patterns of these two groups is given by the following general observation.

Proposition G. *Let* $\mathfrak{G}$ *be a group containing only finitely many subgroups of index* n *for every positive integer* n, *and let* $\mathfrak{H}$ *be a subgroup of index* 2 *in* $\mathfrak{G}$. *Then*

$$\Pi(\mathfrak{G}) = \Big(\Pi(\mathfrak{H}) \cap (\mathbb{N} - 2\mathbb{N})\Big) \cup 2\,\Pi(\mathfrak{H}). \tag{5}$$

This rather surprising relationship allows us in particular to translate results concerning the parity patterns of the groups $\widetilde{\mathfrak{H}}(q)$ into results for the patterns $\Pi(\mathfrak{H}(q))$. It also bears on the conjecture of Grady and Newman concerning groups of the form $\mathfrak{G} = C_2 * \ldots * C_2$ mentioned earlier in this introduction. At first sight, replacing the investigation of the parity pattern $\Pi(\mathfrak{H}(q))$ by the corresponding problem for the group $\widetilde{\mathfrak{H}}(q)$ might not strike one as a particularly promising reduction, since for $q > 5$ both groups have so far entirely resisted all attempts at such an analysis, due to the enormous complexity of the arithmetic functions involved, which becomes uncontrollable as $q \to \infty$. A breakthrough for the groups $\widetilde{\mathfrak{H}}(q)$ – and hence via (5) for Hecke groups – is made possible by (i) divising an approach concentrating the main complexity of the problem in the series

$$\mathcal{R}_q^{\mu,\nu}(z) := \frac{d^\nu}{dz^\nu}\left\{\sum_{n=0}^{\infty} h_n(q)h_{n+\mu}(q)z^n/n!\right\}\Big/\left\{\sum_{n=0}^{\infty} h_n^2(q)z^n/n!\right\},$$

where $h_n(q) := |\operatorname{Hom}(C_q, S_n)|$, (ii) showing that for $q \geq 3$, $\mu \geq 1$, and $\nu \geq 0$ the series $\mathcal{R}_q^{\mu,\nu}(z)$ are in fact *integral* power series, and (iii) the fact that in the critical equation (44) governing the generating function $\sum_{n=0}^{\infty} s_{n+1}(\widetilde{\mathfrak{H}}(q))\, z^n$ the series $\mathcal{R}_q^{\mu,\nu}(z)$, apart from shift factors, always occur with even coefficient, and thus, given (ii), can be ignored modulo 2 (this is the advantage of working with the groups $\widetilde{\mathfrak{H}}(q)$). It is remarkable that, while an exact computation of the remainder terms $\mathcal{R}_q^{\mu,\nu}(z)$ (which all previous attempts have, at least implicitly, aimed at) appears entirely out of reach for large q, their mere integrality can, by a fairly subtle argument occupying the whole of Section 7, be established for arbitrary q. As a consequence of these developments we obtain the differential equation

$$\widetilde{X}_q(z) = \sum_{d|q}\sum_{\mu\geq 0} z^{2d-3}\left(\widetilde{X}_q'(z)\right)^\mu \left(\widetilde{X}_q(z)\right)^{d-2(\mu+1)} \times\left[\binom{d-1}{2\mu} z\,\widetilde{X}_q(z) + (d-1)\binom{d-2}{2\mu}\right] \tag{6}$$

for the mod 2 projection $\widetilde{X}_q(z)$ of the generating function $\sum_{n=0}^{\infty} s_{n+1}(\widetilde{\mathfrak{H}}(q))\, z^n$. A detailed analysis of equation (6) then leads to a rather complete set of results for the parity patterns $\Pi(\widetilde{\mathfrak{H}}(q))$, which are translated into results concerning the patterns $\Pi(\mathfrak{H}(q))$ in Section 8, following the proof of Proposition G.

Acknowledgement. I would like to thank Christian Krattenthaler for a number of valuable remarks and comments concerning an earlier version of this paper.

2. Some preliminaries concerning virtually free groups

2.1. The structure theorem and some invariants

Let $\mathfrak{G}$ be a finitely generated virtually free group and denote by $m_{\mathfrak{G}}$ the least common multiple of the orders of the finite subgroups in $\mathfrak{G}$. By Stallings' structure theorem on groups with infinitely many ends and the subsequent work of Karrass, Pietrowski, and Solitar, $\mathfrak{G}$ can be presented as the fundamental group of a finite graph of groups $(\mathfrak{G}(-), Y)$ in the sense of Bass and Serre with finite vertex groups $\mathfrak{G}(v)$; cf. [37] and [17], or [5, Sect. IV.1.9]. The fact that, conversely, the fundamental group of a finite graph of finite groups is always virtually free of finite rank is more elementary, and can be found for instance in [35, Sect. II.2.6]. It follows in particular from this characterization and the universal covering construction in the category of graphs of groups that a torsion-free subgroup of a finitely generated virtually free group is in fact free (which was the original contribution of Stallings' work to the structure theory of virtually free groups).

If $\mathfrak{F}$ is a free subgroup of finite index in $\mathfrak{G}$ then, following an idea of Wall, one defines the rational Euler characteristic $\chi(\mathfrak{G})$ of $\mathfrak{G}$ as

$$\chi(\mathfrak{G}) = -\frac{\mathrm{rk}(\mathfrak{F}) - 1}{(\mathfrak{G} : \mathfrak{F})}. \tag{7}$$

This is well-defined in view of Schreier's theorem [34], and if $\mathfrak{G} \cong \pi_1(\mathfrak{G}(-), Y)$ is a decomposition of $\mathfrak{G}$ in terms of a graph of groups, then we have

$$\chi(\mathfrak{G}) = \sum_{v \in V(Y)} \frac{1}{|\mathfrak{G}(v)|} - \sum_{e \in E(Y)} \frac{1}{|\mathfrak{G}(e)|}, \tag{8}$$

where $V(Y)$ and $E(Y)$ denote respectively the set of vertices and (geometric) edges of Y.The latter formula reflects the fact that in our situation the Euler characteristic in the sense of Wall coincides with the equivariant Euler characteristic $\chi_T(\mathfrak{G})$ of $\mathfrak{G}$ relative to the tree T canonically associated with $\mathfrak{G}$ in the sense of Bass and Serre; cf. [1, Chap. IX, Prop. 7.3] or [36, Prop. 14]. Define the (free) rank $\mu(\mathfrak{G})$ of $\mathfrak{G}$ to be the rank of a free subgroup of index $m_{\mathfrak{G}}$ in $\mathfrak{G}$. The existence of such a subgroup follows from [35, Lemmas 8 and 10] or formulae (11) and (12) below. Observe that, in view of (7), $\mu(\mathfrak{G})$ is connected with the Euler characteristic of $\mathfrak{G}$ via

$$\mu(\mathfrak{G}) + m_{\mathfrak{G}}\chi(\mathfrak{G}) = 1, \tag{9}$$

which shows in particular that $\mu(\mathfrak{G})$ is well-defined.

The type $\tau(\mathfrak{G})$ of a finitely generated virtually free group $\mathfrak{G} \cong \pi_1(\mathfrak{G}(-), Y)$ is defined as the tuple

$$\tau(\mathfrak{G}) = \big(m_{\mathfrak{G}}; \zeta_1(\mathfrak{G}), \ldots, \zeta_\kappa(\mathfrak{G}), \ldots, \zeta_{m_{\mathfrak{G}}}(\mathfrak{G})\big),$$

where the $\zeta_\kappa(\mathfrak{G})$ are integers indexed by the divisors of $m_{\mathfrak{G}}$, given by

$$\zeta_\kappa(\mathfrak{G}) = \Big|\Big\{e \in E(Y) : |\mathfrak{G}(e)| \,\Big|\, \kappa\Big\}\Big| - \Big|\Big\{v \in V(Y) : |\mathfrak{G}(v)| \,\Big|\, \kappa\Big\}\Big|.$$

It can be shown that the type $\tau(\mathfrak{G})$ is in fact an invariant of the group $\mathfrak{G}$, i.e., independent of the particular decomposition of $\mathfrak{G}$ in terms of a graph of groups $(\mathfrak{G}(-), Y)$, and that two virtually free groups $\mathfrak{G}_1$ and $\mathfrak{G}_2$ contain the same number of free subgroups of index n for each $n \in \mathbb{N}$ if and only if $\tau(\mathfrak{G}_1) = \tau(\mathfrak{G}_2)$; cf. [25, Theorem 2]. Note that as a consequence of (8) the Euler characteristic of $\mathfrak{G}$ can be expressed in terms of the type $\tau(\mathfrak{G})$ via

$$\chi(\mathfrak{G}) = -m_{\mathfrak{G}}^{-1} \sum_{\kappa | m_{\mathfrak{G}}} \varphi(m_{\mathfrak{G}}/\kappa)\, \zeta_\kappa(\mathfrak{G}), \tag{10}$$

where φ is Euler's totient function. It follows in particular that if two virtually free groups have the same number of free index n subgroups for every n, then their Euler characteristics must coincide.

2.2. The functions $a_\lambda(\mathfrak{G})$ and $b_\lambda(\mathfrak{G})$

The total information on the number of free subgroups of given finite index in a finitely generated virtually free group $\mathfrak{G}$ is concentrated in the constant $m_{\mathfrak{G}}$ and the function $b(\mathfrak{G}) : \mathbb{N} \to \mathbb{N}$ given by

$$b_\lambda(\mathfrak{G}) = \text{number of free subgroups of index } \lambda m_{\mathfrak{G}} \text{ in } \mathfrak{G}.$$

Our approach to the function $b_\lambda(\mathfrak{G})$ is to relate it to another arithmetic function $a_\lambda(\mathfrak{G})$ which turns out to be easier to compute. Define a torsion-free $\mathfrak{G}$-action on a set Ω to be a $\mathfrak{G}$-action on Ω which is free when restricted to finite subgroups. For a finite set Ω to admit a torsion-free $\mathfrak{G}$-action it is necessary and sufficient that $|\Omega|$ be divisible by $m_{\mathfrak{G}}$. For $\lambda \in \mathbb{N}_0$ define $a_\lambda(\mathfrak{G})$ by the condition that

$$\begin{aligned}&(\lambda m_{\mathfrak{G}})!\, a_\lambda(\mathfrak{G})\\ &\quad = \text{number of torsion-free } \mathfrak{G}\text{-actions on a set with } \lambda m_{\mathfrak{G}} \text{ elements},\end{aligned}$$

in particular, $a_0(\mathfrak{G}) = 1$. Then the arithmetic functions $a_\lambda(\mathfrak{G})$ and $b_\lambda(\mathfrak{G})$ are related via the transformation formula[4]

$$\sum_{\mu=1}^{\lambda} a_{\lambda-\mu}(\mathfrak{G})\, b_\mu(\mathfrak{G}) = m_{\mathfrak{G}}\, \lambda\, a_\lambda(\mathfrak{G}), \quad \lambda \geq 1. \tag{11}$$

Moreover, a careful analysis of the universal mapping property associated with the presentation $\mathfrak{G} \cong \pi_1(\mathfrak{G}(-), Y)$ of $\mathfrak{G}$ in terms of a graph of groups $(\mathfrak{G}(-), Y)$ leads to the explicit formula

$$a_\lambda(\mathfrak{G}) = \frac{\prod\limits_{e \in E(Y)} \left[(\lambda m_{\mathfrak{G}}/|\mathfrak{G}(e)|)!\; |\mathfrak{G}(e)|^{\lambda m_{\mathfrak{G}}/|\mathfrak{G}(e)|} \right]}{\prod\limits_{v \in V(Y)} \left[(\lambda m_{\mathfrak{G}}/|\mathfrak{G}(v)|)!\; |\mathfrak{G}(v)|^{\lambda m_{\mathfrak{G}}/|\mathfrak{G}(v)|} \right]}, \quad \lambda \geq 0 \tag{12}$$

for $a_\lambda(\mathfrak{G})$; compare [25, Prop. 3]. From this formula it can be deduced that the sequence $a_\lambda(\mathfrak{G})$ is of hypergeometric type and that the generating function $\alpha_{\mathfrak{G}}(z) := \sum_{\lambda=0}^{\infty} a_\lambda(\mathfrak{G}) z^\lambda$ satisfies a homogeneous linear differential equation

$$A_0(\mathfrak{G})\, \alpha_{\mathfrak{G}}(z) + (A_1(\mathfrak{G})\, z - m_{\mathfrak{G}})\, \alpha'_{\mathfrak{G}}(z) + \sum_{\mu=2}^{\mu(\mathfrak{G})} A_\mu(\mathfrak{G})\, z^\mu\, \alpha_{\mathfrak{G}}^{(\mu)}(z) = 0 \tag{13}$$

of order $\mu(\mathfrak{G})$ with integral coefficients

$$A_\mu(\mathfrak{G}) = \frac{1}{\mu!} \sum_{j=0}^{\mu} (-1)^{\mu-j} \binom{\mu}{j} m_{\mathfrak{G}}\, (j+1) \prod_{\kappa \mid m_{\mathfrak{G}}} \prod_{\substack{1 \leq k \leq m_{\mathfrak{G}} \\ (m_{\mathfrak{G}}, k) = \kappa}} (j m_{\mathfrak{G}} + k)^{\zeta_\kappa(\mathfrak{G})},$$

$$0 \leq \mu \leq \mu(\mathfrak{G}); \tag{14}$$

cf. [25, Prop. 5].

2.3. A recurrence relation for $b_\lambda(\mathfrak{G})$

Inserting formula (12) into (11) yields a recursive description of the arithmetic function $b_\lambda(\mathfrak{G})$ attached to a finitely generated virtually free group $\mathfrak{G}$. However, formulae obtained in this way (referred to as being of Hall type) usually turn out to be quite unsatisfactory when dealing with number-theoretic aspects of the sequence $b_\lambda(\mathfrak{G})$ such as divisibility properties. Instead, we will derive a recurrence relation for $b_\lambda(\mathfrak{G})$ which differs considerably in form from (11) and

[4] See for instance [25, Cor. 1].

is rather well adapted for studying divisibility properties of the function $b_\lambda(\mathfrak{G})$. The following result will be the starting point of our present investigations.

Proposition 1. *Let $\mathfrak{G}$ be a finitely generated virtually free group. Then the function $b_\lambda(\mathfrak{G})$ satisfies the recursion*

$$b_{\lambda+1}(\mathfrak{G}) = \sum_{\mu=1}^{\mu(\mathfrak{G})} \sum_{\substack{\lambda_1,\dots,\lambda_\mu>0\\ \lambda_1+\cdots+\lambda_\mu=\lambda}} (\mu!\, m_{\mathfrak{G}}^{\mu})^{-1}\, \mathcal{B}_\mu^{(\mathfrak{G})}(\lambda_1,\dots,\lambda_\mu) \prod_{j=1}^{\mu} b_{\lambda_j}(\mathfrak{G})$$

$$(\lambda \geq 1,\ b_1(\mathfrak{G}) = A_0(\mathfrak{G})) \quad (15)$$

with coefficients

$$\mathcal{B}_\mu^{(\mathfrak{G})}(\lambda_1,\dots,\lambda_\mu) := \sum_{\nu=\mu}^{\mu(\mathfrak{G})} A_\nu(\mathfrak{G})\, B_{\mu,\nu}(\lambda_1,\dots,\lambda_\mu),$$

where

$$B_{\mu,\nu}(\lambda_1,\dots,\lambda_\mu) := \nu! \sum_{\substack{\nu_1,\dots,\nu_\mu\geq 0\\ \nu_1+\cdots+\nu_\mu=\nu-\mu}} \prod_{j=1}^{\mu} \left[\binom{\lambda_j-1}{\nu_j}\Big/(\nu_j+1)\right]$$

and $A_\nu(\mathfrak{G})$ is as in (14).

Proof. Introduce the generating function

$$\beta_{\mathfrak{G}}(z) := \sum_{\lambda=0}^{\infty} b_{\lambda+1}(\mathfrak{G})\, z^\lambda.$$

Then, in view of equation (11), $\beta_{\mathfrak{G}}(z)$ is related to $\alpha_{\mathfrak{G}}(z)$ via the identity

$$\beta_{\mathfrak{G}}(z) = m_{\mathfrak{G}} \frac{d}{dz}\big(\log \alpha_{\mathfrak{G}}(z)\big). \quad (16)$$

We will use Bell's formula[5]

$$\frac{d^\mu}{dz^\mu} f(g(z)) = \sum_{\pi\vdash\mu} \frac{\mu!}{\prod_{j=1}^{\infty}\pi_j!} \left[\prod_{j=1}^{\infty}\left(\frac{g^{(j)}(z)}{j!}\right)^{\pi_j}\right] f^{(\|\pi\|)}(g(z)) \quad (17)$$

[5] By a partition π we mean any sequence $\pi = \{\pi_j\}_{j\geq 1}$ of non-negative integers, such that $\pi_j = 0$ for all but finitely many j. The integer $|\pi| = \sum_{j=1}^{\infty} j\,\pi_j$ is called the weight of π, and $\|\pi\| = \sum_{j=1}^{\infty}\pi_j$ is the norm or length of the partition π. If $|\pi| = 0$, π is called the empty partition, otherwise π is non-empty. As usual, we also write $\pi \vdash \mu$ for $|\pi| = \mu$, and say that π is a partition of μ.

for the derivatives of a composite function to compute the higher derivatives of $\alpha_{\mathfrak{G}}(z) = \exp(m_{\mathfrak{G}}^{-1} \int \beta_{\mathfrak{G}}(z)\, dz)$. Applying (17) with $f(t) = \mathrm{e}^t$ and $g(z) = m_{\mathfrak{G}}^{-1} \int \beta_{\mathfrak{G}}(z)\, dz$ we find after some routine manipulations that

$$\alpha_{\mathfrak{G}}^{(\mu)}(z) = \mu!\, \alpha_{\mathfrak{G}}(z) \sum_{\nu=1}^{\mu} \sum_{\substack{\mu_1,\dots,\mu_\nu>0 \\ \mu_1+\cdots+\mu_\nu=\mu}} (\nu!\, m_{\mathfrak{G}}^{\nu})^{-1} \prod_{j=1}^{\nu} \frac{\beta_{\mathfrak{G}}^{(\mu_j-1)}(z)}{\mu_j!}, \quad \mu \geq 1.$$

Combining these identities for $1 \leq \mu \leq \mu(\mathfrak{G})$ with (13) we obtain for $\beta_{\mathfrak{G}}(z)$ the differential equation

$$\beta_{\mathfrak{G}}(z) = A_0(\mathfrak{G}) + \sum_{\mu=1}^{\mu(\mathfrak{G})} \sum_{\nu=1}^{\mu} \sum_{\substack{\mu_1,\dots,\mu_\nu>0 \\ \mu_1+\cdots+\mu_\nu=\mu}} \binom{\mu}{\mu_1, \dots, \mu_\nu} (\nu!\, m_{\mathfrak{G}}^{\nu})^{-1}$$

$$\times A_\mu(\mathfrak{G})\, z^\mu \prod_{j=1}^{\nu} \beta_{\mathfrak{G}}^{(\mu_j-1)}(z) \tag{18}$$

with $A_\mu(\mathfrak{G})$ as in (14). Our claim (15) follows now by comparing coefficients in (18). □

Remark 1. *Since the numbers $A_\nu(\mathfrak{G})$ as well as the $B_{\mu,\nu}(\lambda_1, \dots, \lambda_\mu)$ are integral, so are the numbers $\mathcal{B}_\mu^{(\mathfrak{G})}(\lambda_1, \dots, \lambda_\mu)$, and the recursion (15) comes rather close to having all its coefficients $(\mu!\, m_{\mathfrak{G}}^{\mu})^{-1} \mathcal{B}_\mu^{(\mathfrak{G})}(\lambda_1, \dots, \lambda_\mu)$ in $\mathbb{Z}$. However, if $\mathfrak{G} = F_r$ is the free group of rank r, then $m_{\mathfrak{G}} = 1$, and $A_\nu(\mathfrak{G}) = S(r+1, \nu+1)$ is a Stirling number of the second kind; and if $r \geq 2$, the coefficient for $\mu = r-1$ turns out to be*

$$\big((r-1)!\big)^{-1} \mathcal{B}_{r-1}^{(F_r)}(\lambda_1, \dots, \lambda_{r-1}) = \frac{r(\lambda+2)}{2},$$

which is not in general integral. Also, for $\mathfrak{G} = F_4$, the coefficients for $\mu = 2$ are

$$\frac{\mathcal{B}_2^{(F_4)}(\lambda_1, \lambda_2)}{2} = 2(\lambda+1)(\lambda+2) - (\lambda_1+1)(\lambda_2+1) + 5,$$

which depends on the variables λ_1 and λ_2, and not only on their sum λ.

3. A recursive description of the function $\chi_q^*(\lambda)$

Write $\chi_q^*(\lambda)$ for $b_q(\lambda) := b_\lambda(\mathfrak{H}(q))$ considered modulo 2, so that χ_q^* is nothing but the characteristic function of the set $\Pi_q^* := \Pi^*(\mathfrak{H}(q)) \subseteq \mathbb{N}$. The purpose of this section is to establish a recursive description of the function $\chi_q^*(\lambda)$, which in turn will be used in the next section to prove our main results concerning the parity patterns Π_q^*.

3.1. Some computations and a lemma

Concerning the invariants considered in the last section we find for $\mathfrak{G} = \mathfrak{H}(q)$ the following.

$$m_q := m_{\mathfrak{H}(q)} = \begin{cases} q, & q \text{ even} \\ 2q, & q \text{ odd}; \end{cases} \tag{19}$$

in particular, m_q is always even. Indeed, if $\mathfrak{G}$ is any finitely generated virtually free group and $(\mathfrak{G}(-), Y)$ is a Stallings decomposition for $\mathfrak{G}$, then we have

$$m_{\mathfrak{G}} = \big[|\mathfrak{G}(v)| \,:\, v \in V(Y)\big],$$

i.e., $m_{\mathfrak{G}}$ coincides with the least common multiple of the orders of the vertex groups in this decomposition. This follows from the universal covering construction and the fact that a finite group has to fix a vertex when acting on a tree.

For $\kappa \mid m_q$,

$$\zeta_\kappa^{(q)} := \zeta_\kappa(\mathfrak{H}(q)) = \begin{cases} 1, & 2 \nmid \kappa \text{ and } \kappa < q \\ -1, & \kappa = m_q \\ 0, & \text{otherwise.} \end{cases} \tag{20}$$

By (8), (9), and (19) we have

$$\chi(\mathfrak{H}(q)) = -\frac{q-2}{2q}$$

and

$$\mu_q := \mu(\mathfrak{H}(q)) = \begin{cases} q-1, & q \text{ odd} \\ q/2, & q \text{ even.} \end{cases}$$

Write $A_\mu^{(q)}$ for $A_\mu(\mathfrak{H}(q))$. The following result describes an important divisibility property of the numbers $A_\mu^{(q)}$.

Lemma 1. *For $0 \leq \mu \leq \mu_q$ the integer $A_\mu^{(q)}$ is divisible by m_q^μ, and*

$$m_q^{-\mu} A_\mu^{(q)} \equiv \binom{\mu_q}{\mu} \quad \text{mod } 2.$$

Proof. Denote by K_q the set of all integers k satisfying $1 \leq k \leq m_q$ and $2 \nmid (k, m_q) < q$. This set has

$$\sum_{\substack{\kappa \mid q \\ 2 \nmid \kappa < q}} \varphi(m_q/\kappa) = \mu_q$$

elements. Here, we have used (20) plus equations (9) and (10) to evaluate the left-hand sum. We have

$$A_\mu^{(q)} = \frac{1}{\mu!} \sum_{j=0}^{\mu} (-1)^{\mu-j} \binom{\mu}{j} \prod_{k \in K_q} (k + j\, m_q) \tag{21}$$

and

$$\prod_{k \in K_q} (k + j\, m_q)$$

$$= \prod_{k \in K_q} k + \sum_{\nu=1}^{\mu_q} (j\, m_q)^\nu \Bigg(\prod_{k \in K_q} k \Bigg) \Bigg(\sum_{\substack{k_1, \ldots, k_\nu \in K_q \\ k_1 < \ldots < k_\nu}} \frac{1}{k_1 \ldots k_\nu} \Bigg). \tag{22}$$

Since

$$A_0^{(q)} = \prod_{k \in K_q} k \equiv 1 \mod 2,$$

we may assume that $\mu > 0$. Inserting (22) into (21) and using the facts that

$$\sum_{j=0}^{\mu} (-1)^{\mu-j} \binom{\mu}{j} = 0 \quad (\mu > 0),$$

and that

$$\frac{1}{\mu!} \sum_{j=1}^{\mu} (-1)^{\mu-j} \binom{\mu}{j} j^\nu = S(\nu, \mu) \quad (\mu, \nu > 0)$$

is a Stirling number of the second kind, we find that for $\mu > 0$

$$A_\mu^{(q)} = m_q^\mu \Bigg[\Bigg(\prod_{k \in K_q} k \Bigg) \Bigg(\sum_{\substack{k_1, \ldots, k_\mu \in K_q \\ k_1 < \ldots < k_\mu}} \frac{1}{k_1 \ldots k_\mu} \Bigg)$$

$$+ \sum_{\nu=\mu+1}^{\mu_q} m_q^{\nu-\mu}\, S(\nu, \mu) \Bigg(\prod_{k \in K_q} k \Bigg) \Bigg(\sum_{\substack{k_1, \ldots, k_\nu \in K_q \\ k_1 < \ldots < k_\nu}} \frac{1}{k_1 \ldots k_\nu} \Bigg) \Bigg].$$

This shows that $A_{\mu}^{(q)}$ is divisible by m_q^{μ} for all μ with $0 < \mu \leq \mu_q$, and that for each such μ

$$m_q^{-\mu}\, A_{\mu}^{(q)} \equiv \Bigg(\prod_{k \in K_q} k\Bigg)\Bigg(\sum_{\substack{k_1,\ldots,k_\mu \in K_q \\ k_1 < \ldots < k_\mu}} \frac{1}{k_1 \ldots k_\mu}\Bigg) \mod 2.$$

The right-hand side of the latter congruence is a sum of $\binom{\mu_q}{\mu}$ odd integers, whence the lemma. □

3.2. The function $\chi_q^*(\lambda)$

We shall require one further piece of preparation.

Lemma 2. *Let $\pi = \{\pi_j\}_{j \geq 1}$ be a partition. Then we have*

$$\nu_2\Bigg(\prod_{j \geq 1} j^{\pi_j}\Bigg) \leq \nu_2\big((2(|\pi| - ||\pi||))!\big) \tag{23}$$

with equality occurring if and only if $|\pi| \leq ||\pi|| + 1$.

Proof. We may assume that $||\pi|| > 0$, i.e., that π is a non-empty partition. Write $|\pi| = ||\pi|| + m$ with some integer $m \geq 0$. By the inequality relating arithmetic and geometric mean, we have

$$2^{\nu_2(\prod_{j \geq 1} j^{\pi_j})} \leq \prod_{j \geq 1} j^{\pi_j} \leq \left(1 + \frac{m}{||\pi||}\right)^{||\pi||}.$$

On the other hand,

$$\nu_2\big((2(|\pi| - ||\pi||))!\big) \geq m + \frac{m-1}{2} + \frac{m-3}{4} = \frac{7m-5}{4}.$$

Since the sequence $\{(1 + \frac{m}{\ell})^{\ell}\}_{\ell \geq 1}$ is increasing for each fixed $m \geq 0$ and converges to e^m, validity of the inequality $\mathrm{e}^{4m} < 2^{7m-5}$ for some m implies (23) with strict inequality for all partitions π such that $|\pi| - ||\pi|| = m$ with this particular m. The latter inequality holds for all

$$m > \frac{5\log(2)}{7\log(2) - 4} \approx 4.0676207,$$

and a check of the remaining cases where $|\pi| - ||\pi|| \leq 4$ completes the proof. □

We can now proceed with our analysis of the function $\chi_q^*(\lambda)$. In Proposition 1 put $\mathfrak{G} = \mathfrak{H}(q)$, and write $\mathcal{B}_{\mu}^{(q)}(\lambda_1, \ldots, \lambda_\mu)$ for $\mathcal{B}_{\mu}^{(\mathfrak{H}(q))}(\lambda_1, \ldots, \lambda_\mu)$. Multiply both sides of (15) by a sufficiently large odd number B depending only on q to obtain

$$B\, b_q(\lambda+1) = \sum_{\mu=1}^{\mu_q} 2^{-(\nu_2(\mu!)+\mu\nu_2(m_q))} \sum_{\substack{\lambda_1,\dots,\lambda_\mu>0\\ \lambda_1+\cdots+\lambda_\mu=\lambda}} \tilde{\mathcal{B}}_\mu^{(q)}(\lambda_1,\dots,\lambda_\mu)$$

$$\times \prod_{j=1}^{\mu} b_q(\lambda_j), \quad \lambda \geq 1, \tag{24}$$

with integers $\tilde{\mathcal{B}}_\mu^{(q)}(\lambda_1,\dots,\lambda_\mu)$ satisfying

$$\nu_2\big(\tilde{\mathcal{B}}_\mu^{(q)}(\lambda_1,\dots,\lambda_\mu)\big) \;=\; \nu_2\big(\mathcal{B}_\mu^{(q)}(\lambda_1,\dots,\lambda_\mu)\big).$$

Decompose $\mathcal{B}_\mu^{(q)}(\lambda_1,\dots,\lambda_\mu)$ as

$$\mathcal{B}_\mu^{(q)}(\lambda_1,\dots,\lambda_\mu) = \mu!\, A_\mu^{(q)} \,+\, \Theta_\mu^{(q)}(\lambda) \,+\, R_\mu^{(q)}(\lambda_1,\dots,\lambda_\mu),$$

where

$$\Theta_\mu^{(q)}(\lambda) := \begin{cases} (\mu+1)!\,(\lambda-\mu)\,A_{\mu+1}^{(q)}/2, & \mu<\mu_q\\ 0, & \mu=\mu_q \end{cases}$$

and

$$R_\mu^{(q)}(\lambda_1,\dots,\lambda_\mu) := \sum_{\nu=\mu+2}^{\mu_q} \sum_{\substack{\nu_1,\dots,\nu_\mu\geq 0\\ \nu_1+\cdots+\nu_\mu=\nu-\mu}} A_\nu^{(q)} \frac{\nu!}{(\nu_1+1)\dots(\nu_\mu+1)}$$

$$\times \binom{\lambda_1-1}{\nu_1}\cdots\binom{\lambda_\mu-1}{\nu_\mu}.$$

By Lemmas 1 and 2 we have for $\nu \geq \mu \geq 1$

$$\nu_2\Big(\frac{\nu!\,A_\nu^{(q)}}{(\nu_1+1)\dots(\nu_\mu+1)}\Big) \;\geq\; \nu_2(\nu!) \,+\, \nu\,\nu_2(m_q) \;-\; \nu_2((2(\nu-\mu))!)$$

with equality occurring at most in the cases when $\nu - \mu \leq 1$. It follows that for $\nu \geq \mu + 2$

$$\begin{aligned}
&\nu_2\left(\frac{2^{-(\nu_2(\mu!)+\mu\nu_2(m_q))}\,\nu!\,A_\nu^{(q)}}{(\nu_1+1)\dots(\nu_\mu+1)}\right)\\
&\quad > \nu\,\nu_2(m_q) + \nu_2(\nu!) - \nu_2((2(\nu-\mu))!) - \nu_2(\mu!) - \mu\,\nu_2(m_q)\\
&\quad \geq \nu - \mu + \nu_2(\nu!) - \nu_2(\mu!) - \nu_2((2(\nu-\mu))!) = \nu_2\left(\binom{2\nu}{2\mu}\right) \geq 0.
\end{aligned}$$

Hence, for each $\mu \in [\mu_q]$, the term $R_\mu^{(q)}(\lambda_1, \dots, \lambda_\mu)$ is divisible by $2^{\nu_2(\mu!)+\mu\nu_2(m_q)}$, and

$$R_\mu^{(q)}(\lambda_1, \dots, \lambda_\mu)/2^{\nu_2(\mu!)+\mu\nu_2(m_q)}$$

is even. Furthermore, we claim that $\Theta_\mu^{(q)}(\lambda)$ is also divisible by $2^{\nu_2(\mu!)+\mu\nu_2(m_q)}$ for all $\mu \in [\mu_q]$, and that

$$\Theta_\mu^{(q)}(\lambda)/2^{\nu_2(\mu!)+\mu\nu_2(m_q)} \equiv \begin{cases} \binom{q/2}{\mu+1}\lambda, & \mu \equiv 0\,(2) \text{ and } \nu_2(q) = 1 \\ 0, & \text{otherwise} \end{cases} \mod 2. \tag{25}$$

This is certainly true if $\mu = \mu_q$. Hence, we may suppose that $\mu < \mu_q$ and consider divisibility of $(\mu+1)!\,A_{\mu+1}^{(q)}$ by $2^{1+\nu_2(\mu!)+\mu\nu_2(m_q)}$. Now, if μ is odd, then $\nu_2((\mu+1)!) > \nu_2(\mu!)$, and our claim follows from the fact that, by Lemma 1, $2^{(\mu+1)\nu_2(m_q)}$ divides $A_{\mu+1}^{(q)}$. If, on the other hand, μ is even, then $\nu_2((\mu+1)!) = \nu_2(\mu!)$, and Lemma 1 tells us that $(\mu+1)!\,A_{\mu+1}^{(q)}$ is divisible by $2^{1+\nu_2(\mu!)+\mu\nu_2(m_q)}$, and that

$$(\mu+1)!\,A_{\mu+1}^{(q)}/2^{1+\nu_2(\mu!)+\mu\nu_2(m_q)} \equiv 2^{\nu_2(m_q)-1}\binom{\mu_q}{\mu+1} \mod 2. \tag{26}$$

For $\nu_2(q) > 1$ the right-hand side of (26) vanishes modulo 2. Similarly, if q is odd, then $\mu_q = q - 1$ is even, and $\binom{\mu_q}{\mu+1} \equiv 0\ (2)$. However, if $\nu_2(q) = 1$, then $m_q = q$, and $\mu_q = q/2$ is odd. Thus, our claim (25) is proved. Finally, again by Lemma 1, $2^{\nu_2(\mu!)+\mu\nu_2(m_q)}$ divides $\mu!\,A_\mu^{(q)}$ for every $\mu \in [\mu_q]$, and

$$\mu!\,A_\mu^{(q)}/2^{\nu_2(\mu!)+\mu\nu_2(m_q)} \equiv \binom{\mu_q}{\mu} \mod 2.$$

We conclude that $\tilde{\mathcal{B}}_\mu^{(q)}(\lambda_1, \ldots, \lambda_\mu)$ is divisible by $2^{\nu_2(\mu!)+\mu\nu_2(m_q)}$ and that

$$\tilde{\mathcal{B}}_\mu^{(q)}(\lambda_1, \ldots, \lambda_\mu)/2^{\nu_2(\mu!)+\mu\nu_2(m_q)}$$
$$\equiv \begin{cases} 0, & \nu_2(q) = 1 \,\&\, \mu \equiv 0\,(2) \,\&\, \lambda \equiv 1\,(2) \\ \binom{\mu_q}{\mu}, & \text{otherwise} \end{cases} \quad \text{mod } 2. \tag{27}$$

Evaluating (24) modulo 2 in the light of (27) we find for the function $\chi_q^*(\lambda)$ the $GF(2)$-recurrence relation

$$\chi_q^*(\lambda+1) = \sum_{\mu=1}^{\mu_q} \sum_{\substack{\lambda_1,\ldots,\lambda_\mu>0 \\ \lambda_1+\cdots+\lambda_\mu=\lambda}} \Big[1 + (1+\mu)\,\lambda\,\delta_{1,\nu_2(q)}\Big]$$
$$\times \binom{\mu_q}{\mu} \chi_q^*(\lambda_1)\ldots\chi_q^*(\lambda_\mu) \qquad (\lambda \geq 1,\ \chi_q^*(1) = 1). \tag{28}$$

Introduce the generating series

$$X_q^*(z) := 1 + \sum_{\lambda=1}^{\infty} \chi_q^*(\lambda)\, z^\lambda \ \in\ GF(2)[[z]].$$

Multiplying both sides of (28) by z^λ and summing over $\lambda \geq 1$, the left-hand side becomes

$$\sum_{\lambda=1}^{\infty} \chi_q^*(\lambda+1)\, z^\lambda = z^{-1}\Big[1 + z + X_q^*(z)\Big],$$

while the corresponding right-hand side is the sum of

$$\Sigma_1 := \sum_{\lambda=1}^{\infty} \sum_{\mu=1}^{\mu_q} \sum_{\substack{\lambda_1,\ldots,\lambda_\mu>0 \\ \lambda_1+\cdots+\lambda_\mu=\lambda}} \binom{\mu_q}{\mu} \chi_q^*(\lambda_1)\ldots\chi_q^*(\lambda_\mu)\, z^\lambda$$

and

$$\Sigma_2 := \sum_{\lambda=1}^{\infty} \sum_{\mu=1}^{\mu_q} \sum_{\substack{\lambda_1,\ldots,\lambda_\mu>0 \\ \lambda_1+\cdots+\lambda_\mu=\lambda}} (1+\mu)\,\lambda\,\delta_{1,\nu_2(q)} \binom{\mu_q}{\mu} \chi_q^*(\lambda_1)\ldots\chi_q^*(\lambda_\mu)\, z^\lambda.$$

By the binomial law in the ring $GF(2)[[z]]$

$$\Sigma_1 = \sum_{\mu=1}^{\mu_q} \binom{\mu_q}{\mu} \Big(1 + X_q^*(z)\Big)^\mu = 1 + \big(X_q^*(z)\big)^{\mu_q}.$$

Also,

$$\Sigma_2 = \delta_{1,\nu_2(q)}\, z \sum_{\mu=1}^{\mu_q} (1+\mu) \binom{\mu_q}{\mu} \left[\left(1 + X_q^*(z)\right)^{\mu} \right]' = 0.$$

Summarizing the preceding discussion we have the following result.

Proposition 2. *For every integer $q \geq 3$ the generating function $X_q^*(z)$ satisfies the identity*

$$X_q^*(z) = 1 + z\left(X_q^*(z)\right)^{\mu_q}. \tag{29}$$

Equivalently, for every $q \geq 3$ the characteristic function $\chi_q^*(\lambda)$ satisfies the $GF(2)$-recursion

$$\chi_q^*(\lambda+1) = \sum_{\substack{\pi \vdash \lambda \\ ||\pi|| \leq \mu_q}} \binom{\mu_q}{||\pi||} \frac{||\pi||!}{\prod_{j=1}^{\infty} \pi_j!} \prod_{j=1}^{\infty} \left(\chi_q^*(j)\right)^{\pi_j}, \quad \lambda \geq 1 \tag{30}$$

starting from $\chi_q^*(1) = 1$.

Identity (29) can be interpreted as providing a recursive reconstruction of the set Π_q^*. Define $\Pi_q^*(\lambda) := \Pi_q^* \cap [\lambda]$. Rewriting (29) in the form

$$\sum_{\lambda \in \Pi_q^*} z^{\lambda} \;=\; z \left[1 \;+\; \sum_{\lambda \in \Pi_q^*} z^{\lambda} \right]^{\mu_q}$$

and comparing coefficients we find that $\Pi_q^*(1) = \{1\}$, and, assuming $\Pi_q^*(\lambda)$ to be known for some $\lambda \geq 1$, we see that $\Pi_q^*(\lambda+1)$ is determined by the condition that $\lambda + 1 \in \Pi_q^*$ if and only if the number of representations of λ in the form $\lambda = \lambda_1 + \cdots + \lambda_{\mu_q}$ with $\lambda_j \in \Pi_q^*(\lambda) \cup \{0\}$ is odd.

4. The free parity pattern of a Hecke group

4.1. The results

In this section we will exploit the identity (29) to obtain results concerning the parity patterns Π_q^*. Equation (30) tells us in particular that Π_q^* is already determined by the free rank μ_q, and the question arises whether, conversely, the parity pattern Π_q^* also determines μ_q, or whether there exist Hecke groups having different free ranks while exhibiting the same free parity pattern. Our first result shows that the latter situation cannot arise.

Theorem 1. *Let $q_1, q_2 \geq 3$ be integers. Then we have $\Pi^*_{q_1} = \Pi^*_{q_2}$ if and only if $\mu_{q_1} = \mu_{q_2}$.*

Proof. We observed already that $\mu_{q_1} = \mu_{q_2}$ implies $\Pi^*_{q_1} = \Pi^*_{q_2}$. Conversely, let $q_1, q_2 \geq 3$ be integers with $\Pi^*_{q_1} = \Pi^*_{q_2} =: \Pi^*$. Then $X^*_{q_1}(z) = X^*_{q_2}(z) =: X^*(z)$, and (29) implies that $(X^*(z))^{\mu_{q_1}} = (X^*(z))^{\mu_{q_2}}$. Suppose without loss of generality that $\mu_{q_1} \leq \mu_{q_2}$, and rewrite the last equation as

$$(X^*(z))^{\mu_{q_1}}\big[(X^*(z))^{\vartheta} - 1\big] = 0, \tag{31}$$

where $\vartheta := \mu_{q_2} - \mu_{q_1} \geq 0$. Since $X^*(z) \neq 0$ and $GF(2)[[z]]$ has no zero divisors, we find from the binomial law in $GF(2)[[z]]$ the $GF(2)$-relation

$$1 = (X^*(z))^{\vartheta} = \sum_{\nu=0}^{\vartheta} \binom{\vartheta}{\nu} \bigg(\sum_{\lambda\in\Pi^*} z^{\lambda}\bigg)^{\nu}. \tag{32}$$

Using the fact that $\chi^*_q(1) = 1$ for every $q \geq 3$ and comparing coefficients we deduce that $\binom{\vartheta}{\nu} \equiv 0\ (2)$ for all $\nu \in [\vartheta]$, which is impossible for $\vartheta > 0$. Hence, we must have $\vartheta = 0$. □

Next, we prove that the series $X^*_q(z)$ is never rational over $GF(2)$. This implies in particular that the sets Π^*_q are always infinite, and that certain infinite sets of positive integers – for instance the sets $\mathcal{C}_m = 1 + m\mathbb{N}_0$ with $m \in \mathbb{N}$ – cannot occur as free parity pattern of a Hecke group. Moreover, we determine the second entry $\alpha^*_q := \min_{\lambda\in\Pi^*_q\setminus\{1\}} \lambda$ of Π^*_q, and we show that $\Pi^*_q \subseteq 1 + 2^{\nu_2(\mu_q)}\mathbb{N}_0$.

Theorem 2. *Let $q \geq 3$ be an integer. Then we have the following.*

(i) *Every entry of Π^*_q is congruent to 1 modulo $2^{\nu_2(\mu_q)}$.*
(ii) *The series $X^*_q(z)$ is not rational over $GF(2)$; in particular, the set Π^*_q is infinite.*
(iii) *We have $\alpha^*_q = 2^{\nu_2(\mu_q)} + 1$; in particular, given an integer $\alpha^* \geq 2$, the set $\{1, \alpha^*\}$ can be extended to a free parity pattern of some Hecke group if and only if $\alpha^* - 1$ is a 2-power.*

Proof. (i) Let $\ell := \nu_2(\mu_q)$. We establish the implication

$$\chi^*_q(\lambda + 1) = 1 \Rightarrow \lambda \equiv 0 \mod 2^{\ell} \tag{33}$$

for all $\lambda \in \mathbb{N}_0$ by induction on λ. The implication (33) holds trivially if $\lambda = 0$. Suppose that (33) holds for all non-negative integers $\lambda < L$ with some integer

$L \geq 1$, and that $\chi_q^*(L+1) = 1$. By (30) and our inductive hypothesis,

$$\chi_q^*(L+1) = \sum_{\substack{\pi \vdash L \\ \pi_j > 0 \Rightarrow j \equiv 1\,(2^\ell) \\ \binom{\mu_q}{||\pi||} \equiv 1\,(2)}} \frac{||\pi||!}{\prod_{j\geq 1} \pi_j!} \prod_{j \geq 1} \left(\chi_q^*(j)\right)^{\pi_j}.$$

If the right-hand side of the latter equation is to be non-zero, then in particular there must exist a partition π of L all of whose parts are congruent to 1 modulo 2^ℓ, and such that $\binom{\mu_q}{||\pi||}$ is odd. The first of these conditions implies that $L \equiv ||\pi||$ mod 2^ℓ, while the second condition, in view of Lucas' formula,[6] forces $||\pi||$ to be divisible by 2^ℓ, hence $L \equiv 0\ (2^\ell)$ as claimed.

(ii) Let $X_q^*(z) = \varphi^*(z)/\psi^*(z)$ with relatively prime polynomials $\varphi^*(z), \psi^*(z) \in GF(2)[z]$, and let $v = \deg(\varphi^*(z)) - \deg(\psi^*(z))$ be the (total) degree of $X_q^*(z)$. Multiplying both sides by $(\psi^*(z))^{\mu_q}$, equation (29) takes the form

$$\varphi^*(z)\big(\psi^*(z)\big)^{\mu_q - 1} = \big(\psi^*(z)\big)^{\mu_q} + z\,\big(\varphi^*(z)\big)^{\mu_q}. \tag{34}$$

Since $\mu_q \geq 2$, $\psi^*(z)$ must divide the right-hand side of (34), as it divides the left-hand side, hence $\psi^*(z) \mid z\,(\varphi^*(z))^{\mu_q}$. Thus, as $(\varphi^*(z), \psi^*(z)) = 1$, it follows that $\psi^*(z) \mid z$, and hence that $\psi^*(z) = 1$, since $\psi^*(z)$ must have a non-zero constant term; in particular, $v = \deg(\varphi^*(z)) \geq 0$. Comparing degrees on both sides of (29) now gives

$$(\mu_q - 1)v + 1 = 0,$$

which is impossible for $v \geq 0$.

(iii) Let $\nu_2(\mu_q) = \ell$, i.e., $\mu_q = 2^\ell m$ with some odd integer m. Recall that if p is a prime and $a = \sum_{j\geq 0} a_j^{(p)}\, p^j$ and $b = \sum_{j\geq 0} b_j^{(p)}\, p^j$ are non-negative integers written in base p, then, by Lucas' Theorem[7]

$$\binom{a}{b} \equiv \prod_{j \geq 0} \binom{a_j^{(p)}}{b_j^{(p)}} \quad \text{mod } p; \tag{35}$$

in particular, $\binom{a}{b} \equiv 1\ (2)$ if and only if $a_j^{(2)} \geq b_j^{(2)}$ for all $j \geq 0$. Write $m = \sum_{j\geq 0} m_j\, 2^j$ with $m_j \in \{0, 1\}$ and $m_0 = 1$. Then

$$\mu_q = 2^\ell\, m = 2^\ell + \sum_{j>0} m_j\, 2^{\ell+j},$$

[6] See formula (35) below.
[7] Cf. for instance [2, Theorem 3.4.1].

and Lucas' Theorem implies that the summation in (30) can be restricted to partitions $\pi \vdash \lambda$ such that $2^\ell \leq ||\pi|| \leq \mu_q$. This observation yields in particular that $\chi_q^*(2) = \cdots = \chi_q^*(2^\ell) = 0$. On the other hand, we see that $\chi_q^*(2^\ell + 1) = \chi_q^*(1) = 1$, hence $\alpha_q^* = 2^\ell + 1$ as claimed. □

We shall now obtain a more explicit description of the parity patterns Π_q^*. Using the recurrence relation (30), viewed over $\mathbb{Z}$, define an integral sequence $\widehat{\chi}_q^*(\lambda)$ starting from $\widehat{\chi}_q^*(1) = 1$, and let $\widehat{X}_q^*(z) := 1 + \sum_{\lambda \geq 1} \widehat{\chi}_q^*(\lambda)\, z^\lambda \in \mathbb{Z}[[z]]$. Then $\widehat{X}_q^*(z) \equiv X_q^*(z) \mod 2$, and the series $\widehat{X}_q^*(z)$ satisfies the functional equation

$$\widehat{X}_q^*(z) = 1 + z\left(\widehat{X}_q^*(z)\right)^{\mu_q}. \tag{36}$$

Let $\widehat{X}_q^*(z) = 1 + \widehat{F}(z)$. Then equation (36) takes the form

$$\widehat{F}(z) = z\,\Phi\big(\widehat{F}(z)\big),$$

where

$$\Phi(\zeta) := (1 + \zeta)^{\mu_q}.$$

By Lagrange inversion,

$$\begin{aligned}\left\langle z^\lambda,\ \widehat{F}(z)\right\rangle &= \frac{1}{\lambda}\left\langle \zeta^{\lambda-1},\ \left(\Phi(\zeta)\right)^\lambda\right\rangle\\ &= \frac{1}{\mu_q\lambda + 1}\binom{\mu_q\lambda + 1}{\lambda},\end{aligned}$$

i.e.,

$$\widehat{X}_q^*(z) = \sum_{\lambda \geq 0} \frac{1}{\mu_q\lambda + 1}\binom{\mu_q\lambda + 1}{\lambda} z^\lambda. \tag{37}$$

By Kummer's formula[8] for the p-adic norm of binomial coefficients we have

$$\nu_2\binom{a}{b} = \mathfrak{s}(b) + \mathfrak{s}(a - b) - \mathfrak{s}(a),$$

where $\mathfrak{s}(x)$ denotes the sum of digits in the binary representation of x. Hence,

$$\begin{aligned}&\nu_2\left(\frac{1}{\mu_q\lambda + 1}\binom{\mu_q\lambda + 1}{\lambda}\right)\\ &= \nu_2\left(\frac{1}{\lambda}\binom{\mu_q\lambda}{\lambda - 1}\right) = \mathfrak{s}(\lambda) + \mathfrak{s}((\mu_q - 1)\lambda + 1) - \mathfrak{s}(\mu_q\lambda) - 1,\end{aligned}$$

and we obtain the following explicit description of the patterns Π_q^*.

[8] Cf. [18, pp. 115–116].

Theorem 3. *For every $q \geq 3$,*

$$\Pi_q^* = \Big\{\lambda \in \mathbb{N} : \ \mathfrak{s}(\lambda) + \mathfrak{s}((\mu_q - 1)\lambda + 1) - \mathfrak{s}(\mu_q \lambda) = 1\Big\}.$$

As is apparent from Theorem 3, the parity patterns Π_q^* will not in general lend themselves to a straightforward explicit characterization as in the case of the modular group; instead, Π_q^* generically tends to inherit the well-known kind of fractal behaviour observed in Pascal's triangle when evaluated modulo 2. There is however one special case where we can describe the patterns Π_q^* in a completely explicit way, namely when μ_q is a 2-power. Hence, while a canonical generalization of the free parity pattern met in the modular group $\mathfrak{H}(3)$ and in $\mathfrak{H}(4)$ to all Hecke groups does not exist, this type of pattern precisely characterizes two infinite series of Hecke groups. For an integer $q \geq 3$ define

$$\Lambda_q^* := \left\{\frac{\mu_q^\sigma - 1}{\mu_q - 1} : \ \sigma = 1, 2, \ldots\right\},$$

i.e., Λ_q^* is the set of partial sums of the geometric series $\sum_{\sigma \geq 0} \mu_q^\sigma$ generated by the free rank μ_q of $\mathfrak{H}(q)$.

Theorem 4. *Let $q \geq 3$ be an integer. Then the following assertions are equivalent:*

(i) $\Pi_q^* = \Lambda_q^*$.
(ii) $\chi_q^*(\lambda) = 0$ *for* $2 \leq \lambda \leq \mu_q$.
(iii) q *or* $q - 1$ *is a 2-power.*

Proof. Since (i) clearly implies (ii), it suffices to prove the implications (ii) $\Rightarrow$ (iii) and (iii) $\Rightarrow$ (i). Suppose first that neither q nor $q - 1$ is a 2-power. Then μ_q is not a 2-power, i.e., $\mu_q = 2^\ell m$ with $\ell \geq 0$ and some odd integer $m > 1$, and, by the third part of Theorem 2,

$$2 \leq \alpha_q^* = 2^\ell + 1 \leq \mu_q,$$

contradicting (ii). This proves the implication (ii) $\Rightarrow$ (iii). Now suppose that q or $q - 1$ is a 2-power. Then μ_q is a 2-power, say $\mu_q = 2^\ell$ with some $\ell \geq 1$. But then $\mathfrak{s}(\mu_q \lambda) = \mathfrak{s}(\lambda)$, and the condition on λ in Theorem 3 simplifies to

$$\mathfrak{s}((\mu_q - 1)\lambda + 1) = 1,$$

that is,

$$\lambda = \frac{2^\alpha - 1}{2^\ell - 1} \quad \text{with} \ \ \ell \mid \alpha \ \ \text{and} \ \ \alpha \geq 1.$$

Assertion (i) follows now from Theorem 3, and the proof of Theorem 4 is complete. □

By specializing Theorem 4 to the case when q is a prime number, we obtain a characterization of Fermat primes among the set of all odd primes.

Corollary 1. *Let $q > 2$ be a prime. Then the following assertions are equivalent:*

(i) $\Pi_q^* = \Lambda_q^* = \left\{ \frac{(q-1)^\sigma - 1}{q-2} : \sigma = 1, 2, \ldots \right\}$.
(ii) $\chi_q^*(\lambda) = 0$ *for* $2 \leq \lambda \leq q-1$.
(iii) q *is a Fermat prime.*

4.2. An example

If q is such that μ_q is not a 2-power, then we are outside of the scope of Theorem 4, and cannot hope to describe the parity pattern Π_q^* by means of a closed formula of Stothers' type. Nevertheless, Theorem 3 still provides a useful and fairly explicit description of Π_q^*. As an example, consider the groups $\mathfrak{H}(q)$ with $q = 2(2^\rho + 1)$ and $\rho \geq 1$. These are the Hecke groups for which $\mu_q - 1$ is a 2-power. For such q, equation (29) gives the relation

$$X_q^*(z) = \frac{1}{1 + zX_q^*(z^{2^\rho})}. \tag{38}$$

Putting $\beta := 2^\rho$ and iterating (38), we find for the generating function $X_q^*(z)$ the expansion

$$X_q^*(z) = \cfrac{1}{1 + \cfrac{z}{1 + \cfrac{z^\beta}{1 + \cfrac{z^{\beta^2}}{1 + \ddots}}}} \tag{39}$$

as a continued fraction over $GF(2)$, which exhibits $X_q^*(z)$ as the quotient $X_q^*(z) = A(z)/B(z)$ of two formal power series $A(z) = \lim_{\nu\to\infty} A_\nu(z)$ and $B(z) = \lim_{\nu\to\infty} B_\nu(z)$, where $A_\nu(z)$ and $B_\nu(z)$ are the approximands of numerator and denominator in (39). These approximands are given by the recursions

$$A_{\nu+1}(z) = A_\nu(z) + z^{\beta^{\nu-1}} A_{\nu-1}(z) \quad (\nu \geq 1;\ A_0 = 0,\ A_1 = 1)$$

respectively

$$B_{\nu+1}(z) = B_\nu(z) + z^{\beta^{\nu-1}} B_{\nu-1}(z) \quad (\nu \geq 1;\ B_0 = B_1 = 1),$$

which in turn are equivalent to the functional equations

$$F(x, z) + F(x^\beta, z) + F(x^{\beta^2} z, z) + x^\beta = 0$$

respectively

$$G(x, z) + G(x^\beta, z) + G(x^{\beta^2} z, z) + x = 0$$

for the generating functions

$$F(x, z) := \sum_{\nu=0}^{\infty} A_\nu(z) x^{\beta^\nu} \quad \text{and} \quad G(x, z) := \sum_{\nu=0}^{\infty} B_\nu(z) x^{\beta^\nu}.$$

The reader may find it instructive to try to use these functional equations to obtain an explicit formula for the power series $F(x, z)$ and $G(x, z)$, and hence for the series $A(z)$ and $B(z)$. Equation (39) is an interesting and, in a sense, explicit (though not particularly illuminating) description of the patterns $\Pi^*_{2(2^\rho+1)}$. Turning to Theorem 3, we see that

$$\mathfrak{s}((\mu_q - 1)\lambda + 1) = \mathfrak{s}(2^\rho\lambda + 1) = \mathfrak{s}(\lambda) + 1,$$

and our condition on λ becomes

$$2\,\mathfrak{s}(\lambda) = \mathfrak{s}(2^\rho\lambda + \lambda).$$

The latter condition holds if and only if the binary representations $\lambda = \sum_{j\geq 0} \lambda_j 2^j$ of λ respectively $2^\rho\lambda = \sum_{j\geq 0} \lambda_j 2^{j+\rho}$ of $2^\rho\lambda$ do not overlap, i.e., if and only if $\lambda_j = 1$ always implies $\lambda_{j+\rho} = 0$. Hence, we find from Theorem 3 that

$$\Pi^*_{2(2^\rho+1)} = \left\{\lambda = \sum_{j\geq 0} \lambda_j 2^j \in \mathbb{N} :\ \lambda_j = 1 \Rightarrow \lambda_{j+\rho} = 0 \ \text{ for all } \ j \geq 0\right\},\ \rho \geq 1. \tag{40}$$

This is a much more useful description of Π^*_q for these q; in particular, we immediately infer from (40) that

$$\Pi^*_{2(2^\rho+1)} \cap [2^\rho + 1] = [2^\rho], \quad \rho \geq 1.$$

5. The parity patterns of the groups $\widetilde{\mathfrak{H}}(q)$

For an integer $q \geq 3$ let $\widetilde{\mathfrak{H}}(q) := C_q * C_q$, and denote by $\widetilde{\chi}_q(n)$ the number $s_n(\widetilde{\mathfrak{H}}(q))$ of index n subgroups in $\widetilde{\mathfrak{H}}(q)$ evaluated modulo 2, so that $\widetilde{\chi}_q$ is simply the characteristic function of the set $\widetilde{\Pi}_q := \Pi(\widetilde{\mathfrak{H}}(q)) \subseteq \mathbb{N}$. The purpose of this section is to establish a differential equation for the generating function $\widetilde{X}_q(z) := \sum_{n\geq 0} \widetilde{\chi}_q(n+1)\, z^n$, and to exploit this identity to investigate the parity patterns $\widetilde{\Pi}_q$. The results of this section depend upon certain properties of the infinite rectangular array

$$\Big(|\operatorname{Hom}(C_q, S_\mu)| \cdot |\operatorname{Hom}(C_q, S_\nu)|\Big)_{\mu,\nu\geq 0},$$

which will be established in Section 7.

5.1. A functional equation for $\widetilde{X}_q(z)$

Put $h_n(q) := |\operatorname{Hom}(C_q, S_n)|$ with the convention that $h_0(q) = 1$ and $h_n(q) = 0$ for $n < 0$. Our starting point is the recurrence relation[9]

$$h_n(q) = \sum_{d|q} (n-1)_{d-1}\, h_{n-d}(q), \quad (n \geq 1,\ h_0(q) = 1), \tag{41}$$

which follows from the identity

$$\sum_{n=0}^{\infty} h_n(q) z^n/n! = \exp\Big(\sum_{d|q} z^d/d\Big) \tag{42}$$

due to Chowla, Herstein, and Scott [3] by differentiating and comparing coefficients.[10] Squaring (41), multiplying both sides by $z^{n-1}/(n-1)!$, and summing

[9] For a ring R with identity element 1, an element $r \in R$, and an integer k, we set $(r)_k := \prod_{\nu=0}^{k-1} (r - \nu)$, with the usual convention that an empty product should equal 1. This is the falling factorial r of order k. Its analogue $\langle r\rangle_k = \prod_{\nu=0}^{k-1} (r + \nu)$, the rising factorial r of order k, is sometimes called a Pochhammer symbol.

[10] More general results concerning the enumeration of permutation representations respectively wreath product representations of (arbitrary) groups, which contain formulae (11), (42), and (43) below as special cases, can be found in [6] and [29].

over $n \geq 1$ gives

$$\sum_{n=1}^{\infty} h_n^2(q)\, z^{n-1}/\,(n-1)! = \sum_{n=1}^{\infty} \sum_{d|q} (n-1)_{d-1}^2\, h_{n-d}^2(q)\, z^{n-1}/\,(n-1)!$$

$$+\,2 \sum_{n=1}^{\infty} \sum_{\substack{d_1,d_2|q\\ d_1<d_2}} (n-1)_{d_1-1}\,(n-1)_{d_2-1}\, h_{n-d_1}(q)\, h_{n-d_2}(q)\, z^{n-1}/\,(n-1)!.$$

Interchanging the summations over n and d, respectively n and the pairs (d_1, d_2), introducing the series

$$\widetilde{H}_q(z) = \sum_{n=0}^{\infty} h_n^2(q)\, z^n/\,n! = \sum_{n=0}^{\infty} |\operatorname{Hom}(\widetilde{\mathfrak{H}}(q), S_n)|\, z^n/\,n!,$$

and rewriting the individual terms occurring in the last equation, we obtain the relation

$$\widetilde{H}_q'(z) = \sum_{d|q} z^{d-1}\big(z^{d-1}\,\widetilde{H}_q(z)\big)^{(d-1)}$$
$$+\,2 \sum_{\substack{d_1,d_2|q\\ d_1<d_2}} z^{d_1-1}\Big(z^{d_2-1} \sum_{n=0}^{\infty} h_n(q) h_{n+d_2-d_1}(q) z^n/n!\Big)^{(d_1-1)}.$$

Applying Leibniz's formula for the higher derivatives of a product function to the latter equation, dividing both sides of the resulting identity by $\widetilde{H}_q(z)$, and using the fact that, by Dey's formula [4, Theorem 6.10],

$$\widetilde{S}_q(z) := \sum_{n=0}^{\infty} s_{n+1}(\widetilde{\mathfrak{H}}(q))\, z^n = \widetilde{H}_q'(z)/\widetilde{H}_q(z), \tag{43}$$

we find that

$$\widetilde{S}_q(z) = \sum_{d|q} \sum_{\nu=0}^{d-1} \binom{d-1}{\nu} (d-1)_{d-\nu-1}\, z^{d+\nu-1}\, \widetilde{H}_q^{(\nu)}(z)/\widetilde{H}_q(z)$$
$$+\,2 \sum_{\substack{d_1,d_2|q\\ d_1<d_2}} \sum_{\nu=0}^{d_1-1} \binom{d_1-1}{\nu} (d_2-1)_{d_1-\nu-1}\, z^{d_2+\nu-1}\, \mathcal{R}_q^{d_2-d_1,\nu}(z), \tag{44}$$

where, for $q \geq 3$, $\mu > 0$, and $\nu \geq 0$

$$\mathcal{R}_q^{\mu,\nu}(z) := \left(\sum_{n\geq 0} h_n(q)\, h_{n+\mu}(q)\, z^n/n! \right)^{(\nu)} / \widetilde{H}_q(z).$$

We now make use of the facts, to be proved among other things in Section 7, that the series $\widetilde{H}_q^{(\nu)}(z)/\widetilde{H}_q(z)$ is an integral power series for every $\nu \geq 0$, and that

$$\widetilde{H}_q^{(\nu)}(z)/\widetilde{H}_q(z) \equiv \sum_{\mu=0}^{\lfloor \nu/2 \rfloor} \binom{\nu}{2\mu} \left(\widetilde{S}_q'(z)\right)^{\mu} \left(\widetilde{S}_q(z)\right)^{\nu-2\mu} \mod 2, \quad \nu \geq 0. \tag{45}$$

Also, at this stage the following result comes into play.

For every $q \geq 3$, each $0 < \mu < q$, and every $0 \leq \nu \leq q-2$, the series $\mathcal{R}_q^{\mu,\nu}(z)$, viewed as a power series, has integral coefficients. (46)

Again, this will be established in Section 7. Taking into account (44), (45), and (46), we find for the series $\widetilde{X}_q(z)$ the differential equation

$$\widetilde{X}_q(z) = \sum_{d|q} \sum_{\mu \geq 0} z^{2d-3} \left(\widetilde{X}_q'(z)\right)^{\mu} \left(\widetilde{X}_q(z)\right)^{d-2(\mu+1)} \times \left[\binom{d-1}{2\mu} z\, \widetilde{X}_q(z) + (d-1) \binom{d-2}{2\mu} \right] \tag{47}$$

over $GF(2)$. Equation (47), which describes the dynamics of the parity pattern $\widetilde{\Pi}_q$, is the counterpart, for the group $\widetilde{\mathfrak{H}}(q)$ and the arithmetic function $s_n(\widetilde{\mathfrak{H}}(q))$, of the algebraic identity (29) governing the series $X_q^*(z)$. Comparing coefficients in (47), one obtains a recurrence relation for the function $\widetilde{\chi}_q(n)$, which we do not bother to write down explicitly, but whose existence serves to show that equation (47) uniquely determines its solution $\widetilde{X}_q(z)$. The first conclusion to be drawn from equation (47) is that for q odd we have $\widetilde{X}_q'(z) = 0$, i.e., the parity pattern $\widetilde{\Pi}_q$ consists only of odd numbers. Hence, for such q, equation (47) degenerates to the algebraic identity

$$\widetilde{X}_q(z) = \sum_{d|q} z^{2(d-1)} \left(\widetilde{X}_q(z)\right)^{d-1}, \tag{48}$$

and the corresponding recurrence relation for the characteristic function $\widetilde{\chi}_q(n)$ is given by

$$\widetilde{\chi}_q(n+1) = \sum_{1<d|q} \sum_{\substack{\pi \vdash n-2(d-1) \\ ||\pi|| \leq d-1}} \binom{d-1}{||\pi||} \frac{||\pi||!}{\prod_{j=1}^{\infty} \pi_j!} \prod_{j=1}^{\infty} \left(\widetilde{\chi}_q(j+1)\right)^{\pi_j}$$

$$(n \geq 1,\ \widetilde{\chi}_q(1) = 1,\ q \text{ odd}). \quad (49)$$

An immediate consequence of (49) is that for q odd we have

$$\widetilde{\chi}_q(2) = \widetilde{\chi}_q(3) = \cdots = \widetilde{\chi}_q(2(p(q)-1)) = 0 \quad \text{and} \quad \widetilde{\chi}_q(2p(q)-1) = 1,$$

where $p(q)$ denotes the smallest prime divisor of q.

5.2. The parity patterns $\widetilde{\Pi}_q$

Here we exploit the differential equation (47) to obtain results concerning the parity patterns of the groups $\widetilde{\mathfrak{H}}(q)$. We begin by settling the case when q is even, after which we can dispose of (47) in favour of the simpler identity (48).

Theorem 5. *If $q > 2$ is even, then $\widetilde{\Pi}_q = \mathbb{N}$.*

Proof. In (47) put $\widetilde{X}_q(z) = (1+z)^{-1}$. Upon multiplication with $(1+z)^{q-1}$ the resulting equation becomes

$$(1+z)^{q-2} = \sum_{(d,\mu)} z^{2d-3}(1+z)^{q-d}\left[dz + d - 1\right], \quad (50)$$

where the right-hand sum extends over all pairs of non-negative integers (d, μ) such that $d \mid q$ and $\binom{d-1}{2\mu} \equiv 1\ (2)$. As the terms of this sum are independent of μ, we can simplify (50) by determining for which divisors d of q the number of integers μ with $\mu \geq 0$ and $\binom{d-1}{2\mu} \equiv 1\ (2)$ is odd. Using (35) we find that if d is odd this happens exactly for $d = 1$, whereas if d is even, we must have $d = 2$. Thus, since q is even, equation (50) (and hence (47) with $\widetilde{X}_q(z) = (1+z)^{-1}$) is equivalent to the $GF(2)$-relation

$$(1+z)^{q-2} = (1+z)^{q-1} + z(1+z)^{q-2},$$

which obviously holds. Since the differential equation (47) has a unique solution, we conclude that for q even indeed $\widetilde{\Pi}_q = \mathbb{N}$ as claimed. □

As we have seen in Section 4, the free parity pattern Π_q^* determines the free rank μ_q of the associated Hecke group $\mathfrak{H}(q)$, and hence determines the structure of $\mathfrak{H}(q)$ up to finitely many (in fact at most two) isomorphism types. The corresponding statement for the parity patterns $\widetilde{\Pi}_q$ fails to hold in the light of

Theorem 5. However, if we restrict attention to the case when q is odd, then the following analogue of Theorem 1 holds.

Theorem 6. *Let $q_1, q_2 \geq 3$ be odd integers, and suppose that $\widetilde{\Pi}_{q_1} = \widetilde{\Pi}_{q_2}$. Then we have $q_1 = q_2$.*

Proof. Put $\widetilde{X}_{q_1}(z) = \widetilde{X}_{q_2}(z) =: \widetilde{X}(z)$, and consider the equation

$$\sum_{d \in D(q_1) \Delta D(q_2)} z^{2(d-1)} \left(\widetilde{X}(z)\right)^{d-1} = 0, \tag{51}$$

which follows from (48). Here, $D(q)$ is the set of divisors of a positive integer q, and Δ denotes the symmetric difference of sets. For $q_1 \neq q_2$ the left-hand side Σ of (51) would be of order $\mathrm{ord}(\Sigma) = 2(d_1 - 1)$, where $d_1 := \min_{d \in D(q_1) \Delta D(q_2)} d$, in particular we would have $\Sigma \neq 0$. Hence, we must have $q_1 = q_2$. □

For an integer $q \geq 3$ define $\widetilde{\alpha}_q := \min_{n \in \widetilde{\Pi}_q \setminus \{1\}} n$, and as before let $p(q)$ be the smallest prime divisor of q. Our next result, which is an analogue of Theorem 2, summarizes some general properties of the parity patterns $\widetilde{\Pi}_q$ in the case where q is odd.

Theorem 7. *Let $q \geq 3$ be an odd integer. Then we have the following.*

(i) *Every member of the set $\widetilde{\Pi}_q$ is congruent to 1 modulo 4.*
(ii) *The generating function $\widetilde{X}_q(z)$ is not rational over $GF(2)$; in particular, the set $\widetilde{\Pi}_q$ is infinite.*
(iii) *We have $\widetilde{\alpha}_q = 2\,p(q) - 1$.*

Proof. Property (iii) has already been observed (cf. the remark following equation (49)).
(i) We prove the implication

$$\widetilde{\chi}_q(n+1) = 1 \Rightarrow n \equiv 0 \mod 4 \tag{52}$$

for all $n \in \mathbb{N}_0$ by induction on n. This implication holds trivially if $n = 0$. Suppose that (52) holds for all non-negative integers $n < N$ with some integer $N \geq 1$, and that $\widetilde{\chi}_q(N+1) = 1$. By (49) and our inductive hypothesis,

$$\widetilde{\chi}_q(N+1) = \sum_{1<d|q} \sum_{\substack{\pi \vdash N-2(d-1) \\ \pi_j > 0 \Rightarrow j \equiv 0(4)}} \binom{d-1}{||\pi||} \frac{||\pi||!}{\prod_{j \geq 1} \pi_j!} \prod_{j \geq 1} \left(\widetilde{\chi}_q(j+1)\right)^{\pi_j},$$

and if the right-hand side of the latter equation is to be non-zero, then in particular there must exist a divisor $1 < d \mid q$ and a partition π of $N - 2(d-1)$ all of whose parts are divisible by 4. Since q is odd, these conditions force N to be divisible by 4.

(ii) Suppose that $\widetilde{X}_q(z) = \widetilde{\varphi}(z)/\widetilde{\psi}(z)$ with relatively prime polynomials $\widetilde{\varphi}(z), \widetilde{\psi}(z) \in GF(2)[z]$, and let $v = \deg(\widetilde{\varphi}(z)) - \deg(\widetilde{\psi}(z))$ be the total degree of $\widetilde{X}_q(z)$. Multiplying both sides by $(\widetilde{\psi}(z))^{q-1}$, equation (48) takes the form

$$\widetilde{\varphi}(z)\left(\widetilde{\psi}(z)\right)^{q-2} = \sum_{d|q} z^{2(d-1)}\left(\widetilde{\varphi}(z)\right)^{d-1}\left(\widetilde{\psi}(z)\right)^{q-d}. \tag{53}$$

Since $q > 2$, $\widetilde{\psi}(z)$ must divide the right-hand side of (53), as it divides the left-hand side , and, as in the proof of Theorem 2 (ii), we conclude that $\widetilde{\psi}(z) = 1$. This implies in particular, that $v = \deg(\widetilde{\varphi}(z)) \geq 0$, and comparing degrees on both sides of (48) gives

$$v = (v + 2)(q - 1),$$

which is impossible. □

The parity patterns $\widetilde{\Pi}_q$ also admit of an analogue of Theorem 3, which we describe next. For an odd integer $q \geq 3$, let $1 = d_0 < d_1 < \cdots < d_r = q$ be the set of divisors of q in increasing order, and let

$$\underline{d}_q := \left(\frac{d_1 - 1}{2}, \frac{d_2 - 1}{2}, \ldots, \frac{d_r - 1}{2}\right) \in \mathbb{N}_0^r.$$

As usual, we define the norm of a vector $\underline{v} = (v_1, \ldots, v_r) \in \mathbb{N}_0^r$ as $||\underline{v}|| = \sum_{j=1}^r v_j$, and if $\underline{u} = (u_1, \ldots, u_r)$ and $\underline{v} = (v_1, \ldots, v_r)$ are two such vectors, then their scalar product is given by $\underline{u} \cdot \underline{v} = \sum_{j=1}^r u_j v_j$. Using the recurrence relation (49), viewed as a $\mathbb{Z}$-relation, define an integral sequence $\widehat{\widetilde{\chi}}_q(n)$ starting from $\widehat{\widetilde{\chi}}_q(1) = 1$, and let $\widehat{\widetilde{X}}_q(z) := \sum_{n\geq 0} \widehat{\widetilde{\chi}}_q(n+1)\, z^n$. Then

(i) $\widehat{\widetilde{X}}_q(z) \equiv \widetilde{X}_q(z) \mod 2$,
(ii) the series $\widehat{\widetilde{X}}_q(z)$ satisfies the functional equation

$$\widehat{\widetilde{X}}_q(z) = \sum_{d|q} z^{2(d-1)}\left(\widehat{\widetilde{X}}_q(z)\right)^{d-1},$$

(iii) $\widehat{\widetilde{\chi}}_q(n+1) = 0$ for all $n \geq 0$ such that $4 \nmid n$.

The first two assertions are clear, the third follows by an argument similar to the proof of Theorem 7 (i). By assertion (iii),

$$\begin{aligned} z^2\, \widehat{\widetilde{X}}_q(z) &= \sum_{\mu\geq 0} \widehat{\widetilde{\chi}}_q(4\mu + 1)\, z^{4\mu+2} \\ &= F(z^2), \end{aligned}$$

where $F(t) := \sum_{\mu \geq 0} \widehat{\widetilde{X}}_q(4\mu+1)\, t^{2\mu+1}$. Writing $\widehat{\widetilde{X}}_q(z)$ as $\widehat{\widetilde{X}}_q(z) = z^{-2}\, F(z^2)$ and substituting $t = z^2$, the functional equation in (ii) becomes

$$F(t) = t\, \Phi(F(t)),$$

where

$$\Phi(\zeta) := \sum_{d|q} \zeta^{d-1}.$$

By Lagrange inversion, we have for $n \geq 1$

$$\begin{aligned}\left\langle t^n,\ F(t)\right\rangle &= \frac{1}{n}\left\langle \zeta^{n-1},\ \left(\Phi(\zeta)\right)^n\right\rangle\\ &= \frac{1}{n} \sum_{\substack{\underline{n} \in \mathbb{N}_0^r\\ \underline{d}_q \cdot \underline{n} = \frac{n-1}{2}}} \binom{n}{\underline{n},\, n - \|\underline{n}\|},\end{aligned}$$

and hence

$$\widehat{\widetilde{X}}_q(z) = \sum_{\mu \geq 0} \left[\frac{1}{2\mu+1} \sum_{\substack{\underline{n} \in \mathbb{N}_0^r\\ \underline{d}_q \cdot \underline{n} = \mu}} \binom{2\mu+1}{\underline{n},\, 2\mu+1-\|\underline{n}\|}\right] z^{4\mu}. \tag{54}$$

Taking into account the first part of Theorem 7, assertion (i), equation (54), and the fact that the 2-adic norm of a multinomial coefficient is given by[11]

$$\nu_2\binom{n}{n_1, \ldots, n_r} = \sum_{j=1}^{r} \mathfrak{s}(n_j) - \mathfrak{s}(n),$$

we now obtain the following explicit description of the parity patterns $\widetilde{\Pi}_q$ in the case where q is odd.

Theorem 8. *Let $q \geq 3$ be an odd integer. Then*

(i) $\widetilde{\Pi}_q \subseteq 1 + 4\mathbb{N}_0$,
(ii) *given an integer $\mu \geq 0$ we have $1 + 4\mu \in \widetilde{\Pi}_q$ if and only if the set*

$$\Bigg\{\underline{n} \in \mathbb{N}_0^r :\ \underline{d}_q \cdot \underline{n} = \mu \quad \text{and} \quad \sum_{j=1}^{r} \mathfrak{s}(n_j) + \mathfrak{s}(2\mu+1-\|\underline{n}\|) - \mathfrak{s}(2\mu) = 1\Bigg\}$$

has odd cardinality.

[11] This follows immediately from Legendre's formula for the p-adic norm of factorials.

The description of the parity pattern $\widetilde{\Pi}_q$ given in Theorem 8 simplifies considerably if q is a prime number.

Corollary 2. *Let $q > 2$ be a prime. Then*

$$\widetilde{\Pi}_q = \Big\{2(q-1)n + 1 : \ n \in \mathbb{N}_0 \ \text{ and } \ \mathfrak{s}(n) + \mathfrak{s}((q-2)n+1) - \mathfrak{s}((q-1)n) = 1\Big\}.$$

Our next result records some consequences of Corollary 2.

Corollary 3. *Let $q > 2$ be a prime. Then*

(i) $\widetilde{\Pi}_q \backslash \{1\} \subseteq 1 + 2(q-1) + 4(q-1)\mathbb{N}_0$,
(ii) *we have $1 + 2(q-1)(2^\alpha + 1) \in \widetilde{\Pi}_q$ for some $\alpha \in \mathbb{N}$ if and only if $a_\alpha = 1$, where $q - 1 = \sum_{j \geq 1} a_j 2^j$ with $a_j \in \{0, 1\}$.*

Proof. (i) Let $n = \sum_{j \geq 0} n_j 2^j$ with $n_j \in \{0, 1\}$, and suppose that $n_0 = 0$ and that $\binom{(q-1)n+1}{n} \equiv 1 \ (2)$. Assuming inductively that $n_j = 0$ for all j with $0 \leq j < J$ and some $J \in \mathbb{N}$, we find that

$$1 + (q-1)n = 1 + \sum_{j \geq J} \frac{q-1}{2} n_j 2^{j+1},$$

in particular $n_J = 0$ by Lucas' Theorem. Hence, we must have $n = 0$, and our claim follows from Corollary 2.
(ii) Write $q - 1 = \sum_{j \geq 1} a_j 2^j$ with $a_j \in \{0, 1\}$, and put $n = 2^\alpha + 1$ with some $\alpha \geq 1$. Then

$$1 + (q-1)n = 1 + \sum_{1 \leq j \leq \alpha} a_j 2^j + \sum_{j > \alpha} (a_j + a_{j-\alpha}) 2^j,$$

hence, by Lucas' Theorem,

$$\binom{(q-1)n+1}{n} \equiv 1 \ (2) \ \Leftrightarrow \ a_\alpha = 1.$$

Our claim follows now from Corollary 2. □

As is apparent from Theorem 8 and Corollary 2, the parity patterns $\widetilde{\Pi}_q$ with q odd can in general not be expected to admit of a characterization in terms of closed formulae. Instead, they generically tend to exhibit the type of fractal behaviour peculiar to parametrized binomial or multinomial coefficients when evaluated modulo 2. Indeed, the only exception occurs when q is a Fermat prime. For an integer $q \geq 3$ define

$$\Lambda_q := \left\{ \frac{2(q-1)^\sigma - q}{q-2} : \ \sigma = 1, 2, \dots \right\},$$

i.e., Λ_q is the set of partial sums of the series $1 + 2\sum_{\sigma\geq 1}(q-1)^\sigma$.

Theorem 9. *Let $q \geq 3$ be an odd integer. Then the following assertions are equivalent:*

(i) $\widetilde{\Pi}_q = \Lambda_q$.
(ii) $\widetilde{\chi}_q(n) = 0$ *for* $n \in ([2q(q-1)] - \{1, 2q-1\}) \cap (\mathbb{N} - 2\mathbb{N})$.
(iii) q *is a Fermat prime.*

Proof. Since (i) clearly implies (ii), it suffices to prove the implications (ii)⇒(iii) and (iii)⇒(i). Assume first that q is not a Fermat prime. If q is not a prime, then $2 \leq 2p(q) - 1 < 2q - 1$, and, by the third part of Theorem 7, we have $\widetilde{\chi}_q(2p(q) - 1) = 1$. Suppose on the other hand that q is a prime, and let

$$\nu_0 := \min \left\{ k \in [q-1] : \ \binom{q-1}{k} \equiv 1 \ (2) \right\}.$$

Since q is not a Fermat prime, we certainly have $1 \leq \nu_0 < q - 1$. Define $n_0 := 2(\nu_0 + 1)(q-1)$. Then $2q - 1 < n_0 + 1 \leq 2q(q-1)$, and from (49), the fact that $\widetilde{\chi}_q(2) = \cdots = \widetilde{\chi}_q(2(q-1)) = 0$, and the definition of ν_0 we find that

$$\begin{aligned} \widetilde{\chi}_q(n_0 + 1) &= \sum_{\substack{\pi \vdash 2\nu_0(q-1) \\ \nu_0 \leq ||\pi|| \leq q-1 \\ \pi_j > 0 \Rightarrow j \geq 2(q-1)}} \binom{q-1}{||\pi||} \frac{||\pi||!}{\prod_{j=1}^\infty \pi_j!} \prod_{j=1}^\infty \left(\widetilde{\chi}_q(j+1)\right)^{\pi_j} \\ &= \widetilde{\chi}_q(2q-1) = 1. \end{aligned}$$

In both cases we obtain a contradiction to (ii), thus proving the implication (ii)⇒(iii). Now suppose that q is a Fermat prime, say $q = 2^{2^\lambda} + 1$. Then $\mathfrak{s}((q-1)n) = \mathfrak{s}(n)$, and the condition on n in Corollary 2 becomes

$$\mathfrak{s}((2^{2^\lambda} - 1)n + 1) = 1,$$

that is,

$$n = \frac{2^\alpha - 1}{2^{2^\lambda} - 1} \quad \text{with} \quad 2^\lambda \mid \alpha \quad \text{and} \quad \alpha \geq 0.$$

Assertion (i) follows now from Corollary 2, and the proof of Theorem 9 is complete. □

Remark 2. *Let $q \geq 3$ be an odd integer, but not a Fermat prime. Then, by the previous theorem, we have $\widetilde{\Pi}_q \neq \Lambda_q$ and, more precisely, there exists a deviation of $\widetilde{\Pi}_q$ from the pattern Λ_q within the set of integers $[2q(q-1)] - \{1, 2q-1\}$. In fact, our proof of the implication* (ii)⇒(iii) *of Theorem* 9 *yields a slightly sharper result. Let $n_q := \min_{n \in \widetilde{\Pi}_q \,\Delta\, \Lambda_q} n$, i.e., n_q is the smallest integer for which $\widetilde{\Pi}_q$ and Λ_q differ as to their containing or not containing this number. Then we have*

$$n_q = \begin{cases} 2\,(2^{\nu_2(q-1)} + 1)\,(q-1) + 1, & q \text{ a prime} \\ 2\,p(q) - 1, & \text{otherwise.} \end{cases} \tag{55}$$

Furthermore, in the first case, i.e., q a prime but not a Fermat prime, we have $2 \leq 2^{\nu_2(q-1)} \leq \frac{q-1}{3}$, which when combined with (55) *yields the estimate*

$$6q - 5 \;\leq\; n_q \;\leq\; \frac{2}{3}\,(q-1)\,(q+2) + 1 \tag{56}$$

in terms of q alone. For $q = 7$ both these bounds are sharp.

6. Some combinatorial interpretations of the series $\widehat{X}_q^*(z)$ and $\widehat{\widetilde{X}}_q(z)$

In Section 4.1 we associated, for every integer $q \geq 3$, a canonical lifting $\widehat{X}_q^*(z) \in \mathbb{Z}[[z]]$ to the $GF(2)$-series $X_q^*(z) = 1 + \sum_{\lambda \in \Pi_q^*} z^\lambda$. Similarly, in Section 5.2, a lifting $\widehat{\widetilde{X}}_q(z)$ of the $GF(2)$-series $\widetilde{X}_q(z) = \sum_{n \in \widetilde{\Pi}_q} z^{n-1}$ was defined and explicitly computed for every odd integer $q > 1$. Here, we shall describe a number of combinatorial interpretations for the coefficients of these liftings $\widehat{X}_q^*(z)$ and $\widehat{\widetilde{X}}_q(z)$. Let $S \subseteq \mathbb{N}$ be a set of positive integers. By a *plane S-tree* we mean a plane tree with the property that every non-terminal vertex has (outer) degree an element of S. Given $S \subseteq \mathbb{N}$ and non-negative integers m, n, we denote by $T_S(m, n)$ the number of plane S-trees having m terminal vertices and a total of n vertices. Let

$$U_S = U_S(t, z) := \sum_{n \geq 0} \sum_{m \geq 0} T_S(m, n)\, t^m\, z^n .$$

Then U_S satisfies the functional equation[12]

$$U_S = t\,z + z \sum_{\sigma \in S} U_S^{\sigma}. \tag{57}$$

In order to establish a connection with the series $\widehat{X}_q^*(z)$, put $S = \{\mu_q\}$ and $t = 1/z$. Then equation (57) becomes

$$U_{\{\mu_q\}}(1/z, z) = 1 + z\left(U_{\{\mu_q\}}(1/z, z)\right)^{\mu_q},$$

and, in view of (36), we must have $U_{\{\mu_q\}}(1/z, z) = \widehat{X}_q^*(z)$. Consequently, for $q \geq 3$ and $\lambda \geq 0$,

$$\frac{1}{\mu_q\lambda + 1}\binom{\mu_q\lambda + 1}{\lambda} = \left\langle z^{\lambda},\, \widehat{X}_q^*(z)\right\rangle = \left\langle z^{\lambda},\, U_{\{\mu_q\}}(1/z, z)\right\rangle$$

equals

(a) the number of plane trees with exactly λ non-terminal vertices, each of which having (outer) degree precisely μ_q.

These tree numbers in turn can be reinterpreted in terms of other combinatorial objects. For $\lambda \geq 0$, the coefficient $\left\langle z^{\lambda},\, \widehat{X}_q^*(z)\right\rangle$ equals

(b) the number of sequences $i_1 i_2 \ldots i_{\lambda\mu_q}$ with $i_j \in \{-1, \mu_q - 1\}$ for all $j \in [\lambda\mu_q]$, such that (i) there are a total of $(\mu_q - 1)\lambda$ values of j for which $i_j = -1$, and (ii) we have $i_1 + i_2 + \cdots + i_j \geq 0$ for all j,
(c) the number of bracketings of a word of length $\lambda(\mu_q - 1) + 1$ subject to λ μ_q-ary operations,
(d) the number of paths p in the (x, y)-plane starting in the origin $(0, 0)$ and terminating in the point $(\lambda\mu_q, 0)$, using steps $(1, \sigma)$ with $\sigma \in \{-1, \mu_q - 1\}$, such that p never passes below the x-axis,
(e) the number of paths p in the (x, y)-plane from $(0, 0)$ to $((\mu_q - 1)\lambda, (\mu_q - 1)\lambda)$, using steps $(\mu_q - 1, 0)$ or $(0, 1)$, such that p never passes above the line $x = y$,
(f) the number of ways of dissecting a convex $(\lambda(\mu_q - 1) + 2)$-gon into λ convex $(\mu_q + 1)$-gons, by drawing diagonals which do not intersect in their interiors;

cf. [38, Prop. 6.2.1]. If we deform the (x, y)-plane by means of the transformation

$$x' = (\mu_q - 1)\lambda - y, \quad y' = \lambda - \frac{x}{\mu_q - 1},$$

[12] Cf. [38, Prop. 6.2.4].

then we find from (e) that $\left\langle z^\lambda,\ \widehat{X}_q^*(z)\right\rangle$ also equals

(g) the number of lattice paths p in the 2-dimensional integral lattice $\mathbb{Z}^2$ starting in the origin $(0, 0)$ and terminating in the lattice point $((\mu_q - 1)\lambda, \lambda)$, such that (i) p consists only of positive horizontal and vertical unit steps, and (ii) p never passes above the line $x = (\mu_q - 1)\,y$.[13]

Let $q \geq 3$ be an odd integer. Putting $S = \{d - 1 : 1 < d \mid q\}$ and $t = 1$, equation (57) becomes

$$U_S(1, z) = z \sum_{d \mid q} \big(U_S(1, z)\big)^{d-1},$$

which is precisely the kind of equation determining the series $F(z)$ occurring in Section 5.2. Proceeding as before, we find that for $\mu \geq 0$

$$\frac{1}{2\mu + 1} \sum_{\substack{\underline{n} \in \mathbb{N}_0^r \\ \underline{d}_q \cdot \underline{n} = \mu}} \binom{2\mu + 1}{\underline{n},\ 2\mu + 1 - ||\underline{n}||}$$

$$= \left\langle z^{4\mu},\ \widehat{\widehat{X}}_q(z)\right\rangle = \left\langle z^{2\mu+1},\ F(z)\right\rangle = \left\langle z^{2\mu+1},\ U_S(1, z)\right\rangle$$

equals

(a) the number of plane trees on $2\mu + 1$ vertices, such that every non-terminal vertex has (outer) degree an element of the set $S = \{d - 1 : 1 < d \mid q\}$,
(b) the number of sequences $i_1 i_2 \ldots i_{2\mu}$, where $i_j \in \{d - 2 : d \mid q\}$ for all $j \in [2\mu]$, such that $i_1 + i_2 + \cdots + i_j \geq 0$ for all j, and $i_1 + i_2 + \cdots + i_{2\mu} = 0$,
(c) the number of bracketings of a word of some length m, $1 \leq m \leq 2\mu + 1$, subject to $2\mu - m + 1$ σ-ary operations, where $\sigma \in \{d - 1 : 1 < d \mid q\}$,
(d) the number of paths p in the (x, y)-plane from $(0, 0)$ to $(2\mu, 0)$ using steps $(1, d - 2)$, where $d \mid q$, such that p never passes below the x-axis,
(e) the number of paths p in the (x, y)-plane from $(0, 0)$ to $(m - 1, m - 1)$ for some m, $1 \leq m \leq 2\mu + 1$, using steps $(d - 2, 0)$ or $(0, 1)$ with $1 < d \mid q$, having a total of 2μ steps, such that p never passes above the line $x = y$,
(f) the number of dissections of a convex $(m + 1)$-gon for some m, $1 \leq m \leq 2\mu + 1$, into $2\mu - m + 1$ regions, each a convex d-gon with some $1 < d \mid q$, by drawing diagonals which do not intersect in their interiors.

[13] Cf. [24, pp. 8 – 9]. See also [33, Chap. I] for related results concerning the enumeration of lattice paths.

7. Square decomposition of subdiagonals and the series $\mathcal{R}_q^{\mu,\nu}(z)$

The proof of formula (47) – and hence that of its consequences Theorems 5–9 – given in Section 5 rests on two statements of a rather technical nature, left unproven at that stage so as not to disrupt the discussion too much. These statements are: (i) formula (45) computing the quotients $\widetilde{H}_q^{(\nu)}(z)/\widetilde{H}_q(z)$ modulo 2 in terms of the series $\widetilde{S}_q(z)$, and (ii) assumption (46) concerning the integrality of the series $\mathcal{R}_q^{\mu,\nu}(z)$. The purpose of this section is to establish these two facts, in this way completing the proofs of the results in Section 5. We begin by showing the following.

Lemma 3. *For every* $\nu \in \mathbb{N}_0$ *the series* $\widetilde{H}_q^{(\nu)}(z)/\widetilde{H}_q(z)$ *is an integral power series, and satisfies the congruence* (45).

Proof. Rewrite (43) in the form $\widetilde{H}_q(z) = \exp\left(\int \widetilde{S}_q(z)\,dz\right)$ and apply Bell's formula (17) to obtain

$$\widetilde{H}_q^{(\nu)}(z)/\widetilde{H}_q(z) = \sum_{\pi \vdash \nu} \frac{\nu!}{\prod_{j=1}^{\infty}(j!)^{\pi_j}\,\pi_j!} \prod_{j=1}^{\infty}\left(\widetilde{S}_q^{(j-1)}(z)\right)^{\pi_j}, \quad \nu \geq 0. \tag{58}$$

The coefficients occurring in (58) have a natural combinatorial interpretation. For a partition π consider the set $\mathcal{M}(\pi)$ of all maps $[||\pi||] \to 2^{[|\pi|]}$, denoted by $i \mapsto N_i$, such that

(i) $N_i \cap N_j = \emptyset$ for $1 \leq i, j \leq ||\pi||$ and $i \neq j$,
(ii) $\bigcup_{i=1}^{||\pi||} N_i = [|\pi|]$,
(iii) $|N_{i_j}| = j$ for $i_j \in [\pi_j] + \sum_{k=1}^{j-1} \pi_k$ and all $j \geq 1$.

We have

$$|\mathcal{M}(\pi)| = |\pi|!/\prod_{j=1}^{\infty}(j!)^{\pi_j}.$$

Let $G(\pi)$ be the permutation group consisting of all permutations of $S_{||\pi||}$ leaving invariant the decomposition

$$[||\pi||] = \coprod_{j\geq 1}\left([\pi_j] + \textstyle\sum_{k=1}^{j-1}\pi_k\right)$$

of $[||\pi||]$, and let $G(\pi)$ act (from the right say) on $\mathcal{M}(\pi)$ in the natural way. Clearly, this action of $G(\pi)$ on $\mathcal{M}(\pi)$ is free, and $G(\pi) \cong \prod_{j\geq 1} S_{\pi_j}$. Hence,

$$\frac{|\pi|!}{\prod_{j\geq 1}(j!)^{\pi_j}\,\pi_j!} = |\mathcal{M}(\pi)/G(\pi)|;$$

in particular these numbers are integers. Since the series $\widetilde{S}_q(z)$ is integral, our first claim follows from (58). Moreover, using the fact that $\widetilde{S}_q^{(\mu)}(z) \equiv 0\ (2)$ for $\mu \geq 2$, we have modulo 2

$$\widetilde{H}_q^{(\nu)}(z)/\widetilde{H}_q(z) \equiv \sum_{\substack{\pi \vdash \nu \\ \pi_j = 0\,(j>2)}} \frac{\nu!}{\prod_{j\geq 1} (j!)^{\pi_j}\, \pi_j!} \prod_{j \geq 1} \left(\widetilde{S}_q^{(j-1)}(z)\right)^{\pi_j}$$

$$\equiv \sum_{\mu=0}^{\lfloor \nu/2 \rfloor} \binom{\nu}{2\mu} \left(\widetilde{S}_q'(z)\right)^{\mu} \left(\widetilde{S}_q(z)\right)^{\nu - 2\mu},$$

which is (45). □

We now come to the main topic of this section, namely the verification of assumption (46) concerning the integrality of the series $\mathcal{R}_q^{\mu,\nu}(z)$. Given any $q \geq 3$, define a system of polynomials $\mathfrak{p}_q^{\ell,\mu}(t) \in \mathbb{Z}[t]$ indexed by two extra parameters $\ell \in \mathbb{Z}$ and $\mu \in \mathbb{N}_0$ via the equations

$$\mathfrak{p}_q^{\ell,0}(t) = \delta_{\ell,0} \quad (\ell \in \mathbb{Z}), \tag{59}$$

$$\mathfrak{p}_q^{\ell,\mu}(t) = 0 \quad (\ell < 0,\ \mu > 0), \tag{60}$$

$$\mathfrak{p}_q^{\ell,\mu}(t) = \sum_{\substack{d \mid q \\ d \leq \mu}} (t + \mu - 1)_{d-1}\, \mathfrak{p}_q^{\ell,\mu-d}(t) + \sum_{\substack{d \mid q \\ d > \mu}} (t + \mu - 1)_{\mu-1} \times \mathfrak{p}_q^{\ell+\mu-d,d-\mu}(t + \mu - d) \qquad (\ell \geq 0,\ \mu > 0). \tag{61}$$

An immediate induction on ℓ, followed by induction on μ, shows that (59) – (61) uniquely define a system $\{\mathfrak{p}_q^{\ell,\mu}(t)\}_{(q,\ell,\mu)}$ of integral polynomials parametrized by the triples

$$(q, \ell, \mu) \in \left(\mathbb{N} - \{1, 2\}\right) \times \mathbb{Z} \times \mathbb{N}_0.$$

As our next result shows, these polynomials $\mathfrak{p}_q^{\ell,\mu}(t)$, for every $q \geq 3$, relate the subdiagonals of the array $\left(h_\mu(q)\, h_\nu(q)\right)_{\mu,\nu \geq 0}$ to the terms of its main diagonal. It is this important observation which underlies our proof of hypothesis (46).

Lemma 4. *For every $q \geq 3$, $\mu \geq 0$, and $n \geq 0$, and with $\mathfrak{p}_q^{\ell,\mu}(t)$ as defined above, we have*

$$h_n(q)\, h_{n+\mu}(q) = \sum_{\ell=0}^{n} \mathfrak{p}_q^{\ell,\mu}(n)\, (n)_\ell\, h_{n-\ell}^2(q). \tag{62}$$

Proof. We will fix $q \geq 3$, and prove (62) for all $\mu, n \geq 0$ by induction on n, followed by induction on μ. Suppose first that $n = 0$. In this case we have to show that

$$h_\mu(q) = \mathfrak{p}_q^{0,\mu}(0), \quad \mu \geq 0. \tag{63}$$

This holds for $\mu = 0$ in view of (59). Fix an integer $M > 0$, and assume that (63) holds for all μ such that $0 \leq \mu < M$. Then, by (41), our inductive hypothesis, (60), and (61) with $\mu = M$ and $\ell = t = 0$,

$$\begin{aligned} h_M(q) &= \sum_{\substack{d \mid q \\ d \leq M}} (M-1)_{d-1}\, h_{M-d}(q) = \sum_{\substack{d \mid q \\ d \leq M}} (M-1)_{d-1}\, \mathfrak{p}_q^{0,M-d}(0) \\ &= \mathfrak{p}_q^{0,M}(0). \end{aligned}$$

Hence, (62) holds for $n = 0$, all $\mu \geq 0$, and $q \geq 3$ as fixed above. Now let $N > 0$ be an integer, and suppose that (62) holds for all n with $0 \leq n < N$, all $\mu \geq 0$, and our fixed value of q. In order to complete the induction step for our induction on n, we have to show that

$$h_N(q)\, h_{N+\mu}(q) = \sum_{\ell=0}^{N} \mathfrak{p}_q^{\ell,\mu}(N)\,(N)_\ell\, h_{N-\ell}^2(q) \tag{64}$$

holds for all $\mu \geq 0$. To see this, we proceed again by induction on μ. In view of (59), (64) is true for $\mu = 0$. Suppose then that (64) holds true for all μ with $0 \leq \mu < M$ and some positive integer M. Then, by (41), the inductive hypotheses, (60), and (61) with $\mu = M$ and $t = N$,

$$\begin{aligned} &h_N(q)\, h_{N+M}(q) \\ &\quad = \sum_{d \mid q} (N+M-1)_{d-1}\, h_N(q)\, h_{N+M-d}(q) \\ &\quad = \sum_{\substack{d \mid q \\ d \leq M}} (N+M-1)_{d-1} \sum_{\ell=0}^{N} \mathfrak{p}_q^{\ell,M-d}(N)\,(N)_\ell\, h_{N-\ell}^2(q) \\ &\quad\quad + \sum_{\substack{d \mid q \\ M<d\leq N+M}} (N+M-1)_{d-1} \sum_{\ell=0}^{N+M-d} \mathfrak{p}_q^{\ell,d-M}(N+M-d) \\ &\quad\quad \times (N+M-d)_\ell\, h_{N+M-d-\ell}^2(q) \end{aligned}$$

$$= \sum_{\ell=0}^{N} \Bigg[\sum_{\substack{d|q \\ d \le M}} (N+M-1)_{d-1}\, \mathfrak{p}_q^{\ell, M-d}(N)$$

$$+ \sum_{\substack{d|q \\ d > M}} (N+M-1)_{M-1}\, \mathfrak{p}_q^{\ell+M-d, d-M}(N+M-d) \Bigg] (N)_\ell\, h_{N-\ell}^2(q)$$

$$= \sum_{\ell=0}^{N} \mathfrak{p}_q^{\ell,M}(N)\,(N)_\ell\, h_{N-\ell}^2(q).$$

This completes the proof of (64), and hence of the lemma. □

We now come to the main result of this section.

Proposition 3. *For every* $(q, \mu, \nu) \in (\mathbb{N} - \{1, 2\}) \times \mathbb{N} \times \mathbb{N}_0$, *the series* $\mathcal{R}_q^{\mu,\nu}(z)$ *is an integral power series; in particular, assumption* (46) *holds true.*

Proof. With the convention that the zero polynomial has degree -1, decompose each $\mathfrak{p}_q^{\ell,\mu}(t)$ according to [25, Lemma 5] in the form

$$\mathfrak{p}_q^{\ell,\mu}(t) = \sum_{\kappa=0}^{d_q^{\ell,\mu}} a_q^{\ell,\mu}(\kappa)\,(t)_\kappa, \quad (q, \ell, \mu) \in (\mathbb{N} - \{1, 2\}) \times \mathbb{Z} \times \mathbb{N}_0, \quad (65)$$

where $\deg\left(\mathfrak{p}_q^{\ell,\mu}(t)\right) = d_q^{\ell,\mu}$, and with $a_q^{\ell,\mu}(\kappa) \in \mathbb{Z}$ for all q, ℓ, μ, and κ. Then, by Lemma 4 and (65), we have for $q \ge 3$ and $\mu > 0$ that

$$\sum_{n=0}^{\infty} h_n(q)\, h_{n+\mu}(q)\, z^n/n! = \sum_{n=0}^{\infty} \sum_{\ell=0}^{n} \sum_{\kappa=0}^{d_q^{\ell,\mu}} a_q^{\ell,\mu}(\kappa)\,(n)_\kappa\,(n)_\ell\, h_{n-\ell}^2(q)\, z^n/n!$$

$$= \sum_{\ell=0}^{\infty} \sum_{\kappa=0}^{d_q^{\ell,\mu}} a_q^{\ell,\mu}(\kappa) \sum_{n=0}^{\infty} (n+\ell)_\kappa\, h_n^2(q)\, z^{n+\ell}/n!$$

$$= \sum_{\ell=0}^{\infty} \sum_{\kappa=0}^{d_q^{\ell,\mu}} a_q^{\ell,\mu}(\kappa)\, z^\kappa \left(z^\ell\, \widetilde{H}_q(z)\right)^{(\kappa)}.$$

Invoking Leibniz's formula and introducing the numbers

$$A_{\ell,i}^{q,\mu} := \sum_{\kappa=0}^{d_q^{\ell,\mu}} \binom{\kappa}{i} (\ell)_{\kappa-i}\, a_q^{\ell,\mu}(\kappa)$$

this becomes

$$\sum_{n=0}^{\infty} h_n(q)\, h_{n+\mu}(q)\, z^n/n! = \sum_{\ell=0}^{\infty} \sum_{i=0}^{\infty} A_{\ell,i}^{q,\mu}\, z^{\ell+i}\, \widetilde{H}_q^{(i)}(z).$$

Taking the ν-th derivative, applying Leibniz's formula again, and dividing throughout by the series $\widetilde{H}_q(z)$ then gives

$$\mathcal{R}_q^{\mu,\nu}(z) = \sum_{\ell=0}^{\infty} \sum_{i=0}^{\infty} \sum_{j=0}^{\infty} \binom{\nu}{j} (\ell+i)_{\nu-j}\, A_{\ell,i}^{q,\mu}\, z^{\ell+i+j-\nu}\, \widetilde{H}_q^{(i+j)}(z)/\widetilde{H}_q(z).$$

Since the coefficients $A_{\ell,i}^{q,\mu}$ are well-defined integers, the family of series

$$\left\{ \binom{\nu}{j} (\ell+i)_{\nu-j}\, A_{\ell,i}^{q,\mu}\, z^{\ell+i+j-\nu}\, \widetilde{H}_q^{(i+j)}(z)/\widetilde{H}_q(z) \right\}_{\ell,i,j}$$

is summable (because of the factor $z^{\ell+i+j-\nu}$), and the series $\widetilde{H}_q^{(i+j)}(z)/\widetilde{H}_q(z)$ are integral for all $i, j \geq 0$ by Lemma 3, we conclude that for every $q \geq 3$, $\mu > 0$, and $\nu \geq 0$ the remainder series $\mathcal{R}_q^{\mu,\nu}(z)$ is indeed integral, as claimed. □

With Lemma 3 and Proposition 3 in hand, the proof of formula (47), and hence of the results in Section 5.2 is complete.

8. The parity pattern of a Hecke group

In this final section we establish a relationship between the parity pattern of an arbitrary group and that of one of its index 2 subgroups. Once established, this relationship will allow us in particular to translate the results obtained for the patterns $\widetilde{\Pi}_q$ into results concerning the parity patterns of Hecke groups.

8.1. Index 2 descent

As a rule, divisibility properties of subgroup counting functions (unlike their growth behaviour) tend to react extremely sensitively to movements within a commensurability class; in particular, when passing from a group to one of its finite index subgroups, arithmetic structure of this kind is usually severely deformed if not completely destroyed. The following result is a non-trivial exception to this rule.

Proposition 4. *Let $\mathfrak{G}$ be a group containing only finitely many subgroups of index n for every positive integer n, and let $\mathfrak{H}$ be a subgroup of index 2 in $\mathfrak{G}$. Then*

$$\Pi(\mathfrak{G}) = \Big(\Pi(\mathfrak{H}) \cap (\mathbb{N} - 2\mathbb{N})\Big) \cup 2\,\Pi(\mathfrak{H}). \tag{66}$$

Proof. Every subgroup $\mathfrak{G}'$ of index n in $\mathfrak{G}$, which is not contained in $\mathfrak{H}$, intersects $\mathfrak{H}$ in a subgroup $\mathfrak{H}'$ with $(\mathfrak{H} : \mathfrak{H}') = n$. Hence, each such $\mathfrak{G}'$ is contained in the set $\bigcup_{(\mathfrak{H}:\mathfrak{H}')=n} \mathfrak{S}(\mathfrak{H}')$, where

$$\mathfrak{S}(\mathfrak{H}') := \Big\{\mathfrak{G}' \leq \mathfrak{G} : \ \mathfrak{G}' \cap \mathfrak{H} = \mathfrak{H}' \ \text{ and } \ \mathfrak{G}'\mathfrak{H} = \mathfrak{G}\Big\},$$

and, conversely, each $\mathfrak{G}' \in \bigcup_{(\mathfrak{H}:\mathfrak{H}')=n} \mathfrak{S}(\mathfrak{H}')$ is of index n in $\mathfrak{G}$ and not contained in $\mathfrak{H}$. It follows that

$$s_n(\mathfrak{G}) = \sum_{(\mathfrak{H}:\mathfrak{H}')=n} |\mathfrak{S}(\mathfrak{H}')| + \begin{cases} s_{n/2}(\mathfrak{H}), & 2 \mid n \\ 0, & 2 \nmid n. \end{cases} \tag{67}$$

Fix an element ζ with $\mathfrak{G} = \langle \mathfrak{H}, \zeta \rangle$. Given a subgroup $\mathfrak{H}'$ of index n in $\mathfrak{H}$ and a right transversal $1 = \mathfrak{h}_1, \mathfrak{h}_2, \ldots, \mathfrak{h}_n$ for $\mathfrak{H}'$ in $\mathfrak{H}$, the elements $\mathfrak{g}_{\mu,\nu} := \mathfrak{h}_\mu \zeta^\nu$ with $(\mu, \nu) \in [n] \times \{0, 1\}$ form a right transversal for $\mathfrak{H}'$ in $\mathfrak{G}$. A subgroup $\mathfrak{G}' \in \mathfrak{S}(\mathfrak{H}')$ must contain $\mathfrak{H}'$ as a subgroup of index 2, and an element $\mathfrak{g}_{\mu,1}$ for some μ. Hence, the sets

$$\mathfrak{G}'_\mu := \mathfrak{H}' \cup \mathfrak{H}'\mathfrak{g}_{\mu,1}, \quad \mu \in [n]$$

exhaust all possibilities for a subgroup $\mathfrak{G}' \in \mathfrak{S}(\mathfrak{H}')$. Clearly, such a set $\mathfrak{G}'_\mu$ is contained in $\mathfrak{S}(\mathfrak{H}')$ if and only if $\mathfrak{G}'_\mu$ is a subgroup of $\mathfrak{G}$. The necessary and sufficient condition for the latter to hold is that $\mathfrak{g}_{\mu,1}\mathfrak{H}'\mathfrak{g}_{\mu,1} = \mathfrak{H}'$, i.e., $\mathfrak{g}_{\mu,1}$ has to represent an involution in the group $N_{\mathfrak{G}}(\mathfrak{H}')/\mathfrak{H}'$. By a well-known theorem of Frobenius [9],[14] if d divides the order of a finite group $\mathfrak{G}$, then the number of solutions in $\mathfrak{G}$ of the equation $x^d = 1$ is divisible by d. Applying this result with $d = 2$ to the group $N_{\mathfrak{G}}(\mathfrak{H}')/\mathfrak{H}'$ we see that $|N_{\mathfrak{G}}(\mathfrak{H}')/\mathfrak{H}'|$ is congruent modulo 2 to the number of solutions in $N_{\mathfrak{G}}(\mathfrak{H}')/\mathfrak{H}'$ of the equation $x^2 = 1$. Those among these solutions which are represented by elements $\mathfrak{g}_{\mu,0}$ constitute the totality of solutions of this equation in the subgroup $N_{\mathfrak{H}}(\mathfrak{H}')/\mathfrak{H}'$; hence, applying Frobenius' result again, their number is congruent modulo 2 to $|N_{\mathfrak{H}}(\mathfrak{H}')/\mathfrak{H}'|$. Thus,

$$|N_{\mathfrak{G}}(\mathfrak{H}')/\mathfrak{H}'| \equiv |N_{\mathfrak{H}}(\mathfrak{H}')/\mathfrak{H}'| + |\mathfrak{S}(\mathfrak{H}')| \bmod 2. \tag{68}$$

It follows from (68) that $|\mathfrak{S}(\mathfrak{H}')|$ is odd if and only if $(N_{\mathfrak{G}}(\mathfrak{H}') : \mathfrak{H}') \equiv 0\ (2)$ and $(N_{\mathfrak{H}}(\mathfrak{H}') : \mathfrak{H}') \equiv 1\ (2)$. Consequently, in view of this and equation (67),

$$s_n(\mathfrak{G}) \equiv |\Omega_n| + \begin{cases} s_{n/2}(\mathfrak{H}), & 2 \mid n \\ 0, & 2 \nmid n \end{cases} \bmod 2, \tag{69}$$

[14] See also [16].

where

$$\Omega_n := \Big\{\mathfrak{H}' \leq \mathfrak{H}: \ (\mathfrak{H}:\mathfrak{H}') = n, \ (N_{\mathfrak{H}}(\mathfrak{H}'):\mathfrak{H}') \equiv 1\,(2), \ \text{and} \\ (N_{\mathfrak{G}}(\mathfrak{H}'):\mathfrak{H}') \equiv 0\,(2)\Big\}.$$

Denote by $\mathfrak{U}_n(\mathfrak{H})$ the set of all index n subgroups in $\mathfrak{H}$. Since $\mathfrak{H}$ is normal in $\mathfrak{G}$, $\mathfrak{G}$ acts on $\mathfrak{U}_n(\mathfrak{H})$ by conjugation, and this action restricts to an action of $\mathfrak{G}$ (and hence of $\mathfrak{H}$) on the set Ω_n. Therefore, if n is even and $\mathfrak{H}' \in \Omega_n$, then $(\mathfrak{H} : N_{\mathfrak{H}}(\mathfrak{H}')) \equiv 0\,(2)$, and Ω_n decomposes into classes of even length under $\mathfrak{H}$. Hence, in this case $|\Omega_n| \equiv 0\,(2)$, and, by (69),

$$s_n(\mathfrak{G}) \equiv s_{n/2}(\mathfrak{H}) \bmod 2, \quad 2 \mid n. \tag{70}$$

Suppose on the other hand that n is odd, and consider the action of $\mathfrak{G}$ on the set

$$\mathfrak{U}_n(\mathfrak{H}) - \Omega_n = \Big\{\mathfrak{H}' \leq \mathfrak{H}: \ (\mathfrak{H}:\mathfrak{H}') = n \ \text{ and } \ (N_{\mathfrak{G}}(\mathfrak{H}'):\mathfrak{H}') \equiv 1\,(2)\Big\}.$$

Then, if $\mathfrak{H}' \in \mathfrak{U}_n(\mathfrak{H}) - \Omega_n$, we have $(\mathfrak{G} : N_{\mathfrak{G}}(\mathfrak{H}')) \equiv 0\,(2)$, i.e., $\mathfrak{U}_n(\mathfrak{H}) - \Omega_n$ decomposes into classes of even length under the action of $\mathfrak{G}$. Hence in this case

$$s_n(\mathfrak{H}) = |\mathfrak{U}_n(\mathfrak{H})| \equiv |\Omega_n| \bmod 2,$$

and by (69)

$$s_n(\mathfrak{G}) \equiv s_n(\mathfrak{H}) \bmod 2, \quad 2 \nmid n. \tag{71}$$

Statements (70) and (71) can be rephrased as

$$\Pi(\mathfrak{G}) \cap 2\,\mathbb{N} = 2\,\Pi(\mathfrak{H}) \tag{72}$$

respectively

$$\Pi(\mathfrak{G}) \cap (\mathbb{N} - 2\,\mathbb{N}) = \Pi(\mathfrak{H}) \cap (\mathbb{N} - 2\,\mathbb{N}). \tag{73}$$

Taking the union of (72) and (73) yields (66). □

As an illustration, let $\mathfrak{G} = C_2^{*r}$ be the free product of $r \geq 2$ copies of the cyclic group of order 2. Then we have $m_{\mathfrak{G}} = 2$ and $\mu(\mathfrak{G}) = r - 1$, hence $\mathfrak{G}$ contains an index 2 subgroup which is free of rank $r - 1$. Since a finitely generated infinite free group has all its subgroup numbers odd,[15] we find from Proposition 4 that $\Pi(\mathfrak{G}) = \mathbb{N}$, i.e., $s_n(\mathfrak{G})$ is odd for all $n \geq 1$. This had been conjectured in [10], and proved for the case when $r \geq 4$ in a subsequent paper; cf. [11, Theorem 1].

[15] This follows immediately from M. Hall's recursion formula [14, Theorem 5.2].

8.2. The parity patterns Π_q

Denote by $\chi_q(n)$ the number $s_n(\mathfrak{H}(q))$ of index n subgroups in the Hecke group $\mathfrak{H}(q)$ evaluated modulo 2, so that χ_q is nothing but the characteristic function of the set $\Pi_q := \Pi(\mathfrak{H}(q)) \subseteq \mathbb{N}$, and let $X_q(z) := \sum_{n=0}^{\infty} \chi_q(n+1)\, z^n$ be the associated generating function. Note that for every $q \geq 3$ the group $\mathfrak{H}(q)$ is a split extension of $\widetilde{\mathfrak{H}}(q)$ by C_2; indeed, if $\mathfrak{H}(q) = \langle \sigma, \tau \,|\, \sigma^2 = \tau^q = 1 \rangle$, then the subgroup generated by τ and τ^σ is isomorphic to $\widetilde{\mathfrak{H}}(q)$ with transversal $\{1, \sigma\}$. Hence, Proposition 4 applies to the effect that (i) for q even, $\Pi_q = \mathbb{N}$ (by Theorem 5), and (ii) for q odd, $\Pi_q = \widetilde{\Pi}_q \cup 2\,\widetilde{\Pi}_q$ (making use of Theorem 7 (i)). Given the latter equation our results concerning the groups $\widetilde{\mathfrak{H}}(q)$ for q odd translate into results for Hecke groups, and we find the following.

Theorem 10. *Let $q \geq 3$ be an integer.*

(a) *If q is even, then $\Pi_q = \mathbb{N}$.*
(b) *If q is odd, then*
 (i) $\Pi_q \subseteq (1 + 4\,\mathbb{N}_0) \cup (2 + 8\,\mathbb{N}_0)$,
 (ii) *the series $X_q(z)$ is not rational over $GF(2)$; in particular, the set Π_q is infinite,*
 (iii) *the first entries of Π_q are 1, 2, and $2\,p(q) - 1$,*
 (iv) *we have $X_q(z) = \widetilde{X}_q(z) + z\,\widetilde{X}_q(z^2)$.*

Proof. Assertion (a) has already been observed, parts (i), (iii), and (iv) of (b) are immediate from parts (i) and (iii) of Theorem 7 and the relationship between the patterns Π_q and $\widetilde{\Pi}_q$. To prove the second part of (b), note that

$$z\,X_q'(z) = \sum_{n \in \Pi_q \cap 2\mathbb{N}} z^{n-1},$$

and hence that

$$X_q(z) + z\,X_q'(z) = \sum_{n \in \Pi_q \cap (\mathbb{N} - 2\mathbb{N})} z^{n-1} = \widetilde{X}_q(z).$$

Consequently, if $X_q(z)$ were rational over $GF(2)$, so would be $\widetilde{X}_q(z)$, contradicting Theorem 7 (ii). □

Theorem 11. *Let $q_1, q_2 \geq 3$ be odd integers, and suppose that $\Pi_{q_1} = \Pi_{q_2}$. Then we have $q_1 = q_2$.*

This is an immediate consequence of Theorem 6. Furthermore, the explicit characterization of the patterns $\widetilde{\Pi}_q$ given in Theorem 8 translates into the following description of the parity patterns Π_q.

Theorem 12. *Let $q \geq 3$ be odd, let $r + 1 = \sigma_0(q)$ be the number of divisors of q, and let $\underline{d}_q$ be as in Theorem 8. Then we have*[16]

$$\Pi_q = \Theta_q \ \cup\ 2\,\Theta_q,$$

where Θ_q consists of all positive integers $n \equiv 1$ (4) such that the set

$$\Big\{ \underline{n} \in \mathbb{N}_0^r : \ \underline{d}_q \cdot \underline{n} = \frac{n-1}{4} \quad \text{and}$$
$$\sum_{j=1}^{r} \mathfrak{s}(n_j) + \mathfrak{s}(\frac{n+1}{2} - ||\underline{n}||) - \mathfrak{s}(\frac{n-1}{2}) = 1 \Big\}$$

has odd cardinality.

The description of the parity pattern Π_q given in Theorem 12 simplifies considerably if q is a prime number.

Corollary 4. *Let $q > 2$ be a prime. Then $\Pi_q = \Theta_q \cup 2\,\Theta_q$, where*

$$\Theta_q = \Big\{ 2\,(q-1)\,n + 1 : \ n \in \mathbb{N}_0 \ \text{ and}$$
$$\mathfrak{s}(n) + \mathfrak{s}((q-2)\,n + 1) - \mathfrak{s}((q-1)\,n) = 1 \Big\}.$$

Our next result is the analogue of Corollary 3 for the patterns Π_q.

Corollary 5. *Let $q > 2$ be a prime. Then*

(a) $\Pi_q \backslash \{1, 2\} \subseteq 1 + 2(q-1) + 4(q-1)\mathbb{N}_0 \ \cup\ 2 + 4(q-1) + 8(q-1)\mathbb{N}_0$,
(b) *for every $\alpha \in \mathbb{N}$ the following assertions are equivalent:*
 (i) $1 + 2\,(q-1)\,(2^\alpha + 1) \in \Pi_q$,
 (ii) $2 + 4\,(q-1)\,(2^\alpha + 1) \in \Pi_q$,
 (iii) $a_\alpha = 1$,
 where $q - 1 = \sum_{j \geq 1} a_j\, 2^j$ with $a_j \in \{0, 1\}$.

Just as in the case of the patterns $\widetilde{\Pi}_q$ we can, in general, not expect the parity patterns Π_q to admit of a characterization in terms of closed formulae. Again, the only exception occurs when q is a Fermat prime.

Theorem 13. *Let $q \geq 3$ be an odd integer. Then the following assertions are equivalent:*

(i) $\Pi_q = \Lambda_q \ \cup\ 2\,\Lambda_q$.
(ii) $\chi_q(n) = 0$ *for* $n \in ([2q\,(q-1)] - \{1, 2q-1\}) \cap (\mathbb{N} - 2\mathbb{N})$.
(iii) *q is a Fermat prime.*

[16] We use the notation of Theorem 8.

If we restrict Theorem 13 to the case when q is a prime, then, in view of Remark 2, we can be slightly more precise.

Corollary 6. *Let $q > 2$ be a prime. Then the following assertions are equivalent:*

(i) $\Pi_q = \Lambda_q \cup 2\,\Lambda_q$.
(ii) $\chi_q(n) = 0$ *for* $n = 2(2^\nu + 1)(q - 1) + 1$ *and* $1 \leq \nu \leq \log_2(\frac{q-1}{3})$.
(iii) *q is a Fermat prime.*

References

[1] K. S. Brown, *Cohomology of groups*, Springer, New York, 1982.
[2] P. J. Cameron, *Combinatorics*, Cambridge University Press, 1994.
[3] S. Chowla, I. N. Herstein, and W. R. Scott, The solutions of $x^d = 1$ in symmetric groups, *Norske Vid. Selsk.* **25** (1952), 29–31.
[4] I. M. S. Dey, Schreier systems in free products, *Proc. Glasgow Math. Soc.* **7** (1965), 61–79.
[5] W. Dicks and M. J. Dunwoody, *Groups acting on graphs*, Cambridge University Press, 1989.
[6] A. Dress and T. Müller, Decomposable functors and the exponential principle, *Adv. in Math.* **129** (1997), 188–221.
[7] M. du Sautoy, Finitely generated groups, p–adic analytic groups and Poincaré series, *Ann. of Math.* **137** (1993), 639–670.
[8] M. du Sautoy and F. Grunewald, Analytic properties of zeta functions and subgroup growth, *Ann. of Math.* **152** (2000), 793–833.
[9] G. Frobenius, Über einen Fundamentalsatz der Gruppentheorie, *Berl. Sitz.* (1903), 987–991.
[10] M. Grady and M. Newman, Counting subgroups of given index in Hecke groups, Contemporary Mathematics Vol. 143, Am. Math. Society, 1993, 431–436.
[11] M. Grady and M. Newman, Some divisibility properties of the subgroup counting function for free products, *Math. Comp.* **58** (1992), 347–353.
[12] M. Gromov, Groups of polynomial growth and expanding maps, *Publ. Math. IHES* **53** (1981), 53–78.
[13] M. Hall, Coset representations in free groups, *Trans. Amer. Math. Soc.* **67** (1949), 421–432.
[14] M. Hall, Subgroups of finite index in free groups, *Can. J. Math.* **1** (1949), 187–190.
[15] M. Hall and T. Radó, On Schreier systems in free groups, *Trans. Amer. Math. Soc,* **64** (1948), 386–408.
[16] P. Hall, On a theorem of Frobenius, *Proc. London Math. Soc.* **40** (1936), 468–501.
[17] A. Karrass, A. Pietrowski, and D. Solitar, Finite and infinite cyclic extensions of free groups, *J. Austral. Math. Soc.* **16** (1973), 458–466.
[18] E. Kummer, Über die Ergänzungssätze zu den allgemeinen Reciprocitätsgesetzen, *J. reine u. angew. Math.* **44** (1852), 93–146. Reprinted in Collected Papers (edited by A. Weil), Vol. I, 485–538, Springer, New York, 1975.

[19] A. Lubotzky, *Subgroup growth*, lecture notes prepared for the conference 'Groups 1993 Galway/St. Andrews', University College, Galway.

[20] A. Lubotzky, Counting finite index subgroups, *Proceedings of the conference 'Groups 1993 Galway/St. Andrews'*, 368–404.

[21] A. Lubotzky, Subgroup growth and congruence subgroups, *Invent. Math.* **119** (1995), 267–295.

[22] A. Lubotzky, A. Mann, and D. Segal, Finitely generated groups of polynomial subgroup growth, *Israel J. Math.* **82** (1993), 363–371.

[23] A. Meyer, Divisibility properties of subgroup counting functions associated with finitely generated virtually free groups, PhD thesis, University of Bielefeld, 1998.

[24] S. G. Mohanti, *Lattice path counting and applications*, Academic Press, New York, 1979.

[25] T. Müller, Combinatorial aspects of finitely generated virtually free groups, *J. London Math. Soc.* (2) **44** (1991), 75–94.

[26] T. Müller, Counting free subgroups of finite index, *Archiv d. Math.* **59** (1992), 525–533.

[27] T. Müller, Subgroup growth of free products, *Invent. Math.* **126** (1996), 111–131.

[28] T. Müller, Combinatorial classification of finitely generated virtually free groups, *J. Algebra* **195** (1997), 285–294.

[29] T. Müller, Enumerating representations in finite wreath products, *Adv. in Math.* **153** (2000), 118–154.

[30] T. Müller, Monomial representations and subgroup growth, in preparation.

[31] T. Müller, Poincaré's problem for free products, in preparation.

[32] T. Müller, Remarks on the PhD thesis of A. Meyer, unpublished manuscript, 1998.

[33] T. V. Narayana, *Lattice path combinatorics with statistical applications*, Math. Expositions No. 23, University of Toronto Press, London, 1979.

[34] O. Schreier, Die Untergruppen der freien Gruppen, *Abh. Math. Sem. Univ. Hamburg* **5** (1927), 161–183.

[35] J.–P. Serre, *Trees*, Springer, Berlin–Heidelberg–New York, 1980.

[36] J.–P. Serre, Cohomologie des groupes discrets, in: *Prospects in Mathematics*, Ann. of Math. Stud. vol. 70, Princeton Univerity Press, 1971, 77–169.

[37] J. Stallings, On torsion-free groups with infinitely many ends, *Ann. of Math.* **88** (1968), 312–334.

[38] R. Stanley, *Enumerative Combinatorics*, Vol. 2, Cambridge University Press, New York, 1999.

[39] W. Stothers, The number of subgroups of given index in the modular group, *Proc. Royal Soc. Edinburgh* **78A** (1977), 105–112.

13

Automorphisms of the binary tree: state-closed subgroups and dynamics of $1/2$-endomorphisms

V. Nekrashevych and S. Sidki*

1. Introduction

Automorphisms of regular 1-rooted trees of finite valency have been the subject of vigorous investigations in recent years as a source of remarkable groups which reflect the recursiveness of these trees (see [S1], [G2]). It is not surprising that the recursiveness could be interpreted in terms of automata. Indeed, the automorphisms of the tree have a natural interpretation as input-output automata where the states, finite or infinite in number, are themselves automorphisms of the tree. On the other hand input-output automata having the same input and output alphabets can be seen as endomorphisms of a 1-rooted tree indexed by finite sequences from this alphabet. It is to be noted that the set of automorphisms having a finite number of states and thus corresponding to finite automata, form an enumerable group called the *group of finite-state automorphisms*. The calculation of the product of two automorphisms of the tree involve calculating products between their states which are not necessarily elements of the group generated by the two automorphisms. In order to remain within the same domain of calculation we have defined a group G as *state-closed* provided the states of its elements are also elements of G [S2]. Among the outstanding examples of state-closed groups are the classes of self-reproducing (fractal-like) groups constructed in [G1, GS, BSV] which are actually generated by automorphisms with finite number of states, or equivalently, generated by finite automata. The state-closed condition has allowed the use of induction on the length function to prove detailed properties of these groups. Of course, if the group is not state-closed one may take its state-closure. In doing so, properties such as finite generation, may not be conserved. At any rate, state-closed groups which are finitely generated yet not necessarily finite-state are subgroups of another

* The second author acknowledges support from FAPDF of Brazil.

enumerable group called the *group of functionally recursive automorphisms* [BS1].

An important set of examples of state-closed groups of classical nature are the m-dimensional affine groups $\mathbb{Z}^m \cdot \mathrm{GL}(m, \mathbb{Z})$. It was shown in [BS2] that for every m the corresponding affine group is faithfully represented as a state-closed group of automorphisms of the 2^m-ary tree. The purpose of the present paper is to investigate state-closed groups of automorphisms of the 2-tree with emphasis on subgroups of m-dimensional affine groups.

Given an abstract group G with a subgroup H of index 2, we call a homomorphism $\rho : H \to G$ *a* $1/2$-*endomorphism* of G. The ρ−core(H) is the maximal subgroup K of H which is normal in G and is ρ-invariant (that is, $\rho(K) \leq K$), and ρ is called *simple* provided ρ−core(H) is trivial. The concept of $1/2$-endomorphism is intimately related to that state-closed group. For if G is a non-trivial state-closed group, then the stabilizer subgroup G_1 of the first level vertices of the tree is of index 2 in G and the restriction of the action of elements of G_1 to one of the two maximal subtrees provides us with such a map. On the other hand, as we will show, a group G with a $1/2$-endomorphism $\rho : H \to G$ admits representations into the automorphism group of the binary tree as a state-closed group. These representations are faithful if and only if ρ is simple. Certain classes of groups which have faithful representations as groups of automorphisms of the binary tree fail to have a faithful representation as a state-closed groups. One instance of this breakdown occurs in finitely generated free non-abelian nilpotent groups. We show that these groups cannot admit a faithful state-closed representation on the binary tree. The situation for torsion-free non-abelian polycyclic groups is mixed and the problem of describing the polycyclic state-closed groups is open. Another open problem in this context concerns the existence of non-cyclic state-closed free groups. It is to be noted that homomorphisms $\rho : H \to G$ where H is a subgroup of finite index in G (so called virtual endomorphisms of G) have been the subject of recent studies (see [GM] and [Nek]). These works as well as ours represent first explorations of a new topic in combinatorial group theory.

We call torsion-free abelian groups of finite rank m which are state-closed subgroups of the automorphism group of the binary tree *m-dimensional (binary) lattices*. A large part of our work is devoted to the classification of these lattices. We prove that the $1/2$-endomorphism associated to each such group is the restriction of an irreducible linear transformation defined on a rational vector space of dimension m. When the lattice is generated by finite-state automorphisms of the tree, we prove that its corresponding linear transformation is necessarily a contracting map, in the sense that the roots of the characteristic polynomial has absolute value less than 1. A classification of these polynomials having degree less than 6 reveal the surprising fact that the class number corresponding to each is 1. The connection between the finite-state condition for the lattice with the dynamical behavior of the associated linear transformation and the number theoretic observations about the characteristic polynomials of these transformations confirms the wide scope of interaction between these notions about tree automorphisms and other topics in mathematics. In particular, the relationship with dynamical systems as expounded in [B] is strongly enhanced.

A state-closed group G whose associated $1/2$-endomorphism is onto is called *recurrent*. We describe the topological closure of a recurrent lattice G in terms of a ring of certain infinite series which generalize the ring of dyadic integers. Here, the group of automorphisms of the tree is considered as a profinite topological group with respect to the pro-2 topology; in this setting, the topological closure of an abelian subgroup is again abelian. We also prove that the elements of the group G act on such series as *generalized adding machines*. This action may be viewed as numeration systems for abelian groups. In the case of rank 1 we get usual dyadic numeration system (or "nega-dyadic"). For rank 2 we get numeration systems similar to the numeration systems for complex numbers. For more on numeration systems one can read in [K] and [Sa].

One of the nice properties of recurrent lattices is that they are topologically determined by their $1/2$-endomorphisms. Indeed, we prove that any two recurrent lattices with the same associated $1/2$-endomorphism have equal topological closures.

One special class of m-dimensional recurrent lattices admits for every m a "large" linear group normalizer within the finite-state group of automorphisms of the binary tree. More precisely, let $\mathbb{Z}^m$ denote the m-dimensional lattice, $\varsigma : \mathrm{GL}(m, \mathbb{Z}) \to \mathrm{GL}(m, \mathbb{Z}_2)$ the natural "modulo 2" epimorphism and $B(m, \mathbb{Z})$ the pre-image of the Borel subgroup of $\mathrm{GL}(m, \mathbb{Z}_2)$. We prove that the affine group $\mathbb{Z}^m \cdot B(m, \mathbb{Z})$ admits a faithful representation as a state-closed, finite-state group of automorphisms of the binary tree. This result is optimal in the sense that $[\mathrm{GL}(m, \mathbb{Z}) : B(m, \mathbb{Z})] = (2^m - 1)(2^{m-1} - 1) \cdots (2^2 - 1)$ is the maximal odd factor of $|\mathrm{GL}(m, \mathbb{Z}_2)|$ and $B(m, \mathbb{Z})$ is a maximal subgroup of $\mathrm{GL}(m, \mathbb{Z})$ with respect to avoiding having elements of odd order. It also puts in perspective the main result in [BS2] that the affine group $\mathbb{Z}^m \cdot \mathrm{GL}(m, \mathbb{Z})$ has a faithful, finite-state, state-closed representation on a 2^m-ary tree. One needs exponentially high valency for the tree since the minimum degree of a transitive representation of $\mathrm{GL}(m, \mathbb{Z}_2)$ is $2^m - 1$ for $m > 4$ [KL].

The paper was developed during a visit of the first author to Universidade de Brasilia in February of 1999. He is very grateful for the hospitality and fruitful collaboration.

2. Tree automorphisms, automata

We present below definitions and preliminary notions about the binary tree, its automorphisms and their interpretation as automata. The one-rooted regular binary tree $\mathcal{T}_2$ may be identified with the monoid $\mathcal{M}$ freely generated by a set $Y = \{0, 1\}$ and ordered by the relation

$$v \leq u \text{ if and only if } u \text{ is a prefix of } v;$$

the identity element of $\mathcal{M}$ is the empty sequence $\emptyset$ (the root of the tree).

Let $\mathcal{A} = Aut(\mathcal{T}_2)$ be the automorphism group of the tree. The group of permutations $P(Y)$ is the cyclic group of order 2 generated by the transposition $\sigma = (0, 1)$. This permutation is extended "rigidly" to an automorphism of $\mathcal{A}$ by

$$(y \cdot u)^\sigma = y^\sigma \cdot u, \forall y \in Y, \forall u \in \mathcal{M}.$$

An automorphism $\alpha \in \mathcal{A}$ induces $\sigma^{i_\emptyset}$ where $i_\emptyset = 0, 1$, on the set $Y \subset \mathcal{M}$. Therefore the automorphism affords the representation $\alpha = \alpha' \sigma^{i_\emptyset}$, where α' fixes Y point-wise. Furthermore, α' induces for each $y \in Y$ an automorphism α'_y of the subtree whose vertices form the set $y \cdot \mathcal{M}$. On using the canonical isomorphism $yu \mapsto u$ between this subtree and the tree $\mathcal{T}$, we may consider (or, renormalize) α' as a function from Y into $\mathcal{A}$; in notational form, $\alpha' \in \mathcal{F}(Y, \mathcal{A})$. Thus, $\alpha = (\alpha_0, \alpha_1)\sigma^{i_\emptyset}$ and the group $\mathcal{A}$ is an infinitely iterated wreath product

$$\mathcal{A} = \mathcal{A} \wr \langle \sigma \rangle.$$

It is convenient to denote α by $\alpha_\emptyset$ and α'_y by α_y. In order to describe α_y we use the same procedure as in the case of α. Successive applications produce the set

$$\Sigma(\alpha) = \{\sigma^{i_u} \mid u \in \mathcal{M}\}$$

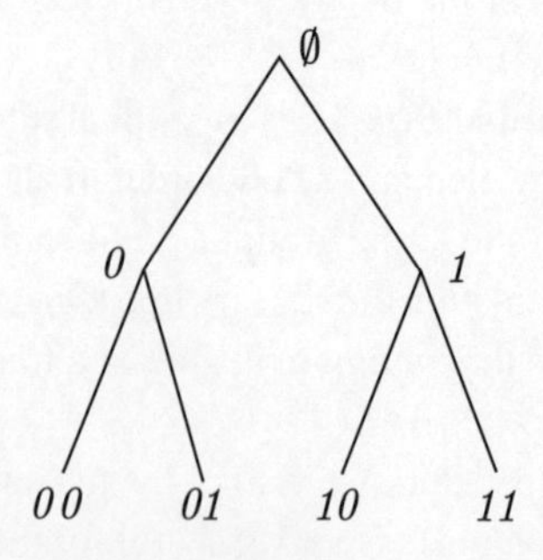

Figure 13.1 Binary tree

of permutations of Y which describes faithfully the automorphism α. Another by-product of the procedure is the set *states* of α,

$$Q(\alpha) = \{\alpha_u \mid u \in \mathcal{M}\}.$$

The definition of the product of automorphisms implies the following important properties of the function Q

$$Q(\alpha^{-1}) = Q(\alpha)^{-1},$$

$$Q(\alpha\beta) \subseteq Q(\alpha)Q(\beta), \forall \alpha, \beta \in \mathcal{A}.$$

If $Q(\alpha)$ is finite then α is said to be a *finite-state* automorphism. The set of finite-state automorphisms form the enumerable subgroup $\mathcal{F}$ of $\mathcal{A}$. The notion of finite-state automorphism is a special case of the more general functionally recursive automorphism. A finite set of automorphisms S is *functionally recursive* provided for each $\gamma \in S$, its states γ_0, γ_1 are group words in the elements of S. An automorphism α is functionally recursive provided α is an element of some functionally recursive set. The set of functionally recursive automorphisms form an enumerable group $\mathcal{R}$.

The interpretation of α as an automaton proceeds as follows: the input and output alphabets are the same set $Y = \{0, 1\}$; the set of states is $Q(\alpha)$; the initial state is α; let $y \in Y, \alpha_u \in Q(\alpha)$ and z the image of y under σ^{i_u}, then the state-transition function is $y : \alpha_u \mapsto \alpha_{uy}$ and the output function is $\alpha_u : y \mapsto z$. Thus, a finite-state automorphism corresponds to a finite automaton. A finite automaton is usually depicted by a directed graph called *the Moore diagram.* The vertices of the diagram correspond to the states of the automaton; and the arrows correspond to the transitions. For every $y \in Y$ and every state α_u we draw an arrows from α_u to α_{uy} and label it by $(y|z)$, where $z = y^{\sigma^{i_u}}$ as above. Then the arrows correspond to the transitions while the labels show the output. The set of infinite sequences $c = (c_0, c_1, c_2, \dots)$ with $c_i \in \{0, 1\}$ correspond to ends or boundary points of the tree. The action of an automorphism extends naturally to the boundary. Let $c = (c_0, c_1, c_2, \dots)$. For $\alpha = (\alpha_0, \alpha_1)\sigma$, we have

$$c^\alpha = \begin{cases} (1, (c_1, c_2, \dots)^{\alpha_0}), & \text{if } c_0 = 0, \\ (0, (c_1, c_2, \dots)^{\alpha_1}), & \text{if } c_0 = 1 \end{cases}$$

and for $\alpha = (\alpha_0, \alpha_1)$,

$$c^\alpha = \begin{cases} (0, (c_1, c_2, \dots)^{\alpha_0}), & \text{if } c_0 = 0, \\ (1, (c_1, c_2, \dots)^{\alpha_1}), & \text{if } c_0 = 1. \end{cases}$$

The boundary points $c = (c_0, c_1, c_2, \dots)$ also correspond to a dyadic integer $\xi = c_0 + c_1 2 + c_2 2^2 + \cdots + c_i 2^i + \cdots$ and the action of the tree

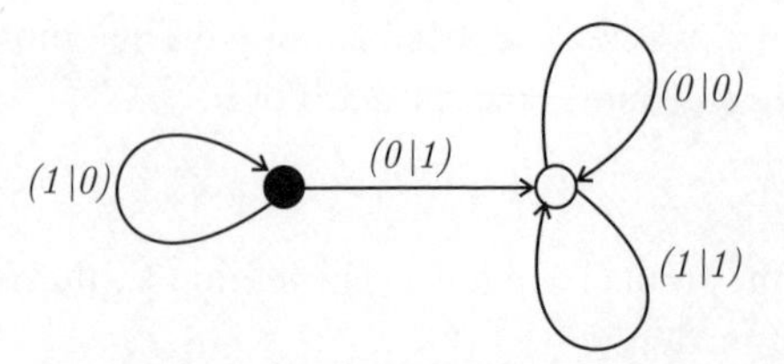

Figure 13.2 Adding machine

automorphism α can thus be translated to an action on the ring of dyadic integers. For example, consider the automorphism $\tau = (e, \tau)\sigma$. Then

$$c^\tau = \begin{cases} (1, c_1, c_2, \dots), & \text{if } c_0 = 0, \\ (0, (c_1, c_2, \dots)^\tau), & \text{if } c_0 = 1, \end{cases}$$

which translates to the binary addition

$$\xi^\tau = 1 + \xi,$$

and this fact justifies referring to τ as the *binary adding machine*. The diagram of the automaton corresponding to the binary adding machine is shown in Figure 13.2.

We have developed sufficient language to give the definition of three examples of self reproducing groups.

(i) Let $\alpha = (e, \alpha_1)$ where $\alpha_1 = (\sigma, \alpha_{11})$ and $\alpha_{11} = (\sigma, \alpha)$. The group $\langle \alpha, \alpha_1, \alpha_{11}, \sigma \rangle$ is a state-closed infinite 2-group of intermediate growth [G1].

(ii) Let $\alpha = (\alpha_0, \alpha)$, $\alpha_0 = (\alpha_{00}, e)$, $\alpha_{00} = (\sigma, \sigma)$. Then the group $\langle \alpha, \alpha_0, \alpha_{00}, \sigma \rangle$ is a state-closed infinite 2-group [S2].

(iii) Let $\tau = (e, \tau)\sigma$, $\mu = (e, \mu^{-1})\sigma$. Then $\langle \tau, \mu \rangle$ is state-closed, torsion-free, and is just-nonsolvable [BSV].

3. State-closed groups

Let G be a non-trivial state-closed group. Since G is non-trivial, some of states of its elements are active, and so the state-closed condition implies that there exists an element of G that is active; that is, G is transitive on the first level of the binary tree. The first level stabilizer G_1 is a subgroup of index 2 in the group G and $G = G_1 \cup G_1 a$ for some choice $a \in G \setminus G_1$. Then in the action of G on the tree, $a = (a_0, a_1)\sigma$ and $h = (h_0, h_1)$ for all $h \in G_1$. Since G is state-closed, $a_0, a_1 \in G$, and likewise, $h_0, h_1 \in G$ for all $h \in G_1$. Therefore the projections π_0, π_1 of G_1 on its first and second coordinates are $1/2$-endomorphisms from the subgroup G_1 into the group G. We note that in case G is recurrent, $\pi_0(G_1) = G$

and therefore G is transitive on all levels of the binary tree. Now, since $a^2 = (a_0a_1, a_1a_0) \in G_1$ we have $a_1 = a_0^{-1}\pi_0(a^2)$. Also, we have for every $h \in G_1$, $h^a = \sigma(a_0^{-1}, a_1^{-1})(h_0, h_1)(a_0, a_1)\sigma = (h_1^{a_1}, h_0^{a_0})$; thus $\pi_1(h^a) = h_0^{a_0} = \pi_0(h)^{a_0}$. Hence, the projections π_0, π_1 satisfy the following conditions

$$a_1 = a_0^{-1}\pi_0(a^2), \quad \pi_1(h) = \pi_0(h^{a^{-1}})^{a_0}.$$

In case G is abelian, the second condition simplifies to $\pi_1(h) = \pi_0(h)$ for all $h \in G_1$.

Examples. (1) Define the following sequence of elements of $Aut(\mathcal{T}_2)$, $\sigma_\mathbf{0} = \sigma$, $\sigma_\mathbf{1} = (e, \sigma_\mathbf{0})$ and for $i \geq 1$, $\sigma_\mathbf{i} = (e, \sigma_\mathbf{i-1})$. Define also for $0 \leq n \leq \infty$ the subgroups $P_n = \langle \sigma_\mathbf{i} \mid 0 \leq i \leq n \rangle$. Then P_n is isomorphic to the wreath product, iterated n times, of cyclic groups of order 2. It is easy to see from the definition of the generators $\sigma_\mathbf{i}$ that P_n is state-closed and moreover, P_∞ is recurrent.

(2) Let $\tau = (e, \tau)\sigma$ be the binary adding machine. Define $\tau_\mathbf{0} = \tau$ and $\tau_\mathbf{i} = (e, \tau_\mathbf{i-1})$ for $i \geq 1$. Define also the subgroups $\Upsilon_n = \langle \tau_i \mid 0 \leq i \leq n \rangle$. Then Υ_n is state-closed. When n is finite this group factors as $\Upsilon_n = N \cdot P_{n-1}$, where N is the normal closure of $\langle \tau_\mathbf{n} \rangle$ in Υ_n and is free abelian group of rank $n + 1$. We note that $\Upsilon_0, \dots, \Upsilon_\infty$ are recurrent groups.

Proposition 3.1. *Let G be a group of automorphisms of the binary tree, generated by a finite set S. Then, G is state-closed if and only if S is functionally recursive.*

Proof. Let $S = \{\alpha, \beta, \dots, \gamma\}$. Suppose G is state-closed. Then as $\alpha_0 \in G$, it is a word $\alpha_0 = \alpha_0(\alpha, \beta, \dots, \gamma)$ in the elements of S, and so every state of every element in S is also a word in the elements of S. Thus, S is functionally recursive. On the other hand, if S is functionally recursive, then by definition, every state δ of every element of S is some word $\delta = \delta(\alpha, \beta, \dots, \gamma)$ in the elements of S; therefore, $\delta \in G$. □

Lemma 3.1. *Let G be a state-closed group and $\widehat{G}$ its topological closure. Then $\widehat{G}$ is also state-closed.*

Proof. An automorphism $\omega \in \mathcal{A}$ belongs to $\widehat{G}$ if and only if for every $n \in \mathbb{N}$, the action of ω on the first n levels of the tree coincides with an action of some $g \in G$ on these levels (g depends on n). If $\omega = (\omega_0, \omega_1)\sigma^i$, $i \in \{0, 1\}$ then $g = (g_0, g_1)\sigma^i$ for some $g_0, g_1 \in G$ and then the action of ω_0 on first $n - 1$ levels of the tree coincides with the action of g_0; the same is true for ω_1 and g_1 respectively. Thus the action of ω_0 and ω_1 on any finite number of levels coincides with actions of some elements of G, thus they belong to $\widehat{G}$. □

Definition 1. Let $qN_\mathcal{A}(G) = \{\alpha \in \mathcal{A} \mid G^\alpha \leq G\}$ denote the *semi-normalizer* of G in the group $\mathcal{A}$ of automorphisms of the tree and let $C_\mathcal{A}(G)$ denote the centralizer of G in $\mathcal{A}$. For any element $\alpha \in \mathcal{A}$ we denote (α, α) by $\alpha^{(1)}$ and inductively, for any $n \geq 0$, $(\alpha^{(n)}, \alpha^{(n)})$ by $\alpha^{(n+1)}$.

We are able to produce information about the form of elements of the semi-normalizer and centralizer of a recurrent group.

Proposition 3.2. *Let G be a recurrent group, $\widehat{G}$ its topological closure and a an active element of G.*

(i) *Given $\alpha \in qN_{\mathcal{A}}(G)$, there exist $i \in \{0, 1\}$, $\beta \in qN_{\mathcal{A}}(G)$, $u \in G$ such that*

$$\alpha = \beta^{(1)}(e, u)a^i \text{ and } (a_0^\beta u a_0^{-1}, u^{-1}a_1^\beta a_1^{-1}) \in G.$$

(ii) *Suppose G is also abelian. Then u in the above formula satisfies*

$$u^2 = \left(a_0^{-1}a_1\right)^{-1}\left(a_0^{-1}a_1\right)^\beta.$$

In addition, $C_{\mathcal{A}}(G)$ coincides with $\widehat{G}$.

Proof. (i) Let $a = (a_0, a_1)\sigma \in G$, $h = (h_0, h_1) \in G_1$, $\alpha \in qN_{\mathcal{A}}(G)$. There exists a unique $i \in \{0, 1\}$ such that $\alpha' = \alpha a^{-i}$ is inactive; clearly, $\alpha' \in qN_{\mathcal{A}}(G)$. Thus we may assume $\alpha = (\alpha_0, \alpha_1)$. Now, $a^\alpha = (\alpha_0^{-1}a_0\alpha_1, \alpha_1^{-1}a_1\alpha_0)\sigma \in G$, $h^\alpha = (h_0^{\alpha_0}, h_1^{\alpha_1}) \in G_1$. Since G is state-closed, $\alpha_0^{-1}a_0\alpha_1 = k, \alpha_1^{-1}a_1\alpha_0 = k' \in G$ and also $h_0^{\alpha_0}, h_1^{\alpha_1} \in G$. Since G is recurrent, h_0 can be equal to any element of G, thus $\alpha_0 \in qN_{\mathcal{A}}(G)$. We find that $\alpha_1 = a_0^{-1}\alpha_0 k$. Thus, $\alpha = \beta^{(1)}(e, u)$ for $u = \left(a_0^{-1}\right)^{\alpha_0} k \in G$ and $\beta = \alpha_0 \in qN_{\mathcal{A}}(G)$. The second part follows from computing the commutator $[\beta^{(1)}(e, u), a^{-1}]$.

(ii) As G is abelian, then for all $h \in G_1$, $h = (h_0, h_0)$ and so, $a_0^\beta u a_0^{-1} = u^{-1}a_1^\beta a_1^{-1}$; thus, $u^2 = \left(a_0^{-1}a_1\right)^{-1}\left(a_0^{-1}a_1\right)^\beta$ follows. Again, as G is abelian then so is $\widehat{G}$; thus, $\widehat{G} \leq C_{\mathcal{A}}(G)$. Now let $\alpha \in C_{\mathcal{A}}(G)$. Then from part (i), $\alpha = \beta^{(1)}(e, u)a^i$ and clearly, $\alpha' = \beta^{(1)}(e, u) \in C_{\mathcal{A}}(G)$. On applying α' to G_1, we conclude that $\beta \in C_{\mathcal{A}}(G)$, since $\pi_0(G_1) = G$. Now $a^{\alpha'} = (\beta^{-1}a_0\beta u, u^{-1}\beta^{-1}a_1\beta)\sigma = (a_0 u, u^{-1}a_1)\sigma = a$ implies $u = e$. Hence, $\alpha = \beta^{(1)}a^i$, $\beta \in C_{\mathcal{A}}(G)$. Successive developments of α yield $\alpha \in \widehat{G}$. □

3.1. State-closed representations

Given a group G, we describe below all the state-closed representations of G on the binary tree. Consider a subgroup H of G among the subgroups of index 2 in G. This subgroup contains the subgroup G^2 generated by the squares of the elements of G and G^2 itself contains the commutator subgroup G' of G. Given such a subgroup H, we choose a $1/2$-endomorphism $\rho : H \to G$, $a \in G \setminus H$ and $a_0 \in G$. We will prove that the quadruple (H, ρ, a, a_0) defines uniquely a state-closed representation of G on the binary tree.

Theorem 3.1. *Let G be a group, (H, ρ, a, a_0) a quadruple as defined above and $a_1 = a_0^{-1}\rho(a^2)$. Also, let σ be the rigid extension of the transposition* $(0, 1)$ *to an automorphism of the binary tree $\mathcal{T}_2$. Then the map $\varphi : G \to Aut(\mathcal{T}_2)$ defined recursively by the rules:*

$$(ha)^\varphi = (\rho(h)^\varphi a_0^\varphi, \rho\left(h^{a^{-1}}\right)^{a_0\varphi} a_1^\varphi)\sigma$$

$$h^\varphi = (\rho(h)^\varphi, \rho\left(h^{a^{-1}}\right)^{a_0\varphi})$$

is a homomorphism such that the first level stabilizer of G^φ coincides with H^φ. The kernel of φ is equal to $\rho-\mathrm{core}(H)$.

Proof. It follows from the definition of the map φ that $(ha)^\varphi = h^\varphi a^\varphi$ for every $h \in H$. Thus, in order to prove that φ is a homomorphism it is sufficient to check the following equalities:

$$\left(a^2\right)^\varphi = (a^\varphi)^2,$$
$$(h_1h_2)^\varphi = h_1^\varphi h_2^\varphi,$$
$$\left(h^a\right)^\varphi = (h^\varphi)^{a^\varphi}.$$

We prove them by induction on the tree level.

$$1)\ \left(a^2\right)^\varphi (a^\varphi)^{-2} = \left(\rho\left(a^2\right)^\varphi, \rho\left(\left(a^2\right)^{a^{-1}}\right)^{a_0\varphi}\right)\left(\left(a_0^\varphi, a_1^\varphi\right)\sigma\right)^{-2}$$
$$= \left((a_0a_1)^\varphi, (a_1a_0)^\varphi\right)\left(a_0^\varphi a_1^\varphi, a_1^\varphi a_0^\varphi\right)^{-1}$$
$$= \left((a_0a_1)^\varphi\left(a_0^\varphi a_1^\varphi\right)^{-1}, (a_1a_0)^\varphi\left(a_1^\varphi a_0^\varphi\right)^{-1}\right).$$

The verification of the homomorphism condition clearly reduces to the next level.

$$2)\ (h_1h_2)^\varphi\left(h_1^\varphi h_2^\varphi\right)^{-1} = \left(\rho\left(h_1h_2\right)^\varphi, \rho\left(h_1^{a^{-1}}h_2^{a^{-1}}\right)^{a_0\varphi}\right)\cdot$$
$$\left(\rho\left(h_1\right)^\varphi\rho\left(h_2\right)^\varphi, \rho\left(h_1^{a^{-1}}\right)^{a_0\varphi}\rho\left(h_2^{a^{-1}}\right)^{a_0\varphi}\right)^{-1}$$
$$= \left(\left(\rho\left(h_1\right)\rho\left(h_2\right)\right)^\varphi\left(\rho\left(h_1\right)^\varphi\rho\left(h_2\right)^\varphi\right)^{-1},\right.$$
$$\left.\left(\rho\left(h_1^{a^{-1}}\right)^{a_0}\rho\left(h_2^{a^{-1}}\right)^{a_0}\right)^\varphi\left(\rho\left(h_1^{a^{-1}}\right)^{a_0\varphi}\rho\left(h_2^{a^{-1}}\right)^{a_0\varphi}\right)^{-1}\right).$$

Again, the reduction in this case is clear.

$$3)\ (h^a)^\varphi \left((h^\varphi)^{a^\varphi}\right)^{-1} = \left(\rho\left(h^a\right)^\varphi, \rho\left(h\right)^{a_0\varphi}\right)\cdot$$

$$\left(\sigma\left(\left(a_0^\varphi\right)^{-1}, \left(a_1^\varphi\right)^{-1}\right)\left(\rho\left(h\right)^\varphi, \rho\left(h^{a^{-1}}\right)^{a_0\varphi}\right)\left(a_0^\varphi, a_1^\varphi\right)\sigma\right)^{-1}$$

$$= \left(\rho\left(h^a\right)^\varphi\left(\rho\left(h^{a^{-1}}\right)^{a_0\varphi a_1^\varphi}\right)^{-1}, \rho\left(h\right)^{a_0\varphi}\left(\rho\left(h\right)^{\varphi a_0^\varphi}\right)^{-1}\right)$$

$$= \left(\rho\left(h^a\right)^\varphi\left(\rho\left(h^a\right)^{a_1^{-1}\varphi a_1^\varphi}\right)^{-1}, \rho\left(h\right)^{a_0\varphi}\left(\rho\left(h\right)^{\varphi a_0^\varphi}\right)^{-1}\right);$$

the last equality follows from

$$\rho\left(h^{a^{-1}}\right) = \rho\left(h^a\right)^{\rho(a^{-2})} = \rho\left(h^a\right)^{a_1^{-1}a_0^{-1}}.$$

Now we consider whether the first coordinate of $(h^a)^\varphi\left((h^\varphi)^{a^\varphi}\right)^{-1}$ is trivial. The following sequence of equivalent statements lead to the desired reduction

$$\rho\left(h^a\right)^\varphi\left(\rho\left(h^a\right)^{a_1^{-1}\varphi a_1^\varphi}\right)^{-1} = e,$$

$$\rho\left(h^a\right)^{a_1^{-1}\varphi a_1^\varphi} = \rho\left(h^a\right)^\varphi,$$

$$a_1^\varphi\rho\left(h^a\right)^\varphi\left(a_1^\varphi\right)^{-1} = \left(\rho\left(h^a\right)^{a_1^{-1}}\right)^\varphi.$$

The question of triviality of the second coordinate of $(h^a)^\varphi\left((h^\varphi)^{a^\varphi}\right)^{-1}$ reduces more simply to the next level. Now let N be the kernel of the homomorphism φ. Then for every $h \in N$, h^φ is inactive, and consequently, $h \in H$. Thus $h^\varphi = e = (e, e) = (\rho(h)^\varphi, \rho(h^{a^{-1}})^{a_0\varphi})$. Hence $\rho(h) \in N$ and N is ρ-invariant. On the other hand, if M is a subgroup of H which is normal in G and is also ρ-invariant then for every $h \in M$ the elements $\rho(h)$ and $\rho(h^{a^{-1}})^{a_0}$ belong to M. It follows inductively from the representation $h^\varphi = (\rho(h)^\varphi, \rho(h^{a^{-1}})^{a_0\varphi})$ that h^φ is trivial. Hence, $M \leq N$ and kernel of φ is $\rho-\mathrm{core}(H)$. □

Remarks. We maintain the above notation.

(1) The relationship between the $\ker(\rho)$ and $\rho-\mathrm{core}(H)$ is not totally clear. Yet, it is obvious that $D = \ker(\rho) \cap \ker(\rho)^a$ is contained in $\rho-\mathrm{core}(H)$. Thus if ρ is simple then $D = \{e\} = [\ker(\rho), \ker(\rho)^a]$. We conclude that in case the abelian subgroups are cyclic (for instance, when G is a free group), then the condition ρ is simple implies ρ is a monomorphism. In the other direction, if G

is abelian then $\ker(\rho) = \ker(\rho)^a = D$, and all simple $1/2$-endomorphisms are also monomorphisms.

(2) Suppose that G is a normal subgroup of some group F and suppose that ρ is a restriction of an endomorphism $\widehat{\rho}$ of F. Suppose in addition that G is of finite index k in F, an let F^k be the subgroup of F generated by the k-th powers of its elements. Then $L = \left(F^k\right)^2 \leq H$ and so, $L^\rho = L^{\widehat{\rho}} \leq L$; thus, $L \leq \rho\text{-core}\,(H)$. It follows then that if $F = G$ and ρ is simple then G is an elementary abelian 2-group.

3.2. Extensions and restrictions of $1/2$-endomorphisms

The next results concern manners of producing simple $1/2$-endomorphisms.

Lemma 3.2. *Let G be a group and H a subgroup of index 2. Suppose that $\rho : H \to G$ is a simple $1/2$-endomorphism of G. Then the restriction of ρ to $H \cap H^\rho$ is a simple $1/2$-endomorphism of H^ρ.*

Proof. Since H is not ρ-invariant, there exists an element b in $\rho(H)$ outside H. Therefore, $G = H\langle b\rangle$. Let K be a subgroup of $H \cap H^\rho$, which is normal in H^ρ and is ρ-invariant. Then K is contained in $K^{\rho^{-1}}$ which is a normal subgroup of H. Therefore, $K^{\rho^{-1}b}$ is normal in H and $K = K^b \leq K^{\rho^{-1}b}$. Now let $M = K^{\rho^{-1}} \cap K^{\rho^{-1}b}$. Then M is a normal subgroup of G and $K \leq M$. Also, we have

$$K \leq M \leq K^{\rho^{-1}},\ M^\rho \leq K$$

and thus M is ρ-invariant. Since ρ is simple, it follows that both M and K are trivial. □

Proposition 3.3. *Let G be a group and H a subgroup of index 2. Suppose that $\rho : H \to G$ is a simple homomorphism. Let $\widetilde{G} = G \times G$ be the direct product of G with itself, $\widetilde{H} = H \times G$, and $\widetilde{\rho} : H \times G \to G \times G$ the map defined by $\widetilde{\rho} : (h, x) \to (x, h^\rho)$ for all $h \in H, x \in G$. Then $\widetilde{H}$ is a subgroup of index 2 in $\widetilde{G}$ and $\widetilde{\rho}$ is a simple homomorphism.*

Proof. Let $\widetilde{K}$ be a subgroup of $\widetilde{H}$, normal in $\widetilde{G}$ and $\widetilde{\rho}$-invariant. Then the projection of $\widetilde{K}$ on its first coordinate produces a subgroup K of H, normal in G. Furthermore, if $y = (h, x) \in \widetilde{K}$ then we have $h \in K$, $y^{\widetilde{\rho}}(x, h^\rho) \in \widetilde{K}$, $x \in H$, $y^{\widetilde{\rho}^2} = (h^\rho, x^\rho) \in \widetilde{K}$ and so $h^\rho \in K$; that is, K is a ρ-invariant. We conclude that K is trivial and thus $y = (e, x)$, $y^{\widetilde{\rho}} = (x, e) = (e, e)$, $y = e$; hence $\widetilde{K}$ is trivial. □

A direct application of this proposition is

Corollary 3.1. *Let $k \geq 0$, $m = 2^k$. Consider the free-abelian group $G = \mathbb{Z}^m$ of rank m, its subgroup $H = 2\mathbb{Z} \times \mathbb{Z}^{m-1}$ and the rational vector space $V = \mathbb{Q}^m$. Define the following linear transformations of V represented by the matrices*

$$\mathbf{A}_0 = \frac{1}{2}, \mathbf{A}_1 = \begin{pmatrix} 0 & 1 \\ \frac{1}{2} & 0 \end{pmatrix}, \ldots, \mathbf{A}_k = \begin{pmatrix} 0 & \mathbf{I}_{2^{k-1}} \\ \mathbf{A}_{k-1} & 0 \end{pmatrix}$$

with respect to the canonical basis. Then $\mathbf{A}_k$ defines a simple homomorphism from H into G, for all $k \geq 0$.

The next result will be used in the final section of the paper in order to extend certain simple $1/2$-endomorphisms associated to lattices to their affine groups.

Lemma 3.3. *Let G be a group which admits a factorization $G = MH$ where H is a subgroup of index 2 and M a normal subgroup such that $C_G(M) \leq M$. Furthermore, let $\rho : H \to G$ be a homomorphism, $N = H \cap M$ and η the restriction of ρ to N. Suppose $\eta(N) \leq M$ and η simple. Then ρ is also simple.*

Proof. Suppose η is simple and define $K = \rho-\text{core}(H)$, $D = K \cap M$. Then D is a normal subgroup of G contained in N and

$$D^\eta = (K \cap N)^\eta \leq K^\rho \cap N^\eta \leq K \cap M = D,$$

is a subgroup of the $\eta-\text{core}(M)$. Thus D is trivial and as both K and M are normal subgroups of G, K centralizes M. We conclude that $K \leq M$, $K \leq \eta-\text{core}(M)$ and $K = \{e\}$. □

3.3. State-closed solvable groups

We present in this section a number of results, some positive and others negative, for state-closed representations of finitely generated solvable groups.

Proposition 3.4. *Let G be an abelian state-closed group of automorphisms of the binary tree. Then G is either torsion-free or an elementary abelian 2-group. If G is an elementary abelian 2-group then it is a subgroup of the topological closure of $\langle \sigma, \sigma^{(1)}, \sigma^{(2)}, \ldots \rangle$.*

Proof. Suppose the group is not torsion-free. Then the set of involutions $\Omega_2(G)$ is not contained in the stabilizer subgroup G_1, for otherwise this set would be ρ-invariant and therefore trivial. Let a be an involution such that $G = G_1 \oplus \langle a \rangle$. But then, $G_1^2 = G^2$ and thus $G^2 = \{e\}$. As a is active, $a = (a', a')\sigma$, and the elements of G_1 have the form $h = (x, x)$, with $a', x \in G$. It becomes clear on developing the elements of G that these belong to the topological closure of $\langle \sigma, \sigma^{(1)}, \sigma^{(2)}, \ldots \rangle$. □

Theorem 3.2. *Let G be a finitely generated free nilpotent group of class k, H a subgroup of G of finite index containing the commutator subgroup G' and $\rho : H \to G$ a homomorphism. Then G' is ρ-invariant. Therefore, G is a state closed group acting on the binary tree if and only if $k = 1$.*

Proof. We proceed by induction on the nilpotency class of k of G and may assume $k \geq 2$.

There exists a free generating set $S = \{a_1, a_2, \ldots, a_m\}$ of G modulo G' and positive integers $n_1, n_2, \ldots, n_m$ such that $U = \{a_1^{n_1}, a_2^{n_2}, \ldots, a_m^{n_m}\}$ is a free generating set for H modulo G'.

First will show that $\gamma_k(G)$ is ρ-invariant. Note that

$$\gamma_k(H) \leq \gamma_k(G) \leq H,$$
$$\gamma_k(H)^\rho = \gamma_k(H^\rho) \leq \gamma_k(G).$$

It is well-known that $\gamma_k(G)$ is generated by $a = [a_{i_1}, a_{i_2}, \ldots, a_{i_k}]$ where $a_{i_s} \in S$. Likewise, $\gamma_k(H)$ is generated by $b = [a_{i_1}^{n_{i_1}}, a_{i_2}^{n_{i_2}}, \ldots, a_{i_k}^{n_{i_k}}]$. Note that $b = [a_{i_1}, a_{i_2}, \ldots, a_{i_k}]^n = a^n$ where $n = n_{i_1}, n_{i_2} \ldots n_{i_k}$. Now, $b^\rho = [a_{i_1}^{n_{i_1}\rho}, a_{i_2}^{n_{i_2}\rho}, \ldots, a_{i_k}^{n_{i_k}\rho}] = (a^\rho)^n$. Since $G/\gamma_k(G)$ is torsion-free it follows that $a^\rho \in \gamma_k(G)$ and $\gamma_k(G)$ is ρ-invariant. Therefore ρ induces a homomorphism $\overline{\rho} : H/\gamma_k(G) \to G/\gamma_k(G)$ and the proof of the first statement follows by induction. The second statement is an immediate conclusion. □

Remark. In the above proposition, the hypothesis that the subgroup H contain G' is necessary. For let G be the free nilpotent class group of nilpotency class 2, freely generated by a, b, and let $z = [a, b]$. Consider the subgroup $H = \langle a^k, b\rangle$ of G where $k \geq 2$. Then $[G : H] = k^2$ and H is not a normal subgroup of G. The map $\rho : a^k \to b, b \to a^{-1}$ extends to an epimorphism $\rho : H \to G$ and $\rho(z^k) = \rho[a^k, b] = [b, a^{-1}] = z$. If K is a nontrivial subgroup of H and is normal in G, then K contains z^{ik} for some $i \geq 1$; choose ik to be minimal. Then on applying ρ to K we produce z^i which shows that K cannot be ρ-invariant and therefore it follows that ρ is simple.

Theorem 3.3. *Let G be a finitely generated nilpotent group and suppose G is a state-closed group of automorphisms of the binary tree. Then G is torsion-free or a finite 2-group.*

Proof. It is well-known that the set $T(G)$ of torsion elements of G is indeed a finite subgroup. We proceed by induction on $|T(G)|$. Suppose $T(G)$ is nontrivial. Choose an involution a in G outside G_1; therefore, $G = G_1\langle a\rangle$. Recall that π_0 is the projection map of the stabilizer subgroup G_1 on its first coordinate. The group G is described by the quadruple (G_1, π_0, a, a_0). We may obtain other faithful state-closed representations of G by different choices of a_0. Let $a_0 = e$. Then we have $a = (e, a_1)\sigma$, and since $o(a) = 2$, we find that $a_1 = e$ and $a = \sigma$. By Lemma 3.5, π_0 is a simple 1/2-endomorphism of $(G_1)^{\pi_0}$. As $|T(G_1)| < |T(G)|$, we conclude by induction that $(G_1)^{\pi_0}$ is either torsion-free or a finite 2-group. Since $a = \sigma$, we have that $G_1 \leq (G_1)^{\pi_0} \times (G_1)^{\pi_0}$ and so, G_1 is either torsion-free or a finite 2-group. We have to discuss the first alternative only. In this case, $T(G) = \langle\sigma\rangle$ and is central. We conclude that $G^2 = G_1^2$ and thus G is an elementary abelian 2-group. □

The question as to which finitely generated solvable groups admit faithful state-closed representations is open. We give below some examples of non-abelian groups with such representations.

Examples. We have shown in Lemma 3.3 of [BS2] that if ξ is a dyadic unit then $\lambda = \lambda^{(1)}(e, \tau^{(\xi-1)/2})$ conjugates $\tau = (e, \tau)\sigma$ to τ^ξ. The affine group of the dyadic integers is a metabelian group and is state-closed.

If $\xi = -1$ then $\lambda = \lambda^{(1)}(e, \tau^{-1})$ inverts τ and the group $G = \langle\tau, \lambda\rangle$ is state-closed and polycyclic. If $\xi = 3$, then $\lambda = \lambda^{(1)}(e, \tau)$ conjugates $\tau = (e, \tau)\sigma$ to τ^3 and $G = \langle\tau, \lambda\rangle$ is a metabelian torsion-free recurrent group but is not polycyclic.

More polycyclic examples will be constructed in the last section of this paper.

4. Lattices of finite rank

4.1. Generating pairs

Let G be an m-dimensional lattice. We recall the formulas in Section 3,

$$a_1 = a_0^{-1}\pi_0(a^2),$$

$$\pi_1(h) = \pi_0(h^{a^{-1}})^{a_0} = \pi_0(h).$$

Therefore the elements of G have the following developments:

$$a = (a_0, a_0^{-1}\pi_0(a^2))\sigma,$$

$$h = (\pi_0(h), \pi_0(h)), \quad h \in H.$$

Any subgroup of G of finite index also has rank m, and thus so does the first level stabilizer G_1. We choose a free generating set $\{\mathbf{v_1}, \mathbf{v_2}, \ldots, \mathbf{v_m}\}$ of G such that G_1 is freely generated by $\{2\mathbf{v_1}, \mathbf{v_2}, \ldots, \mathbf{v_m}\}$ and let $a = \mathbf{v_1}$. Consider the vector space $V = \mathbb{Q} \otimes G$ and denote the extension of π_0 to V by A. Then A is an invertible linear transformation of V and the elements of G are represented in additive notation by

$$v_1 = (a_0, -a_0 + 2A(v_1))\sigma,$$
$$h = (A(h), A(h)) \text{ for all } h \in G_1.$$

On choosing $r = -a_0 + A(v_1)$ in V, v_1 may be re-written as

$$v_1 = (A(v_1) - r, A(v_1) + r)\sigma.$$

In this form the development of the elements of G simplify to

$$\begin{cases} u = (A(u) - r, A(u) + r)\sigma & \text{for all } u \in G \setminus G_1, \\ h = (A(h), A(h)) & \text{for all } h \in G_1. \end{cases}$$

Note that $r \in G + A(G) \setminus G$. The matrix representation of A with respect to the basis $\{v_1, v_2, \ldots, v_m\}$ has the form

$$\mathbf{A} = \begin{pmatrix} \frac{a_{11}}{2} & a_{12} & \cdots & a_{1m} \\ \frac{a_{21}}{2} & a_{22} & \cdots & a_{2m} \\ \vdots & \vdots & \ddots & \vdots \\ \frac{a_{m1}}{2} & a_{m2} & \cdots & a_{mm} \end{pmatrix}, \tag{1}$$

where all the a_{ij}'s are integers; this is so since A maps the group $\langle 2v_1, v_2, \ldots, v_m\rangle$ into the group $\langle v_1, v_2, \ldots, v_m\rangle$. Considering that G_1 is not A-invariant, the first column $\mathbf{A}(v_1)$ is not an integral vector, yet $2\mathbf{A}(v_1)$ is integral. The same observation holds for $r = -a_0 + \mathbf{A}(v_1)$. The characteristic polynomial $f(x)$ of A has the form $f(x) = x^m + \frac{1}{2}g(x)$ where $g(x)$ is an integral polynomial of degree $m - 1$.

Definition 2. Let $\mathbf{A}$ be an invertible $m \times m$ matrix with rational coefficients. Then $W = \mathbf{A}^{-1}(\mathbb{Z}^m) \cap \mathbb{Z}^m$ is called the $\mathbb{Z}$-*domain* of $\mathbf{A}$. If W has index n in $\mathbb{Z}^m$ then $\mathbf{A}$ is called $1/n$-*integral*. If the restriction of $\mathbf{A}$ to W defines a simple homomorphism then $\mathbf{A}$ is said to be *simple*.

For an arbitrary basis of the group G, the matrix $\mathbf{A}$ associated to the $1/2$-endomorphism π_0 from the stabilizer subgroup G_1 into the group G must be $1/2$-integral. It is clear that a matrix is $1/2$-integral if and only if the sum of any two (not necessary distinct) of its non-integral columns is integral. It follows from the arguments above that any $1/2$-integral matrix is conjugate to a matrix of the type (1). The following proposition, which is an analogue of [BJ, Prop. 10.1], gives a criterion for a $1/n$-integral matrix $\mathbf{A}$ to be simple, when considered as a homomorphism from its $\mathbb{Z}$-domain into $\mathbb{Z}^m$.

Proposition 4.1. *Let* $\mathbf{A}$ *be an* $m \times m$ *matrix over* $\mathbb{Q}$. *Suppose* $\mathbf{A}$ *is an* $1/n$-*integral matrix. Then* $\mathbf{A}$ *is simple if and only if its characteristic polynomial is not divisible by a monic polynomial with integral coefficients. Furthermore, if* n *is a prime number, then* $\mathbf{A}$ *is simple if and only if its characteristic polynomial of* $\mathbf{A}$ *is irreducible.*

Proof. Let $W = \mathbf{A}^{-1}(\mathbb{Z}^m) \cap \mathbb{Z}^m$ be the $\mathbb{Z}$-domain of $\mathbf{A}$.

(i) Suppose $\mathbf{A}$ is not simple and let $\{0\} \neq U \leq W$ be such that $\mathbf{A}(U) \leq U$ and $\mathbf{C}$ be the restriction of $\mathbf{A}$ to U. Then the characteristic polynomial of $\mathbf{C}$ is a monic polynomial with integral coefficients and is a factor of the characteristic polynomial of $\mathbf{A}$. In the other direction, suppose $f(x) = x^k + a_1 x^{k-1} + \cdots + a_k \in \mathbb{Z}[x]$ is an irreducible factor of the characteristic polynomial of $\mathbf{A}$. Let $\widehat{U} \leq \mathbb{Q}^m$ be the kernel of the operator $f(\mathbf{A})$. Then for arbitrary nonzero element $v \in \widehat{U}$ the vectors $v, \mathbf{A}(v), \mathbf{A}^2(v), \ldots \mathbf{A}^{k-1}(v)$ form a basis of the space $\widehat{U}$ and the matrix of the operator $\mathbf{A}|_{\widehat{U}}$ in this basis is obviously integral. Therefore there exists a nonzero integer q such that all the vectors $qv, q\mathbf{A}(v), q\mathbf{A}^2(v), \ldots q\mathbf{A}^{k-1}(v)$ are integral and form a basis of the space $\widehat{U}$ and the matrix of $\mathbf{A}|_{\widehat{U}}$ is integral. Thus $U = \widehat{U} \cap \mathbb{Z}^m$ is a nontrivial invariant group, and $\mathbf{A}$ is not simple.

(ii) Suppose n is a prime number. Then any $1/n$-integral matrix is similar to a matrix of the type

$$\mathbf{A} = \begin{pmatrix} \frac{a_{11}}{n} & a_{12} & \cdots & a_{1m} \\ \frac{a_{21}}{n} & a_{22} & \cdots & a_{2m} \\ \vdots & \vdots & \ddots & \vdots \\ \frac{a_{m1}}{n} & a_{m2} & \cdots & a_{mm} \end{pmatrix}.$$

Let $f(x)$ be the characteristic polynomial of $\mathbf{A}$, then the polynomial $nf(x)$ has integral coefficients. It is clear that in any nontrivial decomposition of $nf(x)$ into a product of two polynomials with integral coefficients one of the factors will have a leading coefficient equal to 1. Thus the irreducibility of $f(x)$ is equivalent to the simplicity of $\mathbf{A}$. □

Corollary 4.1. *Let* G *be an* m-*dimensional lattice and* K *a sub-lattice of* G. *Then* K *is also* m-*dimensional.*

Proof. Suppose K is a proper non-trivial subgroup of G, which is state-closed. Since $\mathbb{Q} \otimes K_1 = \mathbb{Q} \otimes K$ and is A-invariant and as A is irreducible, we get that $\mathbb{Q} \otimes K = \mathbb{Q} \otimes G$ and therefore K and G have equal ranks. □

Example. We cannot conclude in the above corollary that $K = G$. For, let G be the 2-dimensional lattice generated by $\alpha = \left(e, \alpha\beta^2\right)\sigma, \beta = (\alpha, \alpha)$. Then $K = \langle \alpha, \beta^2 \rangle$ is a proper sub-lattice of G.

We have obtained the following variant of Theorem 3.1 for torsion-free abelian groups.

Proposition 4.2. *Let $G = \mathbb{Z}^m$ and let $\mathbf{A}$ be an $m \times m$ invertible rational matrix which is $1/2$-integral and simple. Let $H = \mathbf{A}^{-1}(G) \cap G$ and $r \in G + \mathbf{A}(G) \setminus G$. Then the pair $(\mathbf{A}, r)$ determines uniquely a representation φ of the group G as an m-dimensional lattice, by the rules*

$$v^{\varphi} = \begin{cases} \left((\mathbf{A}(v) - r)^{\varphi}, (\mathbf{A}(v) + r)^{\varphi}\right)\sigma & \textit{if } v \in G \setminus H \\ \left(\mathbf{A}(v)^{\varphi}, \mathbf{A}(v)^{\varphi}\right) & \textit{if } v \in H. \end{cases}$$

Let us start again with an abstract torsion-free abelian group G of rank m and a faithful state-closed representation φ of G described by the quadruple (H, ρ, a, a_0). The following steps lead to the description of the representation φ by the pair of parameters $(\mathbf{A}, r)$.

(i) Identify G with the additive group $\mathbb{Z}^m$, fix the canonical basis $\{e_1, e_2, \ldots, e_m\}$. Identify a with e_1 and identify the basis of H with $\{2e_1, e_2, \ldots, e_m\}$.
(ii) Define the vector space $V = \mathbb{Q}^m$ and extend ρ to a linear transformation A of V which is represented in the canonical basis as an irreducible $1/2$-integral matrix $\mathbf{A}$.
(iii) Define $r = -a_0 + \mathbf{A}(e_1)$. Then $H = \mathbf{A}^{-1}(G) \cap G$, $r \in G + \mathbf{A}(G) \setminus G$. Hence the representation φ is described simply by the *generating pair* $(\mathbf{A}, r)$.

If $\mathbf{A}$ is an invertible integral matrix, then for $\mathbf{A}' = \mathbf{T}\mathbf{A}\mathbf{T}^{-1}$ and $r' = \mathbf{T}r$, the pair $(\mathbf{A}', r')$ defines a state-closed representation φ' of G and $\varphi = T\varphi'$; in particular, $G^{\varphi} = G^{\varphi'}$. Therefore, in classifying the generating pairs $(\mathbf{A}, r)$, we may restrict $\mathbf{A}$ to representatives of the similarity classes of the $1/2$-integral irreducible matrices under conjugation by $\mathbf{T} \in \mathrm{GL}(m, \mathbb{Z})$.

4.2. 1-dimensional lattices

We give below a description of the 1-dimensional lattices.

Theorem 4.1. *Let $G = \langle g \rangle$ be a cyclic group of infinite order, let $\varphi : G \to Aut(\mathcal{T}_2)$ be a faithful state-closed representation of G and denote g^{φ} by α. Furthermore, let $\tau = (e, \tau)\sigma$, $\mu = (e, \mu^{-1})\sigma$. Then there exist integers c, d such that $\alpha = \left(\alpha^c, \alpha^d\right)\sigma$ and $c + d$ odd. If $\alpha = \left(\alpha^c, \alpha^{1-c}\right)\sigma$, then $\tau = \alpha^{-2c+1}$. If $\alpha = \left(\alpha^c, \alpha^{-1-c}\right)\sigma$, then $\mu = \alpha^{2c+1}$. In addition, any $\alpha = \left(\alpha^c, \alpha^d\right)\sigma$ with $c + d$ odd is conjugate to $\tau = (e, \tau)\sigma$ by a functionally recursive automorphism of the tree.*

Proof. Let $(\mathbf{A}, r)$ be the generating pair of φ. Obviously, $H = \langle g^2 \rangle$. On denoting g^φ by α we have $\alpha = (\alpha^c, \alpha^d)\sigma$, for some integers c, d. In additive notation, $\rho : 2\alpha \mapsto (c+d)\alpha$ and so $\mathbf{A} = (\frac{c+d}{2})$ and $r = \frac{d-c}{2}$. Since $\mathbf{A}$ is simple, $f = c + d$ is odd. Let $c + d = \pm 1$. When $\alpha = \left(\alpha^c, \alpha^{1-c}\right)\sigma$, we have

$$\alpha^2 = (\alpha, \alpha), \alpha^{-2c} = \left(\alpha^{-c}, \alpha^{-c}\right), \alpha^{-2c+1} = \left(e, \alpha^{1-2c}\right)\sigma$$

and thus $\tau = \alpha^{-2c+1}$. Similarly, if $\alpha = \left(\alpha^c, \alpha^{-1-c}\right)\sigma$, we have $\mu = \alpha^{2c+1}$. Let $w = (c + d - 1)/2$. Define the automorphisms of the tree, $\lambda = (\lambda, \lambda\alpha^{(-c+d)w})$ and $\gamma = (\lambda^{-1}\gamma, \alpha^{-c}\lambda^{-1}\gamma)$. Then $\{\alpha, \lambda, \gamma\}$ is a functionally recursive set and it can be verified that $\lambda^{-1}\alpha\lambda = \alpha^{c+d}$ and $\gamma^{-1}\alpha\gamma = \tau$. □

5. Finite-state lattices

5.1. Contracting maps

If A is a linear transformation of a vector space over a subfield of the complex numbers, then the *spectral radius* of A, denoted by $\kappa(A)$, is the largest absolute value of the eigenvalues of A. If $\kappa(A) < 1$ then A is a *contracting map*.

Theorem 5.1. *Suppose G is an m-dimensional lattice defined by (A, r). Then G is finite-state if and only if A is a contracting map.*

Proof. Since G is a lattice, the characteristic polynomial $f(x)$ of A is irreducible. Therefore the eigenvalues λ_i of A are all distinct and thus A is diagonalizable over the complex numbers. We will fix a basis $\{e_1, e_2, ...e_m\}$ for G, which will be identified with the Euclidean basis for $\mathbb{R}^m$. Let $\{\varepsilon_1, \varepsilon_2, \dots, \varepsilon_m\}$ be the basis of $\mathbb{C} \otimes G$, formed by the eigenvectors of A such that $A(\varepsilon_i) = \lambda_i \varepsilon_i$. Let $(\xi_1(v), \xi_2(v), \dots, \xi_m(v))$ be the coordinate vector of $v \in G$ with respect to this basis.

(i) Suppose by contradiction that G is finite state and $\kappa(A) \geq 1$. Let $\lambda = \lambda_1$, $|\lambda| \geq 1$. Since the group G has rank m there exists a vector $v \in G$ with a nonzero first coordinate $\xi_1(v)$. Let us fix this v. We are going to find a sequence $\{v_n\}$ of states of v such that the sequence $\{|\xi_1(v_n)|\}$ is nondecreasing. Using this we will prove that v has infinitely many states. The sequence will be defined inductively. Set $v_0 = v$. For every $n \geq 0$, v_{n+1} is a state of v_n and is defined by the rules below.

If v_n is inactive then $v_{n+1} = A(v_n)$. Then v_{n+1} is a state of v_n and $\xi_1(v_{n+1}) = \lambda\xi_1(v_n)$ with $|\ \xi_1(v_{n+1})\ | = |\ \lambda\ | \cdot |\ \xi_1(v_n)\ | \geq |\ \xi_1(v_n)\ |$.

If v_n is active then the vectors $A(v_n) + r$ and $A(v_n) - r$ are states of v_n. If $\xi_1(r) \neq 0$ then either $|\xi_1(A(v_n) + r)| > |\xi_1(A(v_n))| \geq |\xi_1(v_n)|$ or $|\xi_1(A(v_n) - r)| > |\xi_1(A(v_n))| \geq |\xi_1(v_n)|$. Then we choose v_{n+1} to be equal to one of the vectors $A(v_{\mathbf{n}}) + r$, $A(v_{\mathbf{n}}) - r$ so that $|\xi_1(v_{\mathbf{n+1}})| > |\xi_1(v_{\mathbf{n}})|$. If $\xi_1(r) = 0$ then we choose $v_{\mathbf{n+1}} = A(v_{\mathbf{n}}) + r$. Then $|\xi_1(v_{\mathbf{n+1}})| = |\lambda| \cdot |\xi_1(v_{\mathbf{n}})| \geq |\xi_1(v_{\mathbf{n}})|$.

An infinite number of elements from the sequence $\{v_{\mathbf{n}}\}$ are active, otherwise all the vectors $A^n(v)$ are eventually integral which would contradict the simplicity of the matrix $\mathbf{A}$. Thus in the case where $\xi_1(r) \neq 0$, the sequence $\{|\xi_1(v_{\mathbf{n}})|\}$ is nondecreasing and has infinitely many different elements. If $\xi_1(r) = 0$ then $\xi_1(v_{\mathbf{n}}) = \lambda^n \xi_1(v)$. Therefore the sequence $\{\xi_1(v_{\mathbf{n}})\}$ may contain a finite number of different elements only when λ is a root of unity. But by Proposition 4.1 this contradicts the simplicity of $\mathbf{A}$. Hence in all cases, v has an infinite number of different states.

(ii) Suppose $\kappa = \kappa(A) < 1$. We will prove that every element $v \in G$ has finite number of states. Define the max-norm of a vector by $\|u\| = \max\{|\xi_1(u)|, \ldots, |\xi_m(u)|\}$. Then $\|A^n(u)\| \leq \kappa^n \|u\|$ for all $n \in \mathbb{N}$ and $u \in G$. Any state of v, seen as a tree automorphism, is equal to v or to a vector of the type $A^n(v) \pm A^{n_1}(r) \pm A^{n_2}(r) \pm \cdots \pm A^{n_l}(r)$, where $n > n_1 > n_2 > \ldots > n_l > 0$ is a decreasing sequence of positive integers. The norms of the latter states can be estimated as follows:

$$
\begin{aligned}
&\|A^n(v) \pm A^{n_1}(r) \pm A^{n_2}(r) \pm \cdots \pm A^{n_l}(r)\| \leq \\
&\qquad \|A^n(v)\| + \|A^{n_1}(r)\| + \|A^{n_2}(r)\| + \cdots + \|A^{n_l}(r)\| \leq \\
&\qquad \|A^n(v)\| + \|A^{n-1}(r)\| + \|A^{n-2}(r)\| + \cdots + \|A(r)\| + \|r\| \leq \\
&\qquad \kappa^n \|v\| + \left(\kappa^{n-1} + \kappa^{n-2} + \cdots + \kappa + 1\right) \|r\| \leq \\
&\qquad \kappa^n \|v\| + \sum_{s=0}^{\infty} \kappa^s \|r\| \leq \|v\| + (1-\kappa)^{-1} \|r\|.
\end{aligned}
$$

Therefore all the states of the tree automorphism v lie inside some finite ball with respect to the max-norm which itself is contained in a finite Euclidean ball. As the states are integral vectors, there exists only a finite number of them. □

Example. Let $\mathbf{A}$ be the $m \times m$ matrix

$$
\begin{pmatrix}
0 & 1 & 0 & \ldots & 0 \\
0 & 0 & 1 & \ldots & 0 \\
\vdots & \vdots & \vdots & \ddots & \vdots \\
0 & 0 & \ldots & \ldots & 1 \\
1/2 & 0 & \ldots & \ldots & 0
\end{pmatrix}.
$$

Then the characteristic polynomial of $\mathbf{A}$ is $f(x) = x^m - 1/2$ and $\mathbf{A}$ is an irreducible 1/2-integral matrix with spectral radius $\sqrt[m]{1/2}$. If we choose $r = \frac{1}{2} e_m$ then the generating pair $(\mathbf{A}, r)$ defines a group G with the following set of free

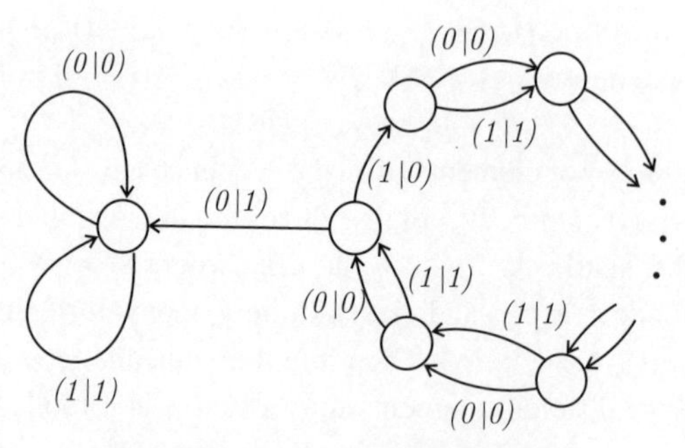

Figure 13.3 Automaton generating $\mathbb{Z}^n$

generators

$$\alpha = (e, \alpha^{(m-1)})\sigma, \alpha^{(1)}, \alpha^{(2)}, \ldots, \alpha^{(m-1)}.$$

Thus, G is state-closed and finite-state. Indeed G is recurrent, as the first level stabilizer G_1 is freely generated by $\alpha^{(1)}, \alpha^{(2)}, \ldots, \alpha^{(m-1)}, \alpha^2 = (\alpha^{(m-1)}, \alpha^{(m-1)}) = \alpha^{(m)}$ whose projection on the first coordinate produces G. The automata corresponding to the generators of G are obtained by choosing different initial states in the Moore diagram shown on Figure 13.3.

The group G is a minimal lattice. To prove this, it is sufficient to show that the state-closure of any non-trivial cyclic subgroup generated by $\beta = \alpha^{i_0}\alpha^{(1)i_1}\alpha^{(2)i_2}\ldots\alpha^{(m-1)i_{m-1}}$ is the whole group G. We define a norm of β by

$$\mid \beta \mid = \mid i_0 \mid + \mid i_1 \mid + \cdots + \mid i_{m-1} \mid .$$

Consider a minimal counterexample; that is, β has minimal non-zero norm such that the set of states $Q(\beta)$ does not generate G. We may choose β such that $i_0 \geq 0$. If $i_0 = 2i'$ then $\beta = (\beta_0, \beta_0) \in G_1$ and $\beta_0 = \alpha^{(m-1)i'}\alpha^{i_1}\alpha^{(1)i_2}\ldots\alpha^{(m-2)i_{m-1}}$. If $i_0 = 2i' + 1$ then $\beta = (\beta_0, \beta_1)\sigma$ where again $\beta_0 = \alpha^{(m-1)i'}\alpha^{i_1}\alpha^{(1)i_2}\ldots\alpha^{(m-2)i_{m-1}}$ which has norm $\mid i' \mid + \mid i_1 \mid + \cdots + \mid i_{m-1} \mid$. Thus by the minimality of β, we have $i_0 = 0$. A repetition of this argument leads to a contradiction.

Theorem 5.2. *Let G be finite-state lattice of rank m defined by the pair (A, r). Then G is a recurrent lattice and the characteristic polynomial of A^{-1} is an integral monic irreducible polynomial. Furthermore, for a fixed m, there exist only finitely many* $\mathrm{GL}(m, \mathbb{Z})$*-similarity classes of linear transformations A which are $1/2$-integral, irreducible and contracting.*

Proof. Let $f(x) = x^m + \frac{1}{2}(a_{m-1}x^{m-1} + \ldots + a_0)$ be the characteristic polynomial of A. Then as the spectral radius of A is less than 1, and since the coefficients

$\frac{1}{2}a_i$ of $f(x)$ are symmetric polynomials in the roots, we get that $|a_i| < 2\binom{m}{i}$. In particular, $1 \le |a_0| < 2$ and $|a_0| = 1$. Since the a_i's are integers, the number of possible characteristic polynomials $f(x)$ is finite. As the matrix $\mathbf{A}$ is conjugate to a matrix of the type (1) and has determinant $\pm\frac{1}{2}$, the image of G_1 under A is G; that is, A is recurrent. Now, the characteristic polynomial of A^{-1} is the integral irreducible polynomial $h(x) = a_0x^m + a_1x^{m-1} + \cdots + a_{m-1}x + 2$, where $|a_0| = 1$. By a theorem of Latimer and MacDuffee [N] the number of $\mathrm{GL}(m, \mathbb{Z})$-similarity classes of linear transformations with characteristic polynomial $h(x)$ is equal to the number of ideal classes in the order $\mathbb{Z}[\alpha]$, where α is a root of $h(x)$, and therefore is finite. We conclude that there is a finite number of similarity classes of A^{-1} and therefore of A as well. □

Remarks. (i) Although a finite-state lattice is recurrent, the converse is not necessarily true. For example, let G be determined by $(\mathbf{A}, \mathbf{r})$ where $\mathbf{A} = \begin{pmatrix} 0 & 1 \\ 1/2 & 1 \end{pmatrix}$. Then, since $\det(\mathbf{A}) = -1/2$, the group G is recurrent. However, it is not finite-state since $\kappa(\mathbf{A}) > 1$.

(ii) If a lattice G is finite-state we cannot conclude that it is a minimal lattice. For, let $\mathbf{A} = \begin{pmatrix} 1 & -1 \\ 5/2 & -2 \end{pmatrix}$ and let G bc the group generated by $\alpha = (e, \alpha^2\beta^5)\sigma$, $\beta = (\alpha^{-1}\beta^{-2}, \alpha^{-1}\beta^{-2})$. We verify directly that the states of α are contained in $K = \langle \alpha, \beta^5 \rangle$ and therefore G is not a minimal lattice.

5.2. Finite-state lattices of small rank

Theorem 5.3. *Consider the automorphisms of the binary tree $\tau = (e, \tau)\sigma$, $\mu = (e, \mu^{-1})\sigma$. A subgroup G of automorphisms of the binary tree is a 1-dimensional finite-state lattice if and only if it is generated by an l-th root of τ or μ for some odd integer l.*

Proof. Let $G = \langle \alpha \rangle$ be a 1-dimensional finite-state lattice with generating pair (A, r). By Proposition 5.1, $A = \pm 1/2$. When $(A, r) = (1/2, 1/2)$, we get a cyclic group generated by the adding machine $\tau = (e, \tau)\sigma$ and when $(A, r) = (-1/2, -1/2)$ we get a cyclic group generated by $\mu = (e, \mu^{-1})\sigma$. By Theorem 4.1, α is an l-th root of τ or μ for some odd integer l. On the other hand it can be checked directly that a group generated by an l-th root of τ or μ for an odd integer l is state-closed and finite-state. □

Theorem 5.4. *Let $\mathbf{A}$ be a 2×2 rational $1/2$-integral matrix. Then $(\mathbf{A}, r)$ is a generating pair for a 2-dimensional lattice if and only if $\mathbf{A}$ is $\mathrm{GL}(2, \mathbb{Z})$-similar to* *the companion matrix of one of the six polynomials* $x^2 \pm \frac{1}{2}$, $x^2 \pm \frac{1}{2}x + \frac{1}{2}$, $x^2 \pm x + \frac{1}{2}$.

Proof. The matrix $\mathbf{A}$ is equivalent to $\begin{pmatrix} \frac{1}{2}a_{11} & a_{12} \\ \frac{1}{2}a_{21} & a_{22} \end{pmatrix}$ where the a_{ij}'s are integers. By Theorem 5.2 its characteristic polynomial is $f(x) = x^2 + \frac{1}{2}a_1 x + \frac{1}{2}a_0$ with $|a_0| < 2$, $|a_1| < 4$. Thus, $|a_0| = 1$, $|a_1| = 0, 1, 2, 3$. The roots of $f(x)$ are $z_1 = \frac{1}{4}(-a_1 + \sqrt{a_1^2 - 8a_0})$, $z_2 = \frac{1}{4}(-a_1 - \sqrt{a_1^2 - 8a_0})$. If $\Delta = a_1^2 - 8a_0 \leq 0$ then $a_0 = 1$, $|a_1| = 0, 1, 2$ follow. On the other hand, suppose $\Delta = a_1^2 - 8a_0 > 0$; that is, $a_1^2 \geq 8a_0$. If $a_0 = 1$ then $a_1 = \pm 3$, $\Delta = 1$, and so $|z_1|$ or $|z_2| = 1$. Thus, $a_0 = -1$, $\Delta = a_1^2 + 8$. On substituting the possible values of a_1 in z_i we find that the only possibility for a_1 is 0. Hence the pair (a_0, a_1) varies over the set $\{(1, 0), (1, \pm 1), (1, \pm 2), (-1, 0)\}$.

Now we try to determine the possible reductions of the matrix $\mathbf{A}$ modulo conjugations by invertible integral matrices $\begin{pmatrix} s_{11} & s_{12} \\ s_{21} & s_{22} \end{pmatrix}$. On conjugating $\mathbf{A}$ by $\begin{pmatrix} -1 & 0 \\ 0 & 1 \end{pmatrix}$, we may assume $a_{21} \geq 0$. We have $\text{trace}(\mathbf{A}) = \frac{1}{2}a_{11} + a_{22} = -\frac{1}{2}a_1$, $\det(\mathbf{A}) = \frac{1}{2}a_{11}a_{22} - \frac{1}{2}a_{21}a_{12} = \frac{1}{2}a_0$. Therefore, $a_{22} = -\frac{1}{2}a_1 - \frac{1}{2}a_{11}$, $a_{21}a_{12} = -(\frac{1}{2}a_{11}^2 + \frac{1}{2}a_1 a_{11} + a_0)$. On conjugating $\mathbf{A}$ by $\begin{pmatrix} 1 & 0 \\ s & 1 \end{pmatrix}$ we transform $\frac{1}{2}a_{11}$ into $\frac{1}{2}a_{11} + sa_{12}$, and on conjugating by $\begin{pmatrix} 1 & s \\ 0 & 1 \end{pmatrix}$, we transform $\frac{1}{2}a_{11}$ into $\frac{1}{2}(a_{11} - sa_{21})$. Thus, we can reduce the absolute value of $\frac{1}{2}a_{11}$ unless possibly when $|a_{11}| \leq |a_{12}|, \frac{1}{2}|a_{21}|$. Hence we may assume, $2a_{11}^2 \leq \left|\frac{1}{2}a_{11}^2 + \frac{1}{2}a_1 a_{11} + a_0\right|$. The possible solutions (a_0, a_1, a_{11}) are contained in the table

a_0	1	1	1	1	1	−1
a_1	0	−1	1	2	−2	0
a_{11}	0	−1, 0	0, 1	0, 1	−1, 0	0

On using the fact that $a_{22} = -\frac{1}{2}(a_1 + a_{11})$ is an integer, the possibilities for a_{11} are reduced further and on using the formula $a_{12}a_{21} = -(\frac{1}{2}a_{11}^2 + \frac{1}{2}a_1 a_{11} + a_0)$, we obtain the following table

a_0	1	1	1	1	1	−1
a_1	0	−1	1	2	−2	0
a_{11}	0	−1	1	0	0	0
a_{22}	0	1	−1	−1	1	0
$a_{12}a_{21}$	−1	−2	−2	−1	−1	1

The different columns of the table correspond to different characteristic polynomials. As we may always choose $a_{21} > 0$, there are at most 8 classes. Further equivalences may occur only within the same column of the table. Suppose a_{21} is an even integer $2a'_{21}$, as may occur in the second and third columns. Then $\begin{pmatrix} a_{22} & -2a_{12} \\ -a'_{21} & a_{11} \end{pmatrix}$ conjugates $\mathbf{A}$ into $\begin{pmatrix} \frac{1}{2}a_{11} & 2a_{12} \\ \frac{1}{2}a'_{21} & a_{22} \end{pmatrix}$. Thus, in the third and fourth columns a_{21} may be chosen to be equal to 1. Considering that $|a_{11}| = 1$ in these columns, we may conjugate the corresponding matrix into one where $a_{11} = 0$. Hence, the equivalence classes of $\mathbf{A}$ are represented uniquely by the companion matrix of the characteristic polynomial of $\mathbf{A}$. □

Remark. The polynomials $f(x) = x^m + \frac{1}{2}(a_{m-1}x^{m-1} + \ldots + a_0)$ such that the absolute value of each of their roots is less than 1 present some remarkable features. As we showed in Theorem 5.2, there is a finite number of such polynomials for each fixed degree m. The proof of the above theorem shows that the number for $m = 2$ is 6, and that the class number corresponding to each polynomial is 1. We note that Pavel Guerzhoy has investigated with the use of the Number Theoretic package PARI-GP ([PG] polynomials of higher degree. He reproduced our 6 polynomials of degree 2, produced a complete list of 14 polynomials of degree 3, 36 polynomials of degree 4 and 58 of degree 5. In addition, it turns out that the class number in all these cases is 1. We note that though the class number eventually increases for higher degrees, still 1 seems to be quite predominant.

6. Recurrent lattices

6.1. Adding machines

As we commented earlier, the automorphism $\tau = (e, \tau)\sigma$ which generates a lattice of rank 1 represents binary addition in its action on the dyadic integers and was thus called an adding machine. We will show below that the notion of an adding machine can be generalized to lattices of arbitrary rank. Let G be an m-dimensional recurrent lattice defined by the generating pair (A, r). Also let $A^{-k}(G) = \{v \in G \mid A^k(v) \in G\}$ and recall that G_k is the pointwise stabilizer of the k-th level of the tree.

Lemma 6.1. *The group $A^{-n}(G)$ coincides with G_n for every n.*

Proof. We proceed by induction on n. We have $A^0(G) = G = G_0$. Assume $n \geq 1$. The group $A^{-n}(G)$ stabilizes the first level. For every $g \in A^{-n}(G)$, we have $g = (A(g), A(g))$ and $A(g) \in A^{-n+1}(G)$ which by the inductive hypothesis is the stabilizer of the $(n-1)$-st level of the tree. Thus $g \in G_n$ and $A^{-n}(G) \leq G_n$. On the other hand, we have $A(G_1) = G, A(G_2) \leq G_1, \ldots, A(G_n) \leq G_{n-1}$, and thus $G_n \leq A^{-n}(G)$. Hence $G_n = A^{-n}(G)$. □

Since $A(G_1) = G$, it follows that there exists $d \in G \setminus G_1$ such that $r = A(d)$; we fix such a d. Therefore, $A^{-1}(d) \in G_1 \setminus G_2$ and for all $n \geq 0$, $A^{-n}(d) \in G_n \setminus G_{n+1}$. We define an A-adic number as the series $c_0 d + c_1 A^{-1}(d) + \cdots + c_n A^{-n}(d) + \cdots$ where the c_n's are 0 or 1. These A-dic numbers belong to the closure of $\widehat{G}$ of the group G in the automoprhism group of the tree with respect to the 2-adic topology. We will show that the group G acts as an adding machine on these numbers.

Lemma 6.2. *The set of all A-adic numbers coincides with the topological closure $\widehat{G}$ of the group G. The map Ψ from the boundary of the binary tree to $\widehat{G}$, defined by*

$$\Psi(c_0, c_1, \ldots, c_n, \ldots) = c_0 d + c_1 A^{-1}(d) + \cdots + c_n A^{-n}(d) + \cdots$$

is well-defined and bijective.

Proof. We have to prove that the series $c_0 d + c_1 A^{-1}(d) + c_2 A^{-2}(d) + \cdots + c_n(A^{-n} d) + \cdots$ is convergent; equivalently, the sequence of its partial sums is a Cauchy sequence. This follows directly from the fact that $G_n = A^{-n}(G)$. Thus the map Ψ is well defined. In order to prove that every element of $\widehat{G}$ can be uniquely expanded in such a way, it is sufficient to prove that for every n and every $g \in G$ there exists a unique element of the form $g_n = c_0 d + c_1 A^{-1}(d) + \cdots + c_n A^{-n}(d)$ where $c_i \in \{0, 1\}$ such that $g - g_n \in G_n$. Let us prove this by induction on n. If $g \in G_1$ then $g - d \notin G_1$. On the other hand, if $g \notin G_1$ then $g - d \in G_1$. Thus the assertion is true for $n = 1$. Suppose it is true for $n = k-1, k \geq 2$. Since $g - g_{k-1} \in G_{k-1} = A^{-k+1}(G)$ and A is injective, there exists unique $h \in G$ such that $g - g_{k-1} = A^{-k+1}(h)$. If $h \in G_1 = A^{-1}(G)$ then $g - g_{k-1} \in A^{-k}(G) = G_k$ and we put $g_k = g_{k-1}$. Then $g - g_k \in G_k$ but $g - \left(g_{k-1} + A^{-k+1}(d)\right) = A^{-k+1}(h) - A^{-k+1}(d) = A^{-k+1}(h-d) \notin G_k$. If $h \notin G_1$ then $h - d \in G_1 = A^{-1}(G)$ and if we put $g_k = g_{k-1} + A^{-k+1}(d)$ then $g - g_k = A^{-k+1}(h-d) \in A^{-k}(G) = G_k$ but $g - g_k = A^{-k+1}(h) \notin A^{-k}(G) = G_k$. Thus, in any case, g_k can be chosen uniquely. □

The following proposition shows that the above identification Ψ is compatible with the action of G both on $\widehat{G}$ and on the boundary of the tree.

Proposition 6.1. *Let Ψ be the bijection from the boundary of the tree to $\widehat{G}$. Then for every element w of the boundary and $g \in G$ we have*

$$w^g = \Psi^{-1}\left(\Psi(w) + g\right).$$

Proof. Let $g \in G$, $w = (c_0, c_1, \dots)$ be arbitrary infinite path of the binary tree ($c_i \in \{0, 1\}$) and $w' = (c_1, c_2, \dots)$. Denote by $\psi(g)$ the tree automorphism defined by the rule

$$w^{\psi(g)} = \Psi^{-1}\left(\Psi\left(w\right) + g\right).$$

We have to prove that $\psi(g) = g$. If $g \in G_1$ then

$$g + \Psi(w) = g + c_0 d + c_1 A^{-1}(d) + \cdots = c_0 d \\ + A^{-1}\left(A\left(g\right) + c_1 d + c_2 A^{-1}\left(d\right) + \cdots\right).$$

Therefore $w^{\psi(g)} = \Psi^{-1}\left(\Psi\left(w\right) + g\right)$ is the sequence $\left(c_0, \left(w'\right)^{\psi(g_0)}\right)$ where, $g_0 = A(g)$. Thus, in this case, $\psi(g) = \left(\psi\left(A\left(g\right)\right), \psi\left(A\left(g\right)\right)\right)$. If $g \notin G_1$ and $c_0 = 0$ then

$$g + \Psi(w) = g + c_1 A^{-1}(d) + \cdots = d + (g - d) + c_1 A^{-1}(d) + \cdots \\ = 1 \cdot d + A^{-1}(A\left(g - d\right) + c_1 d + +c_2 A^{-1}\left(d\right) + \cdots).$$

If $g \notin G_1$ and $c_0 = 1$ then

$$g + \Psi(w) = g + d + c_1 A^{-1}(d) + c_2 A^{-2}\left(d\right) \cdots = 0 \cdot d \\ +A^{-1}(A\left(g + d\right) + c_1 + c_2 A^{-1}\left(d\right) + c_3 A^{-2}\left(d\right) + \cdots).$$

Therefore in this case

$$\psi(g) = \left(\psi\left(A\left(g - d\right)\right), \psi\left(A\left(g + d\right)\right)\right)\sigma \\ = \left(\psi\left(A\left(g\right) - r\right), \psi\left(A\left(g\right) + r\right)\right)\sigma.$$

We conclude that the map ψ satisfies the same recurrence as that which defines the group G and therefore $\psi(g) = g$ for every $g \in G$. □

Now it follows from Proposition 6.1 that the natural action of $\widehat{G}$ on itself is identified, by using the bijection Ψ, with the action of $\widehat{G}$ on the tree.

Example. Let $\mathbf{A} = \begin{pmatrix} 0 & 1 \\ 1/2 & 0 \end{pmatrix}$, $r = \begin{pmatrix} 0 \\ 1/2 \end{pmatrix}$. Then $\mathbf{A}^{-1} = \begin{pmatrix} 0 & 2 \\ 1 & 0 \end{pmatrix}$ and the multiplicative group G is freely generated by $\alpha = (e, \alpha^{(1)})\sigma, \alpha^{(1)}$. In additive notation, G has the basis $\alpha = v_1 = \begin{pmatrix} 1 \\ 0 \end{pmatrix}$, $\alpha^{(1)} = v_2 = \begin{pmatrix} 0 \\ 1 \end{pmatrix}$. We compute

$$d = \mathbf{A}^{-1} r = \begin{pmatrix} 1 \\ 0 \end{pmatrix} = v_1, \mathbf{A}^{-1} d = \begin{pmatrix} 0 \\ 1 \end{pmatrix} = v_2, \dots,$$

$$\mathbf{A}^{-2i} d = 2^i v_1, \mathbf{A}^{-(2i+1)} d = 2^i v_2, \dots$$

An element $w \in \widehat{G}$, in additive notation, has the form

$$w = \xi_1 v_1 + \xi_2 v_2,$$

where $\xi_1 = \sum c_i 2^i$, $\xi_2 = \sum d_i 2^i$ are dyadic integers. The element of the boundary which corresponds to w is $\Psi^{-1}(w) = (f_0, f_1, \dots, f_{2i}, f_{2i+1}, \dots)$ where $f_{2i} = c_i$, $f_{2i+1} = d_i$. Now, it can be checked that applying α to $f = \sum f_i 2^i$ corresponds to calculating $\Psi^{-1}(w + v_1)$.

6.2. Topological closure of recurrent lattices

Let G be a recurrent lattice and (A, r) its generating pair. We will prove in this section that A determines the topological closure $\widehat{G}$ of G, independently of r. Recall from Proposition 3.2 that the centralizer of a recurrent abelian group G is equal to $\widehat{G}$. As was mentioned earlier, there exists $d \in G \setminus G_1$ such that $r = A(d)$. Then the closure $\widehat{G}$, by Proposition 6.1, can be naturally identified with the group of formal series of the form $c_0 d + c_1 A^{-1}(d) + c_2 A^{-2}(d) + \cdots$, $c_i \in \{0, 1\}$.

Lemma 6.3. *For any formal power series $f(x) = 1 + b_1 x + b_2 x^2 + \cdots + b_n x^n + \cdots \in \mathbb{Z}[[x]]$ the operator $f(A^{-1}) = I + b_1 A^{-1} + b_2 A^{-2} + \cdots + b_n A^{-n} + \cdots$ is a well-defined continuous automorphism of the group $\widehat{G}$.*

Proof. For any $g \in \widehat{G}$, the sequence of partial sums of the series $g + b_1 A^{-1}(g) + b_2 A^{-2}(g) + \cdots + b_n A^{-n}(g) + \cdots$ is a Cauchy sequence and thus the series is convergent to an element of the group $\widehat{G}$. Thus $f(A^{-1})$ is an endomorphism of the group $\widehat{G}$ which leaves invariant $G_n = A^{-n}(G)$ for all $n \geq 0$. Hence, $f(A^{-1})$ is a continuous function. Now, since every series $f(x) = 1 + b_1 x + \cdots + b_n x^n + \cdots \in \mathbb{Z}[[x]]$ is a unit in this ring, the endomorphism $f(A^{-1})$ is an automorphism of the group $\widehat{G}$. □

Theorem 6.1. *Let L and M be two recurrent m-dimensional lattices with the respective generating pairs $(A, A(d_1))$ and $(A, A(d_2))$. Then their closures $\widehat{L}$ and $\widehat{M}$ are equal.*

Proof. Since d_2 is an integral vector, there exists a sequence $(c_0, c_1, \dots, c_n, \dots)$ with $c_n \in \{0, 1\}$ such that $d_2 = c_0 d_1 + c_1 A^{-1}(d_1) + \cdots + c_n A^{-n}(d_1) + \cdots$. Since $A(d_2)$ is not integral, $c_0 = 1$. By Lemma 6.3, the sum $B = 1 + c_1 A^{-1} + c_2 A^{-2} + \cdots + c_n A^{-n} + \cdots$ defines an automorphism of the group $\widehat{L}$. Obviously, A commutes with B and thus the generating pair of the group $B^{-1}(L)$ is $\left(BAB^{-1}, BA(d_1)\right) = (A, A(d_2))$. We conclude that $B^{-1}(L) = M$ and $\widehat{M} = \widehat{L}$. □

We conclude that recurrent lattices defined by matrices with the same characteristic polynomial have equal completions, irrespective of the vector r.

7. Finite-state representations of affine groups

Let V be the free abelian group of rank m generated by the canonical basis of column vectors $\{v_1, v_2, \ldots, v_m\}$ and let W be the subgroup of V generated by $\{2v_1, v_2, \ldots, v_m\}$. Consider the matrix defined on the vector space $\mathbb{Q} \otimes V$,

$$\mathbf{A} = \begin{pmatrix} 0 & 1 & 0 & \ldots & 0 \\ 0 & 0 & 1 & \ldots & 0 \\ \vdots & \vdots & \vdots & \ddots & \vdots \\ 0 & 0 & \ldots & \ldots & 1 \\ 1/2 & 0 & \ldots & \ldots & 0 \end{pmatrix}$$

Conjugation of elementary transformations $E_{i,j}(t)$ by $\mathbf{A}^{-1}$ has the following effect:

$$\mathbf{A}E_{i,j}(t)\mathbf{A}^{-1} = E_{i-1,j-1}(\delta t),$$

where the indices are written modulo m and

$$\delta = \begin{cases} 1 & \text{if } i = j, \text{ or } 1 < i, j, \\ 2 & \text{if } 1 = j < i, \\ 1/2 & \text{if } 1 = i < j. \end{cases}$$

Let Γ be the subgroup of $\mathrm{GL}(m, \mathbb{Z})$ consisting of matrices $\mathbf{B} = (b_{ij})$ where b_{ij} is even for $i < j$. Then Γ is a maximal subgroup of $\mathrm{GL}(m, \mathbb{Z})$ with respect to the property of not containing non-trivial elements of odd order. The group Γ is generated by the set of elementary matrices

$$S = \left\{E_{i,j}(1) \text{ for } i > j,\ E_{i,j}(2) \text{ for } i < j,\ E_{i,i}(-1) \text{ for all } i\right\}.$$

We verify the following conditions:

$$W \text{ is } \Gamma\text{-invariant}, \mathbf{A}(W) = V, \mathbf{A}\Gamma\mathbf{A}^{-1} = \Gamma.$$

Let G be the semi-direct product of V by Γ. The elements of G are written as $(v, \mathbf{B})$, where $v \in V$, $\mathbf{B} \in \Gamma$ and the product is defined by $(v, \mathbf{B})(v', \mathbf{B}') = (v + \mathbf{B}(v'), \mathbf{B} \cdot \mathbf{B}')$. We simplify the notation $(v, \mathbf{B})$ as $v \cdot \mathbf{B}$. Then $H = W \cdot \Gamma$ is a subgroup of index 2 in G. Also, $G = VH$ and, clearly, V is self-centralizing in G. Define the $1/2$-endomorphism $\rho : H \to G$ by

$$\rho : w \cdot \mathbf{B} \mapsto \mathbf{A}(w) \cdot \mathbf{ABA}^{-1}.$$

Then ρ is an isomorphism from H onto G and so by using it we may define a recurrent representation φ of G on the binary tree. Since ρ restricted to W is simple we conclude from Lemma 3.3 that ρ is also simple and so, the representation φ is faithful.

7.1. Polycyclic examples

Using the affine group, we can prove the existence of subgroups of G which are torsion-free polycyclic and non-abelian. To this effect let $\mathbf{C} = \mathbf{A}^{-1} - \mathbf{I}$. Then $\mathbf{C} \in \Gamma$; this follows from the fact that $x^n - 2 = (x-1)(x^{n-1} + \cdots + x + 1) - 1$. Since $\mathbf{A}^{-1}$ is irreducible on $\mathbb{Q} \otimes V$ then so is $\mathbf{C}$. Let $M = V\langle \mathbf{C} \rangle$. Then by Lemma 3.3, the map ρ induces a simple $1/2$-endomorphism on M. One may generalize this construction by considering the group $\mathbf{U}$ of units in the ring of algebraic integers of $\mathbb{Q}[\mathbf{C}]$. Then $\mathbf{U}$ is a subgroup of Γ. Dirichlet's Unit Theorem [BSh] provides us with a torsion-free abelian subgroup F of D of higher rank for polynomials of higher degrees m. Then $M = VF$ is a state-closed metabelian polycyclic group.

7.2. Finite-state state-closed representation of affine groups

We proceed to construct a concrete representation $\varphi : G(= V\Gamma) \to Aut(\mathcal{T}_2)$ and show that the image is generated by finite-state automorphisms of the binary tree.

Choose $\left(v_1^{\varphi}\right)_0 = e$. Then,

$$\begin{aligned} v_1^{\varphi} &= (e, 2\mathbf{A}(v_1)^{\varphi})\sigma = (e, v_m^{\varphi})\sigma, \\ v_2^{\varphi} &= (v_1^{\varphi}, v_1^{\varphi}), \ldots, \\ v_i^{\varphi} &= (v_{i-1}^{\varphi}, v_{i-1}^{\varphi}), \ldots, \\ v_m^{\varphi} &= (v_{m-1}^{\varphi}, v_{m-1}^{\varphi}). \end{aligned}$$

Drop φ from the notation; so,

$$\begin{aligned} v_1 &= (e, v_m)\sigma, \\ v_i &= (v_{i-1}, v_{i-1}) \text{ for } 2 \leq i \leq m-1, \\ v_m &= (v_{m-1}, v_{m-1}). \end{aligned}$$

The representation of $\mathbf{B} \in \Gamma$ is given by

$$\mathbf{B}^\varphi = (\rho(\mathbf{B})^\varphi, \rho(\mathbf{B}^{v_1^{-1}})^\varphi) = (\rho(\mathbf{B})^\varphi, (A(I - \mathbf{B})v_1)) \cdot \rho(\mathbf{B})^\varphi);$$

and so, in particular, for the generators of Γ in S,

$$E_{i,j}(t)^\varphi = (E_{i-1,j-1}(\delta t)^\varphi, \big(\mathbf{A}(I - E_{i,j}(t))v_1)\big) \cdot E_{i-1,j-1}(\delta t)^\varphi).$$

Note that

$$\mathbf{A}(I - E_{i,j}(t))v_1 = \begin{cases} 0 & \text{if } j \neq 1, \\ -tv_{i-1} & \text{if } i \neq 1,\ j = 1, \\ \frac{1}{2}(1-t)\,v_m & \text{if } i = 1,\ j = 1. \end{cases}$$

Hence, on removing the φ from the notation, we have

$$E_{i,j}(t) = \begin{cases} (E_{i-1,j-1}(\delta t), E_{i-1,j-1}(\delta t)) & \text{if } j \neq 1, \\ (E_{i-1,m}(\delta t), (-tv_{i-1}) \cdot E_{i-1,m}(\delta t)) & \text{if } i \neq 1,\ j = 1, \\ (E_{m,m}(\delta t), \big(\frac{1-t}{2} v_m\big) \cdot E_{m,m}(\delta t)) & \text{if } i = 1,\ j = 1. \end{cases}$$

We have arrived finally at the form the generators take in this representation:

$$\begin{aligned} E_{1,1}(-1) &= (E_{m,m}(-1), v_m \cdot E_{m,m}(-1)), \\ E_{j,j}(-1) &= (E_{j-1,j-1}(-1), E_{j-1,j-1}(-1)) && \text{if } 1 < j, \\ E_{1,j}(2) &= (E_{m,j-1}(1), E_{m,j-1}(1)) && \text{if } 1 < j, \\ E_{i,j}(2) &= (E_{i-1,j-1}(2), E_{i-1,j-1}(2)) && \text{if } 1 < i < j, \\ E_{i,1}(1) &= (E_{i-1,m}(\delta), (-v_{i-1}) \cdot E_{i-1,m}(\delta)) && \text{if } 1 < i, \\ E_{i,j}(1) &= (E_{i-1,j-1}(1), E_{i-1,j-1}(1)) && \text{if } 1 < j < i, \end{aligned}$$

The representation of G is finite-state. This is so because the set of states of each v_j is equal to $\{e, v_i \mid 1 \leq i \leq m\}$ and the states of the given generators $E_{i,j}(t)$ of Γ are contained in the product of sets $Y \cdot S$ where $Y = \{e, \pm v_i \mid 1 \leq i \leq m\}$.

References

[B] H. Bass, M. Otero-Espinar, D. Rockmore, C. P. L. Tresser, *Cyclic Renormalization and the Automorphism Groups of Rooted Trees*, Lecture Notes in Mathematics vol. 1621, Springer, Berlin, 1995.

[BJ] O. Bratelli and P.E.T. Jorgensen, *Iterated function systems and permutation representations of the Cuntz algebra*, vol. 139, Memoirs of the Amer. Math. Soc., no. 663, Providence, Rhode Island, 1999.

[BSh] Z. I. Borevich, I. R. Shafarevich, *Number Theory*, Academic Press, New York, 1966.

[BS1] A. M. Brunner, S. Sidki, On the automorphism group of the one-rooted binary tree, *J. Algebra* **195** (1997), 465–486.

[BS2] A. M. Brunner, S. Sidki, The generation of GL$(n, \mathbb{Z})$ by finite state automata, *Int. J. of Algebra and Computation* **8** (1998), 127–139.

[BSV] A. M. Brunner, S. Sidki, A. C. Vieira, A just-nonsolvable torsion-free group defined on the binary tree, *J. Algebra* **211** (1999), 99–144.

[G1] R. I. Grigorchuk, On the Burnside problem for periodic groups, *Functional Anal. Appl.* **14** (1980), 41–43.

[G2] R. Grigorchuk, Just infinite branch groups, in: *New Horizons in pro-p Groups* (M. P. F. du Sautoy, A. Shalev, D. Segal eds.), Progress in Mathematics vol. 184, Birkhauser, Boston, 2000, 121–179.

[GM] R. I. Grigorchuk, H. J. Mamaghani, On the use of iterates of endomorphisms for constructing groups with specific properties, *Math Studii, Lviv* **8** (1997), 198–206.

[GS] N. Gupta, S. Sidki, Extensions of groups by tree automorphisms, in: *Contributions to Group Theory*, Contemp. Math. 33 (1984), 232–246.

[KL] P. Kleidman, M. Liebeck, *The Subgroup Structure of the Finite Classical Groups*, LMS Lecture Notes vol. 129, Cambridge University Press, Cambridge, 1990.

[K] D. Knuth, *The Art of Computer Programming vol. 2 (Seminumerical Algorithms)*, Addison-Wesley, 1969.

[Nek] V. Nekrashevych, Stabilizers of transitive actions on locally finite graphs, *Int. J. of Algebra and Computation* **10** (2000), 591–602.

[N] M. Newman, *Integral Matrices*, Academic Press, 1972.

[PG] PARI-GP, C. Batut et.al. *User's Guide*, Laboratoire A2X, U.M.R. 9936 du C.N.R.S., Universitie Bordeaux I. France, 1998.

[Sa] T. Safer, Polygonal Radix Representations of Complex Numbers, *Theoret. Comput. Sci.* **210** (1999), 159–171.

[S1] S. Sidki, *Regular Trees and their Automorphisms*, Monografias de Matematica vol. 56, IMPA, Rio de Janeiro, 1998.

[S2] S. Sidki, Automorphisms of one-rooted trees: growth, circuit structure and acyclicity, *J. of Mathematical Sciences* **100** (2000), 1925–1943.

[V] A. Vince, *Digit tiling of Euclidian space*, in: Directions in mathematical quasicrystals, Amer. Math. Soc., Providence, Rhode Island, 2000, 329–370.

14

The mapping class group of the twice punctured torus

J. R. Parker and C. Series

Introduction

Let Σ be a (possibly punctured) surface of negative Euler characteristic, and let $C(\Sigma)$ be the set of isotopy classes of families of disjoint simple closed curves on Σ. When Σ is a once punctured torus Σ_1, there is a well known recursive structure on $C(\Sigma_1)$ which arises from the relationships between $C(\Sigma_1)$ (identified with the extended rational numbers), continued fractions, and $\mathrm{PSL}(2, \mathbb{Z})$ (the mapping class group of Σ_1) [11], [20]. The results in this paper arose out of a search for an analogous structure on $C(\Sigma_2)$, where Σ_2 is a torus with two punctures. Masur and Minsky [14], [15] have recently described an alternative approach. Our method is motivated by the Bowen-Series construction [4], [21] of Markov maps for Fuchsian groups. This generalised the relationship between $\mathrm{PSL}(2, \mathbb{Z})$ (now thought of as a Fuchsian group acting in the hyperbolic plane) and continued fractions (now thought of as points in the limit set of $\mathrm{PSL}(2, \mathbb{Z})$), to a large class of Fuchsian groups Γ. The Markov map was a map on the boundary at infinity, in other words the limit set $\Lambda(\Gamma)$, which generated continued fraction expansions for points in $\Lambda(\Gamma)$, and whose admissible sequences simultaneously gave an elegant solution to the word problem in Γ [21] (see Section 1.1 below).

The idea behind this paper rests on the analogy between Γ acting on the hyperbolic plane and the mapping class group $\mathcal{MCG}(\Sigma)$ acting on Teichmüller space $\mathcal{T}(\Sigma)$. In this analogy, the boundary S^1 of the hyperbolic plane (or the limit set of Γ) is replaced by a suitable boundary of $\mathcal{T}(\Sigma)$. We use the *Thurston boundary*, namely the space $\mathcal{PML}(\Sigma)$ of projective measured laminations on Σ (see Section 1.4 below). The mapping class group $\mathcal{MCG}(\Sigma)$ (see Section 1.5 below) acts on both $\mathcal{T}(\Sigma)$ and $\mathcal{PML}(\Sigma)$. By analogy with the Bowen-Series construction, we define a Markov map f on $\mathcal{PML}(\Sigma)$ which has the same relation to the action of $\mathcal{MCG}(\Sigma)$ on $\mathcal{PML}(\Sigma)$ as the Bowen-Series map has to the action of Γ on $\Lambda(\Gamma) = S^1$. Thurston's well known theory of train tracks [17], [22] gives $\mathcal{ML}(\Sigma)$ a piecewise linear cone structure. Here we use the variant of π_1-train tracks introduced by Birman and Series [2]. This special class of train tracks is defined relative to a fixed choice of fundamental domain and associated geometric generators for $\pi_1(\Sigma)$, in such a way that an integer weighting yields not only a (multiple) simple loop but simultaneously allows one to read off a shortest representative as a cyclic word in $\pi_1(\Sigma)$. Thus the space $\mathcal{ML}(\Sigma)$ is partitioned into finitely many maximal cells corresponding to weightings on the (finitely many) possible π_1-train tracks associated to a given fundamental domain for Σ_2.

In this paper we study the special case of the twice punctured torus Σ_2. We use this structure to construct a Markov map f on $\mathcal{ML}(\Sigma_2)$. The Markov partition is essentially the set of maximal cells and the restriction of f to each cell is a specific (rather simple) element of $\mathcal{MCG}(\Sigma_2)$ which acts linearly on the set of weights. Labelling the cells by the corresponding elements of $\mathcal{MCG}(\Sigma_2)$, we show that the f-expansions (that is the labelled orbit paths of f, see Section 1.1) give a unique normal form for the elements of $\mathcal{MCG}(\Sigma_2)$. In particular, the labels are a set of generators for $\mathcal{MCG}(\Sigma_2)$, and comparison of normal forms for nearby elements allows us to find a presentation for $\mathcal{MCG}(\Sigma_2)$. Since the map f is Markov, the f-expansions lie in a subshift of finite type which is in fact close to geodesic with respect to the set of generators in question. In the language of automatic groups (see Section 1.2 below), these normal forms for elements of $\mathcal{MCG}(\Sigma_2)$ allow us to construct a word acceptor. If we can show that these normal forms satisfy the fellow traveller property then this gives an explicit automatic structure on $\mathcal{MCG}(\Sigma_2)$. We conclude the paper by showing that this is indeed the case. Our method is rather similar to that given by Mosher [16] who shows that any mapping class group has an automatic structure.

The dimension of $\mathcal{PML}(\Sigma)$ is necessarily odd. We choose to study the case of the twice punctured torus because it is one of the few three dimensional examples. The details of the construction are rather special; we conjecture that the underlying principles are not. One of the main obstacles to finding a complete generalisation of these techniques is the difficulty of finding a map f and a Markov partition of $\mathcal{PML}(\Sigma)$ suitably related to the piecewise linear structure on $\mathcal{PML}(\Sigma)$.

An illuminating discussion can be made for the once punctured torus Σ_1, where the dimension of $\mathcal{PML}(\Sigma_1)$ is 1. Here the Teichmüller space is the upper half plane and the mapping class group is PSL$(2, \mathbb{Z})$. Although the final results are familiar, the methods may be of interest, and we begin by presenting this example in some detail to explain our ideas. Presumably similar methods would prove the automaticity result for Bowen-Series expansions in the Fuchsian group case.

The outline of the paper is as follows. Section 1 draws together the necessary background material from a variety of sources. Section 2 gives the construction for the once punctured torus. Sections 3, 4 and 5 extend this construction to the twice punctured torus. Section 3 contains an explicit development of the Birman-Series construction for Σ_2. Section 4 gives the action of the mapping class group and the construction of the Markov map f. In Section 5 we construct a word difference machine which shows that the f-expansions satisfy the fellow traveller property.

The results in this paper arose out of discussions between the authors and Linda Keen as part of a project to understand the Maskit embedding of the twice punctured torus, [12]. We would also like to thank David Epstein and Sarah Rees for helpful discussions about automatic groups. Part of this research was carried out while the first author was supported by a S.E.R.C./E.P.S.R.C. Research Fellowship held at the University of Warwick in the period 1992 – 1994.

Index of symbols used

δ_j	a Dehn twist on the once or twice punctured torus: Sections 1.5, 2.2, 3.2.
ι_j	a symmetry of the once or twice punctured torus: Sections 2.2, 3.2.
ρ_j	the composition of Dehn twists $\delta_j \delta_0 \delta_j$ for $j = 1, 2$: Sections 4.6, 5.3, 5.4.
ϕ	an element of the mapping class group $\mathcal{MCG}(\Sigma_j)$.
ψ	a word difference: Sections 1.2, 2.5, 5.
e	the identity element of a group.
$\mathbf{e}_j$, $\mathbf{e}^i_j$	irreducible loops on the once or twice punctured torus: Sections 2.1, 3.1.
f_j	a Markov map on $\mathcal{ML}(\Sigma_j)$.
I_j	maximal cells for $\mathcal{ML}(\Sigma_1)$: Section 2.1.
Δ_j	maximal cells for $\mathcal{ML}(\Sigma_2)$: Section 3.1.
Σ_j	the j times punctured torus, for $j = 1, 2$.
A_j, B_j, ...	regions in $\mathcal{ML}(\Sigma_2)$ or $\mathcal{F}$: Sections 4.1, 4.5.
Q_j, R_j, ...	the union of several regions in $\mathcal{F}$: Sections 2.4, 5.2.
X°	the interior of a set X.
$\mathcal{A}$	the alphabet for a word acceptor: Sections 1.2, 2.4, 4.7.
$\mathcal{D}$	the collection of word differences: Sections 1.2, 2.5, 5.1, 5.5.
$\mathcal{F}$	the set of Farey blocks (pairs) for Σ_2 (or Σ_1): Sections 2.3, 4.3.

1. Background

In this section we gather together all of the background material we need. This is taken from a variety of different areas. Much of the material is expository in nature.

1.1. Markov Maps and the Bowen-Series construction

A *Markov map* on a space X is a map $f : X \to X$, together with a finite (or in certain cases infinite) partition of X into sets X_i, such that $f(X_i)$ is an exact union of sets X_j. We say that f satisfies the *Markov property*: if $f(X_i^\circ) \cap X_j^\circ$ is non empty then $X_j \subset f(X_i)$. For $\xi \in X$, let $p(\xi) = j$ if $\xi \in I_j$. The sequence $p(\xi), p(f(\xi)), p(f^2(\xi)), p(f^3(\xi)), \ldots$ is called the *f-expansion* of ξ. Associated to f is a transition matrix of zeros and ones recording which transitions between states can occur. Since f is Markov, all infinite sequences with allowable transitions occur. The finite blocks which occur in these expansions are called *admissible*. Often a Markov map is required to be expanding or to have other differentiability properties. Such questions will not concern us here.

The Bowen-Series construction was modelled on the relationship between $\mathrm{PSL}(2, \mathbb{Z})$ acting in the upper half plane model of the hyperbolic plane $\mathbb{H}^2$ and the continued fraction transformation acting on the extended real line $\mathbb{R} \cup \{\infty\}$. The continued fraction map

$$f(x) = \begin{cases} x - 1 & \text{if } x \geq 1 \\ x + 1 & \text{if } x \leq -1 \\ \frac{-1}{x} & \text{if } |x| \leq 1 \end{cases}$$

can be regarded as an example of a Markov map with the partition Π into intervals $[\infty, -1], [-1, 0], [0, 1], [1, \infty]$. (For simplicity here and in what follows we omit details about endpoints. As defined above, the map f is 2-valued at the endpoints.) The f-expansion of a point $\xi \in \mathbb{R}$ is essentially the same as its continued fraction expansion. We note that the restriction of f to each element of Π belongs to the finite subset $\Gamma_0 = \{x \mapsto x - 1, x \mapsto x + 1, x \mapsto -1/x\} \subset \mathrm{PSL}(2, \mathbb{Z})$. (The well known fact that Γ_0 is a generating set for $\mathrm{PSL}(2, \mathbb{Z})$ may be proved using these expansions [21].) Via f-expansions, $\mathbb{R} \cup \{\infty\}$ may be mapped in an obvious way into $\Pi_{n=0}^{\infty}\Gamma_0$, giving an alternative viewpoint in which points in $\mathbb{R}$ are regarded as infinite words in Γ_0 [19]. Furthermore, the finite admissible blocks which occur in these f-expansions give an elegant and well known solution to the word problem in $\mathrm{PSL}(2, \mathbb{Z})$ [19]: each finite admissible block is a shortest word relative to the generators Γ_0 and every element in PSL(2, $\mathbb{Z}$) occurs as an admissible block in precisely one way.

This construction was generalised in [4] to the case of an arbitrary Fuchsian group Γ acting in the disc model of the hyperbolic plane with a given geometric set of generators Γ_0. (See [21] for the best exposition.) This involves the construction of a Markov map f on the boundary at infinity, the unit circle S^1. The elements of the partition were intervals I_j, and, for each j, the restriction $f|_{I_j}$ was in Γ_0. The f-expansions carried full information about the Γ-action on S^1, in the sense that two points were in the same Γ orbit if and only if the "tails" of their f-expansions agreed. Furthermore, and this is the point of interest here, these f-expansions simultaneously generate a most elegant solution to the word problem in Γ [21]. If to each partition interval is associated $f|_{I_j} \in \Gamma_0$, the f-expansions map to a set of infinite sequences in Γ_0. The finite admissible blocks in the f-expansion, give unique shortest representatives for words in Γ relative to the generators Γ_0. Clearly, this comes very close to saying they generate an automatic structure for Γ.

1.2. Automatic groups

In this section we give the properties of automatic groups that will be used later. More general references to this and related material are the books of Epstein et al [7] and Holt [9], to which the reader is referred for more details. See also [10], [18].

An *alphabet* $\mathcal{A}$ is a finite set. A *language* $\mathcal{L}$ over an alphabet $\mathcal{A}$ is a collection of finite sequences of elements of $\mathcal{A}$ (called *words* or *strings*). The *length* of a string $w = (a_1, \dots , a_n)$ is $|w| = n$.

For the purpose of this paper a *finite state automaton* over an alphabet $\mathcal{A}$ is a finite, directed, edge labelled graph whose vertices are called *states* and whose directed edges are called *arrows*. There is a specified state called the *start state* and a partition of the states into two disjoint sets, the *accept states* and the *non-accept states*. Every arrow from a state is labelled with a symbol from $\mathcal{A}$ and no two arrows from the same state have the same label. Given any string $w = (a_1, \dots , a_n)$ over $\mathcal{A}$ and any state s there is at most one path of arrows starting at s so that the jth arrow is labelled with a_j. This path terminates at some state s'. We say that w goes from s to s'.

The language *accepted* by this automaton is the collection of strings w over $\mathcal{A}$ which go from the start state to some accept state. A language $\mathcal{L}$ over $\mathcal{A}$ is called *regular* if it is accepted by some finite state automaton over $\mathcal{A}$ and this automaton is said to *recognise* the language $\mathcal{L}$.

Let G be a group with identity element e. Consider an alphabet $\mathcal{A}$ and a map $\mathcal{A} \to G$ denoted by $a \mapsto \overline{a}$. This extends to a map from the collection of strings over $\mathcal{A}$ to G by $w = (a_1, \dots, a_n) \mapsto \overline{w} = \overline{a}_1 \dots \overline{a}_n$, the product of the images of the letters in w. If every element of G can be described in this way we call $\mathcal{A}$ a finite generating set for G. A language $\mathcal{L} = \mathcal{L}(G)$ over $\mathcal{A}$ is called an *automatic structure* for G if two conditions are satisfied. First, $\mathcal{L}(G)$ is a regular language which maps onto G. That is, there is a finite state automaton so that every element of G may be described by (at least) one path through this automaton. This automaton is called the *word acceptor* $\mathcal{W}(G)$. The second property is known as the fellow traveller property which we explain below. If a group G has an automatic structure then G is called an *automatic group*.

For a group G with finite generating set $\mathcal{A}$, the *word length* of $g \in G$ denoted $|g|$ with respect to $\mathcal{A}$ is the shortest length of any word in $\mathcal{A}$ representing g. The *word metric* on G is $d(g, h) = |g^{-1}h|$. Given a word $w = (a_1, \dots, a_n)$ over $\mathcal{A}$, for each integer $0 \le t \le n$, denote by $w(t) = (a_1, \dots, a_t)$ the prefix of w of length t, and for integers $t \ge n$ denote $w(t) = w$. Given a constant k, two words w, v over $\mathcal{A}$ are *k-fellow travellers* if $d\big(w(t), v(t)\big) \le k$ for all $t \ge 0$. Also k is called the fellow traveller constant for w and v. The group G satisfies the *fellow traveller property* if there is a constant k such that for any words $w, v \in \mathcal{L}$ with $d(\overline{w}, \overline{v}) \le 1$ then w and v are k-fellow travellers.

Let w and v be a pair of words as above and $a \in \mathcal{A} \cup \{e\}$ so that $\overline{wa} = \overline{v}$. If w and v are k-fellow travellers then $\psi(t) = w(t)^{-1}v(t)$ has length at most k for all $t \ge 0$. Thus for all choices of w and v with $d(\overline{w}, \overline{v}) \le 1$ the $\psi(t)$ lie in a finite set $\mathcal{D}$, the collection of *word differences* and $\mathcal{A} \cup \{e\} \subset \mathcal{D} \subset \mathcal{L}$. Knowledge of the word differences allows us to reconstruct the multiplicative structure of G in an automated way.

More precisely, the fellow traveller property is equivalent to the existence of *multiplier automata* $\mathcal{M}_a$ for each $a \in \mathcal{A} \cup \{\$\}$ for G [7], [10]. Each $\mathcal{M}_a$ is a 2-*stringed automaton* whose alphabet is $\mathcal{A}' \times \mathcal{A}'$, where $\mathcal{A}'$ is the *padded alphabet* $\mathcal{A} \cup \{\$\}$. It accepts the *padded pair* (w^+, v^+) for strings (w, v) over $\mathcal{A}$ whenever w, v are accept states of $\mathcal{W}(G)$ and $\overline{wa} = \overline{v}$. Here the symbols w^+, v^+ indicate that the padding symbol \$, which maps to the identity in G, may be added to the shorter of w, v to make them have equal length. The automaton $\mathcal{M}_\$$ recognises identity in G, replacing the condition $\overline{wa} = \overline{v}$ by $\overline{w} = \overline{v}$.

The multiplier automata $\mathcal{M}_a$ for $a \in \mathcal{A} \cup \{\$\}$ may be constructed by means of a *word-difference machine*, clearly explained in [10] and summarised here. This is really a collection of new automata, all of which have the same state space, namely the set of triples (s_1, s_2, ψ) such that s_1, s_2 are states of $\mathcal{L}(G)$ and $\psi \in \mathcal{D}$. The start state is (s_0, s_0, e) where s_0 is the start state of $\mathcal{L}(G)$ and e is the identity of G. For $a, b \in \mathcal{A}$ there is an arrow from (s_1, s_2, ψ) to (s_1', s_2', ψ') if and only if there are arrows $s_1 \xrightarrow{a} s_1'$ and $s_2 \xrightarrow{b} s_2'$ in the word acceptor and if $\overline{\psi'} = \overline{x^{-1}\psi y}$. In the automaton $\mathcal{M}_a$, the state (s_1, s_2, d) is a success state if s_1, s_2 are in $\mathcal{L}$ and if $\bar{\psi} = \bar{a}$.

There is an extra technicality needed to deal with the padding symbol. Namely, we have to add an extra state to the word acceptor $\mathcal{W}(G)$ which is reached when $\mathcal{W}(G)$ is in an accept state and the padding symbol is read. If either of s_1 or s_2 is this extra state, then one or other of x, y as above will be replaced by the padding symbols \$ and the condition $\overline{\psi'} = \overline{x^{-1}\psi y}$ will be replaced by $\overline{\psi'} = \overline{\psi y}$ or $\overline{\psi'} = \overline{x^{-1}\psi}$. Frequently we shall think of these conditions as commutative squares or triangles of relations between elements in the group.

If we can construct a word difference machine using a finite set of word differences $\mathcal{D}'$, then we have clearly verified the fellow traveller property and can use the above process to simultaneously construct all the multiplicative automata $\mathcal{M}_a$. The process can be seen as concatenating squares or triangles to yield a collection of cross paths in $\mathcal{D}$ between all the prefixes $w(t)$ and $wa(t)$ occurring in the normal forms of any two words w and wa for $a \in \mathcal{A}$. The collection of all those squares and triangles which arise is easily seen to give a presentation for the group.

We can often use a Markov map to construct a word acceptor. This is analogous to Mosher's construction of a word acceptor by reversing the combing process [16]. Suppose that we have a (fixed point) free action of a group G on a space X and a Markov map f defined with respect to a partition $\{X_i\}$ of X so that on X_i the map f is some element of G. Suppose that there exists a particular $x \in X$ so that for each $g \in G$ there is a non-negative integer n so that $f^n(gx) = x$, and so that $n = 0$ if and only if g is the identity e.[1] Then we can use f to define a word acceptor for G as follows. There is a special start state corresponding to x and there is one state for each X_i in the Markov partition. For each X_i suppose that $f|_{X_i} = \alpha_i$ and $f|_{X_i} = \alpha_i : X_i \to X_j \cup \cdots \cup X_k$. We draw an arrow from the each of the states $X_j, \ldots, X_k$ to X_i with the label $\alpha_i{}^{-1}$. This means that all the arrows arriving in each state have the same label and all arrows leaving each state have different labels. (Strictly speaking, there should be arrows leaving each state with every label in the alphabet. If there are letters in the alphabet that do not occur as labels leaving a particular state X_i then we draw arrows from X_i with these labels to a new state called the *fail state*. All arrows leaving the fail state return there. In practice we do not use the fail state and will omit all arrows leading there.) In order to read a normal form

[1] This is very close to the property of *orbit equivalence*: f is said to be orbit equivalent to G on X if for any $x, y \in X$ then $x = gy$ for some $g \in G$ if and only if $f^n x = f^m y$ for some $m, n \geq 0$.

for $g \in G$ we consider the word in G obtained by inverting the composition of the particular values of f arising from $f^n(gx) = x$. This is the same as path through the word acceptor corresponding to g.

We shall find a suitable set of word differences $\mathcal{D}$ by starting from the generating set $\mathcal{A}$ and successively adding more words ψ' as dictated by the conditions $\overline{\psi'} = \overline{x^{-1}\psi y}$ until the collection we arrive at becomes closed under further moves of this kind. The method is similar to Mosher's construction of "raising bems" [16]. Since the states of the word acceptor are elements of the partition of X, the states of the difference machine are elements (X_i, X_j, ψ) for X_i, X_j in the Markov partition and $\psi \in \mathcal{D}$. The new relations will be of the form $\overline{\psi'} = \overline{\alpha_i{}^{-1}\psi\alpha_j}$ where $\alpha_i = f|_{X_i}$. We shall also allow degenerate squares or triangles corresponding to pairs of states (X_i, X_j, ψ), (X_i', X_j, e) with $\psi = f|_{X_i}$, giving the trivial relation $\overline{e} = \overline{\psi^{-1}\psi}$. During this process of adding new word differences, it will unfortunately sometimes be necessary to subdivide some of the states X_i. This is because the various word differences $\psi \in \mathcal{D}$ may map the state X_i to a number of different states on which the definition of f varies, thus possibly introducing several different variants of the relation $\overline{\psi'} = \overline{\alpha_i^{-1}\psi\alpha_j}$. Technically, this means that we have to add new states (Y_{i_k}, Y_{j_l}, ψ) to the word difference automaton, where the new sets Y_{i_k}, Y_{j_l} are certain subsets of X_i and X_j. However, these difficulties are also resolved after a finite number of steps, and it should be clear that from the resulting collection of squares and triangles we can construct automata as required.

1.3. Multiple simple loops and π_1-train tracks

In what follows we do not use the conventional Thurston theory of train tracks (for which see [17], [22]) but a variant due to Birman and Series [2]. We will only be concerned with punctured surfaces and there the theory is much easier. Thus we restrict our attention to this case. For details see also [12].

A *loop* on a surface Σ is a closed curve. A loop is called *simple* if it has no self intersections. A loop is *boundary parallel* or *peripheral* if it is homotopic to a loop around a puncture. A *multiple simple loop* is a collection of pairwise disjoint simple loops none of which is either homotopically trivial or boundary parallel. For the p times punctured torus Σ_p the maximal number of non-trivial homotopy classes of disjoint, non-boundary parallel curves is p. Thus a multiple simple loop γ on Σ_p can be written as $m_1\gamma_1 + \cdots + m_p\gamma_p$ where m_j is a non-negative integer and the γ_j are distinct homotopy classes of simple closed curves on Σ_p.

For definiteness, fix a choice of hyperbolic structure on Σ and let $R \subset \mathbb{H}^2$ be a fundamental region for our surface whose vertices are all at punctures of the surface.[2] Suppose R has sides σ_k and side pairing maps $\mu_k : \sigma_k \mapsto \sigma_{k'}$ where $\mu_{k'} = \mu_k{}^{-1}$ for each k. Let $\overline{R}$ be the closure of R in $\mathbb{H}^2$. A π_1-*train track* τ is a collection of pairwise disjoint arcs $\alpha_j : [0, 1] \to \overline{R}$ so that

(i) $\alpha_j(0) \in \sigma_k$ and $\alpha_j(1) \in \sigma_l$,
(ii) $\alpha_j(\lambda) \in R^\circ$ for $\lambda \in (0, 1)$,
(iii) at most one arc joins each pair of sides,
(iv) no arc goes from one side to itself. That is, if k and l are as in (i) then $k \neq l$.

An arc of τ is called a *corner arc* if it joins adjacent sides of R. Each corner arc faces a particular vertex of R and for each vertex cycle in the side pairing of R we have the corresponding *corner cycle* consisting of all corner branches corresponding to the same puncture. A *weighting* w on a π_1-train track τ is an assignment of a non-negative number $w(\alpha_j)$ to each arc α_j of τ. A weighting is *integral* if each weight is a (non-negative) integer. We define the *length* of w, denoted $|w|$, as $|w| = \sum w(\alpha_j)$ where the sum is over all arcs α_j of τ.

We now explain how to collapse a multiple simple loop to obtain a π_1-train track with an integral weighting. We begin by lifting the multiple simple loop γ to the fundamental region R. The multiple simple loop becomes a collection of arcs, called *strands*, joining sides of R. We say that a multiple simple loop γ is *supported* on a π_1-train track τ if, for every strand of γ there is an arc of τ joining the same pair of sides. If γ is supported on τ we may give τ an integral weighting w_γ by assigning to each arc of τ the number of strands of γ joining that pair of sides. This weighting has the following properties (see [2], [12]):

(i) For each side pairing $\mu_k : \sigma_k \to \sigma_{k'}$, the sum of the weights of arcs with endpoints on σ_k is the same as the sum of the weights of arcs with endpoints on $\sigma_{k'}$.
(ii) At least one arc in each corner cycle must have weight zero.

The first condition holds because when we perform the gluing coming from μ_k each endpoint of a strand of γ on σ_k is identified with the endpoint of a strand on $\sigma_{k'}$. Thus the total numbers of endpoints on this pair of sides are the same. The second condition holds because, if not, the strands in corner cycle would join up to give a peripheral loop in γ. If a (non-negative but not necessarily integral) weighting satisfies (i) and (ii) we call it a *proper weighting*.

[2] Our results are combinatorial in nature and hence independent of the particular hyperbolic structure chosen. Nevertheless, since the theory of π_1-train tracks involves hyperbolic geometry, some choice needs to be made.

Conversely, every proper integral weighting w on a π_1-train track τ gives rise to a multiple simple loop γ. This means that in order to study multiple simple loops it is sufficient to study proper integral weightings on π_1-train tracks. Let $W(\tau)$ denote the collection of all proper weightings on the π_1-train track τ and $W_{\mathcal{O}}(\tau)$ the collection of proper integral weightings on τ (see [12]).

1.4. Irreducible loops and $\mathcal{PML}$

A π_1-train track τ is said to be *recurrent* (see [22]) if there exists a proper integral weighting $w \in W_{\mathcal{O}}(\tau)$ so that $w(\alpha_j)$ is non-zero for all branches α_j of τ. Such a π_1-train track τ is said to be *maximal* if there does not exist a recurrent π_1-train track τ' so that τ is properly contained in τ' in the obvious sense. It follows from Thurston's theory, or as one can directly verify in the special cases of concern to us here, that if τ is a maximal recurrent train track then the dimension of $W(\tau)$ is $6g - 6 + 2p$ where Σ is a surface of genus g with p points removed. We call the collection of all proper weightings $W(\tau)$ on a maximal recurrent π_1-train track τ a *maximal cell*. Any simple loop γ defines a recurrent π_1-train track $\tau(\gamma)$ with weights $w(\gamma)$ as above. A simple loop γ is said to be *irreducible* if $w(\gamma) \neq w_1 + w_2$ for any $w_1, w_2 \in W_{\mathcal{O}}\big(\tau(\gamma)\big)$ and $w_j \neq 0$ for $j = 1, 2$. Clearly there are only finitely many maximal recurrent π_1-train tracks. We shall see below that if Σ_p is the p-times punctured torus, for $p = 1, 2$, each maximal cell is the linear span of $2p$ irreducible loops.

We denote the collection of all homotopy classes of multiple simple, non-boundary parallel loops on Σ by $\mathcal{ML}_{\mathcal{O}}(\Sigma)$ and the collection of all measured geodesic laminations on Σ by $\mathcal{ML}(\Sigma)$. It is a theorem of Birman and Series [2] that $\mathcal{ML}_{\mathcal{O}}(\Sigma)$ and $\mathcal{ML}(\Sigma)$ can be identified with the collections of proper integral weightings and proper weightings respectively on π_1-train tracks on Σ. For $\mathcal{ML}_{\mathcal{O}}(\Sigma)$ the proof of this follows the outline given above. If $w \in \mathcal{ML}(\Sigma)$ then clearly w is contained in some maximal cell $W(\tau)$. Thus $\mathcal{ML}(\Sigma)$ is the union of maximal cells $W(\tau_i)$ where τ_i runs over all maximal recurrent π_1-train tracks on Σ. This gives $\mathcal{ML}(\Sigma)$ a natural cell structure. In the cases we are interested in, namely Σ_1 and Σ_2, we shall prove (Propositions 2.1.1 and 3.1.2) the following result.

Proposition 1.4.1. *For $p = 1, 2$ let Σ_p denote the p times punctured torus. There are finitely many irreducible loops $\mathbf{e}_1, \dots, \mathbf{e}_k$ on Σ_p so that for each maximal π_1-train track τ the corresponding maximal cell $W(\tau)$ is the positive linear span of $2p$ irreducible loops: $W(\tau) = \mathrm{sp}^+\{\mathbf{e}_{i_1}, \dots, \mathbf{e}_{i_{2p}}\}$, where $i_j \in \{1, \dots, k\}$. Also the intersection of two cells is*

$$W(\tau) \cap W(\tau') = \mathrm{sp}^+\left(\{\mathbf{e}_{i_1}, \dots, \mathbf{e}_{i_{2p}}\} \cap \{\mathbf{e}_{i'_1}, \dots, \mathbf{e}_{i'_{2p}}\}\right),$$

where

$$W(\tau) = \mathrm{sp}^{+}\{\mathbf{e}_{i_1}, \ldots, \mathbf{e}_{i_{2p}}\} \text{ and } W(\tau') = \mathrm{sp}^{+}\{\mathbf{e}_{i'_1}, \ldots, \mathbf{e}_{i'_{2p}}\}.$$

The space $W(\tau)$ may be projectivised in a natural way to obtain $\mathcal{P}W(\tau)$ and similarly $W_{\mathcal{O}}(\tau)$ can be projectivised to obtain the set of rational weightings $\mathcal{P}W_{\mathcal{O}}(\tau)$. The space $\mathcal{PML} = \mathcal{PML}(\Sigma)$ is the union over all τ of the corresponding cones $\mathcal{P}W(\tau)$, which we call π_1-cones, glued along their lower dimensional common simplices as in the proposition. We denote the union of all rational weightings $\mathcal{P}W_{\mathcal{O}}(\tau)$ by $\mathcal{PML}_{\mathcal{O}}(\Sigma)$. Using the Birman-Series identification, the space $\mathcal{PML}(\Sigma)$ can be naturally identified with the space of projective measured laminations on Σ, shown by Thurston to be a sphere of dimension $6g - 7 + 2p$ [17]. Thus in our two examples, we expect $\mathcal{PML}(\Sigma_1)$ and $\mathcal{PML}(\Sigma_2)$ to be S^1 and S^3 respectively. In each example we shall first determine the maximal cones. Gluing up using Proposition 1.4.1 will allow us to see explicitly how the spheres S^1 and S^3 are formed.

1.5. Dehn twists and the mapping class group

The (orientation preserving) *mapping class group* $\mathcal{MCG} = \mathcal{MCG}(\Sigma)$ of Σ is the group of isotopy classes of (orientation preserving) automorphisms of Σ, see [1] or [16] for example. That is, an element of $\mathcal{MCG}$ is an (orientation preserving) homeomorphism of Σ to itself and two such homeomorphisms give the same element of $\mathcal{MCG}$ if one can be deformed to the other isotopically along a continuous path of homeomorphisms of Σ to itself. There is a natural action of the mapping class group on Teichmüller space of Σ as the Teichmüller modular group. This action can be extended to $\mathcal{PML}(\Sigma)$, [23], and it is this action we will consider here.

Let w be a simple closed curve on a surface Σ parametrised by $\xi \in [0, l_w]$ where l_w is the length of w. Consider a small tubular neighbourhood around w in Σ and denote this by $N_w = [0, 1] \times w$. We define a homeomorphism of Σ called (left) *Dehn twist about* w (see [1] for example) denoted δ_w as the identity on $\Sigma - N_w$ and by requiring that $(\eta, \xi) \in [0, 1] \times w$ is mapped by δ_w to $(\eta, \xi - \eta l_w)$ where $\eta \in [0, 1]$ and $\xi - \eta l_w$ is defined mod l_w. Observe that if $\eta = 0$ or 1 then δ_w is the identity. By a well known result of Dehn [6], the (orientation preserving) mapping class group is generated by Dehn twists.

We will produce a set of Dehn twists which we shall show are generators for the mapping class groups of Σ_1 and Σ_2. We will then investigate the action of these Dehn twists on the piecewise linear structure on $\mathcal{PML}(\Sigma_p)$ given by π_1-train tracks. In particular we show that the Dehn twists act piecewise linearly on $\mathcal{PML}(\Sigma_p)$ with respect to this piecewise linear structure. Moreover, the action also restricts to an action on $\mathcal{PML}_{\mathcal{O}}(\Sigma_p)$. This gives a piecewise linear action of $\mathcal{MCG}$ on $\mathcal{PML}$ and $\mathcal{PML}_\mathcal{O}$ respectively. This action is not free in the sense that there are elements of $\mathcal{PML}_\mathcal{O}$ which have non-trivial stabilisers in $\mathcal{MCG}$. For example, performing a Dehn twist about w fixes w and any curve disjoint from w. In order to construct a Markov map whose orbits describe $\mathcal{MCG}$ we need to find a (fixed point) free action of $\mathcal{MCG}$ on a suitable space Y. For the case of the once punctured torus, Y will be the space of (ordered) Farey neighbours, that is, pairs of curves which intersect exactly once. For the twice punctured torus we will generalise this idea by defining quadruples of curves in a special topological configuration which we call Farey blocks. The space Y of Farey blocks will admit a free action of $\mathcal{MCG}(\Sigma_2)$.

2. The mapping class group of the once punctured torus

In this section we carry out the construction of π_1-train tracks, a Markov map and an automatic structure for the once punctured torus. Much of the material in this section is, to some extent, well known. However, we shall adopt a non-standard view point. The reasons for including this section are two-fold. First, the main structure of the argument is the same as for the twice punctured torus. Thus it will serve as motivation and a guide for what follows, explaining the main ideas with computations of a much more manageable scale. Secondly, when we are dealing with the twice punctured torus there are several steps in the construction of the Markov map and automatic structure. One of these steps is essentially the construction we present in this section. This will save us considerable effort later on. In Sections 2.1 and 2.2 we show that certain elementary Dehn twists act on $\mathcal{ML}(\Sigma_1)$ exactly like the continued fraction map on $\mathbb{R} \cup \{\infty\}$. In Section 2.3 we construct the Markov map and in Sections 2.4 and 2.5 we explain how it gives the automatic structure for $\mathcal{MCG}(\Sigma_1) = \mathrm{PSL}(2, \mathbb{Z})$.

2.1. π_1-train tracks and the cell structure of $\mathcal{ML}(\Sigma_1)$

As remarked in Section 1.3, we start by fixing a definite hyperbolic structure for Σ_1 and a fundamental domain $R_1 \subset \mathbb{H}^2$ for the action of $\pi_1(\Sigma)$ on the hyperbolic plane.

The fundamental domain R_1 we choose is the standard rectangular one with opposite sides identified by side pairings which match the midpoints of the sides. The region R_1 has four vertices all of which project to the puncture of Σ_1 (see Fig. 2.1.1). We label these v_1, v_2, v_3, v_4 in clockwise order. Writing $v_i v_j$ for the side joining v_i to v_j, the side pairings will be S carrying $v_1 v_2$ to $v_4 v_3$ and T carrying $v_1 v_4$ to $v_2 v_3$. The maps S and T correspond to homotopy

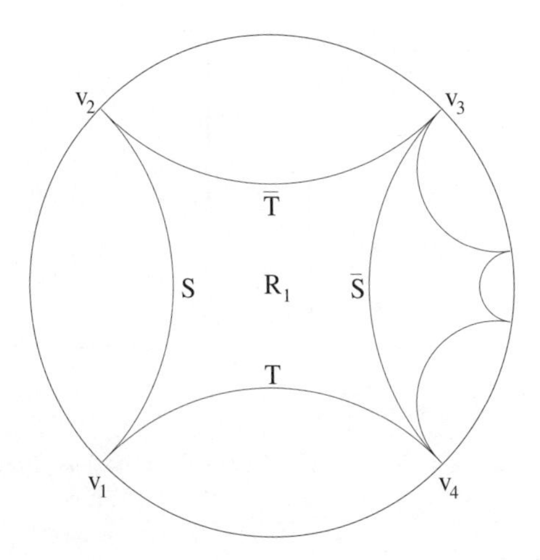

Figure 2.1.1 A hyperbolic fundamental domain R_1 for Σ_1, where $\overline{S}$, $\overline{T}$ denote S^{-1}, T^{-1}.

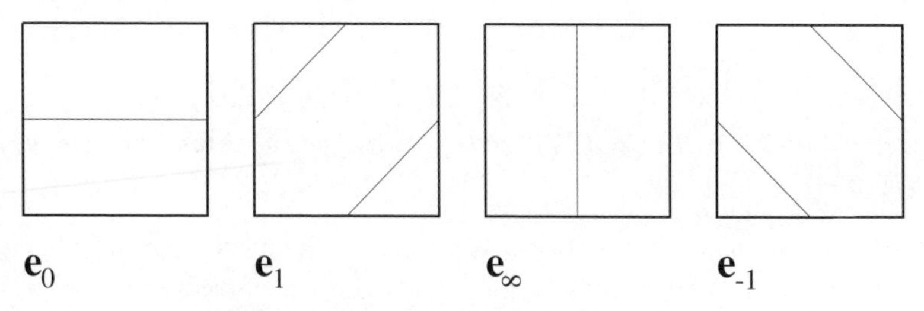

Figure 2.1.2 The elementary π_1-train tracks.

classes of simple closed curves which freely generate the fundamental group. We now introduce the irreducible loops that will form the basis for $\mathcal{ML}(\Sigma_1)$ as explained in Proposition 1.4.1. They are defined as follows (see Fig. 2.1.2.):[3]

- $\mathbf{e}_0$ consists of a single arc joining v_1v_2 and v_3v_4;
- $\mathbf{e}_1$ consists of an arc joining v_1v_2 and v_2v_3 and an arc joining v_3v_4 and v_4v_1;
- $\mathbf{e}_\infty$ consists of a single arc joining v_2v_3 and v_4v_1;
- $\mathbf{e}_{-1}$ consists of an arc joining v_2v_3 and v_3v_4 and an arc joining v_4v_1 and v_1v_2.

We mention in passing that one may also define these loops in terms of the cutting sequences as discussed in [3] or [21]: $\mathbf{e}_0 = S$, $\mathbf{e}_1 = ST$, $\mathbf{e}_\infty = T$ and $\mathbf{e}_{-1} = S^{-1}T$. (Since the loops are un-oriented, strictly speaking $\mathbf{e}_0 = S$ or S^{-1} and so on.)

[3] For simplicity we draw R_1 as a Euclidean rectangle in what follows.

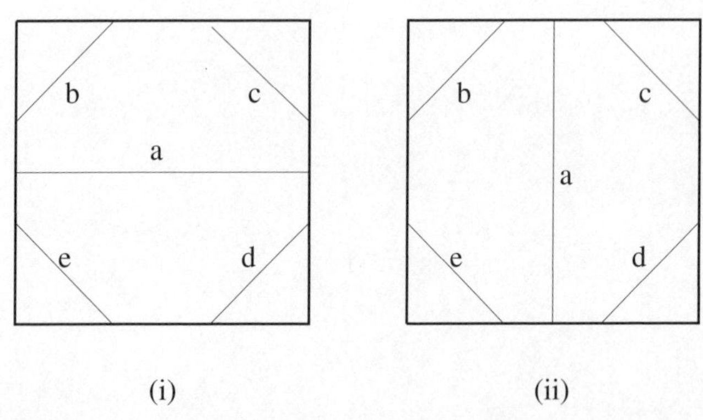

Figure 2.1.3 The two possible configurations for a maximal weighted π_1-train track on R_1.

Next, we define cells in $\mathcal{ML}(\Sigma_1)$. We show below that these are maximal. The cells are:

$$I_0 = \mathrm{sp}^+\{\mathbf{e}_0, \mathbf{e}_1\}, \quad I_1 = \mathrm{sp}^+\{\mathbf{e}_0, \mathbf{e}_{-1}\},$$

$$I_2 = \mathrm{sp}^+\{\mathbf{e}_\infty, \mathbf{e}_{-1}\}, \quad I_3 = \mathrm{sp}^+\{\mathbf{e}_\infty, \mathbf{e}_1\}.$$

Proposition 2.1.1. *The cells I_0, I_1, I_2, I_3 are maximal and their union is $\mathcal{ML}(\Sigma_1)$.*

Proof. It is sufficient to show that no extra arcs can be added to any of these four π_1-train tracks, and that any loop is supported on one of them. This is carried out in the appendix to [2]. For convenience we reproduce it here. Clearly, any maximal π_1-train track on Σ_1 must have one of the two forms illustrated in Fig. 2.1.3. Summing the weights over the two pairs of identified sides and cancelling a we obtain two equations

$$b + e = c + d, \quad b + c = e + d.$$

We may solve these to obtain $b = d$ and $c = e$. Now we know that on the corner cycle we cannot have all the weights non-zero. Thus $b = 0$ or $c = 0$. Since all the weights are non-negative, this means that there are four configurations of π_1-train track corresponding to non-boundary parallel, simple loops on Σ_1. □

Notation 2.1.2. By Proposition 2.1.1 any simple closed curve γ may be represented as $a\mathbf{e}_i + b\mathbf{e}_j$ where $w(\gamma) \in I_k = \mathrm{sp}^+\{\mathbf{e}_i, \mathbf{e}_j\}$ for some $k = 0, 1, 2, 3$. We always write the ordered pair (a, b) to represent $a\mathbf{e}_0 + b\mathbf{e}_1$ if $w(\gamma) \in I_0$, $a\mathbf{e}_0 + b\mathbf{e}_{-1}$ if $w(\gamma) \in I_1$, $a\mathbf{e}_\infty + b\mathbf{e}_{-1}$ if $w(\gamma) \in I_2$ or $a\mathbf{e}_\infty + b\mathbf{e}_1$ if $w(\gamma) \in I_3$.

The notation for the $\mathbf{e}_j$ has the following rationale. Regard R_1 as a square with v_1 in the bottom left hand corner, and side pairings which are Euclidean

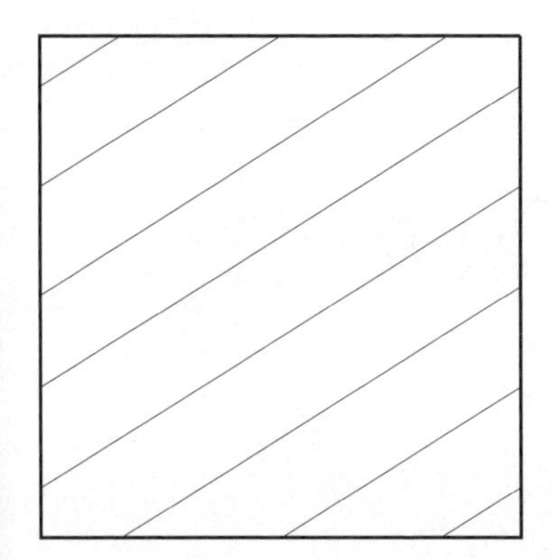

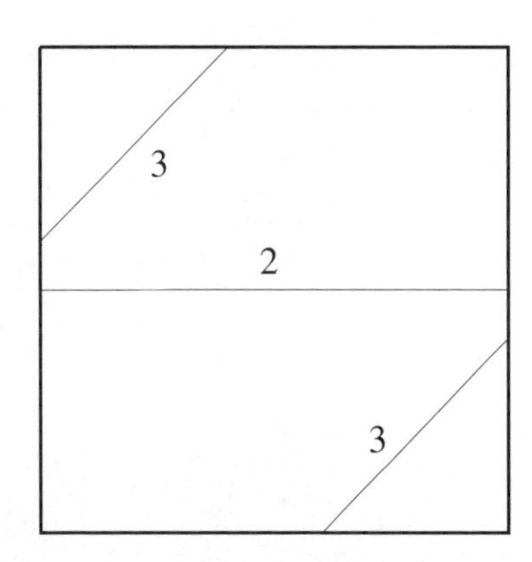

Figure 2.1.4 The line of slope $3/5$ drawn on R_1 and as a weighted π_1-train track.

translations. By Proposition 2.1.1 above, any simple closed curve γ on Σ_1 is supported on one of the four maximal cells I_k. Thus, up to homotopy, γ is equivalent to a family of parallel Euclidean straight lines across R_1. Lifting to the Euclidean universal cover of R_1, that is $\mathbb{R}^2$, such a family of lines links to form a line of rational slope on the plane. With this identification, it is clear that the curve we have labelled $\mathbf{e}_j$ has slope j.

More generally, we obtain an identification of $\mathcal{PML}(\Sigma_1)$ with the extended real line $\mathbb{R} \cup \{\infty\}$ by mapping $(a, b) \in I_0$ to the point $b/(a+b)$, $(a, b) \in I_1$ to $-b/(a+b)$, $(a, b) \in I_2$ to $-(a+b)/b$ and $(a, b) \in I_3$ to $(a+b)/b$. An example, the curve represented by $(2, 3) = 2\mathbf{e}_0 + 3\mathbf{e}_1 \in I_0$, is shown in Fig. 2.1.4. This corresponds to the line of slope $3/5$ in $\mathbb{R}^2$.

The maximal cells I_0, I_1, I_2, I_3 have their boundaries identified as in Proposition 1.4.1. In this case it is easy to see that $I_0 \cap I_1 = \mathrm{sp}^+\{\mathbf{e}_0\}$, $I_1 \cap I_2 = \mathrm{sp}^+\{\mathbf{e}_{-1}\}$, $I_2 \cap I_3 = \mathrm{sp}^+\{\mathbf{e}_\infty\}$ and $I_3 \cap I_0 = \mathrm{sp}^+\{\mathbf{e}_0\}$. The other two intersections are empty. This is illustrated in Fig. 2.1.5, from which one clearly sees that $\mathcal{PML}(\Sigma_1) \sim S^1$. We remark that in the appendix to [2], Birman and Series considered oriented curves and so found a different cell structure for $\mathcal{ML}(\Sigma_1)$.

2.2. Dehn twists and the mapping class group

Let the Dehn twists about $\mathbf{e}_\infty$ and $\mathbf{e}_0$ be denoted by δ_0 and δ_1 respectively. We now want to investigate the effects of these twists on the projective structure on the space of train tracks constructed in the previous section. In order to simplify things we will make use of some natural symmetries of R_1 and the π_1 train tracks we constructed above. These symmetries are defined as follows:

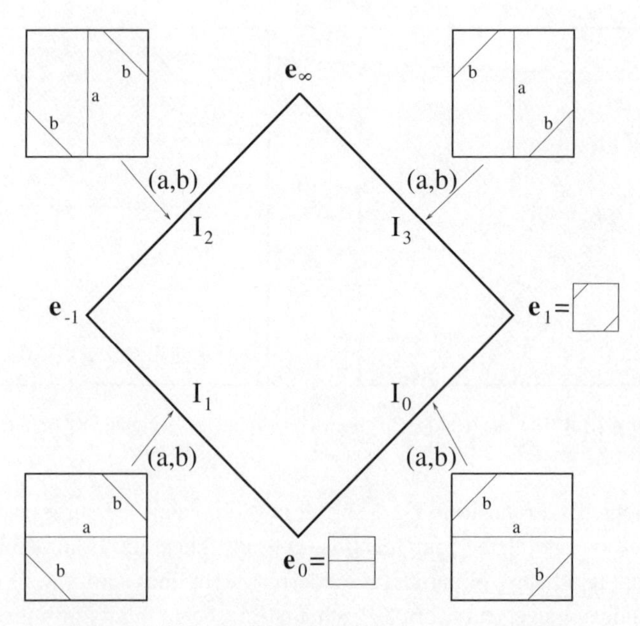

Figure 2.1.5 The partition of $\mathcal{ML}(\Sigma_1)$ into maximal cells.

- ι_1 interchanges the pairs (v_1, v_4), (v_2, v_3);
- ι_2 cyclically permutes the vertices sending v_1 to v_2, v_2 to v_3 and so on;
- ι_3 fixes v_1 and v_3 and interchanges (v_2, v_4).

In addition, we will sometimes write ι_0 for the identity map. Observe that applying ι_2 twice we get a rotation of R_1 by 180° which interchanges v_1, v_3 and v_2, v_4. Even though this is not the identity on R_1 it does act as the identity on each of the I_j. (This map is just the map which sends any curve to itself with the opposite orientation.) The ι_j act on $\mathcal{PML}(\Sigma_1)$ as the Klein 4-group. On these irreducible loops this action is given by:

$$\iota_1 : \mathbf{e}_0 \mapsto \mathbf{e}_0, \quad \mathbf{e}_1 \mapsto \mathbf{e}_{-1}, \quad \mathbf{e}_\infty \mapsto \mathbf{e}_\infty, \quad \mathbf{e}_{-1} \mapsto \mathbf{e}_1,$$

$$\iota_2 : \mathbf{e}_0 \mapsto \mathbf{e}_\infty, \quad \mathbf{e}_1 \mapsto \mathbf{e}_{-1}, \quad \mathbf{e}_\infty \mapsto \mathbf{e}_0, \quad \mathbf{e}_{-1} \mapsto \mathbf{e}_1,$$

$$\iota_3 : \mathbf{e}_0 \mapsto \mathbf{e}_\infty, \quad \mathbf{e}_1 \mapsto \mathbf{e}_1, \quad \mathbf{e}_\infty \mapsto \mathbf{e}_0, \quad \mathbf{e}_{-1} \mapsto \mathbf{e}_{-1}.$$

Note that the action of the symmetries extends naturally to the cells I_j. Moreover, the action is given by $\iota_j(I_0) = I_j$. The benefit of applying these symmetries is that we only need consider the action of δ_0 on $\mathcal{ML}(\Sigma_1)$. The action of $\delta_0{}^{-1}$ and $\delta_1{}^{\pm 1}$ will follow by symmetry as follows. We claim that

$$\iota_1\delta_0\iota_1 = \delta_0{}^{-1} \qquad \iota_2\delta_0\iota_2 = \delta_1 \qquad \iota_3\delta_0\iota_3 = \delta_1{}^{-1}$$

$$\iota_1\delta_1\iota_1 = \delta_1{}^{-1} \qquad \iota_2\delta_1\iota_2 = \delta_0 \qquad \iota_3\delta_1\iota_3 = \delta_0{}^{-1}.$$

This is because ι_1 and ι_3 reverse orientation and so conjugate right Dehn twists to left Dehn twists, while ι_2 preserves orientation but interchanges $\mathbf{e}_0$ and $\mathbf{e}_\infty$.

Applying a Dehn twist to a weighted π_1-train track sometimes results in an *unreduced* π_1-train track, that is a π_1-train track which may have arcs with both ends on the same edge of the fundamental domain. An unreduced π_1-train track satisfies conditions (i) – (iii) given in Section 1.3 but fails to satisfy (iv). The process of converting an unreduced (weighted) π_1-train track into a (reduced) π_1-train track is called *pulling tight*. Suppose that the unreduced π_1-train track τ has an arc α from the side σ_k to itself and that this arc has weight $w(\alpha)$. Suppose also that τ has a proper integral weighting. We begin by converting it into a multiple simple loop γ on R_1. This means that we replace each arc α_j with weight $w(\alpha_j)$ by $w(\alpha_j)$ strands joining the same pair of sides as α_j. In particular we have $w(\alpha)$ strands from σ_k to itself. We now perform a homotopy of Σ which will remove the intersections of all these strands with σ_k. This is done as follows. We can always choose an innermost strand β, which together with an arc of σ_k bounds a disc in R_1 containing no other strands. Suppose that the endpoints of β on σ_k are x_+ and x_-. Now consider the images of x_+ and x_- under the side pairing map μ_k. These are points of $\sigma_{k'}$ that are ends of strands β_+ and β_- respectively. (To find out which strands, put an orientation on σ_k and $\sigma_{k'}$ consistent with μ_k and then count endpoints from the corresponding ends of σ_k and $\sigma_{k'}$.) The other endpoints of β_+ and β_- are points y_+ and y_- on sides σ_+ and σ_-. We replace β, β_+ and β_- by a single strand β' from y_+ to y_-. It is clear that this strand can be drawn disjoint from the other strands of γ. We have reduced the number of strands by two. This process clearly terminates after a finite number of applications, giving a multiple simple loop which has no strands with both endpoints on σ_k. Repeating for all k gives a reduced π_1-train track on R_1 with a proper integral weighting.

Proposition 2.2.1. *Let $(a, b) \in I_0$. The Dehn twists act on I_0 as follows:*

$$\delta_0(a, b) = (b, a + b) \in I_3$$
$$\delta_0^{-1}(a, b) = (b, a) \in I_1$$
$$\delta_1(a, b) = \begin{cases} (a - b, b) \in I_0 \text{ if } a \geq b \\ (b - a, a) \in I_3 \text{ if } a \leq b \end{cases}$$
$$\delta_1^{-1}(a, b) = (a + b, b) \in I_0.$$

Proof. This follows directly from the linearity theorems in [2]. We include an alternative proof as an illustration of how we manipulate π_1-train tracks. It is illustrated in Fig. 2.2.1 below. We begin with the train track for a general integral point $(a, b) \in I_0$. We want to perform the Dehn twist $\delta_j^{\pm 1}$, for $j = 0, 1$, about the

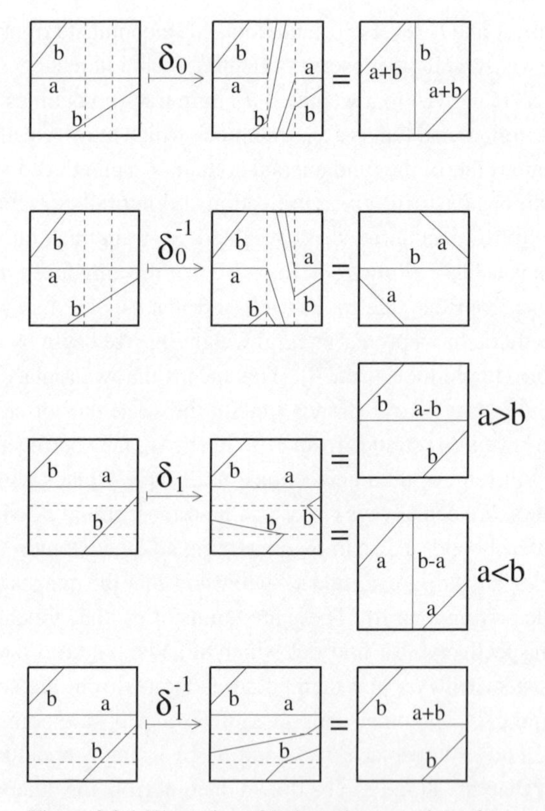

Figure 2.2.1 The action of Dehn twists on points of I_0.

curve γ_j which is either $\mathbf{e}_\infty$ or $\mathbf{e}_0$ respectively. We draw a tubular neighbourhood about γ_j as a strip going from one side to the opposite side. This strip is bounded by dotted lines in the figure. The Dehn twist is the identity outside this strip and inside the strip fixes one boundary component of the cylinder about γ_j while rotating the other component one whole turn. In between we interpolate linearly so that an arc of a train track which went straight across is now wrapped once around the cylinder before emerging on the other side. It still carries the same weight which represents the number of strands in the corresponding multiple simple loop. In the cases illustrated in the top or bottom diagrams in Fig. 2.2.1, all that remains is to gather together arcs whose endpoints lie on the same sides and add their weights. In the middle two cases in Fig. 2.2.1 we need an intermediate step. Namely the image train track is unreduced so that we need to pull tight. In the first case, that is $\delta_0{}^{-1}(a, b)$, the unreduced arc joins the bottom side to itself and has weight b. Convert the train track into a multiple simple loop by drawing $w(\alpha)$ strands joining the same pair of sides as α for each arc α with weight $w(\alpha)$. This gives b strands joining the bottom side to itself. Reading from the left, the first b endpoints of strands along the bottom edge are joined to the next b strands in the reverse order. Likewise reading from the left, the first b strands on the top side are ends of strands all joining the left hand side and the next b strands all join the right hand side. Thus we may pull all b simple loops tight at once by replacing all $3b$ of these strands by b strands joining the left and right sides. The result after converting back to a π_1-train track is shown in the right hand column. For $\delta_1(a, b)$ we perform the same process but the result is slightly more complicated. Now we have an arc of weight b joining the right hand side to itself. When we convert to a multiple simple loop this arc becomes b strands which, reading from the bottom, are the first b strands on the right hand side. These are joined to the next b strands on the same side. On the left hand side the first b strands from the bottom have their other endpoint on the bottom side. However we need to take care when finding the next b strands. If $a \geq b$ these next b strands join the right hand side and we may pull these loops tight to obtain b strands joining the bottom side to the right hand side. There were $a - b$ strands joining the left and right sides which we have not used and these remain after pulling tight. In the case where $a \leq b$ we can pull a strands tight in this way. There remain $b - a$ strands joining the right hand side to itself. Their other ends join the left and top sides. Thus after pulling tight we obtain $b - a$ strands joining the top and bottom sides and a remain joining the left and top sides. After reconverting to π_1-train tracks one obtains the result, again shown in the right hand column. To complete the proof for general weightings (a, b), note that we can clearly obtain the result for proper rational weightings by clearing denominators in $\mathcal{ML}_{\mathcal{O}}(\Sigma_1)$. The result for general $w(\alpha)$ follows by continuity. □

It is well known that the Dehn twists δ_0 and δ_1 generate the mapping class group $\mathcal{MCG}(\Sigma_1)$. In fact we have $\iota_2 = \delta_1\delta_0\delta_1$. Together with $\iota_2\delta_1\iota_2 = \delta_0$ and the fact that ι_2 has order 2 this immediately gives

$$\delta_1\delta_0\delta_1 = \delta_0\delta_1\delta_0, \quad (\delta_0\delta_1)^3 = e.$$

It turns out that this gives a presentation of $\mathcal{MCG}(\Sigma_1)$:

$$\mathcal{MCG}(\Sigma_1) = \langle \delta_0, \delta_1 \,|\, \delta_1\delta_0\delta_1 = \delta_0\delta_1\delta_0, (\delta_0\delta_1)^3 = e \rangle.$$

This may be seen either using standard facts about the modular group or can be deduced from the automatic structure given below, see Section 2.5. In order to obtain the identification of $\mathcal{MCG}(\Sigma_1)$ with PSL(2, $\mathbb{Z}$), consider the identification of $\mathcal{PML}(\Sigma_1)$ with $\mathbb{R} \cup \{\infty\}$ given in Section 2.1. It is easy to see that after making this identification, the action of δ_0 and δ_1 on $\mathbb{R} \cup \{\infty\}$ is given by

$$\delta_0 : x \mapsto x/(-x + 1), \quad \delta_1 : x \mapsto x + 1.$$

Notice that this is essentially the same as the continued fraction map explained in Section 1.1. We see that $\iota_2 : x \mapsto -1/x$. If we had been considering oriented curves as in the appendix of [2] it is clear that $\iota_2{}^2$ would fix each non-trivial, non-peripheral curve but reverse its orientation. This corresponds to the matrix $-I$ and we would have obtained an action of SL(2, $\mathbb{Z}$) rather than PSL(2, $\mathbb{Z}$).

2.3. The Markov map and Farey pairs

We will now define a Markov map f_1 on $\mathcal{ML}(\Sigma_1)$ from which we shall construct the desired automatic structure for $\mathcal{MCG}(\Sigma_1)$. The Markov partition of $\mathcal{ML}(\Sigma_1)$ will consist of the four maximal cells I_0, I_1, I_2 and I_3. These cells are closed and therefore intersect along their boundaries. This gives rise to ambiguities, but this will not present a problem. The map $f_1|_{I_j}$ will be chosen from $\{\delta_0{}^{\pm 1}, \delta_1{}^{\pm 1}\}$ in such a way that f_1 has the required Markov property.

Lemma 2.3.1. $\delta_1(I_0) = I_0 \cup I_3$.

Proof. By Proposition 2.2.1 we see that $\delta_1(I_0) \subset I_0 \cup I_3$. Also $\delta_1{}^{-1}(I_0) \subset I_0$ and $\delta_1{}^{-1}(I_3) = \iota_3\delta_0\iota_3(\iota_3 I_0) = \iota_3\delta_0(I_0) \subset \iota_3(I_3) = I_0$. This gives the result. □

We now define $f_1|_{I_0} = \delta_1$ and $f_1|_{I_j}$ by symmetry. In summary, f_1 is defined as

$$\begin{aligned}
f_1|_{I_0} &= \delta_1 &&: I_0 \mapsto I_0 \cup I_3\\
f_1|_{I_1} &= \delta_1{}^{-1} &&: I_1 \mapsto I_1 \cup I_2\\
f_1|_{I_2} &= \delta_0 &&: I_2 \mapsto I_1 \cup I_2\\
f_1|_{I_3} &= \delta_0{}^{-1} &&: I_3 \mapsto I_0 \cup I_3.
\end{aligned}$$

On a boundary $I_i \cap I_j$ the map is considered to be two valued. By Lemma 2.3.1 the map f satisfies the Markov property of Section 1.1. The following lemma will be crucial for constructing the automaton. Recall the definition of length given in Section 1.3. In this case we can see by direct inspection that if $w = (a, b) \in I_j$ then $|w| = a + 2b$ for $j = 0, 1, 2, 3$.

Lemma 2.3.2. *Let $w \in \mathcal{ML}_{\mathcal{O}}(\Sigma_1)$ be a proper integral weighting on a π_1-train track τ. Then $|f_1(w)| \leq |w|$ with equality if and only if $w = a\mathbf{e}_0$ or $a\mathbf{e}_\infty$ for $a \in \mathbb{N}$.*

Proof. This is easy to check from Proposition 2.2.1. □

The rough idea of the construction of the word acceptor for $\mathcal{MCG}(\Sigma_1)$ is to use the four cells I_0, I_1, I_2, I_3 as states and to define arrows using the transition matrix associated to the Markov map f_1. However there is a problem with this idea, namely $\mathcal{MCG}(\Sigma_1)$ does not act freely on $\mathcal{ML}(\Sigma_1)$. In other words, $\phi\gamma = \gamma$ for $\phi \in \mathcal{MCG}(\Sigma_1)$ and γ a simple loop on Σ_1 does not imply that ϕ is the identity. We therefore need to consider the action of $\mathcal{MCG}(\Sigma_1)$ on a space of slightly more elaborate objects on which the action is fixed point free. To this end, we introduce the notation of a Farey pair.

Two (homotopy classes of) simple closed curves on Σ_1 are called *Farey neighbours* if they (have representatives that) intersect exactly once. Notice that this condition automatically implies that these curves do not divide the punctured torus so neither of them can be boundary parallel or homotopically trivial. We consider ordered pairs of Farey neighbours (γ_1, γ_2) which we refer to as *Farey pairs*. It is clear that $(\mathbf{e}_0, \mathbf{e}_\infty)$ and $(\mathbf{e}_\infty, \mathbf{e}_0)$ are both Farey pairs. We denote the set of all Farey pairs by $\mathcal{F}$. It is easy to see the Farey pair $(\mathbf{e}_0, \mathbf{e}_\infty)$ has trivial stabiliser in $\mathcal{MCG}(\Sigma_1)$ and that, for any other Farey pair (γ_1, γ_2), there is an element ϕ of $\mathcal{MCG}(\Sigma_1)$ sending it to $(\mathbf{e}_0, \mathbf{e}_\infty)$. As $(\mathbf{e}_0, \mathbf{e}_\infty)$ has trivial stabiliser this element is unique. In particular, ι_2 sends $(\mathbf{e}_\infty, \mathbf{e}_0)$ to $(\mathbf{e}_0, \mathbf{e}_\infty)$. Fortunately, the notion of Farey neighbours is compatible with the cell structure of $\mathcal{ML}(\Sigma_1)$ in the following sense.

Proposition 2.3.3. Let (γ_1, γ_2) be a pair of Farey neighbours. If $\{\gamma_1, \gamma_2\} \neq \{\mathbf{e}_0, \mathbf{e}_\infty\}$ then γ_1 and γ_2 are both contained in I_j for some $j = 0, 1, 2, 3$.

Proof. The easiest way to see this is to use the well known fact that, using the identification of $\mathcal{PML}(\Sigma_1)$ with $\mathbb{R} \cup \{\infty\}$ given in Section 2.1, a pair of Farey neighbours corresponds to a pair of rational numbers p/q and r/s with $ps - qr = \pm 1$, see [19]. Provided we have $\{\pm p/q, \pm r/s\} \neq \{0 = 0/1, \infty = 1/0\}$, it is clear that p/q and r/s are both contained in one of the intervals $[-\infty, -1]$, $[-1, 0]$, $[0, 1]$ or $[1, \infty]$. The result follows from the discussion in Section 2.1. □

On a cell I_j, the map f_1 is constantly equal to a fixed element α_j of $\mathcal{MCG}(\Sigma_1)$ with possible ambiguity at the endpoints. Proposition 2.3.3 allows us to extend the action of f_1 to $\mathcal{F} - \big\{(\mathbf{e}_0, \mathbf{e}_\infty), (\mathbf{e}_\infty, \mathbf{e}_0)\big\}$ by defining $f_1(\gamma_1, \gamma_2) = \big(\alpha_j(\gamma_1), \alpha_j(\gamma_2)\big)$ whenever γ_1 and γ_2 are both in I_j. (Notice that this automatically takes care of the ambiguities at the endpoints.) Since the mapping class group preserves Farey neighbours, it follows that $\big(\alpha_j(\gamma_1), \alpha_j(\gamma_2)\big) \in \mathcal{F}$ and we can continue to iterate f_1 until possibly $f_1{}^n(\gamma_1, \gamma_2) \in \big\{(\mathbf{e}_0, \mathbf{e}_\infty), (\mathbf{e}_\infty, \mathbf{e}_0)\big\}$. The following shows that the iteration process will always terminate in this way. It is an immediate consequence of Lemma 2.3.2.

Lemma 2.3.4 Suppose $(\gamma_1, \gamma_2) \in \mathcal{F}$ and $\{\gamma_1, \gamma_2\} \neq \{\mathbf{e}_0, \mathbf{e}_\infty\}$. Then

$$|f_1(\gamma_1)| + |f_1(\gamma_2)| < |\gamma_1| + |\gamma_2|.$$

Suppose now that $\phi \in \mathcal{MCG}(\Sigma_1)$. Since the condition of being Farey neighbours is topological, the pair $\big(\phi(\mathbf{e}_0), \phi(\mathbf{e}_\infty)\big)$ is always a Farey pair. Our normal form results from the following proposition.

Proposition 2.3.5 Let $\phi \in \mathcal{MCG}(\Sigma_1)$. Then there exists a non-negative integer n so that $\iota_2{}^{\varepsilon} f_1{}^n\big(\phi(\mathbf{e}_0), \phi(\mathbf{e}_\infty)\big) = (\mathbf{e}_0, \mathbf{e}_\infty)$ where $\varepsilon = 0$ or 1.

Proof. This follows immediately from the above discussion and Lemma 2.3.4. □

Remark. This proposition shows that the actions of f_1 and $\mathcal{MCG}(\Sigma_1)$ on the space of Farey neighbours are *orbit equivalent*. In other words, for any pairs of Farey neighbours $\{\gamma_1, \gamma_2\}$ and $\{\gamma_1', \gamma_2'\}$ we have

$$\{\gamma_1', \gamma_2'\} = \big\{\phi(\gamma_1), \phi(\gamma_2)\big\}$$

for $\phi \in \mathcal{MCG}(\Sigma_1)$ if and only if there exist non-negative integers m, n so that

$$f_1{}^n\{\gamma_1, \gamma_2\} = f_1{}^m\{\gamma_1', \gamma_2'\}.$$

The concept of orbit equivalence is of considerable importance in ergodic theory since any properties depending only on the orbit structure, for example invariant measures, can now be studied relative to f_1 rather than the group $\mathcal{MCG}(\Sigma_1)$. In particular, we have shown that the action on the space of Farey neighbours is *hyperfinite*, see [5]. This should be compared with the analogous results for Fuchsian groups in [4], [21].

2.4. The normal form and the word acceptor

Let us denote the exceptional Farey pairs $(\mathbf{e}_0, \mathbf{e}_\infty)$ and $(\mathbf{e}_\infty, \mathbf{e}_0)$ by K_0 and K_2. We extend the definition of f_1 by setting $f_1|_{K_0} = \iota_0 = e$ and $f_1|_{K_2} = \iota_2 = \delta_1\delta_0\delta_1$. Thus we can write $\mathcal{F} = I_0 \cup I_1 \cup I_2 \cup I_3 \cup K_0 \cup K_2$. The point of Proposition 2.3.5 is that it allows us to define normal forms for elements of $\mathcal{MCG}(\Sigma_1)$ in the following way. For any $\phi \in \mathcal{MCG}(\Sigma_1)$, the pair $\big(\phi(\mathbf{e}_0), \phi(\mathbf{e}_\infty)\big)$ lies in some cell, U_n say, where U_n is one of $I_0, \dots, I_3, K_0, K_2$. As we apply the map f_1 we move through a sequence of cells

$$U_n \longrightarrow U_{n-1} \longrightarrow \cdots \longrightarrow U_1 \longrightarrow U_0 = K_0.$$

Here each cell U_j for $j \geq 2$ is one of the four cells $I_0, \dots, I_3$ and U_1 is one of the five cells $I_0, \dots, I_3, K_2$. At each stage, $f_1|_{U_j} = \alpha_j$, a fixed element in the set $\{\delta_0{}^{\pm 1}, \delta_1{}^{\pm 1}\}$ (or possibly ι_2 if $U_1 = K_2$). Thus we have

$$f_1{}^n\big(\phi(\mathbf{e}_0), \phi(\mathbf{e}_\infty)\big) = \big(\alpha_1\alpha_2\cdots\alpha_n\phi(\mathbf{e}_0), \alpha_1\alpha_2\cdots\alpha_n\phi(\mathbf{e}_\infty)\big) = (\mathbf{e}_0, \mathbf{e}_\infty).$$

Since $\mathcal{MCG}(\Sigma_1)$ acts freely on the space $\mathcal{F}$ this shows that $\alpha_1\alpha_2\cdots\alpha_n\phi = e$ giving the normal form $\phi = \alpha_n{}^{-1}\cdots\alpha_1{}^{-1}$. In particular, we have shown that $\{\delta_0{}^{\pm 1}, \delta_1{}^{\pm 1}\}$ generate $\mathcal{MCG}(\Sigma_1)$.

For example, suppose that $\phi = {\delta_0}^2\delta_1$. We claim that the normal form for ϕ is $\delta_0{\delta_1}^{-1}\iota_2$. From Proposition 2.2.1 we see that $\phi(\mathbf{e}_0) = (1, 1) \in I_3$ and $\phi(\mathbf{e}_\infty) = (0, 1) \in I_3$. Thus we need to find the f_1-expansion for the Farey pair $\big(\phi(\mathbf{e}_0), \phi(\mathbf{e}_\infty)\big) \in I_3$. Applying $f_1|_{I_3} = {\delta_0}^{-1}$ we obtain $f_1\phi(\mathbf{e}_0) = (0, 1) \in I_0$ and $f_1\phi(\mathbf{e}_\infty) = (1, 0) \in I_0$. Applying $f_1|_{I_0} = \delta_1$ we obtain K_2. Applying $f_1|_{K_2} = \iota_2$ brings us back to K_0. Thus

$$f_1{}^3\big(\phi(\mathbf{e}_0), \phi(\mathbf{e}_\infty)\big) = \big(\iota_2\delta_1{\delta_0}^{-1}\phi(\mathbf{e}_0), \iota_2\delta_1{\delta_0}^{-1}\phi(\mathbf{e}_\infty)\big) = (\mathbf{e}_0, \mathbf{e}_\infty).$$

From Proposition 2.3.5 we see that $\iota_2\delta_1{\delta_0}^{-1}\phi = e$ and hence $\phi = \delta_0{\delta_1}^{-1}\iota_2$ as claimed.

Now we follow the procedure outlined in Section 1.2 and construct a finite state automaton that recognises our normal form. We have just extended the definition of f_1 to the $\mathcal{F}$, the set of Farey pairs:

$$\begin{aligned}
f_1|_{I_0} &= \delta_1 &&: I_0 \mapsto I_0 \cup I_3 \cup K_0 \cup K_2\\
f_1|_{I_1} &= {\delta_1}^{-1} &&: I_1 \mapsto I_1 \cup I_2 \cup K_0 \cup K_2\\
f_1|_{I_2} &= \delta_0 &&: I_2 \mapsto I_1 \cup I_2 \cup K_0 \cup K_2\\
f_1|_{I_3} &= {\delta_0}^{-1} &&: I_3 \mapsto I_0 \cup I_3 \cup K_0 \cup K_2\\
f_1|_{K_0} &= e &&: K_0 \mapsto K_0\\
f_1|_{K_2} &= \iota_2 &&: K_2 \mapsto K_0.
\end{aligned}$$

In order to define the word acceptor, we define six states labelled I_0, I_1, I_2, I_3, K_0, K_2 and we draw an arrow from state U to state V labelled by α if $U \subset f_1(V)$ and $f_1|_V = \alpha^{-1}$. The start state is K_0. It is clear that there is at most one arrow with a given label from each state and that all arrows ending at a particular state have the same label. Any path in this graph beginning at K_0 and following arrows in the given direction gives the normal form for an element of $\mathcal{MCG}(\Sigma_1)$ by reading the labels on the arrows in the order given by the path. Moreover any ϕ has a unique normal form given in this way. The word acceptor is shown in Fig. 2.4.1. We remark that because we are dealing with composition of functions we read all strings from right to left. Perhaps it is worth pointing out that the normal forms that this word acceptor produces all have the form $W(\delta_0, {\delta_1}^{-1})\iota_2^\varepsilon$ or $W({\delta_0}^{-1}, \delta_1)\iota_2^\varepsilon$ where $W(\alpha, \beta)$ is any string in the letters α and β and ε is either 0 or 1.

For example the element $\phi = {\delta_0}^2\delta_1$ considered in the example above corresponds to the following path in the word acceptor

$$K_0 \xrightarrow{\iota_2} K_2 \xrightarrow{{\delta_1}^{-1}} I_0 \xrightarrow{\delta_0} I_3.$$

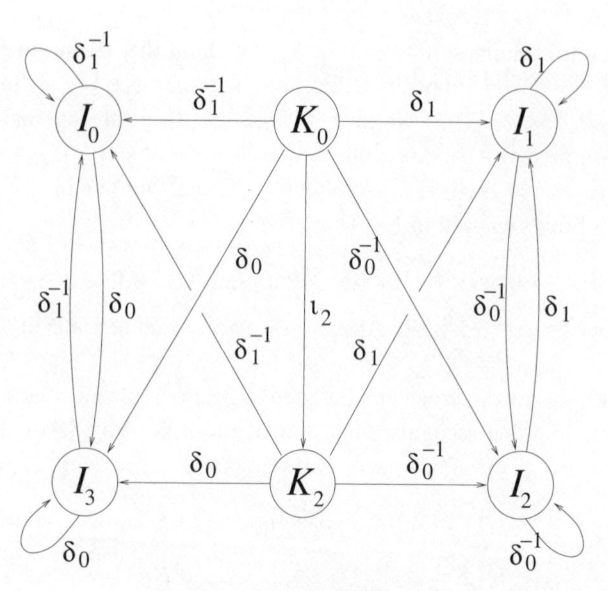

Figure 2.4.1 The word acceptor.

2.5. The word difference machine

In order to produce an automatic structure for $\mathcal{MCG}(\Sigma_1)$ we now explain how to construct the word difference machine as explained in Section 1.2. We must find a finite set of words $\mathcal{D}$ in $\mathcal{MCG}(\Sigma_1)$ called the *word differences*. This set should have the following properties: first $\mathcal{D}$ should contain the identity e and all the letters of the alphabet $\mathcal{A}$. The second property is slightly more complicated. Suppose that $\psi \in \mathcal{D}$ and U, V are two subsets of the I_j for $j = 0, 1, 2, 3$ or K_j for $j = 0, 2$ with the property that $\psi(U) = V$. Let $f_1|_U = \alpha$ and $f_1|_V = \beta$ denote the restriction of f_1 to U and V. Here α and β are particular elements of the alphabet $\mathcal{A}$. Then $\psi' = \beta\psi\alpha^{-1}$ sends $\alpha(U)$ to $\beta(V)$.[4] Our second requirement on $\mathcal{D}$ is that for all $\psi \in \mathcal{D}$ we should have $\psi' \in \mathcal{D}$ for all choices of U and V. This means that we get a commutative diagram which we call a *square*:

$$\begin{array}{ccc} U & \xrightarrow{\psi} & V \\ \downarrow{\scriptstyle\alpha} & & \downarrow{\scriptstyle\beta} \\ \alpha(U) & \xrightarrow{\psi'} & \beta(V) \end{array} \qquad (*)$$

This square corresponds to the relation $\psi' = \beta\psi\alpha^{-1}$ in G. We want to be able to concatenate squares vertically. In general $\alpha(U)$ and $\beta(V)$ will contain points in

[4] This differs from the expression in Section 1.2 as we are now reading strings from right to left.

several elements of the partition of $\mathcal{F}$ into I_j and K_j. This means that f_1 may not be a fixed element of $\mathcal{MCG}(\Sigma_1)$ on $\alpha(U)$ and $\beta(V)$. Let U' and V' be subsets of $\alpha(U)$ and $\beta(V)$ which each lie in a single set in the partition and satisfy $\psi'(U') = V'$. We are always able to subdivide $\alpha(U)$ and $\beta(V)$ into finitely many pieces for which this property holds. Since U' and V' are each contained in a single set of the partition, the restriction of f_1 to each of these two sets is a fixed element of $\mathcal{MCG}(\Sigma_1)$. In this way we can now construct several squares

$$\begin{array}{ccc} U' & \xrightarrow{\psi'} & V' \\ {\scriptstyle \alpha'}\downarrow & & \downarrow{\scriptstyle \beta'} \\ \alpha'(U') & \xrightarrow{\psi''} & \beta'(V') \end{array} \qquad (*)$$

each of which may be placed below $(*)$. In other words, we may concatenate squares vertically. A special case is where the word difference is the identity e

$$\begin{array}{ccc} U & \xrightarrow{e} & U \\ {\scriptstyle \alpha}\downarrow & & \downarrow{\scriptstyle \alpha} \\ \alpha(U) & \xrightarrow{e} & \alpha(U) \end{array}$$

For example, if $U = I_0$, $V = I_1$ and $\psi = \delta_0{}^{-1}$ we can construct the following square. This may be verified using Proposition 2.2.1 and the discussion in Section 2.4.

$$\begin{array}{ccc} I_0 & \xrightarrow{\delta_0{}^{-1}} & I_1 \\ {\scriptstyle f_1|_{I_0}=\delta_1}\downarrow & & \downarrow{\scriptstyle f_1|_{I_1}=\delta_1{}^{-1}} \\ I_0 \cup I_3 \cup K_0 \cup K_2 & \xrightarrow[\iota_2]{} & I_1 \cup I_2 \cup K_0 \cup K_2 \end{array}$$

In order to concatenate vertically, we need to subdivide the sets in the bottom line of this square. It could be followed by squares whose top lines are one of

$$I_0 \xrightarrow{\iota_2} I_2, \qquad I_3 \xrightarrow{\iota_2} I_1, \qquad K_0 \xrightarrow{\iota_2} K_2, \qquad K_2 \xrightarrow{\iota_2} K_0.$$

For example, it could be followed by the following square with $U' = I_0$ and $V' = I_2$:

$$\begin{array}{ccc} I_0 & \xrightarrow{\iota_2} & I_2 \\ {\scriptstyle f_1|_{I_0}=\delta_1}\downarrow & & \downarrow{\scriptstyle f_1|_{I_2}=\delta_0} \\ I_0 \cup I_3 \cup K_0 \cup K_2 & \xrightarrow[\iota_2]{} & I_1 \cup I_2 \cup K_0 \cup K_2 \end{array}$$

In order to simplify such diagrams, we make the following definitions

$$Q_0 = I_0 \cup I_3 \cup K_0 \cup K_2, \qquad Q_2 = I_1 \cup I_2 \cup K_0 \cup K_2.$$

If $\psi = \alpha$ or $\psi = \beta^{-1}$ we define degenerate squares, or *triangles* as follows. In the case where $\psi = \alpha$, we do not apply f_1 to V. This means that V may contain points from several sets in the partition, indeed we can take $V = \alpha(U)$. Likewise if $\psi = \beta^{-1}$ we define a triangle by not applying f_1 to U. Again U may contain points from several sets in the partition and we may take $U = \beta(V)$. In both cases ψ' is the identity map e:

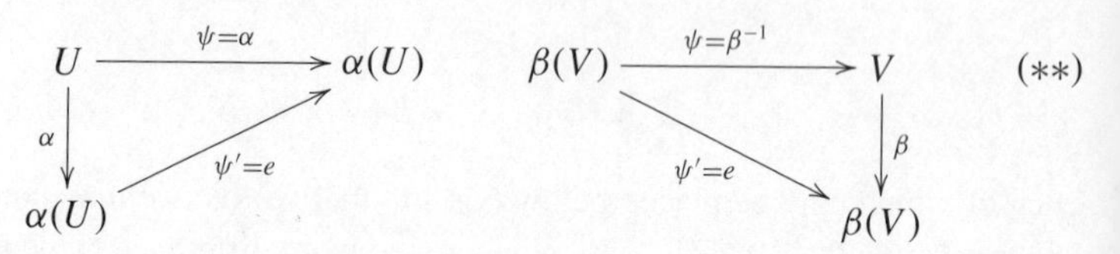

We now show that we can take the set of word differences to be

$$\mathcal{D} = \{e, \delta_0, {\delta_0}^{-1}, \delta_1, {\delta_1}^{-1}, \iota_2\}.$$

We first display all squares and triangles for which U is (a subset of) I_0. In the cases of triangles where $\psi = \beta^{-1}$ then we replace U by $\beta(V)$ and include the cases for which $I_0 \subset \beta(V)$ in our list.

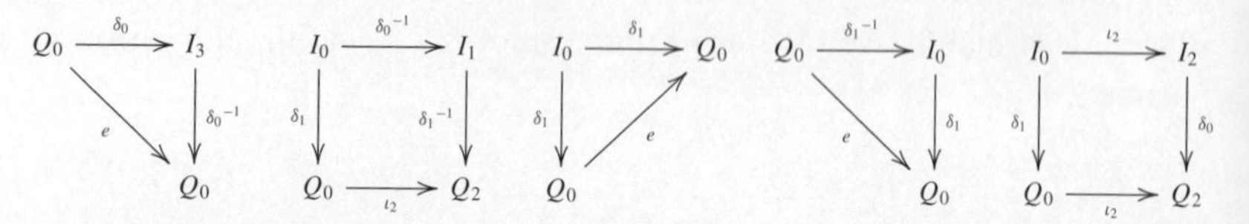

The squares and triangles where U is (a subset of) I_1, I_2 or I_3 or where $\beta(V) = Q_2$ may be obtained from these by symmetry. Finally, suppose that U is either K_0 or K_2. If the word difference ψ is ${\delta_j}^{\pm 1}$ for $j = 0, 1$ the relevant squares and triangles have already been included in the above list. Moreover, for such word differences the new word difference ψ' is e. On the other hand, if the word difference is ι_2 we obtain triangles

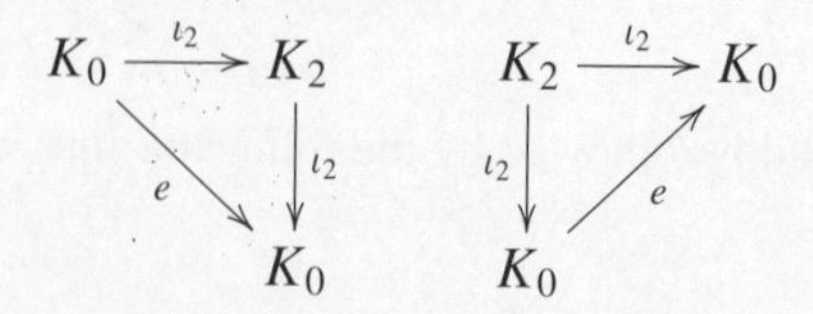

We now explain how to construct the word difference machine from these squares and triangles. Following the outline in Section 1.2, the states of the machine should be any triple (U, V, ψ) which appears as the top line of one of

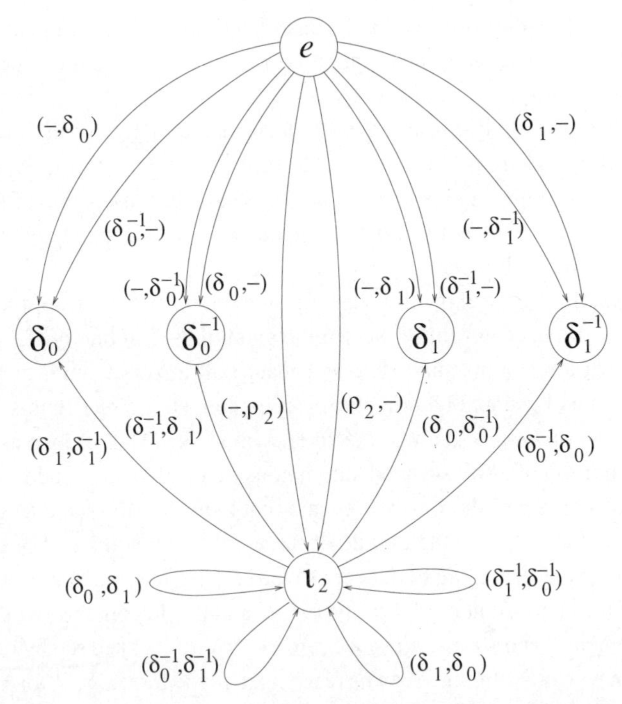

Figure 2.5.1 The (asynchronous) word difference machine. (In addition there are edges from e to itself labelled (δ_0, δ_0), $(\delta_0^{-1}, \delta_0^{-1})$, (δ_1, δ_1), $(\delta_1^{-1}, \delta_1^{-1})$ and (ι_2, ι_2).)

our squares or triangles. However, since the only function of the choice of U, V is to determine the value of the map f, we may as well take the states of the word difference machine to be the elements of $\mathcal{D}$. The arrows will be ordered pairs $(x, y) \in \mathcal{A} \cup \{-\}$ where x and y are essentially the inverses of α and β. In other words, given a square of the form $(*)$ we draw an arrow from ψ' to ψ and label it $(\alpha^{-1}, \beta^{-1})$. Similarly the triangles $(**)$ correspond to arrows from e to ψ labelled $(\alpha^{-1}, -)$ and $(-, \beta^{-1})$ respectively. We illustrate this in Fig. 2.5.1. In addition there should be arrows labelled (α, α) from e to itself for all $\alpha \in \mathcal{A}$. It is automatic from our construction that for any path in the word difference machine with strings of labels $(\alpha_j{}^{-1}, \beta_j{}^{-1})$ the strings of labels $\alpha_j{}^{-1}$ and $\beta_j{}^{-1}$ are both paths through the word acceptor. A result of this construction is that we have verified the presentation for $\mathcal{MCG}(\Sigma_1)$ given in section 2.2 (compare the proof of Theorem 2.3.12 of [7], page 51). In order to see this, observe that any closed path through the Cayley graph can be decomposed into a union of triangles each of which has one side of length at most one and of which the other two sides are paths in normal form leading back to the identity. This forms a van Kampen diagram for the closed path by covering it with squares and triangles of the form we have constructed above. One can easily verify that each of these squares and triangles corresponds to a relation which may be derived from $\delta_1\delta_0\delta_1 = \delta_0\delta_1\delta_0$ or $(\delta_0\delta_1)^3 = e$.

This has essentially constructed a 2-stringed automaton for the word difference machine. There is still a technical problem to be overcome. Namely, the word difference machine is asynchronous. This is because some of the labels have the form $(\alpha^{-1}, -)$ or $(-, \beta^{-1})$. In fact, this will occur exactly once when we are dealing with pairs of elements of the group which differ by a word of length exactly one. Specifically, the first time the normal forms of the prefixes from the word acceptor differ we see the symbol "–" in one of the strings in the word difference machine. This is because all arrows from e to any other state have this form and no other arrow does. In order to rectify this difficulty we need to *synchronise* the word difference machine. This is done as follows. In the definition of a two stringed automaton, we need to put a padding symbol \$ at the end of one of the words to ensure that they have the same length. Thus we need to move "–" in the middle of the word to a \$ at the end of the word. This is achieved by adding to our set of word differences the diagonals in each square. That is, for squares of the type (∗) we add the diagonal word differences $\beta^{-1}\psi' = \psi\alpha^{-1}$ and $\beta\psi = \psi'\alpha$. (We remark that it is easy to see by inspection that this new word difference can be rearranged to the form $\iota_2\delta_j{}^{\pm 1}$ for $j = 0, 1$.) This has the following effect. Suppose the normal forms from the word acceptor for words differing by ψ are $\alpha_1{}^{-1}\alpha_2{}^{-1}\alpha_3{}^{-1}\alpha_4{}^{-1}$ and $\beta_1{}^{-1}\beta_2{}^{-1}\alpha_4{}^{-1}$. Below we give the path in the word difference machine above (read from bottom to top), the corresponding squares, the amended squares and finally the path in the synchronised difference machine.

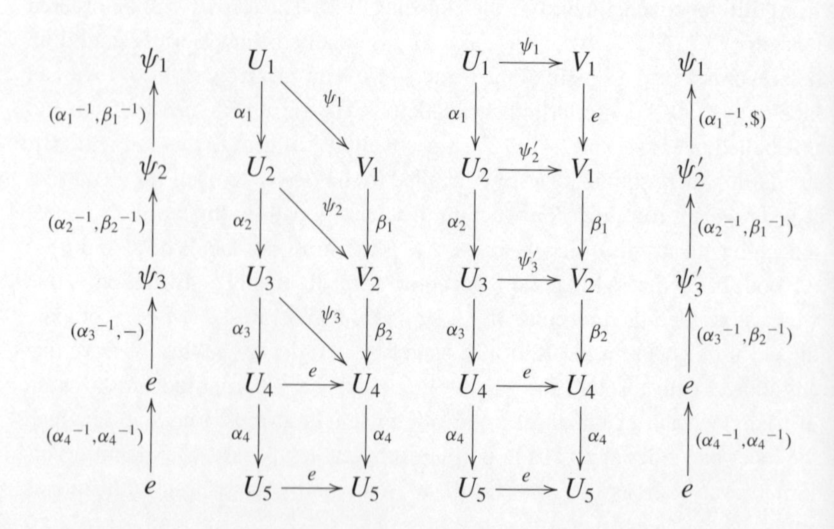

Here ψ_2, ψ_3, ψ_2' and ψ_3' are chosen so that the middle two diagrams commute. It is clear how to change the word difference machine in the light of this example. Of course there are now rather more states and arrows in the synchronised difference machine. In particular, the new states are

$$\mathcal{D}' = \left\{e, \delta_0, {\delta_0}^{-1}, \delta_1, {\delta_1}^{-1}, \iota_2, \iota_2\delta_0, \iota_2{\delta_0}^{-1}, \iota_2\delta_1, \iota_2{\delta_1}^{-1}\right\}.$$

3. Train tracks for the twice punctured torus

We now turn our attention to the twice punctured torus Σ_2. We want to mimic the constructions of the previous section. As we shall see, at every stage the basic ideas are the same but the implementation is considerably more complex.

3.1. The cell structure of $\mathcal{ML}(\Sigma_2)$

We briefly go through the construction of π_1-train tracks for the twice punctured torus; see also [12].

Once again, we fix a hyperbolic structure on Σ_2 by specifying a fundamental polygon for the action of $\pi_1(\Sigma_2)$ on $\mathbb{H}^2$. The fundamental domain R_2 that we choose to work with has six vertices, all of which project to punctures of Σ_2.

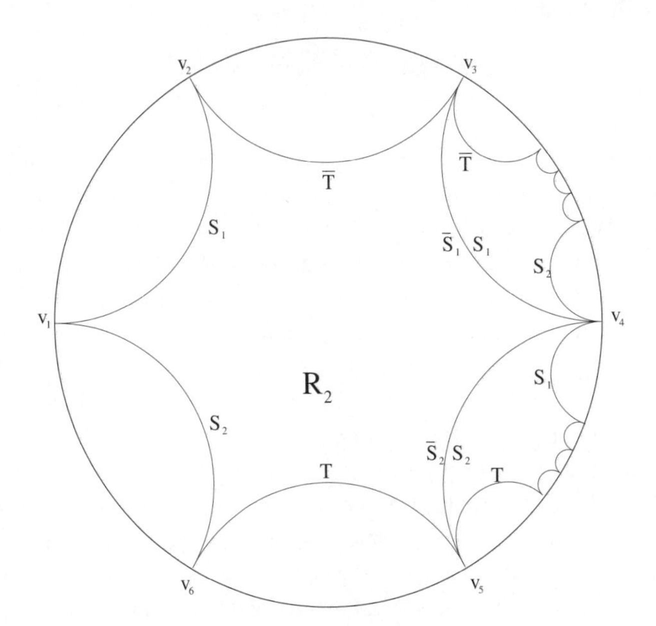

Figure 3.1.1 A hyperbolic fundamental domain R_2 for Σ_2.

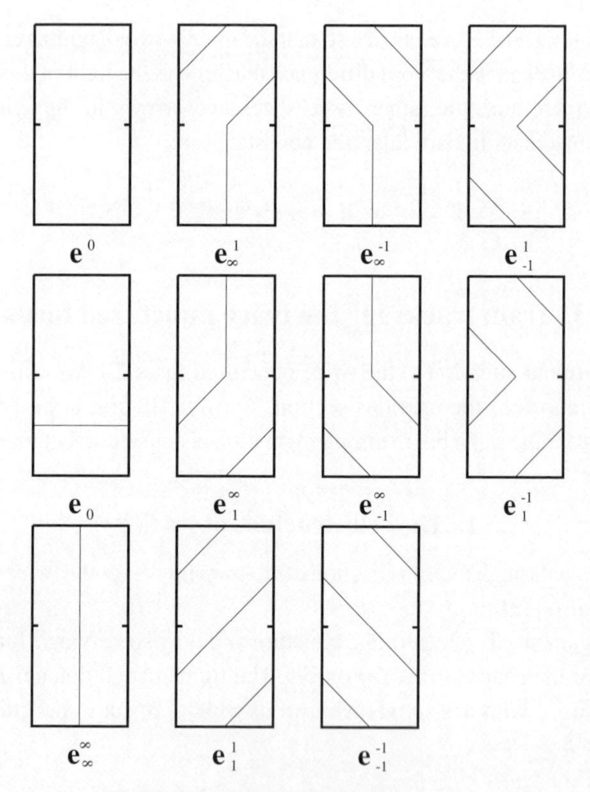

Figure 3.1.2 The irreducible loops on Σ_2.

We label these $v_1, \ldots, v_6$ in clockwise order. The side pairings will be S_1 identifying v_1v_2 to v_4v_3, S_2 identifying v_6v_1 with v_5v_4 and T identifying v_5v_6 with v_3v_2. We assume that S_1, S_2 and T match the endpoints of the respective sides. It is clear that v_1 and v_4 project to one puncture and the other four vertices project to the other. The maps S_1, S_2 and T correspond to homotopy classes of simple closed curves that generate the fundamental group $\pi_1(\Sigma_2)$.

We now introduce the irreducible loops that will form the basis of $\mathcal{ML}(\Sigma_2)$. Fig. 3.1.2 is a schematic picture of the eleven loops as they appear on the fundamental domain R_2. The end of a strand on one side of R_2 is glued by a side-pairing transformation to the corresponding end of the paired side. Thus shortest words representing these loops can be either computed directly or read off using the method of cutting sequences, see [2], [3] or [12]. For example, in the loop $\mathbf{e}_1^1$ there are three strands. The end of the strand on v_4v_3 is glued to the end on v_1v_2; the end on v_2v_3 is glued to the end on v_5v_6, and the end on v_5v_4 is

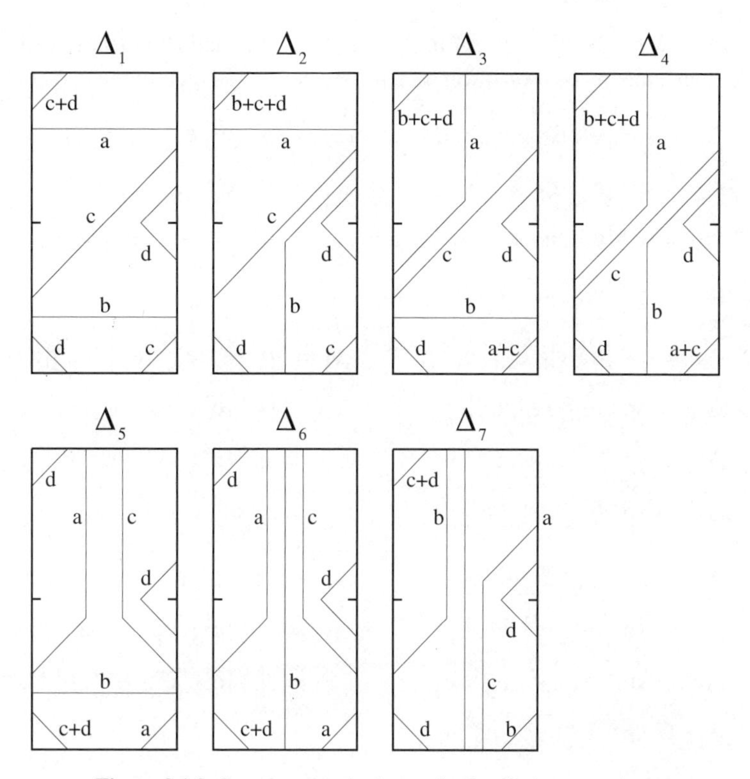

Figure 3.1.3 Generic points in the maximal cells $\Delta_1, \ldots, \Delta_7$.

glued to the end on $v_1 v_6$. Thus the cutting sequence is $S_1 T S_2$, which as one may easily verify represents this loop in $\pi_1(\Sigma_2)$. Since we are only interested in the un-oriented loop up to free homotopy, any cyclic permutation of this sequence or its inverse would work just as well. The full list of cutting sequences for loops is

$$\mathbf{e}^0 = S_1, \quad \mathbf{e}^1_\infty = S_1 T, \quad \mathbf{e}^{-1}_\infty = S_1{}^{-1} T, \quad \mathbf{e}^1_{-1} = S_1 T S_2{}^{-1},$$

$$\mathbf{e}_0 = S_2, \quad \mathbf{e}^\infty_1 = T S_2, \quad \mathbf{e}^\infty_{-1} = T S_2{}^{-1}, \quad \mathbf{e}^{-1}_1 = S_1{}^{-1} T S_2,$$

$$\mathbf{e}^\infty_\infty = T, \quad \mathbf{e}^1_1 = S_1 T S_2, \quad \mathbf{e}^{-1}_{-1} = S_1{}^{-1} T S_2{}^{-1}.$$

The reason for our notation is the following. If we split R_2 into two boxes, the upper one with vertices v_1, v_2, v_3, v_4 and the lower with vertices v_1, v_4, v_5, v_6 (see [12]) then $\mathbf{e}^i_j$ has gradient i in the upper box and j in the lower box. Where there is no superscript (subscript) then the relevant loop has no arcs in the upper (respectively lower) box. This idea is developed further in [12].

We now define 28 cells Δ_j in $\mathcal{ML}(\Sigma_2)$. As we shall show, these cells are maximal, meeting only on lower dimensional faces, and their union is $\mathcal{ML}(\Sigma_2)$.

$$\Delta_1 = \mathrm{sp}^+\{\mathbf{e}^0, \mathbf{e}_0, \mathbf{e}^1_1, \mathbf{e}^1_{-1}\}, \qquad \Delta_2 = \mathrm{sp}^+\{\mathbf{e}^0, \mathbf{e}^1_\infty, \mathbf{e}^1_1, \mathbf{e}^1_{-1}\},$$

$$\Delta_3 = \mathrm{sp}^+\{\mathbf{e}^\infty_1, \mathbf{e}_0, \mathbf{e}^1_1, \mathbf{e}^1_{-1}\}, \qquad \Delta_4 = \mathrm{sp}^+\{\mathbf{e}^\infty_1, \mathbf{e}^1_\infty, \mathbf{e}^1_1, \mathbf{e}^1_{-1}\},$$

$$\Delta_5 = \mathrm{sp}^+\{\mathbf{e}^\infty_1, \mathbf{e}_0, \mathbf{e}^\infty_{-1}, \mathbf{e}^1_{-1}\}, \qquad \Delta_6 = \mathrm{sp}^+\{\mathbf{e}^\infty_1, \mathbf{e}^\infty_\infty, \mathbf{e}^\infty_{-1}, \mathbf{e}^1_{-1}\},$$

$$\Delta_7 = \mathrm{sp}^+\{\mathbf{e}^\infty_\infty, \mathbf{e}^\infty_1, \mathbf{e}^1_\infty, \mathbf{e}^1_{-1}\}.$$

$$\Delta_8 = \mathrm{sp}^+\{\mathbf{e}^0, \mathbf{e}_0, \mathbf{e}^{-1}_{-1}, \mathbf{e}^{-1}_1\}, \qquad \Delta_9 = \mathrm{sp}^+\{\mathbf{e}^0, \mathbf{e}^{-1}_\infty, \mathbf{e}^{-1}_{-1}, \mathbf{e}^{-1}_1\},$$

$$\Delta_{10} = \mathrm{sp}^+\{\mathbf{e}^\infty_{-1}, \mathbf{e}_0, \mathbf{e}^{-1}_{-1}, \mathbf{e}^{-1}_1\}, \qquad \Delta_{11} = \mathrm{sp}^+\{\mathbf{e}^\infty_{-1}, \mathbf{e}^{-1}_\infty, \mathbf{e}^{-1}_{-1}, \mathbf{e}^{-1}_1\},$$

$$\Delta_{12} = \mathrm{sp}^+\{\mathbf{e}^\infty_{-1}, \mathbf{e}_0, \mathbf{e}^\infty_1, \mathbf{e}^{-1}_1\}, \qquad \Delta_{13} = \mathrm{sp}^+\{\mathbf{e}^\infty_{-1}, \mathbf{e}^\infty_\infty, \mathbf{e}^\infty_1, \mathbf{e}^{-1}_1\},$$

$$\Delta_{14} = \mathrm{sp}^+\{\mathbf{e}^\infty_\infty, \mathbf{e}^\infty_{-1}, \mathbf{e}^{-1}_\infty, \mathbf{e}^{-1}_1\}.$$

$$\Delta_{15} = \mathrm{sp}^+\{\mathbf{e}_0, \mathbf{e}^0, \mathbf{e}^1_1, \mathbf{e}^{-1}_1\}, \qquad \Delta_{16} = \mathrm{sp}^+\{\mathbf{e}_0, \mathbf{e}^\infty_1, \mathbf{e}^1_1, \mathbf{e}^{-1}_1\},$$

$$\Delta_{17} = \mathrm{sp}^+\{\mathbf{e}^1_\infty, \mathbf{e}^0, \mathbf{e}^1_1, \mathbf{e}^{-1}_1\}, \qquad \Delta_{18} = \mathrm{sp}^+\{\mathbf{e}^1_\infty, \mathbf{e}^\infty_1, \mathbf{e}^1_1, \mathbf{e}^{-1}_1\},$$

$$\Delta_{19} = \mathrm{sp}^+\{\mathbf{e}^1_\infty, \mathbf{e}^0, \mathbf{e}^{-1}_\infty, \mathbf{e}^{-1}_1\}, \qquad \Delta_{20} = \mathrm{sp}^+\{\mathbf{e}^1_\infty, \mathbf{e}^\infty_\infty, \mathbf{e}^{-1}_\infty, \mathbf{e}^{-1}_1\},$$

$$\Delta_{21} = \mathrm{sp}^+\{\mathbf{e}^\infty_\infty, \mathbf{e}^1_\infty, \mathbf{e}^\infty_1, \mathbf{e}^{-1}_1\}.$$

$$\Delta_{22} = \mathrm{sp}^+\{\mathbf{e}_0, \mathbf{e}^0, \mathbf{e}^{-1}_{-1}, \mathbf{e}^1_{-1}\}, \qquad \Delta_{23} = \mathrm{sp}^+\{\mathbf{e}_0, \mathbf{e}^\infty_{-1}, \mathbf{e}^{-1}_{-1}, \mathbf{e}^1_{-1}\},$$

$$\Delta_{24} = \mathrm{sp}^+\{\mathbf{e}^{-1}_\infty, \mathbf{e}^0, \mathbf{e}^{-1}_{-1}, \mathbf{e}^1_{-1}\}, \qquad \Delta_{25} = \mathrm{sp}^+\{\mathbf{e}^{-1}_\infty, \mathbf{e}^\infty_{-1}, \mathbf{e}^{-1}_{-1}, \mathbf{e}^1_{-1}\},$$

$$\Delta_{26} = \mathrm{sp}^+\{\mathbf{e}^{-1}_\infty, \mathbf{e}^0, \mathbf{e}^1_\infty, \mathbf{e}^1_{-1}\}, \qquad \Delta_{27} = \mathrm{sp}^+\{\mathbf{e}^{-1}_\infty, \mathbf{e}^\infty_\infty, \mathbf{e}^1_\infty, \mathbf{e}^1_{-1}\},$$

$$\Delta_{28} = \mathrm{sp}^+\{\mathbf{e}^\infty_\infty, \mathbf{e}^{-1}_\infty, \mathbf{e}^\infty_{-1}, \mathbf{e}^1_{-1}\}.$$

The statement that Δ_j is a cell should be interpreted in the following way. One needs to check that the four irreducible loops defining Δ_j are all supported on a common π_1-train track τ_j. This is immediate since one checks that, in each case, all four loops can be drawn in R_2 in such a way that they intersect only on the boundary ∂R_2. The arcs may be homotoped so that their endpoints are at the midpoints of the sides of R_2. Since the midpoints are identified by the side pairings, this exactly gives a π_1-train track in the sense of [2]. The cell Δ_j consists of all proper weightings on the π_1-train track τ_j. Fig. 3.1.4 shows the π_1-train track τ_1 which supports for the cell Δ_1. We normally draw this as in the top left hand corner of Fig. 3.1.3 where it is clearer that the weighting shown is $a\mathbf{e}^0 + b\mathbf{e}_0 + c\mathbf{e}^1_1 + d\mathbf{e}^1_{-1}$. It is easy to check that this is a proper weighting as defined in Section 3.1.

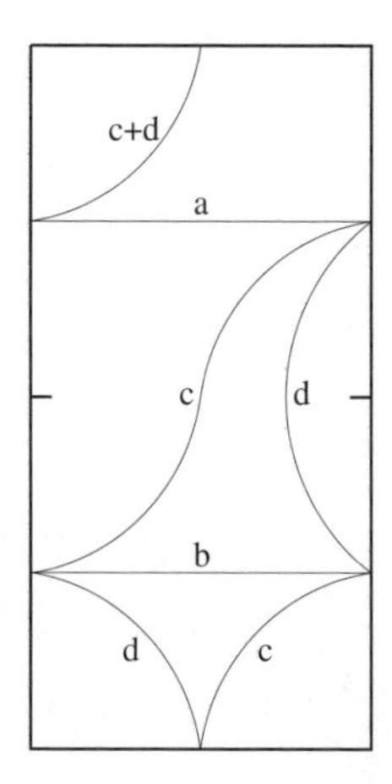

Figure 3.1.4 The π_1-train track τ_1 corresponding to Δ_1.

Notation 3.1.1. When we want to speak of a point of one of these cells we write it as an ordered quadruple (a, b, c, d) to represent $a\mathbf{e}_i + b\mathbf{e}_j + c\mathbf{e}_k + d\mathbf{e}_k \in \mathrm{sp}^+\{\mathbf{e}_i, \mathbf{e}_j, \mathbf{e}_k, \mathbf{e}_l\}$ where the irreducible loops are taken in the order given above. Thus, for example $(a, b, c, d) \in \Delta_1$ means $a\mathbf{e}^0 + b\mathbf{e}_0 + c\mathbf{e}_1^1 + d\mathbf{e}_{-1}^1$.

Proposition 3.1.1. *The cells $\Delta_1, \ldots, \Delta_{28}$ are maximal and their union is $\mathcal{ML}(\Sigma_2)$.*

Proof. (Outline) The idea is similar to the proof of Proposition 2.1.1. We will sketch the idea and then illustrate it by performing the computation in one case. All other cases are similar, straightforward, and left to the reader. The idea is the following. It is clear that any multiple simple loop may be homotoped so that it runs along a collection of arcs in R_2 joining midpoints of distinct sides, and which meet only on ∂R_2. Collapsing all arcs joining the same pair of sides yields a properly weighted π_1-train track on R_2. We now reverse this process and investigate what the possibilities for maximal weightings of this kind are. Take a copy of R_2 and draw strands joining the midpoints of pairs of distinct sides in such a way that no two strands intersect, that no two strands join the same pair of sides and that no more strands can be added without violating the previous two conditions. Now put a weight on each strand. In order to be proper, the weights must satisfy the following conditions as outlined in Section 1.3:

(i) all weights should be non-negative and not all zero,
(ii) either the weight on the corner strand separating v_1 from the rest of R should be zero or else the weight on the corner strand separating v_4 should be zero,

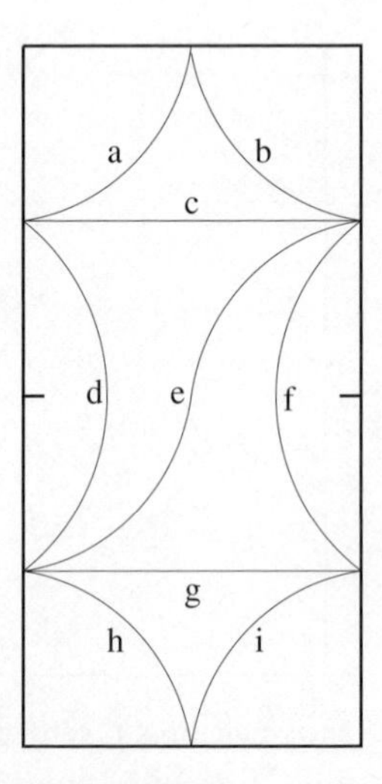

Figure 3.1.5 Possible weightings on a maximal configuration.

(iii) the weight on at least one of the four corner strands separating v_2, v_3, v_5, v_6 from the rest of R_2 should be zero,

(iv) the sum of the weights of all strands ending on a given side should be equal to the sum of weights on the side it is identified with.

Condition (iv) puts three linear relations between the weights. Solving these relations and inserting conditions (i), (ii) and (iii), we see that we must be in one of the 28 maximal cells defined above depending on our initial configuration of strands. Detailed computations for the configuration of Fig. 3.1.5 are carried out below. □

An Example. We now perform the computations in the configuration given Fig. 3.1.5 which is a maximal diagram of the type described. By equating the weights on each side we see that

$$a + b = h + i, \quad a + c + d = b + c + e + f,$$
$$d + e + g + h = f + g + i.$$

At least one of d and f must vanish by condition (ii). (Otherwise there would be a loop homotopic to one of the punctures.) Without loss of generality we suppose $d = 0$. This means

$$a + b = h + i, \quad a = b + e + f, \quad e + g + h = f + g + i.$$

Substituting for a in the first equation we obtain

$$2b + e + f = h + i, \quad e + h = f + i.$$

Adding these and cancelling $f + h$ from each side we get $i = b + e$. Substituting and cancelling once again we find that $h = b + f$. To summarise:

$$a = b + e + f, \quad d = 0, \quad h = b + f, \quad i = b + e.$$

Now by (iii), at least one of a, b, h, i must vanish. By inspection, if any of a, h, i vanish then so must b. (Remember all weights are non-negative.) This means that b must be zero and the train track is:

$$c\mathbf{e}^0 + g\mathbf{e}_0 + e\mathbf{e}_1^1 + f\mathbf{e}_{-1}^1 = (c, g, e, f) \in \Delta_1,$$

corresponding, after changing labels, to the picture shown in Fig. 3.1.4.

We now indicate how the lower dimensional facets in the boundaries of the maximal cells Δ_j for $j = 1, \dots, 28$ fit together in such a way that the resulting manifold is a 3-sphere. To do this, observe that there are fourteen maximal cells containing the irreducible loop $\mathbf{e}_{-1}^1$ and fourteen containing $\mathbf{e}_1^{-1}$. Moreover, these two irreducible loops never occur together in one of the cells (or else there would be loops around both punctures). Thus each maximal cell is a cone with apex $\mathbf{e}_{-1}^1$ or $\mathbf{e}_1^{-1}$ over the cell spanned by the other three irreducible loops. One can verify that there are fourteen possibilities for these cells spanned by three loops and that each one arises. Moreover, these fourteen cells may be glued together to form a 2-sphere as indicated in Fig. 3.1.6. Thus the fourteen maximal cells involving $\mathbf{e}_{-1}^1$ form a cone over the 2-sphere, that is a 3-ball. Similarly the other fourteen maximal cells also give a 3-ball. When the boundaries of these two balls are glued together in the obvious manner they form a 3-sphere. We show which maximal cells intersect to give the three-cells on the 2-sphere in Fig. 3.1.7.

3.2. Dehn twists and a presentation for $\mathcal{MCG}(\Sigma_2)$

Let the Dehn twists about $\mathbf{e}_\infty^\infty$, $\mathbf{e}^0$ and $\mathbf{e}_0$ be denoted by δ_0, δ_1 and δ_2 respectively. We now want to investigate the action of these Dehn twists on the cell structure of $\mathcal{ML}(\Sigma_2)$ given in the previous section. Again we begin by introducing some symmetries that will simplify matters. We are only interested in symmetries which fix the punctures. The symmetry group will be isomorphic to Klein's four group and we describe its non-trivial elements by their action on the vertices of R_2:

- ι_1 interchanges the pairs (v_1, v_4), (v_2, v_3), (v_5, v_6);
- ι_2 interchanges the pairs (v_1, v_4), (v_2, v_5), (v_3, v_6);
- ι_3 fixes v_1, v_4 and interchanges the pairs (v_2, v_6), (v_3, v_5).

When necessary we shall denote the identity by ι_0. It is clear that ι_1 and ι_3 are orientation reversing homeomorphisms of Σ_2 and that ι_2 is orientation preserving.

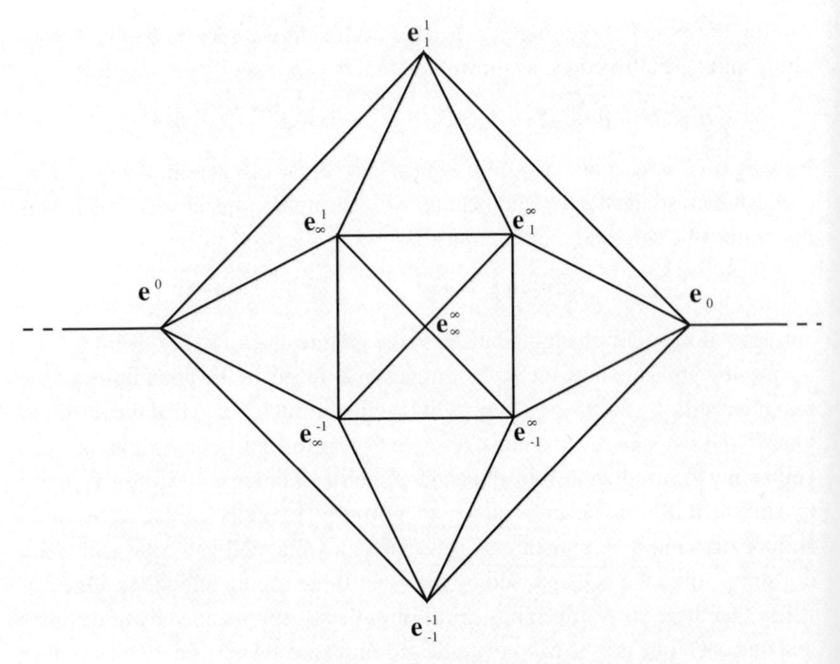

Figure 3.1.6 Dividing the 2−sphere into 14 three-cells.

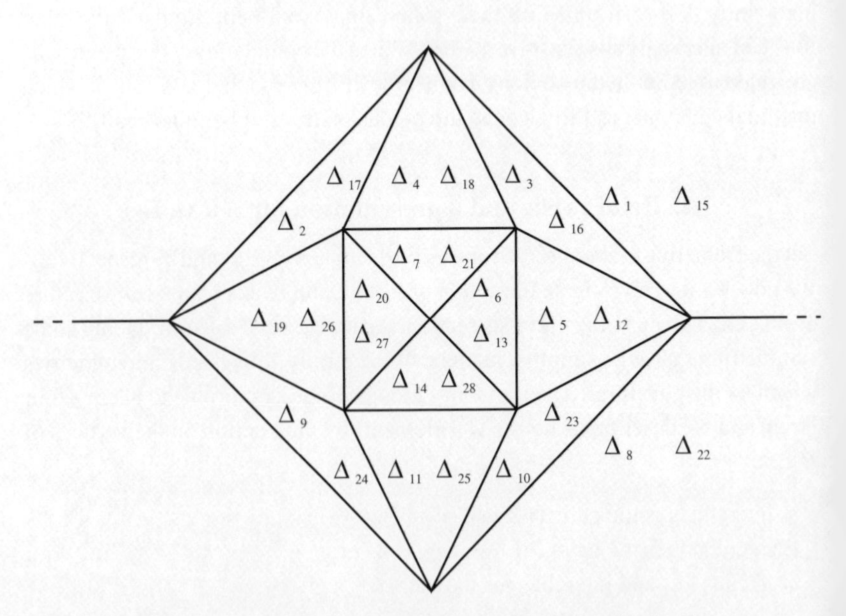

Figure 3.1.7 Fig. 3.1.6 repeated but indicating the maximal cells.

We easily see that ι_j has the following effect on the eleven irreducible loops:

$$\begin{aligned}
\iota_1 &: \mathbf{e}^i_j \longmapsto \mathbf{e}^{-i}_{-j}, & \mathbf{e}^0 &\longmapsto \mathbf{e}^0, & \mathbf{e}_0 &\longmapsto \mathbf{e}_0,\\
\iota_2 &: \mathbf{e}^i_j \longmapsto \mathbf{e}^j_i, & \mathbf{e}^0 &\longmapsto \mathbf{e}_0, & \mathbf{e}_0 &\longmapsto \mathbf{e}^0,\\
\iota_3 &: \mathbf{e}^i_j \longmapsto \mathbf{e}^{-j}_{-i}, & \mathbf{e}^0 &\longmapsto \mathbf{e}_0, & \mathbf{e}_0 &\longmapsto \mathbf{e}^0
\end{aligned}$$

where $i, j \in \{\infty, \pm 1\}$ and $-\infty = \infty$. Thus these actions clearly respect the cell structure of $\mathcal{ML}(\Sigma_2)$. In particular, the maximal cells $\Delta_8, \ldots, \Delta_{28}$ can be expressed as $\Delta_{j+7k} = \iota_k(\Delta_j)$ for $j = 1, \ldots, 7$ and $k = 1, 2, 3$. The symmetries ι_k conjugate the Dehn twists δ_j to one another. It is easy to check that

$$\begin{aligned}
\iota_1\delta_0\iota_1 &= \delta_0{}^{-1}, & \iota_1\delta_1\iota_1 &= \delta_1{}^{-1}, & \iota_1\delta_2\iota_1 &= \delta_2{}^{-1},\\
\iota_2\delta_0\iota_2 &= \delta_0, & \iota_2\delta_1\iota_2 &= \delta_2, & \iota_2\delta_2\iota_2 &= \delta_1,\\
\iota_3\delta_0\iota_3 &= \delta_0{}^{-1}, & \iota_3\delta_1\iota_3 &= \delta_2{}^{-1}, & \iota_3\delta_2\iota_3 &= \delta_1{}^{-1}.
\end{aligned}$$

We can also express ι_2 in terms of the δ_j as

$$\iota_2 = \delta_2\delta_0\delta_2\delta_1\delta_0\delta_2 = (\delta_0\delta_1\delta_2)^2.$$

The proof of this identity will be an easy exercise once the effect of the Dehn twists on $\mathcal{ML}(\Sigma_2)$ has been investigated in the next section. This proof will be left to the reader.

Before we go on to investigate the effect of Dehn twists on π_1-train tracks we will discuss the presentation of $\mathcal{MCG}$ in terms of Dehn twist generators. The first claim is that δ_0, δ_1 and δ_2 generate the (orientation preserving) mapping class group of Σ_2. As in section 2, this will be a consequence of our construction of the Markov map, which will once again produce a unique normal form for every element of $\mathcal{MCG}(\Sigma_2)$ in terms of δ_0, δ_1 and δ_2 (and their inverses).

There are certain relations in G which arise by inspection on Σ_2. For example, since $\mathbf{e}^0$ and $\mathbf{e}_0$ are disjoint, the Dehn twists δ_1 and δ_2 commute. Also, since $\mathbf{e}^0$ and $\mathbf{e}_0$ each intersect $\mathbf{e}^\infty_\infty$ exactly once, for $j = 1, 2$ the Dehn twists δ_0 and δ_j satisfy the braid relation $\delta_0\delta_j\delta_0 = \delta_j\delta_0\delta_j$. Finally, ι_2 is an involution so its square is the identity. Using the form for ι_2 constructed above, we can write $\iota_2 = (\delta_0\delta_1\delta_2)^2$. This gives the relation $(\delta_0\delta_1\delta_2)^4 = e$ (compare [1]).

It turns out that these are all the relations we need to give a presentation of G. This will follow from our construction of the word difference machine (section 5) along the same lines as explained in section 2.5. Hence we obtain the following presentation for $\mathcal{MCG}(\Sigma_2)$. This resembles the presentations for other mapping class groups given by Birman in [1] and could have been derived using methods similar to hers.

Theorem 3.2.1. *The mapping class group of the twice punctured torus admits a presentation:*

$$\begin{aligned}\mathcal{MCG}(\Sigma_2) = \big\langle \delta_0, \delta_1 \delta_2 \big| \delta_1\delta_2 = \delta_2\delta_1,\ \delta_1\delta_0\delta_1 = \delta_0\delta_1\delta_0, \\ \delta_2\delta_0\delta_2 = \delta_0\delta_2\delta_0,\ (\delta_0\delta_1\delta_2)^4 = e\big\rangle.\end{aligned}$$

In section 5 of [13] Magnus gives the following presentation of $\mathcal{MCG}(\Sigma_2)$. The generators are r, s, ρ, σ, τ subject to the following relations:

$$\begin{array}{llll} s^2 = (r^{-1}s)^3, & s\tau s^{-1} = \rho, & s\rho s^{-1} = \rho\tau^{-1}\rho^{-1}, & \\ r\tau r^{-1} = \tau, & r\rho r^{-1} = \rho\tau^{-1}, & s^{-4}\rho\tau^{-1}\rho^{-1}\tau = 1, & \\ \sigma^2 = s^{-4}, & \sigma r\sigma^{-1} = r, & \sigma s\sigma^{-1} = s, & \sigma\tau\sigma^{-1} = \tau^{-1}s^4. \end{array}$$

One may pass from our presentation to Magnus' presentation via the substitution

$$\begin{aligned} r = \delta_1, \quad s = \delta_1\delta_0\delta_1, \quad \rho = \delta_1\delta_0\delta_1{\delta_2}^{-1}{\delta_0}^{-1}{\delta_1}^{-1}, \\ \sigma = {\delta_1}^{-1}{\delta_0}^{-1}{\delta_1}^{-1}{\delta_1}^{-1}{\delta_0}^{-1}{\delta_1}^{-1}, \quad \tau = \delta_1{\delta_2}^{-1}. \end{aligned}$$

The proof of this is straightforward and is left to the reader.

3.3. The effect of Dehn twists on $\mathcal{PML}$

We now investigate the effect of the Dehn twists on the maximal cells. Using general results of Birman-Series [2] and Hamidi-Tehrani-Chen [8] we know that these maps are piecewise linear. In fact using reductions similar to those in section 2 we will show this directly. We begin by summarising the results.

Proposition 3.3.1. *The Dehn twist δ_1 has the following effect on $\Delta_1, \dots, \Delta_7$:*

(i) *δ_1 maps Δ_1 piecewise linearly to $\Delta_1 \cup \Delta_3 \cup \Delta_5$ as follows:*

$$\delta_1(a,b,c,d) = \begin{cases} (a-c-d, b, c, d) \in \Delta_1 & \text{if } c+d \le a \\ (c+d-a, b, a-d, d) \in \Delta_3 & \text{if } d \le a \le c+d \\ (c, b, d-a, a) \in \Delta_5 & \text{if } a \le d. \end{cases}$$

(ii) δ_1 *maps* Δ_2 *piecewise linearly to* $\Delta_2 \cup \Delta_4 \cup \Delta_7 \cup \Delta_6$ *as follows:*

$$\delta_1(a, b, c, d) = \begin{cases} (a-b-c-d, b, c, d) \in \Delta_2 & \text{if } b+c+d \le a \\ (b+c+d-a, b, a-b-d, d) \in \Delta_4 & \text{if } b+d \le a \le b+c+d \\ (b+d-a, c, a-d, d) \in \Delta_7 & \text{if } d \le a \le b+d \\ (c, b, d-a, a) \in \Delta_6 & \text{if } a \le d. \end{cases}$$

(iii) δ_1 *maps* Δ_3 *to* Δ_{12}, Δ_4 *to* Δ_{13}, Δ_5 *to* Δ_{10}, *and* Δ_6 *to* Δ_{11} *as follows:*

$$\delta_1(a, b, c, d) = (d, b, c, a).$$

(iv) δ_1 *maps* Δ_7 *to* Δ_{14} *as follows:*

$$\delta_1(a, b, c, d) = (c, d, a, b).$$

We remark that inverting the maps in Propositions 3.3.1(i) and (ii) shows that ${\delta_1}^{-1}$ maps $\bigcup_{j=1}^{7} \Delta_j$ to $\Delta_1 \cup \Delta_2$. We now investigate the action of δ_2.

Proposition 3.3.2. *The Dehn twist* δ_2 *has the following effect on* $\Delta_1, \dots, \Delta_7$

(i) δ_2 *maps* Δ_1 *to* $\Delta_1 \cup \Delta_2$ *as follows:*

$$\delta_2(a, b, c, d) = \begin{cases} (a, b+d-c, c, d) \in \Delta_1 & \text{if } c \le b+d \\ (a, c-b-d, b+d, d) \in \Delta_2 & \text{if } b+d \le c. \end{cases}$$

(ii) δ_2 *maps* Δ_2 *to* $\Delta_1 \cup \Delta_2$ *as follows:*

$$\delta_2(a, b, c, d) = \begin{cases} (a, d-c, c, b+d) \in \Delta_1 & \text{if } c \le d \\ (a, c-d, d, b+d) \in \Delta_2 & \text{if } d \le c. \end{cases}$$

(iii) δ_2 *maps* Δ_3 *to* $\Delta_4 \cup \Delta_3 \cup \Delta_7$ *as follows:*

$$\delta_2(a, b, c, d) = \begin{cases} (a, b+d-a-c, c, d) \in \Delta_3 & \text{if } a+c \le b+d \\ (a, a+c-b-d, b+d-a, d) \in \Delta_4 & \text{if } a \le b+d \le a+c \\ (a-b-d, b+d, c, d) \in \Delta_7 & \text{if } b+d \le a. \end{cases}$$

(iv) δ_2 *maps* Δ_4 *to* $\Delta_4 \cup \Delta_3 \cup \Delta_7$ *as follows:*

$$\delta_2(a, b, c, d) = \begin{cases} (a, d-a-c, c, b+d) \in \Delta_3 & \text{if } a+c \leq d \\ (a, a+c-d, d-a, b+d) \in \Delta_4 & \text{if } a \leq d \leq a+c \\ (a-d, d, c, b+d) \in \Delta_7 & \text{if } \leq a. \end{cases}$$

(v) δ_2 *maps* Δ_5 *to* $\Delta_5 \cup \Delta_6$ *as follows:*

$$\delta_2(a, b, c, d)$$
$$= \begin{cases} (a, b+c+d-a, c, d) \in \Delta_5 & \text{if } a \leq b+c+d \\ (b+c+d, a-b-c-d, c, d) \in \Delta_6 & \text{if } b+c+d \leq a. \end{cases}$$

(vi) δ_2 *maps* Δ_6 *to* $\Delta_5 \cup \Delta_6$ *as follows:*

$$\delta_2(a, b, c, d) = \begin{cases} (a, c+d-a, b+c, d) \in \Delta_5 & \text{if } a \leq c+d \\ (c+d, a-c-d, b+c, d) \in \Delta_6 & \text{if } c+d \leq a. \end{cases}$$

(vii) δ_2 *maps* Δ_7 *to* $\Delta_5 \cup \Delta_6$ *as follows:*

$$\delta_2(a, b, c, d) = \begin{cases} (b, d-b, a, c+d) \in \Delta_5 & \text{if } b \leq d \\ (d, b-d, a, c+d) \in \Delta_6 & \text{if } d \leq b. \end{cases}$$

We remark that inverting these maps shows that ${\delta_2}^{-1}$ maps $\Delta_1 \cup \Delta_2$ to $\Delta_1 \cup \Delta_2$; $\Delta_3 \cup \Delta_4 \cup \Delta_7$ to $\Delta_3 \cup \Delta_4$, and $\Delta_5 \cup \Delta_6$ to $\Delta_5 \cup \Delta_6 \cup \Delta_7$. We now turn our attention to ${\delta_0}^{-1}$.

Proposition 3.3.3. *The Dehn twist* ${\delta_0}^{-1}$ *has the following effect on* $\Delta_1, \dots, \Delta_7$ *:*

(i) ${\delta_0}^{-1}$ *maps* Δ_1 *to* Δ_{25}, Δ_2 *to* Δ_{24}, Δ_3 *to* Δ_{23}, *and* Δ_4 *to* Δ_{22} *as follows:*

$${\delta_0}^{-1}(a, b, c, d) = (a, b, c, d).$$

(ii) ${\delta_0}^{-1}$ *maps* Δ_5 *to* $\Delta_5 \cup \Delta_6$ *as follows:*

$${\delta_0}^{-1}(a, b, c, d) = \begin{cases} (c, a-c, b+c, d) \in \Delta_5 & \text{if } c \leq a \\ (a, c-a, b+c, d) \in \Delta_6 & \text{if } a \leq c. \end{cases}$$

(iii) ${\delta_0}^{-1}$ *maps* Δ_6 *to* $\Delta_5 \cup \Delta_6$ *as follows:*

$${\delta_0}^{-1}(a, b, c, d) = \begin{cases} (b+c, a-b-c, c) \in \Delta_5 & \text{if } b+c \leq a \\ (a, b+c-a, c) \in \Delta_6 & \text{if } a \leq b+c. \end{cases}$$

(iv) $\delta_0{}^{-1}$ *maps* Δ_7 *onto* $\Delta_7 \cup \Delta_4 \cup \Delta_3 \cup \Delta_2 \cup \Delta_1$ *as follows:*

$$\delta_0{}^{-1}(a, b, c, d)$$
$$= \begin{cases} (a - b - c, b, c, d) \in \Delta_7 & \text{if } b + c \leq a \\ (a - c, a - b, b + c - a, d) \in \Delta_4 & \text{if } b, c \leq a \leq b + c \\ (a - c, b - a, c, d) \in \Delta_3 & \text{if } c \leq a \leq b \\ (c - a, a - b, b, d) \in \Delta_2 & \text{if } b \leq a \leq c \\ (c - a, b - a, a, d) \in \Delta_1 & \text{if } a \leq b, c. \end{cases}$$

We remark that inverting these maps we see that δ_0 maps $\Delta_1 \cup \Delta_2 \cup \Delta_4 \cup \Delta_3 \cup \Delta_7$ to Δ_7 and $\Delta_5 \cup \Delta_6$ to $\Delta_5 \cup \Delta_6$.

Remark. *Points in the images of the Dehn twists are in fact well defined. In other words, if equality holds in one of the conditions above, we are on the common boundary of two maximal cells. For example if* $(a, b, c, d) \in \Delta_1$ *with* $a = c + d$ *then*

$$\delta_1(c + d, b, c, d) = b\mathbf{e}_0 + c\mathbf{e}_1^1 + d\mathbf{e}_{-1}^1 \in \Delta_1 \cap \Delta_3.$$

Proof of Propositions 3.3.1, 3.3.2, and 3.3.3. The method will be the same for each of the propositions and is completely analogous to the proof of Proposition 2.2.1. The proof of Proposition 3.3.1(i) follows from an analysis of Fig. 3.3.1, Proposition 3.3.1(ii) from Fig. 3.3.2 and Proposition 3.3.3(iv) from Fig. 3.3.3. The proofs of the other parts follow along similar lines but are easier. To avoid repetition we will discuss the proof of Proposition 3.3.1(i) and leave the rest to the reader. These proofs should be compared to the very similar discussion in section 3.2 of [8].

In Fig. 3.3.1, we start off with the π_1-train track for a general point $(a, b, c, d) \in \Delta_1$. We want to perform a Dehn twist δ_1 about $\mathbf{e}^0$. A tubular neighbourhood about $\mathbf{e}^0$ is represented by the pair of dotted lines in the figure. The Dehn twist is the identity everywhere outside these dotted lines and so everything we see here remains the same. In the cylinder between the dotted lines, we do a whole turn to the left. This means that a line crossing this cylinder is wrapped once around the cylinder before exiting in the same place. In the diagram this is represented by the diagonal lines inside the strip (in the train track on the top right). The new train track is a representation of the original train track after the Dehn twist has taken place. Unfortunately, this train track is unreduced. That is, it has strands from one side to itself (the upper right hand side v_4v_3). We need to remove these loops by pulling them tight as explained in section 2.2.

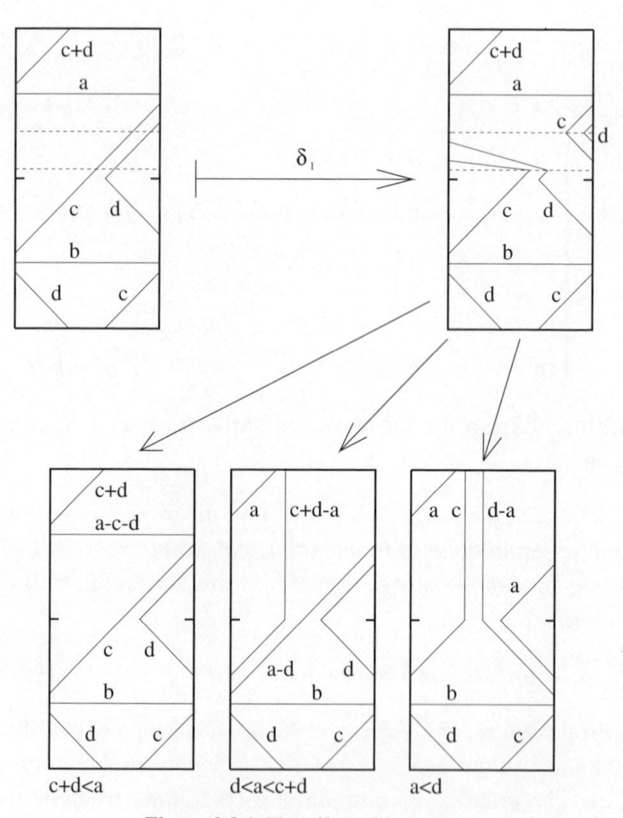

Figure 3.3.1 The effect of δ_1 on Δ_1.

The details of how to pull tight are as follows. The results are shown in the lower part of Fig. 3.3.1. As usual, it is enough to assume that all the weights are integers. Recall that an integral weight m on a strand means that m strands join the same pair of sides of R_2. The ends of these strands are identified, preserving order, with the strands on the paired side. In the top right hand diagram of Fig. 3.3.1, there are $c + d$ strands joining the side v_4v_3 to itself. The side pairing $S_1{}^{-1}$ takes the $2(c + d)$ ends of these loops to the lowest $2(c + d)$ ends of strands emanating from v_1v_2. We begin with the innermost of these loops, that is the strand that, together with an arc of v_4v_3, bounds a disc containing no strands. The endpoints of this loop are the $(c + d)$th and $(c + d + 1)$th ends from the bottom of v_4v_3. These are identified by the side pairing $S_1{}^{-1}$ with the $(c + d)$th and $(c + d + 1)$th ends from the bottom of v_1v_2. Providing a and d are non-zero these are ends of strands joining v_1v_2 with v_5v_4 and v_4v_3 respectively. When we pull this loop tight these three strands become a single strand from v_5v_4 to v_4v_3. Doing this $\min\{a, d\}$ times we get this number of strands joining v_5v_4 and v_4v_3.

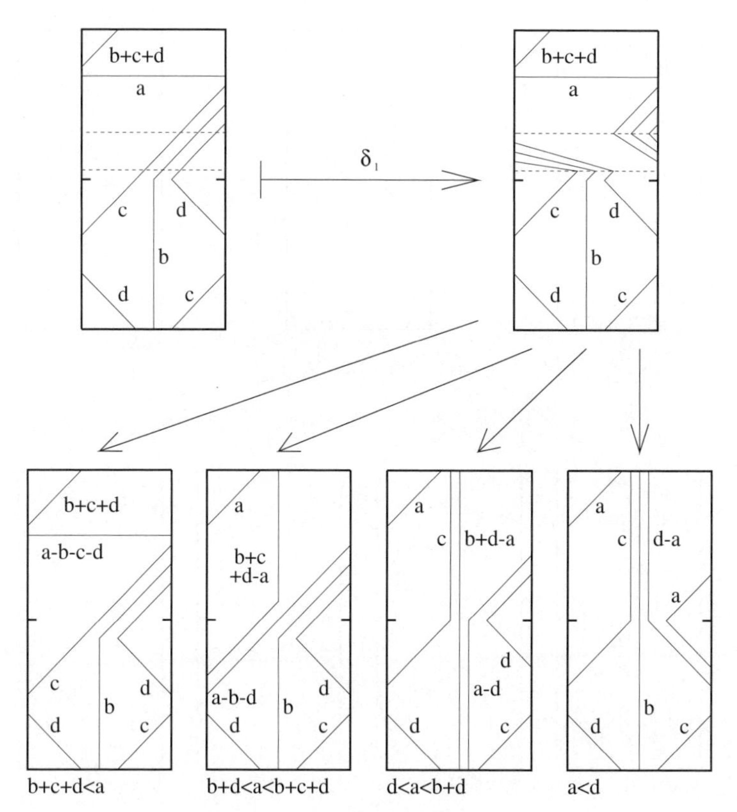

Figure 3.3.2 The effect of δ_1 on Δ_2.

In the case where $a \leq d$ we have exhausted all a strands joining the sides v_1v_2 and v_4v_3. We therefore continue using the $c + d$ strands joining v_3v_2 and v_1v_2. After pulling each of the next $d - a$ loops tight we obtain strands from v_3v_2 to v_5v_4. Finally each of the remaining c loops gives a strand from to v_3v_2 to v_6v_1. There are a strands from v_1v_2 to v_3v_2 remaining that have not been changed. Putting all this information together we obtain the π_1-train track in the bottom right of Fig. 3.3.1. It is then clear that this train track is the point $(c, b, d - a, a) \in \Delta_5$.

The case $a \geq d$ is similar but with further sub-cases $c + d \leq a$ and $d \leq a \leq c + d$. These give the other train tracks on the lower line of Fig. 3.3.1 and the points of Δ_1 and Δ_3 listed in the statement of Proposition 3.3.1. The rest of the propositions follow similarly. □

The following corollary is an immediate consequence of these results. It may be verified by considering the image of each of the $\mathbf{e}^i_j$ and extending linearly to the whole of $\mathcal{ML}(\Sigma_2)$.

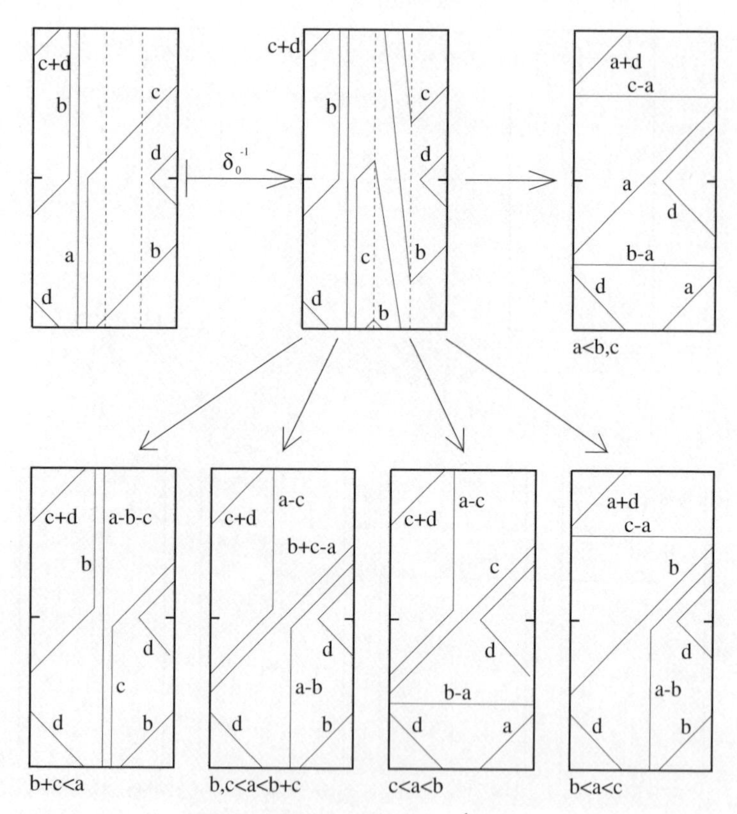

Figure 3.3.3 The effect of $\delta_0{}^{-1}$ on Δ_7.

Corollary 3.3.4 We have $\iota_2 = (\delta_0\delta_1\delta_2)^2$.

4. The Markov map

In this section we construct the Markov map which will be the key to constructing our normal form for elements of $\mathcal{MCG}(\Sigma_2)$.

4.1. Canonical Dehn twists for maximal cells

For each maximal cell Δ_j we now define a *canonical Dehn twist* η_j which is one of $\delta_0{}^{\pm 1}$, $\delta_1{}^{\pm 1}$, $\delta_2{}^{\pm 1}$. For simplicity we work with $\Delta_1, \ldots, \Delta_7$ and then define canonical Dehn twists for the other cells by symmetry. These twists will map maximal cells onto the union of other cells and are the essential step for defining our Markov map.

Definition. *The canonical Dehn twist on* Δ_1, Δ_2, Δ_5, *and* Δ_6 *is* δ_1. *The canonical Dehn twist on* Δ_3, Δ_4, *and* Δ_7 *is* ${\delta_0}^{-1}$.

From Propositions 3.3.1, 3.3.2 and 3.3.3 we see that the images of the canonical Dehn twists are as follows

$$\begin{aligned}
\delta_1 \quad &: \Delta_1 \longrightarrow \Delta_1 \cup \Delta_3 \cup \Delta_5 \\
\delta_1 \quad &: \Delta_2 \longrightarrow \Delta_2 \cup \Delta_4 \cup \Delta_6 \cup \Delta_7 \\
{\delta_0}^{-1} &: \Delta_3 \longrightarrow \Delta_{23} = \iota_3(\Delta_2) \\
{\delta_0}^{-1} &: \Delta_4 \longrightarrow \Delta_{22} = \iota_3(\Delta_1) \\
\delta_1 \quad &: \Delta_5 \longrightarrow \Delta_{10} = \iota_1(\Delta_3) \\
\delta_1 \quad &: \Delta_6 \longrightarrow \Delta_{11} = \iota_1(\Delta_4) \\
{\delta_0}^{-1} &: \Delta_7 \longrightarrow \Delta_1 \cup \Delta_2 \cup \Delta_3 \cup \Delta_4 \cup \Delta_7.
\end{aligned}$$

We now show that the canonical Dehn twists map maximal cells onto the union of other maximal cells listed above. The *interior* of a maximal cell is defined to be Δ_j°, the collection of points $(a, b, c, d) \in \Delta_j$ with a, b, c, d all positive.

Proposition 4.1.1. *Denote the canonical Dehn twist on the cell* Δ_k *by* η_k. *For each* $j, k \in \{1, \ldots, 28\}$, *if* $\Delta_j^\circ \cap \eta_k(\Delta_k^\circ)$ *is non-empty then* $\Delta_j \subset \eta_k(\Delta_k)$.

Proof. Since the maximal cells only overlap on their boundaries and we are assuming that $\Delta_j^\circ \cap \eta_k(\Delta_k^\circ)$ is non-empty it is sufficient to consider only the maximal cells which appear in the images of canonical Dehn twists listed above. In order to check that $\Delta_j \subset \eta_k(\Delta_k)$ we need only show ${\eta_k}^{-1}(\Delta_j) \subset \Delta_k$. This may be checked using the propositions of section 3.3 by inspection on a case by case basis. □

4.2. The Markov map on $\mathcal{ML}(\Sigma_2)$

We are going to define the Markov map on the partition of $\mathcal{ML}(\Sigma_2)$ into maximal cells by taking the shortest word in canonical Dehn twists that maps the maximal cell Δ_j onto a union of at least two maximal cells. This will ensure that the resulting mapping has enough "expansion" to strictly reduce length. We will see that composition of at most three canonical Dehn twists has this property. As usual there will be an ambiguity as to how the Markov

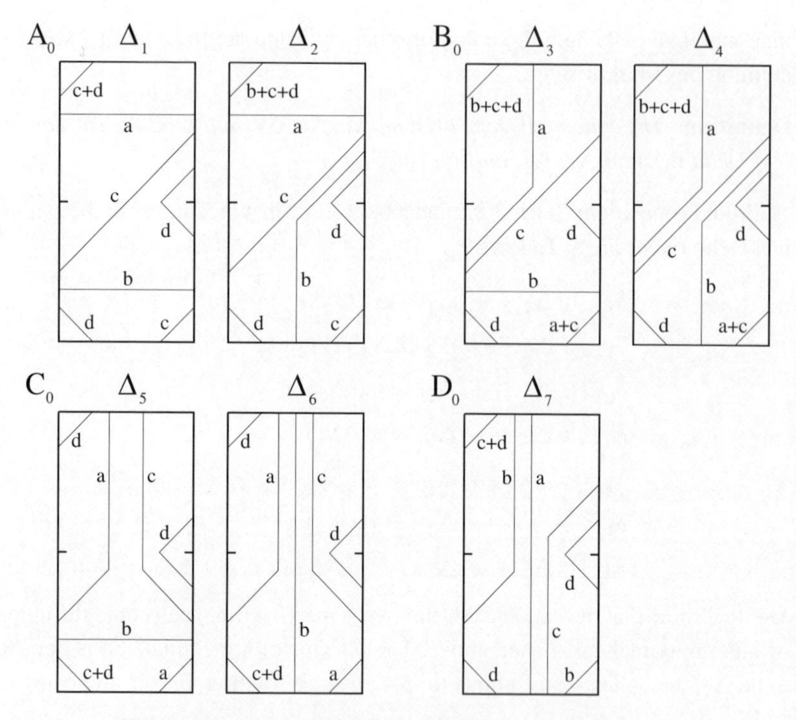

Figure 4.2.1 Generic points in the regions A_0, B_0, C_0 and D_0.

map is defined on the intersections of maximal cells (that is on their common boundary faces).

It turns out that certain pairs of maximal cells, for example $\Delta_1 \cup \Delta_2$, have the same canonical Dehn twist and always occur together in the image of a canonical twist. Others, for example Δ_7, occur on their own in the image. This means that we can make our partition of $\mathcal{ML}(\Sigma_2)$ coarser. That is, we group the maximal cells Δ_j in pairs or on their own according to how they behave under these canonical Dehn twists. These groupings will be called *regions*. The regions are defined as follows:

$$
\begin{aligned}
A_0 &= \Delta_1 \cup \Delta_2, & B_0 &= \Delta_3 \cup \Delta_4, & C_0 &= \Delta_5 \cup \Delta_6, & D_0 &= \Delta_7,\\
A_1 &= \Delta_8 \cup \Delta_9, & B_1 &= \Delta_{10} \cup \Delta_{11}, & C_1 &= \Delta_{12} \cup \Delta_{13}, & D_1 &= \Delta_{14},\\
A_2 &= \Delta_{15} \cup \Delta_{16}, & B_2 &= \Delta_{17} \cup \Delta_{18}, & C_2 &= \Delta_{19} \cup \Delta_{20}, & D_2 &= \Delta_{21},\\
A_3 &= \Delta_{22} \cup \Delta_{23}, & B_3 &= \Delta_{24} \cup \Delta_{25}, & C_3 &= \Delta_{26} \cup \Delta_{27}, & D_3 &= \Delta_{28}.
\end{aligned}
$$

There are four different types of region which we call A_j, B_j, C_j, D_j where $j = 0, 1, 2, 3$ is determined by $A_j = \iota_j(A_0)$ and so on.

Lemma 4.2.1. *The canonical Dehn twists map regions onto unions of other regions. For A_0, B_0, C_0 and D_0 these regions are given below.*

$$\begin{aligned}
\delta_1 &: A_0 \longmapsto A_0 \cup B_0 \cup C_0 \cup D_0 \\
\delta_0^{-1} &: B_0 \longmapsto A_3 = \iota_3(A_0) \\
\delta_1 &: C_0 \longmapsto B_1 = \iota_1(B_0) \\
\delta_0^{-1} &: D_0 \longmapsto A_0 \cup B_0 \cup D_0.
\end{aligned}$$

The results for the other regions may be obtained by symmetry.

We are now ready to define the Markov map f_2 on $\mathcal{ML}(\Sigma_2)$. This is defined to be the shortest word in the canonical Dehn twists that maps each region onto at least two other regions. It is given as follows:

$$\begin{aligned}
f_2|_{A_0} &= \delta_1 : A_0 \longmapsto A_0 \cup B_0 \cup C_0 \cup D_0 \\
f_2|_{B_0} &= \delta_2^{-1}\delta_0^{-1} : B_0 \longmapsto A_3 \cup B_3 \cup C_3 \cup D_3 \\
f_2|_{C_0} &= \delta_2\delta_0\delta_1 : C_0 \longmapsto A_2 \cup B_2 \cup C_2 \cup D_2 \\
f_2|_{D_0} &= \delta_0^{-1} : D_0 \longmapsto A_0 \cup B_0 \cup D_0 \\
f_2|_{A_1} &= \delta_1^{-1} : A_1 \longmapsto A_1 \cup B_1 \cup C_1 \cup D_1 \\
f_2|_{B_1} &= \delta_2\delta_0 : B_1 \longmapsto A_2 \cup B_2 \cup C_2 \cup D_2 \\
f_2|_{C_1} &= \delta_2^{-1}\delta_0^{-1}\delta_1^{-1} : C_1 \longmapsto A_3 \cup B_3 \cup C_3 \cup D_3 \\
f_2|_{D_1} &= \delta_0 : D_1 \longmapsto A_1 \cup B_1 \cup D_1 \\
f_2|_{A_2} &= \delta_2 : A_2 \longmapsto A_2 \cup B_2 \cup C_2 \cup D_2 \\
f_2|_{B_2} &= \delta_1^{-1}\delta_0^{-1} : B_2 \longmapsto A_1 \cup B_1 \cup C_1 \cup D_1 \\
f_2|_{C_2} &= \delta_1\delta_0\delta_2 : C_2 \longmapsto A_0 \cup B_0 \cup C_0 \cup D_0 \\
f_2|_{D_2} &= \delta_0^{-1} : D_2 \longmapsto A_2 \cup B_2 \cup D_2 \\
f_2|_{A_3} &= \delta_2^{-1} : A_3 \longmapsto A_3 \cup B_3 \cup C_3 \cup D_3 \\
f_2|_{B_3} &= \delta_1\delta_0 : B_3 \longmapsto A_0 \cup B_0 \cup C_0 \cup D_0 \\
f_2|_{C_3} &= \delta_1^{-1}\delta_0^{-1}\delta_2^{-1} : C_3 \longmapsto A_1 \cup B_1 \cup C_1 \cup D_1 \\
f_2|_{D_3} &= \delta_0 : D_3 \longmapsto A_3 \cup B_3 \cup D_3.
\end{aligned}$$

It is clear from Proposition 4.1.1 and Lemma 4.2.1 that f_2 satisfies the Markov property with respect to the partition of $\mathcal{ML}(\Sigma)$ into the sixteen regions $A_0, \ldots, D_3$.

Recall that the *length* $|\gamma|$ of a multiple simple loop γ supported on a weighted π_1-train track τ is the sum of the weights on all its strands. For a proper integral weighting this is a positive integer. Let $(a, b, c, d) \in \Delta_j$ with the standard bases. We now compare the length of the π_1-train track represented by this point with the length of its image under the Markov map f_2 as defined above. We have

Proposition 4.2.2. The Markov map f_2 does not increase length. Moreover f_2 strictly decreases length for every train track except those of the form:

$$a\mathbf{e}^0 + b\mathbf{e}_0,\ a\mathbf{e}^1_\infty + b\mathbf{e}^\infty_1,\ a\mathbf{e}^{-1}_\infty + b\mathbf{e}^\infty_{-1},\ a\mathbf{e}^\infty_\infty + b\mathbf{e}^1_{-1},\ a\mathbf{e}^\infty_\infty + b\mathbf{e}^{-1}_1, \quad (1)$$

$$a\mathbf{e}^0 + b\mathbf{e}^1_\infty + c\mathbf{e}^{-1}_\infty,\ a\mathbf{e}^\infty_\infty + b\mathbf{e}^1_\infty + c\mathbf{e}^{-1}_\infty,\ a\mathbf{e}_0 + b\mathbf{e}^\infty_1 + c\mathbf{e}^\infty_{-1},$$
$$a\mathbf{e}^\infty_\infty + b\mathbf{e}^\infty_1 + c\mathbf{e}^\infty_{-1}. \quad (2)$$

Proof. This follows from an analysis of lengths for the action of f_2 on each maximal cell Δ_j:

cell	$\|(a, b, c, d)\|$	f_2	$\|f_2(a, b, c, d)\|$	$\|(a, b, c, d)\| - \|f_2(a, b, c, d)\|$
Δ_1	$a + b + 3c + 3d$	δ_1	$a + b + 2c + 2d$	$c + d$
Δ_2	$a + 2b + 3c + 3d$	δ_1	$a + b + 2c + 2d$	$b + c + d$
Δ_3	$2a + b + 3c + 3d$	${\delta_2}^{-1}{\delta_0}^{-1}$	$a + b + 2c + 2d$	$a + c + d$
Δ_4	$2a + 2b + 3c + 3d$	${\delta_2}^{-1}{\delta_0}^{-1}$	$a + b + 2c + 3d$	$a + b + c$
Δ_5	$2a + b + 2c + 3d$	$\delta_2\delta_0\delta_1$	$2a + b + 2c + d$	$2d$
Δ_6	$2a + b + 2c + 3d$	$\delta_2\delta_0\delta_1$	$2a + b + 2c + d$	$2d$
Δ_7	$a + 2b + 2c + 3d$	${\delta_0}^{-1}$	$a + b + c + 3d$	$b + c.$

□

In what follows we will be particularly interested in the four train tracks (2) (as illustrated in the lower line of Fig. 4.2.2) which are $\Delta_{19} \cap \Delta_{26}$, $\Delta_{20} \cap \Delta_{27}$ (which together form $C_2 \cap C_3$) and $\Delta_5 \cap \Delta_{12}$, $\Delta_6 \cap \Delta_{13}$ (which together form $C_0 \cap C_1$).

4.3. The space of Farey blocks

As for the once punctured torus, in order to construct a normal form and word acceptor from the Markov map we constructed in the previous section, we need to find a space on which $\mathcal{MCG}(\Sigma_2)$ acts without fixed points. For the once punctured torus we could take the space of Farey pairs. For the twice punctured torus we will generalise this to the space of *Farey blocks*. A Farey block is an ordered quadruple of (homotopy classes of) simple loops which lie in a certain topological configuration on Σ_2.

Let γ_i and γ_j be (homotopy classes of) simple closed curves on Σ_2. Define the intersection number $i(\gamma_i, \gamma_j)$ to be the minimal number of points in $\gamma_i \cap \gamma_j$

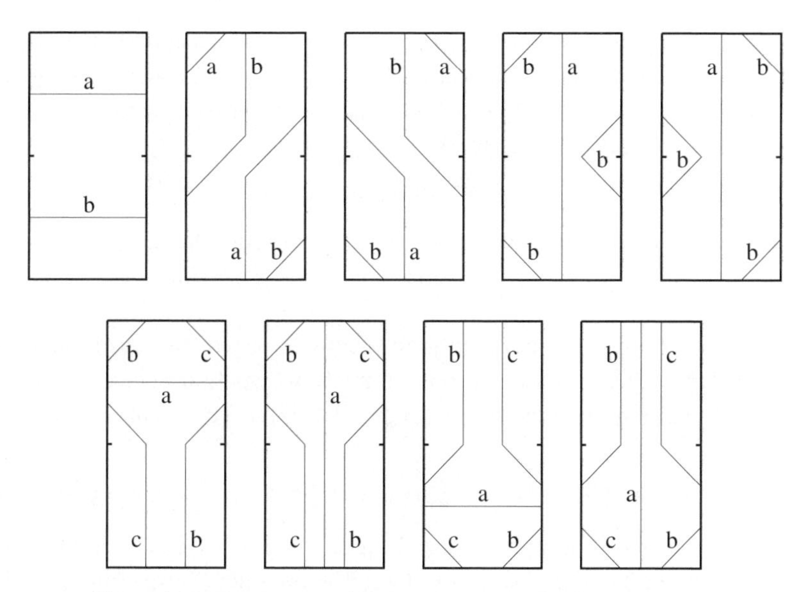

Figure 4.2.2 The nine exceptional configurations of Proposition 4.2.2.

as γ_i, γ_j vary through all elements of their free homotopy class. The idea behind the construction of Farey blocks is the following. Consider a pair of Farey neighbours on the twice punctured torus, that is an ordered pair of curves (γ_1, γ_2) that intersect exactly once, so that $i(\gamma_1, \gamma_2) = 1$. Both of these curves are necessarily non-dividing and there exists a curve β that separates $\gamma_1 \cup \gamma_2$ from the punctures and which is unique up to isotopy. The curve β is the commutator $[\gamma_1, \gamma_2]$. On an unpunctured or once punctured torus β would be homotopically trivial or peripheral respectively. Moreover, in that case (γ_1, γ_2) would have trivial stabiliser in the mapping class group. However, for the twice punctured torus this is not the case. There is a non-trivial homeomorphism which preserves γ_1, γ_2 and β and which interchanges the punctures. The square of this map is the Dehn twist about β. We get around this difficulty by considering an extra curve γ_3 that is disjoint from γ_1 and intersects γ_2 and β once and twice respectively. Here is the precise definition.

Definition. *A* **Farey block** *is an ordered quadruple of (homotopy classes of) curves* $(\gamma_1, \gamma_2, \beta; \gamma_3)$ *with the properties that:*

(i) γ_j *is non-dividing for* $j = 1, 2, 3$ *and* β *is dividing,*
(ii) $i(\gamma_j, \beta) = 0$ *for* $j = 1, 2$ *and* $i(\gamma_3, \beta) = 2$,
(iii) $i(\gamma_2, \gamma_j) = 1$ *for* $j = 1, 3$ *and* $i(\gamma_1, \gamma_3) = 0$.

The collection of all Farey blocks will be denoted $\mathcal{F}$. Since Farey blocks are only defined using topological data, namely the intersection number and the separation properties of (homotopy class of) simple closed curves, the image of a Farey block under an element of $\mathcal{MCG}(\Sigma_2)$ is also a Farey block. This defines an action of $\mathcal{MCG}(\Sigma_2)$ on $\mathcal{F}$. We claim that this action is free. In order to see this, consider the Farey block

$$\big(\mathbf{e}_0, \mathbf{e}_\infty^\infty, \mathbf{e}_1^\infty + \mathbf{e}_{-1}^\infty; \mathbf{e}^0\big).$$

It is clear that this has trivial stabiliser. (It is easy to see that this block is mapped to itself by ι_1 but we are only considering orientation preserving automorphisms of Σ_2.)

The idea behind the normal form for the twice punctured torus is similar to that for the once punctured torus. Namely we apply a general element $\phi \in \mathcal{MCG}(\Sigma_2)$ to the Farey block $\big(\mathbf{e}_0, \mathbf{e}_\infty^\infty, \mathbf{e}_1^\infty + \mathbf{e}_{-1}^\infty; \mathbf{e}^0\big)$ to obtain a new Farey block

$$\phi\big(\mathbf{e}_0, \mathbf{e}_\infty^\infty, \mathbf{e}_1^\infty + \mathbf{e}_{-1}^\infty; \mathbf{e}^0\big) = \big(\phi(\mathbf{e}_0), \phi(\mathbf{e}_\infty^\infty), \phi(\mathbf{e}_1^\infty + \mathbf{e}_{-1}^\infty); \phi(\mathbf{e}^0)\big).$$

The idea is to apply the Markov map f_2 repeatedly to $\phi\big(\mathbf{e}_0, \mathbf{e}_\infty^\infty, \mathbf{e}_1^\infty + \mathbf{e}_{-1}^\infty; \mathbf{e}^0\big)$ until we get back our original Farey block. The resulting f_2-expansion should be the normal form for ϕ. In practice it is slightly more complicated than this.

Crucial to the construction for the once punctured torus, in section 2, was the following fact. If (γ_1, γ_2) was a Farey pair, then γ_1 and γ_2 both lay in the same cell I_j in $\mathcal{ML}(\Sigma_1)$ so that we could take $f_1(\gamma_1) = \alpha_j(\gamma_1)$, $f_1(\gamma_2) = \alpha_j(\gamma_2)$ for the same element $\alpha_j \in \mathcal{MCG}(\Sigma_2)$. In consequence, $\big(f_1(\gamma_1), f_1(\gamma_2)\big)$ was again a Farey pair and applying f_1 to a sequence of such pairs gave a well-defined sequence of elements in $\mathcal{MCG}(\Sigma_1)$, which defined our normal forms.

The analogous statement about Farey blocks is almost, but unfortunately not quite, correct. In fact we can observe that if $(\gamma_1, \gamma_2, \beta; \gamma_3)$ is a Farey block and if β is in the interior of a region R then γ_1 and γ_2 are also in R. This is because γ_1 and γ_2 are disjoint from β and so the π_1-train tracks representing these γ_i and β are both contained in the same maximal cell (compare this with Proposition 2.3.3). In addition, if γ_1 is in the R° then γ_3 is also in R. If, however, γ_1 is on the boundary of R then γ_3 may be in an adjacent region. In the next section we characterise those weighted π_1-train tracks which represent dividing curves. This will enable us to determine those exceptional cases which cause difficulty in extending the map f_2.

4.4. Dividing curves

The following lemma will allow us to determine which dividing curves do not lie in the interior of any region.

Lemma 4.4.1. *If a proper integral weighting w on a π_1-train track τ on R_2 represents a connected dividing loop β on Σ_2 then*

(i) *for each side* σ *of* R_2, *the sum of the weights on arcs with endpoints on* σ *is even,*

(ii) *the weights of* w *have no common factor.*

Proof. Let τ be a π_1-train track carrying weight $w(\alpha)$ on strand α. As usual, the weighting can be expanded to the loop β by replacing each strand α with $w(\alpha)$ disjoint strands joining the same pair of sides and gluing their ends together with side pairings. Since β is connected, condition (ii) is clear.

Any dividing curve β separates Σ_2 into a torus with a hole and a sphere with 2 punctures and a hole. Colouring the components of $\Sigma_2 - \beta$ with distinct colours, we see that both punctures lie in a region of the same colour. The colouring lifts to a colouring of $R_2 - \beta$ in such a way that colours alternate along the sides of R_2, changing each time an endpoint of a strand of β meets the side. Since β avoids the punctures, the two segments of side which meet in a puncture have the same colour. The condition (i) is now clear. □

Proposition 4.4.2. Every dividing curve is in the interior of a region with the following six exceptions, see Fig. *4.4.1:*

$$\chi_0 = \mathbf{e}_1^\infty + \mathbf{e}_{-1}^\infty \in C_0 \cap C_1,\ \ \beta_0 = \mathbf{e}_1^1 + \mathbf{e}_{-1}^1 \in A_0 \cap B_0,$$
$$\beta_1 = \mathbf{e}_1^{-1} + \mathbf{e}_{-1}^{-1} \in A_1 \cap B_1,$$
$$\chi_2 = \mathbf{e}_\infty^1 + \mathbf{e}_\infty^{-1} \in C_2 \cap C_3,\ \ \beta_2 = \mathbf{e}_1^1 + \mathbf{e}_1^{-1} \in A_2 \cap B_2,$$
$$\beta_3 = \mathbf{e}_{-1}^1 + \mathbf{e}_{-1}^{-1} \in A_3 \cap B_3.$$

Proof. Any weighting w representing a dividing curve must satisfy conditions (i) and (ii) of Lemma 4.4.1. Recall that in section 3.1 we mentioned the idea of splitting R_2 into two boxes by drawing a horizontal line from v_1 to v_4. We claim that if w represents a dividing curve then, in at least one of these boxes, there are arcs with non-zero weights across each of the four corners. We say that such a box *contains a cross*. (This condition plays a crucial role in [12]. As can be seen in Fig. 4.4.1, the curves χ_0, β_0, β_1 contain a cross in the lower box and χ_2, β_2, β_3 have one in the upper box.) Let w be a weighting representing a connected dividing curve where neither box contains a cross. If a box does not contain a cross, then one can see by inspection of the irreducible loops in Fig. 3.1.2 (or the generic π_1-train tracks in Fig. 3.1.3), that it contains a certain number a of corner strands across one pair of opposite corners and a number b strands going across a pair of opposite sides. Thus the number of strands ending at each pair of opposite sides is a and $a + b$ respectively. By (i) both a and $a + b$ are even. Thus so is b. Doing this for both boxes we obtain a contradiction to (ii).

Now we refer to Fig. 4.2.1 which shows the generic configurations for weights in any of the four regions A_0, B_0, C_0, D_0. Let us take the region

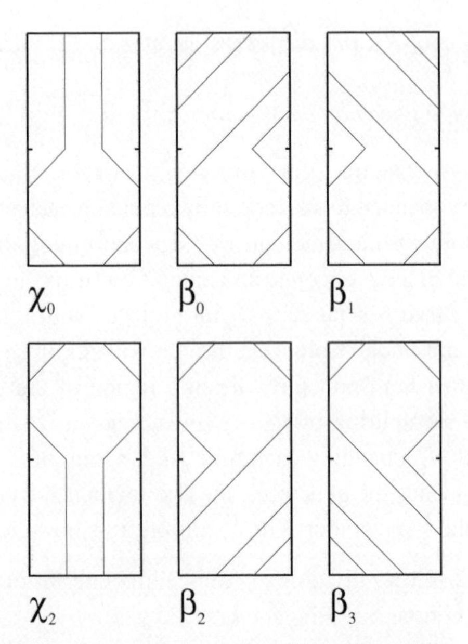

Figure 4.4.1 The six curves of Proposition 4.4.2.

$A_0 = \Delta_1 \cup \Delta_2$; the other cases are similar. The observation that w contains a cross translates into the statement that $c > 0$ and $d > 0$, where, as shown in the top left two diagrams, c and d are the weights on $\mathbf{e}_1^1$ and $\mathbf{e}_{-1}^1$ respectively. Since Δ_1 and Δ_2 are glued across the face $b = 0$ to form the region A_0, any such weighting is in the interior of A_0 unless a (the coefficient of $\mathbf{e}^0$) vanishes. It is however easy to see that any weight $b\mathbf{e}_0 + c\mathbf{e}_1^1 + d\mathbf{e}_{-1}^1$ or $b\mathbf{e}_\infty^1 + c\mathbf{e}_1^1 + d\mathbf{e}_{-1}^1$ represents a multiple loop, one of whose components is $\mathbf{e}_1^1 + \mathbf{e}_{-1}^1$. This contradicts the hypothesis that w represents a connected curve unless $b = 0$ and $c = 1 = d$, in which case $w = \beta_0$.

For $B_0 = \Delta_3 \cup \Delta_4$, the weighting w contains a cross provided $a + c > 0$ and $d > 0$. Again $b = 0$ gives $\Delta_3 \cap \Delta_4$ in the interior of B_0. Thus we have to check two cases, $c = 0$ and $a = 0$. When $a = 0$ we obtain a contradiction as above unless $w = \beta_0$. When $c = 0$, Lemma 4.4.1(i) shows that all the weights are even, in contradiction to Lemma 4.4.1(ii).

The cases of C_0 and D_0 are similar and can be left to the reader. □

The following lemma will be needed in the next section.

Lemma 4.4.3. Let β be a simple closed dividing curve on Σ_2.

(i) If β is in the interior of A_0 then $f_2|_{A_0}(\beta) = \delta_1(\beta)$ is either in the interior of one of A_0, B_0, C_0, D_0 or else $\delta_1(\beta) = \beta_0$.

(ii) If β is in the interior of D_0 then $f_2|_{D_0}(\beta) = \delta_0{}^{-1}(\beta)$ is either in the interior of one of A_0, B_0, D_0 or else $\delta_0{}^{-1}(\beta) = \beta_0$.

Proof. Both parts are similar. We only consider (i). Consider δ_1 acting on $\mathcal{ML}(\Sigma_2)$. We know that β is in the interior of A_0 and $\delta_1(A_0) = A_0 \cup B_0 \cup C_0 \cup D_0$. Therefore $\delta_1(\beta)$ is in the interior of $A_0 \cup B_0 \cup C_0 \cup D_0$. In other words, $\delta_1(\beta)$ is in the interior of one of A_0, B_0, C_0, D_0 or else it is in the common boundary of at least 2 of these regions. By Proposition 4.4.1 we see that the only possibility is that $\delta_1(\beta) = \beta_0$. □

We will need to characterise the curve γ_3 in the Farey block $(\gamma_1, \gamma_2, \beta; \gamma_3)$ when $(\gamma_1, \gamma_2, \beta)$ is either $(\mathbf{e}_0, \mathbf{e}_\infty^\infty, \chi_0)$ or $(\mathbf{e}_\infty^\infty, \mathbf{e}_0, \chi_0)$.

Proposition 4.4.4. If $(\gamma_1, \gamma_2, \beta; \gamma_3)$ is a Farey block with $(\gamma_1, \gamma_2, \beta) = (\mathbf{e}_0, \mathbf{e}_\infty^\infty, \chi_0)$ then, for some non-negative integer a, γ_3 has one of the following forms:

(i) $\mathbf{e}^0$,
(ii) $\mathbf{e}_1^\infty + \mathbf{e}_{-1}^1 + a\chi_0 = (a+1, 0, a, 1) \in \Delta_5 \cap \Delta_6$,
(iii) $\mathbf{e}_{-1}^\infty + \mathbf{e}_1^{-1} + a\chi_0 = (a+1, 0, a, 1) \in \Delta_{12} \cap \Delta_{13}$.

If $(\gamma_1, \gamma_2, \beta) = (\mathbf{e}_\infty^\infty, \mathbf{e}_0, \chi_0)$ then, for some non-negative integer a, γ_3 has one of the following forms:

(iv) $\mathbf{e}_{-1}^1 + a\chi_0 = (a, 0, a, 1) \in \Delta_5 \cap \Delta_6$,
(v) $\mathbf{e}_1^{-1} + a\chi_0 = (a, 0, a, 1) \in \Delta_{12} \cap \Delta_{13}$.

Remark. *Observe that, apart from the case $(\gamma_1, \gamma_2, \beta; \gamma_3) = (\mathbf{e}_0, \mathbf{e}_\infty^\infty, \chi_0; \mathbf{e}^0)$, in this situation all four curves $\gamma_1, \gamma_2, \gamma_3, \beta$ are in either $\Delta_5 \cup \Delta_6 = C_0$ or $\Delta_{12} \cup \Delta_{13} = C_1$.*

Proof. We begin with the case $(\gamma_1, \gamma_2, \beta) = (\mathbf{e}_0, \mathbf{e}_\infty^\infty, \chi_0)$. We know γ_3 must be disjoint from $\gamma_1 = \mathbf{e}_0$. There is a one parameter family of simple closed curves on Σ_2 disjoint from $\mathbf{e}_0$. (To see this, observe that $\Sigma_2 - \{\mathbf{e}_0\}$ is topologically a four times punctured sphere. It is well known that simple closed curves on the four punctured sphere are parametrised by $\mathbb{Q} \cup \{\infty\}$.) It is not hard to show that all curves disjoint from $\mathbf{e}_0$ must have one of the following six types (see Fig. 4.4.2):

$$(a, 0, b, b) \in \Delta_1, \quad (a, 0, b, a+b) \in \Delta_3, \quad (a+b, 0, a, b) \in \Delta_5,$$
$$(a+b, 0, a, b) \in \Delta_{12}, \quad (a, 0, b, a+b) \in \Delta_{10}, \quad (a, 0, b, b) \in \Delta_8.$$

It is easy to see that the intersection numbers of these curves with $\gamma_2 = \mathbf{e}_\infty^\infty$ are: $a+2b$, $a+2b$, b, b, $a+2b$, $a+2b$ respectively. Since γ_3 should intersect $\gamma_2 = \mathbf{e}_\infty^\infty$ exactly once we obtain the result.

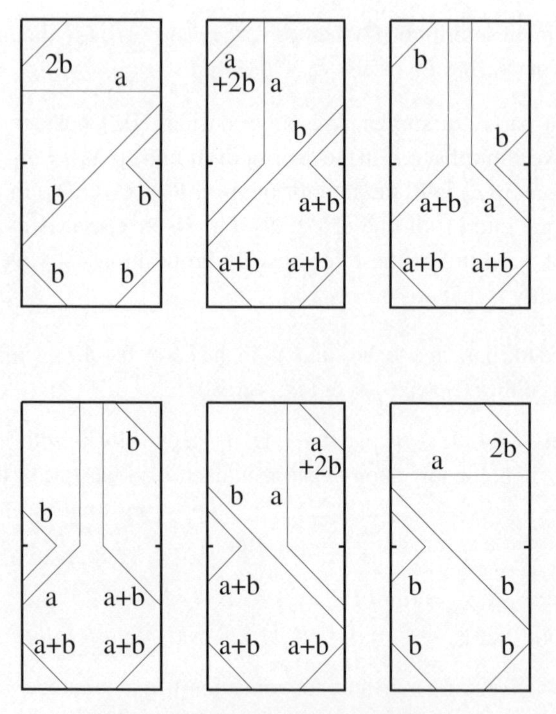

Figure 4.4.2 The curves disjoint from $\mathbf{e}_0$.

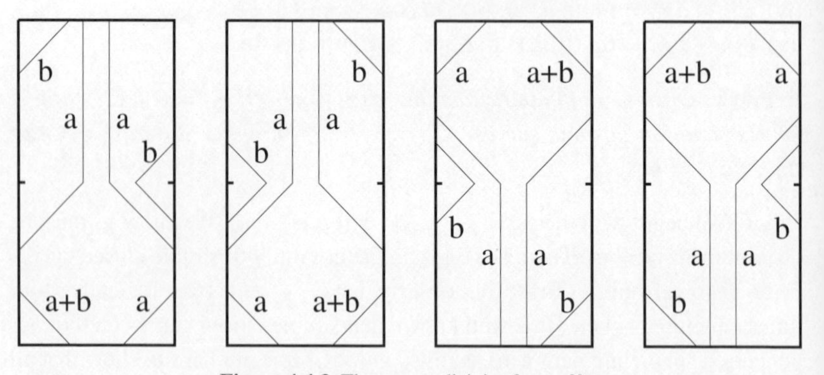

Figure 4.4.3 The curves disjoint from $\mathbf{e}_\infty^\infty$.

Now we turn our attention to the case $(\gamma_1, \gamma_2, \beta) = (\mathbf{e}_\infty^\infty, \mathbf{e}_0, \chi_0)$. All curves disjoint from $\gamma_1 = \mathbf{e}_\infty^\infty$ must have one of the following four forms (see Fig. 4.4.3):

$$(a, 0, a, b) \in \Delta_6, \quad (a, 0, a, b) \in \Delta_{13}, \quad (a, 0, a, b) \in \Delta_{20}, \quad (a, 0, a, b) \in \Delta_{27}.$$

The intersection numbers of these curves with $\gamma_2 = \mathbf{e}_0$ are b, b, $2a+b$, $2a+b$ respectively. The result follows as above. $\square$

The next proposition is a similar characterisation of γ_3 when $(\gamma_1, \gamma_2, \beta)$ is $(\mathbf{e}^0, \mathbf{e}_\infty^\infty, \chi_2)$ or $(\mathbf{e}_\infty^\infty, \mathbf{e}^0, \chi_2)$.

Proposition 4.4.5. If $(\gamma_1, \gamma_2, \beta; \gamma_3)$ is a Farey block with $(\gamma_1, \gamma_2, \beta) = (\mathbf{e}^0, \mathbf{e}_\infty^\infty, \chi_2)$ then, for some non-negative integer a, γ_3 has one of the following forms:

(i) $\mathbf{e}_0$,
(ii) $\mathbf{e}_\infty^1 + \mathbf{e}_1^{-1} + a\chi_2 = (a+1, 0, a, 1) \in \Delta_{19} \cap \Delta_{20}$,
(iii) $\mathbf{e}_\infty^{-1} + \mathbf{e}_{-1}^1 + a\chi_2 = (a+1, 0, a, 1) \in \Delta_{26} \cap \Delta_{27}$.

If $(\gamma_1, \gamma_2, \beta) = (\mathbf{e}_\infty^\infty, \mathbf{e}^0, \chi_2)$ then, for some non-negative integer a, γ_3 has one of the following forms:

(iv) $\mathbf{e}_1^{-1} + a\chi_2 = (a, 0, a, 1) \in \Delta_{19} \cap \Delta_{20}$,
(v) $\mathbf{e}_{-1}^1 + a\chi_2 = (a, 0, a, 1) \in \Delta_{26} \cap \Delta_{27}$.

Proof. This follows by applying ι_2 to the results in Proposition 4.4.4. $\square$

Observe that in this case we have $(\gamma_1, \gamma_2, \beta; \gamma_3) = (\mathbf{e}^0, \mathbf{e}_\infty^\infty, \chi_2; \mathbf{e}_0)$ or else all four curves are in $\Delta_{19} \cup \Delta_{20} = C_2$ or $\Delta_{26} \cup \Delta_{27} = C_3$.

4.5. Subdividing $\mathcal{F}$: the states of the word acceptor

In the next section we shall extend the Markov map f_2 to a map on the space of Farey blocks $\mathcal{F}$. Inverting this map will give the word acceptor. This construction will resemble that given in section 2.3 but will involve some extra steps. The first part of this process is to divide $\mathcal{F}$ into subsets which will form the states of the word acceptor. We begin by stratifying $\mathcal{F}$ into three subsets. Define

$$\begin{aligned}\mathcal{F}_0 = \big\{(\gamma_1, \gamma_2, \beta; \gamma_3) \in \mathcal{F} : \beta = \chi_0,\ \{\gamma_1, \gamma_2\} = \{\mathbf{e}_0, \mathbf{e}_\infty^\infty\} \text{ or}\\ \beta = \chi_2,\ \{\gamma_1, \gamma_2\} = \{\mathbf{e}^0, \mathbf{e}_\infty^\infty\}\big\},\end{aligned}$$

$$\begin{aligned}\mathcal{F}_1 = \big\{(\gamma_1, \gamma_2, \beta; \gamma_3) \in \mathcal{F} : \beta = \chi_0,\ \{\gamma_1, \gamma_2\} \neq \{\mathbf{e}_0, \mathbf{e}_\infty^\infty\} \text{ or}\\ \beta = \chi_2,\ \{\gamma_1, \gamma_2\} \neq \{\mathbf{e}^0, \mathbf{e}_\infty^\infty\}\big\},\end{aligned}$$

$$\mathcal{F}_2 = \big\{(\gamma_1, \gamma_2, \beta; \gamma_3) \in \mathcal{F} : \beta \notin \{\chi_0, \chi_2\}\big\}.$$

Our goal is to define a map f on $\mathcal{F}$ so that for each Farey block $(\gamma_1, \gamma_2, \beta; \gamma_3)$ there is a non-negative integer n so that

$$f^n(\gamma_1, \gamma_2, \beta; \gamma_3) = (\mathbf{e}_0, \mathbf{e}_\infty^\infty, \chi_0; \mathbf{e}^0).$$

At each stage, we want f to equal a specific one of the sixteen possible values of f_2, so that for any Farey block we have

$$f(\gamma_1, \gamma_2, \beta; \gamma_3) = \big(f_2(\gamma_1), f_2(\gamma_2), f_2(\beta); f_2(\gamma_3)\big)$$
$$= \big(\phi(\gamma_1), \phi(\gamma_2), \phi(\beta); \phi(\gamma_3)\big)$$

for some $\phi \in \mathcal{MCG}(\Sigma_2)$. The difficulty is that, if the members of the block $(\gamma_1, \gamma_2, \beta; \gamma_3)$ lie in different regions, it is not clear which value for ϕ to pick. This happens, for example, in the case of the Farey block $(\mathbf{e}^1_{-1}, \mathbf{e}_0, \beta_0; \mathbf{e}^\infty_\infty)$.

We resolve this difficulty by using the results of the previous section and the stratification above. Roughly speaking, f will be defined as follows. On $\mathcal{F}_2$, the map f will take the value of f_2 on the region which contains the dividing curve β. Applying f will decrease the length $|\beta|$ of β. We continue applying f until $\beta = \chi_0$ or χ_2. Thus, there will be a non-negative integer n_2 so that $f^{n_2}(\gamma_1, \gamma_2, \beta; \gamma_3) \in \mathcal{F}_1 \cup \mathcal{F}_0$. On $\mathcal{F}_1$ the map f will be the Markov map f_1 on $\mathcal{ML}(\Sigma_1)$ constructed in section 2.3. This map will fix β and reduce $|\gamma_1| + |\gamma_2|$. Thus, there will be a non-negative integer n_1 so that $f^{n_1+n_2}(\gamma_1, \gamma_2, \beta; \gamma_3) \in \mathcal{F}_0$. Finally, on $\mathcal{F}_0$ the map f will fix $|\gamma_1| + |\gamma_2| + |\beta|$ and decrease $|\gamma_3|$. There will be a non-negative integer n_0 so that $f^{n_0+n_1+n_2}(\gamma_1, \gamma_2, \beta; \gamma_3) = (\mathbf{e}_0, \mathbf{e}^\infty_\infty, \chi_0; \mathbf{e}^0)$.

In order to carry out the details we shall introduce a number of extra regions which will become the states of the word acceptor. These regions will all be subsets of the space $\mathcal{F}$ of Farey blocks.

We begin by considering $\mathcal{F}_2$. By definition, if the Farey block $(\gamma_1, \gamma_2, \beta; \gamma_3)$ is in $\mathcal{F}_2$ then the dividing curve β is neither χ_0 nor χ_2. Using Proposition 4.4.2, we see that either β is in the interior of one of the sixteen regions $A_0, \ldots, D_3$ or else β is one of the four dividing curves $\beta_0, \beta_1, \beta_2, \beta_3$. (Recall that for these four curves $\beta_j \in A_j \cap B_j$). We divide $\mathcal{F}_2$ into twenty regions as follows. We call these $\mathbf{A}_j, \mathbf{B}_j, \mathbf{C}_j, \mathbf{D}_j, \mathbf{E}_j$ for $j = 0, 1, 2, 3$. The sixteen subsets $\mathbf{A}_0, \ldots, \mathbf{D}_3$ of $\mathcal{F}_2$ are defined to consist of all Farey blocks for which β is in the interior of the region $A_0, \ldots, D_3$ in $\mathcal{ML}(\Sigma_2)$, respectively. The four subsets $\mathbf{E}_j$ of $\mathcal{F}_2$ are defined to consist of all Farey blocks for which $\beta = \beta_j$ where $j = 0, 1, 2, 3$. That is when $j = 0$:

$$\mathbf{A}_0 = \big\{(\gamma_1, \gamma_2, \beta; \gamma_3) \in \mathcal{F} : \beta \in (\Delta_1 \cup \Delta_2)^\circ\big\},$$

$$\mathbf{B}_0 = \big\{(\gamma_1, \gamma_2, \beta; \gamma_3) \in \mathcal{F} : \beta \in (\Delta_3 \cup \Delta_4)^\circ\big\},$$

$$\mathbf{C}_0 = \big\{(\gamma_1, \gamma_2, \beta; \gamma_3) \in \mathcal{F} : \beta \in (\Delta_5 \cup \Delta_6)^\circ\big\},$$

$$\mathbf{D}_0 = \big\{(\gamma_1, \gamma_2, \beta; \gamma_3) \in \mathcal{F} : \beta \in \Delta_7^\circ\big\},$$

$$\mathbf{E}_0 = \big\{(\gamma_1, \gamma_2, \beta; \gamma_3) \in \mathcal{F} : \beta = \beta_0\big\}.$$

For $k = 1, 2, 3$ apply the symmetry ι_k to the above five regions in order to obtain $\mathbf{A}_k$, $\mathbf{B}_k$, $\mathbf{C}_k$, $\mathbf{D}_k$ and $\mathbf{E}_k$.

By abuse of notation we will frequently drop the bold face notation $\mathbf{A}_j$ and simply write A_j when the meaning is clear from the context. Thus the reader should keep clearly in mind that A_j may either denote a region of $\mathcal{ML}(\Sigma_2)$ as defined in section 4.2 or the subset $\mathbf{A}_j \subset \mathcal{F}_2$ of Farey blocks. The subsets $\mathbf{A}_j, \mathbf{B}_j, \ldots, \mathbf{E}_j$ will be states in the word acceptor we are aiming to construct. Before defining the map f on each of these twenty regions (whose inverse will give the arrows in the word acceptor), we will proceed to define the states for the strata $\mathcal{F}_1$ and $\mathcal{F}_0$.

We consider first the Farey blocks in $\mathcal{F}_1$. Here we have $\beta = \chi_0$ or χ_2. For the sake of definiteness we will describe the situation for χ_0 in detail. In order to perform the same constructions for χ_2 it is necessary to apply the symmetry ι_2. The curve χ_0 divides Σ_2 into two components, one of which is a twice punctured disc and the other is a torus with a hole. All homotopically non-trivial, non-peripheral simple closed curves on Σ_2 that are disjoint from χ_0 are contained in the one holed torus component of $\Sigma_2 - \chi_0$. In particular, this is true for γ_1 and γ_2. The stabiliser of χ_0 in $\mathcal{MCG}(\Sigma_2)$ is the group generated by δ_0 and δ_2. The action of the group they generate on the one-holed-torus component of $\Sigma_2 - \chi_0$ exactly corresponds to the action of $\mathcal{MCG}(\Sigma_1)$ on Σ_1 considered in section 2. Therefore, we divide $\mathcal{F}_1$ into states which correspond to the intervals I_0, I_1, I_2 and I_3 for the space of Farey pairs on the once punctured torus. This is done as follows:

$$F_0 = \{(\gamma_1, \gamma_2, \beta; \gamma_3) \in \mathcal{F} : \beta = \chi_0;\ \gamma_1, \gamma_2 \in \mathrm{sp}^+\{\mathbf{e}_0, \mathbf{e}_1^\infty\}\},$$

$$G_0 = \{(\gamma_1, \gamma_2, \beta; \gamma_3) \in \mathcal{F} : \beta = \chi_0;\ \gamma_1, \gamma_2 \in \mathrm{sp}^+\{\mathbf{e}_\infty^\infty, \mathbf{e}_1^\infty\}\},$$

$$F_1 = \{(\gamma_1, \gamma_2, \beta; \gamma_3) \in \mathcal{F} : \beta = \chi_0;\ \gamma_1, \gamma_2 \in \mathrm{sp}^+\{\mathbf{e}_0, \mathbf{e}_{-1}^\infty\}\},$$

$$G_1 = \{(\gamma_1, \gamma_2, \beta; \gamma_3) \in \mathcal{F} : \beta = \chi_0;\ \gamma_1, \gamma_2 \in \mathrm{sp}^+\{\mathbf{e}_\infty^\infty, \mathbf{e}_{-1}^\infty\}\},$$

$$F_2 = \{(\gamma_1, \gamma_2, \beta; \gamma_3) \in \mathcal{F} : \beta = \chi_2;\ \gamma_1, \gamma_2 \in \mathrm{sp}^+\{\mathbf{e}^0, \mathbf{e}_\infty^1\}\},$$

$$G_2 = \{(\gamma_1, \gamma_2, \beta; \gamma_3) \in \mathcal{F} : \beta = \chi_2;\ \gamma_1, \gamma_2 \in \mathrm{sp}^+\{\mathbf{e}_\infty^\infty, \mathbf{e}_\infty^1\}\},$$

$$F_3 = \{(\gamma_1, \gamma_2, \beta; \gamma_3) \in \mathcal{F} : \beta = \chi_2;\ \gamma_1, \gamma_2 \in \mathrm{sp}^+\{\mathbf{e}^0, \mathbf{e}_\infty^{-1}\}\},$$

$$G_3 = \{(\gamma_1, \gamma_2, \beta; \gamma_3) \in \mathcal{F} : \beta = \chi_2;\ \gamma_1, \gamma_2 \in \mathrm{sp}^+\{\mathbf{e}_\infty^\infty, \mathbf{e}_\infty^{-1}\}\}.$$

Finally, we consider $\mathcal{F}_0$. Here either $\beta = \chi_0$ and $\{\gamma_1, \gamma_2\} = \{\mathbf{e}_0, \mathbf{e}_\infty^\infty\}$ or $\beta = \chi_2$ and $\{\gamma_1, \gamma_2\} = \{\mathbf{e}^0, \mathbf{e}_\infty^\infty\}$. In Propositions 4.4.4 and 4.4.5 we analysed the different possibilities for γ_3. We divide $\mathcal{F}_3$ into states according to these different possibilities:

$$H_0 = \{(\mathbf{e}_\infty^\infty, \mathbf{e}_0, \chi_0; \gamma_3) \in \mathcal{F} : \gamma_3 = (a, 0, a, 1) \in \Delta_5 \cap \Delta_6 \text{ where } a \in \mathbb{N} \cup \{0\}\},$$

$$I_0 = \{(\mathbf{e}_0, \mathbf{e}_\infty^\infty, \chi_0; \gamma_3) \in \mathcal{F} : \gamma_3 = (a+1, 0, a, 1) \in \Delta_5 \cap \Delta_6 \text{ where } a \in \mathbb{N} \cup \{0\}\},$$

$$H_1 = \{(\mathbf{e}_\infty^\infty, \mathbf{e}_0, \chi_0; \gamma_3) \in \mathcal{F} : \gamma_3 = (a, 0, a, 1) \in \Delta_{12} \cap \Delta_{13} \text{ where } a \in \mathbb{N} \cup \{0\}\},$$

$$I_1 = \{(\mathbf{e}_0, \mathbf{e}_\infty^\infty, \chi_0; \gamma_3) \in \mathcal{F} : \gamma_3 = (a+1, 0, a, 1) \in \Delta_{12} \cap \Delta_{13} \text{ where } a \in \mathbb{N} \cup \{0\}\},$$

$$H_2 = \{(\mathbf{e}_\infty^\infty, \mathbf{e}^0, \chi_2; \gamma_3) \in \mathcal{F} : \gamma_3 = (a, 0, a, 1) \in \Delta_{19} \cap \Delta_{20} \text{ where } a \in \mathbb{N} \cup \{0\}\},$$

$$I_2 = \{(\mathbf{e}^0, \mathbf{e}_\infty^\infty, \chi_2; \gamma_3) \in \mathcal{F} : \gamma_3 = (a+1, 0, a, 1) \in \Delta_{19} \cap \Delta_{20} \text{ where } a \in \mathbb{N} \cup \{0\}\},$$

$$H_3 = \{(\mathbf{e}_\infty^\infty, \mathbf{e}^0, \chi_2; \gamma_3) \in \mathcal{F} : \gamma_3 = (a, 0, a, 1) \in \Delta_{26} \cap \Delta_{27} \text{ where } a \in \mathbb{N} \cup \{0\}\},$$

$$I_3 = \{(\mathbf{e}^0, \mathbf{e}_\infty^\infty, \chi_2; \gamma_3) \in \mathcal{F} : \gamma_3 = (a+1, 0, a, 1) \in \Delta_{26} \cap \Delta_{27} \text{ where } a \in \mathbb{N} \cup \{0\}\},$$

$$J_0 = \{(\mathbf{e}_0, \mathbf{e}_\infty^\infty, \chi_0; \mathbf{e}^0) \in \mathcal{F}\},$$

$$J_2 = \{(\mathbf{e}^0, \mathbf{e}_\infty^\infty, \chi_2; \mathbf{e}_0) \in \mathcal{F}\}.$$

In each case, notice that $(a, 0, a, 0)$ corresponds to the loop β, that is χ_0 or χ_2.

4.6. The definition of the map f

Having defined the states which partition the space $\mathcal{F}$ of Farey blocks, we now turn our attention to the definition of the map f. First we consider states in $\mathcal{F}_2$. For Farey blocks in $\mathbf{A}_j, \ldots, \mathbf{D}_j$ we define f to be the same as the Markov map f_2 on the corresponding regions $A_j, \ldots, D_j$ in $\mathcal{ML}(\Sigma_2)$. We need to be slightly more careful in computing the images of these blocks.

Lemma 4.6.1. *On $\mathcal{F}_2$ the Dehn twist δ_1 maps A_0 onto $A_0 \cup B_0 \cup C_0 \cup D_0 \cup E_0$ and δ_0^{-1} maps D_0 onto $A_0 \cup B_0 \cup D_0 \cup E_0$.*

Proof. This is an immediate consequence of Lemma 4.4.3. □

To define f in $\mathbf{E}_j$ we have a choice since, as in Proposition 4.4.2, $\beta_j \in A_j \cap B_j$. We choose $f|_{\mathbf{E}_j} = f_2|_{A_j}$ so that, for example, $f|_{\mathbf{E}_0} = f_2|_{A_0} = \delta_1$. It is easy to check that $\delta_1(\beta_0) = \chi_0$ so that, with this definition, f maps $\mathbf{E}_0$ into $\mathcal{F}_1 \cup \mathcal{F}_0$. (We remark that $f_2|_{B_0}(\beta_0) = \delta_2{}^{-1}\delta_0{}^{-1}(\beta_0) = \chi_2$ so this would also be true if we had made the other choice.) Since $\delta_1{}^{-1}(\chi_0) = \beta_0$ it is easy to see that f maps $\mathbf{E}_0$ onto that subset of $\mathcal{F}_1 \cup \mathcal{F}_0$ consisting of all Farey blocks with $\beta = \chi_0$. Applying symmetries, corresponding results are true for $f(\mathbf{E}_j)$ for $j = 1, 2, 3$. We can now summarise the effect of the map f on all states in $\mathcal{F}_2$. In each case, the arrow indicates that f maps the gives state onto the union of the states listed on the right. Note that for simplicity, we have now replaced the bold $\mathbf{A}_j, \dots, \mathbf{E}_j$ with $A_j, \dots, E_j$ for $j = 0, 1, 2, 3$.

$$
\begin{aligned}
f|_{A_0} &= \delta_1 : A_0 \longmapsto A_0 \cup B_0 \cup C_0 \cup D_0 \cup E_0,\\
f|_{B_0} &= \delta_2{}^{-1}\delta_0{}^{-1} : B_0 \longmapsto A_3 \cup B_3 \cup C_3 \cup D_3 \cup E_3,\\
f|_{C_0} &= \delta_2\delta_0\delta_1 : C_0 \longmapsto A_2 \cup B_2 \cup C_2 \cup D_2 \cup E_2,\\
f|_{D_0} &= \delta_0{}^{-1} : D_0 \longmapsto A_0 \cup B_0 \cup D_0 \cup E_0,\\
f|_{E_0} &= \delta_1 : E_0 \longmapsto F_0 \cup G_0 \cup H_0 \cup I_0 \cup J_0 \cup F_1 \cup G_1\\
&\qquad \cup H_1 \cup I_1,\\
f|_{A_1} &= \delta_1{}^{-1} : A_1 \longmapsto A_1 \cup B_1 \cup C_1 \cup D_1 \cup E_1,\\
f|_{B_1} &= \delta_2\delta_0 : B_1 \longmapsto A_2 \cup B_2 \cup C_2 \cup D_2 \cup E_2,\\
f|_{C_1} &= \delta_2{}^{-1}\delta_0{}^{-1}\delta_1{}^{-1} : C_1 \longmapsto A_3 \cup B_3 \cup C_3 \cup D_3 \cup E_3,\\
f|_{D_1} &= \delta_0 : D_1 \longmapsto A_1 \cup B_1 \cup D_1 \cup E_1,\\
f|_{E_1} &= \delta_1{}^{-1} : E_1 \longmapsto F_0 \cup G_0 \cup H_0 \cup I_0 \cup J_0 \cup F_1 \cup G_1\\
&\qquad \cup H_1 \cup I_1,\\
f|_{A_2} &= \delta_2 : A_2 \longmapsto A_2 \cup B_2 \cup C_2 \cup D_2 \cup E_2,\\
f|_{B_2} &= \delta_1{}^{-1}\delta_0{}^{-1} : B_2 \longmapsto A_1 \cup B_1 \cup C_1 \cup D_1 \cup E_1,\\
f|_{C_2} &= \delta_1\delta_0\delta_2 : C_2 \longmapsto A_0 \cup B_0 \cup C_0 \cup D_0 \cup E_0\\
f|_{D_2} &= \delta_0{}^{-1} : D_2 \longmapsto A_2 \cup B_2 \cup D_2 \cup E_2,\\
f|_{E_2} &= \delta_2 : E_2 \longmapsto F_2 \cup G_2 \cup H_2 \cup I_2 \cup J_2 \cup F_3 \cup G_3\\
&\qquad \cup H_3 \cup I_3,\\
f|_{A_3} &= \delta_2{}^{-1} : A_3 \longmapsto A_3 \cup B_3 \cup C_3 \cup D_3 \cup E_3,
\end{aligned}
$$

$$
\begin{aligned}
f|_{B_3} &= & \delta_1\delta_0 &: B_3 \longmapsto A_0 \cup B_0 \cup C_0 \cup D_0 \cup E_0,\\
f|_{C_3} &= & \delta_1{}^{-1}\delta_0{}^{-1}\delta_2{}^{-1} &: C_3 \longmapsto A_1 \cup B_1 \cup C_1 \cup D_1 \cup E_1,\\
f|_{D_3} &= & \delta_0 &: D_3 \longmapsto A_3 \cup B_3 \cup D_3 \cup E_3,\\
f|_{E_3} &= & \delta_2{}^{-1} &: E_3 \longmapsto F_2 \cup G_2 \cup H_2 \cup I_2 \cup J_2 \cup F_3 \cup G_3\\
&&&\qquad\quad \cup H_3 \cup I_3.
\end{aligned}
$$

Having defined f on all of $\mathcal{F}_2$, we now consider $\mathcal{F}_1$. As we described in the previous section, we may regard (γ_1, γ_2) as lying on a one-holed torus (one of the components of $\Sigma_2 - \beta$). Moreover, as γ_1 and γ_2 intersect exactly once they correspond to Farey neighbours. Therefore, we define f to agree with the map f_1 on the space of Farey pairs and described in section 2.4. Results about the image of this map follow as in that section. Thus we have:

$$
\begin{aligned}
f|_{F_0} &= \delta_2 : F_0 \longmapsto F_0 \cup G_0 \cup H_0 \cup I_0 \cup J_0 \cup H_1 \cup I_1,\\
f|_{G_0} &= \delta_0{}^{-1} : G_0 \longmapsto F_0 \cup G_0 \cup H_0 \cup I_0 \cup J_0 \cup H_1 \cup I_1,\\
f|_{F_1} &= \delta_2{}^{-1} : F_1 \longmapsto F_1 \cup G_1 \cup H_0 \cup I_0 \cup J_0 \cup H_1 \cup I_1,\\
f|_{G_1} &= \delta_0 : G_1 \longmapsto F_1 \cup G_1 \cup H_0 \cup I_0 \cup J_0 \cup H_1 \cup I_1,\\
f|_{F_2} &= \delta_1 : F_2 \longmapsto F_2 \cup G_2 \cup H_2 \cup I_2 \cup J_2 \cup H_3 \cup I_3,\\
f|_{G_2} &= \delta_0{}^{-1} : G_2 \longmapsto F_2 \cup G_2 \cup H_2 \cup I_2 \cup J_2 \cup H_3 \cup I_3,\\
f|_{F_3} &= \delta_1{}^{-1} : F_3 \longmapsto F_3 \cup G_3 \cup H_2 \cup I_2 \cup J_2 \cup H_3 \cup I_3,\\
f|_{G_3} &= \delta_0 : G_3 \longmapsto F_3 \cup G_3 \cup H_2 \cup I_2 \cup J_2 \cup H_3 \cup I_3.
\end{aligned}
$$

Finally, we consider Farey blocks in $\mathcal{F}_0$. If we were in the case of the once punctured torus we would need to apply a power of the involution $\delta_1\delta_0\delta_1$. For the twice punctured torus, this element has infinite order (its fourth power is Dehn twist about χ_2). Therefore we need to investigate the effect of powers of $\delta_1\delta_0\delta_1$ and $\delta_2\delta_0\delta_2$. For simplicity we denote $\delta_j\delta_0\delta_j$ by ρ_j for $j = 1, 2$. We have the following lemma.

Lemma 4.6.2. *If the Farey block $(\gamma_1, \gamma_2, \beta; \gamma_3)$ is in H_0 then its image under $\rho_2{}^{-1}$ is in I_0 if $a \geq 1$ or is in J_0 if $a = 0$. If $(\gamma_1, \gamma_2, \beta; \gamma_3)$ is in I_0 then its image under $\rho_2{}^{-2}$ is in I_0 if $a \geq 1$ or is in J_0 if $a = 0$. Moreover any Farey block in I_0 or J_0 arises as the image of a Farey block in H_0 or I_0 in this way.*

Proof. The curve χ_0 is fixed under application of δ_0 or δ_2 and so under any power of ρ_2. It follows by a similar argument to those given in section 2.1 that ρ_2 interchanges $\mathbf{e}_0$ and $\mathbf{e}_\infty^\infty$. It remains to check the effect of ρ_2 on the possible curves γ_3. If the Farey block $(\gamma_1, \gamma_2, \beta; \gamma_3)$ is in I_0 then $\gamma_3 = (a+1, 0, a, 1) \in$

$\Delta_5 \cap \Delta_6$. We claim that $\rho_2{}^{-1}$ sends $(a+1, 0, a, 1) \in \Delta_5 \cap \Delta_6$ to $(a, 0, a, 1) \in \Delta_5 \cap \Delta_6$. Likewise, if $(\gamma_1, \gamma_2, \beta; \gamma_3)$ is in H_0 then $\gamma_3 = (a, 0, a, 1) \in \Delta_5 \cap \Delta_6$ and we claim that $\rho_2{}^{-1}$ sends $(a, 0, a, 1) \in \Delta_5 \cap \Delta_6$ to $(a, 0, a-1, 1) \in \Delta_5 \cap \Delta_6$ if $a \geq 1$ or to $\mathbf{e}^0$ if $a = 0$. This claim is proved using Propositions 3.3.2 and 3.3.3 as follows:

$$\begin{array}{ccc}
(a+1, 0, a, 1) \in \Delta_5 \cap \Delta_6 & \qquad & (a, 0, a, 1) \in \Delta_5 \cap \Delta_6 \\
\Big\downarrow {\scriptstyle \delta_2{}^{-1}} & & \Big\downarrow {\scriptstyle \delta_2{}^{-1}} \\
(a+1, 0, a, 1) \in \Delta_5 \cap \Delta_6 & & (a, 1, a-1, 1) \in \Delta_6 \\
\Big\downarrow {\scriptstyle \delta_0{}^{-1}} & & \Big\downarrow {\scriptstyle \delta_0{}^{-1}} \\
(a, 1, a, 1) \in \Delta_5 & & (a, 0, a-1, 1) \in \Delta_5 \cap \Delta_6 \\
\Big\downarrow {\scriptstyle \delta_2{}^{-1}} & & \Big\downarrow {\scriptstyle \delta_2{}^{-1}} \\
(a, 0, a, 1) \in \Delta_5 \cap \Delta_6 & & (a, 0, a-1, 1) \in \Delta_5 \cap \Delta_6
\end{array}$$

where we assume $a \geq 1$ in the right hand column. If $a = 0$ then

$$\rho_2{}^{-1}(\mathbf{e}^1_{-1}) = \delta_2{}^{-1}\delta_0{}^{-1}\delta_2{}^{-1}(\mathbf{e}^1_{-1}) = \delta_2{}^{-1}\delta_0{}^{-1}(\mathbf{e}^1_{\infty}) = \delta_2{}^{-1}(\mathbf{e}^0) = \mathbf{e}^0. \qquad \square$$

Corollary 4.6.3. Applying $\rho_2{}^{-j}$ to Farey blocks in H_0, I_0 decreases the length of γ_3 by $2j$. Therefore we define f on $\mathcal{F}_0$ as follows.

$$\begin{array}{llll}
f|_{H_0} = \rho_2{}^{-1} : H_0 \longrightarrow I_0 \cup J_0, & & f|_{I_0} = \rho_2{}^{-2} : I_0 \longrightarrow I_0 \cup J_0, \\
f|_{H_1} = \rho_2 \quad : H_1 \longrightarrow I_1 \cup J_0, & & f|_{I_1} = \rho_2{}^{2} \quad : I_1 \longrightarrow I_1 \cup J_0, \\
f|_{H_2} = \rho_1{}^{-1} : H_2 \longrightarrow I_2 \cup J_2, & & f|_{I_2} = \rho_1{}^{-2} : I_2 \longrightarrow I_2 \cup J_2, \\
f|_{H_3} = \rho_1 \quad : H_3 \longrightarrow I_3 \cup J_2, & & f|_{I_3} = \rho_1{}^{2} \quad : I_3 \longrightarrow I_3 \cup J_2, \\
f|_{J_0} = e \quad : J_0 \longrightarrow J_0, & & f|_{J_2} = \iota_2 \quad : J_2 \longrightarrow J_0.
\end{array}$$

This completes the definition of the map f.

4.7. The word acceptor

We have now subdivided the space $\mathcal{F}$ of Farey blocks into states and defined the map f on each state, in such a way that

(i) for each state U, we have $f|_U \equiv \phi$ for some $\phi \in \mathcal{MCG}(\Sigma_2)$, and
(ii) for any states U and V if $f(U) \cap V^\circ \neq \emptyset$ then $V \subset f(U)$.

In order for f to define a normal form leading to a suitable word acceptor, we now only need to verify that successive applications of the map f always eventually terminate in the end state J_0. Of course, this requirement was central to our choice of definition of the map f. In analogy with the case of the once punctured torus Σ_1 we have:

Proposition 4.7.1. *Let $(\gamma_1, \gamma_2, \beta; \gamma_3)$ be any Farey block in $\mathcal{F}$. There exists a non-negative integer n so that $f^n(\gamma_1, \gamma_2, \beta; \gamma_3) = (\mathbf{e}_0, \mathbf{e}_\infty^\infty, \chi_0; \mathbf{e}^0)$.*

Proof. This is similar to Proposition 2.3.5. First, if $(\gamma_1, \gamma_2, \beta; \gamma_3) \in \mathcal{F}_2$ it follows from Proposition 4.2.2 and Proposition 4.4.2. that f strictly decreases $|\beta|$. Thus there is a non-negative integer n_2 so that $f^{n_2}(\gamma_1, \gamma_2, \beta; \gamma_3)$ is in $\mathcal{F}_1$ or $\mathcal{F}_0$. For any Farey block in $\mathcal{F}_1$ the map f strictly decreases $|\gamma_1| + |\gamma_2|$. This follows from Lemma 2.3.4. Thus there is a non-negative integer n_1 so that $f^{n_2+n_1}(\gamma_1, \gamma_2, \beta; \gamma_3)$ is in $\mathcal{F}_0$. For any Farey block in $\mathcal{F}_0$ other than $(\mathbf{e}_0, \mathbf{e}_\infty^\infty, \chi_0; \mathbf{e}^0)$ or $(\mathbf{e}^0, \mathbf{e}_\infty^\infty, \chi_2; \mathbf{e}_0)$, using Corollary 4.6.3 we see that the map strictly decreases $|\gamma_3|$. Thus there is a non-negative integer n_0 so that $f^{n_2+n_1+n_0}(\gamma_1, \gamma_2, \beta; \gamma_3)$ is either $(\mathbf{e}_0, \mathbf{e}_\infty^\infty, \chi_0; \mathbf{e}^0)$ or $(\mathbf{e}^0, \mathbf{e}_\infty^\infty, \chi_2; \mathbf{e}_0)$. Finally,

$$f|_{J_2}(\mathbf{e}^0, \mathbf{e}_\infty^\infty, \chi_2; \mathbf{e}_0) = \iota_2(\mathbf{e}^0, \mathbf{e}_\infty^\infty, \chi_2; \mathbf{e}_0) = (\mathbf{e}_0, \mathbf{e}_\infty^\infty, \chi_0; \mathbf{e}^0). \qquad \square$$

Remark. *Just as in Section* 2, *this proposition proves that the actions of f and $\mathcal{MCG}(\Sigma_2)$ are* orbit equivalent, *in other words, for any $(\gamma_1, \gamma_2, \beta; \gamma_3)$ and $(\gamma_1', \gamma_2', \beta'; \gamma_3')$ in $\mathcal{F}$ we have*

$$(\gamma_1', \gamma_2', \beta'; \gamma_3') = \big(\phi(\gamma_1), \phi(\gamma_2), \phi(\beta); \phi(\gamma_3)\big)$$

for $\phi \in \mathcal{MCG}(\Sigma_2)$ if and only if there exist non negative integers m, n so that

$$f^n(\gamma_1, \gamma_2, \beta; \gamma_3) = f^m(\gamma_1', \gamma_2', \beta'; \gamma_3').$$

See the remark following Proposition 2.3.5 *for the significance of this observation.*

Proposition 4.7.1 allows us to construct a normal form for elements of $\mathcal{MCG}(\Sigma_2)$. Namely, for any $\phi \in \mathcal{MCG}(\Sigma_2)$, we determine the map f^n for which

$$f^n\big(\phi(\mathbf{e}_0), \phi(\mathbf{e}_\infty^\infty), \phi(\chi_0); \phi(\mathbf{e}^0)\big) = (\mathbf{e}_0, \mathbf{e}_\infty^\infty, \chi_0; \mathbf{e}^0).$$

Since at each stage f is a fixed element of $\mathcal{MCG}(\Sigma_2)$ this, together with the fact that $\mathcal{MCG}(\Sigma_2)$ acts freely on $\mathcal{F}$, gives a unique expression for ϕ. The details of the normal form are now rather complicated and are best described in terms of a *word acceptor* for $\mathcal{MCG}(\Sigma_2)$. The states of the word acceptor are obviously

$A_j, \dots, I_j$ for $j = 0, 1, 2, 3$ and J_0, J_2. To get the arrows we need to invert the map f on each separate state. The alphabet $\mathcal{A}$ which labels these arrows will consist of all possible values of f^{-1}, namely:

$$\mathcal{A} = \left\{ \begin{array}{c} e, \iota_2, \delta_0{}^{\pm 1}, \delta_1{}^{\pm 1}, \delta_2{}^{\pm 1}, \delta_1\delta_0, \delta_1{}^{-1}\delta_0{}^{-1}, \delta_2\delta_0, \delta_2{}^{-1}\delta_0{}^{-1}, \\ (\delta_1\delta_0\delta_2)^{\pm 1}, (\delta_2\delta_0\delta_1)^{\pm 1}, \rho_1{}^{\pm 1}, \rho_2{}^{\pm 1}, \rho_1{}^{\pm 2}, \rho_2{}^{\pm 2}, \end{array} \right\}.$$

We list the labelled arrows leading from each state. We begin with J_0. It has the following arrows which may be read off from the definition of f. The arrows from J_0 are.

$$J_0 \xrightarrow{\delta_1{}^{-1}} E_0, \quad J_0 \xrightarrow{\delta_2{}^{-1}} F_0, \quad J_0 \xrightarrow{\delta_0} G_0, \quad J_0 \xrightarrow{\rho_2} H_0, \quad J_0 \xrightarrow{\rho_2{}^2} I_0,$$
$$J_0 \xrightarrow{\delta_1} E_1, \quad J_0 \xrightarrow{\delta_2} F_1, \quad J_0 \xrightarrow{\delta_0{}^{-1}} G_1, \quad J_0 \xrightarrow{\rho_2{}^{-1}} H_1, \quad J_0 \xrightarrow{\rho_2{}^{-2}} I_1,$$
$$J_0 \xrightarrow{\iota_2} J_2.$$

The arrows from J_2 may be found from those from J_0 by applying the symmetry ι_2, with one exception: there is no arrow from J_2 to J_0. The arrows from I_j may be found by applying ι_j to the following arrows from I_0:

$$I_0 \xrightarrow{\delta_1{}^{-1}} E_0, \quad I_0 \xrightarrow{\delta_2{}^{-1}} F_0, \quad I_0 \xrightarrow{\delta_0} G_0, \quad I_0 \xrightarrow{\rho_2} H_0, \quad I_0 \xrightarrow{\rho_2{}^2} I_0,$$
$$I_0 \xrightarrow{\delta_1} E_1, \quad I_0 \xrightarrow{\delta_2} F_1, \quad I_0 \xrightarrow{\delta_0{}^{-1}} G_1.$$

The arrows from H_j may be found by applying ι_j to the following arrows from H_0:

$$H_0 \xrightarrow{\delta_1{}^{-1}} E_0, \quad H_0 \xrightarrow{\delta_2{}^{-1}} F_0, \quad H_0 \xrightarrow{\delta_0} G_0,$$
$$H_0 \xrightarrow{\delta_1} E_1, \quad H_0 \xrightarrow{\delta_2} F_1, \quad H_0 \xrightarrow{\delta_0{}^{-1}} G_1.$$

The arrows from F_j and G_j are very similar. They may be found by applying ι_j to the following arrows from F_0 and G_0:

$$F_0 \xrightarrow{\delta_1{}^{-1}} E_0, \quad F_0 \xrightarrow{\delta_1} E_1, \quad F_0 \xrightarrow{\delta_2{}^{-1}} F_0, \quad F_0 \xrightarrow{\delta_0} G_0,$$
$$G_0 \xrightarrow{\delta_1{}^{-1}} E_0, \quad G_0 \xrightarrow{\delta_1} E_1, \quad G_0 \xrightarrow{\delta_2{}^{-1}} F_0, \quad G_0 \xrightarrow{\delta_0} G_0.$$

The arrows from A_0, B_0, D_0 and E_0 are all similar. They are

$$A_0 \xrightarrow{\delta_0} D_0, \quad A_0 \xrightarrow{\delta_1{}^{-1}} A_0, \quad A_0 \xrightarrow{\delta_0{}^{-1}\delta_1{}^{-1}} B_3, \quad A_0 \xrightarrow{\delta_2{}^{-1}\delta_0{}^{-1}\delta_1{}^{-1}} C_2,$$

$$B_0 \xrightarrow{\delta_0} D_0, \quad B_0 \xrightarrow{\delta_1{}^{-1}} A_0, \quad B_0 \xrightarrow{\delta_0{}^{-1}\delta_1{}^{-1}} B_3, \quad B_0 \xrightarrow{\delta_2{}^{-1}\delta_0{}^{-1}\delta_1{}^{-1}} C_2,$$

$$D_0 \xrightarrow{\delta_0} D_0, \quad D_0 \xrightarrow{\delta_1{}^{-1}} A_0, \quad D_0 \xrightarrow{\delta_0{}^{-1}\delta_1{}^{-1}} B_3, \quad D_0 \xrightarrow{\delta_2{}^{-1}\delta_0{}^{-1}\delta_1{}^{-1}} C_2,$$

$$E_0 \xrightarrow{\delta_0} D_0, \quad E_0 \xrightarrow{\delta_1{}^{-1}} A_0, \quad E_0 \xrightarrow{\delta_0{}^{-1}\delta_1{}^{-1}} B_3, \quad E_0 \xrightarrow{\delta_2{}^{-1}\delta_0{}^{-1}\delta_1{}^{-1}} C_2.$$

Finally, the arrows from C_0 are

$$C_0 \xrightarrow{\delta_1{}^{-1}} A_0, \quad C_0 \xrightarrow{\delta_0{}^{-1}\delta_1{}^{-1}} B_3, \quad C_0 \xrightarrow{\delta_2{}^{-1}\delta_0{}^{-1}\delta_1{}^{-1}} C_2.$$

The arrows from A_j, B_j, C_j, D_j and E_j for $j = 1, 2, 3$ may be found by applying the symmetry ι_j to A_0, B_0, C_0, D_0 or E_0 respectively.

In Fig. 4.7.1 we have given a schematic representation of the arrows from $A_0, \ldots, J_0$ listed above. In order to simplify the diagram we have drawn a single arrow to represent several between different pairs of states. In order to reconstruct the word acceptor, the diagram should be reproduced with all suffices

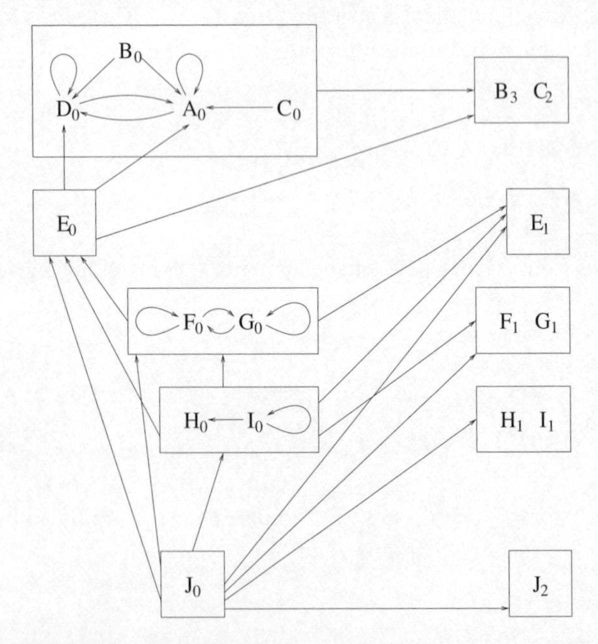

Figure 4.7.1 A diagram of the arrows from states $A_0, \ldots, J_0$.
(An arrow from (or to) a box means that there should be an arrow from (or to) each state in that box. Each arrow should be labelled with the inverse of f on the state it points towards.)

$j = 0, 1, 2, 3$ by applying symmetries ι_j. The arrows to the right hand columns indicate how these four different diagrams are linked. An arrow between two of the rectangular boxes should be replaced with arrows between all the states in each of the two boxes. Finally, the arrows from state U_j to V_k should be labelled with the inverse of $f|_{V_k}$. Observe that all arrows between boxes either go upwards or across but never downwards. This gives the word acceptor the structure of a partially ordered set.

5. The word difference machine

In this final section, following the procedures of sections 1.2 and 2.5, we construct a word difference machine for the word acceptor of $\mathcal{MCG}(\Sigma_2)$.

5.1. Outline of the construction

The construction of the word difference machine is very similar to the construction for the once punctured torus given earlier. Our notation will follow that established in the introduction to section 2.5. As before, the word difference machine is a 2-stringed finite state automaton. Its states are the elements of a set of word differences $\mathcal{D}$. As before, the basic building blocks are squares

$$\begin{array}{ccc} U & \xrightarrow{\psi} & V \\ {\scriptstyle\alpha}\downarrow & & \downarrow{\scriptstyle\beta} \\ \alpha(U) & \xrightarrow{\psi'} & \beta(V) \end{array} \qquad (*)$$

where now U, V are subsets of the states $A_j, \ldots, I_j, J_0, J_2 \subset \mathcal{F}$. As usual, in such a square $\psi, \psi' \in \mathcal{D}$, $\psi(U) = V$, $\alpha = f|_U$, $\beta = f|_V$ and $\psi' = \beta\psi\alpha^{-1}$. In addition to degenerate squares, or triangles, of type $(**)$

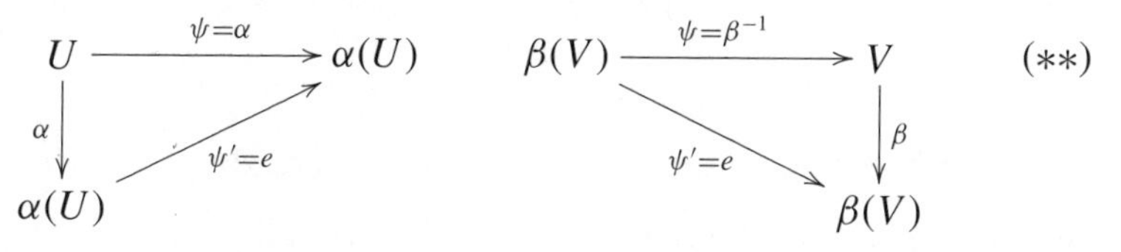

used in section 2.5 we also introduce further degenerate squares where we only apply f to either U or to V. If we do not apply f to U, then we may relax the requirement that U be contained in a single state in the partition of $\mathcal{F}$. In this case we write $\psi^{-1}(V)$ in place of U. Likewise when we do not apply f to V we write $\psi(U)$ instead of V and allow it to contain points in more than one state.

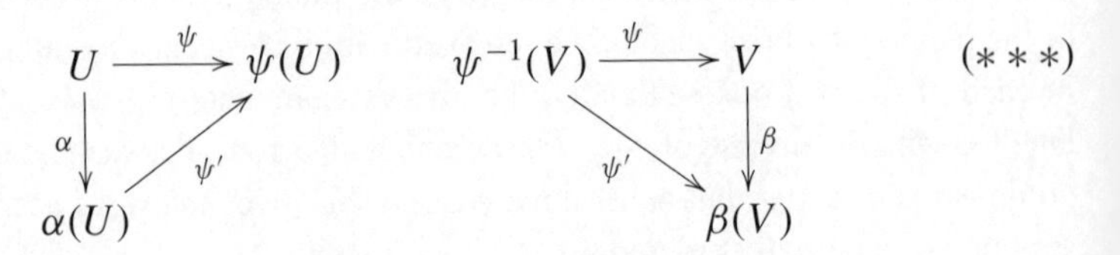

We require that $\psi' = \psi\alpha^{-1}$ or $\psi' = \beta\psi$ respectively is in $\mathcal{D}$. We need to be careful that only finitely many such triangles occur in each path through the word difference machine. This is a key point which we will discuss later. The arrows in the word difference machine will consist of pairs $(\alpha^{-1}, \beta^{-1})$ corresponding to the diagram $(*)$ or $(\alpha^{-1}, -)$, $(-, \beta^{-1})$ corresponding to $(**)$ or $(* * *)$. For example, a square of type $(*)$ will give an arrow $(\alpha^{-1}, \beta^{-1})$ from $\psi' \in \mathcal{D}$ to $\psi \in \mathcal{D}$. The set $\mathcal{D}$ should contain all elements of the alphabet $\mathcal{A}$ constructed in the previous section. In fact we will begin by considering the set

$$\mathcal{D}_0 = \left\{e, \iota_2, \delta_0{}^{\pm 1}, \delta_1{}^{\pm 1}, \delta_2{}^{\pm 1}\right\}.$$

As all the elements of $\mathcal{A}$ have length at most six in these letters we can break squares involving word differences in $\mathcal{A}$ down into at most six squares (placed horizontally) involving word differences in $\mathcal{D}_0$ (see section 5.5 below). During our construction we will add to the list $\mathcal{D}_0$. Recall that in Section 3.3 we found various relations in $\mathcal{MCG}(\Sigma_2)$. We will use these when constructing the squares. The fact that we need to use no more relations is the proof that the presentation for $\mathcal{MCG}(\Sigma_2)$ given in Theorem 3.2.1 works. Because of the stratification of states in $\mathcal{F}$ described in section 4.5, the normal form given by the word acceptor for each element ϕ in $\mathcal{MCG}(\Sigma_2)$ may be broken down as $\phi = \phi_2\phi_1\phi_0$ where $\phi_j(\mathbf{e}_0, \mathbf{e}_\infty^\infty, \chi_0; \mathbf{e}^0)$ is in $\mathcal{F}_j$ for $j = 0, 1, 2$. We will break the word difference machine into subgraphs which correspond to these pieces. There will additionally be a fourth subgraph which will correspond to certain special word differences, which we call exceptional (see section 5.4 below).

5.2. Squares and triangles arising from states in $\mathcal{F}_2$

We begin by constructing squares and triangles where U and V are contained in $\mathcal{F}_2$ and where $\psi \in \mathcal{D}_0 = \{\delta_j{}^{\pm 1} | j = 0, 1, 2\} \cup \{\iota_2\}$. By use of the symmetries ι_j for $j = 1, 2, 3$ we may restrict our attention to the case where U and V are subsets of $\mathbf{A}_0$, $\mathbf{B}_0$, $\mathbf{C}_0$, $\mathbf{D}_0$ and $\mathbf{E}_0$. As usual, we drop the distinction between $\mathbf{A}_0, \ldots, \mathbf{E}_0$ and $A_0, \ldots, E_0$. Since $\delta_1 = f|_{A_0}$ we already know its effect on

A_0. Thus we can write down the triangle:

$$\begin{array}{ccc} A_0 & \xrightarrow{\delta_1} & A_0 \cup B_0 \cup C_0 \cup D_0 \cup E_0 \\ \downarrow{\scriptstyle \delta_1} & \nearrow{\scriptstyle e} & \\ A_0 \cup B_0 \cup C_0 \cup D_0 \cup E_0 & & \end{array}$$

In order to simplify things further, we define $Q_j = A_j \cup B_j \cup C_j \cup D_j \cup E_j$ for $j = 0, 1, 2, 3$. Now suppose that U is one of B_0 or C_0. Using Proposition 3.3.1, we obtain two squares:

$$\begin{array}{ccc} B_0 & \xrightarrow{\delta_1} & C_1 \\ {\scriptstyle \delta_2{}^{-1}\delta_0{}^{-1}}\downarrow & & \downarrow{\scriptstyle \delta_2{}^{-1}\delta_0{}^{-1}\delta_1{}^{-1}} \\ Q_3 & \xrightarrow[e]{} & Q_3 \end{array} \qquad \begin{array}{ccc} C_0 & \xrightarrow{\delta_1} & B_1 \\ {\scriptstyle \delta_2\delta_0\delta_1}\downarrow & & \downarrow{\scriptstyle \delta_2\delta_0} \\ Q_2 & \xrightarrow[e]{} & Q_2 \end{array}$$

When considering D_0 and E_0 we make the following definitions which again simplify the notation.

$$\begin{aligned} R_j &= A_j \cup B_j \cup D_j \cup E_j && \text{for } j = 0, 1, 2, 3; \\ S_j &= F_j \cup G_j && \text{for } j = 0, 1, 2, 3; \\ T_j &= H_j \cup I_j \cup J_j \cup H_{j+1} \cup I_{j+1} && \text{for } j = 0, 2. \end{aligned}$$

We obtain

$$\begin{array}{ccc} D_0 & \xrightarrow{\delta_1} & D_1 \\ {\scriptstyle \delta_0{}^{-1}}\downarrow & & \downarrow{\scriptstyle \delta_0} \\ R_0 & \xrightarrow{\delta_1\delta_0\delta_1} & R_1 \end{array} \qquad \begin{array}{ccc} E_0 & \xrightarrow{\delta_1} & S_0 \cup S_1 \cup T_0 \\ {\scriptstyle \delta_1}\downarrow & \nearrow{\scriptstyle e} & \\ S_0 \cup S_1 \cup T_0 & & \end{array}$$

Now we do the same for $\psi = \delta_1{}^{-1}$. We obtain one triangle:

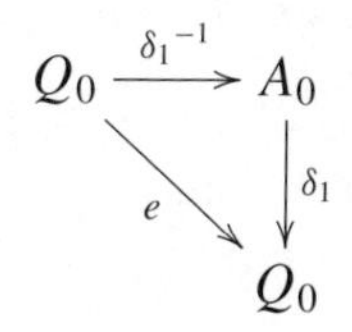

Now we consider squares where $\psi = \delta_2$. From Proposition 3.3.2 we know the effect of δ_2 on $\Delta_1, \ldots, \Delta_7$. We know that $\delta_2 : A_0 \longrightarrow A_0$ and $\delta_2 : B_0 \longrightarrow B_0 \cup D_0$. We divide $U = B_0$ into maximal subsets $U_1 = \delta_2{}^{-1}(B_0)$ or $U_2 = \delta_2{}^{-1}(D_0)$.

This gives:

$$\begin{array}{ccc} A_0 & \xrightarrow{\delta_2} & A_0 \\ \downarrow{\scriptstyle \delta_1} & & \downarrow{\scriptstyle \delta_1} \\ Q_0 & \xrightarrow{\delta_2} & Q_0 \end{array} \qquad \begin{array}{ccc} \delta_2{}^{-1}(B_0) & \xrightarrow{\delta_2} & B_0 \\ \downarrow{\scriptstyle \delta_2{}^{-1}\delta_0{}^{-1}} & & \downarrow{\scriptstyle \delta_2{}^{-1}\delta_0{}^{-1}} \\ C_3 \cup D_3 & \xrightarrow{\delta_0} & Q_3 \end{array} \qquad \begin{array}{ccc} \delta_2{}^{-1}(D_0) & \xrightarrow{\delta_2} & D_0 \\ \downarrow{\scriptstyle \delta_2{}^{-1}\delta_0{}^{-1}} & & \downarrow{\scriptstyle \delta_0{}^{-1}} \\ A_3 \cup B_3 \cup E_3 & \xrightarrow{\delta_2\delta_0} & Q_0 \end{array}$$

Now $\delta_2(C_0 \cup D_0) = C_0$. Therefore we divide $V = C_0$ into subsets $V_1 = \delta_2(C_0)$ and $V_2 = \delta_2(D_0)$.

$$\begin{array}{ccc} C_0 & \xrightarrow{\delta_2} & \delta_2(C_0) \\ \downarrow{\scriptstyle \delta_2\delta_0\delta_1} & & \downarrow{\scriptstyle \delta_2\delta_0\delta_1} \\ Q_2 & \xrightarrow{\delta_0} & C_2 \cup D_2 \end{array} \qquad \begin{array}{ccc} D_0 & \xrightarrow{\delta_2} & \delta_2(D_0) \\ \downarrow{\scriptstyle \delta_0{}^{-1}} & & \downarrow{\scriptstyle \delta_2\delta_0\delta_1} \\ R_0 & \xrightarrow{\iota_2\delta_2{}^{-1}} & A_2 \cup B_2 \cup E_2 \end{array} \qquad \begin{array}{ccc} E_0 & \xrightarrow{\delta_2} & E_0 \\ \downarrow{\scriptstyle \delta_1} & & \downarrow{\scriptstyle \delta_1} \\ S_0 \cup S_1 \cup T_0 & \xrightarrow{\delta_2} & S_0 \cup S_1 \cup T_0 \end{array}$$

We remark that it is not immediately obvious that the bottom lines of these squares are as claimed. We now do an example which illustrates how the bottom line is found. The rest are simple and are left to the reader. The example we choose is

$$\begin{array}{ccc} \delta_2{}^{-1}(B_0) \cap B_0 & \xrightarrow{\delta_2} & B_0 \\ \downarrow{\scriptstyle \delta_2{}^{-1}\delta_0{}^{-1}} & & \downarrow{\scriptstyle \delta_2{}^{-1}\delta_0{}^{-1}} \\ C_3 \cup D_3 & \xrightarrow{\delta_0} & Q_3 \end{array}$$

We do this using the following result.

Proposition 5.2.1. *If* $(a, b, c, d) \in \Delta_3 \cup \Delta_4 = B_0$ *and* $\delta_2(a, b, c, d) \in \Delta_3 \cup \Delta_4 = B_0$ *then*

$$\delta_2{}^{-1}\delta_0{}^{-1}(a, b, c, d) \in \Delta_{26} \cup \Delta_{27} \cup \Delta_{28} = C_3 \cup D_3$$

and

$$\delta_2{}^{-1}\delta_0{}^{-1}\delta_2(a, b, c, d) = \delta_0\delta_2{}^{-1}\delta_0{}^{-1}(a, b, c, d) \in Q_3.$$

Moreover, any point of Q_3 *arises in this way.*

Proof. The word difference can of course be found without reference to the states on the bottom line of the diagram. It is clear that in the present case the new word difference ψ' is

$$\left(\delta_2{}^{-1}\delta_0{}^{-1}\right)\delta_2\left(\delta_2{}^{-1}\delta_0{}^{-1}\right)^{-1} = \delta_2{}^{-1}\delta_0{}^{-1}\delta_2\delta_0\delta_2 = \delta_0$$

using the relation $\delta_2\delta_0\delta_2 = \delta_0\delta_2\delta_0$. We now justify the claims about the regions involved in this diagram. We begin by identifying $U_1 = B_0 \cap \delta_2{}^{-1}(B_0)$.

We know that $B_0 = \Delta_3 \cup \Delta_4$. By Proposition 3.3.2 (iii) and (iv) we see that $(a, b, c, d) \in \Delta_3$ is sent to $\Delta_3 \cup \Delta_4 = B_0$ provided $a \leq b + d$ and that $(a, b, c, d) \in \Delta_4$ is sent to B_0 provided $a \leq d$. Thus,

$$U_1 = \{(a, b, c, d) \in \Delta_3 : a \leq b + d\} \cup \{(a, b, c, d) \in \Delta_4 : a \leq d\}.$$

By definition $\delta_2(U_1) \subset B_0$, and it is not hard to check that $\delta_2(U_1) = B_0$. We now investigate the effect of ${\delta_2}^{-1}{\delta_0}^{-1}$ on U. We know from Proposition 3.4.3 (i) that ${\delta_0}^{-1}$ sends Δ_3 and Δ_4 to Δ_{23} and Δ_{22} respectively by ${\delta_0}^{-1}(a, b, c, d) = (a, b, c, d)$. Thus U is sent to the appropriate subset of A_3. Applying ι_3 to Proposition 3.3.1 (i) we see that if $(a, b, c, d) \in \Delta_{22}$ with $a \leq d$ then ${\delta_2}^{-1}(a, b, c, d)$ is in Δ_{26}. Similarly, if $(a, b, c, d) \in \Delta_{23}$ with $a \leq b + d$ then ${\delta_2}^{-1}(a, b, c, d)$ is in Δ_{27} or Δ_{28}. It is not hard to show that this map is surjective and so ${\delta_2}^{-1}{\delta_0}^{-1}(U) = C_3 \cup D_3$. Finally, we know that ${\delta_2}^{-1}{\delta_0}^{-1}$ maps B_0 onto Q_3. □

By similar reasoning we construct all the squares and triangles for which $\psi = {\delta_2}^{-1}$.

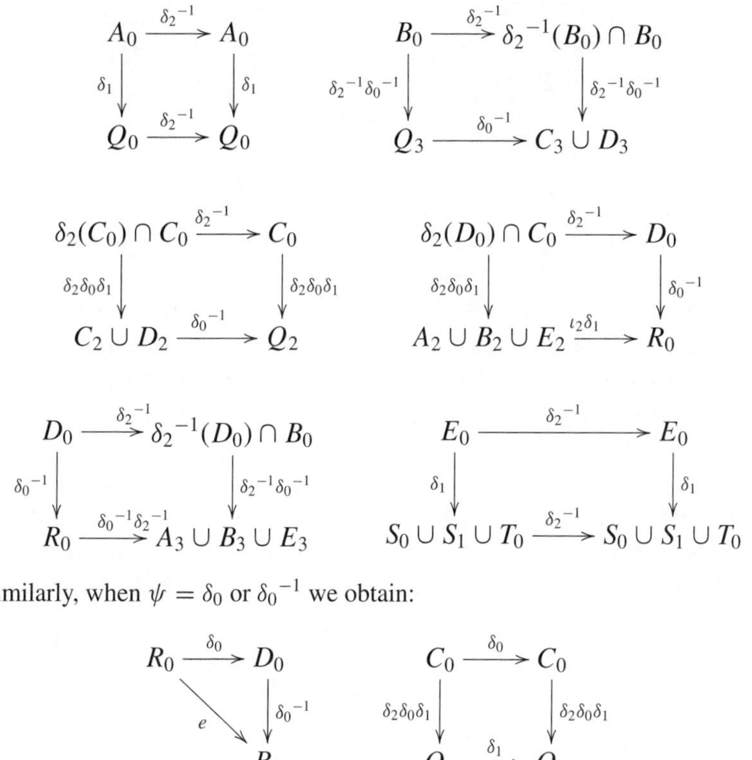

$$\begin{CD} A_0 @>{\delta_2^{-1}}>> A_0 \\ @V{\delta_1}VV @VV{\delta_1}V \\ Q_0 @>{\delta_2^{-1}}>> Q_0 \end{CD} \qquad \begin{CD} B_0 @>{\delta_2^{-1}}>> {\delta_2}^{-1}(B_0) \cap B_0 \\ @V{\delta_2^{-1}\delta_0^{-1}}VV @VV{\delta_2^{-1}\delta_0^{-1}}V \\ Q_3 @>{\delta_0^{-1}}>> C_3 \cup D_3 \end{CD}$$

$$\begin{CD} \delta_2(C_0) \cap C_0 @>{\delta_2^{-1}}>> C_0 \\ @V{\delta_2\delta_0\delta_1}VV @VV{\delta_2\delta_0\delta_1}V \\ C_2 \cup D_2 @>{\delta_0^{-1}}>> Q_2 \end{CD} \qquad \begin{CD} \delta_2(D_0) \cap C_0 @>{\delta_2^{-1}}>> D_0 \\ @V{\delta_2\delta_0\delta_1}VV @VV{\delta_0^{-1}}V \\ A_2 \cup B_2 \cup E_2 @>{\iota_2\delta_1}>> R_0 \end{CD}$$

$$\begin{CD} D_0 @>{\delta_2^{-1}}>> {\delta_2}^{-1}(D_0) \cap B_0 \\ @V{\delta_0^{-1}}VV @VV{\delta_2^{-1}\delta_0^{-1}}V \\ R_0 @>{\delta_0^{-1}\delta_2^{-1}}>> A_3 \cup B_3 \cup E_3 \end{CD} \qquad \begin{CD} E_0 @>{\delta_2^{-1}}>> E_0 \\ @V{\delta_1}VV @VV{\delta_1}V \\ S_0 \cup S_1 \cup T_0 @>{\delta_2^{-1}}>> S_0 \cup S_1 \cup T_0 \end{CD}$$

Similarly, when $\psi = \delta_0$ or ${\delta_0}^{-1}$ we obtain:

$$\begin{array}{ccc} R_0 & \xrightarrow{\delta_0} & D_0 \\ & \searrow^{e} & \Big\downarrow {\scriptstyle \delta_0^{-1}} \\ & & R_0 \end{array} \qquad \begin{CD} C_0 @>{\delta_0}>> C_0 \\ @V{\delta_2\delta_0\delta_1}VV @VV{\delta_2\delta_0\delta_1}V \\ Q_2 @>{\delta_1}>> Q_2 \end{CD}$$

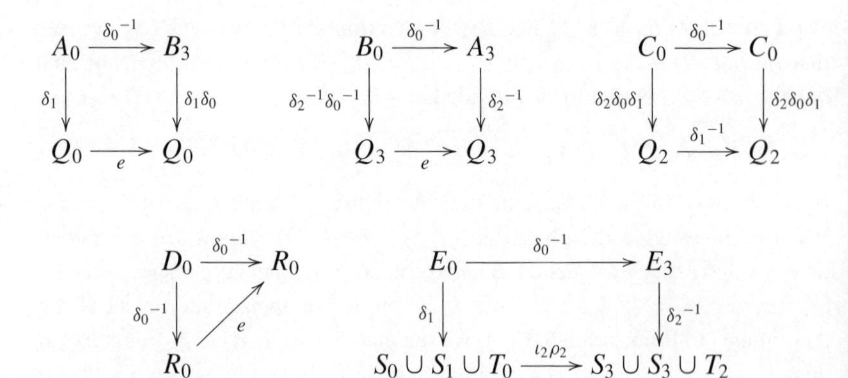

Finally we consider the word difference ι_2.

$$
\begin{array}{ccc}
A_0 & \xrightarrow{\iota_2} & A_2 \\
\delta_1 \big\downarrow & & \big\downarrow \delta_2 \\
Q_0 & \xrightarrow{\iota_2} & Q_2
\end{array}
\qquad
\begin{array}{ccc}
B_0 & \xrightarrow{\iota_2} & B_2 \\
\delta_2{}^{-1}\delta_0{}^{-1} \big\downarrow & & \big\downarrow \delta_1{}^{-1}\delta_0{}^{-1} \\
Q_3 & \xrightarrow{\iota_2} & Q_1
\end{array}
\qquad
\begin{array}{ccc}
C_0 & \xrightarrow{\iota_2} & C_2 \\
\delta_2\delta_0\delta_1 \big\downarrow & & \big\downarrow \delta_1\delta_0\delta_2 \\
Q_2 & \xrightarrow{\iota_2} & Q_0
\end{array}
$$

$$
\begin{array}{ccc}
D_0 & \xrightarrow{\iota_2} & D_2 \\
\delta_0{}^{-1} \big\downarrow & & \big\downarrow \delta_0{}^{-1} \\
R_0 & \xrightarrow{\iota_2} & R_2
\end{array}
\qquad
\begin{array}{ccc}
E_0 & \xrightarrow{\iota_2} & E_2 \\
\delta_1 \big\downarrow & & \big\downarrow \delta_2 \\
S_0 \cup S_1 \cup T_0 & \xrightarrow{\iota_2} & S_2 \cup S_3 \cup T_2
\end{array}
$$

Now consider the diagrams we have constructed above. If the bottom line consists of a word difference between regions we have already constructed then we have no more work. However, there are some diagrams for which this is not the case. First, there are those diagrams where the bottom line involves the regions making up S_j or T_j. We will consider these in the next section. Secondly, there are word differences between regions in Q_j or R_j which we have not yet considered. We consider these individually.

The easiest case is where we have a word difference of $\iota_2\psi$ where ψ is one of the word differences that we can already deal with. These can be analysed as follows. For each of the squares we have constructed with word difference ψ and written in the form $(*)$ we add a new square with word difference $\iota_2\psi$:

$$
\begin{array}{ccccc}
U & \xrightarrow{\psi} & V & \xrightarrow{\iota_2} & W \\
\alpha \big\downarrow & & \big\downarrow \beta & & \big\downarrow \gamma \\
U' & \xrightarrow{\psi'} & V' & \xrightarrow{\iota_2} & W'
\end{array}
$$

where $W = \iota_2(V)$, $W' = \iota_2(V')$ and $\gamma = \iota_2\beta\iota_2$ is $f|_W$. If ψ' has the form $\iota_2\psi''$ for some ψ'', then we use $\iota_2{}^2 = e$ to get a word difference of ψ''. This is illustrated in the example given in section 5.5 below.

The remaining word differences we have to consider are

$$
\begin{array}{rrcll}
\delta_2\delta_0 : & A_3 \cup B_3 \cup E_3 & \longrightarrow & R_0, & (1)\\
\delta_0{}^{-1}\delta_2{}^{-1} : & R_0 & \longrightarrow & A_3 \cup B_3 \cup E_3, & (2)\\
\rho_1 = \delta_1\delta_0\delta_1 : & R_0 & \longrightarrow & R_1. & (3)
\end{array}
$$

The word difference (3) is slightly more complicated than the others. We will treat this word difference and those arising from it separately in section 5.4. These will be called *exceptional word differences* and will constitute a separate subgraph of the word difference machine.

We now consider the word difference (1). By examining which point goes to which, we see that this may be broken into four new arrows. Namely

$$
\begin{array}{ll}
\delta_2\delta_0 : \delta_0{}^{-1}\delta_2{}^{-1}(B_0) \cap A_3 \longrightarrow B_0, & \delta_2\delta_0 : \delta_0{}^{-1}\delta_2{}^{-1}(D_0) \cap A_3 \longrightarrow D_0,\\
\delta_2\delta_0 : B_3 \longrightarrow A_0, & \delta_2\delta_0 : E_3 \longrightarrow E_0.
\end{array}
$$

Now $\delta_0{}^{-1}\delta_2{}^{-1}(B_0) \subset A_3$ and $\delta_0{}^{-1}\delta_2{}^{-1}(D_0) \subset A_3$. Thus we obtain the following squares

$$
\begin{array}{ccc}
\delta_0{}^{-1}\delta_2{}^{-1}(B_0) & \xrightarrow{\delta_2\delta_0} & B_0\\
\downarrow{\scriptstyle \delta_2{}^{-1}} & & \downarrow{\scriptstyle \delta_2{}^{-1}\delta_0{}^{-1}}\\
C_3 \cup D_3 & \xrightarrow{\delta_0} & Q_3
\end{array}
\qquad
\begin{array}{ccc}
\delta_0{}^{-1}\delta_2{}^{-1}(D_0) & \xrightarrow{\delta_2\delta_0} & D_0\\
\downarrow{\scriptstyle \delta_2{}^{-1}} & & \downarrow{\scriptstyle \delta_0{}^{-1}}\\
A_3 \cup B_3 \cup E_3 & \xrightarrow{\delta_2\delta_0} & R_0
\end{array}
$$

$$
\begin{array}{ccc}
B_3 & \xrightarrow{\delta_2\delta_0} & A_0\\
\downarrow{\scriptstyle \delta_1\delta_0} & & \downarrow{\scriptstyle \delta_1}\\
Q_0 & \xrightarrow{\delta_2} & Q_0
\end{array}
\qquad
\begin{array}{ccc}
E_3 & \xrightarrow{\delta_2\delta_0} & E_0\\
\downarrow{\scriptstyle \delta_2{}^{-1}} & & \downarrow{\scriptstyle \delta_1}\\
S_2 \cup S_3 \cup T_2 & \xrightarrow{\iota_2\delta_0{}^{-1}\delta_1{}^{-1}} & S_0 \cup S_1 \cup T_0
\end{array}
$$

The word differences on the bottom lines of these diagrams either have been considered above or else involve S_j and T_j. In the latter case we will consider them in the next section.

We now consider the word difference (2). It is rather similar to (1), and splits as

$$
\begin{array}{ll}
\delta_0{}^{-1}\delta_2{}^{-1} : B_0 \longrightarrow \delta_0{}^{-1}\delta_2{}^{-1}(B_0) \cap A_3, & \delta_0{}^{-1}\delta_2{}^{-1} : A_0 \longrightarrow B_3,\\
\delta_0{}^{-1}\delta_2{}^{-1} : D_0 \longrightarrow \delta_0{}^{-1}\delta_2{}^{-1}(D_0) \cap A_3, & \delta_0{}^{-1}\delta_2{}^{-1} : E_0 \longrightarrow E_3.
\end{array}
$$

$$\begin{array}{ccc} B_0 & \xrightarrow{\delta_0{}^{-1}\delta_2{}^{-1}} & \delta_0{}^{-1}\delta_2{}^{-1}(B_0) \\ \downarrow{\scriptstyle \delta_2{}^{-1}\delta_0{}^{-1}} & & \downarrow{\scriptstyle \delta_2{}^{-1}} \\ Q_3 & \xrightarrow{\delta_0{}^{-1}} & C_3 \cup D_3 \end{array} \qquad \begin{array}{ccc} D_0 & \xrightarrow{\delta_0{}^{-1}\delta_2{}^{-1}} & \delta_0{}^{-1}\delta_2{}^{-1}(D_0) \\ \downarrow{\scriptstyle \delta_0{}^{-1}} & & \downarrow{\scriptstyle \delta_2{}^{-1}} \\ R_0 & \xrightarrow{\delta_0{}^{-1}\delta_2{}^{-1}} & A_3 \cup B_3 \cup E_3 \end{array}$$

$$\begin{array}{ccc} A_0 & \xrightarrow{\delta_0{}^{-1}\delta_2{}^{-1}} & B_3 \\ \downarrow{\scriptstyle \delta_1} & & \downarrow{\scriptstyle \delta_1\delta_0} \\ Q_0 & \xrightarrow{\delta_2{}^{-1}} & Q_0 \end{array} \qquad \begin{array}{ccc} E_0 & \xrightarrow{\delta_0{}^{-1}\delta_2{}^{-1}} & E_3 \\ \downarrow{\scriptstyle \delta_1} & & \downarrow{\scriptstyle \delta_2{}^{-1}} \\ S_0 \cup S_1 \cup T_0 & \xrightarrow{\iota_2\delta_2\delta_0} & S_2 \cup S_3 \cup T_2 \end{array}$$

The word differences on the bottom lines of these diagrams either have been considered above or else involve S_j and T_j, again to be treated in the next section.

Fig. 5.2.1 shows the non-exceptional arrows constructed above where U is one of A_0, B_0, C_0 or D_0. The labels on the arrows may be obtained from the

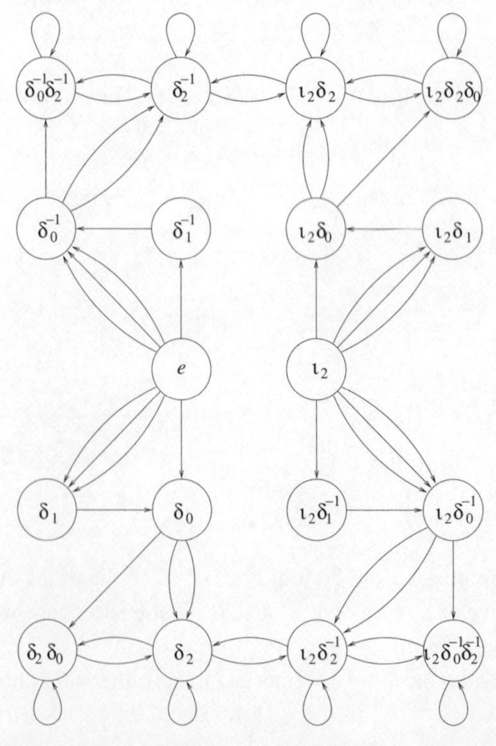

Figure 5.2.1 Non-exceptional arrows in the word difference machine where U is one of A_0, B_0, C_0, D_0. (Each arrow is labelled by a pair $(\alpha^{-1}, \beta^{-1})$ as described in Section 5.1.)

squares listed above. For example, the three arrows from e to δ_1 are labelled $({\delta_1}^{-1}, -)$, $(\delta_0\delta_2, \delta_1\delta_0\delta_2)$ and $({\delta_1}^{-1}{\delta_0}^{-1}{\delta_2}^{-1}, {\delta_0}^{-1}{\delta_2}^{-1})$.

5.3. Squares and triangles arising from states in $\mathcal{F}_0$ and $\mathcal{F}_1$

We proceed along the lines of the previous section. What we do now is essentially the same as in section 2.5. Recall that $S_j = F_j \cup G_j$ for $j = 0, 1, 2, 3$ and $T_j = H_j \cup I_j \cup J_j \cup H_{j+1} \cup I_{j+1}$ for $j = 0, 2$. We begin with the word diffences δ_1 and δ_1^{-1}. We have:

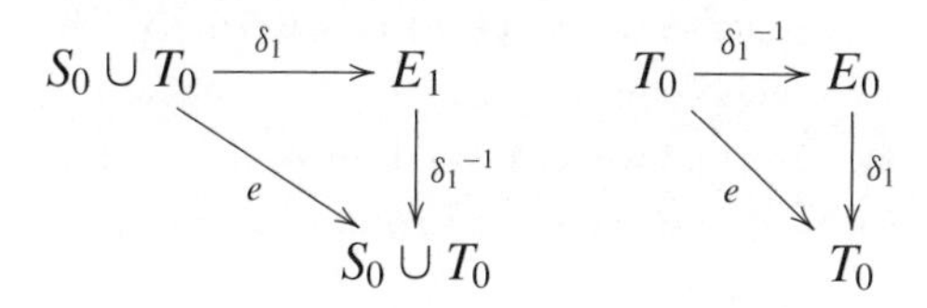

We now consider δ_2:

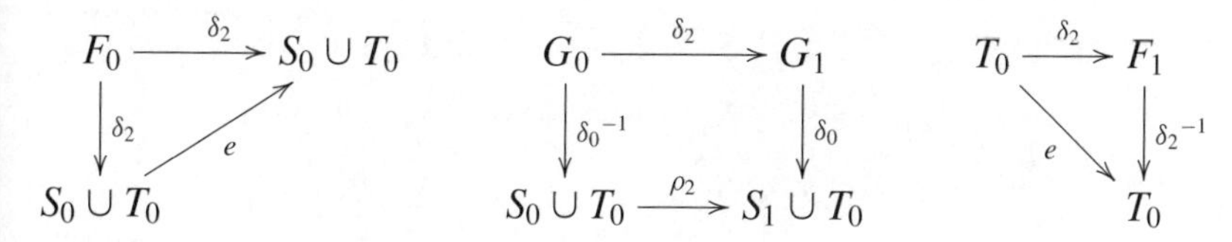

(Recall that $\rho_2 = \delta_2\delta_0\delta_2$.) We now consider ${\delta_2}^{-1}$ and δ_0:

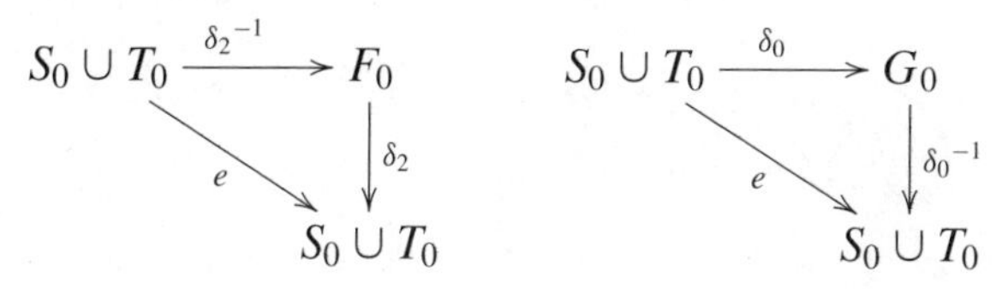

We now consider ${\delta_0}^{-1}$:

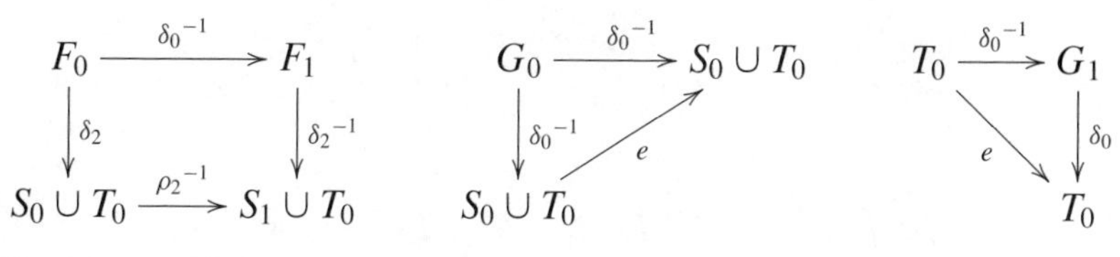

We now consider ι_2:

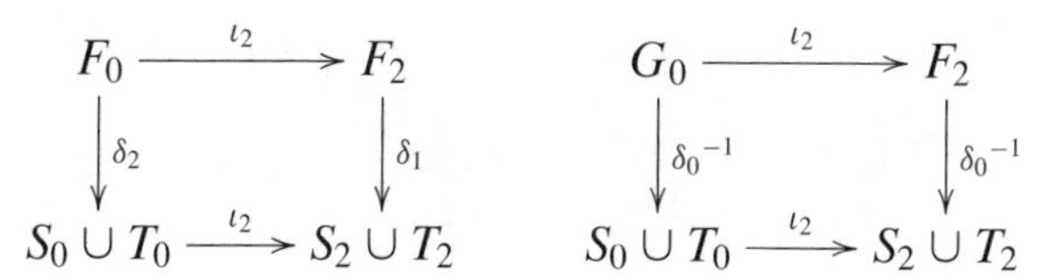

As for word differences between states in $\mathcal{F}_2$, if we have a square for the word difference ψ of the form $(*)$ then we add a new square with the word difference $\iota_2\psi$:

$$\begin{array}{ccccc}
U & \xrightarrow{\psi} & V & \xrightarrow{\iota_2} & W \\
\alpha\big\downarrow & & \big\downarrow\beta & & \big\downarrow\gamma \\
U' & \xrightarrow{\psi'} & V' & \xrightarrow{\iota_2} & W'
\end{array}$$

where $W = \iota_2(V)$, $W' = \iota_2(V')$ and $\gamma = \iota_2\beta\iota_2$ is $f|_W$. Now if ψ' were $\iota_2\psi''$ then we have $\iota_2\psi' = {\iota_2}^2\psi'' = \psi''$ since ${\iota_2}^2$ is the identity. This has completed the construction of squares and triangles for the standard word differences. There remain a few cases that we have not dealt with. These arise is the bottom line in some of the diagrams we have found. The relevant word differences are

$$\begin{aligned}
\rho_2 :\quad & S_0 \cup T_0 \longrightarrow S_1 \cup T_0, && (1)\\
{\rho_2}^{-1} :\quad & S_0 \cup T_0 \longrightarrow S_1 \cup T_0, && (2)\\
{\delta_0}^{-1}{\delta_1}^{-1} : S_2 \cup S_3 \cup T_2 & \longrightarrow S_2 \cup S_3 \cup T_2, && (3)\\
\delta_2\delta_0 : S_0 \cup S_1 \cup T_0 & \longrightarrow S_0 \cup S_1 \cup T_0 && (4)
\end{aligned}$$

where, as in section 4.6, $\rho_j = \delta_j\delta_0\delta_j$ for $j = 1, 2$. We consider them separately.

The word difference (1) splits as

$$\rho_2 : F_0 \longrightarrow G_1, \quad \rho_2 : G_0 \longrightarrow F_1,$$
$$\rho_2 : H_0 \cup J_0 \longrightarrow I_0, \quad \rho_2 : I_0 \longrightarrow H_0,$$

as well as ρ_2 acting on H_1 and I_1. Applying the symmetry ι_1 this is equivalent to ${\rho_2}^{-1}$ acting on H_0 and I_0 which we consider under (2). Thus we get diagrams

$$\begin{array}{ccc}
F_0 & \xrightarrow{\rho_2} & G_1 \\
\big\downarrow\delta_2 & & \big\downarrow\delta_0 \\
S_0 \cup T_0 & \xrightarrow{\rho_2} & S_1 \cup T_0
\end{array}
\qquad
\begin{array}{ccc}
G_0 & \xrightarrow{\rho_2} & F_1 \\
\big\downarrow{\delta_0}^{-1} & & \big\downarrow{\delta_2}^{-1} \\
S_0 \cup T_0 & \xrightarrow{\rho_2} & S_1 \cup T_0
\end{array}$$

$$\begin{array}{ccc}
H_0 \cup J_0 & \xrightarrow{\rho_2} & I_0 \\
 & \searrow^{e} & \big\downarrow{\rho_2}^{-1} \\
 & & H_0 \cup J_0
\end{array}
\qquad
\begin{array}{ccc}
I_0 & \xrightarrow{\rho_2} & H_0 \\
{\rho_2}^{-1}\big\downarrow & & \big\downarrow{\rho_2}^{-2} \\
H_0 \cup J_0 & \xrightarrow[e]{} & H_0 \cup J_0
\end{array}$$

Similarly for (2). We get

$$\rho_2{}^{-1} : F_0 \longrightarrow G_1, \quad \rho_2{}^{-1} : G_0 \longrightarrow F_1,$$
$$\rho_2{}^{-1} : H_0 \longrightarrow I_0, \quad \rho_2{}^{-1} : I_0 \longrightarrow H_0 \cup J_0.$$

In addition, there is $\rho_2{}^{-1}$ acting on H_1 and I_1. Applying the symmetry ι_1, we can obtain these word differences from ρ_2 acting on H_0 and I_0. This was done above. The remaining word differences give the following diagrams

$$\begin{array}{ccc} F_0 & \xrightarrow{\rho_2{}^{-1}} & G_1 \\ \downarrow{\scriptstyle \delta_2} & & \downarrow{\scriptstyle \delta_0} \\ S_0 \cup T_0 & \xrightarrow{\rho_2{}^{-1}} & S_1 \cup T_0 \end{array} \qquad \begin{array}{ccc} G_0 & \xrightarrow{\rho_2{}^{-1}} & F_1 \\ \downarrow{\scriptstyle \delta_0{}^{-1}} & & \downarrow{\scriptstyle \delta_2{}^{-1}} \\ S_0 \cup T_0 & \xrightarrow{\rho_2{}^{-1}} & S_1 \cup T_0 \end{array}$$

$$\begin{array}{ccc} H_0 & \xrightarrow{\rho_2{}^{-1}} & I_0 \\ {\scriptstyle \rho_2{}^{-2}}\downarrow & & \downarrow{\scriptstyle \rho_2{}^{-1}} \\ H_0 \cup J_0 & \xrightarrow[e]{} & H_0 \cup J_0 \end{array} \qquad \begin{array}{ccc} I_0 & \xrightarrow{\rho_2{}^{-1}} & H_0 \cup J_0 \\ {\scriptstyle \rho_2{}^{-1}}\downarrow & \nearrow{\scriptstyle e} & \\ H_0 \cup J_0 & & \end{array}$$

We now consider the word differences (3) and (4). The only way that these word differences can occur is from one of the following diagrams

$$\begin{array}{ccc} E_3 & \xrightarrow{\iota_2\delta_2\delta_0} & E_2 \\ \downarrow{\scriptstyle \delta_2{}^{-1}} & & \downarrow{\scriptstyle \delta_2} \\ S_2 \cup S_3 \cup T_2 & \xrightarrow{\delta_0{}^{-1}\delta_1{}^{-1}} & S_2 \cup S_3 \cup T_2 \end{array} \qquad \begin{array}{ccc} E_0 & \xrightarrow{\iota_2\delta_0{}^{-1}\delta_2{}^{-1}} & E_1 \\ \downarrow{\scriptstyle \delta_1} & & \downarrow{\scriptstyle \delta_1{}^{-1}} \\ S_0 \cup S_1 \cup T_0 & \xrightarrow{\delta_2\delta_0} & S_0 \cup S_1 \cup T_0 \end{array}$$

Applying the symmetries we see that it is sufficient to consider the following word differences

$$\delta_0\delta_2 : S_0 \cup T_0 \longrightarrow S_0 \cup S_1 \cup T_0, \quad (5)$$
$$\delta_0{}^{-1}\delta_2{}^{-1} : S_0 \cup T_0 \longrightarrow F_1, \quad (6)$$
$$\delta_2\delta_0 : S_0 \cup T_0 \longrightarrow G_1, \quad (7)$$
$$\delta_2{}^{-1}\delta_0{}^{-1} : S_0 \cup T_0 \longrightarrow S_0 \cup S_1 \cup T_0. \quad (8)$$

These break down into diagrams as follows

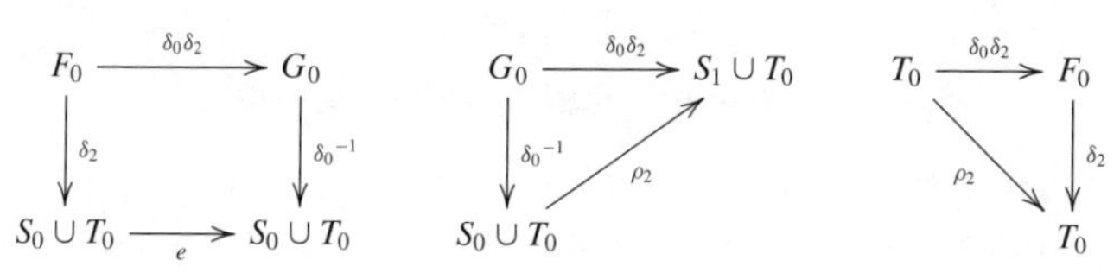

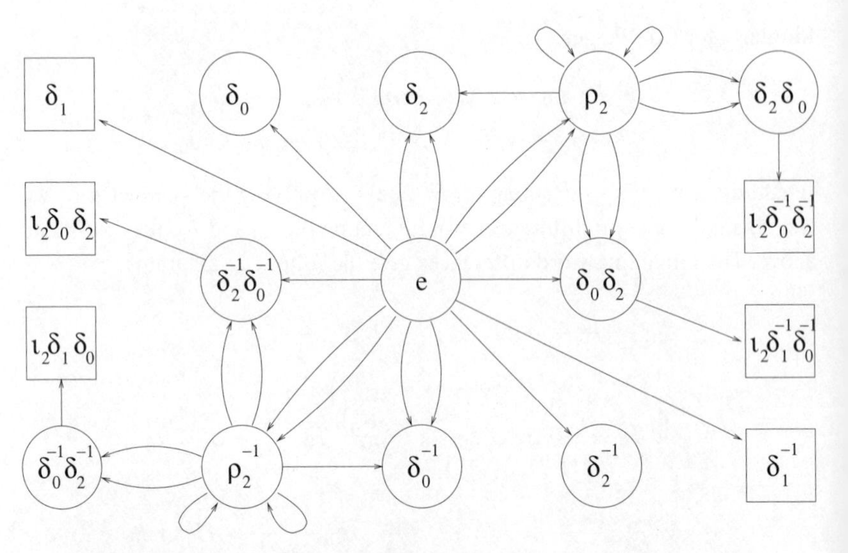

Figure 5.3.1 Arrows in the word difference machine where U is one of F_0, G_0, H_0, I_0, J_0. (In addition there should be word differences $\iota_2\psi$. The square boxes denote word differences where V is E_j.)

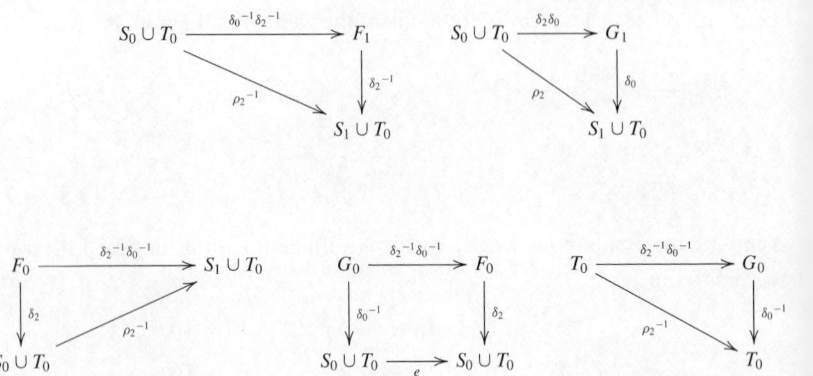

We remark that all these squares have introduced no new word differences in their bottom lines.

5.4. The exceptional word differences

We have now completed the word difference machine except that we have not dealt with word differences arising from the following square (which was called

(3) in section 5.2):

$$\begin{array}{ccc} D_0 & \xrightarrow{\delta_1} & D_1 \\ {\scriptstyle \delta_0^{-1}}\downarrow & & \downarrow{\scriptstyle \delta_0} \\ R_0 & \xrightarrow{\rho_1} & R_1 \end{array}$$

and its symmetric images. Recall that $R_j = A_j \cup B_j \cup D_j \cup E_j$. Observe that $\delta_1 : D_0 \longrightarrow D_1$ and its symmetric images do not arise in the bottom row of any of the squares or triangles we have constructed so far. We call them *exceptional initial state*. There are some states which only arise in paths beginning with an exceptional initial states. We call these *exceptional states*. Because they never occur in a bottom row, once we have gone from an exceptional state to a non-exceptional state (that is any of the states considered in sections 5.2 and 5.3) we can never return to an exceptional state. In order to get from an exceptional to a non-exceptional state it is usually necessary to pass through a triangle. Thus a triangle of this special type can only occur once in any path through the word difference machine. The exceptional states constitute a separate sub-graph of the word difference machine. We now construct the exceptional states arising from the map: $\rho_1 : R_0 \longrightarrow R_1$. Intersecting this with the states of $\mathcal{F}_2$ gives:

$$\begin{array}{lrll} \rho_1 : & \rho_1{}^{-1}(D_1) \cap A_0 & \longrightarrow D_1, & (1) \\ \rho_1 : & \rho_1{}^{-1}(B_1) \cap A_0 & \longrightarrow B_1, & (2) \\ \rho_1 : & D_0 & \longrightarrow \rho_1(D_0) \cap A_1, & (3) \\ \rho_1 : & B_0 & \longrightarrow \rho_1(B_0) \cap A_1, & (4) \\ \rho_1 : & E_0 & \longrightarrow E_1. & (5) \end{array}$$

Each of these five maps is the top line in a square or triangle of the form $(*)$, $(**)$ or $(***)$. We claim that $\rho_1{}^{-1}(D_1)$ and $\rho_1{}^{-1}(B_1)$ are subsets of A_0 and that $\rho_1(D_0)$ and $\rho_1(B_0)$ are subsets of A_1. This may be checked using Propositions 3.3.1, and 3.3.3. The maps (1) and (3) above give rise to squares of the form $(*)$ for which the new word difference ψ' is again ρ_1. They are

$$\begin{array}{ccc} \rho_1{}^{-1}(D_1) & \xrightarrow{\rho_1} & D_1 \\ {\scriptstyle \delta_1}\downarrow & & \downarrow{\scriptstyle \delta_0} \\ R_0 & \xrightarrow{\rho_1} & R_1 \end{array} \qquad \begin{array}{ccc} D_0 & \xrightarrow{\rho_1} & \rho_1(D_0) \\ {\scriptstyle \delta_0^{-1}}\downarrow & & \downarrow{\scriptstyle \delta_1^{-1}} \\ R_0 & \xrightarrow{\rho_1} & R_1 \end{array}$$

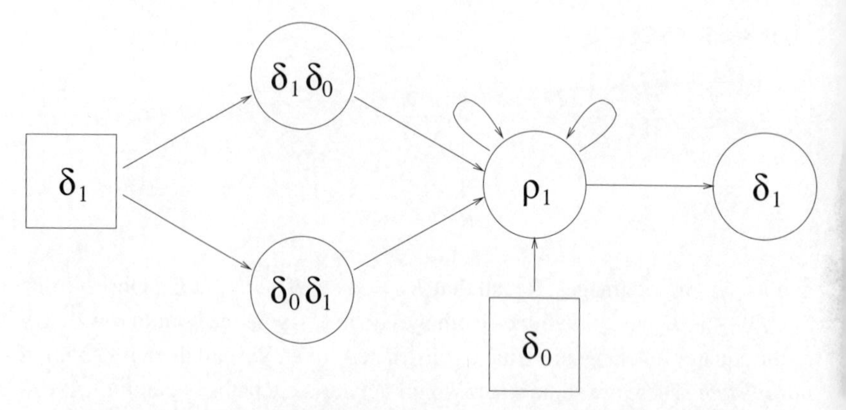

Figure 5.4.1 Arrows between exceptional word differences ending at the initial difference δ_1. (The square boxes denote non-exceptional word differences.)

The map (5) gives a square which leads directly to a non-exceptional state involving $\mathcal{F}_1$ and $\mathcal{F}_0$. It is

$$\begin{array}{ccc} E_0 & \xrightarrow{\ \rho_1\ } & E_1 \\ \downarrow{\scriptstyle \delta_1} & & \downarrow{\scriptstyle \delta_1^{-1}} \\ S_0 \cup S_1 \cup T_0 & \xrightarrow{\ \delta_0\ } & S_0 \cup S_1 \cup T_0 \end{array}$$

The maps (2) and (4) are immediately followed by a triangle and a square.

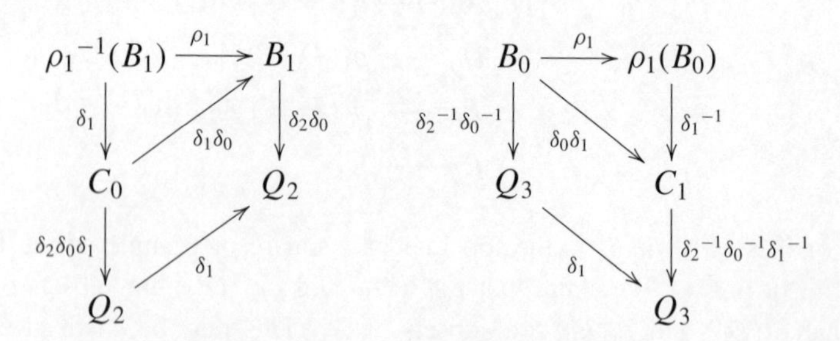

When we form arrows in the word difference machine, we reverse the arrows in each of the squares and triangles listed above. Recall that exceptional initial states never occur as the bottom line in any of the squares we have constructed. This means that there are no arrows leading out of exceptional initial states. In other words they are dead ends in the word difference machine. Moreover, there are no arrows to non-exceptional states from the subgraph of the word difference machine consisting of exceptional states.

5.5. Synchronising the word difference machine

Sections 5.2, 5.3 and 5.4 contain a full list of all word differences. There is one last technical problem because, as in section 2.5, the word difference machine we have constructed using this process is not synchronised. This is because of the presence of triangles rather than squares. We need to check that only finitely many triangles can occur in any path through the difference machine and to then compensate for this by adding padding symbols \$.

the previous sections we have constructed, up to symmetry, all the squares triangles that give rise to arrows in the difference machine. As before, we se the following notation for triangles:

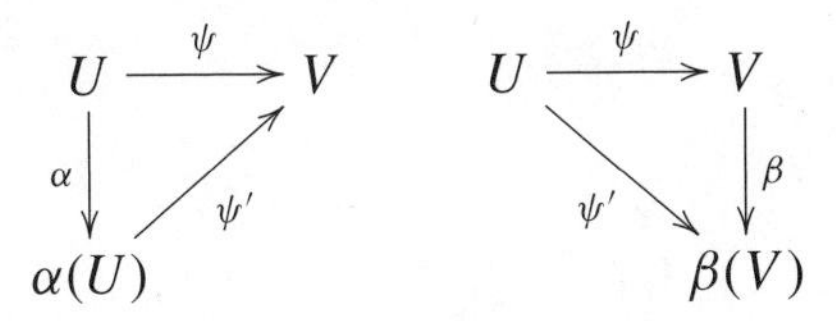

There are exactly three ways that triangles can arise. First, there are triangles where ψ' is the identity. Clearly, this can occur at most once in any path through the difference machine. This is equivalent to saying that the only arrows leading to the state e in the difference machine also start at e.

Secondly, there are triangles between non-exceptional states where ψ is $(\delta_j\delta_0)^{\pm 1}$ or $(\delta_0\delta_j)^{\pm 1}$ and ψ' is $\rho_j{}^{\pm 1}$ for $j = 1, 2$. For example

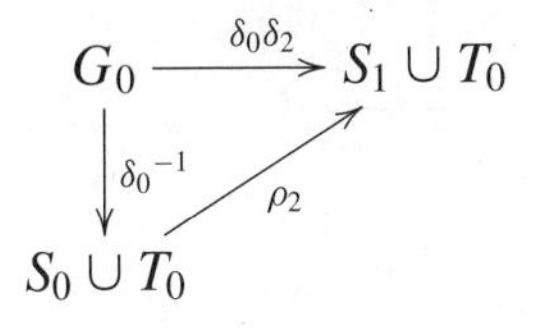

Any subsequent squares have the word difference e or $\rho_j{}^{\pm 1}$. Therefore this type of triangle can occur at most once. This is equivalent to saying that, in Fig. 5.3.1, the only arrows leading to the state $\rho_2{}^{\pm 1}$ begin either at $\rho_2{}^{\pm 1}$ or at e.

Thirdly, there are triangles occurring during transition from exceptional states to non-exceptional states. Once we leave exceptional states we never return and so this can occur at most once. In other words, in Fig. 5.4.1 there are no arrows from exceptional states to non-exceptional states and the triangles occur on the arrows from $\delta_1\delta_0$ and $\delta_0\delta_1$ to ρ_1. We now give an example which contains all three types of triangles. This is the worst possible case we must deal with. This is the word difference δ_1 for the following Farey block in D_0:

$$\gamma_1 = (0, 1, 2, 3) \in \Delta_7,\ \gamma_2 = (1, 1, 5, 8) \in \Delta_7,$$
$$\beta = (3, 3, 11, 15) \in \Delta_7;\ \gamma_3 = (1, 1, 4, 6) \in \Delta_7.$$

The squares associated with reducing this Farey block back to J_0 are:

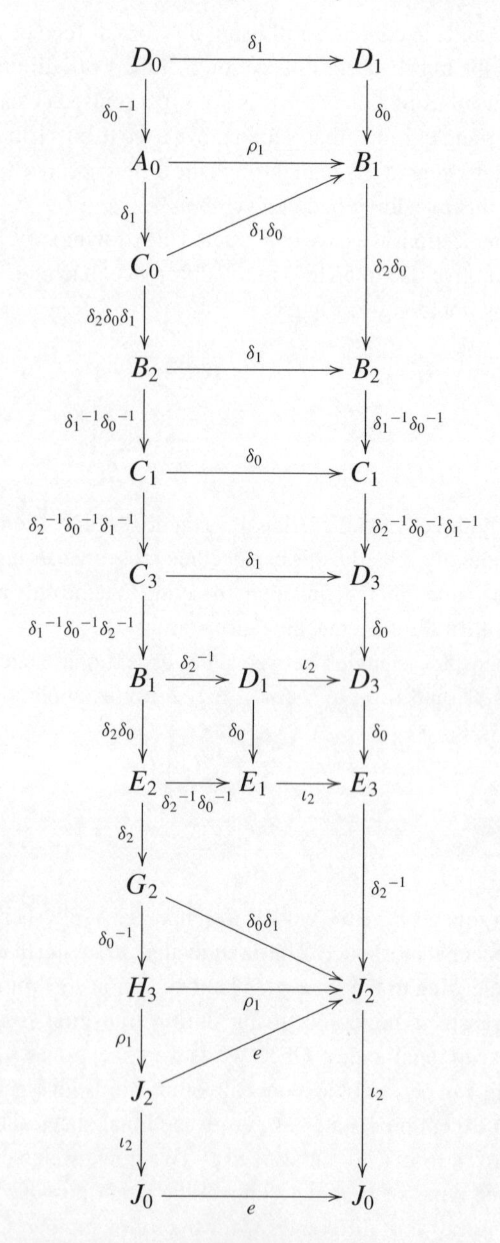

In order to synchronise the word difference machine, as we saw in section 2.5, it is necessary to add to the set of word differences by adding in diagonals to squares. Because we can have more than one triangle these diagonals may carry over several squares. In our example, we need word differences $\delta_2\delta_1\delta_0\delta_1$ between H_3 and E_3; ${\delta_0}^{-1}\delta_2\delta_0\delta_1$ between G_2 and D_3 and so on. Adding all possible diagonals in groups of one, two and three squares adds considerably to the possible word differences. Finally, we need to make further changes in the collection of word differences. This is because we need to consider initial word differences for all elements of the alphabet $\mathcal{A}$. So far we have only considered initial word differences in $\mathcal{D}_0$ as indicated in section 5.1. Because all words in $\mathcal{A}$ have length at most six in the letters of $\mathcal{D}_0$ this means we need to concatenate up to six word differences. For example, the word difference $\delta_1\delta_0\delta_2$ on the state D_0 gives rise to three squares concatenated horizontally:

$$\begin{array}{ccccccc}
D_0 & \xrightarrow{\delta_2} & \delta_2(D_0)\cap C_0 & \xrightarrow{\delta_0} & \delta_0\delta_0(D_0)\cap C_0 & \xrightarrow{\delta_1} & \delta_1\delta_0\delta_2(D_0)\cap B_0 \\
{\scriptstyle {\delta_0}^{-1}}\downarrow & & \downarrow{\scriptstyle \delta_2\delta_0\delta_1} & & \downarrow{\scriptstyle \delta_2\delta_0\delta_1} & & \downarrow{\scriptstyle \delta_2\delta_0} \\
R_0 & \xrightarrow[\iota_2{\delta_2}^{-1}]{} & A_2\cup B_2\cup E_2 & \xrightarrow[\delta_1]{} & R_2 & \xrightarrow[e]{} & R_2
\end{array}$$

In fact, this may be simplified to give the square

$$\begin{array}{ccc}
D_0 & \xrightarrow{\delta_1\delta_0\delta_2} & \delta_1\delta_0\delta_2(D_0)\cap B_0 \\
{\scriptstyle {\delta_0}^{-1}}\downarrow & & \downarrow{\scriptstyle \delta_2\delta_0} \\
R_0 & \xrightarrow{\iota_2} & R_2
\end{array}$$

All these changes make the final collection of word differences $\mathcal{D}$ rather large but it is still finite. Thus we have a word difference machine in the usual sense. This makes $\mathcal{MCG}(\Sigma_2)$ automatic. It seems likely that the same structure shows that $\mathcal{MCG}(\Sigma_2)$ is biautomatic (in other words generators can be added at either end) but we have not explored this possibility.

References

[1] J. S. Birman, Braids, *Links and Mapping Class Groups*, Annals of Math. Studies vol. 82, Princeton University Press, 1974.

[2] J. S. Birman and C. Series, Algebraic linearity for an automorphism of a surface group, *J. Pure and Applied Algebra* **52** (1988), 227–275.

[3] J. S. Birman and C. Series, Dehn's algorithm revisited, with applications to simple curves on surfaces, in: *Combinatorial Group Theory and Topology* (S. Gersten and J. Stallings eds.), Annals of Math. Studies vol. 111, Princeton University Press 1987, 451–478.

[4] R. Bowen and C. Series, Markov maps associated to Fuchsian groups, *Inst. Hautes Études Sci. Publ. Math.* **50** (1979), 153–170.

[5] A. Connes, J. Feldman and B. Weiss, An amenable equivalence relation is generated by a single transformation, *Ergodic Theory and Dynamical Systems* **1** (1981), 435–450.

[6] M. Dehn, Die Gruppe der Abbildungklassen, *Acta Math.* **69** (1938), 135–206.

[7] D. B. A. Epstein, J. W. Cannon, D. F. Holt, S. V. F. Levy, M. S. Paterson and W. P. Thurston, *Word Processing in Groups*, Jones and Bartlett, Boston, 1992.

[8] H. Hamidi-Tehrani and Z.-H. Chen, Surface diffeomorphisms via train-tracks, *Topology and its Applications* **20** (1996), 1–27.

[9] D. F. Holt, *The Warwick Automatic Groups Software*, DIMACS Series in Mathematics and Computer Science vol. 25, 1996.

[10] D. F. Holt, Automatic groups, subgroups and cosets, in: *The Epstein Birthday Schrift* (I. Rivin, C. Rourke and C. Series eds.), Geometry & Topology Monographs vol. 1, 1998, 249–260.

[11] L. Keen and C. Series, Pleating coordinates for the Maskit embedding of the Teichmüller space of punctured tori, *Topology* **32** (1993), 719–749.

[12] L. Keen, J. R. Parker and C. Series, Combinatorics of simple closed curves on the twice punctured torus, *Israel J. Math.* **112** (1999), 29–60.

[13] W. Magnus, Über Automorphismen von Fundamentalgruppen berandeter Flächen, *Math. Annalen* **109** (1934), 617–646.

[14] H. A. Masur and Y. N. Minsky, Geometry of the complex of curves I: Hyperbolicity, *Invent. Math.* **138** (1999), 103–149.

[15] H. A. Masur and Y. N. Minsky, Geometry of the complex of curves II: Hierarchical structure, *Geom. Funct. Anal.* **10** (2000), 902–974.

[16] L. Mosher, Mapping class groups are automatic, *Ann. of Math.* **142** (1995), 303–384.

[17] R. C. Penner with J. L. Harer, *Combinatorics of Train Tracks*, Annals of Math. Studies vol. 125, Princeton University Press, 1992.

[18] S. Rees, Hairdressing in groups: a survey of combings and formal languages, in: *The Epstein Birthday Schrift* (I. Rivin, C. Rourke and C. Series eds.) Geometry & Topology Monographs vol. 1, 1998, 493–509.

[19] C. Series, The modular surface and continued fractions, *J. London Math. Soc.* **31** (1985), 69–85.

[20] C. Series, The geometry of Markoff numbers, *Math. Intelligencer* **7** (1985), 20–29.

[21] C. Series, Geometrical methods of symbolic coding, in: *Ergodic Theory, Symbolic Dynamics and Hyperbolic Spaces* (T. Bedford, M. Keane and C. Series eds.), Oxford University Press 1991, 125–151.

[22] W. P. Thurston, *The Geometry and Topology of Three-Manifolds*, Princeton University Press, 1997.

[23] W. P. Thurston, On the geometry and dynamics of diffeomorphisms of surfaces, *Bull. Amer. Math. Soc.* **19** (1988), 417–431.

15

Kac–Moody groups: split and relative theories. Lattices

B. Remy

Introduction

Historical sketch of Kac–Moody theory. — Kac–Moody theory was initiated in 1968, when V. Kac and R. Moody independently defined infinite-dimensional Lie algebras generalizing complex semi-simple Lie algebras. Their definition is based on Serre's presentation theorem describing explicitly the latter (finite-dimensional) Lie algebras [Hu1, 18.3]. A natural question then is to integrate Kac–Moody Lie algebras as Lie groups integrate real Lie algebras, but this time in the infinite-dimensional setting. This difficult problem led to several propositions. In characteristic 0, a satisfactory approach consists in seeing them as subgroups in the automorphisms of the corresponding Lie algebras [KP1,2,3]. This way, V. Kac and D. Peterson developed the structure theory of Kac–Moody algebras in complete analogy with the classical theory: intrinsic definition and conjugacy results for Borel (resp. Cartan) subgroups, root decomposition with abstract description of the root system... Another aspect of this work is the construction of generalized Schubert varieties. These algebraic varieties enabled O. Mathieu to get a complete generalization of the character formula in the Kac–Moody framework [Mat1]. To this end, O. Mathieu defined Kac–Moody groups over arbitrary fields in the formalism of ind-schemes [Mat2].

Combinatorial approach. — Although the objects above – Kac–Moody groups and Schubert varieties – can be studied in a nice algebro-geometric context, we will work with groups arising from another, more combinatorial viewpoint. All of this work is due to J. Tits [T4,5,6,7], who of course contributed also to the previous problems. The aim is to get a much richer combinatorial structure for these groups. This led J. Tits to the notion of " Root Group Datum" axioms [T7] whose geometric counterpart is the theory of Moufang twin buildings. The starting point of the construction of Kac–Moody groups [T4] is a generalization of Steinberg's presentation theorem [Sp, Theorem 9.4.3] which concerns simply connected semi-simple algebraic groups. In this context, the groups $\mathrm{SL}_n(\mathbb{K}[t, t^{-1}])$ are Kac–Moody groups obtained via Tits' construction.

Relative Kac–Moody theory in characteristic 0. — So far, the infinite-dimensional objects alluded to were analogues of split Lie algebras and split algebraic groups. Still, it is known that by far not all interesting algebraic groups are covered by the split theory – just consider the simplest case of the multiplicative group of a quaternion skew field. This is the reason why Borel-Tits theory [BoT] is so important: it deals with algebraic groups over arbitrary fields $\mathbb{K}$, and the main results provide a combinatorial structure theorem for $\mathbb{K}$-points, conjugacy theorems for minimal $\mathbb{K}$-parabolic subgroups (resp. for maximal $\mathbb{K}$-split tori). This theory calls for a generalization in the Kac–Moody setting: this work was achieved in the characteristic 0 case by G. Rousseau [Rou1,2,3; B_3R].

Two analogies. — The example of the groups $\mathrm{SL}_n(\mathbb{K}[t, t^{-1}])$ is actually a good guideline since it provides at the same time another analogy for Kac–Moody groups. Indeed, over finite fields $\mathbb{K} = \mathbb{F}_q$, the former ones are arithmetic groups in the function field case. Following the first analogy – a Kac–Moody group is an infinite-dimensional reductive group, a large part of the present paper describes the author's thesis [Ré2] whose basic goal is to define $\mathbb{K}$-forms of Kac–Moody groups with no assumption on the ground field $\mathbb{K}$. This prevents us from using the viewpoint of automorphisms of Lie algebras as in the works previously cited, but the analogues of the main results of Borel-Tits theory are proved. This requires first to reconsider Tits' construction of Kac–Moody groups and to prove results in this (split) case, interesting in their own right. On the other hand, the analogy with arithmetic groups will also be discussed so as to see Kac–Moody groups as discrete groups – lattices for $\mathbb{F}_q$ large enough – of their geometries.

Tools. — Let us talk about tools now, and start with the main difficulty: no algebro-geometric structure is known for split Kac–Moody groups as defined by J. Tits. The idea is to replace this structure by two well-understood actions. The first one is not so mysterious since it is linear: it is in fact the natural generalization of the adjoint representation of algebraic groups. It plays a crucial role because it enables to endow a large family of subgroups with a structure of algebraic group. The second kind of action involves buildings: it is a usual topic in group theory to define a suitable geometry out of a given group to study it (use of Cayley graphs, of boundaries...) Kac–Moody groups are concerned by building techniques thanks to their nice combinatorial properties, refining that of BN-pairs (for an account on the general use of buildings in group theory, see [T2]). There exists some kind of (non-unique) correspondence between buildings and

BN-pairs [Ron §5], but it is too general to provide precise information in specific situations. Some refinements are to be adjusted accordingly – see the example of Euclidean buildings and " Valuated Root Data" in Bruhat-Tits theory [BrT1]. As already said, in the Kac–Moody setting the refinements consist in requiring the " Root Group Datum" axioms at the group level and in working with twin buildings at the geometric level. Roughly speaking, a twin building is the datum of two buildings related by opposition relations between chambers and apartments.

Organization of the paper. — This article is divided into four parts. Part 1 deals with group combinatorics in a purely abstract context. Since the combinatorial axioms will be satisfied by both split and non-split groups, it is an efficient way to formalize properties shared by them. We describe the geometry of twin buildings and the corresponding group theoretic axioms. The aim is to obtain two kinds of Levi decomposition which will be of interest later on. Part 2 describes the split theory of Kac–Moody groups. In particular, we explain why the adjoint representation can be seen as a substitute for a global algebro-geometric structure. An illustration of this is a repeatedly used argument combining negative curvature and algebraic groups arguments. Part 3 presents the relative theory of almost split Kac–Moody groups. A sketch of the proof of the structure theorem for rational points is given. The particular case of a finite ground field is considered, as well as a classical example of a twisted group leading to a semi-homogeneous twin tree. Finally, part 4 adopts the viewpoint of discrete groups. We first show that Kac–Moody theory enables to produce hyperbolic buildings (among many other possibilities) and justify why these geometries are particularly interesting. Then we show that Kac–Moody groups or their spherical parabolic subgroups over a finite ground field are often lattices of their buildings. This leads to an analogy with arithmetic lattices over function fields. The assumed knowledge for this article consists of general facts from building theory and algebraic groups. References for buildings are K. Brown's book [Br2] for the apartment systems viewpoint and M. Ronan's book [Ron] for the chamber systems one. Concerning algebraic groups, the recent books [Bo] and [Sp] are the main references dealing with relative theory.

Acknowledgements. — This work presents a Ph.D. prepared under the supervision of G. Rousseau. It is a great pleasure to thank him, as well as M. Bourdon who drew my attention to the discrete groups viewpoint. I am very grateful to H. Abels, P. Abramenko, and Th. Müller for their kindness. It was a pleasure and a nice experience to take part in the conference " Groups : geometric and combinatorial aspects" organized by H. Helling and the latter, and to be welcomed at the SFB 343 (University of Bielefeld).

1. Abstract group combinatorics and twin buildings

The aim of this section is to provide all the abstract background we will need to study split and almost split Kac–Moody groups. Subsection 1.1 introduces the root system of a Coxeter group. These roots will index the group combinatorics of the " Twin Root Datum" axioms presented in Subsection 1.2. In the next subsection, we describe the geometric side of the (TRD)-groups, that is the twin buildings on which they operate. Subsection 1.4 is dedicated to the geometric notions and realizations to be used later. At last, Subsection 1.5 goes back to group theory providing some semi-direct decomposition results for distinguished classes of subgroups.

1.1. Root systems and realizations of a Coxeter group

The objects we define here will be used to index group theoretic axioms (1.2.A) and to describe the geometry of buildings (1.3.B).

A. Coxeter complex. Root systems. Let $M = [M_{st}]_{s,t\in S}$ denote a Coxeter matrix with associated Coxeter system (W, S). For our purpose, it is sufficient to suppose the canonical generating set S finite. The group W admits the following presentation :

$$W = \langle s \in S \mid (st)^{M_{st}} = 1 \text{ whenever } M_{st} < \infty \rangle.$$

We shall use the *length function* $\ell : W \to \mathbb{N}$ defined w.r.t. S. The existence of an abstract simplicial complex acted upon by W is the starting point of the definition of buildings of type (W, S) in terms of apartment systems. This complex is called the *Coxeter complex* associated to W, and will be denoted by $\Sigma(W, S)$ or Σ. It describes the combinatorial geometry of " slices" in a building of type (W, S) – the so-called *apartments*. The abstract complex Σ is made of translates of the *special subgroups* $W_J := \langle J \rangle$, $J \subset S$, ordered by inclusion [Br2, p.58-59]. The *root system* of (W, S) is defined by means of the length function ℓ [T4, §5]. The set W admits a W-action via left translations. *Roots* are distinguished halves of this W-set, whose elements will be called *chambers*.

Definition. (i) *The* simple root *of index s is the half* $\alpha_s := \{w \in W \mid \ell(sw) > \ell(w)\}$.

(ii) *A* root *of W is a half of the form $w\alpha_s$, $w \in W$, $s \in S$. The set of roots will be denoted by Φ; it admits an obvious W-action.*

(iii) *A root is called* positive *if it contains* 1; *otherwise, it is called* negative. *Denote by Φ_+ (resp. Φ_-) the set of positive (resp. negative) roots of Φ.*

(iv) *The* opposite *of a root is its complement.*

Remark. *The opposite of the root $w\alpha_s$ is indeed a root since it is $ws\alpha_s$.*

The next definitions are used for the group combinatorics presented in 1.2.A.

Definition. (i) *A pair of roots $\{\alpha;\beta\}$ is called* prenilpotent *if both intersections $\alpha\cap\beta$ and $(-\alpha)\cap(-\beta)$ are non-empty.*

(ii) *Given a prenilpotent pair of roots $\{\alpha;\beta\}$, the* interval $[\alpha;\beta]$ *is by definition the set of roots γ with $\gamma\supset\alpha\cap\beta$ and $(-\gamma)\supset(-\alpha)\cap(-\beta)$. We also set $]\alpha;\beta[:=[\alpha;\beta]\setminus\{\alpha;\beta\}$.*

B. The Tits cone. Linear refinements. We introduce now a fundamental realization of a Coxeter group : the *Tits cone*. It was first defined in [Bbk, V.4] to which we refer for proofs. This approach will also allow us to consider roots as linear forms [Hu2, II.5]. We keep the Coxeter system (W,S) with Coxeter matrix $M=[M_{st}]_{s,t\in S}$. Consider the real vector space V over the symbols $\{\alpha_s\}_{s\in S}$ and define the *cosine matrix* A of W by $A_{st}:=-\cos(\pi/M_{st})$: this is the matrix of a bilinear form B (w.r.t. the basis $\{\alpha_s\}_{s\in S}$). To each s of S is associated the involution $\sigma_s:\lambda\mapsto\lambda-2B(\alpha_s,\lambda)\alpha_s$, and the assignment $s\mapsto\sigma_s$ defines a faithful representation of W. The *(positive) half-space* in V^* of an element λ in V is denoted by $D(\lambda)$: $D(\lambda):=\{x\in V^*\mid\lambda(x)>0\}$. We denote its boundary – the kernel of λ – by $\partial\lambda$.

Definition. (i) *The* standard chamber c *is the simplicial cone $\bigcap_{s\in S}D(\alpha_s)$ of elements of V^* on which all linear forms α_s are positive. A* chamber *is a W-translate of c.*

(ii) *The* standard facet of type J, $J\subset S$, *is $F_I:=\bigcap_{s\in I}\partial\alpha_s\cap\bigcap_{s\in S\setminus I}D(\alpha_s)$. A* facet of type J *is a W-translate of F_J.*

(iii) *The* Tits cone $\overline{\mathcal{C}}$ *of W is the union of the closures of all chambers $\bigcup_{w\in W}w\overline{c}=\bigsqcup_{w\in W,J\subset S}wF_J$.*

A study of the action of W on $\overline{\mathcal{C}}$ shows that the type is well-defined. A facet is *spherical* if it is of type J with W_J finite. Facets of all types are represented here, as simplicial cones. The simplicial complex so obtained is not locally finite in general, but its interior $\mathcal{C}$ is. In fact, a facet is of non-spherical type if and only if it lies in the boundary of the Tits cone. Further properties of the cone $\overline{\mathcal{C}}$ are available in [V].

Example. *In the case of an affine reflexion group, the Tits cone is made up of the union of an open half-space and the origin, which is the only nonspherical* *facet. The affine space in which the standard representation of the group is defined is just the affinisation of this cone* [Hu2, II.6].

The viewpoint of linear forms for roots enables us to introduce another – more restrictive – notion of interval of roots.

Definition. *Given two roots α and β, the* linear interval *they define is the set $[\alpha; \beta]_{lin}$ of positive linear combinations of them, seen as linear forms on V.*

Figure. —

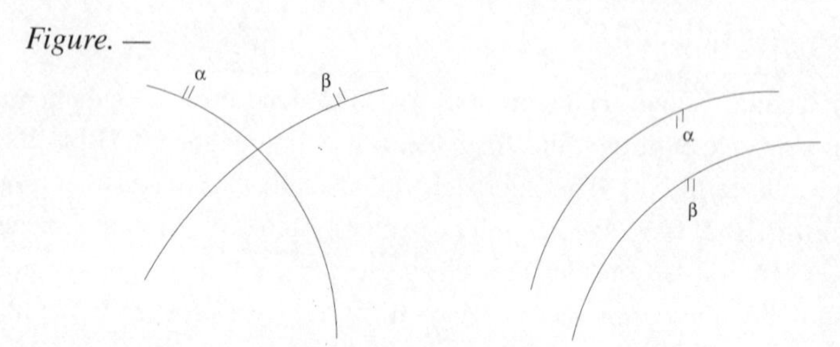

The picture above illustrates a general fact: there are two ways to be prenilpotent. Either the walls of the roots intersect along a spherical facet – and the four pairs of roots of the form $\{\pm\alpha; \pm\beta\}$ are prenilpotent, or a root contains the other – and only two pairs among the four are prenilpotent.

Remark. *For a prenilpotent pair of roots, one has $[\alpha; \beta]_{lin} \subset [\alpha; \beta]$, still the notions do not coincide, even in the Kac–Moody situation. G. Rousseau indicated an example of strict inclusion provided by a tesselation of the hyperbolic plane $\mathbb{H}^2$* [Ré2, 5.4.2].

C. Geometric realizations. We will try to use geometry as much as possible instead of abstract set-theoretic structures. As an example, we will often represent Coxeter complexes by *polyhedral complexes* [BH]. We are interested in such spaces with the following additional properties.

(i) The complex is labelled by a fixed set of subsets of S – the *types*. We call *facets* the polyhedra inside. Codimension 1 facets are called *panels*, maximal facets are called *chambers*.

(ii) There is a countable family of codimension 1 subcomplexes – the *walls* – w.r.t. which are defined involutions – the *reflections*. A reflection fixes its wall and stabilizes the whole family of them.

(iii) For a chamber c, there is a bijection between the set of generators S and the walls supporting the panels of c. The corresponding reflections define a faithful representation of W by label-preserving automorphisms of the complex.

(iv) The W-action is simply transitive on chambers.

Remark. (1) *The Tits cone satisfies all the conditions above, with simplicial cones as facets instead of polyhedra.*

(2) *We may use geometric realizations where only spherical types appear. This is the case in the examples below.*

Example. (1) *The simplest example with infinite Coxeter group is given by the real line and its tesselation by the segments defined by consecutive integers. It is acted upon by the infinite dihedral group D_∞, and the corresponding buildings are trees.*

(2) *Another famous example associated to an affine reflection group comes from the tesselation of the Euclidean plane by equilateral triangles. Buildings with this geometry as apartments are called $\widetilde{\mathrm{A}}_2$- or triangle buildings. They are interesting because, even if they belong to the well-known class of Euclidean buildings, many of them do not come from Bruhat-Tits theory.*

(3) *A well-known way to concretely construct a Coxeter group with a realization of its Coxeter complex is to apply Poincaré polyhedron theorem* [Mas, IV.H.11]. *This works in the framework of spherical, Euclidean or hyperbolic geometry; we just have to consider reflections w.r.t. to a suitable polyhedron. For a tiling of a hyperbolic space $\mathbb{H}^n$, the corresponding buildings are called hyperbolic, Fuchsian in the two-dimensional case.*

Figure. —

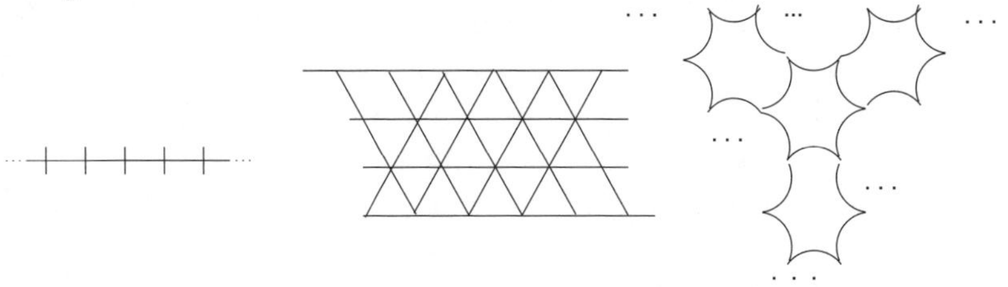

1.2. Axioms for group combinatorics

We can now give axioms refining BN-pairs and adapted to the Kac–Moody situation.

A. The (TRD) axioms. The axioms listed below are indexed by the set of roots Φ of the Coxeter system (W, S). It is a slight modification of axioms proposed in [T7]. That it implies the group combinatorics introduced by V. Kac and D. Peterson [KP2] follows from [Ré2, Théorème 1.5.4], which elaborates on [Ch] and [T3].

Definition. *Let G be an abstract group containing a subgroup H. Suppose G is endowed with a family $\{U_\alpha\}_{\alpha\in\Phi}$ of subgroups indexed by the set of roots*

Φ, and define the subgroups $U_+ := \langle U_\alpha \mid \alpha \in \Phi_+ \rangle$ and $U_- := \langle U_\alpha \mid \alpha \in \Phi_- \rangle$. Then, the triple $\big(G, \{U_\alpha\}_{\alpha\in\Phi}, H\big)$ is said to satisfy the (TRD) axioms *if the following conditions are satisfied.*

(TRD0) *Each U_α is non-trivial and normalized by H.*
(TRD1) *For each prenilpotent pair of roots $\{\alpha; \beta\}$, the commutator subgroup $[U_\alpha, U_\beta]$ is contained in the subgroup $U_{]\alpha,\beta[}$ generated by the U_γ's, with $\gamma \in]\alpha, \beta[$.*
(TRD2) *For each s in S and u in $U_{\alpha_s} \setminus \{1\}$, there exist uniquely defined u' and u'' in $U_{-\alpha_s} \setminus \{1\}$ such that $m(u) := u'uu''$ conjugates U_β onto $U_{s\beta}$ for every root β. Besides, it is required that for all u and v in $U_{\alpha_s} \setminus \{1\}$, one should have $m(u)H = m(v)H$.*
(TRD3) *For each s in S, $U_{\alpha_s} \not\subset U_-$ and $U_{-\alpha_s} \not\subset U_+$.*
(TRD4) *G is generated by H and the U_α's.*

Such a group will be referred to as a (TRD)-group. *It will be called a* $(\mathrm{TRD})_{\mathrm{lin}}$-group *or said to satisfy the* $(\mathrm{TRD})_{\mathrm{lin}}$ axioms *if* (TRD1) *is still true after replacing intervals by linear ones.*

Remark. (1) *A consequence of Borel-Tits theory* [BoT, Bo, Sp] *is that isotropic reductive algebraic groups satisfy $(TRD)_{lin}$ axioms. We will see in* 2.3.A *that so do split Kac–Moody groups. The case of non-split Kac–Moody groups is the object of the Galois descent theorem – see* 3.2.A.
(2) *The case of algebraic groups suggests to take into account more carefully proportionality relations between roots seen as linear forms. It is indeed possible to formalize the difference between reduced and non-reduced infinite root systems, and to derive refined $(TRD)_{lin}$ axioms* [Ré2, 6.2.5].

B. Main consequences. We can derive a first list of properties for a (TRD)-group G.

Two BN-pairs. — The main point is the existence of two BN-pairs in the group G. Define the *standard Borel subgroup of sign ϵ* to be $B_\epsilon := HU_\epsilon$. The subgroup $N < G$ is by definition generated by H and the $m(u)$'s of axiom (TRD2). Then, one has

$$H = \bigcap_{\alpha\in\Phi} N_G(U_\alpha) = B_+ \cap N = B_- \cap N,$$

and (G, B_+, N, S) and (G, B_-, N, S) are BN-pairs sharing the same Weyl group $W = N/H$. As B_+ and B_- are not conjugate, the positive and the negative

BN-pairs do not carry the same information. A conjugate of B_+ (resp. B_-) will be called a *positive (resp. negative) Borel subgroup.*

Refined Bruhat and Birkhoff decompositions. — A formal consequence of the existence of a BN-pair is a Bruhat decomposition for the group. In our setting, the decomposition for each sign can be made more precise. For each $w \in W$, define the subgroups $U_w := U_+ \cap wU_-w^{-1}$ and $U_{-w} := U_- \cap wU_+w^{-1}$. The *refined Bruhat decompositions* are then [KP2, Proposition 3.2] :

$$G = \bigsqcup_{w \in W} U_w w B_+ \quad \text{and} \quad G = \bigsqcup_{w \in W} U_{-w} w B_-,$$

with uniqueness of the first factor. A third decomposition involves both signs and will be used to define the twinned structures (1.3.A). More precisely, the *refined Birkhoff decompositions* are [KP2, Proposition 3.3]:

$$G = \bigsqcup_{w \in W} (U_+ \cap wU_+w^{-1}) w B_- = \bigsqcup_{w \in W} (U_- \cap wU_-w^{-1}) w B_+,$$

once again with uniqueness of the first factors.

Other unique writings. — Another kind of unique writing result is valid for the groups $U_{\pm w}$. For each $z \in W$, define the (finite) sets of roots $\Phi_z := \Phi_+ \cap z^{-1}\Phi_-$ and $\Phi_{-z} := \Phi_- \cap z^{-1}\Phi_+$. Then, the group U_w (resp. U_{-w}) is in bijection with the set-theoretic product of the root groups indexed by $\Phi_{w^{-1}}$ (resp. $\Phi_{-w^{-1}}$) for a suitable (cyclic) ordering on the latter set [T4, proposition 3 (ii)], [Ré2, 1.5.2].

Two buildings. — Let us describe now how to construct a building out of each BN-pair, the twin structure relating them being defined in 1.3.A. Fix a sign ϵ and consider the corresponding BN-pair (G, B_ϵ, N, S). Let $d_\epsilon : G/B_\epsilon \times G/B_\epsilon \to W$ be the application which associates to $(gB_\epsilon, hB_\epsilon)$ the element w such that $B_\epsilon g^{-1} h B_\epsilon = B_\epsilon w B_\epsilon$. Then, d_ϵ is a W*-distance* making $(G/B_\epsilon, d_\epsilon)$ a building [Ron, 5.3]. The *standard apartment of sign* ϵ is $\{wB_\epsilon\}_{w \in W}$, the relevant apartment system being the set of its G-transforms. A *facet* of type J is a translate $gP_{\epsilon,J}$ of a *standard parabolic subgroup* $P_{\epsilon,J} := B_\epsilon W_J B_\epsilon$.

C. *Examples.* — The most familiar examples of groups enjoying the properties above are provided by Chevalley groups over Laurent polynomials. These groups are Kac–Moody groups of affine type. We briefly describe the case of the special linear groups $\mathrm{SL}_n(\mathbb{K}[t, t^{-1}])$, $n \geq 2$. From the Kac–Moody viewpoint, the ground field is $\mathbb{K}$. The buildings involved are Bruhat-Tits. They are associated to the p-adic Lie groups $\mathrm{SL}_n\big(\mathbb{F}_q(\!(t)\!)\big)$ and $\mathrm{SL}_n\big(\mathbb{F}_q(\!(t^{-1})\!)\big)$ respectively in the case of a finite ground field $\mathbb{F}_q$. The Weyl group is the affine reflection group $\mathcal{S}_n \ltimes \mathbb{Z}^{n-1}$. The Borel subgroups are

$$B_+ := \left\{ M \in \begin{pmatrix} \mathbb{K}[t] & & \mathbb{K}[t] \\ & \ddots & \\ t\mathbb{K}[t] & & \mathbb{K}[t] \end{pmatrix} \middle| \det M = 1 \right\};$$

$$B_- := \left\{ M \in \begin{pmatrix} \mathbb{K}[t^{-1}] & & t^{-1}\mathbb{K}[t^{-1}] \\ & \ddots & \\ \mathbb{K}[t^{-1}] & & \mathbb{K}[t^{-1}] \end{pmatrix} \middle| \det M = 1 \right\}.$$

As subgroup H, we take the standard Cartan subgroup T of $\mathrm{SL}_n(\mathbb{K})$ made up of diagonal matrices with coefficients in $\mathbb{K}^\times$ and determinant 1. For $1 \leq i \leq n-1$, the monomial matrices with $\begin{pmatrix} 0 & 1 \\ -1 & 0 \end{pmatrix}$ in position $(i, i+1)$ on the diagonal (and 1's everywhere else on it) lift the $n-1$ simple reflexions generating the finite Weyl group of $\mathrm{SL}_n(\mathbb{K})$. The last reflexion "responsible for the affinisation" is lifted by

$$N_n := \begin{pmatrix} 0 & 0 & -t^{-1} \\ 0 & \begin{pmatrix} 1 & & 0 \\ & \ddots & \\ 0 & & 1 \end{pmatrix} & 0 \\ t & 0 & 0 \end{pmatrix}.$$

The situation is the same for root groups: besides the simple root groups of $\mathrm{SL}_n(\mathbb{K})$, one has to add

$$U_n := \left\{ u_n(k) := \begin{pmatrix} 1 & 0 & 0 \\ & \ddots & \\ 0 & 1 & 0 \\ & & \ddots \\ kt & 0 & 1 \end{pmatrix} \middle| k \in \mathbb{K} \right\},$$

to get the complete family of subgroups indexed by the simple roots.

1.3. Moufang twin buildings

The theory of *Moufang twin buildings* is the geometric side of the group combinatorics above. A good account of their general theory is [A1]. It was initiated in [T6], [T7] and *a posteriori* in [T1], whereas [RT2,3] deals with the special case of twin trees. References for the classification problem are [MR] for the unicity step, and then B. Mühlherr's work, in particular [Mü] and his forthcoming Habilitationschrift.

A. Twin buildings. The definition of a *twin building* is quite similar to that of a building in terms of W-distance [T7, A1 §2].

Definition. *A* twin building of type (W, S) *consists of two buildings* $(\mathcal{I}_+, d_+)$ *and* $(\mathcal{I}_-, d_-)$ *of type* (W, S) *endowed with a* (W-)codistance. *By definition, the latter is a map* $d^* : (\mathcal{I}_+ \times \mathcal{I}_-) \cup (\mathcal{I}_- \times \mathcal{I}_+) \to W$ *satisfying the following conditions for each sign* ϵ *and all chambers* x_ϵ *in* $\mathcal{I}_\epsilon$ *and* $y_{-\epsilon}$, $y'_{-\epsilon}$ *in* $\mathcal{I}_{-\epsilon}$.

(TW1) $d^*(y_{-\epsilon}, x_\epsilon) = d^*(x_\epsilon, y_{-\epsilon})^{-1}$.
(TW2) *If* $d^*(x_\epsilon, y_{-\epsilon}) = w$ *and* $d_{-\epsilon}(y_{-\epsilon}, y'_{-\epsilon}) = s \in S$ *with* $\ell(ws) < \ell(w)$, *then* $d^*(x_\epsilon, y'_{-\epsilon}) = ws$.
(TW3) *If* $d^*(x_\epsilon, y_{-\epsilon}) = w$ *then for each* $s \in S$, *there exists* $z_{-\epsilon} \in \mathcal{I}_{-\epsilon}$ *with* $d_{-\epsilon}(y_{-\epsilon}, z_{-\epsilon}) = s$ *and* $d^*(x_\epsilon, z_{-\epsilon}) = ws$.

From this definition can be derived two *opposition relations*. Two chambers are *opposite* if they are at codistance 1. Given an apartment $\mathbb{A}_\epsilon$ of sign ϵ, an *opposite* of it is an apartment $A_{-\epsilon}$ such that each chamber of $\mathbb{A}_\epsilon$ admits exactly one opposite in $\mathbb{A}_{-\epsilon}$. In this situation, the same assertion is true after inversion of signs. An apartment admits at most one opposite, and the set of apartments having an opposite forms an apartment system in the building [A1 §2].

Remark. *We defined two opposition relations, but we have to be careful with them. Whereas an element of the apartment system defined above admits by definition exactly one opposite, a chamber admits many opposites in the building of opposite sign. In the sequel, the bold letter* $\mathbb{A}$ *will refer to a pair* $(\mathbb{A}_+, \mathbb{A}_-)$ *of opposite apartments.*

The connection with group combinatorics is folklore [Ré2, 2.6.4].

Proposition. *Let* G *be a* (TRD)-*group. Then the buildings associated to the two* BN-*pairs of* G *are Moufang and belong to a twin building structure. In particular, two facets of opposite signs are always contained in a pair of opposite apartments. Moreover,* G *is transitive on pairs of opposite chambers.*

So far, we explained how to derive two buildings from (TRD)-groups. The main consequences of Moufang property will be described in the next subsection. So what is left to do is to define the codistance. Whereas W-distance is deduced from Bruhat decomposition, the W-codistance can be made completely explicit thanks to the Birkhoff decomposition 1.2.B. Two chambers of opposit signs gB_+ and hB_- are at *codistance* w if and only if $g^{-1}h$ is in the Birkhoff class B_+wB_-. This definition does not depend on the choice of g and h in their class.

Examples. — 1. The class of twin trees has been studied in full generality in [RT2,3] where many properties are established. For instance, the trees have to be at least semi-homogeneous. The proofs are simpler than in the general case thanks to the use of an integral codistance which faithfully reflects the properties of the W-codistance.

2. For the groups $\mathrm{SL}_n(\mathbb{K}[t, t^{-1}])$ of 1.2.C, the opposition relation can be formulated more concretely. Two chambers gB_+ and hB_- are opposite if and only if their stabilizers – Borel subgroups of opposite signs – intersect along a conjugate of T. Over $\mathbb{F}_q$, this is equivalent to intersecting along a group of (minimal) order $(q-1)^{n-1}$.

B. Geometric description of the group action. Suppose we are given a (TDR)-group G and denote by $\mathbb{A}$ the standard pair of opposite apartments in its twin building, that is $\{wB_+\}_{w\in W} \sqcup \{wB_-\}_{w\in W}$. We want to give a geometric description of the G-action on its twin building.

Kernel of the action. — One has (i) $H = \mathrm{Fix}_G(\mathbb{A}_+) = \mathrm{Fix}_G(\mathbb{A}_-)$.

This result is the geometric formulation of [KP2, Corollary 3.4]; it shows that the kernel of the action of G on its twin building is contained in $H < G$.

Moufang property. — Roughly speaking, requiring the Moufang property for a building is a way to make sure the latter admits a sufficiently large automorphism group, with well-understood local actions. We just state the main consequence of it, which will enable us to compute the number of chambers whose closure contains a given panel F. This integer will be referred to as the *thickness* at F of the building. Let us start with F_s the positive panel of type s in the closure of the standard chamber c. Then [Ron, (MO1) p.74]:

(ii) The root group U_{α_s} fixes α_s and is normalized by H. It is simply transitive on the chambers containing F_s and $\neq c$.

By homogeneity, (ii) is true for every configuration panel-chamber-wall-root as above, and *mutatis mutandis* everything remains true for the negative building of G.

Figure. —

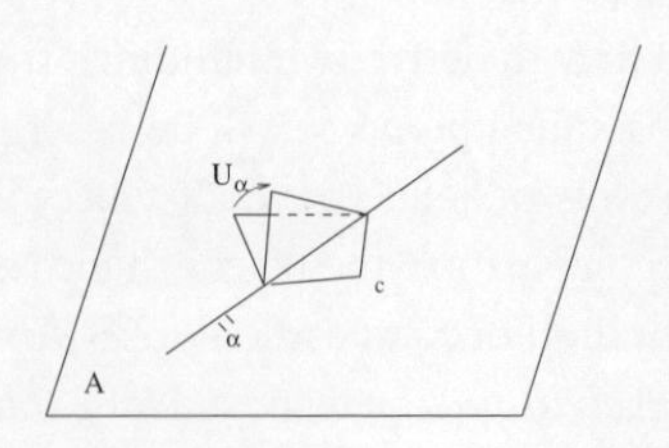

The Weyl group as a subquotient. — The following assertion is just a geometric formulation of axiom (TRD2). Consider the epimorphism $\nu : N \twoheadrightarrow W$ with $\mathrm{Ker}(\nu) = H$.

(iii) The group N stabilizes $\mathbb{A}$; it permutes the U_α's by $nU_\alpha n^{-1} = U_{\nu(n)\alpha}$, and the roots of $\mathbb{A}$ accordingly.

1.4. A deeper use of geometry

In this section, we exploit different geometric notions to study in further detail (TRD)-groups.

A. Convexity and negative curvature. To each of these notions corresponds a specific geometric realization, the *conical* and the *metric* realization, respectively. These geometries can be defined for a single building, they basically differ by the way an apartment is represented inside. The representations of the whole building are obtained via standard glueing techniques [D2, §10].

The conical realization of a building. — An apartment of fixed sign is represented here by the Tits cone of the Weyl group (1.1.B). For twin buildings, a pair of opposite apartments is represented by the union of the Tits cone and its opposite in the ambient real vector space. We can then use the obvious geometric opposition because it is the restriction of the abstract opposition relation of facets. The point is that the shape of simplicial cone for each facet fits in convexity arguments.

The metric realization of a building. — This realization was defined by M. Davis and G. Moussong. Its main interest is that it is non-positively curved. Actually, it satisfies the CAT(0) property: geodesic triangles are at least as thin as in the Euclidean plane. This enables one to apply the following [BrT1, 3.2].

Theorem (Bruhat-Tits fixed point theorem)). *Every group of isometries of a* CAT(0)*-space with a bounded orbit has a fixed point.*

This result generalizes a theorem applied by É. Cartan to Riemannian symmetric spaces to prove conjugacy of maximal compact subgroups in Lie groups. The metric realization only represents facets of spherical type. Indeed, consider the partially ordered set of spherical types $J \subset S$. A chamber is represented by the cone over the barycentric subdivision of this poset. G. Moussong [Mou] proved that this leads to a piecewise Euclidean cell complex which is locally non-positively curved. A simple connectivity criterion by M. Davis [D1] proves the global CAT(0) property for Coxeter groups, and the use of retractions proves it at the level of buildings [D2].

B. Balanced subsets. In a single building, to fix a facet is the condition that defines the so-called family of parabolic subgroups. In the twin situation, we can define another family of subsets taking into account both signs. By taking fixators, it will also give rise to an interesting family of subgroups.

Definition. *Call* balanced *a subset of a twin building contained in a pair of opposite apartments, intersecting the building of each sign and covered by a finite number of spherical facets.*

Suppose we are given a balanced subset Ω and a pair of opposite apartments $\mathbb{A}$ containing it. According to the Bruhat decompositions, for each sign ϵ the apartment $\mathbb{A}_\epsilon = \{wB_\epsilon\}_{w\in W}$ is isomorphic to the Coxeter complex of the Weyl group W. We can then define interesting sets of roots associated with the inclusion $\Omega \subset \mathbb{A}$, namely:

$$\Phi^u(\Omega) = \{\alpha \in \Phi \mid \overline{\alpha} \supset \Omega \quad \partial\alpha \not\supset \Omega\}, \quad \Phi^m(\Omega) = \{\alpha \in \Phi \mid \partial\alpha \supset \Omega\},$$
$$\text{and} \quad \Phi(\Omega) = \{\alpha \in \Phi \mid \overline{\alpha} \supset \Omega\} = \Phi^u(\Omega) \sqcup \Phi^m(\Omega).$$

So as to work in an apartment of fixed sign, we shall use the terminology of separation and strong separation. Using the Tits cone realization, we denote by Ω_+ (resp. Ω_-) the subset $\Omega \cap \mathbb{A}_+$ (resp. the subset of opposites of points in $\Omega \cap \mathbb{A}_-$). This enables to work only in $\mathbb{A}_+$. The root α is said to *separate* Ω_+ from Ω_- if $\overline{\alpha} \supset \Omega_+$ while $-\overline{\alpha} \supset \Omega_-$; α *strongly separates* Ω_+ from Ω_- if it separates Ω_+ from Ω_- and if not both of Ω_+ and Ω_- are contained in the wall $\partial\alpha$. Then $\Phi^u(\Omega)$ is the set of roots strongly separating Ω_+ from Ω_-, $\Phi(\Omega)$ that of roots separating Ω_+ from Ω_-. We will often omit the reference to Ω when it is obvious. Here is an important lemma which precisely makes use of a convexity argument [Ré2, 5.4.5].

Lemma. *The sets of roots Φ^u, Φ^m, and Φ are finite.*

Idea of proof. The set Φ^m is obviously stable under opposition, it is finite since there is only a finite number of walls passing through a given spherical facet. Choose a point in each facet meeting Ω_+ (resp. Ω_-) and denote the barycenter so obtained by x_+ (resp. x_-). The point $x_\pm$ is contained in a spherical facet $F_\pm$. Connect the facets by a minimal gallery Γ, and denote by $d_\pm$ the chamber whose closure contains $F_\pm$ and at maximal distance from the chamber of Γ whose closure contains $F_\pm$. By convexity, a root in Φ^u has to contain d_+ but not d_-, so $\#\Phi^u$ is bounded by the length of a gallery connecting these chambers.

Figure. —

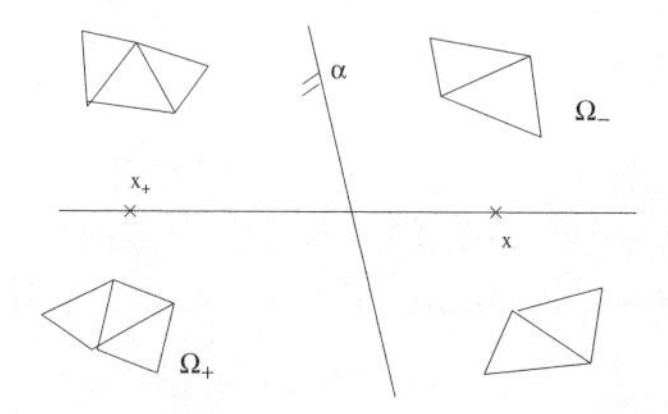

Examples. According to 1.3.A, a pair of spherical points of opposite signs is a simple but fundamental example of balanced subset. Let us describe the sets Φ^m and Φ^u in such a situation. Set $\Omega_\pm := \{x_\pm\} \subset \mathbb{A}_+$ so that $\Omega = \{-x_-; x_+\}$, and draw the segment $[x_-; x_+]$ in the Tits cone. The set Φ^m is empty if $[x_-; x_+]$ intersects a chamber. The roots of Φ^u are the ones that strongly separate x_+ from x_-. Assume first that each point lies in (the interior of) a chamber, which automatically implies that $\Phi^m(\Omega)$ is empty. First, by transitivity of G on the set of pairs of opposite apartments (1.3.A), we may assume that $\mathbb{A}$ is the standard one. Then thanks to the W-action, we can suppose that the positive point x_+ lies in the standard positive chamber c_+; there is a w in W such that x_- lies in wc_+. Choose a minimal gallery between c_+ and wc_+. Then Φ^u is the set $\Phi_{w^{-1}}$ of the $\ell(w)$ positive roots whose wall is crossed by it. The first picture below is an example of type $\widetilde{A}_2$ with $\Phi^m = \varnothing$ and $\#\Phi^u = 5$. Now we keep this $\widetilde{A}_2$ example, fix a wall $\partial\alpha$ and consider pairs of points $\{x_\pm\}$ with $[x_-; x_+] \subset \partial\alpha$. This implies that Φ^m will be always equal to $\{\pm\alpha\}$. Still, as in the previous case, the size of Φ^u will increase when so will the distance between x_- and x_+. So the second picture below is an example of type $\widetilde{A}_2$ with $\#\Phi^m = 2$ and $\#\Phi^u = 8$.

Figure. —

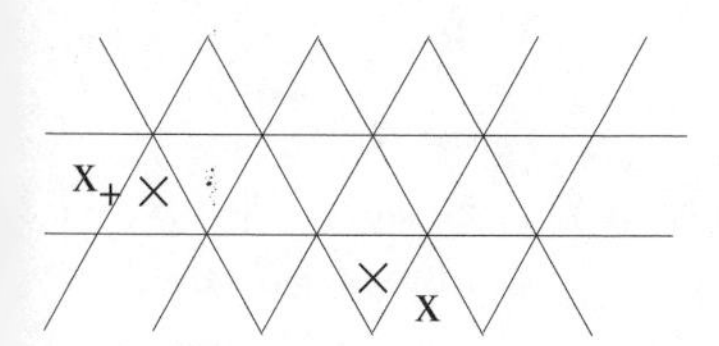

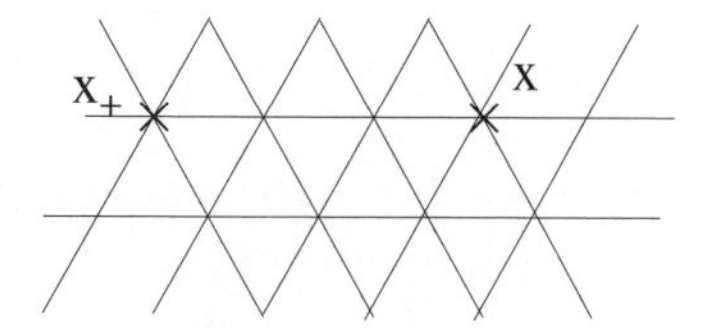

1.5. Levi decompositions

This section is dedicated to the statement of two kinds of decomposition results, each being an abstract generalization of the classical Levi decompositions.

A. Levi decompositions for parabolic subgroups. The first class of subgroups distinguished in a group with a BN-pair is that of parabolic subgroups. The existence of root groups in the (TRD) case suggests to look for more precise decompositions [Ré2, 6.2].

Theorem (Levi decomposition for parabolic subgroups). *Let F be a facet. Suppose F is spherical or G satisfies (TRD)$_{lin}$. Then for every choice of a pair $\mathbb{A}$ of opposite apartments containing F, the corresponding parabolic subgroup* $\mathrm{Fix}_G(F)$ *admits a semi-direct product decomposition*

$$\mathrm{Fix}_G(F) = M(F, \mathbb{A}) \ltimes U(F, \mathbb{A}).$$

The group $M(F, \mathbb{A})$ is the fixator of $F \cup -F$, where $-F$ is the opposite of F in $\mathbb{A}$; it is generated by H and the root groups U_α with $\partial\alpha \supset F$. It satisfies the (TRD) *axioms, and the (TRD)$_{lin}$ refinement if so does G. The group $U(F, \mathbb{A})$ only depends on F, it is the normal closure in* $\mathrm{Fix}_G(F)$ *of the root groups U_α with $\alpha \supset F$.*

Remark. (1) *It is understood in the statement that all the roots considered above are subsets of $\mathbb{A}$. By transitivity of G on pairs of opposite apartments, the root groups are conjugates of that in the (TRD) axioms.*

(2) *In the case of an infinite Weyl group, Levi decompositions show that the action of a (TRD)-group on a single building cannot be discrete since the fixator of a point contains infinitely many root groups.*

B. Levi decompositions for small subgroups. As mentioned in 1.4.B, the twin situation leads to the definition of another class of interesting subgroups.

Definition. *A subgroup of G is called small if it fixes a balanced subset.*

To describe the combinatorial structure of small subgroups, we need to require the additional $(NILP)$ axiom. It is not really useful to state it here explicitly, because our basic object of study is the class of Kac–Moody groups which satisfy it. We have [Ré2, 6.4]:

Theorem (Levi decomposition for small subgroups). *Suppose G is a (TRD)$_{lin}$-group satisfying* (NILP), *and let Ω be a balanced subset. Then for every choice of pair of opposite apartments $\mathbb{A}$ containing Ω, the small subgroup* $\mathrm{Fix}_G(\Omega)$ *admits a semi-direct product decomposition*

$$Fix_G(\Omega) = M(\Omega, \mathbb{A}) \ltimes U(\Omega, \mathbb{A}),$$

where both factors are closely related to the geometry of Ω (and of $\mathbb{A}$, w.r.t. which all roots are defined). Namely, $M(\Omega, \mathbb{A})$ satisfies the (TRD)$_{lin}$ axioms for the root groups U_α with $\partial\alpha \supset \Omega$, that is $\alpha \in \Phi^m(\Omega)$; and $U(\Omega, \mathbb{A})$ is in bijection with the set-theoretic product of the root groups U_α with $\alpha \in \Phi^u(\Omega)$ for any given order.

Remark. *In the Kac–Moody case, it can be shown that $U(\Omega, \mathbb{A})$ only depends on Ω and not on $\mathbb{A}$, so that we will use the notation $U(\Omega)$ instead.*

Example. (1) *As a very special case, consider in the standard pair of opposite apartments the balanced subset $\Omega := \{c_+; wc_-\}$, where $c_\pm$ are the standard chambers. Then, thanks to the first case of example* 1.4.B, *we see that* $\mathrm{Fix}_G(\Omega) = T \ltimes U_w$, *with U_w as defined in* 1.2.B. *Set-theoretically, the latter group is in particular a product of $\ell(w)$ root groups.*

(2) *Another special case is provided by the datum of two opposite spherical facets $\pm F$ in $\mathbb{A}$. Then $U(F \cup -F) = \{1\}$ and the fixator of the pair is at the same time the M-factor of the Levi decomposition above and in the previous sense for the parabolic subgroups* $\mathrm{Fix}_G(F)$ *and* $\mathrm{Fix}_G(-F)$.

C. *Further examples.* The examples below appear in the two simplest cases of special linear groups over Laurent polynomials (1.2.C). They will illustrate both kinds of decomposition. Examples of hyperbolic twin buildings will be treated in Section 4.1.

The case $\mathrm{SL}_2(\mathbb{F}_q[t, t^{-1}])$. The Bruhat-Tits buildings involved are both isomorphic to the homogeneous tree T_{q+1} of valency $q + 1$. A pair of opposite apartments is represented by two parallel real lines divided by the integers. Let us consider first the fixator of a single point. If this point is in an open edge, then its fixator is a Borel subgroup isomorphic to the group $\mathrm{SL}_2\begin{pmatrix}\mathbb{F}_q[t] & \mathbb{F}_q[t]\\ t\mathbb{F}_q[t] & \mathbb{F}_q[t]\end{pmatrix}$. If the point is a vertex, then it is isomorphic to the Nagao lattice $\mathrm{SL}_2(\mathbb{F}_q[t])$. Let us consider now pairs of points $\{ x_+; -x_-\}$ of opposite signs. We use the notation of the examples of 1.4.B. If the points are opposite vertices, then Φ^u is empty and Φ^m consists of two opposite roots: the corresponding small subgroup is equal to the M-factor, a finite group of Lie type and rank one over $\mathbb{F}_q$. If the points are non opposite middles of edges e_+ and $-e_-$, and if the edges e_+ and e_- in the positive straight line are at W-distance w, then the corresponding small fixator is the semi-direct product of T (isomorphic to $\mathbb{F}_q^\times$) by a commutative group isomorphic to the additive group of the $\mathbb{F}_q$-vector space $\mathbb{F}_q^{\ell(w)}$.

Figure. —

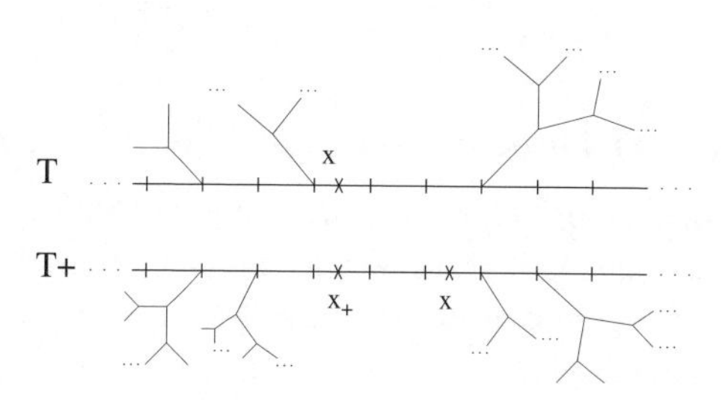

The case $SL_3(\mathbb{F}_q[t, t^{-1}])$. The case of single points is quite similar. For instance, let us fix a vertex x. The set L of chambers whose closure contains x can naturally be seen as the building of $SL_3(\mathbb{F}_q)$. The fixator P of x is isomorphic to $SL_3(\mathbb{F}_q[t])$. The M-factor in the Levi decomposition of P is a finite group of Lie type A_2 over $\mathbb{F}_q$ and its action on the small building L is the natural one. The infinite U-factor fixes L pointwise. Suppose now we are in the second case of example 1.4.B: the pair of points $\{ x_+; -x_- \}$ lies in a wall $\partial\alpha$. We already now that Φ^m contains the two opposite roots $\pm\alpha$, so that the M-factor of $\text{Fix}(\{ x_+; -x_- \})$ is a rank one finite group of Lie type over $\mathbb{F}_q$. If we assume α to be the first simple root of $SL_3(\mathbb{F}_q)$, then M is the group of matrices $\begin{pmatrix} X & 0 \\ 0 & \det X^{-1} \end{pmatrix}$, with $X \in GL_2(\mathbb{F}_q)$. The U-factor is a (commutative) p-group, with $p = \text{char}(\mathbb{F}_q)$. It is isomorphic to the group of matrices $\begin{pmatrix} 1 & 0 & P \\ 0 & 1 & Q \\ 0 & 0 & 1 \end{pmatrix}$, where P and Q are polynomials of $\mathbb{F}_q[t]$ of degree bounded the number of vertices between x_- and x_+.

2. Split Kac–Moody groups

In 1987, J. Tits defined group functors whose values over fields we will call *(split) Kac–Moody groups*. In Subsection 2.1 below, we give the definition of these groups, which involves some basic facts from Kac–Moody algebras to be recalled. Subsection 2.2 relates Kac–Moody groups to the abstract combinatorics previously studied, providing afterwards some more information. The three last subsections then deal with more specific properties. In Subsection 2.3 we present the adjoint representation and Subsection 2.4 combines it with Levi decompositions to endow each small subgroup with the structure of an algebraic group. Finally, Subsection 2.5 presents an argument which will occur several times afterwards. It is used for example to prove a conjugacy theorem for Cartan subgroups.

2.1. Tits functors and Kac–Moody groups

We sum up here the step-by-step definition of split Kac–Moody groups by J. Tits. It is also the opportunity to indicate how the notions defined in Subsection 1.1 for arbitrary Coxeter groups arise in the Kac–Moody context. All the material here is contained in [T4, §2].

A. Definition data. Functorial requirements. The data needed to define a Tits group functor are quite similar to that needed to define a Chevalley-Demazure group scheme.

Definition. (i) *A generalized Cartan matrix is an integral matrix* $A = [A_{s,t}]_{s,t \in S}$ *satisfying:* $A_{s,s} = 2$, $A_{s,t} \leq 0$ *when* $s \neq t$ *and* $A_{s,t} = 0 \Leftrightarrow A_{t,s} = 0$.

(ii) *A* Kac–Moody root datum *is a 5-tuple* $\big(S, A, \Lambda, (c_s)_{s \in S}, (h_s)_{s \in S}\big)$, *where* A *is a generalized Cartan matrix indexed by the finite set* S, Λ *is a free* $\mathbb{Z}$-*module (whose* $\mathbb{Z}$-*dual is denoted by* $\Lambda^{\vee}$*). Besides, the elements* c_s *of* Λ *and* h_s *of* $\Lambda^{\vee}$ *are required to satisfy* $\langle c_s \mid h_t \rangle = A_{ts}$ *for all* s *and* t *in* S.

Thanks to these elementary objects, it is possible to settle abstract functorial requirements generalizing the properties of functors of points of Chevalley-Demazure group schemes [T4, p.545]. The point is that *over fields* a problem so defined does admit a unique solution, concretely provided by a group presentation [T4, Theorem 1]. This is the viewpoint we will adopt in the next subsection and for the rest of our study. Let us first define further objects only depending on a Cartan generalized matrix.

Definition. (i) *The* Weyl group *of a generalized Cartan matrix* A *is the Coxeter group*

$$W = \langle s \in S \mid (st)^{M_{st}} = 1 \ \textit{whenever} \ M_{st} < \infty \rangle$$

with $M_{st} = 2, 3, 4, 6$ *or* ∞ *according to whether* $A_{st}A_{ts} = 0, 1, 2, 3$ *or* > 3.

(ii) *The* root lattice *of* A *is the free* $\mathbb{Z}$-*module* $Q := \bigoplus_{s \in S} \mathbb{Z}a_s$ *over the symbols* a_s, $s \in S$. *We also set* $Q_+ := \bigoplus_{s \in S} \mathbb{N}a_s$.

The group W operates on Q via $s \cdot a_t = a_t - A_{st}a_s$. This allows to define the *root system* Δ^{re} of A by $\Delta^{re} := W \cdot \{a_s\}_{s \in S}$. This set has all the abstract properties of the set of real roots of a Kac–Moody algebra. In particular, one has $\Delta^{re} = \Delta^{re}_+ \sqcup \Delta^{re}_-$, where $\Delta^{re}_\epsilon := \epsilon Q_+ \cap \Delta^{re}$. Let us explain now how to recover the notions of Subsection 1.1. Recall that given a Coxeter matrix M, we defined its cosine matrix so as to introduce the Tits cone on which the group acts. The starting point here is the generalized Cartan matrix which plays the role of the cosine matrix. This allows us to define the Tits cone as in 1.1.B, as well as all the objects related to it. Instead of V, the real vector space we use here is $Q_{\mathbb{R}} := Q \otimes_{\mathbb{Z}} \mathbb{R}$. This way, we get a bijection between Δ^{re} and the root system of the Weyl group as defined in 1.1.A, for which the notions of prenilpotence and intervals coincide. General references for abstract infinite root systems are [H2] and [Ba]. Note that in the Kac–Moody setting, linear combinations of roots give roots only involving integral coefficients. This leads to the following.

Convention. *In Kac–Moody theory, the root system* Δ^{re} *will play the role of the root system of the Weyl group of* A, *and the roots will be denoted by latin letters.*

B. Lie algebras, universal enveloping algebras, Z-forms. As already said, a more constructive approach consists in defining groups by generators and relations. The idea is to mimic a theorem by Steinberg giving a presentation of a split simply connected semi-simple algebraic group over a field. So far, so good. The point is that the relations involving root groups are not easily available. One has to reproduce some kind of Chevalley's construction in the Kac–Moody context. This implies introduction of quite a long series of "tangent" objects such as Lie algebras, universal enveloping algebras and, finally, $\mathbb{Z}$-forms of them.

Kac–Moody algebras. From now on, we suppose we are given a Kac–Moody root datum $\mathcal{D} = \big(S, A, \Lambda, (c_s)_{s\in S}, (h_s)_{s\in S}\big)$.

Definition. *The* Kac–Moody algebra *associated with* $\mathcal{D}$ *is the* $\mathbb{C}$*-Lie algebra* $\mathfrak{g}_{\mathcal{D}}$ *generated by* $\mathfrak{g}_0 := \Lambda^{\vee} \otimes_{\mathbb{Z}} \mathbb{C}$ *and the sets* $\{e_s\}_{s\in S}$ *and* $\{f_s\}_{s\in S}$, *subject to the following relations:*

$$[h, e_s] = \langle c_s, h\rangle e_s \ \textit{and}\ [h, f_s] = -\langle c_s, h\rangle f_s \ \textit{for}\ h\in\mathfrak{g}_0, \quad [\mathfrak{g}_0, \mathfrak{g}_0] = 0,$$

$$[e_s, f_s] = -h_s \otimes 1, \quad [e_s, f_t] = 0 \ \textit{for}\ s \neq t \ \textit{in}\ S,$$

$$(ad\, e_s)^{-A_{st}+1} e_t = (ad\, f_s)^{-A_{st}+1} f_t = 0 \qquad \textit{(Serre relations)}.$$

This Lie algebra admits an abstract Q-gradation by declaring the element e_s (resp. f_s) to be of degree a_s (resp. $-a_s$) and the elements of $\mathfrak{g}_0$ to be of degree 0. Each degree for which the corresponding subspace is non-trivial is called a *root* of $\mathfrak{g}_{\mathcal{D}}$.

For each $s\in S$, the derivations $\mathrm{ad} e_s$ and $\mathrm{ad} f_s$ are locally nilpotent, which enables us to define the automorphisms $\exp(\mathrm{ad} e_s)$ and $\exp(\mathrm{ad} f_s)$ of $\mathfrak{g}_{\mathcal{D}}$. The automorphisms $\exp(\mathrm{ad} e_s).\exp(\mathrm{ad} f_s).\exp(\mathrm{ad} e_s)$ and $\exp(\mathrm{ad} f_s).\exp(\mathrm{ad} e_s).\exp(\mathrm{ad} f_s)$ are equal and will be denoted by s^*. We are interested in s^* because if we define W^* to be the group of automorphisms of $\mathfrak{g}_{\mathcal{D}}$ generated by the s^*'s, then $s^* \mapsto s$ lifts to a homomorphism $\nu : W^* \twoheadrightarrow W$. Considering the W^*-images of the spaces $\mathbb{C}e_s$ and $\mathbb{C}f_s$ makes Δ^{re} appear as a subset of roots of $\mathfrak{g}_{\mathcal{D}}$. A basic fact about real roots a in Kac–Moody theory is that the corresponding homogeneous subspaces $\mathfrak{g}_a \subset \mathfrak{g}_D$ are one-dimensional, so that the W-action on Δ^{re} is lifted by the W^*-action on the homogeneous spaces $\mathfrak{g}_a, a \in \Delta^{re}$. Besides, for $s \in S$ and $w^* \in W^*$, the pair of opposite elements $w^*(\{\pm e_s\})$ only depends on the root $a := \nu(w^*)a_s$; we denote it by $\{\pm e_a\}$.

Divided powers and $\mathbb{Z}$-forms. We want now to construct a $\mathbb{Z}$-form of the universal enveloping algebra $\mathcal{U}\mathfrak{g}_D$ in terms of divided powers. Subrings of this ring will be used for algebraic differential calculus on some subgroups of Kac–Moody groups. For each $u \in \mathcal{U}\mathfrak{g}_D$ and each $n \in \mathbb{N}$, $u^{[n]}$ will denote the divided power $(n!)^{-1}u^n$ and $\binom{u}{n}$ will denote $(n!)^{-1}u(u-1)...(u-n+1)$. For each $s \in S$, $\mathcal{U}_{\{s\}}$ (resp. $\mathcal{U}_{\{-s\}}$) is the subring $\sum_{n\in\mathbb{N}} \mathbb{Z}e_s^{[n]}$ (resp. $\sum_{n\in\mathbb{N}} \mathbb{Z}f_s^{[n]}$) of $\mathcal{U}\mathfrak{g}_D$. We denote by $\mathcal{U}_O$ the subring of $\mathcal{U}\mathfrak{g}_D$ generated by the (degree 0) elements of the form $\binom{h}{n}$, with $h \in \Lambda^\vee$ and $n \in \mathbb{N}$.

Definition. *We denote by $\mathcal{U}_\mathcal{D}$ the subring of $\mathcal{U}\mathfrak{g}_D$ generated by $\mathcal{U}_O$ and the $\mathcal{U}_{\{s\}}$ and $\mathcal{U}_{\{-s\}}$ for s in S. It contains the ideal $\mathcal{U}_\mathcal{D}{}^+ := \mathfrak{g}_D.\mathcal{U}\mathfrak{g}_D \cap \mathcal{U}_\mathcal{D}$.*

The point about $\mathcal{U}_\mathcal{D}$ is the following.

Proposition. *The ring $\mathcal{U}_\mathcal{D}$ is a $\mathbb{Z}$-*form *of $\mathcal{U}\mathfrak{g}_D$ i.e., the natural map $\mathcal{U}_\mathcal{D} \otimes_\mathbb{Z} \mathbb{C} \to \mathcal{U}\mathfrak{g}_D$ is a bijection.*

Remark. *In the finite-dimensional case – when $\mathfrak{g}_D$ is a semi-simple Lie algebra, $\mathcal{U}_\mathcal{D}$ is the ring used to define Chevalley groups by means of formal exponentials* [Hu1, VII].

C. Generators and Steinberg relations. Split Kac–Moody groups. We can now start to work with groups.

First step: root groups and integrality result. Let us consider a root c in Δ^{re}. We denote by $\mathfrak{U}_c$ the $\mathbb{Z}$-scheme isomorphic to the additive one-dimensional group scheme $\mathbf{G}_a$ and whose Lie algebra is the $\mathbb{Z}$-module generated by $\{\pm e_c\}$. For each root c, the choice of a sign defines an isomorphism $\mathbf{G}_a \xrightarrow{\sim} \mathfrak{U}_c$. These choices are analogues of the ones defining an " épinglage" in the finite-dimensional case. We denote for short $\mathfrak{U}_s$ (resp. $\mathfrak{U}_{-s}$) the $\mathbb{Z}$-group scheme $\mathfrak{U}_{a_s}$ (resp. $\mathfrak{U}_{-a_s}$), and x_s (resp. x_{-s}) the isomorphism $\mathbf{G}_a \xrightarrow{\sim} \mathfrak{U}_s$ (resp. $\mathbf{G}_a \xrightarrow{\sim} \mathfrak{U}_{-s}$) induced by the choice of e_s (resp. f_s) in $\{\pm e_s\}$ (resp. $\{\pm f_s\}$). Let us consider now $\{a; b\}$ a prenilpotent pair of roots. The direct sum of the homogeneous subspaces $\mathfrak{g}_c$ of degree $c \in [a; b]_{\text{lin}}$ is a nilpotent Lie subalgebra $\mathfrak{g}_{[a;b]_{\text{lin}}}$ of $\mathfrak{g}_D$, which defines a unique unipotent complex algebraic group $U_{[a;b]_{\text{lin}}}$, by means of formal exponentials.

Proposition. *There exists a unique $\mathbb{Z}$-group scheme $\mathfrak{U}_{[a;b]_{lin}}$ containing the $\mathbb{Z}$-schemes $\mathfrak{U}_c$ for $c \in [a;b]_{lin}$, whose value over $\mathbb{C}$ is the complex algebraic group $U_{[a;b]_{lin}}$ with Lie algebra $\mathfrak{g}_{[a;b]_{lin}}$. Moreover, the product map*

$$\prod_{c\in[a;b]_{lin}} \mathfrak{U}_c \to \mathfrak{U}_{[a;b]_{lin}}$$

is an isomorphism of $\mathbb{Z}$-schemes for every order on $[a;b]_{lin}$.

Remark. *This result is an abstract generalization of an integrality result of Chevalley* [Sp, Proposition 9.2.5] *concerning the commutation constants between root groups of split semi-simple algebraic groups. The integrality here is expressed by the existence of a $\mathbb{Z}$-structure extending the $\mathbb{C}$-structure of the algebraic group. We will make these $\mathbb{Z}$-group schemes a bit more explicit when defining the adjoint representation (in* 2.3.A*).*

Second step: Steinberg functors. Now we see the $\mathbb{Z}$-schemes $\mathfrak{U}_{[a;b]_{\text{lin}}}$ just as group functors over rings. We will amalgamate them so as to obtain a big group functor containing all the relations between the root groups. This is the step of the procedure where algebraic structure is lost. The relations $c \in [a;b]_{\text{lin}}$ give rise to an inductive system of group functors with injective transition maps $\mathfrak{U}_c(R) \hookrightarrow \mathfrak{U}_{[a;b]_{\text{lin}}}(R)$ over every ring R.

Definition. *The* Steinberg functor *– denoted by St_A – is the limit of the inductive system above.*

This group functor only depends on the generalized Cartan matrix A, not on the whole datum $\mathcal{D}$. Besides, the W^*-action on $\mathfrak{g}_{\mathcal{D}}$ extends to a W^*-action on St_A : for each w in W, we will still denote by w^* the corresponding automorphism.

Third step: the tori and the other relations. What is left to do at this level is to introduce a torus, a split one by analogy with Chevalley groups. In fact, it is determined by the lattice Λ of the Kac–Moody root datum $\mathcal{D}$. As in the previous step, we are only interested in group functors. Denote by $\mathcal{T}_\Lambda$ the functor of points $\mathrm{Hom}_{\mathbb{Z}-\mathrm{alg}}(\mathbb{Z}[\Lambda], -) = \mathrm{Hom}_{\mathrm{groups}}(\Lambda, -^\times)$ defined over the "category" $\mathbb{Z}$-alg of rings. For a ring R, r an invertible element in it and h an element of $\Lambda^\vee$, the notation r^h corresponds to the element $\lambda \mapsto r^{\langle\lambda|h\rangle}$ of $\mathcal{T}_\Lambda(R)$. We will also denote $r^{\langle\lambda|h\rangle}$ by $\lambda(r^h)$. This way we can see elements of Λ as characters of $\mathcal{T}_\Lambda$. We finally set $\tilde{s}(r) := x_s(r)x_{-s}(r^{-1})x_s(r)$ for each invertible element r in a ring R, and $\tilde{s} := \tilde{s}(1)$.

Definition. *For each ring R, we define the group $\underline{G}_{\mathcal{D}}(R)$ as the quotient of the free product $St_A(R) * \mathcal{T}_\Lambda(R)$ by the following relations.*

$$tx_s(q)t^{-1} = x_s(c_s(t)q); \quad \tilde{s}(r)t\tilde{s}(r)^{-1} = s(t);$$
$$\tilde{s}(r^{-1}) = \tilde{s}r^{h_s}; \quad \tilde{s}u\tilde{s}^{-1} = s^*(u);$$

with $s \in S$, $c \in \Delta^{re}$, $t \in \mathcal{T}_\Lambda(R)$, $q \in R$, $r \in R^\times$, $u \in \mathfrak{U}_c(R)$. We define this way a group functor denoted by $\underline{G}_{\mathcal{D}}$ and called the Tits functor *associated to $\mathcal{D}$. The value of this functor on the field $\mathbb{K}$ will be called the* split Kac–Moody group *associated to $\mathcal{D}$ and defined over $\mathbb{K}$.*

Remark. *This is the place where we use the duality relations between Λ and $\Lambda^\vee$ required in the definition of $\mathcal{D}$. This duality occurs in the expression $c_s(t)$, and allows to define a W-action on Λ by sending s to the involution of $\Lambda : \lambda \mapsto \lambda - \langle \lambda \mid h_s \rangle c_s$. This yields a W-action on $\mathcal{T}_\Lambda$ thanks to which the expression $s(t)$ makes sense in the defining relations. The expression $s^*(u)$ comes from the W^*-action on the Steinberg functor.*

Example. (1) *As a first example, we can consider a Kac–Moody root datum where A is a Cartan matrix and $\Lambda^\vee$ is the $\mathbb{Z}$-module freely generated by the h_s's. Then what we get over fields is the functor of points of the simply connected Chevalley scheme corresponding to A. This can be derived from the theorem by Steinberg alluded to in* 2.1.A [Sp, Theorem 9.4.3]. *One cannot expect better since as already explained, algebraic structure is lost during the amalgam step defining St_A.*

(2) *More generally, the case where Λ (resp. $\Lambda^\vee$) is freely generated by the c_s's (resp. h_s's) will be referred to as the adjoint (resp. simply connected) case.*

(3) *The groups $\mathrm{SL}_n(\mathbb{K}[t,t^{-1}])$ of* 1.2.C *can be seen as split Kac–Moody groups over $\mathbb{K}$. The generalized Cartan matrix is the matrix $\widetilde{A}_{n-1}$ indexed by $\{0; 1; \ldots n-1\}$, characterized for $n \geq 3$ by $A_{ii} = 2$, $A_{ij} = -1$ for i and j consecutive modulo n and $A_{ij} = 0$ elsewhere. In rank 2, $\widetilde{A}_2$ is just $\begin{pmatrix} 2 & -2 \\ -2 & 2 \end{pmatrix}$. The $\mathbb{Z}$-module Λ (resp. $\Lambda^\vee$) is then the free $\mathbb{Z}$-module of rank $n-1$ generated by the simple roots c_1, ... c_{n-1} of the standard Cartan subgroup T of $\mathrm{SL}_n(\mathbb{K})$ (resp. the corresponding one-parameter multiplicative subgroups). To be complete, one has to add the character $c_0 := -\sum_{1\leq i\leq n-1} c_i$ (resp. the cocharacter $h_0 := -\sum_{1\leq i\leq n-1} h_i$). These examples show in particular that neither Λ nor $\Lambda^\vee$ need to contain the c_s's or the h_s's as free families.*

2.2. Combinatorics of a Kac–Moody group

This section states the connection with the abstract group combinatorics presented in §1, then stresses some specific properties (often due to splitness). We are given here a split Kac–Moody group G defined over $\mathbb{K}$ which acts on its twin building $(\mathcal{I}_+, \mathcal{I}_-, d^*)$.

A. Main statement. As the (TRD) axioms were adjusted to the Kac–Moody situation, it is not surprising to get the following result [Ré2, 8.4.1].

Proposition. *Let G be a Kac–Moody group, namely the value of a Tits group functor over a field $\mathbb{K}$. Then G is a $(TRD)_{lin}$-group for the root groups of its definition, which also satisfies also the axiom* (NILP) *for Levi decomposition of small subgroups* (1.5.B). *The standard Cartan subgroup T plays the role of the normalizer of the U_α's.*

Sketch of proof. In the defining relations of a Kac–Moody group (2.1.C), everything seems to be done so as to conform to the (TRD) axioms. Still, we have to be a bit more careful with the non degeneracy requirements. Indeed, for the first half of axiom (TRD0) and axiom (TRD3), we have to use the adjoint representation defined later. This point is harmless since the definition of this representation does not need group combinatorics.

Remark. *This result can be used to prove the unicity over fields of Kac–Moody groups. Actually, the original group combinatorics involved in the proof by J. Tits* [T4, §5] *was more general.*

B. Specific properties. As Tits functors generalize Chevalley-Demazure schemes, one can expect further properties for at least two reasons. The first one is the "algebraicity" of Kac–Moody groups i.e., the fact that the precise defining relations are analogues of that for algebraic groups. The second reason is splitness.

"Algebraicity" of Tits functors. First, we know that the biggest group normalizing all the root groups is the standard Cartan subgroup T: this is a purely combinatorial consequence of the (TRD) axioms due to the fact that in BN-pairs, Borel subgroups coincide with their normalizer. Besides we have a completely explicit description of the T-action on root groups – by characters. This leads to the following result [Ré2, 10.1.3].

Proposition. *If the ground field $\mathbb{K}$ has more than* 3 *elements, then the fixed points under T in the buildings are exactly the points in $\mathbb{A}$ (the standard pair of opposite apartments).*

Remark. *What this proposition says is that this set is not bigger than* $\mathbb{A}$. *This result has to be related to another one, where the same hypothesis* $|\mathbb{K}| > 3$ *is needed: in this case, N is exactly the normalizer of T in G* [Ré2, 8.4.1].

Another algebraic-like consequence is about Levi decompositions and residues in buildings. Consider the standard facet $\pm F_J$ of spherical type $J \subset S$ and sign $\pm$. Then the M-factor (see 1.5.B, example (2)) is the Kac–Moody group corresponding to the datum where A is restricted to $J \times J$, and only the c_s's and h_s's with $s \in J$ are kept. This is a root datum abstractly defining points of a reductive group. Moreover, the union of the sets of chambers whose closure contains $\pm F_J$ – the *residue* of $\pm F_J$ – is the twin building of the reductive group, on which it operates naturally. The generalization to arbitrary facets is straightforward.

Remark. *The analogy with reductive groups can fail on some points. Consider for instance the case of the centralizer of the standard Cartan subgroup T. It is classical in the algebraic setting that the centralizer of a Cartan subgroup is not bigger than the subgroup itself. Consider once again the groups* $\mathrm{SL}_n(\mathbb{K}[t, t^{-1}])$ *of* 1.2.C. *Then the centralizer of T – the standard Cartan subgroup of* $\mathrm{SL}_n(\mathbb{K})$ *– is infinite-dimensional since it is the group of determinant* 1 *diagonal matrices with monomial coefficients.*

Splitness. The main analogy with split reductive groups is that the torus is a quotient of a finite numbers of copies of $\mathbb{K}^\times$ and each root group is isomorphic to the additive group of the ground field $\mathbb{K}$. Combined with a geometric consequence of the Moufang property, the latter property of root groups says:

Lemma. *Every building arising from a split Kac–Moody group is* homogeneous, *that is thickness is constant* (*equal to* $|\mathbb{K}| + 1$) *over all panels.*

The buildings arising from a split Kac–Moody group will be referred to as *(split) Kac–Moody buildings.*

2.3. The adjoint representation

Besides the geometry of buildings, an important tool to study Kac–Moody groups is a linear representation which generalizes the adjoint representation of algebraic groups. Let us fix a Kac–Moody root datum $\mathcal{D}$ and the corresponding Tits functor $\underline{G}_{\mathcal{D}}$.

A. Formal sums and distribution algebras. Let us go back to the situation of 2.1.C; we work in particular with the prenilpotent pair of roots $\{a; b\}$. In the $\mathbb{Z}$-form $\mathcal{U}_{\mathcal{D}}$, we consider the subring $\mathcal{U}_{[a;b]_{\mathrm{lin}}} := \mathcal{U}_{\mathcal{D}} \cap \mathcal{U}\mathfrak{g}_{[a;b]_{\mathrm{lin}}}$. Let us define the R-algebra $\big(\widehat{\mathcal{U}_{[a;b]_{\text{lin}}}}\big)_R$ of formal sums $\sum_{c\in\mathbb{N}a+\mathbb{N}b} r_c u_c$, with u_c homogeneous of degree c in $\mathcal{U}_\mathcal{D}$ and $r_c \in R$.

Definition. *Denote by* $\mathfrak{U}_{[a;b]_{\text{lin}}}$ *the group functor which to each ring* R *associates the subgroup* $\langle \exp(re_c) \mid r \in R, c \in \Delta^{re} \cap (\mathbb{N}a + \mathbb{N}b)\rangle < \big(\widehat{\mathcal{U}_{[a;b]_{\text{lin}}}}\big)_R^\times$. *Here* $\exp(re_c)$ *is the formal exponential* $\sum_{n\geq 0} r^n e_c^{[n]}$ *and* $\big(\widehat{\mathcal{U}_{[a;b]_{\text{lin}}}}\big)_R^\times$ *is the multiplicative group of the* R*-algebra* $\big(\widehat{\mathcal{U}_{[a;b]_{\text{lin}}}}\big)_R$.

There is a little abuse of notation when denoting by $\mathfrak{U}_{[a;b]_{\text{lin}}}$ a new object, but we will justify it soon. Thanks to a generalized Steinberg commutator formula, one can prove first:

Proposition. *The group functor* $\mathfrak{U}_{[a;b]_{\text{lin}}}$ *is the functor of points of a smooth connected group scheme of finite type over* $\mathbb{Z}$, *with Lie algebra* $\mathfrak{g}_{[a;b]_{\text{lin}}} \cap \mathcal{U}_\mathcal{D}$. *Its value over* $\mathbb{C}$ *is the complex algebraic group* $U_{[a;b]_{\text{lin}}}$, *and we have* $\mathrm{Dist}_e U_{[a;b]_{\text{lin}}} \cong \mathcal{U}\mathfrak{g}_{[a;b]_{\text{lin}}}$.

The notation Dist_e is for the *algebra of distributions supported at unity* of an algebraic group. This proposition is the place where a bit of algebraic differential calculus [DG, II.4] comes into play. We will not detail the use of it, but just say that it allows to prove that the group functors above and of the proposition in 2.1.C coincide [Ré2, 9.3.3]. In other words, this result leads to a concrete description of the group functors occurring in the amalgam defining St_A.

B. The adjoint representation as a functorial morphism. For every ring R, the extension of scalars from $\mathbb{Z}$ to R of $\mathcal{U}_\mathcal{D}$ is denoted by $\big(\mathcal{U}_\mathcal{D}\big)_R$. Denote by $\mathrm{Aut}_{filt}(\mathcal{U}_\mathcal{D})(R)$ the group of automorphisms of the R-algebra $\big(\mathcal{U}_\mathcal{D}\big)_R$ preserving its filtration arising from that of $\mathcal{U}_\mathcal{D}$ and its ideal $\mathcal{U}_\mathcal{D}{}^+ \otimes_\mathbb{Z} R$. The naturality of scalar extension makes the assignement $R \mapsto \mathrm{Aut}_{filt}(\mathcal{U}_\mathcal{D})(R)$ a group functor. We have [Ré2, 9.5.3]:

Theorem. *There is a morphism of group functors* $Ad : \underline{G}_\mathcal{D} \to Aut_{filt}(\mathcal{U}_\mathcal{D})$ *characterized by:*

$$Ad_R\big(x_s(r)\big) = \sum_{n\geq 0} \frac{(ade_s)^n}{n!} \otimes r^n, \quad Ad_R\big(x_{-s}(r)\big) = \sum_{n\geq 0} \frac{(ad f_s)^n}{n!} \otimes r^n,$$

$$Ad_R\big(\mathcal{T}_\Lambda(R)\big) \text{ fixes } \big(\mathcal{U}_0\big)_R \quad \text{and} \quad Ad_R(\mathrm{h})(\mathrm{e_a} \otimes \mathrm{r}) = \mathrm{c_a}(\mathrm{h})(\mathrm{e_a} \otimes \mathrm{r}),$$

for s *in* S, *for each ring* R, *each* h *in* $\mathcal{T}_\Lambda(R)$ *and each* r *in* R.

Sketch of proof. No use of group combinatorics is needed for this result. Roughly speaking, to define the adjoint representation just consists in inserting " ad" in the arguments of the formal exponentials for the root groups, and in making the torus operate diagonally w.r.t. the duality given by $\mathcal{D}$. One has to follow all the steps of the definition of Tits functors, and to verify that all the defining relations involved are satisfied by the partial linear actions of the characterization. Proposition 2.3.B enables to justify that the first formula defines a representation of the Steinberg functor. Then, the torus must be taken into account, but the verification concerning the rest of the defining relations of $\underline{G}_{\mathcal{D}}$ is purely computational. The naturality of the representation is an easy point.

C. The adjoint representation over fields. Restriction to values of Tits functors over fields brings further properties, precisely because this allows the use of group combinatorics according to 2.2.A – see [Ré2, 9.6] for proofs.

Proposition. (i) *Over fields, the kernel of the adjoint representation is the center of the Kac–Moody group G. This center is the intersection of kernels of characters of T, namely the centralizers in T of the root groups.*
(ii) *If the ground field is algebraically closed, the image of Ad is again a Kac–Moody group.*

2.4. Algebraic subgroups

One of the main interests of the adjoint representation is that it provides algebro-geometric structure for the family of small subgroups of a Kac–Moody group.

A. Statement. The following result is, up to technicalities, the combination of the abstract Levi decompositions and restrictions of source and target of the adjoint representation [Ré2, 10.3].

Theorem (Algebraic structure). *For every balanced subset Ω, denote by $G(\Omega)$ the quotient group $\mathrm{Fix}_G(\Omega)/Z\big(\mathrm{Fix}_G(\Omega)\big)$ of the corresponding small subgroup by its center. Then $G(\Omega)$ admits a natural structure of algebraic group, arising from the adjoint representation.*

Remark. *An important question then is to know how far the quotient $G(\Omega)$ is from the small subgroup $\mathrm{Fix}_G(\Omega)$ itself. To be more precise, let us fix a pair $\mathbb{A}$ of opposite apartments containing the balanced subset Ω. Thanks to a suitable g in G, we may assume $\mathbb{A}$ is the standard pair of opposite apartments. At the group level, this corresponds to conjugating the small subgroup so as to make it contain the standard Cartan subgroup T. Then the center of $\mathrm{Fix}_G(\Omega)$ is the intersection of the kernels (in T) of the characters associated to the roots in $\Phi(\Omega)$. (These roots index the root groups generating the small subgroup together with T). This shows that the quotient is always as a group abstractly isomorphic to a diagonalisable algebraic group. This point will be important* *when we handle Cartan subgroups, because preimages of diagonalisable groups will stay diagonalisable.*

Example. *Let us consider some examples from* 1.5.C, *i.e., from some small subgroups of* $\mathrm{SL}_2(\mathbb{K}[t, t^{-1}])$ *and* $\mathrm{SL}_3(\mathbb{K}[t, t^{-1}])$. *There will be no quotient in the* SL_2 *case: the standard Cartan subgroup is one-dimensional. By taking a pair of opposite middles of edges, one gets a small subgroup isomorphic to the M-factor of the last example in* 1.5.C. *This group consists of the matrices* $\begin{pmatrix} X & 0 \\ 0 & \det X^{-1} \end{pmatrix}$, *with* $X \in \mathrm{GL}_2(\mathbb{F}_q)$. *Its center is one-dimensional and this is the corresponding quotient that actually admits an algebraic structure by the procedure above.*

B. *Sketch of proof.* The main tools for this proof are basic results from algebraic geometry or algebraic group theory, such as constructibility of images of morphisms, use of tangent maps to justify that a bijective algebraic morphism is an algebraic isomorphism... Let us make the same reductions as in the remark above. In particular, Ω is in the standard pair of opposite apartments. Recall also the Levi decomposition $\mathrm{Fix}_G(\Omega) = M(\Omega, \mathbb{A}) \ltimes U(\Omega, \mathbb{A})$, with $M(\Omega, \mathbb{A})$ satisfying the abstract (TRD) axioms. The first step is the existence of a finite-dimensional subspace $W(\Omega)_{\mathbb{K}} \subset \big(\mathcal{U}_{\mathcal{D}}\big)_{\mathbb{K}}$ such that the resctriction Ad_Ω of the adjoint representation is the center $Z\big(\mathrm{Fix}_G(\Omega)\big)(< T)$ [Ré2, 10.3.1]. Consequently, Ad_Ω is faithful on $U(\Omega)$ and on the subgroup of $M(\Omega, \mathbb{A})$ generated by a positive half of Φ^m. The latter group is the "unipotent radical" of an abstract Borel subgroup of $M(\Omega, \mathbb{A})$. Then, it is proved that the isomorphic images of these two groups are unipotent closed $\mathbb{K}$-subgroups of $\mathrm{GL}\big(W(\Omega)_{\mathbb{K}}\big)$ [Ré2, 10.3.2]. The next step is to show that the abstract image $M_{\Omega,\mathbb{A}} := \mathrm{Ad}_\Omega\big(M(\Omega, \mathbb{A})\big)$ is also a closed subgroup. This is done by proving that its Zariski closure in $\mathrm{GL}\big(W(\Omega)_{\mathbb{K}}\big)$ is reductive and comparing the classical decompositions so obtained. The proof of this fact works as follows. We know at this level that the image by Ad_Ω of the abstract Borel subgroup of $M(\Omega, \mathbb{A})$ is closed. By Bruhat decomposition and local closedness of orbits, this gives a bound on the dimension of the Zariski closure. The use of Cartier dual numbers allows to compute the Lie algebras involved and to show that the bound is reached. The triviality of the unipotent radical follows from singular tori arguments [Ré2, 10.3.3]. The final step is to show that the semi-direct product so obtained in $\mathrm{GL}\big(W(\Omega)_{\mathbb{K}}\big)$ is an algebraic one: this is proved by Lie algebras (tangent maps) arguments [Ré2, 10.3.4].

2.5. Cartan subgroups

We present now a general argument combining negative curvature and algebraic groups. Then we apply it to give a natural (more intrinsic) definition of *Cartan subgroups*, previously defined by conjugation. Recall that a group of isometries is *bounded* if it acts with bounded orbits [Br2, p.160].

A. A typical argument. We suppose we are in the following setting: a group H acts in a compatible way on a Kac–Moody group G and on its twin building. (For instance, H may be a subgroup of G but this is not the only example.)

Argument. (1) *Justify that H is a bounded group of isometries of both buildings and apply the Bruhat-Tits fixed point theorem to get a balanced H-fixed subset Ω.*

(2) *Apply results from algebraic groups to the small H-stable subgroup* $\mathrm{Fix}_G(\Omega)$.

Remark. *In Subsection* 3.2.B, *the argument is applied to the Galois actions arising from the definition of almost split forms.*

B. Cartan subgroups. Recall that the *standard Cartan subgroup* is the value T of the split torus $\mathcal{T}_\mathcal{D}$ on the ground field $\mathbb{K}$. The provisional definition of Cartan subgroups describes them as conjugates of T. We want to give a more intrinsic characterization of them, by means of their behaviour w.r.t. the adjoint representation Ad. We suppose here that the ground field $\mathbb{K}$ is infinite.

A criterion for small subgroups. This criterion precisely involves the adjoint representation. Let H be a subgroup of the Kac–Moody group G. Call H *Ad-locally finite* if its image by Ad is locally finite, that is the linear span of the H-orbit of each point in $(\mathcal{U}_\mathcal{D})_\mathbb{K}$ is finite-dimensional.

Theorem. *The following assertions are equivalent.*

(i) *H is Ad-locally finite.*
(ii) *H intersects a finite number of Bruhat double classes for each sign $\pm$.*
(iii) *H is contained in the intersection of two fixators of spherical facets of opposite signs.*

Remark. 1. *The importance of this theorem lies in the fact that it is a connection between the two G-actions we chose to study G : the adjoint representation and the action on the twin building.*

2. *This result is due to V. Kac and D. Peterson* [KP3, Theorem 1] *who proved a lemma relating the Q-gradation and the Bruhat decompositions. This lemma was used by them to prove conjugacy of Cartan subgroups over $\mathbb{C}$, with Kac–Moody groups defined as automorphism groups of Lie algebras. No argument of negative curvature appears in their work, they invoke instead convexity properties of a certain distance function on the Tits cone.*

Definition and conjugacy of $\mathbb{K}$-split Cartan subgroups. We say that a subgroup is *Ad-diagonalizable* if its image by the adjoint representation is a diagonalizable group of automorphisms of $(\mathcal{U}_{\mathcal{D}})_{\mathbb{K}}$. Until the end of the section, we will assume the ground field $\mathbb{K}$ to be infinite so as to characterize algebraic groups by their points.

Definition. *A subgroup of G is a* ($\mathbb{K}$-split) Cartan subgroup *if it is Ad-diagonalizable and maximal for this property.*

We will omit the prefix "$\mathbb{K}$-split" when the ground field is separably closed.

Theorem. *The* ($\mathbb{K}$*-split*) *Cartan subgroups are all conjugates by G of the standard Cartan subgroup T.*

Sketch of proof. [Ré2, 10.4.2] — Basically, this is the application of argument 2.5.A, thanks to the smallness criterion above. An Ad-diagonalizable subgroup has to be small and its image in the algebraic subgroup is a Cartan subgroup in the algebraic sense. We have then to use transitivity of G on pairs of opposite apartments and the algebraic conjugacy theorem of Cartan subgroups to send our subgroup in T thanks to a suitable element of G.

C. Connection with geometry. Combining this theorem with the proposition of 2.2.B and the fact that G is transitive on the pairs of opposite apartments, we get [Ré2, 10.4.3 and 10.4.4]:

Corollary. *Suppose the ground field $\mathbb{K}$ is infinite. Then, there is a natural G-equivariant dictionary between the pairs of opposite apartments and the $\mathbb{K}$-split Cartan subgroups of G. Moreover, given a facet F, this correspondence relates pairs of opposite apartments containing F and $\mathbb{K}$-split Cartan subgroups of G contained in the parabolic subgroup* $\mathrm{Fix}_G(F)$.

We saw in the abstract study of (TRD)-groups that parabolic subgroups do not form the only interesting class of subgroups. Let us consider a balanced subset Ω of the twin building of G. A natural question is to know what the correspondence says in this case. First we choose a pair of opposite apartments $\mathbb{A}$ containing Ω. We work in the conical realization, that is, we see $\mathbb{A}$ as the union $\overline{\mathcal{C}} \cup -\overline{\mathcal{C}}$ of a Tits cone and its opposite in the real vector space V^*. Then it makes sense to define the *convex hull* of Ω – denoted by $\mathrm{conv}(\Omega)$ – to be the trace on $\overline{\mathcal{C}} \cup -\overline{\mathcal{C}}$ of the convex hull determined in V^*. The same trick enables us to define the *vectorial extension* $\mathrm{vect}(\Omega, \mathbb{A})$. The difference of notation is justified by [Ré2, 10.4.5]:

Proposition. (i) $\mathbb{K}$*-split Cartan subgroups of G contained in* $\mathrm{Fix}_G(\Omega)$ *are the preimages of the maximal $\mathbb{K}$-split tori of the algebraic group $G(\Omega)$. Besides,* *they are in one-to-one G-equivariant correspondence with the pairs of opposite apartments containing* Ω.

(ii) *The subspace* $\mathrm{conv}(\Omega)$ *and the subgroup* $U(\Omega)$ *only depend on* Ω.

Remark. *The independence of* $\mathbb{A}$ *for the convex hull is actually useful for the Galois descent.*

3. Relative Kac–Moody theory

We describe now an analogue of Borel-Tits theory. Section 3.1 defines the *almost split forms* of Kac–Moody groups. This gives the class of groups which are concerned by the theory. In Section 3.2, we state the structure theorem for rational points and try to give an idea of its quite long proof. The geometric method for the Galois descent follows faithfully the lines of G. Rousseau's work, who used it for the characteristic 0 case [Rou1,2]. Section 3.3 introduces the class of *quasi-split* groups and states that this is the only class of almost split Kac–Moody groups over finite fields. In Section 3.4, we focus on relative links and apartments, so as to compute thicknesses for instance, and, finally, in Secton 3.5 we use the classical example of the unitary group SU_3 to show how things work concretely.

3.1. Definition of forms

The requirements are basically of two kinds. We have to generalize conditions from algebraic geometry, and then to ask for isotropy conditions. We suppose we are given a ground field $\mathbb{K}$, we choose a separable closure $\mathbb{K}_s$ (resp. an algebraic closure $\overline{\mathbb{K}}$) of it. We work with the group $G = \underline{G}_{\mathcal{D}}(\overline{\mathbb{K}})$ determined by $\overline{\mathbb{K}}$ and the Kac–Moody root datum $\mathcal{D}$. To $\mathcal{D}$ is also associated the $\mathbb{Z}$-form $\mathcal{U}_{\mathcal{D}}$; if R is any ring, then the scalar extension of $\mathcal{U}_{\mathcal{D}}$ from $\mathbb{Z}$ to R will be denoted by $\left(\mathcal{U}_{\mathcal{D}}\right)_R$.

A. Functorial forms. Algebraic forms. Here is the part of the conditions aiming at generalizing algebraic ones. Unfortunately, $\mathbb{K}$-forms of Kac–Moody groups are not defined by a list of conditions stated once and for all at the beginning of the study. One has to require a condition, then to derive some properties giving rise to the notions involved in the next requirement, and so on. We sum up here the basic steps of this procedure.

Use of functoriality. A Kac–Moody group is a value of a group functor over a field. Consequently, it is natural to make occur a group functor as main piece of a $\mathbb{K}$-form. A *functorial* $\mathbb{K}$*-form* of G is a group functor defined over field extensions of $\mathbb{K}$ which coincides with the Tits functor $\underline{G}_{\mathcal{D}}$ over extensions of $\overline{\mathbb{K}}$. This is really a weak requirement since in the finite-dimensional case, it just takes into account the functor of points of a scheme. Still, this condition enables us to make the Galois group $\Gamma := \mathrm{Gal}(\mathbb{K}_s/\mathbb{K})$ operate on G [Ré2, 11.1.2]. We explained in 2.2.A that the adjoint representation is a substitute for a global algebro-geometric structure on a Kac–Moody group. It is quite natural then to work with this representation to define forms [Ré2, 11.1.3].

Definition. *A* prealgebraic $\mathbb{K}$-form *of G consists in of a functorial $\mathbb{K}$-form $\underline{G}$ and of a filtered $\mathbb{K}$-form $\mathcal{U}_{\mathbb{K}}$ of the $\overline{\mathbb{K}}$-algebra $(\mathcal{U}_{\mathcal{D}})_{\overline{\mathbb{K}}} := \mathcal{U}_{\mathcal{D}} \otimes_{\mathbb{Z}} \overline{\mathbb{K}}$, both satisfying the following conditions.*

(PREALG1) *The adjoint representation Ad is Galois-equivariant.*

(PREALG2) *The value of $\underline{G}$ on a field extension gives rise to a group embedding.*

Remark. *It is understood in the notion of filtered that $\mathcal{U}_{\mathbb{K}}$ is a direct sum $\mathbb{K} \oplus \mathcal{U}_{\mathbb{K}}{}^{+}$, where $\mathcal{U}_{\mathbb{K}}{}^{+}$ is a $\mathbb{K}$-form of the ideal $(\mathcal{U}_{\mathcal{D}}{}^{+})_{\overline{\mathbb{K}}}$. The existence of a filtered $\mathbb{K}$-form of the $\overline{\mathbb{K}}$-algebra $(\mathcal{U}_{\mathcal{D}})_{\overline{\mathbb{K}}}$ enables us this time to define a Γ-action on $(\mathcal{U}_{\mathcal{D}})_{\overline{\mathbb{K}}}$, and finally on the automorphism group $\mathrm{Aut}_{filt}(\mathcal{U}_{\mathcal{D}})(\overline{\mathbb{K}})$.*

Example. *As a fundamental example, we can consider the datum of the restriction of the Tits functor to the $\mathbb{K}$-extensions (contained in $\overline{\mathbb{K}}$) and the $\mathbb{K}$-algebra $(\mathcal{U}_{\mathcal{D}})_{\mathbb{K}}$. This will be referred to as the split form of G.*

Algebraicity conditions. We suppose now we are given a *splitting extension* $\mathbb{E}/\mathbb{K}$ of the prealgebraic form $(\underline{G}, \mathcal{U}_{\mathbb{K}})$, that is a field over extensions of which the form is the split one. We also assume that *the field $\mathbb{E}$ is infinite* and that *the extension $\mathbb{E}/\mathbb{K}$ is normal*. The definition of Cartan subgroups in terms of Ad and the Γ-equivariance of this representation provides [Ré2, 11.2.2]:

Lemma. *The image of an $\mathbb{E}$-split Cartan subgroup by a Galois automorphism is still an $\mathbb{E}$-split Cartan subgroup.*

In view of the conjugacy of Cartan subgroups, this lemma suggests to rectify Galois automorphisms so as to make them stabilize the standard Cartan subgroup T. If γ denotes a Galois automorphism such that $\gamma T = gTg^{-1}$, then $\bar{\gamma} := \mathrm{int} g^{-1} \circ \gamma$ will denote such a rectified automorphism. At the $\mathbb{E}$-algebras level, the symbol $\bar{\gamma}$ will denote the rectified (semilinear) automorphism $\mathrm{Ad} g^{-1} \circ \gamma$, where γ is the Galois automorphism of $(\mathcal{U}_{\mathcal{D}})_{\mathbb{E}}$ arising from the form. The choice of g modulo N will be harmless. Another class of interesting subgroups is that of root groups, for which things are not so nice. Still, we have [Ré2, 11.2.3]:

Lemma. *If the c_s's are free in the lattice Λ of the Kac–Moody root datum $\mathcal{D}$, then the image of a root group (w.r.t. to T) under a rectified Galois automorphism is still a root group (w.r.t. to T).*

Note that the action on the group G has no reason to respect its nice combinatorics. The additional requirements are then [Ré2, 11.2.5]:

Definition. *An* algebraic $\mathbb{K}$-form *of G is a prealgebraic form $(\underline{G}, \mathcal{U}_{\mathbb{K}})$ such that, for each $\gamma \in \Gamma$, one has:*

(ALG0) *the rectified automorphism $\bar{\gamma}$ stabilizes the family of root groups (w.r.t. to T).*
(ALG1) *$\bar{\gamma}$ respects the abstract Q-gradation of $(\mathcal{U}_{\mathcal{D}})_{\mathbb{E}}$, inducing a permutation of Q still denoted by $\bar{\gamma}$ and satisfying the homogeneity condition $\bar{\gamma}(na) = n\bar{\gamma}(a)$, $a \in \Delta^{re}$, $n \in \mathbb{N}$.*
(ALG2) *$\bar{\gamma}$ respects the algebraic characters and cocharacters of T among all the abstract ones.*

Remark. *The algebraic characters (resp. cocharacters) are the ones arising from Λ (resp. $\Lambda^\vee$):* (ALG2) *is indeed an algebraicity condition. Concerning* (ALG1)*, the justification comes from algebraic differential calculus. If G admitted an algebro-geometric structure – a topological Hopf algebra of regular functions for instance – with U_a as a closed subgroup, then the subring $\bigoplus_{j \leq n} \mathbb{E}\frac{e_a^j}{j!}$ would be the $\mathbb{E}$-algebra of invariant distributions tangent to U_a. So it is natural to require a rectified Galois automorphism (stabilizing T) to respect the order of these distributions.*

Here are now the main consequences of algebraicity [Ré2, 11.2.5 and 11.3.2].

Proposition. (i) (i) *$\bar{\gamma}$ is a group automorphism of Q, it stabilizes Δ^{re} and induces the same permutation as the one defined via the root groups.*
(ii) *$\bar{\gamma}$ stabilizes N inducing an automorphism of W sending the reflexion w.r.t. ∂a to the reflection w.r.t. $\partial\bar{\gamma}(a)$.*
(iii) *If the Dynkin diagram of A is connected, each Galois automorphism sends a Borel subgroup on a Borel subgroup, possibly after opposition of its sign.*

In the non connected case, oppositions of signs can occur componentwise. At this step, there is no reason why the BN-pair of fixed sign should be respected. This is a requirement to be made.

B. Almost splitness. This condition is the "isotropy part" of the requirements [Rou3].

Definition. *The algebraic form* $(\underline{G}, \mathcal{U}_{\mathbb{K}})$ *is* almost split *if the Galois action stabilizes the conjugacy of Borel subgroups of each sign.*

The stability of conjugacy classes of Borel subgroups enables one to rectify each Galois automorphism γ in such a way that $\gamma^* := \mathrm{intg}_\sigma{}^{-1} \circ \gamma$ stabilizes B_+ and B_-. This way, the element g_σ is defined modulo $T = B_+ \cap B_-$. The assignment $\gamma \mapsto \gamma^*$ defines a Γ-action on W stabilizing S and is called the $*$-*action*. The terminology is justified by the analogy with the action considered by J. Tits in the classification problem of semi-simple groups. Now we can turn for the first time to the other interesting G-space, namely the twin building of the group [Ré2, 11.3.3 and 11.3.4].

Proposition. (i) *Via group combinatorics, the Galois group operates on the buildings by automorphisms, up to permutation of types by the* $*$*-action defined above.*

(ii) *The Galois group acts by isometries on the metric realizations of the building of each sign.* Γ*-orbits are finite.*

The last assertion is the starting point of the argument in 2.5.A: there exist Γ-fixed balanced subsets. In fact, we can prove [Ré2, 11.3.5]:

Theorem. *Let us assume now* $\mathbb{E}/\mathbb{K}$ *Galois, and let us consider* Ω *to be balanced and Galois-fixed. Then the subspace* $W(\Omega)_{\mathbb{E}}$ *is defined over* $\mathbb{K}$*, the homomorphism* Ad_Ω *is* Γ*-equivariant, and its image is a closed* $\mathbb{K}$*-subgroup of* $\mathrm{GL}\left(W(\Omega)_{\mathbb{E}}\right)$.

So to speak, this is the first step of the Galois descent: looking at Γ-fixed balanced subsets and at the corresponding small subgroups (resp. algebraic groups).

3.2. The Galois descent theorem

This section states the main structure result about rational points of almost split Kac–Moody $\mathbb{K}$-groups. The geometric side is at the same time the main tool and an interesting result in its own right.

A. Statement [Ré2, 12.4.3]. Here is the result which justifies the analogy between the theory of almost split forms of Kac–Moody groups and Borel-Tits theory for isotropic reductive groups. We keep the almost split $\mathbb{K}$-form $(\underline{G}, \mathcal{U}_{\mathbb{K}})$ of the previous section. We suppose now $\mathbb{E} = \mathbb{K}_s$, i.e., that the form is split over the separable closure $\mathbb{K}_s$. A Kac–Moody group endowed with such a form will be called an *almost split Kac–Moody* $\mathbb{K}$*-group*. By definition, the *group of rational points* of the form is the group of Galois-fixed points $G(\mathbb{K}) := G^\Gamma$.

Theorem (Galois descent). *Let G be a Kac–Moody group, almost split over $\mathbb{K}$. Then*

(i) *The set of Γ-fixed points in the twin building of G is still a twin building $\mathcal{J}^\natural$.*
(ii) *The group of rational points $G(\mathbb{K})$ is a (TRD)$_{\text{lin}}$-group for a natural choice of subgroups suggested by the geometry of a pair of opposite apartments in $\mathcal{J}^\natural$. Indeed, $\mathcal{J}^\natural$ is a geometric realization of the twin building abstractly associated to $G(\mathbb{K})$.*
(iii) *Maximal $\mathbb{K}$-split tori are conjugate in $G(\mathbb{K})$.*

Remark. *The assumption of splitness over $\mathbb{K}_s$ is not necessary in the algebraic case, for which we can apply the combination of a theorem due to Cartier* [Bo, Theorem 18.7] *and a theorem due to Chevalley-Rosenlicht-Grothendieck* [Bo, Theorem 18.2].

B. *Sketch of proof.* Let us start with notation and terminology. The symbol $\Omega^\natural$ will always denote a Γ-fixed balanced subset, and Ω will denote the Γ-stable union of spherical facets covering it. We are working now in the conical realization of the twin building so as to apply convexity arguments. A *generic subspace* is a subspace of a pair of opposite apartments meeting a spherical facet.

Relative geometric objects. We define now relative objects with suggestive terminology. This does not mean that we know at this step that they form a building.

Definition. (i) *A* (spherical) $\mathbb{K}$-facet *is the subset of Γ-fixed points of a (spherical) Γ-stable facet. A* $\mathbb{K}$-chamber *is a spherical $\mathbb{K}$-facet of maximal closure. A* $\mathbb{K}$-panel *is a $\mathbb{K}$-facet of codimension one in the closure of a $\mathbb{K}$-chamber.*
(ii) *A* $\mathbb{K}$-apartment *is a generic Γ-fixed subspace, maximal for these properties.*

Remark. *These geometric definitions also give sense to an opposition relation on $\mathbb{K}$-facets of oppposite signs, which will be called opposite if they form a symmetric subset of a pair of opposite apartments.*

Let us turn now to the other application of argument 2.5.A, with the Galois group as bounded isometry group [Ré2, 12.2.1 and 12.2.3].

Theorem. *For each Γ-fixed balanced subset $\Omega^\natural$, there exists a Γ-stable pair of opposite apartments containing Ω. Since the Galois action preserves barycenters,* $\operatorname{conv}(\Omega^\natural)$ *is Γ-fixed.*

Sketch of proof. Theorem 3.1.B says that $\mathrm{Ad}_\Omega(\mathrm{Fix}_G(\Omega))$ is a $\mathbb{K}$-group, so it admits a Cartan subgroup defined over $\mathbb{K}$. The Γ-equivariant dictionnary of 2.5.B says that this corresponds geometrically to a Γ-stable pair of opposite apartments containing Ω. The last assertion follows from the fact that the convex hull can be determined in any apartment containing $\Omega^\natural$.

The last assertion is the starting point of convexity arguments, which enable one to prove the following facts [Ré2, 12.2.4]:

- Two $\mathbb{K}$-facets of opposite signs are always contained in a $\mathbb{K}$-apartment.
- There is an integer d such that $\mathbb{K}$-apartments and $\mathbb{K}$-chambers are all d-dimensional.
- The convex hull of two opposite $\mathbb{K}$-chambers is a $\mathbb{K}$-apartment, and each $\mathbb{K}$-apartment can be constructed this way. In particular, *$\mathbb{K}$-apartments are symmetric double cones.*
- Each $\mathbb{K}$-facet is in the closure of a spherical one. This enables us to define $\mathbb{K}$-chambers as $\mathbb{K}$-facets (*a priori* not spherical) of maximal closure.

Rational transitivity results. The next step is to study $\mathcal{J}^\natural$ as a $G(\mathbb{K})$-set [Ré2, 12.3.2].

Theorem. (i) *The group of rational points $G(\mathbb{K})$ is transitive on the $\mathbb{K}$-apartments.*
(ii) *Two $\mathbb{K}$-facets of arbitrary signs are always contained in a $\mathbb{K}$-apartment.*

Remark. *The proof of point* (ii) *requires a Levi decomposition argument for a Γ-fixed balanced subset made of three points, so Levi decompositions are useful for balanced subset more general than pairs of spherical points of opposite signs.*

Relative roots. Anisotropic kernel. To introduce a relative (TRD)-structure on the rational points, we need of course to define a relative Coxeter system $(W^\natural, S^\natural)$. We can make the choice of a pair of $\mathbb{K}$-facets $\pm F^\natural$ of opposite signs and of a pair $\pm c$ of opposite chambers in such a way that:

- $F^\natural$ and $-F^\natural$ define a $\mathbb{K}$-apartment $\mathbb{A}_\mathbb{K}$ by $\mathbb{A}_\mathbb{K} := \mathrm{conv}(F^\natural \cup -F^\natural)$;
- the Γ-stable facet $\pm F$ such that $\pm F^\natural = (\pm F)^\Gamma$ is in the closure of the chamber $\pm c$;
- the pair of opposite apartments $\mathbb{A}$ defined by $\pm c$ is Γ-stable.

We see $\mathbb{A}$ as the union of the Tits cone and its opposite in the real vector space V^*. Moreover, we denote by $L^\natural$ the linear span of $\mathbb{A}_\mathbb{K}$ in V^*.

Definition. *The restriction* $a^\natural := a\mid_{L^\natural}$ *of a root* a *of* $\mathbb{A}$ *(seen as a linear form on* V^**) is a* $\mathbb{K}$-root *of* $\mathbb{A}_\mathbb{K}$ *if the trace* $D(a^\natural) := \mathbb{A}_\mathbb{K} \cap D(a)$ *is generic. In this case,* $D(a^\natural)$ *is called the* $\mathbb{K}$-halfspace *of* $a^\natural$.

For a given $\mathbb{K}$-halfspace $D(a^\natural)$, we consider the set $\Phi_{a^\natural}$ of roots b with $D(b) \cap \mathbb{A}_\mathbb{K} = D(a^\natural)$. This set is stable under Γ and so is the group generated by the root groups U_b, $b \in \Phi_{a^\natural}$. Let us denote by $V_{D(a^\natural)}$ the group of its Γ-fixed points.

Definition. (i) *The group* $V_{D(a^\natural)}$ *is the* relative root group *associated to the* $a^\natural$.
(ii) *The fixator* $\mathrm{Fix}_G(\mathbb{A}_\mathbb{K})$, *denoted by* $Z(\mathbb{A}_\mathbb{K})$, *is the* anisotropic kernel *associated to* $\mathbb{A}_\mathbb{K}$.

Rational group combinatorics. The next step is to prove that $G(\mathbb{K})$ admits a nice combinatorial structure, stronger than BN-pairs but weaker than the $(\mathrm{TRD})_{\mathrm{lin}}$-axioms. This group combinatorics is due to V. Kac and D. Peterson: the *refined Tits systems* [KP2]. This is an abstract way to obtain the relative Coxeter system $(W^\natural, S^\natural)$ we are looking for, via BN-pairs. We have [Ré2, 12.4.1 and 12.4.2]:

Lemma. *The group* $W^\natural$ *is a subquotient of* $G(\mathbb{K})$, *namely the quotient of the stabilizer in* $G(\mathbb{K})$ *of* $\mathbb{A}_\mathbb{K}$ *by its fixator. The half of each sign of the double cone* $\mathbb{A}_\mathbb{K}$ *is a geometric realization of the Coxeter complex of* $W^\natural$, *and the latter group is generated by reflections w.r.t. to* $F^\natural$. *The* $\mathbb{K}$-*halfspaces* $D(a^\natural)$ *are in* $W^\natural$-*equivariant bijection with the abstract set of roots* $\Phi_\mathbb{K}$ *of* $(W^\natural, S^\natural)$.

We have now an index set for the $(\mathrm{TRD})_{\mathrm{lin}}$-axioms. What is still unclear is how to define a linear interval of $\mathbb{K}$-halfspaces. For $D(a^\natural)$ and $D(b^\natural)$, we set

$$[D(a^\natural); D(b^\natural)]_{\mathrm{lin.}} := \{D(c^\natural) \mid \exists \lambda, \mu \in \mathbb{R}_+ \quad c^\natural = \lambda a^\natural + \mu b^\natural\}.$$

With all the objects defined above, point (i) of the Galois descent theorem more precisely says that *the group of rational points* $G(\mathbb{K})$ *– endowed with the* $\mathbb{K}$-*points* $Z(\mathbb{K})^\Gamma$ *of the anisotropic kernel and the family* $\{V_{D(a^\natural)}\}_{D(a^\natural)\in\Phi_\mathbb{K}}$ *– satisfies the* $(TRD)_{\mathrm{lin}}$-*axioms*. Point (ii) says that *the set of* Γ-*fixed points in the twin building is the realization of a twin building*. Besides, all the suggestive terminology for relative objects is justified.

Maximal $\mathbb{K}$-*split tori.* In the formulation above, we forgot the maximal $\mathbb{K}$-split tori. According to 2.5.C, it is natural to expect a relative dictionary between $\mathbb{K}$-apartments and maximal $\mathbb{K}$-split tori. A $\mathbb{K}$-*split torus* of G is a $\mathbb{K}$-split subgroup of a Γ-stable Cartan subgroup (this definition makes sense since the latter groups admit natural structures of $\mathbb{K}$-tori). Let us keep the $\mathbb{K}$-apartment $\mathbb{A}_\mathbb{K}$, and $Z(\mathbb{K}) = \mathrm{Fix}_G(\mathbb{A}_\mathbb{K})$.

Proposition. *The anisotropic kernel* $Z(\mathbb{A}_{\mathbb{K}})$ *is the fixator of any pair of opposite* $\mathbb{K}$*-chambers* $\pm F^{\natural}$ *in* $\mathbb{A}_{\mathbb{K}}$. *It is a* Γ*-stable small subgroup and the Levi factor of the* Γ*-stable parabolic subgroup* $\mathrm{Fix}_G(\pm F^{\natural})$. *Its associated algebraic group is semi-simple anisotropic.*

Idea of proof. [Ré2, 12.3.2] — The first assertion follows from example 2 in 1.5.B, and the anisotropy part is justified as in the proof of Lang's theorem below (3.3.A).

As the quotient of a small subgroup defining its associated algebraic group is by a diagonalisable group, the anisotropy assertion gives rise to a $G(\mathbb{K})$-equivariant map

$$\begin{array}{ccc} \{\mathbb{K}\text{-apartments}\} & \longrightarrow & \{\text{Maximal } \mathbb{K}\text{-split tori}\} \\ \mathbb{A}_{\mathbb{K}} & \mapsto & Z_d^0(\mathbb{A}_{\mathbb{K}}), \end{array}$$

where $Z_d^0(\mathbb{A}_{\mathbb{K}})$ is the connected $\mathbb{K}$-split part of the center of $Z(\mathbb{A}_{\mathbb{K}})$. A more precise statement of point (iii) of the Galois descent theorem is that *each* $\mathbb{K}$*-split torus can be conjugated by* $G(\mathbb{K})$ *into* $Z_d^0(\mathbb{A}_{\mathbb{K}})$. In particular, the map above is surjective. In fact, it is bijective for $\mathbb{K}$ large enough.

3.3. A Lang type theorem

We describe here the specific situation in which the ground field is finite.

A. Quasi-splitness. An algebraic $\mathbb{K}$-group is *quasi-split* if it contains a Borel subgroup defined over $\mathbb{K}$, so in the twin situation, it is natural to introduce the following.

Definition. *A Kac–Moody* $\mathbb{K}$*-group is* quasi-split *if it contains two opposite Borel subgroups stable under the Galois group* Γ.

Remark. *Geometrically, this means that two opposite chambers are Galois-stable.*

For comparison, the classical Lang theorem [Bo, 16.6] asserts that over a finite field, an algebraic group has to be quasi-split.

Theorem (Lang's theorem for Kac–Moody groups)**.** *Let* G *be a Kac–Moody group, almost split over a finite field* $\mathbb{F}_q$. *Then* G *is in fact quasi-split over* $\mathbb{F}_q$.

Sketch of proof. [Ré2, 13.2.2] — Actually, the proof of this result makes use of the algebraic one. The argument works as follows. Look at the Levi decomposition of a minimal Γ-stable parabolic subgroup P of G. By minimality, the Levi factor is an algebraic $\mathbb{F}_q$-group which cannot contain a proper parabolic subgroup defined over $\mathbb{F}_q$. Indeed, it would otherwise lead to a Γ-stable parabolic subgroup smaller than P. Consequently, the Levi factor – which is reductive – has no proper parabolic subgroup defined over $\mathbb{F}_q$, but admits an $\mathbb{F}_q$-Borel subgroup: it must be a torus.

B. Down-to-earth constructions of forms. Restriction to quasi-split forms enables one to concretely define twisted Kac–Moody groups. In fact, groups of this kind were already considered by J.-Y. Hée and J. Ramagge, who obtained structure theorems for "rational points" without reference to or use of buildings [H1 and H3], [Ra1 and Ra2]. These groups were studied in the wider context of generalized *Steinberg torsions*, which allowed J.-Y. Hée to define analogues of Ree-Suzuki groups. If we stick to Galois torsion as above, we have the following constructive result [Ré2, 13.2.3].

Proposition. *Consider a generalized Cartan matrix A with Dynkin diagram D. Suppose we are given a Galois extension $\mathbb{E}/\mathbb{K}$ and a morphism $*$: $\mathrm{Gal}(\mathbb{E}/\mathbb{K}) \to$ Aut(D). Then the simply connected (resp. adjoint) split Kac–Moody group defined over $\mathbb{E}$ admits a quasisplit form over $\mathbb{K}$ with $*$-action the morphism above.*

Remark. *The Dynkin diagram of a generalized Cartan matrix is defined in* [K]. *The result is also valid with a suitable notion of automorphism for a Kac–Moody root datum. As we will see it as a procedure to construct twisted geometries which only depends on A, we will not be interested in these refinements.*

3.4. Relative apartments and relative links

We make precise now the determination of the shape of the apartments and of the relative links.

A. Geometry of relative apartments. Our aim here is to derive some more information about the geometry of $\mathbb{K}$-apartments from the $*$-action of the almost split form. This is to be combined with the constructive procedure of quasi-split forms 3.3.B. We are still working with the datum of a pair of $\mathbb{K}$-facets $\pm F^\natural$ of opposite signs and of a pair $\pm c$ of opposite chambers s.t. $F^\natural$ and $-F^\natural$ define a $\mathbb{K}$-apartment $\mathbb{A}_\mathbb{K}$ by $\mathbb{A}_\mathbb{K} := \mathrm{conv}(F^\natural \cup -F^\natural)$; the Γ-stable facet $\pm F$ with $\pm F^\natural = (\pm F)^\Gamma$ is in the closure $\pm\bar{c}$ and the pair of opposite apartments $\mathbb{A}$

defined by $\pm c$ is Γ-stable. Recall that $\mathbb{A}$ is the union of the Tits cone and its opposite in the real vector space V^* and that $L^\natural$ is the linear span of $\mathbb{A}_\mathbb{K}$ in V^*. By transitivity of $G(\mathbb{K})$, all $\mathbb{K}$-chambers are fixed points in Γ-stable facets of fixed type S_0.

Lemma. *Denote by Γ^* the automorphism group of the standard chamber c arising from the $*$-action. Then $L^\natural$ admits the following definition by linear equations:*

$$L^\natural = \{x \in V^* \mid a_s(x) = 0 \ \forall s \in S_0 \text{ and } a_s(x) = a_t(x) \text{ for } \Gamma^* s = \Gamma^* t\},$$

and the $\mathbb{K}$-chamber is a simplicial cone of dimension $d \geq S^\natural$ defined by the same equations for x in c_+.

This result [Ré2, 12.6.1] enables us to determine the shape of a $\mathbb{K}$-chamber or a $\mathbb{K}$-apartment from the datum of the $*$-action. In other words, this is a concrete procedure to determine a part of the geometry of the twisted twin building. To get more precise information, we have to use finer theoretical results concerning infinite Kac–Moody type root systems. Following the abstract results in [Ba], the datum of S_0 and of the $*$-action determines a *relative Kac–Moody matrix*, a relative version of the matrix A. Then, there is a similar rule to deduce a Coxeter from it, and it happens to be naturally isomorphic to the relative Weyl group $W^\natural$. So another piece of the twisted combinatorics is theoretically computable.

B. Geometry of relative residues. Once we have determined the geometry of the slices of the twisted twin building, one may want to know more about the local geometry of it. This we will also accomplish thanks to the knowledge of S_0 and of the $*$-action. In the split case, according to 2.2.B, the fixator of a pair of opposite spherical facets $F_\pm$ is abstractly isomorphic to a split reductive group. Besides, a realization of its twin building is given by the union of the residues of F_+ and F_-. The relative version of this is also valid. More precisely, let us consider a pair of opposite spherical $\mathbb{K}$-facets $\pm F^\natural$ whose union will be denoted by $\Omega^\natural$. It can be shown then that the Galois actions on the Γ-stable residues yields a natural $\mathbb{K}$-structure on the reductive algebraic groups to which the small subgroup $\mathrm{Fix}_G(\Omega^\natural)$ is isomorphic. For instance, if we assume that $\pm F^\natural$ are spherical $\mathbb{K}$-panels, this provides information about rational root groups. In fact, we have [Ré2, 12.5.2 and 12.5.4]:

Lemma. (i) *The union of the sets of $\mathbb{K}$-chambers whose closure contains $\pm F^\natural$ is a geometric realization of the twinning of an algebraic reductive $\mathbb{K}$-group.*

(ii) *Rational root groups are isomorphic to root groups in algebraic reductive $\mathbb{K}$-groups. In particular, the only possible proportionality factor > 1 between $\mathbb{K}$-roots is 2 and rational root groups are metabelian.*

Combined with the geometric consequence of the Moufang property, the second point enables to compute thicknesses.

3.5. A classical twin tree

A. The split situation. Recall that $\mathbf{SL}_3(\mathbb{K}[t,t^{-1}])$ is a split Kac–Moody group defined over the field $\mathbb{K}$ and of affine type $\widetilde{A}_2$. The Dynkin diagram is the simply-laced triangle. The simple roots of the $\widetilde{A}_2$-affine root system are denoted by a_0, a_1 and a_2, for which we choose respectively the parametrizations x_0: $r \mapsto \begin{pmatrix} 1 & 0 & 0 \\ 0 & 1 & 0 \\ -rt & 0 & 1 \end{pmatrix}$, $x_1 : r \mapsto \begin{pmatrix} 1 & r & 0 \\ 0 & 1 & 0 \\ 0 & 0 & 1 \end{pmatrix}$ and $x_2 : r \mapsto \begin{pmatrix} 1 & 0 & 0 \\ 0 & 1 & r \\ 0 & 0 & 1 \end{pmatrix}$.
The standard Cartan subgroup is

$$T = \left\{ D_{u,v} := \begin{pmatrix} u & 0 & 0 \\ 0 & u^{-1}v & 0 \\ 0 & 0 & v^{-1} \end{pmatrix} \middle| \, u, v \in \mathbb{E}^\times \right\}.$$

We work for this example with the Bruhat-Tits realization of the buildings.

B. The involution. Let us consider a separable quadratic extension $\mathbb{E}/\mathbb{K}$. We want to define a quasi-split $\mathbb{K}$-form of $\mathrm{SL}_3(\mathbb{E}[t,t^{-1}])$ so as to produce a semi-homogeneous twin tree by quasi-split Galois descent. From a theoretical point of view (3.3.B), it is enough to find an involution of the Dynkin diagram. We choose the inversion of types 1 and 2. We describe now concretely to which Galois action this choice leads. Let ι (resp. τ) be the inversion (resp. the "antidiagonal" symmetry) of matrices,

$$\tau : \begin{pmatrix} a & b & c \\ d & e & f \\ g & h & i \end{pmatrix} \mapsto \begin{pmatrix} i & f & c \\ h & e & b \\ g & d & a \end{pmatrix}$$

Let $\cdot^\sigma$ be at the same time the non-trivial element of $\mathrm{Gal}(\mathbb{E}/\mathbb{K})$ and this operation on all matrix coefficients. Then the involution $\cdot^* := \tau \circ \iota \circ \cdot^\sigma$ of $\mathbf{SL}_3(\mathbb{E}[t,t^{-1}])$ is the one given by the quasi-split $\mathbb{K}$-form. It is completely determined by the formulae $x_1(r)^* = x_2(r^\sigma)$, $x_2(r)^* = x_1(r^\sigma)$ and $x_0(r)^* = x_0(r^\sigma)$, the analogues for negative simple roots, and $D^*_{u,v} = D_{v^\sigma,u^\sigma}$. The group of monomial matrices lifting modulo T the affine Weyl group $W = \mathcal{S}_3 \ltimes \mathbb{Z}^2$ is also $\cdot^*$-stable. This implies the $\cdot^*$-stability of the standard apartments $\mathbb{A}_+ = \{wB_+\}_{w\in W}$ and $\mathbb{A}_- = \{wB_-\}_{w\in W}$. The set of *-fixed points inside is a $\mathbb{K}$-apartment, since according to the permutation of types, there cannot be a two-dimensional $\mathbb{K}$-chamber. This set $\mathbb{A}_{\mathbb{K}}$ is the median of the standard equilateral triangle (intersecting the edge of type 0) w.r.t. to which $\cdot^*$ operates by symmetry in $\mathbb{A}$. Consequently, each building is a tree and the relative Weyl group $W^\natural$ is the infinite dihedral group D_∞. Translations are lifted by the matrices $\begin{pmatrix} at^n & 0 & 0 \\ 0 & 1 & 0 \\ 0 & 0 & (at^n)^{-1} \end{pmatrix}$; reflexions by the matrices $\begin{pmatrix} 0 & 0 & (at^n)^{-1} \\ 0 & 1 & 0 \\ -at^n & 0 & 0 \end{pmatrix}$, $(n \in \mathbb{Z},\ a \in \mathbb{K}^\times)$. The reason why we call the twin building so obtained a classical twin tree is that, up to sign conventions, the Steinberg torsion we defined is that of the unitary group SU_3 considered in [BrT2, 4.1.9].

C. Semi-homogeneity. What is left to do is to compute valencies. We will get semi-homogeneous trees since there are two shapes of intersections of the $\mathbb{K}$-apartment with walls. We sketch arguments for positive standard vertices and edges, but the situation is completely symmetric w.r.t. signs and can be deduced for every facet by transitivity of the rational points.

The first (simplest) situation is given by the intersection of the standard edge of type 0 with $\mathbb{A}_{\mathbb{K}}$. In this case, only one wall passes through the intersection point and each of the two opposite $\mathbb{K}$-roots is the trace on $\mathbb{A}_{\mathbb{K}}$ of a single root. The corresponding root groups are stable and their groups of rational points are isomorphic to the additive group of $\mathbb{K}$. Hence a valency at all transforms of this $\mathbb{K}$-vertex is equal to $1 + |\mathbb{K}|$.

The other situation corresponds to the fact that $\mathbb{A}_{\mathbb{K}}$ contains the standard vertex of type $\{1; 2\}$ for which 3 roots leave the same $\mathbb{K}$-root as trace on $\mathbb{A}_{\mathbb{K}}$. The small subgroup fixing this vertex and its opposite in $\mathbb{A}$ is stable and abstractly isomorphic to SL_3 over $\mathbb{E}$. According to the action around the vertex, the three positive root groups passing through it are permuted and the small subgroup is abstractly isomorphic to a quasi-split reductive $\mathbb{K}$-group of rank one. The root groups of such groups are described in detail in [BrT2, 4.1.9]; they are rational points of the unipotent radical of a Borel subgroup of SU_3, in bijection with $\mathbb{K}^3$. This can also be seen directly on the explicit description of the Galois action above. Hence a valency at all transforms of this $\mathbb{K}$-vertex is equal to $1 + |\mathbb{K}|^3$. This shows that we built a semi-homogeneous twin tree of valencies $1 + |\mathbb{K}|$ and $1 + |\mathbb{K}|^3$. So in the case of a finite ground field $\mathbb{K} = \mathbb{F}_q$, one gets a locally finite semi-homogeneous twin tree of valencies $1 + q$ and $1 + q^3$.

Figure. — [Here $q = 2$]

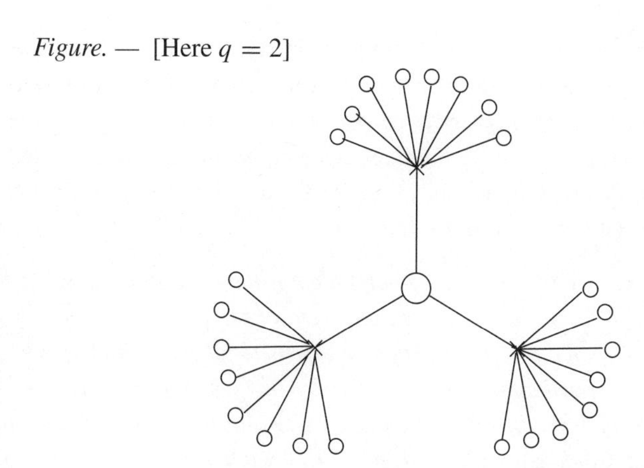

4. Hyperbolic examples. Lattices

So far, we have just considered examples of Chevalley groups over Laurent polynomial rings or twisted versions of them. Section 4.1 deals with split and twisted Kac–Moody groups whose buildings are hyperbolic. In Section 4.2, we show that Kac–Moody theory provides analogues of arithmetic groups over function fields.

4.1. Hyperbolic examples

At last, some more exotic Kac–Moody geometries will be described.

A. Hyperbolic buildings. The definition of such spaces [GP] is the metric version of that of buildings from the apartment system point of view.

Definition. *Let R be a hyperbolic polyhedron providing a Poincaré tiling of the hyperbolic space $\mathbb{H}^n$. A* hyperbolic building of type *R is a piecewise polyhedral cell complex, covered by a family of subcomplexes, the* apartments, *all isomorphic to the tiling above and satisfying the following incidence axioms.*

(i) *Two points are always contained in an apartment.*
(ii) *Two apartments are isomorphic by a polyhedral arrow fixing their intersection.*

A building whose apartments are tilings of the hyperbolic plane $\mathbb{H}^2$ will be called Fuchsian.

The main point about hyperbolic buildings is that we get in this way CAT(-1)-spaces, i.e., metric spaces where geodesic triangles are at least as thin as in $\mathbb{H}^2$. In particular, they are hyperbolic in the sense of Gromov. The additional incidence axioms make these spaces nice geometries where to refine results known for hyperbolic metric spaces, or to generalize results about symmetric spaces with (strict) negative curvature.

Remark. *Note that in contrast with the affine case, for hyperbolic buildings there is usually no relation between the dimension of the apartments and the rank of the Weyl group of the building. The two-dimensional case with a regular right-angled r-gon is particularly striking: $r \geq 5$ may be arbitrary large. Nevertheless, we will sometimes be interested in the case called compact hyperbolic by Bourbaki. It corresponds to a tesselation of $\mathbb{H}^n$ by a hyperbolic simplex. In this particular case, the relation alluded to above is valid; and the Weyl group shares another property with the affine case: every proper subset of canonical generators gives rise to a finite Coxeter subgroup.*

B. Split twin hyperbolic buildings. Our purpose here is to show which kind of hyperbolic buildings can be produced via Kac–Moody machinery. Let us make preliminary remarks, valid in both split and twisted cases. First, the rule of point (i) in the second definition of 2.1.A shows that the Coxeter coefficients involved in Kac–Moody theory are very specific: 2, 3, 4, 6 or ∞. So by far not any Coxeter group appears as the Weyl group of a generalized Cartan matrix. Besides, the thicknesses will always be of the form " 1 + prime power", and will be constant in the split case. These are strong conditions, but not complete obstructions.

Hyperbolic realizations. It is clear that standard glueing techniques [D2, §10] apply in the hyperbolic case once affine tilings are replaced by hyperbolic ones. Given P, a hyperbolic Poincaré polyhedron with dihedral angles $\neq 0$ of the form π/m with $m = 2$, 3, 4 or 6, and which may have vertices at infinity; and given p^k a prime power, there exists a Kac–Moody building of constant thickness $1 + p^k$ admitting a metric realization whose apartments are Poincaré tilings defined by P [Ré1, Sect. 2.3]. This building is a complete geodesic proper CAT(-1) metric space. This is just an existence result, it is expected to be combined with a uniqueness assertion, but the following result due to D. Gaboriau and F. Paulin ([GP], théorème 3.5) is quite disheartening.

Theorem. *Let P be a regular hyperbolic Poincaré polygon with an even number of edges and angles equal to π/m, $m \geq 3$. Let L be a fixed generalized m-gon of classical type and sufficiently large characteristic. Then, there exist uncountably many hyperbolic buildings with apartments the tiling provided by P and links all isomorphic to L.*

The latter theorem is in contrast with the following characterization proved by M. Bourdon [Bou1, 2.2.1], which shows that in the two-dimensional case, there is some kind of dichotomy according to whether $m \geq 3$ or $m = 2$. The latter kind of buildings will be referred to as the class of *right-angled Fuchsian buildings*.

Proposition. *Consider the Poincaré tiling of the hyperbolic plane $\mathbb{H}^2$ by a right-angled r-gon ($r \geq 5$). Then, there exists a unique building $I_{r,q+1}$ satisfying the following conditions:*

- *apartments are all isomorphic to this tesselation;*
- *the link at each vertex is the complete bipartite graph of parameters $(q + 1, q + 1)$.*

Consequently, we know by uniqueness that the building $I_{r,q+1}$ – $r \geq 5$, q a prime power, comes from a Kac–Moody group.

C. Quasi-split Kac–Moody groups and hyperbolic buildings. On the one hand, according to Subsection 2.2.B, there is no possibility to construct buildings with non-constant thickness via split Kac–Moody groups. On the other hand, we know by Section 3.5 that we can obtain semi-homogeneous twin trees from classical groups. Here is an example of what can be done in the context of hyperbolic buildings.

Semi-homogeneous trees in Fuchsian buildings. [Ré2, 13.3] — Definition of $\mathbb{F}_q$-forms relies on the existence of symmetries in the Dynkin diagram. Let us sketch the application of method 3.3.B to the buildings $I_{r,q+1}$. Denote by D the Coxeter diagram of the reflection group associated to the regular right-angled r-gon $R \subset \mathbb{H}^2$. To get it, draw the r-th roots of 1 on the unit circle and connect each pair of non consecutive points by an edge $\overset{\infty}{\text{———}}$. To keep all symmetries, lift this Coxeter diagram into a Dynkin diagram replacing the edges $\overset{\infty}{\text{———}}$ by $\Longleftrightarrow$. The corresponding generalized Cartan matrix A is indexed by $\mathbb{Z}/r$; it is defined by $A_{\bar{i},\bar{i}} = 2$, $A_{\bar{i},\overline{i+1}} = 0$ and -2's elsewhere.

Figure. — [Here $r = 6$]

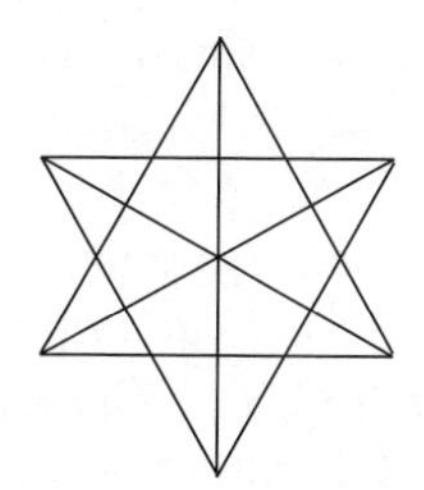

We suppose now that the integer r is odd, so that D is symmetric w.r.t. each axis passing through the middle of two neighbors and the opposite vertex. We fix such a symmetry, and consider the quadratic extension $\mathbb{F}_{q^2}/\mathbb{F}_q$: this defines a quasi-split form as in 3.5.B. Moreover, the arguments to determine the shape of an $\mathbb{F}_q$-chamber or an $\mathbb{F}_q$-apartment are the same. We get a straight line cut into segments and acted upon by the infinite dihedral group D_∞. The twisted buildings are trees.

Figure. — [Here $r = 5$ and $q = 2$]

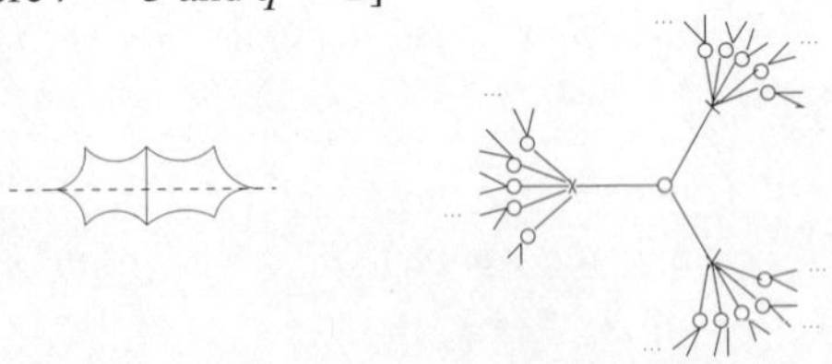

For the valencies, we must once again distinguish two kinds of intersection of the boundary of R with the $\mathbb{F}_q$-apartment, which leads to two different computations. When the $\mathbb{F}_q$-apartment cuts an edge, the rational points of the corresponding root group are isomorphic to $\mathbb{F}_q$, hence a valency equal to $1+q$. The other case is when the $\mathbb{F}_q$-apartment passes through a vertex contained in two orthogonal walls, with two roots leaving the same trace on the Galois-fixed line. The corresponding Galois-stable group is then a direct product of two root groups (each isomorphic to $\mathbb{F}_{q^2}$) acted upon by interversion and Frobenius: the valency at these points is then $1+q^2$. Consequently, we obtain this way a semi-homogeneous twin tree with valencies $1+q$ and $1+q^2$.

Twisted Fuchsian buildings in higher-dimensional hyperbolic buildings. [Ré1, §2] — Galois descent makes the dimension of apartments decrease, so if we want Fuchsian buildings with non constant thickness, we must consider higher dimensional hyperbolic buildings. Let us consider a specific example obtained from a polyhedron in the hyperbolic ball $\mathbf{B}^3$ and admitting many symmetries: it is the truncated regular hyperideal octahedron with dihedral angles $\pi/2$ and $\pi/3$. This polyhedron P is obtained by moving at same speed from the origin eight points on the axis of the standard orthogonal basis of $\mathbb{R}^3$. The condition on the angles impose to move the points outside of $\mathbf{B}^3$. What is left to do then is to cut the cusps by totally geodesic hyperplanes centered at the intersections of $\mathbf{S}^2 = \partial\mathbf{B}^3$ with the axis. A figure of this polyhedron is available in [GP]. The polyhedron P admits a symmetry w.r.t. hyperplanes containing edges and cutting P along an octogon. The residue at such an edge – where the dihedral angle is $\pi/3$ – is the building of SL_3. This enables one to define a quasi-split $\mathbb{F}_q$-form whose building is Fuchsian. The $\mathbb{F}_q$-chambers are octogons and thicknesses at $\mathbb{F}_q$-edges are alternatively $1+q$ and $1+q^3$.

D. Hyperbolic buildings from a geometrical point of view. Here is now a small digression concerning single hyperbolic buildings, without the assumption that they arise from Kac–Moody theory. This often implies to restrict oneself to specific classes of hyperbolic buildings, but provides stimulating results.

Existence, uniqueness and homogeneity. We saw in the previous subsection that unicity results were useful to know which buildings come from Kac–Moody groups. Nevertheless, the "disheartening theorem" above shows that the shape of the chambers and the links are not sufficient in this direction. F. Haglund [Ha] determined conditions to be added so as to obtain characterization results for two-dimensional hyperbolic buildings. His study enables him to know when the (type-preserving) full automorphism group is transitive on chambers.

Automorphism groups. The first step towards group theory then is to consider groups acting on a given hyperbolic building. Two kinds of groups are particularly interesting: full automorphism groups and lattices. Concerning full automorphism groups, F. Haglund and F. Paulin [HP] proved abstract simplicity and non-linearity results for two-dimensional buildings. Besides, É. Lebeau [L] proved that some others of these groups are strongly transitive on their buildings, hence admit a natural structure of BN-pair – see [Ron], §5. Concerning lattices, the first point is the existence problem. In fact, the main construction technique is a generalization to higher dimension of Serre's theory of graph of groups [Se]. D. Gaboriau and F. Paulin [GP] determined the precise local conditions on a *complex of groups* in the sense of A. Haefliger [BH] to have a hyperbolic building as universal covering. This method has the advantage to produce at the same time a uniform lattice with the building. The lattices we will obtain by Kac–Moody theory are *not* uniform.

Boundaries: quasiconformal geometry and analysis. The reason why boundaries of metric hyperbolic metric spaces are interesting is that they can be endowed with a quasiconformal structure – see [GH], §7. In the case of hyperbolic buildings, the structure can be expected to be even richer or better known, in view of the CAT(-1)-property and the incidence axioms. At least for Fuchsian buildings, this is indeed the case as shown by work due to M. Bourdon. In this situation, he relates the precise conformal dimension of the boundary and the growth rate of the Weyl group [Bou2]. In a joint paper with H. Pajot [BP1], the existence of Poincaré inequalities and the structure of Loewner space is proved for the boundaries of the $I_{r,q+1}$'s.

Rigidity. — Rigidity is a vast topic in geometric group theory. In this direction, results were proved by M. Bourdon first [Bou1] and then by M. Bourdon and H. Pajot [BP2]. The first paper deals with Mostow type rigidity of uniform lattices. It concerns right-angled Fuchsian buildings. A basic tool there is a combinatorial cross-ratio constructed thanks to the building structure involved. It enables one to generalize arguments of hyperbolic geometry by D. Sullivan to the singular case of (boundaries of) buildings. The second paper uses techniques from analysis on singular spaces to prove a stronger result implying Mostow rigidity, and in the spirit of results by P. Pansu [P1], B. Kleiner and B. Leeb [KL] for Riemannian symmetric spaces (minus the real and complex hyperbolic spaces) and higher rank Euclidean buildings. More precisely, M. Bourdon and H. Pajot proved [BP2]:

Theorem (Rigidity of quasi-isometries)**.** *Every quasi-isometry of a right-angled Fuchsian building lies within bounded distance from an isometry.*

For the notion of a *quasi-isometry*, see for instance [GH, §5]. Roughly speaking, it is an isometry up to an additive and a multiplicative constant, surjective up to finite distance.

4.2. Analogy with arithmetic lattices

The purpose of this section is to present several arguments to see Kac–Moody groups over finite fields as generalizations of arithmetic groups over function fields.

A. Kac–Moody topological groups and lattices. Let G be a Kac–Moody group, almost split – hence, quasi-split (3.3.A) – over the finite field $\mathbb{F}_q$. We denote by $\Delta_\pm$ the locally finite metric realizations of the buildings associated to G. Recall that all points there are of spherical type, hence the fixator of any point is a spherical parabolic subgroup. By local finiteness, the (isomorphic) corresponding full automorphism groups are locally compact. We use the notations $\mathrm{Aut}_\pm := \mathrm{Aut}(\Delta_\pm)$. The groups Aut_+ and Aut_- are besides unimodular and totally disconnected. We choose a left invariant Haar measure $\mathrm{Vol}_\pm$ and normalize it by $\mathrm{Vol}_\pm\big(\mathrm{Fix}_{\mathrm{Aut}_\pm}(c_\pm)\big) = 1$, where $c_\pm$ is the standard chamber.

Let us turn now to the analogy with arithmetic groups [Mar, p.61]. The affine case is indeed a guideline since Chevalley groups over Laurent polynomials are $\{0;\infty\}$-arithmetic groups over the global field $\mathbb{F}_q(t)$. This is the example of an arithmetic group for two places. If we consider now the fixator of a facet, we get a group commensurable with the value of the latter Chevalley group over the polynomials $\mathbb{F}_q[t]$, so that each fixator of facet is an arithmetic group for one place. Besides, the normalization we chose for the invariant measure consists in the affine case in assigning volume 1 to an Iwahori subgroup. We just replaced here a p-adic Lie group by the full automorphism group of a building.

If we consider now the general case, here is a result supporting the analogy, independently proved by L. Carbone and H. Garland in the split case [CG, Ré3].

Theorem. *Let G be an almost split Kac–Moody group over a finite field $\mathbb{F}_q$. Denote by $\sum_{n\geq 0} d_n t^n$ the growth series of the Weyl group W of G. Assume $\sum_{n\geq 0} \frac{d_n}{q^n} < \infty$. Then:*

(i) *The group G is a lattice of (the full automorphism group of) $\Delta_+ \times \Delta_-$.*
(ii) *For any point in $x_- \in \Delta_-$, $\mathrm{Fix}_G(x_-)$ is a lattice of Δ_+.*
(iii) *These lattices are not uniform.*

Remark. (1) *Non-uniformness is due to J. Tits, who determined fundamental domains for actions of parabolic subgroups in a combinatorial setting more general than the* (TRD) *axioms* [T3, A1 §3].
(2) *Of course, the situation is completely symmetric w.r.t. signs: point* (ii) *is still true after inversion of $+$ and $-$.*
(3) *There is a straightforward generalization of this result to the case of locally finite Moufang twin buildings. The cardinal of the ground field has to be replaced by the maximal cardinal q of root groups. According to the Moufang property* (1.3.B), *this integer is geometrically computed as follows. Consider any chamber, then $q+1$ is the maximal thickness among all the panels in its closure.*

B. *Sketch of proof.* We just consider the case of the fixator of a negative point – spherical negative parabolic subgroup. As we work over a finite field, we can make use of the following facts due to Levi decompositions, 1.5.B and 1.5.A respectively.

(i) The fixator of a pair of (spherical) points of opposite signs is finite.
(ii) Borel subgroups are of finite index in the spherical parabolic subgroups containing them.

Discreteness follows from point (i), and according to point (ii), it is enough to show that the fixator B_- of a negative chamber is of finite covolume. The idea is to adapt Serre's criterion on trees to compute a volume. This requires to look at the series $\sum_{d_+\in C} \frac{1}{|\mathrm{Stab}_{B_-}(d_+)|}$, where C is a complete set of representatives of chambers modulo the discrete group considered. According to Tits' result alluded to above, one has $C = \mathbb{A}_+$. Thanks to a special case of the Levi decompositions (1.5.B), the series is $\leq \frac{1}{|T|}\sum_{w\in W} \frac{1}{q^{\ell(w)}}$. Rearranging elements of W w.r.t. length in the dominating series, one gets the series appearing in the statement.

Remark. *When G is split over $\mathbb{F}_q$, the assumption on the numerical series is sharp.*

C. Classical properties for Kac–Moody lattices. The result above should be seen as a starting point. Indeed, one may wonder whether some classical properties of arithmetic groups are relevant or true for this kind of lattices. Some answers are already known.

Moufang property. The first remark in order for the analogy is to consider Moufang property. Recall that this property is already of interest in classical situations but is not satisfied by all p-adic Lie groups. For instance, it can be proved for many Bruhat-Tits buildings arising from reductive groups over local fields of equal (positive) characteristic, but it is known for no local field of characteristic zero [RT1, §2]. This remark suggests that the analogy with arithmetic groups in characteristic 0 will not be fruitful.

Kazhdan's property (T). This property has proved to be useful in many situations, and makes sense for arbitrary topological groups, endowed with the discrete topology for instance. In the classical case of Lie groups and their lattices, its validity is basically decided by the rank of the ambient Lie group [HV, 2.8]. In rank ≥ 2, all Lie groups are (T) and closed subgroups of finite covolume inherit the property [HV, 3.4]. Over non-archimidean local fields and in rank 1, no group is (T) since a necessary condition is to admit a fixed point for each isometric action on a tree [HV, 6.4]: contradiction with the natural action on the Bruhat-Tits tree. Another necessary condition is for instance compact generation, that is finite generation in the discrete case.

Property (T) is often characterized in terms of unitary representations, but the former necessary condition involving actions on trees suggests to look for geometrical criteria having the advantadge to make generalizations easier. An important example of this approach works as follows. On the one hand, H. Garland used the combinatorial Laplace operator on Bruhat-Tits buildings to obtain vanishing results for cohomology with coefficients in finite-dimensional representations [G]. On the other hand, property (T) is characterized by cohomology vanishing for every unitary representation. The idea to generalize Garland's method to infinite-dimensional representations is due to Pansu-Zuk [Z, P2], then many people elaborated on this. For instance, nice updatings and improvements as well as constructions of exotic discrete groups with property (T) are available in Ballmann-Swiatkowski's paper [BS]. That this concerns Kac–Moody groups with compact hyperbolic Weyl groups was seen independently by Carbone-Garland and Dymara-Januszkiewicz [CG, DJ]. According to the first cohomology vanishing criterion [HV, 4.7], the following result [DJ, Corollary 2] proves more than property (T).

Theorem. *Let G be a Kac–Moody group over $\mathbb{F}_q$ with compact hyperbolic Weyl group. Then for q large enough and $1 \leq k \leq n-1$, the continuous cohomology groups $\mathrm{H}^k_{ct}\big(\mathrm{Aut}(\Delta_\pm), \rho\big)$ with coefficients in any unitary representation ρ vanish.*

Remark. (1) *Note that the case for which we have a positive answer is for a simplex as chamber. Actually, negative results are proved for Fuchsian buildings with non simplicial chambers.*

(2) *Note also the condition on thickness, inside of which is hidden a spectral condition on the Laplace operator. It is actually necessary, in contrast with the classical situation as we will see in the forthcoming subsection.*

Finiteness properties. A way to define these finiteness properties – F_n or FP_n – is to use cohomology or actions on complexes, but for the first two degrees it is closely related to finite generation and finite presentability. A criterion defining the class of non-classical Kac–Moody groups for which positive results are known, involves finiteness of standard subgroups of the Weyl group. More precisely, we say that the Coxeter matrix M is *n-spherical* if $W_J = \langle J \rangle$ is finite for any subset J with $|J| \leq n$. For instance, if M is of irreducible affine or compact hyperbolic type, it is $(r-1)$-spherical but not r-spherical; and 2-spherical means that all coefficients are finite. When results are known, they look like those of arithmetic groups in characteristic p, for which much less is known than in the characteristic 0 case. The abstract framework of group actions on twin buildings in [A1] enables one to formulate and prove results in both classical and Kac–Moody cases. The arguments there rely on homotopy considerations, in particular make use of a deep result by K. Brown [Br1]. If we stick to finite generation or finite presentability, techniques of ordered sets from [T3] can also be used [AM]. Here are more recent results by P. Abramenko [A2]:

Theorem. (i) *If $|S| > 1$ and M is not 2-spherical, then $B_\pm$ is not finitely generated (for arbitrary q).*

(ii) *Suppose that $q > 3$ and that M is 2-spherical but not 3-spherical. Then $B_\pm$ is finitely generated but not finitely presentable.*

(iii) *Suppose that $q > 13$ and M is 3-spherical. Then $B_\pm$ is finitely presentable.*

Remark. (1) *Non finite generation (resp. non finite presentability) is a way to contradict property (T) (resp. Gromov hyperbolicity).*

(2) *The hypothesis "$\mathbb{F}_q$ large enough" once again comes into play. P. Abramenko noticed also that over tiny fields, some groups as in the theorem are lattices of their building but are not finitely generated. This shows that the condition on thickness is necessary for the previous theorem concerning property (T), while it is not so in the classical case (rank argument).*

References

[A1] P. Abramenko, *Twin Buildings and Applications to S-Arithmetic Groups*, Lecture Notes in Mathematics vol. 1641, Springer, 1996.

[A2] P. Abramenko, Finiteness properties of groups acting on twin buildings, this volume.

[AM] P. Abramenko and B. Mühlherr, Présentations de certaines BN-paires jumelées comme sommes amalgamées, *C. R. Acad. Sci. Paris* **325** (1997), 701–706.

[Ba] N. Bardy, *Systèmes de racines infinis*, Mémoire SMF vol. 65, 1996.

[B_3R] V. Back-Valente, N. Bardy-Panse, H. Ben Messaoud, and G. Rousseau, Formes presque déployées des algèbres de Kac–Moody: classification et racines relatives, *J. of Algebra* **171** (1995), 43–96.

[Bbk] N. Bourbaki, *Groupes et algèbres de Lie* IV-VI, Masson, 1981.

[BH] M. Bridson and A. Haefliger, *Metric spaces of nonpositive curvature*, Springer, 1999.

[Bo] A. Borel, *Linear Algebraic Groups*, Graduate Texts in Mathematics vol. 126, Springer, 1990.

[BoT] A. Borel and J. Tits, Groupes réductifs, *Publ. Math. IHÉS* **27** (1965), 55–151.

[Bou1] M. Bourdon, Immeubles hyperboliques, dimension conforme et rigidité de Mostow, *GAFA* **7** (1997), 245–268.

[Bou2] M. Bourdon, Sur les immeubles fuchsiens et leur type de quasi-isométrie, *Erg. Th. and Dynam. Sys.* **20** (2000), 343–364.

[BP1] M. Bourdon and H. Pajot, Poincaré inequalities and quasiconformal structure on the boundary of some hyperbolic buildings, *Proc. AMS* **127** (1999), 2315–2324.

[BP2] M. Bourdon and H. Pajot, Rigidity of quasi-isometries for some hyperbolic buildings, *Comm. Math. Helv.* **75** (2000), 701–736.

[Br1] K. Brown, Finiteness properties of groups, *J. Pure Appl. Algebra* **44** (1987), 45–75.

[Br2] K. Brown, *Buildings*, Springer, 1989.

[BS] W. Ballmann and J. Swiatowski, On L^2-cohomology and property (T) for automorphism groups of polyhedral cell complexes, *GAFA* **7** (1997), 615–645.

[BrT1] F. Bruhat and J. Tits, Groupes réductifs sur un corps local I. Données radicielles valuées, *Publ. Math. IHÉS* **41** (1972), 5–251.

[BrT2] F. Bruhat and J. Tits, Groupes réductifs sur un corps local II. Schémas en groupes. Existence d'une donnée radicielle valuée, *Publ. Math. IHÉS* **60** (1984), 5–184.

[CG] L. Carbone and H. Garland, Existence and deformations of lattices in groups of Kac–Moody type, in preparation.

[Ch] A. Chosson, Présentation des groupes de Kac–Moody, d'après J. Tits, preprint, University of Picardie – Amiens, 1998.

[D1] M. Davis, Groups generated by reflections and aspherical manifolds not covered by Euclidean spaces, *Ann. of Math.* **117** (1983), 293–324.

[D2] M. Davis, Buildings are CAT(0), in: *Geometry and Cohomology in Group Theory* (P.H. Kropholler, G.A. Niblo, and R. Stöhr eds.), LMS Lecture Notes vol. 252, 1997, 108–123.

[DG] M. Demazure and P. Gabriel, *Groupes Algébriques*, Masson/North Holland, 1970.

[DJ] J. Dymara and T. Januszkiewicz, New Kazhdan groups, Geometriæ Dedicata **80** (2000), 311–317.

[G] H. Garland, p-adic curvature and the cohomology of discrete subgroups of p-adic groups, *Ann. of Math.* **97** (1973), 375–423.

[GH] É. Ghys and P. de la Harpe, *Sur les Groupes Hyperboliques d'après Mikhael Gromov*, Progress in Mathematics vol. 83, Birkhäuser, 1990.

[GP] D. Gaboriau and F. Paulin, Sur les immeubles hyperboliques, *Geom. Dedicata* **88** (2001), 153–197.

[H1] J.-Y. Hée, Construction de groupes tordus en théorie de Kac–Moody, *C. R. Acad. Sc. Paris* **310** (1990), 77–80.

[H2] J.-Y. Hée, Systèmes de racines sur un anneau commutatif totalement ordonné, *Geometriæ Dedicata* **37** (1991), 65–102.

[H3] J.-Y. Hée, *Sur la torsion de Steinberg-Ree des Groupes de Chevalley et de Kac–Moody*, Thèse d'Etat University Paris 11, 1993.

[Ha] F. Haglund, Existence, unicité et homogénéité de certains immeubles hyperboliques, preprint, University Paris 11, 1999.

[HP] F. Haglund and F. Paulin, Simplicité de groupes d'automorphismes d'espaces à courbure négative, in: *The Epstein Birthday Schrift*, Geom. Topol. Monogr. vol. 1, Coventry, 1998, 181–248.

[Hu1] J. Humphreys, *Introduction to Lie Algebras and Representation Theory*, Graduate Texts in Mathematics vol. 9, Springer, 1972.

[Hu2] J. Humphreys, *Reflection Groups and Coxeter Groups*, Cambridge University Press, 1990.

[HV] P. de la Harpe and A. Valette, *La Propriété* (T) *de Kazhdan pour les Groupes localement compacts*, Astérisque vol. 175, 1989.

[K] V. Kac, *Infinite-Dimensional Lie Algebras*, Cambridge University Press, 1990.

[KL] B. Kleiner and B. Leeb, Rigidity of quasi-isometries for symmetric spaces and Euclidian buildings, *Publ. Math. IHÉS* **86** (1997), 115–197.

[KP1] V. Kac and D. Peterson, Regular functions on some infinite-dimensional groups, in: *Arithmetic and Geometry* (M. Artin and J. Tate eds.), Progress in Mathematics vol. 36, Birkhäuser, 1983, 141–166.

[KP2] V. Kac and D. Peterson, Defining relations for certain infinite-dimensional groups, Astérisque Hors-Série, 1984, 165–208.

[KP3] V. Kac and D. Peterson, On geometric invariant theory for infinite-dimensional groups, in: Lecture Notes in Mathematics vol. 1271, Springer, 1987, 109–142.

[L] É. Lebeau, *Rigidité et Flexibilité de Complexes polyédraux à Courbure négative*, Thesis ENS Lyon, 1999.

[Mar] G. A. Margulis, *Discrete Subgroups of semisimple Lie Groups*, Springer, 1990.

[Mas] B. Maskit, *Kleinian Groups*, Springer, 1987.

[Mat1] O. Mathieu, Formule des caractères pour les algèbres de Kac–Moody générales, Astérisque vol. 159-160, 1988.

[Mat2] O. Mathieu, Construction d'un groupe de Kac–Moody et applications, *Comp. Math.* **69** (1989), 37–60.

[Mou] G. Moussong, *Hyperbolic Coxeter Groups*, Thesis, Ohio State University, 1988.

[MR] B. Mühlherr and M. Ronan, Local to global structure in twin buildings, *Invent. Math.* **122** (1995), 71–81.

[Mü] B. Mühlherr, Locally split and locally finite buildings of 2-spherical type, *J. Reine u. Angew. Math.* **511** (1999), 119–143.

[P1] P. Pansu, Métriques de Carnot-Carathéodory et quasi-isométries des espaces symétriques de rang 1, *Ann. of Math.* **129** (1989), 1–60.

[P2] P. Pansu, Formules de Matsushima, de Garland et propriété (T) pour des groupes agissant sur des espaces symétriques ou des immeubles, *Bull. Soc. Math. France* **126** (1998), 107–139.

[Ra1] J. Ramagge, A realization of certain affine Kac–Moody groups of types II and III, *J. Algebra* **171** (1995), 713–806.

[Ra2] J. Ramagge, On certain fixed point subgroups of affine Kac–Moody groups, *J. Algebra* **171** (1995), 473–514.

[Ré1] B. Rémy, Immeubles de Kac–Moody hyperboliques, groupes non isomorphes de même immeuble, *Geom. Dedicata* **90** (2002), 29–44.

[Ré2] B. Rémy, Groupes de Kac–Moody déployés et presque déployés, Astérisque vol. 277, Soc. Math. de France, 2002.

[Ré3] B. Remy, Construction de réseaux en théorie de Kac–Moody, *C. R. Acad. Sc. Paris* **329** (1999), 475–478.

[Ron] M. Ronan, *Lectures on Buildings*, Academic Press, 1989.

[Rou1] G. Rousseau, Almost split **K**-forms of Kac–Moody algebras, in: *Infinite-dimensional Lie Algebras and Groups* (Marseille), Adv. Ser. in Math. Physics vol. 7, World Scientific, 1989, 70–85.

[Rou2] G. Rousseau, L'immeuble jumelé d'une forme presque déployée d'une algèbre de Kac–Moody, *Bull. Soc. Math. Belg.* **42** (1990), 673–694.

[Rou3] G. Rousseau, On forms of Kac–Moody algebras, *Proc. of Symp. in Pure Math.* **56** (1994), part 2, 393–399.

[RT1] M. Ronan and J. Tits, Building buildings, *Math. Ann.* **278** (1987), 291–306.

[RT2] M. Ronan and J. Tits, Twin trees. I, *Invent. Math.* **116** (1994), 463–479.

[RT3] M. Ronan and J. Tits, Twin trees. II. Local structure and a universal construction, *Israel J. Math.* **109** (1999), 349–377.

[Se] J.-P. Serre, *Arbres, Amalgames,* SL_2, Astérisque vol. 46, 1977.

[Sp] T. Springer, *Linear Algebraic groups*, Progress in Mathematics vol. 9, second edition, Birkhäuser, 1998.

[T1] J. Tits, *Buildings of Spherical Type and finite BN-Pairs*, Lecture Notes in Mathematics vol. 386, Springer, 1974.

[T2] J. Tits, On buildings and their applications, in: *Proceedings ICM* (Vancouver), 1974, 209–220.

[T3] J. Tits, Ensembles ordonnés, immeubles et sommes amalgamées, *Bull. Soc. Math. Belg.* **38** (1986), 367–387.

[T4] J. Tits, Uniqueness and presentation of Kac–Moody groups over fields, *J. Algebra* **105** (1987), 542–573.

[T5] J. Tits, Groupes associés aux algèbres de Kac–Moody, Bourbaki talk no. 700, Astérisque 177-178, 1989.

[T6] J. Tits, Théorie des groupes, Résumé de cours, annuaire du Collège de France, 1988-89, 81–95, and 1989-90, 87–103.

[T7] J. Tits, Twin buildings and groups of Kac–Moody type, LMS Lecture Notes vol. 165 (1992), 249–286.

[V] E. Vinberg, Discrete linear groups generated by reflections, *Nauk. USSR, Ser. Mat.* **35** (1971), 1083–1119.

[Z] A. Zuk, La propriété (T) de Kazhdan pour les groupes agissant sur les polyèdres, *C. R. Acad. Sci. Paris* **323** (1996), 453–458.

16

On the finite images of infinite groups

D. Segal

Introduction

'There are many well known theorems which assert that one or another kind of infinite group is residually finite. Only recently, however, have results begun to emerge which take residual finiteness as an (explicit or implicit) hypothesis' [S1]. The results referred to concern questions of the type: what can be deduced about a residually finite group if (a) it has relatively few subgroups of finite index, or (b) its finite quotient groups are all relatively small, in some sense? Considerable further progress has been made in the decade since I wrote the above, most of it surveyed in [MS2]. Here I shall discuss two specific problems left open in [MS2], and relations between them.

The first problem concerns *subgroup growth*. We write

$$\mathrm{s}_n(G)$$

to denote the number of subgroups of index at most n in a group G, and say that G has *subgroup growth of type* $\leq f$, for a function f, if

$$\log \mathrm{s}_n(G) = O(\log f(n)).$$

If in fact $\log \mathrm{s}_n(G) = o(\log f(n))$, we say that G has subgroup growth of type *strictly less than* f; and if the type is $\leq f$ but not strictly less than f, then the type is f. (Throughout, logarithms are to base 2.) Thus, for example, the groups with subgroup growth of type $\leq n$ are just the groups of *polynomial subgroup growth*, or *PSG* groups; the finitely generated residually finite groups with this property were characterised in [LMS]: they are precisely the (finitely generated, residually finite) virtually soluble minimax groups. (A soluble group G is *minimax* if there is a finite series $1 = G_0 \lhd G_1 \lhd \cdots \lhd G_h = G$ with each factor G_i/G_{i-1} either cyclic or quasicyclic.) Finitely generated groups with subgroup growth of type

$$f_0(n) := n^{\frac{\log n}{(\log\log n)^2}}$$

exist: they were constructed in [LPS]. Until very recently, no finitely generated group was known having subgroup growth of a type strictly intermediate between n and $f_0(n)$. This led Lubotzky, Pyber and Shalev [LPS] to formulate the following.

Strong Gap Conjecture. *If G is a finitely generated group having subgroup growth of type strictly less than f_0 then G has polynomial subgroup growth.*

If a single counterexample to this were found, the really interesting question would remain of whether, for finitely generated groups, there is *any* gap at all in the possible types of subgroup growth just above PSG; that is, whether the following is true.

Gap Conjecture. *There exists an unbounded increasing function $g : \mathbb{N} \to \mathbb{R}_+$ such that if G is a finitely generated group having subgroup growth of type strictly less than $n^{g(n)}$ then G has polynomial subgroup growth.*

Shortly after the Bielefeld conference, and after writing the first version of this report, I found a family of groups that *refutes* both conjectures. This means that some of the questions raised in my talk have now been answered; most of them are still open, however, and to preserve the balance of the article I shall state all the questions as they arose, giving the answers as far as they are known. The construction of counterexamples to the Gap Conjecture is described in Section 7.

For any group G, the subgroup growth of G is the same as that of its profinite completion $\widehat{G}$ (where by $s_n(\widehat{G})$ one understands the number of *open* subgroups of index at most n in $\widehat{G}$). The question of whether a particular type f of subgroup growth occurs is best approached in two stages: first, what kind (if any) of profinite group Γ can have subgroup growth of type f? And secondly, can such a profinite group Γ be the profinite completion of a finitely generated abstract group G? Of the results stated below, Theorems 1 and 3 belong to the first stage, whereas Theorem 2 and Questions 1 and 2 belong to the second stage.

The Gap Conjecture was for some time known to be *false* for finitely generated *profinite* groups: (easy) soluble counterexamples were constructed in [MS2], and (hard) semisimple ones in [Sh2]. We shall see, however, that groups like those cannot be the profinite completions of finitely generated abstract groups. In Section 6, we discuss the construction of profinite counterexamples to the gap conjecture which *might* be the profinite completions of finitely generated abstract groups; and indeed some of them are, as we shall see in Section 7.

In the opposite direction, we may derive various restrictions on the structure of a profinite group having slow subgroup growth. These amount to the existence of *bounds* for various measures of size for the finite quotients of the group; specifically, on the ranks of abelian sections and on the 'complexity' of the non-abelian chief factors. First of all, we define $\mathrm{r}_p(G)$ to be the maximal rank of any elementary abelian p-section of a finite group G, and for an infinite group G let $\mathrm{ur}_p(G)$ denote the supremum of $\mathrm{r}_p(\overline{G})$ over all finite quotients $\overline{G}$ of G; this is the *upper p-rank* of G. Next, we need to consider non-abelian composition factors of a group. According to the classification of finite simple groups, these will be sporadic simple groups, alternating groups, or groups of Lie type $L(l, p^e)$: here l denotes the Lie rank and p^e the size of the field. I make the following *ad hoc* definition for the 'upper semisimple p-complexity' of G: for a prime p, $\mathrm{ss}_p(G)$ is the supremum of the numbers rle such that some upper chief factor of G is the direct product of r copies of $L(l, p^e)$, for some Lie type L. I also define $\mathrm{ss}_0(G)$ to be the supremum of the numbers rl such that some upper chief factor of G is the direct product of r copies of $\mathrm{Alt}(l)$ or of rl copies of some sporadic group. (An *upper chief factor* of G means a chief factor of some finite quotient of G). The following result, due to Avinoam Mann and stated in [MS2], is proved in Section 1.

Theorem 1. *Let G be a group with subgroup growth of type strictly less than f_0. Then $\mathrm{ss}_p(G)$ is finite for every prime p and for $p = 0$.*

This result is best possible as regards the finiteness of $\mathrm{ss}_0(G)$; for the group with subgroup growth type exactly f_0 constructed in [LPS] has infinitely many alternating groups as upper chief factors. I do not know if it is best possible with regard to the finiteness of the $\mathrm{ss}_p(G)$ for primes p, but this seems likely. Actually, we prove something stronger than Theorem 1: the bounds obtained apply not only to chief factors, but more generally to 'homogeneous upper normal sections' – that is, to sections of the form M/N where $M > N$ are normal subgroups of finite index in G and M/N is a direct product of isomorphic simple groups. It is worth remarking that bounding rle is equivalent to bounding the *rank* of the group $L(l, p^e)^{(r)}$, since the rank of $L(l, p^e)$ can be bounded above and below by simple functions of l and e.

Now a key step in the characterisation of finitely generated groups with PSG was the proof (given in [MS1]) that if a group G has PSG then the numbers $\mathrm{ss}_p(G)$ are not only finite, but *bounded* as p ranges over all primes and zero. This suggests the following.

Question 1. *Does there exist a finitely generated group G such that the numbers $\mathrm{ss}_p(G)$ are finite but unbounded as p ranges over all primes and zero?*

The answer to this question is 'yes'. The finitely generated groups G described in Section 7 have upper chief factors of the form $\mathrm{PSL}_3(p)^{(r_p)}$ for various primes p, where $r_p \to \infty$ with p.

What about groups G for which the numbers $\mathrm{ss}_p(G)$ are uniformly bounded? Let us say that a group G is of *prosoluble type* if every finite quotient of G is soluble – i.e., if $\widehat{G}$ is prosoluble. Combining the argument of [LMS] with Lubotzky's subsequent characterisation of linear groups of slow subgroup growth, one obtains the following (proved in Section 2).

Theorem 2. *Let G be a finitely generated group. If* (i) *the numbers $\mathrm{ss}_p(G)$ are bounded as p ranges over all primes and zero, and* (ii) *G has subgroup growth of type strictly less than $n^{\log n/\log\log n}$, then G is virtually of prosoluble type.*

To say that G is *virtually* of prosoluble type means that G has a normal subgroup of finite index which is of prosoluble type. As the type of subgroup growth is unchanged on passing to a subgroup of finite index, we now restrict attention to groups of prosoluble type. For these we have the following analogue of Theorem 1, proved in Section 3.

Theorem 3. *Let G be a group of prosoluble type. If G has subgroup growth of type strictly less than f_0, then $\mathrm{ur}_p(G)$ is finite for every prime p.*

Now we can ask the following.

Question 2. *Does there exist a finitely generated group G such that the numbers $\mathrm{ur}_p(G)$ are finite but unbounded as p ranges over all primes?*

This is the second of the problems raised in [MS2]. Although not so formulated, it is essentially a question about groups of prosoluble type; for it is shown in [LM2] that every group G with $\mathrm{ur}_2(G)$ finite is virtually of prosoluble type. If the answer to Question 2 is 'yes', then such a group G will again be a candidate for a counterexample to the Strong Gap Conjecture; indeed, in Section 6 we shall see how hypothetical counterexamples to the Gap Conjecture might be obtained, by making the upper p-ranks $\mathrm{ur}_p(G)$ grow very slowly with p. If the answer to Question 2 is 'no', then Theorem 3 implies that the Strong Gap Conjecture is *true* for groups of prosoluble type. Indeed, Theorem B of [MS1] asserts that *every group of finite upper rank has PSG*; the *upper rank* $\mathrm{ur}(G)$ of G is the least integer r such that every subgroup of every finite quotient of G can be generated by r elements (or else ∞). A theorem of Kovács [Ko] shows that $\mathrm{ur}(G) \leq 1 + \sup_p \mathrm{ur}_p(G)$ if G is a group of prosoluble type; so if G is a finitely generated group of prosoluble type and $\mathrm{ur}_p(G)$ is bounded over all primes p then $\mathrm{ur}(G)$ is finite, and hence G has PSG.

In Sections 4 and 5 we consider some special cases of the Gap Conjecture and of Question 2. The first major breakthrough in the study of subgroup growth was the characterisation by Lubotzky and Mann [LM1] of the *residually nilpotent* PSG groups. Using subsequent work of Lubotzky and Shalev, it is not hard to sharpen this as follows.

Theorem 4. *Let G be a finitely generated group which is virtually residually nilpotent. If G has subgroup growth of type strictly less than $n^{\log n/\log\log n}$, then G is virtually a soluble minimax group.*

Such a group G must therefore have PSG, and so the Strong Gap Conjecture is true for residually nilpotent groups. Similarly, Question 2 has a negative answer for such groups.

Theorem 5. *Let G be a finitely generated group which is virtually residually nilpotent. If $\mathrm{ur}_p(G)$ is finite for every prime p then G is virtually a soluble minimax group.*

These results are proved in Section 4. Section 5 deals with soluble groups: Question 2 is reformulated in terms of the module theory of soluble minimax groups, and I report on some partial results obtained in [S3]. These would be superseded if the second part of the following question has a negative answer:

Question 3. (i) *Does there exist a finitely generated soluble group that is residually finite but not virtually residually nilpotent?*
(ii) *Does there exist such a group which is abelian-by-minimax?*

To conclude this introduction, I should mention a striking result obtained by Laci Pyber, since the publication of [MS2]: in the types of subgroup growth for finitely generated groups, there is *no gap* between $n^{\log n}$ and $n!$. The proof develops further the methods introduced in [LPS], and is as yet unpublished. It is shown in [S4] that groups of the kind described in Section 7, below, exhibit a continuous range of subgroup growth types between n and $n^{\log\log n}$. Whether or not there is a genuine gap between types $n^{\log\log n}$ and $n^{\log n}$ remains to be seen.

1. Semisimple chief factors

Theorem 1 follows from the conjunction of the three propositions below. We follow the method of [MS1], Section 4.

Proposition 1.1. *Let G be a group. Suppose that*

$$\log s_n(G) < \frac{1}{49} \log f_0(n) \text{ for all } n \geq k,$$

where $k \in \mathbb{N}$. If $N \cong \mathrm{Alt}(l)^{(r)}$ is an upper normal section of G with $l \geq 5$ then $rl \leq k$.

Proof. Let N be such an upper normal section of G. Then G has a finite quotient $\overline{G}$ such that N is a normal subgroup of $\overline{G}$ and $C_{\overline{G}}(N) = 1$. Let K be the kernel of the permutation action of $\overline{G}$ (given by conjugation) on the set of r simple factors of N. Since $\mathrm{Alt}(l)$ has at most $2l!$ automorphisms we have $|K| \leq (2l!)^r$, while $|\overline{G} : K| \leq r!$. Hence

$$|\overline{G}| \leq r!(2l!)^r < r^r l^{lr} = m, \text{ say.}$$

On the other hand, N contains the direct product of $r[l/4]$ Klein 4-groups. This contains at least $2^{r^2[l/4]^2}$ subgroups; therefore

$$\log s_m(G) \geq r^2[l/4]^2 \geq r^2 l^2/49.$$

Now suppose that $m \geq k$. Then $\log s_m(G) < \log f_0(m)/49$; it follows that

$$r^2 l^2 < \log f_0(m) = \frac{(\log m)^2}{(\log \log m)^2},$$

which implies

$$\log \log m < \frac{r \log r + rl \log l}{rl} \leq \frac{\log r}{5} + \log l.$$

But it is obvious that $\log \log m > \log(lr) = \log l + \log r$, so we have a contradiction. Consequently $m < k$, and the result follows. □

Proposition 1.2. *Let G be a group. Suppose that*

$$\log s_n(G) \leq \frac{1}{5} \log f_0(n) \text{ for all } n \geq k,$$

where $k \in \mathbb{N}$. Then there exists k' such that for every upper normal section N of G, if $N \cong S^{(r)}$ for some sporadic simple group S then $r \leq k'$.

Proof. Let μ be the maximum of $|\mathrm{Aut}(S)|$ as S ranges over the finitely many sporadic groups. Our normal section N contains an elementary abelian 2-group of rank r, hence contains at least $2^{[r/2]^2}$ subgroups. Suppose that $m := (r\mu)^r \geq k$. Arguing as above, we get

$$[\frac{r}{2}]^2 < \log s_m(G) \leq \frac{r^2(\log(r\mu))^2}{5(\log(r \log(r\mu)))^2}.$$

It follows that r is bounded above by a number depending only on μ and k. $\square$

Proposition 1.3. *Let G be a group with subgroup growth of type strictly less than f_0. Let p be a prime. Then there exists $k = k(p)$ such that for every upper normal section N of G, if $N \cong S^{(r)}$ where S is a simple group of Lie type $L(l, p^e)$ then $rle \le k$.*

Proof. Put $q = p^e$. Suppose that S is of classical type, with $l > 8$. Then S has a section isomorphic to $\mathrm{PSL}([(l+1)/2], q)$ ([KL], §4.1), hence has an elementary abelian p-section of rank $[l/4]^2 e$. Also $|\mathrm{Aut}(S)| < q^{2(2l+1)^2}$. In all other cases, we have $l \le 8$, $|\mathrm{Aut}(S)| < q^{2\times 120^2}$, and S contains an elementary abelian p-subgroup of rank e. Let us deal with the classical case where $l > 8$; the other case is similar and slightly simpler. Put

$$m = r^r p^{2(2l+1)^2 er}.$$

As above, we find that

$$s_m(G) > p^{[[l/4]^2 er/2]^2} \ge p^{l^4 e^2 r^2/10000}.$$

Now put $\varepsilon = (10000 \log p)^{-1}$ and let k be so large that $\log s_n(G) < \varepsilon \log f_0(n)$ for all $n \ge k$. Suppose that $m \ge k$. Then

$$l^4 e^2 r^2 \frac{\log p}{100} < \log s_m(G) < \varepsilon \frac{(\log m)^2}{(\log\log m)^2}$$

$$< \varepsilon \frac{r^2(\log r + 10 l^2 e \log p)^2}{(\log r + \log\log(r p^{2(2l+1)^2 e}))^2}.$$

This rearranges to

$$\log r + \log\log(r p^{2(2l+1)^2 e}) < \frac{\log r}{10 l^2 e \log p} + 1;$$

since $2(2l+1)^2 \ge 722$ this implies $\log r + 3 < (\log r)/10 + 1$, a contradiction. It follows that $m < k$, and hence that $rle < k$. $\square$

We conclude this section by justifying a remark in the introduction. The rank $\mathrm{rk}(N)$ of a finite group N is the least d such that every subgroup of N can be generated by d elements.

Proposition 1.4. *Let $N \cong S^{(r)}$ where S is a finite simple group.*
(i) *If S is a sporadic group then*

$$r r_1 \le \mathrm{rk}(N) \le r r_2$$

where r_1, r_2 are respectively the minimal, maximal ranks of sporadic simple groups.

(ii) *If* $S = \mathrm{Alt}(l)$ *then*

$$2r[l/4] \leq \mathrm{rk}(N) \leq rl.$$

(iii) *If* $S = L(l, p^e)$ *then*

$$r[l/4]^2 e \leq \mathrm{rk}(N) \leq re \cdot n(l)^2/2$$

where $n(l) \leq \max\{248, 2l^2\}$ *is the degree of the natural linear representation of* $L(l, p^e)$ *over* $\mathbb{F}_{p^e}$.

Proof. (i) is evident. The lower bounds in (ii) and (iii) were established in the course of proving the preceding propositions. The upper bounds follow from corresponding bounds on the ranks of $\mathrm{Alt}(l)$ and $\mathrm{GL}_n(\mathbb{F}_p)$; see for example [P], Theorem 1.1 and the following discussion. □

2. Groups of bounded complexity

Here we prove Theorem 2. For a group G, let $D(G)$ denote the intersection of the centralisers of all non-abelian upper chief factors of G. The following is a simple observation:

Lemma 2.1. *If* $G/D(G)$ *is virtually soluble then* G *is virtually of prosoluble type.*

Suppose now that the numbers $\mathrm{ss}_p(G)$ are bounded as p ranges over all primes and zero. Then the non-abelian upper chief factors of G that are not products of groups of Lie type have bounded order; so as N ranges over these, the groups $G/\mathrm{C}_G(N)$ also have bounded order, hence may be considered as linear groups of bounded degree. On the other hand, Lemma 4.4 of [MS1] shows that there exists m such that $\mathrm{Aut}(S)$ is a linear group of degree at most m whenever S is a simple factor of Lie type of some upper chief factor of G. If r is an upper bound for the multiplicity of simple factors in any such chief factor N, it follows that $G/\mathrm{C}_G(N)$ is linear of degree at most $r! \cdot rm$. Thus $G/D(G)$ is a sub-Cartesian product of linear groups of bounded degree. Hence if G is also finitely generated, then $G/D(G)$ is a subdirect product of *finitely many* linear groups over fields: this is an application of the Lemma in [LMS], §3. Assume finally that G has subgroup growth of type strictly less than $n^{\log n/\log\log n}$. Then Lubotzky's theorem ([Lu], §5.1) implies that each linear image of G is virtually soluble; consequently $G/D(G)$ is virtually soluble, and Theorem 2 follows by Lemma 2.1.

The structure of *profinite* groups with bounded complexity, in the above sense, is quite straightforward. Suppose G is a profinite group such that the numbers $\mathrm{ss}_p(G)$ are bounded as p ranges over all primes and zero. The proof of Theorem 1.2 in [Sh2] shows that G has closed normal subgroups $R \leq G_0$ such that R is prosoluble, G/G_0 is finite, and G_0/R is a Cartesian product of finite simple groups of Lie type. In [SS2] this description is exploited in order to give a complete characterisation of the profinite groups with PSG: the group G as above has PSG if and only if (i) the prosoluble subgroup R has finite rank and (ii) the orders of the simple factors in G_0/R satisfy a certain arithmetic condition. It would be interesting to explore the arithmetic conditions corresponding to other types of subgroup growth in profinite groups of this kind.

3. Subgroup growth and upper rank

Theorem 3 depends on the following lemma, which is based on the proof of [MS1], Theorem 3.9. We write $\lambda(s)$ to denote the least integer $\geq \log s$ (recall that logarithms are to base 2).

Lemma 3.1. *Let G be a finite soluble group and let p be a prime such that $\mathrm{r}_p(G) \geq 2$. Then there exists $s \in \mathbb{N}$ such that*

$$\mathrm{r}_p(G) \leq s(5 + \lambda(s)) \quad \textit{and} \quad \mathrm{s}_n(G) > p^{s^2/5},$$

where $n = p^{s(5+\lambda(s))}$.

Proof. Assume without loss of generality that $O_{p'}(G) = 1$, and put $N = O_p(G)$. Let $M = \Phi^i(N)$, where i is chosen so as to maximise $\mathrm{d}(M) = s$, say; here, $\Phi^0(N) = N$ and for $j \geq 1$, $\Phi^j(N)$ is the Frattini subgroup of $\Phi^{j-1}(N)$. It follows from [DDMS] Chapter 2, Ex. 6 that $|N : M| \leq p^{s(1+\lambda(s))}$, and from *loc.cit.* Ex. 7 that $\mathrm{rk}(N) \leq s(2 + \lambda(s))$. Also G/N is isomorphic to a completely reducible subgroup of $\mathrm{GL}_{\mathrm{d}(N)}(\mathbb{F}_p)$, so we have $|G : N| \leq p^{3\mathrm{d}(N)} \leq p^{3s}$ [W]. It is easy to see that $\mathrm{r}_p(G) \geq 2$ implies $s \geq 2$. Now put $m = s(5 + \lambda(s))$. The index of $\Phi(M)$ in G is at most p^m, and the $\mathbb{F}_p$-vector space $M/\Phi(M)$ contains more than $p^{s^2/5}$ distinct subspaces, so we have $\mathrm{s}_n(G) > p^{s^2/5}$. On the other hand,

$$\mathrm{r}_p(G) \leq \mathrm{rk}(N) + \log_p |G : N| \leq s(2 + \lambda(s)) + 3s = m. \qquad \square$$

Proof of Theorem 3. Let G be a group with subgroup growth of type strictly less than $f_0(n) = n^{\log n/(\log\log n)^2}$. We show that for each prime p, there is a finite bound for the numbers $\mathrm{r}_p(\overline{G})$ as $\overline{G}$ ranges over the finite soluble quotients of

G. If these account for all the finite quotients of G, then of course this means that $\mathrm{ur}_p(G)$ is finite. Having fixed p, put

$$c = \frac{1}{20 \log p}.$$

According to the hypothesis, there exists $n_0 \in \mathbb{N}$ such that

$$s_n(G) \le n^{c \log n/(\log\log n)^2}$$

for all $n \ge n_0$. I claim that

$$\mathrm{r}_p(\overline{G}) \le \max\{703, \ \log n_0\}$$

for every finite soluble quotient $\overline{G}$ of G. To see this, let $\overline{G}$ be such a quotient, and suppose that $\mathrm{r}_p(\overline{G}) \ge 704$. Let s be the integer given in Lemma 3.1 (with $\overline{G}$ in place of G), and put $m = s(5 + \lambda(s))$. Then $\mathrm{r}_p(\overline{G}) \le m$. Suppose to begin with that $p^m > n_0$. Lemma 3.1 gives

$$\begin{aligned}\frac{s^2}{5} \log p < \log s_{p^m}(G) &\le c(\log p^m)^2/(\log\log p^m)^2 \\ &\le cm^2(\log p)^2/(\log m)^2 \\ &= \frac{m^2 \log p}{20(\log m)^2},\end{aligned}$$

since $p \ge 2$ and $c \log p = 1/20$. Now $5 + \lambda(s) < 2 \log s$, since $m \ge 704$ implies $s \ge 2^6$; so $m < 2s \log s$, and we deduce that

$$s^2(\log m)^2 < \frac{5}{20}(2s \log s)^2 = s^2(\log s)^2,$$

a contradiction since $m > s$. It follows that $p^m \le n_0$, whence

$$\mathrm{r}_p(\overline{G}) \le m \le \log_p n_0 \le \log n_0$$

as claimed. □

4. Residually nilpotent groups

We begin with the

Proof of Theorem 4. Let G be a finitely generated group which is virtually residually nilpotent, having subgroup growth of type strictly less than $n^{\log n/\log\log n}$. We show that G is virtually a soluble minimax group. Replacing G by a suitable subgroup of finite index, we may assume that G is residually nilpotent. Now a theorem of Shalev [Sh1] shows that every pro-p group with subgroup growth of type strictly less than $n^{\log n}$ has finite rank; it follows that, for each prime p, the pro-p completion $\widehat{G}_p$ of G has finite rank, r_p say. Now suppose N is a normal subgroup of G such that G/N is torsion-free and nilpotent, of Hirsch length $\mathrm{h}(G/N) = h$. For each prime p we have

$$h = \dim(\widehat{(G/N)}_p) \leq \mathrm{rk}(\widehat{(G/N)}_p) \leq r_p$$

(see [DDMS], Chapters 3 and 4 for a discussion of the rank and dimension of pro-p groups). Thus r_2, say, is an upper bound for $\mathrm{h}(G/N)$ as G/N ranges over all torsion-free nilpotent quotients of G. We may therefore choose N so that $h = \mathrm{h}(G/N)$ is maximal. For each i put

$$N_i = [N, G, \dots, G] \quad (i \text{ repetitions of } G).$$

Then N/N_i is exactly the torsion subgroup of the finitely generated nilpotent group G/N_i, and N_i/N_{i+1} is an image of $N_{i-1}/N_i \otimes_{\mathbb{Z}} G/G'$. It follows that $\pi(N/N_i) = \pi(N/N_1)$ for all $i \geq 1$, where $\pi(N/N_1) = \pi$, say, is a finite set of primes. Putting

$$s = h + \max_{p \in \pi} r_p,$$

we see that $\mathrm{ur}(G/N_i) \leq s$ for each $i \geq 1$. For each i, the group G/N_i is residually finite, and since G is residually nilpotent we have $\bigcap_{i=1}^{\infty} N_i = 1$. It follows that G is itself residually (finite nilpotent of rank $\leq s$). According to [S2], this implies that G has a nilpotent normal subgroup Q such that G/Q is a subdirect product of finitely many linear groups. Now Lubotzky [Lu], §5.1, has proved that a finitely generated linear group with subgroup growth of type strictly less than $n^{\log n/\log\log n}$ must have PSG; hence G/Q has PSG, and is therefore virtually soluble [LMS]. Thus G is virtually soluble, and the second part of the main result of [S2] now shows that G is virtually nilpotent-by-abelian. An application of Theorem 1.1 of [SS1] now completes the proof (see the Introduction of [SS1]). □

If, in Theorem 4, G is assumed to be soluble, then it is enough to suppose that the subgroup growth is of type strictly less than $n^{\log n}$: the stronger hypothesis was only needed in connection with Lubotzky's theorem on linear groups, and this was used only to show that G is virtually soluble.

Proof of Theorem 5. The group G is as above, except that the hypothesis on subgroup growth is replaced by the hypothesis that $\mathrm{ur}_p(G)$ is finite for every prime p. Since $\mathrm{ur}_2(G)$ is finite, we know from [LM2] that G has a subgroup of finite index that is of prosoluble type; so without loss of generality let us assume that

G is residually nilpotent and that all finite quotients of G are soluble. Arguing as above, we see that G is residually (finite nilpotent of rank $\leq s$) for some finite s. The main result of [S2] now shows that G is virtually nilpotent-by-abelian. Now Lemma 2.2 of [MS1] asserts that a finitely generated nilpotent-by-abelian group of finite upper rank is a minimax group; however, the proof only requires the finiteness of the upper p-ranks, for certain primes p, so we may conclude that G is virtually a minimax group. □

5. Soluble groups

There are some grounds for believing that Question 2 has a negative answer for soluble groups; in [S3] I state this as

Conjecture A. *Let G be a finitely generated soluble group. If* $\mathrm{ur}_p(G)$ *is finite for every prime p then G has finite upper rank.*

(If also G is residually finite, then it follows that G is a minimax group, by Theorem A of [MS1].) This would generalise a theorem of D. J. S. Robinson [R2], which says that a finitely generated soluble group with finite sectional p-rank for every prime p (an $\mathfrak{S}_0$ group) must have finite rank. As we saw in the Introduction, the truth of Conjecture A would imply that of the Strong Gap Conjecture for soluble groups. In fact we can do a lot better.

Proposition 5.1. *Let G be a finitely generated residually finite soluble group with subgroup growth of type strictly less than $n^{\log n}$. If Conjecture A is true, then G is a minimax group.*

Proof. By Theorem A of [MS1], it suffices to show that G has finite upper rank. Arguing by induction on the derived length of G (see below) we reduce to the case where G has an abelian normal subgroup A such that G/A is minimax and residually finite. Suppose that G has infinite upper rank; then Conjecture A says that $\mathrm{ur}_p(G)$ is infinite for some prime p. On the other hand, $\mathrm{ur}_p(G/A) = r$, say, is finite. We shall obtain a contradiction by deriving an upper bound for $\mathrm{ur}_p(G)$. Let q be a large positive integer, and suppose that $\mathrm{ur}_p(G)$ exceeds $q + r$. Then G has a finite quotient $\overline{G}$ such that $\mathrm{r}_p(\overline{G}) \geq q + r$, and we can choose it so that $O_{p'}(\overline{G}) = 1$. Put $N = \mathrm{O}_p(\overline{G})$; note that $N \geq \overline{A}$, the image of A, and that $\mathrm{r}_p(\overline{A}) \geq q$. Now let $M/\overline{A}$ be a normal subgroup of $N/\overline{A}$ such that $\mathrm{d}(M) = s$, say, is as large as possible. Then $s \geq q$ and $s \geq \mathrm{d}(N)$; on the other hand, $\mathrm{rk}(N/M) \leq \mathrm{ur}_p(G/A) = r$. By Lemma 3.18(i) of [DDMS], $M/\Phi(M)$ is self-centralising in $N/\Phi(M)$, and it follows (as in the proof of [*loc.cit.* Lemma

3.18(ii)]) that $|N/M| \leq p^{r\lambda(s)}$. Arguing as in the proof of Lemma 1.1, we now infer that $s_n(\overline{G}) > p^{s^2/5}$, where $n = p^{4s+r\lambda(s)}$. Now there exists n_0 such that

$$\log s_m(G) < (85 \log p)^{-1} (\log m)^2$$

for every $m \geq n_0$. If $n \geq n_0$ this gives

$$(s^2/5) \log p < (85 \log p)^{-1} ((4s + r\lambda(s)) \log p)^2 ,$$

which simplifies to

$$s^2 < 8rs\lambda(s) + r^2\lambda(s)^2$$

and so implies that s is bounded above by some number $f(r)$ depending only on r. It follows that either $n < n_0$ or $s \leq f(r)$; thus we get an upper bound for q, and hence one for $\mathrm{ur}_p(G)$. □

I suspect that the gap in subgroup growth types of finitely generated soluble groups is even larger than that suggested by the above. Examples are constructed in [SS1] of such groups with subgroup growth of type $2^{n^{1/d}}$, for any positive integer d; but the following is open.

Question 4. Does there exist a finitely generated soluble group of infinite rank with subgroup growth of type $\leq 2^{n^{\varepsilon}}$ for every positive ε?

In the paper [S3], Conjecture A is recast as a conjecture about modules over minimax groups, in the following manner. As both hypothesis and conclusion depend only on the finite quotients of G, we may as well assume that G in the conjecture is residually finite. Suppose now that G is a finitely generated, residually finite soluble group of derived length $\ell > 1$, and that $\mathrm{ur}_p(G)$ is finite for every prime p. Let A be a maximal abelian normal subgroup of G containing the last non-trivial term of the derived series of G. Then G/A is again residually finite, by an elementary lemma, and the derived length of G/A is $\ell - 1$. Thus if Conjecture A is true for groups of derived length less than ℓ, then G/A is a minimax group. The conjugation action of G/A on A makes A into a module for the group ring $\mathbb{Z}(G/A) = R$, say. We define

$$\mathrm{ur}_p(A_R) = \sup\{\dim_{\mathbb{F}_p}(\overline{A})\}$$

where $\overline{A}$ runs over all finite R-module images of A/Ap, and

$$\mathrm{ur}(A_R) = \sup_p \mathrm{ur}_p(A_R).$$

It is easy to see that $\mathrm{ur}(G) \leq \mathrm{ur}(G/A) + \mathrm{ur}(A_R)$; on the other hand, if B is an R-submodule of finite index in A then G/B is residually finite, because the class of residually finite minimax groups is closed under extensions [R1, §9.3], and this implies that $\mathrm{ur}_p(A_R) \leq \mathrm{ur}_p(G)$ for each prime p. To establish Conjecture A, then, it would suffice to show that if $\mathrm{ur}_p(A_R)$ is finite for every prime p, then $\mathrm{ur}_p(A_R)$ is uniformly bounded over all p. Thus the conjecture is reduced to a problem about modules for the group ring of a finitely generated minimax group.

Let Γ be a minimax group, and M a $\mathbb{Z}\Gamma$-module. I shall write $\mathrm{ur}_p(M)$ for $\mathrm{ur}_p(M_{\mathbb{Z}\Gamma})$. In general, it is perfectly possible for the upper p-ranks $\mathrm{ur}_p(M)$ to be finite but unbounded as p ranges over all primes: for example M could be the direct sum, over all p, of vector spaces $(\mathbb{F}_p)^p$, with trivial Γ-action. However, this example cannot occur as A_R, above; for A_R satisfies an additional finiteness condition. Let us say that the $\mathbb{Z}\Gamma$-module M is *quasi-finitely generated* if there exists a finitely generated group which is an extension of M by Γ. We can now reformulate our conjecture, as follows.

Conjecture B. *Let Γ be a finitely generated minimax group, and M a quasi-finitely generated $\mathbb{Z}\Gamma$-module. If $\mathrm{ur}_p(M)$ is finite for every prime p then M has finite upper rank.*

Using a considerable amount of heavy machinery, I was able to establish the following special case.

Theorem ([S3, Theorem 3.1]). *Let Γ be a minimax group which is abelian-by-polycyclic, and let M be a finitely generated $\mathbb{Z}\Gamma$-module. If $\mathrm{ur}_p(M)$ is finite for every prime p then M has finite upper rank.*

There is some hope that further work along similar lines may lead to a full proof of Conjecture B, thereby confirming the strong Gap Conjecture for soluble groups.

6. Profinite constructions

The proof of Proposition 3.3 in [MS1] yields the following.

Lemma 6.1. *Let G be a finite soluble group. Then for each integer n,*

$$\mathrm{s}_n(G) \leq n^{r+2},$$

where $r = \max\{\mathrm{r}_p(G) \mid p \leq n\}$.

This may be seen as a sort of converse to Lemma 3.1: together they show that for a group of prosoluble type, slow subgroup growth is more or less equivalent to slow growth of the upper p-ranks as $p \to \infty$. In [MS2], Theorem 1.5, it is used to construct prosoluble groups of arbitrarily slow non-polynomial subgroup growth. However, these groups are metabelian, and so cannot be the profinite completions of finitely generated abstract groups: for it is shown in [SS1] that a finitely generated metabelian group, if its subgroup growth is not polynomial, must have subgroup growth of type at least $2^{n^{1/d}}$ for some positive integer d. (So the Strong Gap Conjecture is true for metabelian groups.)

As indicated in the preceding section, I doubt that there exist soluble counterexamples to the Gap Conjecture. As a first step towards finding counterexamples of prosoluble type, we can use Lemma 6.1 to construct some different prosoluble groups with slow subgroup growth, ones that at least look as if they might be the profinite completions of finitely generated abstract groups. Let $g : \mathbb{N} \to \mathbb{R}_+$ be an unbounded strictly increasing function. We recursively construct a sequence of pairs (M_i, G_i) as follows, with each G_i a finite soluble group and M_i a minimal normal subgroup of G_i. Begin with an arbitrary finite soluble group $G_0 \neq 1$ and minimal normal subgroup M_0 of G_0. Now let $i \geq 1$ and suppose we have defined G_{i-1} and M_{i-1}; put $m_{i-1} = |G_{i-1}|$, let p_i be a prime with

$$p_i > \max\{m_{i-1},\ g^{-1}(m_{i-1})\},$$

and let M_i be a simple $\mathbb{F}_{p_i} G_{i-1}$-module that is non-trivial for M_{i-1}. Then put $G_i = M_i \rtimes G_{i-1}$. Note that

$$m_i = p_i^{d_i} m_{i-1}$$

where $d_i = \dim_{\mathbb{F}_{p_i}} M_i$.

Proposition 6.1. *Let G be the inverse limit of the sequence (G_i), with the obvious maps.*

(i) $\mathrm{d}(G) = \max\{2,\ \mathrm{d}(G_0)\}$;

(ii) *let p be a prime. Then*

$$\mathrm{ur}_p(G) = \begin{cases} \mathrm{r}_p(G_0) & \text{if } p \mid m_0 \\ d_i \leq g(p_i) & \text{if } p = p_i \ \text{ for } i \geq 1; \\ 0 & \text{otherwise} \end{cases}$$

(iii) *for each positive integer n,*

$$\mathrm{s}_n(G) \leq n^{c+g(n)}$$

where $c = 2 + \mathrm{rk}(G_0)$;

(iv) $\mathrm{ur}_p(G)$ *is unbounded as p ranges over all primes;*

(v) *G does not have polynomial subgroup growth.*

Proof. (i) Let $i \geq 1$ and suppose that $x_1, \ldots, x_k$ generate G_{i-1}, where $k \geq 2$. I claim that then G_i is generated by a set of the form $\{a_1x_1, \ldots, a_kx_k\}$ with

$a_1, \ldots, a_k \in M_i$. Indeed, if each such set generates a proper subgroup, then that subgroup is a complement to M_i in G_i; but the number of such complements is just $|M_i|$, by the Schur-Zassenhaus theorem, whereas the number of distinct k-tuples of the given form is $|M_i|^k$, and each k-tuple gives rise to a distinct complement. This is a contradiction since $k \geq 2$. It follows by induction that $\mathrm{d}(G_i) = \max\{2, \mathrm{d}(G_0)\}$ for each $i \geq 1$, and (i) follows.

(ii) Here by $\mathrm{ur}_p(G)$ I mean the maximal p-rank of any quotient of G by an open normal subgroup, which is the same as the maximal p-rank of the groups G_i. So all the statements are obvious except for the claim that $d_i \leq g(p_i)$. However, since g is monotonic this follows from

$$p_i > g^{-1}(m_{i-1}) \geq g^{-1}(d_i),$$

where the second inequality holds because M_i is a 1-generator $\mathbb{F}_{p_i} G_{i-1}$-module.

(iii) This follows from (ii) by Lemma 6.1.

(iv) It suffices to show that d_i is unbounded. Suppose we had $d_i \leq d < \infty$ for all i. Then according to Mal'cev's theorem on soluble linear groups there exists a natural number q such that for each $i \geq 1$, the group $(G_{i-1}^q)'$ acts unipotently on the module M_i. Now choose i large enough to ensure that p_{i-2} and p_{i-1} do not divide q. Then $M_{i-2}M_{i-1} \leq G_{i-1}^q$, so $M_{i-1} = [M_{i-1}, M_{i-2}] \leq (G_{i-1}^q)'$; but M_{i-1} does not act unipotently on M_i since the orders are coprime and the action is non-trivial.

(v) now follows from (iv) and [MS1], Theorem 3.9. □

Question 5. Is there a prosoluble group G of the above form having a dense finitely generated subgroup Γ such that $\widehat{\Gamma} \cong G$, or at least such that the kernel of the natural epimorphism $\widehat{\Gamma} \to G$ is a profinite group of finite rank?

If the answer is 'yes', then we get a positive answer to Question 2; and by choosing the function g to grow arbitarily slowly we obtain counterexamples to the gap conjecture. Groups like G occur as automorphism groups of spherically homogeneous rooted trees, as described in [G]; this may be a good setting in which to seek suitable candidates for Γ.

Next we consider profinite groups of 'semisimple' type. As an analogue to Lemma 6.1 we have the following.

Lemma 6.2. *Let $q = p^e$ where p is a prime, and let $d \geq 2$. Let G be a subgroup of $\mathrm{GL}_d(q)$, of even order g. Then*

$$\mathrm{s}_g(G) \leq g^{4ed^2(3+2\log d)+1} < q^{12ed^5}.$$

Proof. This follows from the first (easier) part of the proof of Lemma 5.1 in [SS2], together with Corollary 3.12 of [MW], which implies that the Fitting length of any soluble subgroup of G is at most $3 + 2\log d$. □

Now let $(q_i = p_i^{e_i})$ be a sequence of prime powers and let (d_i) be a sequence of integers ≥ 2; assume that at least one of these sequences is strictly increasing, and that $(q_i, d_i) \notin \{(2,2), (2,3)\}$ for all i. Put $T_i = \mathrm{PSL}_{d_i}(q_i)$ for each i, and for any subset J of $\mathbb{N}$ let

$$G(J) = \prod_{j \in J} T_j;$$

this is a 2-generator profinite group. Define the function $\eta : \mathbb{N} \to \mathbb{N}$ by setting $\eta(n)$ equal to the largest index j such that $(q_j^{d_j} - 1)/(q_j - 1) \leq n$ if such exists, $\eta(n) = 1$ otherwise, and write

$$h(n) = 1 + 24e_{\eta(n)}d_{\eta(n)}^5.$$

Proposition 6.2. *Let $f : \mathbb{N} \to \mathbb{R}_+$ be a strictly increasing unbounded function. Then there exists an infinite subset J of $\mathbb{N}$ such that the group $G(J)$ satisfies*

$$\mathrm{s}_n(G(J)) \leq n^{f(n)+h(n)} \quad \textit{for all large } n.$$

Proof. Note first that if $j > \eta(n)$ then T_j has no proper subgroup of index $\leq n$ ([KL], Theorem 5.2.2). It follows that $\mathrm{s}_n(G) = \mathrm{s}_n(T_1 \times \cdots \times T_{\eta(n)})$. The result now follows from Lemma 5.2 of [LPS], together with Lemma 6.2 above and the fact that $q_{\eta(n)} < n$. □

This construction is used in [LPS] to produce a group with subgroup growth type exactly $n^{\log n}$. In that case the sequence (q_i) is taken to be constant, and the result depends on Theorem 1.4 of [LPS], which asserts that $a_n(T) \leq n^{c \log n}$ for all n and every finite simple group T (c being an absolute constant); this lies considerably deeper than our Lemma 6.2, but is of no help if we want to construct groups with subgroup growth of type strictly less than $n^{\log n}$. On the other hand, Proposition 4.2 of [LPS] shows that if $q_i = q$ is constant, then the profinite group $G(J)$ contains a dense finitely generated subgroup Γ such that $\widehat{\Gamma} \cong \widehat{\mathbb{Z}} \times G(J)$, and then Γ has subgroup growth of the same type as that of $G(J)$. If the q_i are increasing as well as the d_i such a subgroup Γ may be hard to find in $G(J)$; I give below another construction, analogous to that of Proposition 6.1, that is more amenable.

Given a strictly, but very slowly, increasing unbounded function $f : \mathbb{N} \to \mathbb{R}_+$, let us call the sequence of pairs (q_i, d_i) *suitable* if (i) the corresponding function h defined above satisfies $h = O(f)$, and (ii) at least one of the sequences (e_i), (d_i) tends to ∞. Theorem 4.1 of [MS1] shows that if (ii) holds then for any infinite subset J of $\mathbb{N}$, the profinite group $G(J)$ does *not* have polynomial subgroup growth. With Proposition 6.2 this now gives

Corollary 6.1. *If the sequence (q_i, d_i) is suitable, then the group $G(J)$ has subgroup growth of type $\leq n^f$, but does not have PSG.*

There is no difficulty in constructing suitable sequences. We may for example choose the sequences (e_i) and (d_i) at will, subject to condition (ii), then for each i choose p_i so large that

$$f(p_i^{e_i}) \geq e_i d_i^5.$$

Any such choice then gives rise, according to Corollary 6.1, to a semisimple profinite counterexample to the Gap Conjecture.

Question 6. Given a function f as above, is there a suitable sequence (q_i, d_i) such that the corresponding profinite group $G(J)$ provided by Proposition 6.2 contains a dense finitely generated subgroup Γ such that $\widehat{\Gamma} \cong \Delta \times G(J)$ where Δ is a profinite group of finite rank?

For the second construction, we suppose that each of the simple groups $T_i = \mathrm{PSL}_{d_i}(q_i)$ comes with a faithful permutation representation of degree l_i, and for each k let W_k be the permutational wreath product

$$W_k = T_k \wr T_{k-1} \wr \cdots \wr T_1.$$

Let W be the inverse limit of the groups W_k as $k \to \infty$, relative to the natural epimorphisms

$$W_{k+1} = T_{k+1} \wr W_k \to W_k.$$

Now put $m_k = l_1 \dots l_{k-1}$, and define the function h^* by

$$h^*(n) = 12 e_k d_k^5 m_k^5 \text{ where } k = \eta(n).$$

Lemma 6.3. *Let n be a large natural number and put $k = \eta(n)$. Then*

$$\mathrm{s}_n(W) \leq q_k^{h^*(n)}.$$

Proof. As above, we see that $\mathrm{s}_n(W_j) = \mathrm{s}_n(W_k)$ whenever $j \geq k$. Now W_k is a quotient of $\mathrm{SL}_{d_k}(q_k) \wr W_{k-1}$, which sits naturally inside $\mathrm{GL}_{d_k m_k}(q_k)$. The result therefore follows from Lemma 6.2. □

Now suppose we are given a function $f : \mathbb{N} \to \mathbb{R}_+$ as above. Fix integers $d \geq 2$ and $e \geq 1$, and let (p_i) be a strictly increasing sequence of odd primes such that for each $k > 1$,

$$f(p_k^e) > 12 e d^5 m_k^5,$$

where $m_k = l_1 \dots l_{k-1}$ and $l_i = (p_i^{ed} - 1)/(p_i^e - 1)$ for each i. Then $T_i = \mathrm{PSL}(d, p_i^e)$ has a natural doubly transitive permutation representation on the points of the corresponding projective space over $\mathbb{F}_{p_i^e}$, which has l_i points. Form W as above. I claim that $s_n(W) < n^{f(n)}$ for all $n \geq l_1$; indeed, taking $k = \eta(n)$ we have $q_k = p_k^e < l_k \leq n$ and $h^*(n) = 12ed^5m_k^5 < f(p_k^e) < f(n)$, so the claim follows from Lemma 6.3.

On the other hand, W does not have PSG: for W_k contains an elementary abelian p_k-subgroup of rank $m_k e$, hence contains at least $p_k^{em_k^2/5}$ subgroups, of index less than

$$|W_k| < \prod_{i=1}^{k} p_i^{ed^2m_i} < p_k^{ed^2km_k};$$

so if $s_n(W) \leq n^\alpha$ for all n then $em_k^2/5 \leq ed^2km_k\alpha$, which is false for large k since $m_k \geq p_1^{k-1}$. We have established the following.

Proposition 6.3. *Let $f : \mathbb{N} \to \mathbb{R}_+$ be a strictly increasing unbounded function. Let $d \geq 2$ and $e \geq 1$. Then provided the sequence of primes (p_n) grows sufficiently fast, the profinite group*

$$W = \varprojlim_{n\to\infty} \left(\mathrm{PSL}(d, p_n^e) \wr \mathrm{PSL}(d, p_{n-1}^e) \wr \cdots \wr \mathrm{PSL}(d, p_1^e)\right)$$

has subgroup growth of type $\leq n^f$, but does not have PSG.

The analogue of Question 6 for groups of this nature is answered in the final section.

7. Branch groups with the congruence subgroup property

For each $i \geq 1$ let T_i be a primitive permutation group of degree $l_i \geq 4$. For each k let W_k be the permutational wreath product

$$W_k = T_k \wr T_{k-1} \wr \cdots \wr T_1,$$

and let W be the inverse limit of the groups W_k as $k \to \infty$, relative to the natural epimorphisms

$$W_{k+1} = T_{k+1} \wr W_k \to W_k.$$

Thus W is a profinite group, and a base for the neighbourhoods of 1 in W is provided by the subgroups

$$K_k = \ker(W \to W_k).$$

A dense subgroup G of W is said to have the *congruence subgroup property* if every subgroup of finite index in G contains $G \cap K_k$ for some k. In this case, every finite quotient of G is an image of some W_k and $\widehat{G} \cong W$. In particular, it follows that $s_n(G) = s_n(W)$ for all n.

Theorem 7.1 ([S4]). *Suppose that there exists an r-generator perfect group that has each of the groups T_n as a homomorphic image, and that each T_n is simple. Then W contains a dense $(r+2)$-generator subgroup G that has the congruence subgroup property.*

The profinite group W acts in a natural way on the *spherically homogeneous rooted tree* T of type $(l_1, l_2, \dots)$. In his article [G], Grigorchuk describes the construction of certain finitely generated subgroups of $\mathrm{Aut}(T)$ that he calls *branch groups*; I refer to that article for the definitions of *rooted automorphism* and *directed automorphism*, and the requisite notation. The group G in Theorem 7.1 is generated by the two rooted automorphisms corresponding to the permutations α, β, where $T_1 = \langle \alpha, \beta \rangle$, together with r directed automorphisms $b(1), \dots, b(r)$ of the form

$$b(i) = (\alpha(i)_2, 1, \dots, 1, b(i)_2),$$

where for each n

$$b(i)_n = (\alpha(i)_{n+1}, 1, \dots, 1, b(i)_{n+1}),$$

and $\{\alpha(1)_n, \dots, \alpha(r)_n\}$ is a generating set for T_n that is the image of a generating set for the given perfect group. For details of the proof, see [S4]. Now consider the groups

$$T_n = \mathrm{PSL}(3, p_n),$$

where (p_n) is any sequence of primes. Each of these is generated by the images of the six elementary matrices e_{ij} $(i \neq j \in \{1, 2, 3\})$, which satisfy the relations

$$[e_{ij}, e_{jk}] = e_{ik}.$$

Hence each T_n is an image of the 4-generator perfect group

$$\left\langle X_{ij},\quad i \neq j \in \{1, 2, 3\};\ [X_{ij}, X_{jk}] = X_{ik} \right\rangle.$$

The hypotheses of Theorem 7.1 are therefore satisfied with $r = 4$; and if the sequence (p_n) is chosen to grow fast enough then so are the hypotheses of Proposition 6.3. It follows that the 6-generator group G can be made to have arbitrarily slow non-polynomial subgroup growth. This shows that the Gap Conjecture is false.

References

[DDMS] J. D. Dixon, M. P. F. du Sautoy, A. Mann and D. Segal, *Analytic pro-p Groups, 2nd edition*, Cambridge Studies in Advanced Maths. 61, Cambridge University Press, 1999.

[G] R. I. Grigorchuk, Just infinite branch groups, in: *New horizons in pro-p groups* (du Sautoy, Segal & Shalev eds), Progress in Math. vol. 184, Birkhäuser, Boston, 2000, 121–179.

[KL] P. Kleidman and M. Liebeck, *The subgroup structure of finite classical groups*, LMS Lecture Notes vol. 129, Cambridge University Press, 1990.

[Ko] L.G. Kovács, On finite soluble groups, *Math. Zeit.* **103** (1968), 37–39.

[Lu] A. Lubotzky, Subgroup growth and congruence subgroups, *Invent. Math.* **119** (1995), 267–295.

[LMS] A. Lubotzky, A. Mann and D. Segal, Finitely generated groups of polynomial subgroup growth, *Israel J. Math.* **82** (1993), 363–371.

[LM1] A. Lubotzky and A. Mann, On groups of polynomial subgroup growth, *Invent. Math.* **104** (1991), 521–533.

[LM2] A. Lubotzky and A. Mann, Residually finite groups of finite rank, *Math. Proc. Cambridge Phil. Soc.* **106** (1989), 385–388.

[LPS] A. Lubotzky, L. Pyber and A. Shalev, Discrete groups of slow subgroup growth, *Israel J. Math.* **96** (1996), 399–418.

[MS1] A. Mann and D. Segal, Uniform finiteness conditions in residually finite groups, *Proc. London Math. Soc. (3)* **61** (1990), 529–545.

[MS2] A. Mann and D. Segal, Subgroup growth: some current developments, in: *Infinite Groups 94* (de Giovanni and Newell eds), de Gruyter, 1995.

[MW] O. Manz and T. R. Wolf, *Representations of solvable groups*, LMS Lect. Note Series vol. 185, Cambridge Univ. Press, Cambridge, 1993.

[P] L. Pyber, Asymptotic results for permutation groups, in: *DIMACS Series in discrete mathematics*, vol. 11, 1993, 197–219.

[R1] D. J. S. Robinson, *Finiteness conditions and generalised soluble groups, vols 1 and 2*, Springer, Berlin, 1972.

[R2] D. J. S. Robinson, On the cohomology of soluble groups of finite rank, *J. Pure and Applied Algebra* **6** (1975), 155–164.

[S1] D. Segal, Residually finite groups, in: *Groups–Canberra 1989* (L. G. Kovács ed), Lect. Notes in Maths. vol. 1456, Springer, Berlin, 1990.

[S2] D. Segal, A footnote on residually finite groups, *Israel J. Math.* **94** (1996), 1–5.

[S3] D. Segal, On modules of finite upper rank, *Trans. Amer. Math. Soc.* **353** (2001), 391–410.

[S4] D. Segal, The finite images of finitely generated groups, *Proc. London Math. Soc.* (3) **82** (2001), 597–613.

[SS1] D. Segal and A. Shalev, Groups with fractionally exponential subgroup growth, *J. Pure and Applied Algebra* **88** (1993), 205–223.

[SS2] D. Segal and A. Shalev, Profinite groups with polynomial subgroup growth, *J. London Math. Soc.(2)* **55** (1997), 320–334.

[Sh1] A. Shalev, Growth functions, p-adic analytic groups and groups of finite coclass, *J. London Math. Soc.* **46** (1992), 111–122.

[Sh2] A. Shalev, Subgroup growth and sieve methods, *Proc. London Math. Soc.* **74** (1997), 335–359.

[W] T. R. Wolf, Solvable and nilpotent subgroups of GL(n, q^m), *Canadian Math. J.* **34** (1982), 1097–1111.

Added in proof. Pyber's construction mentioned in the Introduction is in: L. Pyber, Groups of intermediate subgroup growth and a problem of Grothendieck, to appear. For a comprehensive account of subgroup growth see: A. Lubotzky and D. Segal, Subgroup Growth, Birkhäuser-Verlag, Basel, 2003.

17

Pseudo-finite generalized triangle groups

E. B. Vinberg and R. Kaplinsky

A generalized triangle group (g.t.g.) is a group Γ with a fixed presentation of the form

$$\Gamma = \langle x, y \mid x^k = y^l = W(x, y)^m = 1 \rangle, \tag{1}$$

where $k, l, m \geq 2$ and

$$W(x, y) = x^{k_1} y^{l_1} x^{k_2} y^{l_2} \dots x^{k_s} y^{l_s} \tag{2}$$

with $0 < k_i < k$, $0 < l_i < l$, and $s \geq 1$. It is also required that the word W should not be a power of a shorter word. G.t.g. were introduced in [FR] and [BMS]. They have been intensively studied by many authors (see [HMT] and references there). In particular, all finite g.t.g. were found in [HMT] and [LRS].

One of the main tools for studying g.t.g. is constructing their essential homomorphisms to $\mathrm{PSL}_2(\mathbb{C})$. A homomorphism $\varphi : \Gamma \to G$ is called *essential*, if

$$\operatorname{ord} \varphi(x) = k, \quad \operatorname{ord} \varphi(y) = l, \quad \operatorname{ord} \varphi(W(x, y)) = m.$$

It was proved in [BMS] and [FHR] that any g.t.g. admits an essential homomorphism to $\mathrm{PSL}_2(\mathbb{C})$. Most of g.t.g. admit an essential homomorphism to $\mathrm{PSL}_2(\mathbb{C})$ with an infinite image. This is a key step in the classification of finite g.t.g. There are, however, infinite g.t.g. that do not admit such a homomorphism. Let us call a g.t.g. Γ *pseudo-finite* if the image of any essential homomorphisms $\varphi : \Gamma \to \mathrm{PSL}_2(\mathbb{C})$ is finite. In this work we present some partial results on the classification of pseudo-finite g.t.g. This problem was originally motivated by the classification problem for finite groups defined by periodic paired relations (see the definition in [V1]). It seems, however, that it is interesting in its own right. Our results cover the following cases:

1) $m \geq 3$ (see Propositions 4 and 5);

2) $s \leq 3$ (see Propositions 4 and 7 - 10).

The work was completed during our stay at Bielefeld University in August of 1999. We thank this university for its hospitality. The work of the first author was supported by the Humboldt Foundation and by RFBR Grant 98-01-00598.

1. Preliminaries

A pair of matrices $X, Y \in \mathrm{SL}_2(\mathbb{C})$ is called *irreducible*, if they do not have a common eigenvector; or, equivalently, if they generate an irreducible linear group. The following facts can be found, e.g., in [VMH, Appendix] and [V2]. An irreducible pair (X, Y) is defined up to conjugacy by the numbers $\operatorname{tr} X, \operatorname{tr} Y, \operatorname{tr} XY$. Moreover, for any complex numbers a, b, c there are matrices $X, Y \in \mathrm{SL}_2(\mathbb{C})$ such that $\operatorname{tr} X = a, \operatorname{tr} Y = b, \operatorname{tr} XY = c$. A pair (X, Y) is irreducible if and only if the matrix

$$\begin{pmatrix} 2 & \operatorname{tr} X & \operatorname{tr} Y \\ \operatorname{tr} X & 2 & \operatorname{tr} XY \\ \operatorname{tr} Y & \operatorname{tr} XY & 2 \end{pmatrix} \tag{3}$$

is non-degenerate. An irreducible pair is conjugate to a pair of matrices of SU_2 if and only if $\operatorname{tr} X, \operatorname{tr} Y, \operatorname{tr} XY \in \mathbb{R}$ and the (symmetric) matrix (3) is positive definite. The latter means that

$$\begin{aligned} &\operatorname{tr} X = 2\cos\alpha, \quad \operatorname{tr} Y = 2\cos\beta \quad (\alpha, \beta \in (0, \pi)), \\ &\operatorname{tr} XY \in (2\cos(\alpha+\beta),\ 2\cos(\alpha-\beta)). \end{aligned} \tag{4}$$

The boundary cases $\operatorname{tr} XY = 2\cos(\alpha \pm \beta)$ are realized for the pairs of diagonal matrices

$$X = \begin{pmatrix} e^{i\alpha} & 0 \\ 0 & e^{-i\alpha} \end{pmatrix}, \quad Y = \begin{pmatrix} e^{\pm i\beta} & 0 \\ 0 & e^{\mp i\beta} \end{pmatrix}, \tag{5}$$

but also for the pairs of non-commuting matrices

$$X = \begin{pmatrix} e^{i\alpha} & 0 \\ 0 & e^{-i\alpha} \end{pmatrix}, \quad \widetilde{Y} = \begin{pmatrix} e^{\pm i\beta} & 1 \\ 0 & e^{\mp i\beta} \end{pmatrix}. \tag{6}$$

In particular, if X, Y generate an irreducible finite subgroup of $\mathrm{SL}_2(\mathbb{C})$, the conditions (4) hold. Since

$$\operatorname{tr} X^{-1} = \operatorname{tr} X, \quad \operatorname{tr} Y^{-1} = \operatorname{tr} Y, \quad \operatorname{tr} X^{-1}Y^{-1} = \operatorname{tr} YX = \operatorname{tr} XY,$$

any irreducible pair (X, Y) is conjugate to the pair (X^{-1}, Y^{-1}).

For any matrix $X \in \mathrm{SL}_2(\mathbb{C})$ we shall denote by $[X]$ the corresponding element $\{\pm X\}$ of $\mathrm{PSL}_2(\mathbb{C}) = \mathrm{SL}_2(\mathbb{C}) / \{\pm E\}$. The element $[X]$ has order $n \geq 2$

if and only if the eigenvalues of X have the form $e^{\pm\frac{\pi i u}{n}}$ with $(u, n) = 1$ or, equivalently, if

$$\operatorname{tr} X = 2 \cos \frac{\pi u}{n}.$$

A pair of elements $[X]$, $[Y] \in \mathrm{PSL}_2(\mathbb{C})$ is called *irreducible*, if the pair (X, Y) is irreducible in the above sense. An irreducible pair $([X], [Y])$ is still defined up to conjugacy by the numbers $\operatorname{tr} X$, $\operatorname{tr} Y$, $\operatorname{tr} XY$, but these numbers are defined by the elements $[X]$, $[Y]$ only up to multiplying any two of them by -1. This can be applied to constructing essential homomorphisms $\varphi : \Gamma \to \mathrm{PSL}_2(\mathbb{C})$, where Γ is the g.t.g. defined by (1). Set

$$\varphi(x) = [X], \quad \varphi(y) = [Y] \quad (X, Y \in \mathrm{SL}_2(\mathbb{C})). \tag{7}$$

We are to choose X and Y satisfying the conditions

$$\operatorname{tr} X = 2 \cos \frac{\pi u}{k}, \quad \operatorname{tr} Y = 2 \cos \frac{\pi v}{l}, \tag{8}$$

$$\operatorname{tr} W(X, Y) = 2 \cos \frac{\pi w}{m}, \tag{9}$$

where u is prime to k, v is prime to l, and w is prime to m. Since multiplying X (resp. Y) by -1 leads to replacing u (resp. v) with $k - u$ (resp. $l - v$), we may assume that

$$0 < u \leq \frac{k}{2}, \quad 0 < v \leq \frac{l}{2}. \tag{10}$$

For any matrix $Z \in \mathrm{SL}_2(\mathbb{C})$, the Hamilton - Cayley equation gives

$$Z^2 = (\operatorname{tr} Z)\, Z - E.$$

It follows that

$$Z^n = P_n(\operatorname{tr} Z)\, Z - P_{n-1}(\operatorname{tr} Z)\, E,$$

where $P_1(z), P_2(z), \ldots$ are the polynomials defined by

$$P_1(z) = 1, \quad P_2(z) = z, \quad P_{n+1}(z) = z\, P_n(z) - P_{n-1}(z).$$

(These are the Chebyshev polynomials of second kind up to a linear substitution.) Making use of this formula, one can express $\operatorname{tr} W(X, Y)$ as a polynomial in $\operatorname{tr} X$, $\operatorname{tr} Y$, $\operatorname{tr} XY$ (with integral coefficients). Substituting the values of $\operatorname{tr} X$ and $\operatorname{tr} Y$ from (8) and $\operatorname{tr} XY = t$, we obtain a polynomial f of degree s in t

[BMS]. For any (complex) root λ of the algebraic equation

$$f(t) = 2\cos\frac{\pi w}{m} \tag{11}$$

there exists an essential homomorphism $\varphi : \Gamma \to \mathrm{PSL}_2(\mathbb{C})$ satisfying (8) and (9) such that $\operatorname{tr} XY = \lambda$. Moreover, if $\lambda \neq 2\cos(\frac{\pi u}{k} \pm \frac{\pi v}{l})$, the pair (X, Y) is irreducible, so this homomorphism is uniquely defined up to conjugacy. If $\lambda = 2\cos(\frac{\pi u}{k} \pm \frac{\pi v}{l})$, there is an essential homomorphism of Γ to a cyclic group of diagonal matrices. This situation will be investigated in the following section. Another way to find the polynomial f is as follows. Set

$$X = \begin{pmatrix} e^{i\alpha} & 1 \\ 0 & e^{-i\alpha} \end{pmatrix}, \qquad Y = \begin{pmatrix} e^{i\beta} & 0 \\ \tau & e^{-i\beta} \end{pmatrix}$$

with $\alpha = \frac{\pi u}{k}$, $\beta = \frac{\pi v}{l}$. Then

$$\operatorname{tr} X = 2\cos\alpha, \quad \operatorname{tr} Y = 2\cos\beta, \quad \operatorname{tr} XY = \tau + 2\cos(\alpha+\beta).$$

Making use of the formulas

$$X^p = \begin{pmatrix} e^{ip\alpha} & \frac{\sin p\alpha}{\sin\alpha} \\ 0 & e^{-ip\alpha} \end{pmatrix}, \qquad Y^q = \begin{pmatrix} e^{iq\beta} & 0 \\ \frac{\sin q\beta}{\sin\beta}\tau & e^{-iq\beta} \end{pmatrix},$$

one can express $\operatorname{tr} W(X, Y)$ as a polynomial in τ. Substituting

$$\tau = t - 2\cos(\alpha+\beta)$$

we obtain the polynomial $f(t)$. Sometimes we shall extend the notation $[X]$ to any $X \in \mathrm{GL}_2(\mathbb{C})$. More precisely, for any $X \in \mathrm{GL}_2(\mathbb{C})$ we shall denote by $[X]$ the set $\{\lambda X : \lambda \in \mathbb{C}^*\}$ as an element of the group $\mathrm{PGL}_2(\mathbb{C}) = \mathrm{PSL}_2(\mathbb{C})$.

2. G.t.g. admitting an essential homomorphism to a cyclic group

Any finite subgroup of $\mathrm{PSL}_2(\mathbb{C})$ is one of the following groups:

C_n, the cyclic group of order n;
D_n, the dihedral group of order $2n$;
T, the tetrahedral group of order 12;
O, the octahedral group of order 24;
I, the icosahedral group of order 60.

In this section, we consider g.t.g. admitting an essential homomorphism to C_n. Let Γ be the g.t.g. defined by (1).

Proposition 1 ([BMS]). *If there exists an essential homomorphism $\varphi : \Gamma \to C_n$, then there exists an essential homomorphism $\widetilde{\varphi} : \Gamma \to \mathrm{PSL}_2(\mathbb{C})$ with an infinite image.*

Proof. One can interpret φ as a homomorphism of Γ to $\mathrm{PSL}_2(\mathbb{C})$, taking x to $[X]$ and y to $[Y]$, where

$$X = \begin{pmatrix} e^{i\alpha} & 0 \\ 0 & e^{-i\alpha} \end{pmatrix}, \qquad Y = \begin{pmatrix} e^{i\beta} & 0 \\ 0 & e^{-i\beta} \end{pmatrix}$$

with some $\alpha, \beta \in (0, \pi)$. Replacing Y with

$$\widetilde{Y} = \begin{pmatrix} e^{i\beta} & 1 \\ 0 & e^{-i\beta} \end{pmatrix},$$

we obtain the required homomorphism. □

Let us find a criterion for Γ to admit an essential homomorphism to a cyclic group. For any prime p and non-zero integer n, set

$$\nu_p(n) = \max\{\alpha \in \mathbb{Z}_+ : \ p^\alpha \mid n\}.$$

Let n_1, n_2, n_3 be three divisors of n.

Lemma 1. *The congruence*

$$n_1 u_1 + n_2 u_2 + n_3 u_3 \equiv 0 \pmod{n} \tag{12}$$

has a solution with u_1, u_2, u_3 prime to n if and only if the following conditions are satisfied:

(B1′) *for any odd prime p, at least two of the numbers $\nu_p(n_1)$, $\nu_p(n_2)$, $\nu_p(n_3)$ are minimal among them;*

(B2′) *exactly two of the numbers $\nu_2(n_1)$, $\nu_2(n_2)$, $\nu_2(n_3)$ are minimal among them, unless they all equal to $\nu_2(n)$.*

Proof. Decomposing n into primes, we reduce to the case when n is a power of a prime p. Then, cancelling the congruence (12) by a power of p, we reduce to the case, when

$$\min\{\nu_p(n_1), \ \nu_p(n_2), \ \nu_p(n_3)\} = 0,$$

i.e., one of the numbers n_1, n_2, n_3 equals 1. If exactly one of these numbers equals 1 (and two others are non-trivial powers of p), then the congruence (12) has no solutions with u_1, u_2, u_3 prime to p. If at least two of them equal 1, one can easily find a solution of (4) with u_1, u_2, u_3 prime to p, except for the case, when $p = 2, \ \ n_1 = n_2 = n_3 = 1, \ \ n > 1$. This proves the lemma. □

Let us call a triple of non-zero integers $\{n_1, n_2, n_3\}$ *balanced*, if it satisfies the following conditions:

(B1) for any odd prime p, at least two of the numbers $\nu_p(n_1)$, $\nu_p(n_2)$, $\nu_p(n_3)$ are maximal among them;

(B2) exactly two of the numbers $\nu_2(n_1)$, $\nu_2(n_2)$, $\nu_2(n_3)$ are maximal among them, unless they all equal 0.

Set

$$K = k_1 + \cdots + k_s, \qquad L = l_1 + \cdots + l_s. \tag{13}$$

Proposition 2. *The group Γ admits an essential homomorphism to a cyclic group if and only if the triple*

$$\{\frac{k}{(k, K)}, \frac{l}{(l, L)}, m\}$$

is balanced.

Proof. If $\varphi : \Gamma \to C_n$ is an essential homomorphism, then $\varphi(\Gamma) = \langle\varphi(x), \varphi(y)\rangle$ is a cyclic group, whose order is the least common multiple $[k, l]$ of k and l. This shows that we may assume n to be any common multiple of k and l. Let us try to construct an essential homomorphism $\varphi : \Gamma \to \mathbb{Z}_n := \mathbb{Z}/n\mathbb{Z}$, where n is a common multiple of k, l, m. We shall denote a coset $r + n\mathbb{Z}$ by $[r]_n$. If$\varphi(x) = a$ and $\varphi(y) = b$, then $\varphi(W(x, y)) = Ka + Lb$. When a runs over all elements of order k of $\mathbb{Z}_n$, Ka runs over all elements of order $\frac{k}{(k,K)}$, i.e., the elements of the form $[\frac{n(k,K)}{k}u]_n$, where u is prime to n. In an analogous way, when b runs over all the elements of order l, Lb runs over all the elements of the form $[\frac{n(l,L)}{l}v]_n$, where v is prime to n. The order of an element of $\mathbb{Z}_n$ equals m if and only if it has the form $[\frac{n}{m}w]_n$, where w is prime to n. Thus, the group Γ admits an essential homomorphism to $\mathbb{Z}_n$ if and only if the congruence

$$\frac{n\,(k, K)}{k}u + \frac{n\,(l, L)}{l}v \equiv \frac{n}{m}w \pmod{n}$$

has a solution with u, v, w prime to n. According to Lemma 1, this takes place if and only if the triple

$$\{\frac{n(k, K)}{k}, \frac{n(l, L)}{l}, \frac{n}{m}\}$$

satisfies the conditions (B1′) and (B2′) or, equivalently, if the triple

$$\{\frac{k}{(k, K)}, \frac{l}{(l, L)}, m\}$$

satisfies the conditions (B1) and (B2). □

3. Generating pairs of irreducible finite subgroups of $\mathrm{PSL}_2(\mathbb{C})$

Let Γ be the g.t.g. defined by (1) and $\varphi : \Gamma \to \mathrm{PSL}_2(\mathbb{C})$ an essential homomorphism, whose image is an irreducible finite group $F \subset \mathrm{PSL}_2(\mathbb{C})$, i.e., one of the groups D_n, T, O, I. Then

$$\mathrm{ord}\varphi(x) = k, \quad \mathrm{ord}\varphi(y) = l,$$

and $\varphi(x), \varphi(y)$ generate F. It is not difficult to enumerate all generating pairs of each group F of the above list up to conjugacy in $\mathrm{PSL}_2(\mathbb{C})$ or, equivalently, in the normalizer $N(F)$ of F in $\mathrm{PSL}_2(\mathbb{C})$. Note that

$$N(D_n) = D_{2n}, \quad N(T) = N(O) = O, \quad N(I) = I.$$

In order to simplify the task, let us for any generating pair $([X], [Y])$ of a subgroup $F \subset \mathrm{PSL}_2(\mathbb{C})$ consider the element $[Z] \in F$, where $Z \in \mathrm{SL}_2(\mathbb{C})$, satisfying the condition

$$XYZ = 1. \tag{14}$$

Then $([Y], [Z])$ and $([Z], [X])$ will also be generating pairs of F. So any generating triple $([X], [Y], [Z])$ of F satisfying the condition (14), gives rise to 3 generating pairs. (Of course, some of them may be conjugate.) Note that the pair $([Y], [X])$ is conjugate to the pair $([Y]^{-1}, [X]^{-1})$, which is obtained from the "inverse" triple $([Z]^{-1}, [Y]^{-1}, [X]^{-1})$ still satisfying the condition (14). Thus, the problem reduces to a classification of generating triples of irreducible subgroups $F \subset \mathrm{PSL}_2\mathbb{C})$, satisfying the condition (14), up to conjugacy, cyclic permutations, and inversion. Below is a table of all such triples.

We use the following presentation for D_n:

$$D_n = \left\langle\, a, b \mid a^n = b^2 = (ab)^2 = 1 \,\right\rangle,$$

and we identify the groups T, O, I with A_4, S_4, A_5, respectively, via well-known isomorphisms. Note that

$$2\cos\frac{\pi}{5} = \frac{1+\sqrt{5}}{2}, \quad 2\cos\frac{2\pi}{5} = \frac{-1+\sqrt{5}}{2}.$$

In the column "Type" we provide a notation for each type of triples which includes the notation of the corresponding group F. In the column "Orders" the orders of $[X], [Y], [Z]$ are indicated. For $F = D_n$, the number u is prime to n, and one may assume that $0 < u \leqslant \frac{n}{2}$.

It is clear from the very beginning, and it is seen from the table, that the set of all unordered triples $\{\mathrm{tr}X, \mathrm{tr}Y, \mathrm{tr}Z\}$ considered up to multiplying by -1 of

Table 1.

Type	$[X]$	$[Y]$	$[Z]$	Orders	tr X	tr Y	tr Z
$D_n(u)$	a^u	b	$a^u b$	$n, 2, 2$	$2\cos\frac{\pi u}{n}$	0	0
$T(1)$	(123)	(234)	(12)(34)	3, 3, 2	1	1	0
$T(2)$	(123)	(243)	(142)	3, 3, 3	1	1	1
$O(1)$	(1234)	(132)	(14)	4, 3, 2	$\sqrt{2}$	1	0
$O(2)$	(1234)	(1243)	(123)	4, 4, 3	$\sqrt{2}$	$\sqrt{2}$	0
$I(1)$	(12345)	(142)	(15)(34)	5, 3, 2	$\frac{1+\sqrt{5}}{2}$	1	0
$I(2)$	(12354)	(152)	(14)(35)	5, 3, 2	$\frac{1-\sqrt{5}}{2}$	1	0
$I(3)$	(12345)	(132)	(154)	5, 3, 3	$\frac{1+\sqrt{5}}{2}$	1	1
$I(4)$	(12354)	(132)	(145)	5, 3, 3	$\frac{1-\sqrt{5}}{2}$	1	1
$I(5)$	(12345)	(12354)	(13)(24)	5, 5, 2	$\frac{1+\sqrt{5}}{2}$	$\frac{1-\sqrt{5}}{2}$	0
$I(6)$	(12345)	(14352)	(135)	5, 5, 3	$\frac{1+\sqrt{5}}{2}$	$\frac{1+\sqrt{5}}{2}$	1
$I(7)$	(12354)	(15342)	(134)	5, 5, 3	$\frac{1-\sqrt{5}}{2}$	$\frac{1-\sqrt{5}}{2}$	1
$I(8)$	(12345)	(14532)	(145)	5, 5, 3	$\frac{1+\sqrt{5}}{2}$	$\frac{1-\sqrt{5}}{2}$	-1
$I(9)$	(12345)	(12534)	(12453)	5, 5, 5	$\frac{1+\sqrt{5}}{2}$	$\frac{1+\sqrt{5}}{2}$	$\frac{1+\sqrt{5}}{2}$
$I(10)$	(12354)	(12435)	(12543)	5, 5, 5	$\frac{1-\sqrt{5}}{2}$	$\frac{1-\sqrt{5}}{2}$	$\frac{1-\sqrt{5}}{2}$

any two members, is invariant under the Galois group (of a sufficiently large algebraic number field including all the involved numbers).

For any generating pair $([X], [Y])$ of a finite subgroup $F \subset \mathrm{PSL}_2(\mathbb{C})$ and for any $d | \mathrm{ord}[X]$, $e | \mathrm{ord}[Y]$ it is important to know the subgroup generated by $[X]^d$ and $[Y]^e$, and the type of the corresponding triple $([X]^d, [Y]^e, [Y]^{-e}[X]^{-d})$. There are only a few non-trivial cases, which are presented in the following table.

Table 2.

Type of the original triple	Generators of the subgroup	Their orders	Type of the obtained triple
$D_n(u)$	$[X]^d,\ [Y]$	$\frac{n}{d}, 2$	$D_{n/d}(u)$
$O(1)$	$[X]^2,\ [Y]$	2, 3	$T(1)$
$O(1)$	$[X]^2,\ [Z]$	2, 2	$D_4(1)$
$O(2)$	$[X]^2,\ [Y]$	2, 4	$D_4(1)$
$O(2)$	$[X]^2,\ [Z]$	2, 3	$T(1)$
$O(2)$	$[X]^2,\ [Y]^2$	2, 2	$D_2(1)$

4. Admissible transformations. Imprimitive pseudo-finite g.t.g.

Some obvious transformations of the data defining a g.t.g. Γ lead to isomorphic g.t.g. which are, in particular, pseudo-finite if and only if Γ is pseudo-finite. First, one can multiply modulo k the exponents $k_1, \dots, k_s$ by a factor prime to k and, in a similar way, multiply modulo l the exponents $l_1, \dots, l_s$ by a factor prime to l. These transformations can be interpreted as changes of the generators of the cyclic groups $\langle x \rangle$ and $\langle y \rangle$. Second, one can cyclically shift the sequence $(k_1, l_1, \dots, k_s, l_s)$ by an even number. This replaces the relation $W(x, y)^m = 1$ with an equivalent one. Third, one can interchange k and l and simultaneously shift the sequence $(k_1, l_1, \dots, k_s, l_s)$ by an odd number. This can be interpreted as interchanging x and y. Fourth, one can replace the sequence $(k_1, l_1, \dots, k_s, l_s)$ with $(k_s, l_{s-1}, k_{s-1}, \dots, l_1, k_1, l_s)$. This replaces the relation $W(x, y)^m = 1$ with $(W(x^{-1}, y^{-1})^{-1})^m = 1$, which can be interpreted as replacing the relation $(W(x, y))^m = 1$ with the equivalent relation $(W(x, y)^{-1})^m = 1$, combined with changing the generators x and y for x^{-1} and y^{-1}. Transformations of these four types and their combinations are called *admissible*. G.t.g. obtained from each other by admissible transformations are called *equivalent*. It is reasonable to classify pseudo-finite g.t.g. up to equivalence.

A g.t.g. is called *primitive* if

$$(k_1, \dots, k_s, k) = 1, \quad (l_1, \dots, l_s, l) = 1.$$

In the general case, set

$$(k_1, \dots, k_s, k) = d, \quad (l_1, \dots, l_s, l) = e,$$

and consider the primitive g.t.g.

$$\bar{\Gamma} = \left\langle \bar{x}, \bar{y} \mid \bar{x}^{\bar{k}} = \bar{y}^{\bar{l}} = \overline{W}(\bar{x}, \bar{y})^m = 1 \right\rangle,$$

where $\bar{k} = k/d, \bar{l} = l/e$, and $\overline{W}(\bar{x}, \bar{y}) = \bar{x}^{\bar{k}_1} \bar{y}^{\bar{l}_1} \dots \bar{x}^{\bar{k}_s} \bar{y}^{\bar{l}_s}$ with $\bar{k}_i = k_i/d, \bar{l}_i = l_i/e$. There is a natural homomorphism $\pi : \bar{\Gamma} \to \Gamma$, taking $\bar{x}$ to x^d and $\bar{y}$ to y^e. If $\varphi : \Gamma \to G$ is an essential homomorphism, then $\bar{\varphi} = \varphi\pi : \bar{\Gamma} \to G$ is also an essential homomorphism, and $\bar{\varphi}(\bar{\Gamma}) = \langle \varphi(x)^d, \varphi(y)^e \rangle \subset \varphi(\Gamma)$. Conversely, let $\bar{\varphi} : \bar{\Gamma} \to \mathrm{PSL}_2(\mathbb{C})$ be an essential homomorphism. Let $[X]$ be any d-th root of $\bar{\varphi}(\bar{x})$ and $[Y]$ any e-th root of $\bar{\varphi}(\bar{y})$. Then there is a homomorphism $\varphi : \Gamma \to \mathrm{PSL}_2(\mathbb{C})$, taking x to $[X]$ and y to $[Y]$. Obviously, φ is essential and $\bar{\varphi} = \varphi\pi$. It follows that if Γ is pseudo-finite, then $\bar{\Gamma}$ is also pseudo-finite. Let us call a g.t.g. Γ *pseudo-dihedral* (resp. *pseudo-tetrahedral*) if the image of any essential homomorphism $\varphi : \Gamma \to \mathrm{PSL}_2(\mathbb{C})$ is a dihedral (resp. tetrahedral) group.

Proposition 3. *Imprimitive pseudo-finite g.t.g. are (up to equivalence) exactly the groups of the following six types:*

(I1) $\Gamma = \langle x, y \mid x^k = y^2 = (x^{k_1} y \dots x^{k_s} y)^2 = 1 \rangle$,
where $d > 1$ *and the group* $\bar{\Gamma}$ *is pseudo-dihedral;*
(I2) $\Gamma = \langle x, y \mid x^4 = y^4 = (x^{k_1} y^2 \dots x^{k_s} y^2)^2 = 1 \rangle$,
where not all of the exponents $k_1, \dots, k_s$ *are even and the group* $\bar{\Gamma}$ *is pseudo-dihedral;*
(I3) $\Gamma = \langle x, y \mid x^4 = y^4 = (x^2 y^2)^2 = 1 \rangle$;
(I4) $\Gamma = \langle x, y \mid x^3 = y^4 = (x^{k_1} y^2 \dots x^{k_s} y^2)^3 = 1 \rangle$,
where the group $\bar{\Gamma}$ *is pseudo-tetrahedral;*
(I5) $\Gamma = \langle x, y \mid x^3 = y^4 = (x^{k_1} y^2 \dots x^{k_s} y^2)^2 = 1 \rangle$,
where the group $\bar{\Gamma}$ *is pseudo-tetrahedral;*
(I6) $\Gamma = \langle x, y \mid x^2 = y^4 = (x y^2)^2 = 1 \rangle$.

Proof. If an imprimitive g.t.g. Γ is pseudo-finite, then for any essential homomorphism $\bar{\varphi} : \bar{\Gamma} \to \mathrm{PSL}_2(\mathbb{C})$ not only the group $\bar{\varphi}(\bar{\Gamma}) = \langle \bar{\varphi}(\bar{x}), \bar{\varphi}(\bar{y}) \rangle$ is finite, but the group, generated by a d-th root of $\bar{\varphi}(\bar{x})$ and an e-th root of $\bar{\varphi}(\bar{y})$, is still finite. All such possibilities are enumerated in Table 2. Consider them case-by-case. We shall use the notation (7) and set $\bar{X} = X^d$, $\bar{Y} = Y^e$. If $d > 2$ or $k > 4$, then $l = 2$, i.e.,

$$\Gamma = \langle x, y \mid x^k = y^2 = (x^{k_1} y \dots x^{k_s} y)^m = 1 \rangle,$$

and $\mathrm{tr}\bar{X}\bar{Y} = 0$ for any $\bar{\varphi}$. But if $m > 2$, there are at least two possibilities for $\mathrm{tr}\overline{W}(\bar{X}, \bar{Y})$ and, thereby, at least two possibilities for the equation (11) for $\bar{\Gamma}$, which differ only by a constant term. At least one of these polynomials does not vanish at 0. Hence, $m = 2$, and we arrive at case (I1). The cases, when $e > 2$ or $l > 4$, are obtained by interchanging x and y. In all the other cases $d, e \leq 2$ and $k, l \leq 4$. By symmetry, we may (and shall) assume that $e = 2$ and $l = 4$. Under these conditions, if $k = 4$ and $d = 1$, then again $\mathrm{tr}\bar{X}\bar{Y} = 0$ for any $\bar{\varphi}$. Reasoning as above, we can conclude that $m = 2$, which gives the case (I2). If $k = 4$ and $d = 2$, we obtain the case (I3). If $k = 3$, then $d = 1$ and $\bar{\varphi}(\bar{\Gamma})$ must be a tetrahedral group. It follows that $m \leq 3$, so we obtain the case (I4) or (I5). Finally, if $k = 2$, we obtain the case (I6). □

Note that any imprimitive g.t.g. is infinite as a non-trivial amalgamated product. Now we are able to describe all pseudo-finite g.t.g. with $s = 1$.

Proposition 4. . *All the pseudo-finite g.t.g. with $s = 1$ are, up to equivalence, the usual triangle groups*

$$\langle x, y \mid x^k = y^l = (x\,y)^m = 1\rangle$$

with $\frac{1}{k} + \frac{1}{l} + \frac{1}{m} > 1$, *and the following imprimitive g.t.g.:*

1) $\langle x, y \mid x^k = y^2 = (x^d y)^2 = 1\rangle, \quad (d \mid k,\ \ d > 1)$,
2) $\langle x, y \mid x^4 = y^4 = (xy^2)^2 = 1\rangle$,
3) $\langle x, y \mid x^4 = y^4 = (x^2y^2)^2 = 1\rangle$,
4) $\langle x, y \mid x^3 = y^4 = (xy^2)^3 = 1\rangle$,
5) $\langle x, y \mid x^2 = y^4 = (xy^2)^2 = 1\rangle$.

Proof. Changing the generators of the cyclic groups $\langle x\rangle$ and $\langle y\rangle$, we may assume that

$$k_1 = d \mid k, \quad l_1 = e \mid l.$$

If Γ is primitive, i.e., $d = e = 1$, then Γ is a usual triangle group and, as it well-known, it is embedded into $PSL_2(\mathbb{C})$. Therefore, it is pseudo-finite if and only if it is finite, which takes place if and only if $\frac{1}{k} + \frac{1}{l} + \frac{1}{m} > 1$. If Γ is imprimitive, it belongs to one of the types (I1) – (I6) of Proposition 3. The type (I5) is not realized, because the group $\bar{\Gamma}$ in this case is the dihedral group D_3. The other types constitute the above list. □

5. The case $m \geq 3$

Constructing an essential homomorphism of a g.t.g. Γ to $PSL_2(\mathbb{C})$, we can vary the parameters u, v, w in (8) and (9). Let us fix u, v, and vary w. We shall obtain $\varphi(m)$ different algebraic equations of the form (11) with one and the same polynomial f of degree s in the left hand side. Obviously, they do not have common roots. Let N be the total number of their different roots. Then the total number of their roots with multiplicities, which is surely equal to $s\varphi(m)$, does not exceed N plus the number of roots (with multiplicities) of f', whence

$$s(\varphi(m) - 1) \leq N - 1. \tag{15}$$

On the other hand, assuming the group Γ to be pseudo-finite, one can extract from Table 1 all possible values of $\mathrm{tr}XY$ for any fixed values of $\mathrm{tr}X$ and $\mathrm{tr}Y$. They are presented in the following table. It contains all possible values of $\mathrm{tr}X$ and $\mathrm{tr}Y$ up to interchanging them, multiplying by -1, and acting by the Galois group.

Table 3.

k	l	$\operatorname{tr} X$	$\operatorname{tr} Y$	$\operatorname{tr} XY$
≥ 6	2	$2\cos\frac{\pi}{k}$	0	0
5	5	$\frac{1+\sqrt{5}}{2}$	$\frac{1+\sqrt{5}}{2}$	$1,\ \frac{1+\sqrt{5}}{2}$
5	5	$\frac{1+\sqrt{5}}{2}$	$\frac{-1+\sqrt{5}}{2}$	$0, 1$
5	3	$\frac{1+\sqrt{5}}{2}$	1	$0,\ 1,\ \frac{1+\sqrt{5}}{2},\ \frac{-1+\sqrt{5}}{2}$
5	2	$\frac{1+\sqrt{5}}{2}$	0	$0,\ \pm 1,\ \pm\frac{1-\sqrt{5}}{2}$
4	4	$\sqrt{2}$	$\sqrt{2}$	1
4	3	$\sqrt{2}$	1	$0,\ \sqrt{2}$
4	2	$\sqrt{2}$	0	$0, \pm 1$
3	3	1	1	$0,\ 1,\ \frac{1+\sqrt{5}}{2},\ \frac{1-\sqrt{5}}{2}$
3	2	1	0	$0,\ \pm 1,\ \pm\sqrt{2},\ \pm\frac{1+\sqrt{5}}{2},\ \pm\frac{1-\sqrt{5}}{2}$
2	2	0	0	$2\cos\frac{\pi u}{m},\quad (u, m) = 1$

In the last case

$$\Gamma = \langle\, x,\ y \quad|\quad x^2 = y^2 = (x\,y)^m = 1\,\rangle,$$

so $s = 1$ and Γ is a dihedral group. In all the other cases $N \leq 9$ and for $m \geq 3$ the inequality (15) gives an upper bound for s. This bound can be slightly improved with help of the following.

Lemma 2. ([HMT], the proof of Theorem 6.4). *If a g.t.g. Γ admits essential homomorphisms onto D_3 and T, it admits an essential homomorphism onto C_6 and, hence, is not pseudo-finite.*

Proof. Let $\varphi : \Gamma \to D_3$ and $\psi : \Gamma \to T$ be essential homomorphisms. Then the homomorphism

$$\Gamma \to (D_3/C_3) \times (T/D_2) \simeq C_6,$$

$$\gamma \mapsto (\varphi(\gamma)C_3, \psi(\gamma)D_2),$$

is also essential. □

Proposition 5. *All the pseudo-finite g.t.g. with $m \geq 3$ and $s \geq 2$ are, up to equivalence, the following two groups:*

1) $\langle x, y \mid x^3 = y^2 = (xyx^2y)^3 = 1\rangle$;
2) $\langle x, y \mid x^3 = y^2 = (xyxy\,x^2y)^3 = 1\rangle$.

The first of these groups is pseudo-tetrahedral. It gives rise to the following imprimitive pseudo-finite g.t.g. according to the case (I4) of Proposition 3:

$$\langle x, y \mid x^3 = y^4 = (xy^2x^2y^2)^3 = 1\rangle.$$

As it follows from the classification of finite g.t.g., the first of these groups is infinite, while the second one is finite.

Proof. Under our restrictions, the inequality (15) can be satisfied only in five cases of Table 3. In all these cases, the group $\varphi(\Gamma)$ is one of the groups D_3, D_4, D_5, T, O, I. The orders of elements of these groups do not exceed 5. It follows that $m \leq 5$. More precisely, all possible cases are presented in the following table.

Table 4.

k	l	The possible groups $\varphi(\Gamma)$	m	s
5	3	I	3	2, 3
5	2	I	3	2, 3, 4
4	2	D_4, O	4	2
3	3	T, I	3	2, 3
3	2	D_3, T, O, I	3	2 - 7

The case $k = 4, l = 2, m = 3$ is impossible, since in this case $\varphi(\Gamma)$ cannot be the group D_4, which leaves only two possibilities for $\mathrm{tr} XY$. The case $k = 3, l = 2, m = 4$ is impossible, since in this case $\varphi(\Gamma)$ can be only the group O, which again leaves only two possibilities for $\mathrm{tr} XY$. In all the cases, enumerated in Table 4, and all, up to admissible transformations, exponents $k_1, \ldots, k_s, l_1, \ldots, l_s$, we explicitly wrote the equations (11) and found out if all their roots are among the admissible numbers indicated in Table 3. This turned out to be true only in the two cases of the proposition. □

6. The case $m = 2,\ k, l \geq 3$

For $m = 2$, equation (11) takes the form $f(t) = 0$. If the group Γ is pseudo-finite, all the roots of the polynomial f (for any u, v) must be real and lie in the interval

$$\left(2\cos(\frac{\pi u}{k} + \frac{\pi v}{l}),\quad 2\cos(\frac{\pi u}{k} - \frac{\pi v}{l})\right)$$

(see (4)), whence

$$\operatorname{sgn} f(2\cos(\frac{\pi u}{k}+\frac{\pi v}{l}))f(2\cos(\frac{\pi u}{k}-\frac{\pi v}{l}))=(-1)^s.$$

We deduce from this

Proposition 6. *If Γ is a pseudo-finite g.t.g. with $m=2$, then*

$$\operatorname{sgn}(\cos\frac{2\pi Ku}{k}+\cos\frac{2\pi Lv}{l})=(-1)^s \tag{16}$$

for any u prime to k and v prime to l.

(For the notation K and L see (13).)

Proof. Set $\alpha=\frac{\pi u}{k}$ and $\beta=\frac{\pi v}{l}$. Then the commuting matrices X and Y from (5) satisfy the conditions

$$\operatorname{tr}X=2\cos\frac{\pi u}{k},\quad \operatorname{tr}Y=2\cos\frac{\pi v}{l},\quad \operatorname{tr}XY=2\cos(\frac{\pi u}{k}\pm\frac{\pi v}{l}),$$

and, hence,

$$f(2\cos(\frac{\pi u}{k}\pm\frac{\pi v}{l}))=\operatorname{tr}W(X,Y)=2\cos(\frac{\pi Ku}{k}\pm\frac{\pi Lv}{l}).$$

It follows that

$$f(2\cos(\frac{\pi u}{k}+\frac{\pi v}{l}))f(2\cos(\frac{\pi u}{k}-\frac{\pi v}{l}))=$$

$$4\cos(\frac{\pi Ku}{k}+\frac{\pi Lv}{l})\cos(\frac{\pi Ku}{k}-\frac{\pi Lv}{l})=2(\cos\frac{2\pi Ku}{k}+\cos\frac{2\pi Lv}{l}).$$

□

Corollary. *If $k=l=5$, then s is even.*

Proof. Suppose s is odd. Then

$$\cos\frac{2\pi Ku}{5}+\cos\frac{2\pi Lv}{5}<0 \tag{17}$$

for any $u, v\in\{1,2,3,4\}$. If K is divisible by 5, then the first summand is equal to 1 and the inequality (17) cannot hold. Hence, K is not divisible by 5. In the same way, L is not divisible by 5. Consequently, one can choose u, v so that

$$Ku\equiv Lv\equiv 1 \pmod 5.$$

Then (17) does not hold. □

If one of the numbers k, l equals 2 and Γ does not admit an essential homomorphism to a cyclic group, then the condition (16) holds automatically. However, if $k, l \geq 3$, it gives rise to some restrictions on K and L for given k, l and s. They are collected in the following table, containing all possible values of k and l (up to permutation).

Table 5.

k	l	s	Restrictions on K and L
5	5	even	$K \equiv 0 \pmod 5$ or $L \equiv 0 \pmod 5$
5	3	even	$K \equiv 0 \pmod 5$ or $L \equiv 0 \pmod 3$
4	4	even	$K \equiv 0 \pmod 4$ or $L \equiv 0 \pmod 4$; $K, L \not\equiv 2 \pmod 4$
4	3	even	$K \equiv 0 \pmod 4$ or $L \equiv 0 \pmod 3$; $K \not\equiv 2 \pmod 4$
3	3	even	$K \equiv 0 \pmod 3$ or $L \equiv 0 \pmod 3$
5	3	odd	$K \not\equiv 0 \pmod 5$ and $L \not\equiv 0 \pmod 3$
4	4	odd	$K, L \not\equiv 0 \pmod 4$; $K \equiv 2 \pmod 4$ or $L \equiv 2 \pmod 4$
4	3	odd	$K \not\equiv 0 \pmod 4$ and $L \not\equiv 0 \pmod 3$
3	3	odd	$K, L \not\equiv 0 \pmod 3$

For $m = 2$ we did not get an apriori upper bound for s, so we restricted ourselves to the cases $s = 2, 3$. Under this restriction there are only few cases to be checked, taking into account Table 5. The result is contained in the following two propositions.

Proposition 7. *All the primitive pseudo-finite g.t.g. with $m = 2$, $k, l \geq 3$, and $s = 2$ are, up to equivalence, the following seven groups:*

1) $\langle x, y \mid x^5 = y^3 = (xyx^2y^2)^2 = 1 \rangle$;
2) $\langle x, y \mid x^5 = y^3 = (xyx^4y)^2 = 1 \rangle$;
3) $\langle x, y \mid x^4 = y^4 = (xyx^2y^3)^2 = 1 \rangle$;
4) $\langle x, y \mid x^4 = y^3 = (xyx^2y^2)^2 = 1 \rangle$;
5) $\langle x, y \mid x^4 = y^3 = (xyx^3y)^2 = 1 \rangle$;
6) $\langle x, y \mid x^3 = y^3 = (xyx^2y^2)^2 = 1 \rangle$;
7) $\langle x, y \mid x^3 = y^3 = (xyx^2y)^2 = 1 \rangle$.

As it follows from the classification of finite g.t.g., the last two groups are finite, while all the others are infinite.

Remark 1. *It follows from Proposition* 3 *and Proposition* 9 *below, that there are, up to equivalence, exactly two imprimitive pseudo-finite g.t.g. with $m = 2$, $k, l \geq 3$ and $s = 2$, namely, the groups*

$$\langle x, y \mid x^4 = y^4 = (xy^2x^3y^2)^2 = 1 \rangle;$$
$$\langle x, y \mid x^3 = y^4 = (xy^2x^2y^2)^2 = 1 \rangle.$$

Proposition 8. *All the primitive pseudo-finite g.t.g. with $m = 2,\ k, l \geq 3$ and $s = 3$ are, up to equivalence, the following four groups:*

1) $\langle x, y \mid x^5 = y^3 = (xyxyx^4y^2)^2 = 1\rangle$;
2) $\langle x, y \mid x^5 = y^3 = (xyx^2y^2x^3y)^2 = 1\rangle$;
3) $\langle x, y \mid x^4 = y^4 = (xyx^3y^3xy^2)^2 = 1\rangle$;
4) $\langle x, y \mid x^3 = y^3 = (xyxyx^2y^2)^2 = 1\rangle$.

As it follows from the classification of finite g.t.g., all these groups are infinite.

Remark 2. *It follows from Proposition* 3 *and Proposition* 10 *below, that there are no imprimitive pseudo-finite g.t.g. with $m = 2,\ k,\ l \geq 3$ and $s = 3$.*

7. The Case $l = m = 2$

For $l = 2$ we have

$$W(x, y) = x^{k_1}yx^{k_2}y \dots x^{k_s}y.$$

One may assume that $k \geq 3$, otherwise $s = 1$, and the group Γ is dihedral. Admissible transformations in this case reduce to multiplying modulo k the exponents $k_1, \dots, k_s$ by a factor prime to k, cyclic permutations of them, and reversing their order. If, moreover, $m = 2$, the action of the Galois group allows us to restrict the consideration to the case where $u = 1$. Set $\varepsilon = \varepsilon_k = e^{\frac{2\pi i}{k}}$, and choose matrices $X, Y \in \mathrm{SL}_2(\mathbb{C})$ as follows:

$$X = \begin{pmatrix} e^{\frac{\pi i}{k}} & 0 \\ 0 & e^{-\frac{\pi i}{k}} \end{pmatrix} = e^{-\frac{\pi i}{k}} \begin{pmatrix} \varepsilon & 0 \\ 0 & 1 \end{pmatrix}, \quad Y = -i \begin{pmatrix} \tau & 1-\tau \\ 1+\tau & -\tau \end{pmatrix}.$$

Then

$$\operatorname{tr} X = 2\cos\frac{\pi}{k}, \quad \operatorname{tr} Y = 0,$$
$$\operatorname{tr} XY = 2\tau \sin\frac{\pi}{k}, \tag{18}$$

and

$$\operatorname{tr} W(X, Y) = e^{-\pi i\,(\frac{K}{k} + \frac{s}{2})}\, g(\tau),$$

where

$$g(\tau) = \operatorname{tr} \begin{pmatrix} \varepsilon^{k_1} & 0 \\ 0 & 1 \end{pmatrix} \begin{pmatrix} \tau & 1-\tau \\ 1+\tau & -\tau \end{pmatrix} \begin{pmatrix} \varepsilon^{k_2} & 0 \\ 0 & 1 \end{pmatrix} \begin{pmatrix} \tau & 1-\tau \\ 1+\tau & -\tau \end{pmatrix} \dots$$
$$\dots \begin{pmatrix} \varepsilon^{k_s} & 0 \\ 0 & 1 \end{pmatrix} \begin{pmatrix} \tau & 1-\tau \\ 1+\tau & -\tau \end{pmatrix}.$$

Note that multiplying Y by -1 affects tr$X$, tr$Y$ and trXY in the same way as multiplying τ by -1. Obviously,

$$\operatorname{tr} W(X, -Y) = (-1)^s \operatorname{tr} W(X, Y).$$

Hence,

$$g(-\tau) = (-1)^s \, g(\tau),$$

so

$$g(\tau) = \begin{cases} h(\tau^2) & \text{for } s \text{ even}, \\ \tau h(\tau^2) & \text{for } s \text{ odd}, \end{cases}$$

where h is a polynomial of degree $[\frac{s}{2}]$. As it follows from Table 3 and (18), the group Γ is pseudo-finite if and only if all the roots of the polynomial h are among the numbers indicated in the following table.

Table 6.

k	Admissible roots of h
≥ 6	0
5	$0, \ \frac{1+\sqrt{5}}{2\sqrt{5}}, \ \frac{-1+\sqrt{5}}{2\sqrt{5}}$
4	$0, \ \frac{1}{2}$
3	$0, \ \frac{1}{3}, \ \frac{2}{3}, \ \frac{3+\sqrt{5}}{6} \ \frac{3-\sqrt{5}}{6}$

Let us find the polynomial h explicitly. The polynomial $g(\tau)$ is the sum of all the products of entries of the matrices

$$\begin{pmatrix} \varepsilon^{k_1} & 0 \\ 0 & 1 \end{pmatrix}, \begin{pmatrix} \tau & 1-\tau \\ 1+\tau & -\tau \end{pmatrix}, \begin{pmatrix} \varepsilon^{k_2} & 0 \\ 0 & 1 \end{pmatrix}, \begin{pmatrix} \tau & 1-\tau \\ 1+\tau & -\tau \end{pmatrix}, \ldots$$
$$\ldots, \begin{pmatrix} \varepsilon^{k_s} & 0 \\ 0 & 1 \end{pmatrix}, \begin{pmatrix} \tau & 1-\tau \\ 1+\tau & -\tau \end{pmatrix}, \qquad (19)$$

chosen so that the column number of the entry of each matrix equals the row number of the entry of the subsequent matrix, if considering the matrices (19) ordered cyclically. Clearly, one can take only diagonal entries of the matrices

$$\begin{pmatrix} \varepsilon^{k_1} & 0 \\ 0 & 1 \end{pmatrix}, \begin{pmatrix} \varepsilon^{k_2} & 0 \\ 0 & 1 \end{pmatrix}, \ldots, \begin{pmatrix} \varepsilon^{k_s} & 0 \\ 0 & 1 \end{pmatrix}. \qquad (20)$$

Every time, when we retain the number of the diagonal entry passing to the subsequent matrix (20), we have to take an entry $\pm\tau$ of the intermediate factor $\begin{pmatrix} \tau & 1-\tau \\ 1+\tau & -\tau \end{pmatrix}$. Every time, when we switch to another number, we have to take an entry $1\pm\tau$ of the intermediate factor. It follows that each product has the form

$$(-1)^{s-q-r}\,\varepsilon^{k_{i_1}+\ldots+k_{i_s}}\,\tau^{s-2q}\,(1-\tau^2)^q,$$

where $1 \le i_1 < \ldots < i_r \le s$ and the number q is defined as follows:

1) if the set $\{i_1, \ldots, i_r\}$ is a proper subset of $\{1, \ldots, s\}$, then q is equal to the number of its "connected components", where a connected component is a maximal subset of $\{i_1, \ldots, i_r\}$ consisting of consecutive elements of the set $\{1, \ldots, s\}$ considered cyclically ordered;
2) if $\{i_1, \ldots, i_r\} = \{1, \ldots, s\}$, then $q = 0$.

Thus, the polynomial h has the form

$$h(\sigma) = \sum_{q=0}^{[\frac{s}{2}]}(-1)^q h_q\,\sigma^{[\frac{s}{2}]-q}(1-\sigma)^q, \tag{21}$$

where

$$h_0 = \varepsilon^K + (-1)^s \tag{22}$$

and, for $q > 0$,

$$h_q = \sum_{\substack{1\le i_1<\cdots<i_r\le s \\ \{i_1,\ldots,i_r\}\text{ has } q \text{ connected components}}} (-1)^{s-r}\,\varepsilon^{k_{i_1}+\ldots+k_{i_r}}. \tag{23}$$

In particular, for $s = 2$, we have

$$\begin{aligned} h(\sigma) &= (\varepsilon^K + 1)\,\sigma + (\varepsilon^{k_1} + \varepsilon^{k_2})(1-\sigma) \\ &= (\varepsilon^{k_1} - 1)(\varepsilon^{k_2} - 1)\,\sigma + \varepsilon^{k_1} + \varepsilon^{k_2} \end{aligned}$$

Proposition 9. *All the primitive pseudo-finite g.t.g. with $l = m = 2$ and $s = 2$ are, up to equivalence, the following groups:*

1) $\langle x, y \mid x^{2n} = y^2 = (xyx^{n+1}y)^2 = 1\rangle \quad (n \ge 2)$;
2) $\langle x, y \mid x^5 = y^2 = (xyx^2y)^2 = 1\rangle$;
3) $\langle x, y \mid x^3 = y^2 = (xyx^2y)^2 = 1\rangle$.

The groups of the first type are pseudo-dihedral. They give rise to some imprimitive pseudo-finite g.t.g. according to the cases (I1) and (I2) of Proposition

3. The last group is pseudo-tetrahedral. It gives rise to an imprimitive pseudo-finite g.t.g. according to the case (I5) of Proposition 3. As it follows from the classification of finite g.t.g., the groups of the first type are infinite, while the last two groups are finite.

Proof. The root of h equals 0 if and only if $\varepsilon^{k_1} + \varepsilon^{k_2} = 0$, which means that k is even and

$$k_1 - k_2 \equiv \frac{k}{2} \pmod{k}.$$

Under this condition, if Γ is primitive, at least one of the integers k_1, k_2 must be prime to k, and we may assume that it is equal to 1. In this way we obtain the groups of the first type. In all the other cases $k \leq 5$. Only 5 such cases are to be tried. This gives the two last groups of the proposition. □

For $s = 3$ we have

$$\begin{aligned} h(\sigma) = (\varepsilon^K - 1)\sigma + (\varepsilon^{k_2+k_3} + \varepsilon^{k_3+k_1} + \varepsilon^{k_1+k_2} - \varepsilon^{k_1} - \varepsilon^{k_2} - \varepsilon^{k_3}) \\ \times (1-\sigma) = (\varepsilon^{k_1} - 1)(\varepsilon^{k_2} - 1)(\varepsilon^{k_3} - 1)\sigma + \varepsilon^{k_2+k_3} + \varepsilon^{k_3+k_1} \\ + \varepsilon^{k_1+k_2} - \varepsilon^{k_1} - \varepsilon^{k_2} - \varepsilon^{k_3}. \end{aligned}$$

Proposition 10. *All the primitive pseudo-finite g.t.g. with $l = m = 2$ and $s = 3$ are, up to equivalence, the following groups:*

1) $\langle x, y \mid x^{2n} = y^2 = (xyx^{n+1}yx^{n+2}y)^2 = 1\rangle \quad (n \geq 3)$;
2) $\langle x, y \mid x^{3n} = y^2 = (xyx^{n+1}yx^{2n+1}y)^2 = 1\rangle \quad (n \geq 2)$;
3) $\langle x, y \mid x^{6n} = y^2 = (x^d yx^n yx^{5n} y)^2 = 1\rangle \quad (d \mid 6,\ (d, n) = 1)$;
4) $\langle x, y \mid x^{30} = y^2 = (x^2yx^3yx^{26}y)^2 = 1\rangle$;
5) $\langle x, y \mid x^5 = y^2 = (xyx^2yx^3y)^2 = 1\rangle$;
6) $\langle x, y \mid x^5 = y^2 = (xyxyx^4y)^2 = 1\rangle$;
7) $\langle x, y \mid x^4 = y^2 = (xyx^2yx^3y)^2 = 1\rangle$;
8) $\langle x, y \mid x^4 = y^2 = (xyxyx^3y)^2 = 1\rangle$;
9) $\langle x, y \mid x^3 = y^2 = (xyxyx^2y)^2 = 1\rangle$.

The groups of the first three types and the group under 4) are pseudo-dihedral. They give rise to some imprimitive pseudo-finite g.t.g. according to the case (I1) of Proposition 3. As it follows from the classification of finite g.t.g. the groups of the first three types and the groups under 4) and 7) are infinite, while the groups under 5), 6), 8), and 9) are finite. To prove the proposition, we need some lemmas.

Lemma 3. *Let $z_1, z_2, \ldots$ be complex numbers with modulus* 1. *Then*

a) $z_1 + z_2 + z_3 = 0$ *if and only if z_1, z_2, z_3 divide the unit circle into equal parts;*

b) $z_1 + z_2 + z_3 + z_4 = 0$ *if and only if* z_1, z_2, z_3, z_4 *decompose into two pairs of opposite numbers.*

Proof. The lemma is easily proved by a geometrical reasoning. □

Let $\varepsilon_k = e^{\frac{2\pi i}{k}}$ as above, and let φ be a Laurent polynomial with rational coefficients. The following lemma provides an algorithm for finding out if $\varphi(\varepsilon_k) = 0$. Let p be a prime divisor of k.

Lemma 4. a) *If* $p^2 \mid k$, *write the polynomial* φ *in the form*

$$\varphi(z) = \varphi_0(z^p) + \varphi_1(z^p)\, z + \varphi_2(z^p)\, z^2 + \ldots + \varphi_{p-1}(z^p)\, z^{p-1},$$

where $\varphi_0, \varphi_1, \ldots, \varphi_{p-1}$ *are Laurent polynomials (with rational coefficients). Then* $\varphi(\varepsilon_k) = 0$ *if and only if*

$$\varphi_0(\varepsilon_{k/p}) = \varphi_1(\varepsilon_{k/p}) = \ldots = \varphi_{p-1}(\varepsilon_{k/p}) = 0. \qquad (24)$$

b) *If* $p^2 \nmid k$, *write the polynomial* φ *in the form*

$$\varphi(z) = \psi_0(z^p) + \psi_1(z^p)\, z^{\frac{k}{p}} + \psi_2(z^p)\, z^{\frac{2k}{p}} + \ldots + \psi_{p-1}(z^p)\, z^{\frac{(p-1)k}{p}},$$

where $\psi_0, \psi_1, \ldots, \psi_{p-1}$ *are Laurent polynomials (with rational coefficients). Then* $\varphi(\varepsilon_k) = 0$ *if and only if*

$$\psi_0(\varepsilon_{k/p}) = \psi_1(\varepsilon_{k/p}) = \ldots = \psi_{p-1}(\varepsilon_{k/p}). \qquad (25)$$

Proof. a) If $p^2 \mid k$, then $[\mathbb{Q}(\varepsilon_k) : \mathbb{Q}(\varepsilon_{k/p})] = p$, and $1, \varepsilon_k, \varepsilon_k^2, \ldots, \varepsilon_k^{p-1}$ constitute a basis of $\mathbb{Q}(\varepsilon_k)$ over $\mathbb{Q}(\varepsilon_{k/p})$.

b) If $p^2 \nmid k$, then $[\mathbb{Q}(\varepsilon_k) : \mathbb{Q}(\varepsilon_{k/p})] = p - 1$, and $1, \varepsilon^{\frac{k}{p}}, \varepsilon^{\frac{2k}{p}}, \ldots, \varepsilon^{\frac{(p-1)k}{p}}$ linearly span $\mathbb{Q}(\varepsilon_k)$ over $\mathbb{Q}(\varepsilon_{k/p})$ with the only linear dependence

$$1 + \varepsilon^{\frac{k}{p}} + \varepsilon^{\frac{2k}{p}} + \cdots + \varepsilon^{\frac{(p-1)k}{p}} = 0.$$ □

The conditions (24) can be interpreted as follows. Decompose the set of exponents of the polynomial φ into congruence classes modulo p. Then the equalities (24) mean that the sum of terms of φ corresponding to each class vanishes at ε_k. In case b), if not all the residues modulo p are represented by the exponents of non-zero terms of φ (e.g. if the number of these terms is less then p), at least one of the polynomials $\psi_0, \psi_1, \ldots, \psi_{p-1}$ is (identically) equal to 0, and the conditions (25) turn out to be equivalent to the conditions (24).

Corollary. *Assume that among the exponents of non-zero terms of* φ *there is one that is not congruent modulo* p *to any of the others. Let, moreover,* $p^2 \mid k$ *or the number of non-zero terms of* φ *is less than* p. *Then* $\varphi(\varepsilon_k) \neq 0$.

Proof of Proposition 10. The root of the polynomial h equals 0 if and only if

$$\varepsilon^{k_1} + \varepsilon^{k_2} + \varepsilon^{k_3} = \varepsilon^{k_2+k_3} + \varepsilon^{k_3+k_1} + \varepsilon^{k_1+k_2}. \tag{26}$$

Let p be a prime divisor of k. Assume that $p^2 \mid k$ or $p \geq 7$. Due to the preceding corollary the equality (26) can hold only if each of the integers

$$k_1,\ k_2,\ k_3,\ k_2+k_3,\ k_3+k_1,\ k_1+k_2 \tag{27}$$

is congruent modulo p to some of the others. It is easy to see that such a situation takes place only in the following cases, up to permutation of k_1, k_2, k_3:

1) $k_1 \equiv k_2 \pmod{p}, \quad k_3 \equiv k_1 + k_2 \pmod{p}$;
2) $k_1 \equiv k_2 \equiv k_3 \pmod{p}$;
3) $k_2 \equiv k_3 \equiv 0 \pmod{p}$;
4) $k_3 \equiv k_1 + k_2 \equiv 0 \pmod{p}$.

Consider all these cases.

Case 1. In this case, if $p \neq 2$, the decomposition of the set of integers (27) into congruence classes modulo p looks as follows:

$$\{k_1, k_2\} \cup \{k_3, k_1+k_2\} \cup \{k_1+k_3, k_2+k_3\}.$$

By Lemma 4, equality (26) holds only if

$$\varepsilon^{k_1} + \varepsilon^{k_2} = \varepsilon^{k_3} - \varepsilon^{k_1+k_2} = -\varepsilon^{k_1+k_3} - \varepsilon^{k_2+k_3} = 0,$$

which means that k is even and

$$k_1 - k_2 \equiv \frac{k}{2} \pmod{k}, \quad k_3 \equiv k_1 + k_2 \pmod{k}.$$

If the group Γ is primitive, at least one of the integers k_1, k_2 must be prime to k, and we may assume that it is equal to 1. This gives case 1) of the proposition. If $p = 2$, two of the above congruence classes must glue together. It is easy to see that these are the first and the third classes. We get

$$\varepsilon^{k_1} + \varepsilon^{k_2} - \varepsilon^{k_1+k_3} - \varepsilon^{k_2+k_3} = \varepsilon^{k_3} - \varepsilon^{k_1+k_2} = 0.$$

Since

$$\varepsilon^{k_1} + \varepsilon^{k_2} - \varepsilon^{k_1+k_3} - \varepsilon^{k_2+k_3} = (\varepsilon^{k_1} + \varepsilon^{k_2})(1 - \varepsilon^{k_3})$$

and $\varepsilon^{k_3} \neq 1$, we come to the same result as above.

Case 2. The decomposition of the set (27) into congruence classes is

$$\{k_1, k_2, k_3\} \cup \{k_2+k_3, k_3+k_1, k_1+k_2\},$$

whence

$$\varepsilon^{k_1} + \varepsilon^{k_2} + \varepsilon^{k_3} = -\varepsilon^{k_2+k_3} - \varepsilon^{k_3+k_1} - \varepsilon^{k_1+k_2} = 0.$$

Due to Lemma 3, it follows that k is divisible by 3 and $\varepsilon^{k_1}, \varepsilon^{k_2}, \varepsilon^{k_3}$ divide the unit circle into equals parts. At least one of the integers k_1, k_2, k_3 must be prime to k, and we may assume that it is equal to 1. This gives case 2) of the proposition.

Case 3. The decomposition of the set (27) into congruence classes is

$$\{k_1, k_1 + k_2, k_1 + k_3\} \cup \{k_2, k_3, k_2 + k_3\},$$

whence

$$\varepsilon^{k_1} - \varepsilon^{k_1+k_2} - \varepsilon^{k_1+k_3} = \varepsilon^{k_2} + \varepsilon^{k_3} - \varepsilon^{k_2+k_3} = 0.$$

It follows that

$$\varepsilon^{k_2} + \varepsilon^{k_3} = \varepsilon^{k_2+k_3} = 1,$$

which means that k is divisible by 6 and, up to interchanging k_2 and k_3,

$$k_2 \equiv \frac{k}{6} \pmod{k}, \qquad k_3 \equiv \frac{5k}{6} \pmod{k}.$$

Multiplying k_1, k_2, k_3 modulo k by an integer prime to k, one may assume that $k_1 \mid k$. But, if Γ is primitive, $(k_1, \frac{k}{6}) = 1$, whence $k_1 \mid 6$. This gives case 3) of the proposition.

Case 4. If $k_1 \equiv k_2 \pmod{p}$, we come to Case 1. Otherwise, the decomposition of the set (27) into congruence classes is

$$\{k_1, k_1 + k_3\} \cup \{k_2, k_2 + k_3\} \cup \{k_3, k_1 + k_2\},$$

whence

$$\varepsilon^{k_1} - \varepsilon^{k_1+k_3} = \varepsilon^{k_2} - \varepsilon^{k_2+k_3} = \varepsilon^{k_3} - \varepsilon^{k_1+k_2} = 0,$$

which is impossible. If k has no prime divisors satisfying the above conditions, then $k \mid 30$. For all such k and all, up to admissible transformations, k_1, k_2, k_3 we tried the equality (26) with help of a computer. It turned out that it held, beyond the series 1) - 3), only in case 4) of the proposition.

Cases 5–9 *of the proposition.* Finally, if Γ is pseudo-finite, but the root of h does not equal 0, then $k \leq 5$ and the root of h must belong to the numbers indicated in Table 6. There are only few cases to be tried. This gives the remaining 5 cases of the proposition. □

8. Two families of pseudo-dihedral g.t.g.

Finite g.t.g. exist only for $s \leq 8$. The following propositions show that pseudo-finite g.t.g. exist for any s.

Proposition 11. *The group*

$$\Gamma(s, n, c) = \langle x, y \mid x^{sn} = y^2 = (x^c y x^{n+c} y x^{2n+c} y \dots x^{(s-1)n+c} y)^2 = 1 \rangle$$

is pseudo-dihedral (and thereby pseudo-finite) for every choice of s, n, c *with* $0 < c < n$.

Proof. One has to prove that all the roots of the polynomial h (see (21)) equal 0, i.e., that $h_q = 0$ for $q = 1, \dots, [\frac{s}{2}]$. We have (see (23))

$$h_q = \sum_{\substack{1 \leq i_1 < \dots < i_r \leq s \\ \{i_1, \dots, i_r\} \text{ has } q \text{ connected components}}} (-1)^{s-r} \varepsilon^{(i_1 + \dots + i_r - r)n + cr}.$$

We shall prove that the sum $h_{q,r}$ of terms of h_q with fixed r vanishes for each r. Consider the transformation $i \mapsto i + 1$ of the set $\{1, 2, \dots, s\}$ (where $s + 1$ is taken modulo s). It does not change the number of connected components of a subset of $\{1, 2, \dots, s\}$, so $h_{q,r}$ is invariant under this transformation. But, on the other hand, each term of $h_{q,r}$ is multiplied by $\varepsilon^{rn} \neq 1$. Hence, $h_{q,r} = 0$. □

Proposition 12. *The group*

$$\Delta(s, n, c) = \langle x, y \mid x^{2sn} = y^2 = (x^n y x^{3n} y \dots x^{(s-2)n} y x^c y x^{(s+2)n} \\ y \dots x^{(2s-3)n} y x^{(2s-1)n} y)^2 = 1 \rangle$$

is pseudo-dihedral for any choice of s, n, c *with* s *odd and* $0 < c < 2sn$.

Proof. One has to prove that $h_q = 0$ for $q = 1, \dots, [\frac{s}{2}]$. Obviously, $h_q = h'_q \varepsilon^c + h''_q$, where h'_q and h''_q do not depend on c. Hence, it suffices to prove that $h_q = 0$ for $c = 0$ (when $\varepsilon^c = 1$) and for $c = sn$ (when $\varepsilon^c = -1$). For $c = sn$ we have

$$\Delta(s, n, c) = \Gamma(s, 2n, n),$$

so $h_q = 0$ by Proposition 11. For $c = 0$ we have $W(x, y) = y$, so $\operatorname{tr} W(X, Y) = \operatorname{tr} Y = 0$ (identically). □

Note that the above two families cover the series 1) of Proposition 9 and the series 2) and 3) of Proposition 10. Moreover, they cover all the series that we know for $s = 4, 5$.

References

[BMS] G. Baumslag, J. W. Morgan, and P. B. Shalen, Generalized triangle groups, *Math. Proc. Cambridge Philos. Soc.* **102** (1987), 25–31.

[FHR] B. Fine, J. Howie, and G. Rosenberger, One-relator quotients and free products of cyclics, *Proc. Amer. Math. Soc.* **102** (1988), 249–254.

[FR] B. Fine and G. Rosenberger, A note on generalized triangle groups, *Abh. Math. Sem. Univ. Hamburg* **56** (1986), 233–244.

[HMT] J. Howie, V. Metaftsis, and R. M.Thomas, Finite generalized triangle groups, *Trans. Amer. Math. Soc.* **347** (1995), 3613–3623.

[LRS] Lévai, G. Rosenberger, and B. Souvignier, All finite generalized triangle groups, *Trans. Amer. Math. Soc.* **347** (1995), 3625–3627.

[V1] E. Vinberg, Groups defined by periodic paired relations, *Mat. Sbornik* **188** (1997), no. 9, 3–12 (Russian). English translation: Sbornik: Mathematics **188:9**, 1269–1278.

[V2] E. Vinberg, Invariants of 2×2 matrices, Preprint no. 98-113, Universität Bielefeld, SFB 343, 1998, 34 pp.

[VMH] E. Vinberg, J. Mennicke, and H.Helling, On some generalized triangle groups and three-dimensional orbifolds, *Trudy Mosk. Mat. Obshch.* **56** (1995), 5–32 (Russian). English translation: *Trans. Moscow Math. Soc.* (1995), 1–21.